SOFT GROUND ENGINEERING IN COASTAL AREAS

PROCEEDINGS OF THE NAKASE MEMORIAL SYMPOSIUM, 28 NOVEMBER 2002
YOKOSUKA, JAPAN

Soft Ground Engineering in Coastal Areas

Edited by

Takashi Tsuchida
Hiroshima University, Japan

Yoichi Watabe & Minsoo Kang
Port and Airport Research Institute, Yokosuka, Japan

Osamu Kusakabe
Tokyo Institute of Technology, Japan

Masaaki Terashi
Nikken Sekkei Nakase Geotechnical Institute, Kawasaki, Japan

A.A. BALKEMA PUBLISHERS   LISSE / ABINGDON / EXTON (PA) / TOKYO

Published by: A.A. Balkema, a member of Swets & Zeitlinger Publishers
                www.balkema.nl and www.szp.swets.nl

ISBN 90 5809 613 0

Printed in The Netherlands

Soft Ground Engineering in Coastal Areas, Tsuchida et al. (eds)
© 2003 Swets & Zeitlinger, Lisse, ISBN 90 5809 613 0

Table of contents

Estimation of consolidation settlement in the large-scale reclamation

Soil improvement methods in coastal areas

Reuse of dredged soils & behavior of coastal structures under earthquake

Evaluation for the stability of coastal structures

Soft Ground Engineering in Coastal Areas, Tsuchida et al. (eds)
© 2003 Swets & Zeitlinger, Lisse, ISBN 90 5809 613 0

Preface

A symposium was organized to a lasting tribute to Professor Akio Nakase as the Nakase Memorial Symposium, which was the first of this kind in the geotechnical engineering field in Japan. The Memorial Symposium was held at Yokosuka, Japan, from November 28 to 29, 2002 and had a sub-title of "Soft Ground Engineering in Coastal Areas", reflecting Prof. Nakase's major contributions to geotechnical community.

Four organizations and a Technical Committee of International Society for Soil Mechanics and Geotechnical Engineering jointly organized the symposium, which are directly related to Prof. Nakase's career as a geotechnical engineer. These are Port and Airport Research Institute (PARI), National Institute for Land and Infrastructure Management (NILIM), Tokyo Institute of Technology, Nikken Sekkei Nakase Geotechnical Institute and ISSMGE Asian Technical Committee on land Reclamation and Coastal Structure in Asia (ATC12).

From 1961 to 1962, Prof. Nakase studied at the Imperial College of Science and Technology, University of London. He received the Diploma of the Imperial College (DIC) for his dissertation on bearing capacity of clay stratum. His experiences at the Imperial College paved his way to an internationally recognized geotechnical engineer. In fact, 7 overseas friends participated the symposium together with 118 domestic participants. During his time at the Imperial College, he had a close friendship with Professor N.R. Morgenstern and Prof. R.E. Gibson. Prof. Morgenstern delivered the first keynote lecture at the symposium as one of his best friends

Session presentation during the symposium

and Prof. Gibson sent the symposium organizer a letter which moved all the participants. As one of his colleagues during Prof. Nakase's fifteen years time at the Tokyo Institute of Technology, Prof. Y. Yoshimi delivered the second keynote lecture on the 1964 Niigata earthquake in retrospect.

The symposium included two keynote lectures, six sessions and a poster session followed by the memorial party. Topics of the sessions covered a wide range of technical issues, including fundamental characteristics of clayey soils, estimation of consolidation settlement in the large-scale reclamation, soil improvement methods in coastal area, reuse of dredged soils & behavior of coastal structures under earthquake, and evaluation for the stability of coastal structures. Selected twenty papers were presented and discussed at the sessions and twenty-four papers were presented at the poster session. At the memorial party, a string orchestra named Ensemble Civil headed by Mr. Yasuhiro Oh-hashi gave two performances. It was specially arranged because Prof. Nakase's favorite hobby had been playing cello and had been a leader of Ensemble Civil for seven years since its formation.

Performance of Ensemble Civil at the memorial party

This volume contains two keynote lectures and 43 peer-reviewed papers submitted to the symposium, together with the biography and selected papers of Prof. Akio Nakase. These selected papers were appeared in Soils and Foundations (Vol. 7, No. 2, 1967) and the proceedings of the ISSMFE VIII Asian Regional Conference (1987), both of which were edited and published by the Japanese Geotechnical Society. We are grateful for the permission to include these papers in the Proceedings.

We wish to thank all those who contributed to the success of the symposium. The following is the list of executive committee and working committee of symposium.

Executive committee members

Chairperson	Owada, Makoto	President, PARI
Member	Takahashi, Kunio	Executive Director, PARI
	Fukute, Tsutomu	Deputy Director General, NILIM
	Kusakabe, Osamu	Professor, Tokyo Institute of Technology
	Terashi, Masaaki	Director of Nikken Sekkei Nakase Geotechnical Institute
	Kobayashi, Masaki	Kobayashi Softtech Inc.
	Ohta, Hideki	Professor, Tokyo Institute of Technology
	Oh-hashi, Hiroyasu	Representative of Ensemble Civil
	Saito, Kunio	Professor, Chuo University

Working committee members

Secretary	Kusakabe, Osamu	Professor, Tokyo Institute of Technology
Member	Tanaka, Hiroyuki	Director of Geotechnical and Structural Engineering Department, PARI
	Oneda, Hideaki	Director for Special Research on Airport, PARI
	Tsuchida, Takashi	Head of Soil Mechanics and Geo-environment Division, PARI
	Kitazume, Masaki	Head of Soil Stabilization Division, PARI
	Watabe, Yoichi	Senior Research Engineer, PARI
	Kang, Min-Soo	Research Engineer, PARI
	Matsunaga, Yasuo	Head of Planning Division, NILIM
	Kuwano, Jiro	Associate Professor, Tokyo Institute of Technology
	Katagiri, Masaaki	Nikken Sekkei Nakase Geotechnical Institute
	Kamei, Takeshi	Associate Professor, Shimane University

May, 2003

Takashi Tsuchida,
Yoichi Watabe,
Minsoo Kang,
Osamu Kusakabe and
Masaaki Terashi,
Editors.

Akio Nakase   1929–2000

Soft Ground Engineering in Coastal Areas, Tsuchida et al. (eds)
© 2003 Swets & Zeitlinger, Lisse, ISBN 90 5809 613 0

Biography of the late Professor Akio Nakase (1929–2000)

Professor Akio Nakase passed away at his age of 70 years old on November 17, 2000. Professor Nakase was born in Hokkaido on November 3, 1929. He graduated from the University of Tokyo in 1953, then entered the postgraduate course at the same University. On completion of the post graduate course in 1956, he joined the Soil Mechanics Laboratory of the Port and Harbour Research Institute (PHRI), Ministry of Transport. His main subject of research in the Institute was engineering properties of soft marine clays in connection with construction of harbour structures. From 1961 to 1962, he studied at the Imperial College of Science and Technology, University of London, under a scholarship from the United Nations. He received the Diploma of the Imperial College (DIC) for his dissertation on bearing capacity of clay stratum. Later in 1967, the University of Tokyo awarded him the Doctor of Engineering for this work. He became Chief of Soil Mechanics Laboratory in 1964, and Director of Soils Division in 1972. At PHRI, he continued his research on engineering properties of soft marine clays. At the same time, he was busy in consulting on geotechnical problems in harbour and marine works in Japan. He was in charge of compiling the chapters of geotechnical problems in the Technical Standards for Port and Harbour Facilities. For his achievements at PHRI, the Minister of Transport awarded him the Special Prize in 1968. Overseas consulting works were an important part of his activities. He visited the USSR in 1969 as a geotechnical expert of the Japanese Government Mission for surveying ports and transportation problems in the Far East Region of the USSR. He also visited the Southeast Asian countries several times from 1972 to 1974.

In 1974, he moved to the Tokyo Institute of Technology (TIT) as Professor of Geotechnical Engineering and remained this position until his retiring in 1990. During his 16 years' professorship he supervised 26 master thesis's and 11 doctoral thesis's and published a number of technical papers and several books (Stability Chart for Slope and Embankment in 1981, Marine Geotechnology in 1984, for a few.) He served as the Head of Department of Civil Engineering for several terms and was appointed the Director of University Health Centre in 1987 and the Director of University Library in 1988.

At TIT, he continued his research on geotechnical properties of marine clays and intermediate soils. The Japanese Society of Soil Mechanics and Foundation Engineering (JSSMFE) awarded him the prize for research on marine clays and intermediate soils in 1986. In 1978, he visited the United States under the Japan-US Exchange Program of Eminent Scientists. From 1980 to 1985, he served as an adviser to the improvement of soft marine clays in Malaysia. In 1985, he visited Chile as the Chief of the Japanese Government Mission for investigating earthquake damage to harbour structures and for preparing a draft for restoring them. For the past twenty-nine years, he had played an important role in a major project of the Kansai International Airport, and had also been heavily involved in the Offshore Expansion Project of the Tokyo International Airport.

He had undertaken key roles in past International Conferences of ISSMGE. At the 8th ICSMFE in Moscow in 1973, he presented the Special Lecture on Marine Geotechnology. At the 9th ICSMFE in Tokyo in 1977, he was the Acting Secretary General to the Organizing Committee. At the 10th ICSMFE in Stockholm in 1981, he presented the General Report on the special theme of Prediction and Performance. He had been active in various activities of academic societies in Japan. From 1988 to 1989 he served the Vice President of the Japan Society of Civil Engineers (JSCE). In JSSMFE, he was the Vice President for 1988 to 1990 and the President from 1990 to 1992. In May 2000, the Japanese Geotechnical Society (JGS, previously called JSSMFE) and JSCE nominated him as an Honorary Member. JSCE also awarded him the Prize for Distinguished Contributions to Civil Engineering at the same time. Until his death, he was the Chairperson of ISSMGE Technical Committee 30 on Coastal Geotechnical Engineering and the Organizing Committee of International Symposium on Coastal Geotechnical Engineering in Practice held in September 2000. The following are a list of his contributions in reference to the ISSMGE.

1. Sept. 1967	Attended the 3rd Asian Regional Conference on SMFE (Haifa). Served the General-Reporter, Panelist and the Non-Voting Member of the Regional Executive Committee Meeting.
2. April 1973	Presented the Special Lecture " Problems of soil mechanics of the ocean floor" at the 8th Int. Conf. on SMFE (Moscow)
3. April 1975	Attended the Executive Committee Meeting (Istanbul) as the Non-Voting Member.
4. 1975-1978	Member of the Executive Committee for the 9th Int. Conf. on SMFE (Tokyo), substantially served the Acting Secretary General.
5. 1977-1981	Member of the Advisory Committee for the 10th Int. Conf. on SMFE (Stockholm).
6. April 1978	Attended the Advisory Committee Meeting for the 10th ICSMFE in Göteborg.
7. March 1979	Attended the Executive Committee Meeting (Oaxaca) as the Voting-Member.
8. June 1981	Served the General-Reporter at the 10th ICSMFE (Stockholm).
9. Sept. 1983	Served the Panelist at the 7th Asian Regional Conference on SMFE (Haifa), and attended the Executive Committee Meeting as the Non-Voting Member.
10. August 1987	Attended the Executive Committee Meeting (Dublin) as the Voting-Member.
11. July 1987	Presented the Guest Lecture on the Kansai Int. Airport at the 8th Asian Regional Conf. on SMFE (Kyoto).
12. May 1991	Attended the Executive Committee Meeting at the 10th European Regional Conf. on SMFE (Firenze) as the Non-Voting Member.
13. 1992-1994	Chairman of the Asian Technical Committee (AsTC) on Coastal Geotechnical Engineering set up by the Japanese Society
14. March 1994	Attended the 13th ICSMFE (New Delhi) and organized the first Meeting of the AsTC on Coastal Geotechnical Engineering.
15. 1994-1997	Chairman of the Technical Committee 30 (TC 30) on Coastal Geotechnical Engineering.
16. May 1995	Attended the 11th European Regional Conference on SMFE (Copenhagen) and organized the TC 30 Meeting, and also made a presentation at the workshop of the TC 17.
17. 1997-2001	Chairman of the TC 30 for the period of 1997-2001.
18. Sept. 1997	Attended the 14th ICSMGE (Hamburg). And organized the TC 30 Meeting.
19. 1998-2000	Chairman of the Organizing Committee for the International Symposium on Coastal Geotechnical Engineering in Practice (IS - Yokohama 2000).
20. August 1999	Organized the Seminar sponsored by the TC 30 at the 11th Asian Regional Conf. on SMGE (Seoul).
21. Sept. 2000	Organized the International Symposium on Coastal Geotechnical Engineering in Practice (IS - Yokohama 2000).

Professor Akio Nakase with his wife, Kazuko Nakase

After retiring from TIT in 1990, he was appointed Professor Emeritus and became the Special Adviser to the Board of the NIKKEN SEKKEI LTD, one of the largest consulting companies in Japan and established a new geotechnical laboratory with special emphasis on development in waterfront areas in Japan, named NIKKEN SEKKEI Nakase Geotechnical Institute (NNGI). He served as the director of the NNGI until September 2000.

Apart from the engineering profession, he enjoyed various sorts of hobbies. At home he enjoyed gardening, carpentry and model making. His favorite hobby had been playing cello. He continued playing cello for past forty-five years. It was the most enjoyable moment for him to play cello with his wife, Kazuko, playing piano. He had been a member of a non-professional string orchestra for twenty-five years, and enjoyed performing at bi-annual concerts in Tokyo. As part of the JSCE 80th anniversary celebration, JSCE formed a string orchestra by its members named Ensemble Civil. Professor Nakase had been the leader of the Ensemble Civil. This orchestra has given several performances in Japan in connection with special events of JSCE. In July 1995, the Ensemble Civil visited London and gave two performances at the special events of the Institute of Civil Engineers. Some members of the above two string orchestras gave a performance at his funeral.

Professor Nakase was a charming and cultured man. He specially loved the moment when he could drink with young students and engineers, through which he would encourage them to be leading engineers in our profession. He will be greatly missed by his close friends in many parts of the world as well as his students.

Selected papers of Prof. Akio Nakase

Soft Ground Engineering in Coastal Areas, Tsuchida et al. (eds)
© 2003 Swets & Zeitlinger, Lisse, ISBN 90 5809 613 0

THE $\phi_u=0$ ANALYSIS OF STABILITY AND UNCONFINED COMPRESSION STRENGTH**

Akio NAKASE*

ABSTRACT

The accuracy in determining the undrained strength of saturated cohesive soils and its sequence to the result of stability analysis are examined. As for the accuracy in determining the strength, the mechanical disturbance in the whole process from sampling to testing is found to have far greater influence than the effect of stress release in sampling, as far as the unconfined compression strength is concerned.

Three case records of failure in cohesive soils are analyzed by considering respectively the maximum, average and minimum values of measured unconfined compression strength. Results of the analysis show that the occurrence of failure can be explained reasonably by the $\phi_u=0$ analysis, provided that the distribution of the average unconfined compression strength with depth is considered.

1. INTRODUCTION

In most cases the foundation failure takes place in saturated cohesive soils. When a load is applied to failure on saturated cohesive soils of low permeability, the condition of failure may be considered similar to that in routine undrained shear tests, since a loading period is usually too short for the consolidation to occur. Shear stresses in the ground increase during the loading period, however, the strength of soil shows practically no change in this period. Hence the factor of safety against failure becomes minimum at or near end of loading. In such a case, therefore, the strength of soil before the loading is used for the stability analysis, and it is measured in undrained shear tests.

The strength of saturated soil measured in undrained shear tests is called the undrained strength c_u. This strength c_u is independent of a confining pressure, hence the angle of shear resistance ϕ_u is equal to zero. Therefore the stability analysis of cohesive soils, in which the undrained strength c_u is considered, is called the $\phi_u=0$ analysis.

Fellenius was the first to introduce the assumption of $\phi_u=0$ into the stability analysis, while its theoretical basis was put forward by Skempton.[1] Practical limit of the use

---

* Chief of Soil Mechanics Laboratory, Port and Harbour Research Institute, Ministry of Transport, KURIHAMA, JAPAN
**Reprinted from Soils and Foundations, Vol. VII, No.2, 1967

of the $\phi_u=0$ analysis was defined by Bishop and Bjerrum.[2] In their paper Bishop and Bjerrum collected twenty two case records of failure in cohesive soils at end of construction, and stated that an accuracy of $\pm 15\%$ could be expected in the estimate of factor of safety. This statement may indicate that the $\phi_u=0$ analysis is satisfactorily accurate for practical purposes.

It has been recognized that the accuracy of the stability analysis is governed by the accuracy in determining the strength of soil. So far quite a few results of the $\phi_u=0$ analysis of failure have been reported. However, the examination of accuracy in determining the strength of soil for stability analysis does not seem satisfactory. The main subject of the present paper, therefore, is to examine the accuracy in determining the undrained strength of saturated cohesive soil and its sequence to the result of stability analysis.

In site investigations, as far as harbour works are concerned, the unconfined compression test is most commonly used in this country. In what follows, therefore, we shall confine ourselves to the undrained strength obtained from the unconfined compression test.

2. ACCURACY IN DETERMINING UNDRAINED STRENGTH

To the purpose of stability analysis the unconfined compression test is performed on the undisturbed sample taken up from the natural ground. Throughout the whole process from sampling to testing, there may be two factors which are considered to cause any change in the sample condition. One is the release of in-situ confining stress by sampling and the other is the mechanical disturbance to the sample.

The effect of stress release on the soil strength has been studied by Skempton and Sowa,[3] and Noorany and Seed,[4] using triaxial compression apparatus. From the test results on soils of low sensitivity, Skempton and Sowa reported that the decrease in strength due to stress release was in the range of 1 to 2%. On the other hand, Noorany and Seed performed a series of tests on soils of moderate sensitivity, and concluded that the ' perfectly undisturbed ' strength is less than the ' in-situ ' strength by about 6%. This difference was explained by the change in pore pressure set up due to the difference in stress history.

As for the mechanical disturbance to sample, it may be difficult to estimate quantitatively its effect on the strength. The strength of soil decreases with increasing degree of mechanical disturbance. However, the degree of disturbance throughout the whole process from sampling to testing is quite difficult to assess, since the occurrence of mechanical disturbance is largely incidental.

Hvorslev[5] investigated the distribution of unconfined compression strength along the axial direction of sampler tube, and reported that the strengths near the top and bottom of the tube are smaller than those at the middle portion.

Fujishita, Matsumoto and Horie[6] carried out an extensive study of boring and sampling in saturated soft clays. In this study fifteen borings were performed and some two hundred sampler tubes were taken up from a saturated normally consolidated clay. For each bore hole, an undisturbed sample was taken up from every

4

1.5 m. Standard size of the sampler tube was 1 m in length, 75 mm in inner diameter and 1.3 mm in wall thickness. From each sampler tube, several specimens were taken for the unconfined compression test to see the distribution of the strength along the axial direction of tube. Soil in the tube was extruded by pressing the upper end. The test specimen was cylindrical one, 35 mm in diameter and 80 mm in height.

A number of subsoil investigations in the site have shown that the clay is fairly homogeneous. Water content of the clay is 100% to 130%, liquid limit is also 100% to 130% and plasticity index is 65 to 85. Clay fraction smaller than 5 micron is 60% in weight.

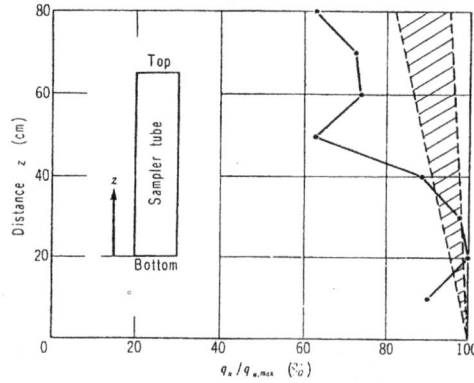

Fig. 1. Distribution of unconfined compression strength in a sampler tube

For each sampler tube, measured unconfined compression strengths were expressed in terms of a ratio to the maximum strength in the tube, which gave a non-dimensional distribution of the strength along the tube. Fig. 1 shows a statistical distribution of the unconfined compression strength along the sampler tube, obtained by summing up the strength distributions of all the sampler tubes. The shaded zone in the figure shows a probable range of strength distribution along the tube, which is estimated from an average distribution of strength with depth. As comparing the obtained strength distribution with the probable range of strength distribution, it may be said that the mechanical disturbance in the upper half of the sampler tube is greater than in the lower half. The strength distribution in Fig. 1 shows a local decrease at the distance of about 50 cm from the bottom. This may be due to the fact that the penetration of sampler tube in the ground was once stopped at about this part.

According to Fig. 1, the relative value of strength is maximum at the distance of 20 to 30 cm from the bottom of sampler tube. Therefore the relationship between the measured maximum strength and depth is to be obtained if the test specimen is taken from each sampler tube at this particular part. However, it is not an actual case, since the degree of mechanical disturbance is not uniform for each sampler tube.

Fig. 2 shows a typical example of the relationship between measured strengths and depth. As seen in the figure, the measured strengths show a considerable scattering, though every step in the whole process has been performed very carefully. Results of the examination at four bore holes showed that the deviation of the measured strengths from their average was 50% at maximum with an average of 36% at the depth of 5 m, and 23% at maximum with an average of 21% at the depth of 15 m.

It is well known that in a normally consolidated clay, the undrained strength increases linearly with depth, provided that the soil density is uniform. Therefore, let us examine a linear relationship between the measured unconfined compression strength q_u and depth z. In this examination five kinds of values were selected from the measured strengths in each sampler tube. They were the maximum strength in a sampler tube $q_{u.max}$, the minimum strength $q_{u.min}$, the strength measured from the

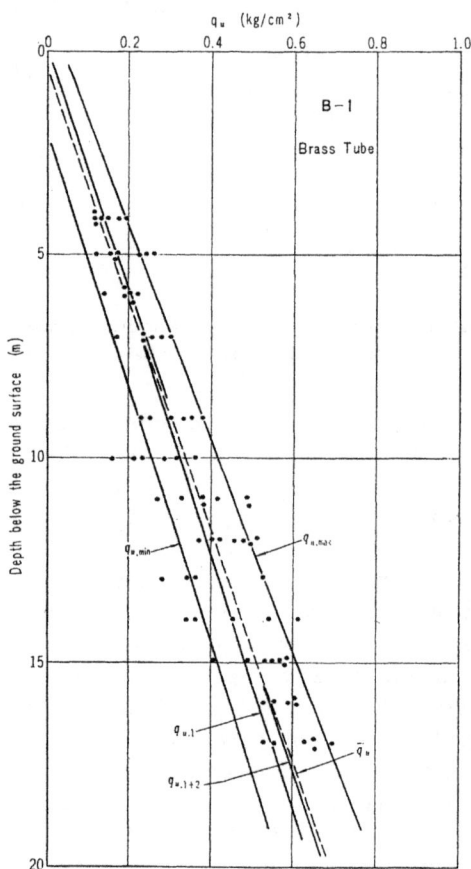

Fig. 2. Typical relationship between measured unconfined
compression strength and depth

specimen at the sampler bottom $q_{u \cdot 1}$, the mean value of strengths measured from the specimens at and next to the bottom $q_{u \cdot 1+2}$ and the average of all the measured strengths $\bar{q}_u$.

In Fig. 2 are shown five straight lines to represent the $q_u - z$ relationships. The linear relationships were determined by the method of the least square. As seen in the figure, the $\bar{q}_u - z$ relationship is very close to the $q_{u \cdot 1} - z$ and $q_{u \cdot 1+2} - z$ relationships. This implies that the $\bar{q}_u$ does not necessarily increase linearly with depth, since, statistically speaking, $q_{u \cdot 1}$ and $q_{u \cdot 1+2}$ are much larger than $\bar{q}_u$ as shown in Fig. 1. Therefore the result of the present examination clearly shows that the degree of disturbance is not uniform for each sampler tube.

As stated before, the decrease in the undrained strength due to stress release on sampling is not much, 1 to 2% being reported by Skempton and Sowa and 6% by Noorany and Seed. Comparing these orders of strength decrease with those of the scattering of measured unconfined compression strength, shown in Fig. 2, it may be concluded that the mechanical disturbance has far greater influence on the accuracy in determining the undrained strength than the effect of stress release does, as far as the unconfined compression strength is concerned.

3. RATE OF INCREASE IN UNDRAINED STRENGTH WITH DEPTH

The undrained strength of normally consolidated cohesive soils is directly proportional to the consolidation pressure. In a homogeneous normally consolidated ground, therefore, the undrained strength is to increase linearly with depth.

From geometry of Mohr's stress diagram, the relationship between undrained strength c_u and effective overburden pressure p is expressed[7]

$$\frac{c_u}{p} = \frac{\{K+(1-K)A_f\}\sin\phi'}{1+(2A_f-1)\sin\phi'} \quad \dots\dots\dots\dots (1)$$

where K : Coefficient of earth pressure at rest
A_f : Pore pressure coefficient at failure
ϕ' : Angle of shear resistance in terms of effective stress.
Coefficient of earth pressure at rest is expressed by an empirical formula[8]

$$K = 1 - \sin\phi' \quad \dots\dots\dots\dots\dots\dots (2)$$

By substituting equation (2) into equation (1) we have

$$\frac{c_u}{p} = \frac{(1-\sin\phi'+A_f\sin\phi')\sin\phi'}{1+(2A_f-1)\sin\phi'} \cdot \quad \dots\dots\dots\dots (3)$$

It has been reported that c_u/p value estimated by equation (3) is in good agreement with the average rate of increase in $c_u = q_u/2$ with effective overburden pressure for cohesive soils in the coastal area in this country. For these soils c_u/p value is reported to be in the range of 0.35 to 0.37.[9]

From results of undrained shear tests on normally consolidated natural clays, Skemp-

7

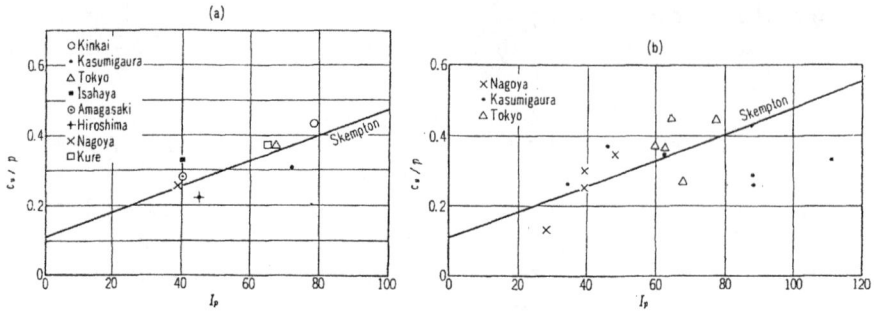

Fig. 3. Relationship between c_u/p and plasticity index

ton found that the c_u/p value was in linear relationship with plasticity index I_p, and expressed[10]

$$\frac{c_u}{p} = 0.11 + 0.0037\, I_p. \quad \dots\dots\dots\dots\dots\dots\dots\dots\dots\dots\dots\dots(4)$$

In Fig. 3(a) measured values of c_u/p are plotted against plasticity index for cohesive soils in this country, which are marine clays except Kasumigaura clay. In the figure is also shown the straight line representing the $c_u/p-I_p$ relationship by equation (4). The c_u/p values in Fig. 3(a) are obtained from relationship between $q_u/2$ and depth, and each plot represents the average of several measurements. For Kinkai clay it has been reported that the c_u/p value observed in consolidation by embankment load is in the range of 0.31 to 0.39.[11] It may be concluded from Fig. 3(a) that the relationship between (c_u/p) value and plasticity index for cohesive soils in this country can be expressed approximately by equation (4).

The plot in Fig. 3(a) is the average of several measurements. In order to show the scattering of original plots, measured values of c_u/p are plotted against plasticity index in Fig. 3 (b) for Nagoya, Tokyo and Kasumigaura clays respectively. For Nagoya clay and Tokyo clay, the scattering seems to be due to a possible error in measurement. In the case of Kasumigaura clay, however, the scattering can not be considered simply due to a usual error in measurement.

For Kasumigaura clay, as shown in Fig. 3 (b), the c_u/p values are practically the same in spite of a great difference, as much as 70, in plasticity index.[12] The present Kasumigaura lake used to be under the sea water several decades ago. And the leaching process is considered to be still going on. In fact, the content of chlorine ion is not uniform along the longitudinal direction of the lake. As a general trend, plasticity index and content of clay fraction increase as the ion content decreases. The case record of Kasumigaura clay would imply that the c_u/p value does not necessarily depend solely on the plasticity index, under an unusual environmental condition.

4. CASE RECORDS OF STABILITY ANALYSIS

4.1 *Method of stability analysis*

Generally speaking there are two methods of stability analysis under an assumption of circular slip surface, i.e., method of slices and friction circle method. The friction circle method, however, is used exclusively for sand or the case in which $\phi \neq 0$. The method of slices is further divided into three types with respect to the difference in practical procedure. These are the USBR method,[13] the Bishop's method[14] and the Tschebotarioff's method.[15] The difference in these three methods lies in the treatment of frictional resistance of soils. In the case of $\phi_u = 0$ analysis, therefore, these three methods are identical.

The factor of safety against failure is expressed in terms of the proportion of shear strength which is mobilized to just maintain the state of critical equilibrium. When the circular slip surface is assumed, however, the above factor of safety can be expressed in terms of the ratio of the restoring moment, which is produced by the shear resistance along an assumed slip surface, and the disturbing moment, which is produced by forces causing the failure.

For obtaining the disturbing moment, two factors, weight of soil and water pressure, are closely interrelated. According to Taylor,[16] the force acting on a soil element under a steady seepage condition is obtained by either of the following two concepts;
(a) vector sum of submerged weight of soil and seepage force, and
(b) vector sum of total weight of soil and resultant water pressure on the whole boundary of the element.

It is rather tedious to draw a flow net for obtaining seepage force, then, the concept (b) is considered advantageous in practice. In the following case records of stability analysis, the concept (b) will be employed.

A key sketch to an actual computation of the disturbing moment is shown in Fig. 4, where γ_1 is a unit weight of partly saturated soil above the phreatic line and γ_2 is a unit weight of saturated soil below the phreatic line, γ_w is a unit weight of water. For convenience of computation, the vertical tension crack at both ends of slip surface and hydrostatic pressure acting on the vertical face are considered, which are based on the above concept (b).

As shown in the preceding section, the undrained strength of soil increases linearly

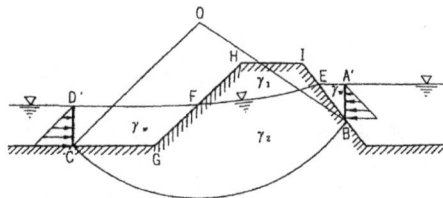

Fig. 4. Key sketch to computation of disturbing moment

9

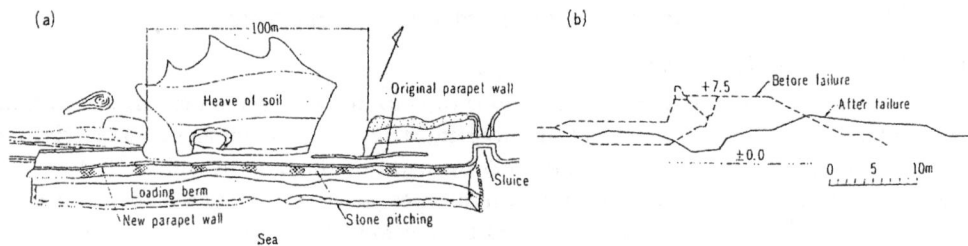

Fig. 5. Plan and cross section of failure (case 1)

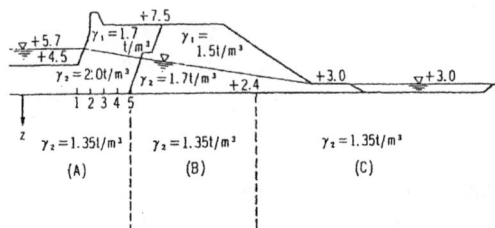

Fig. 6. Assumptions for stability analysis (case 1)

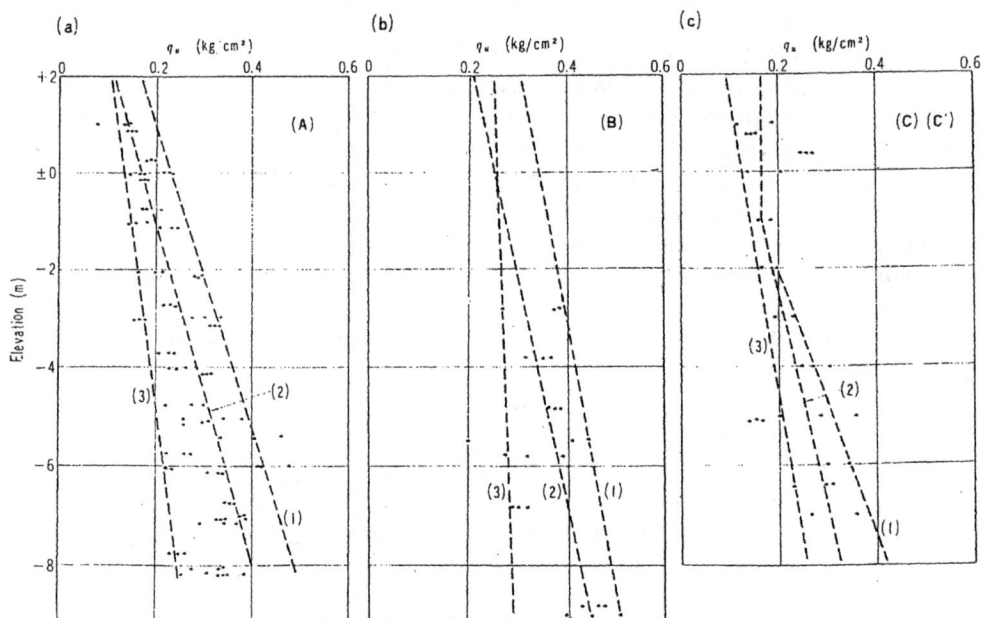

Fig. 7. Relationship between measured q_u and depth (case 1)

with depth, in general.   Therefore this distribution of the undrained strength should be fully taken into consideration for obtaining the restoring moment.[17]

4.2 *Case 1*

4.2.1 *Outline of failure and subsoil conditions*

A part of levee in A port failed and a farm land inside the levee was inundated. Longitudinal length of the failure was about 180 m.   This failure was not eyewitnessed, however, it was supposed to take place at the time of flood tide, at which the sea water level was 5.7 m above the datum.

Upper layer of subsoil is soft silty clay of 15 m thick, which is underlain by a gravel sand layer.   Water content of the clay are between 110% and 150%, which are larger than liquid limit by about 50%.   Void ratio is between 3 and 4.

The levee was constructed about a hundred years ago.   In the course of time, the piping phenomena became appreciable and a new levee widening work had been started. The failure occurred when the widening work was being performed in this particular part of the failed levee.   A plan of the failed levee and its cross sections before and after the failure are shown in Fig. 5.

4.2.2 *Stability analysis*

Cross section of the levee and soil conditions for the stability analysis are shown in Fig. 6.   Since no measurement was made on a phreatic line through the levee, the straight phreatic line was assumed for simplicity as shown in the figure.   This assumption is to result in an error in the safety side.

The foundation clay seems to be divided into three zones with respect to the distribution of strength with depth.   The difference in strength may be due to the difference in stress or environmental conditions such as consolidation by levee load and surface desiccation.

Fig. 7 shows the $q_u - z$ relationships in these three zones, which were measured for the widening work, i.e., shortly before the failure.   In the figure three straight lines are drawn to represent the distribution of q_u with depth.   The lines (1), (2) and (3) represent the distribution of the maximum, average and minimum q_u with depth respectively.   These three lines were determined by eye observation.

In practical computation, the level of EL.$+2.4$ m was assumed to be an imaginary ground surface, and soil above this level was simply assumed to be a load as shown in Fig. 6.   Vertical tension crack was assumed to develop throughout the fill, starting at the intersection point of slip circle and the imaginary ground surface.

The result of stability analysis corresponding to the strength distribution (2) is shown in Fig. 8.   The position of the critical circles seemed to be practically the same for three cases of strength distributions.   The minimum factor of safety was found to be 1.10, 0.93 and 0.80 corresponding to the strength distributions (1), (2) and (3) respectively.   In this analysis the straight phreatic line was assumed as mentioned before, while an actual phreatic line would be of concave shape.   Then the assumption of straight phreatic line would result in too great a fill weight, hence smaller value of factor of safety.   Considering the above situation, it would be said that the stability

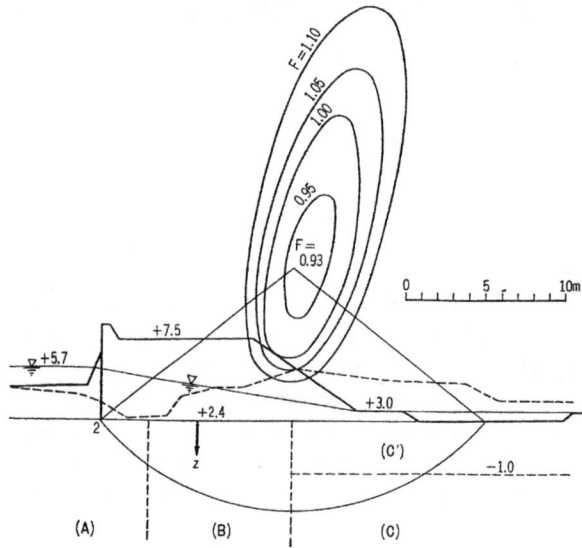

Fig. 8. Result of stability analysis—strength distribution (2) (case 1)

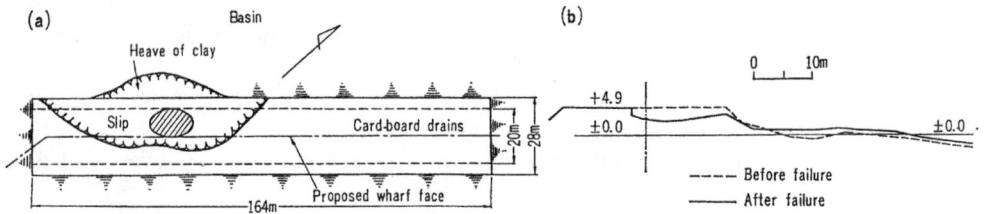

Fig. 9. Plan and cross section of failure (case 2)

analysis based on the average undrained strength, case (2), gave the most realistic result.

4.3 Case 2

4.3.1 Outline of failure and subsoil conditions

A failure took place in an embankment in B port, at which a wharf was being constructed. This failure was eyewitnessed, and it was reported that the failure started about 10 minutes after the cracks had been noticed on the top of embankment, and the main part of failure took place within the following 10 minutes. A longitudinal length of the failure was about 70 m. A plan of the failed embankment and its cross sections before and after the failure are shown in Fig. 9.

Top layer of subsoil consists of sand to silty sand of about 5 m thick, below which silty clay to clay continues to EL.-25 m. Water content of the clay is between 60% and 80%, and void ratio 1.5 and 2.5.

Since the bearing capacity of the foundation soil had been found insufficient, the construction was started with placing a sand fill as the consolidation load for cardboard drains. In the course of placing, local slips took place progressively and a part of the clay was replaced by sand and enfolded within the sand fill.

Consolidation by cardboard drains was to be completed in two steps. The failure took place when, after the first step consolidation was completed, the crown elevation of the embankment was raised from EL.+2.5 m to EL.+5 m.

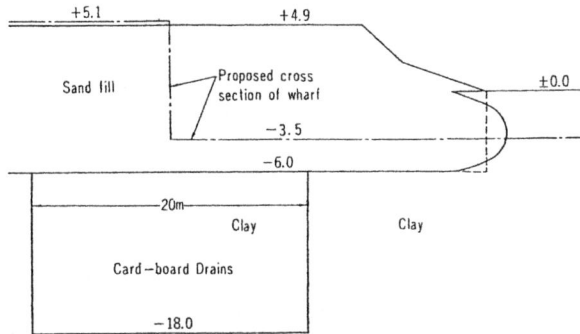

Fig. 10. **Assumptions for stability analysis (case 2)**

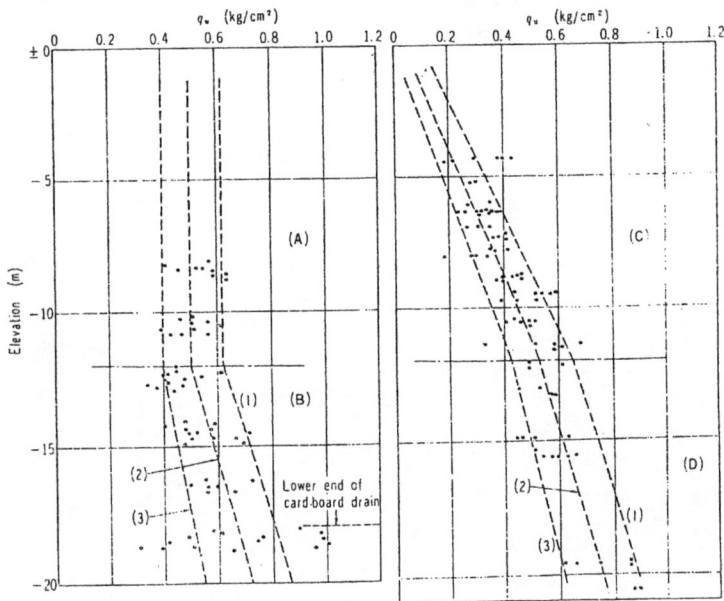

Fig. 11. **Relationship between measured q_u and depth (case 2)**

4.3.2 *Stability analysis*

Cross section and soil conditions for the stability analysis are shown in Fig. 10. With respect to the distribution of undrained strength, the foundation soil was divided into two zones in cross section, i.e., the zones with and without cardboard drains. Fig. 11 shows the q_u-z relationships, which were measured after the first step consolidation was completed, i.e., immediately before the failure. In Fig. 11 the three straight lines (1), (2) and (3) represent the distribution of the maximum, average and minimum strength with depth respectively.

In practical computation the elevation EL.±0 m was considered to be an imaginary ground surface. The sand fill above this level was simply assumed to be a load, and a vertical tension crack was assumed to develop throughout the fill above the elevation EL.±0 m. Restoring moment due to the frictional resistance of sand was worked out by the Tschebotarioff's method.

Fig. 12 shows the result of analysis corresponding to the strength distribution (2). Positions of the critical circle in the three cases of strength distribution were practically identical. The minimum factor of safety was found to be 1.24, 1.01 and 0.78 corresponding to the strength distributions (1), (2) and (3) respectively.

In the above analysis, an angle of shear resistance ϕ' of the sand was assumed to 30°. However the slope angle of the fill was as large as 40°. In addition, it would not be realistic to assume a vertical tension crack in sand fill. Therefore complementary analyses were made on the critical circle shown in Fig. 12, by considering a continuous circular slip surface through sand fill to foundation soil. Results of the analyses showed that in small scale earth structures as in this particular case, the assumption of vertical tension crack in sand filll would not result in any substantial change in the value of factor of safety.

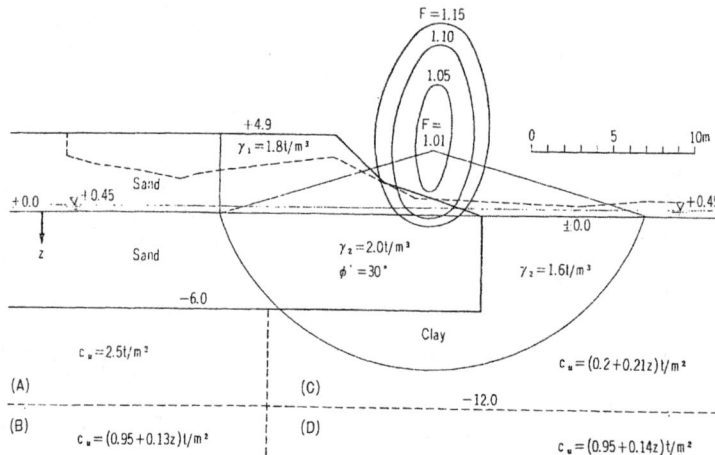

Fig. 12. Result of stability analysis—strength distribution (2) (case 2)

4.4 Case 3

4.4.1 Outline of failure and subsoil conditions

A large scale failure took place in the D berth in C port, which was being constructed. Longitudinal length of the failure was about 200 m. The proposed structure for the D berth was a trestle pier with small sheetpile retaining wall.

The D berth was located on the alluvial deposit, which consisted of silty clay to clay above EL.−30 m and underlain by sand gravel. Water content of the clay was 70% on an average and void ratio about 1.8. The soil between EL.−16 m and EL.−22 m, however, seemed to form a locally soft layer, water content of which was about 100% and void ratio 2.6.

Since the bearing capacity of clay was insufficient to support a pier with 10 m water depth, the soft clay above EL.−16 m had been replaced by sand. At the time of failure a considerable amount of sand was pumped on the backfill from a convenience of construction work in the adjacent pier. A plan of the failed D berth and its cross sections before and after the failure are shown in Fig. 13.

4.4.2 Stability analysis

Cross section and soil conditions for the stability analysis are shown in Fig. 14. It was assumed that the top flow line in the backfill coincided to the surface of backfill, considering the fact that it was immediately before the failure when the pumping was stopped on that day, and also the water was flowing on the top of backfill. The sea water level was taken to that at the ebb tide, at which the failure was considered to occur.

Foundation clay was divided into five zones with respect to the distribution of strength with depth. The boundary of these zones, in cross section, was determined

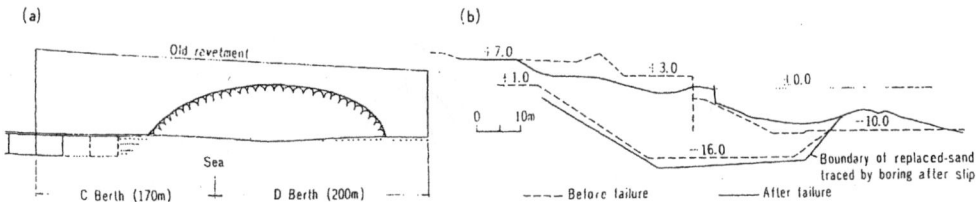

Fig. 13. Plan and cross section of failure (case 3)

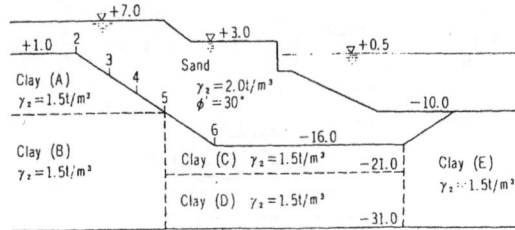

Fig. 14. Assumptions for stability analysis (case 3)

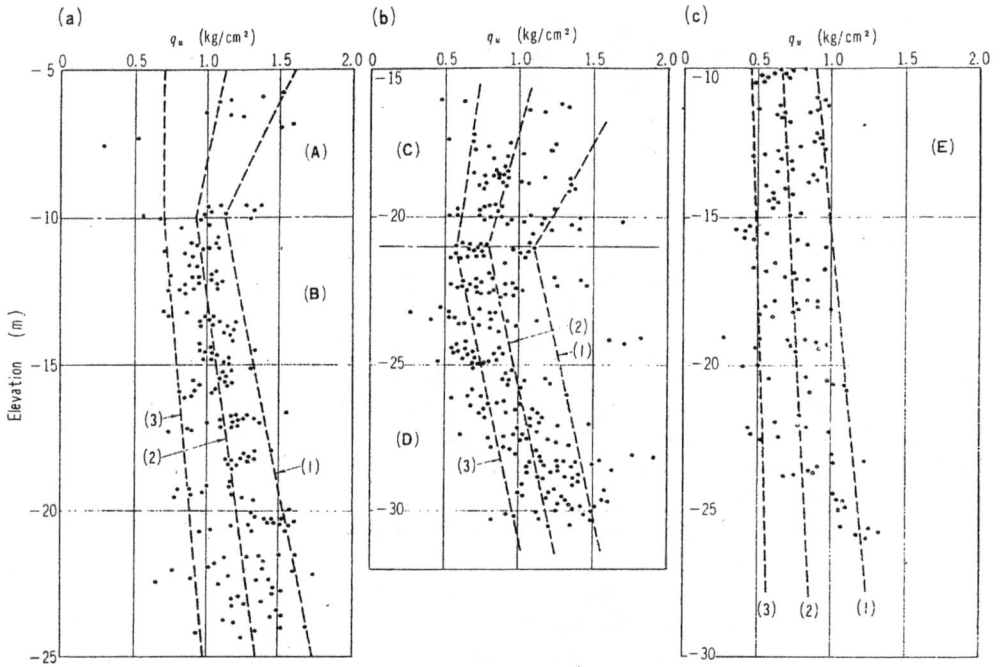

Fig. 15. Relationship between measured q_u and depth (case 3)

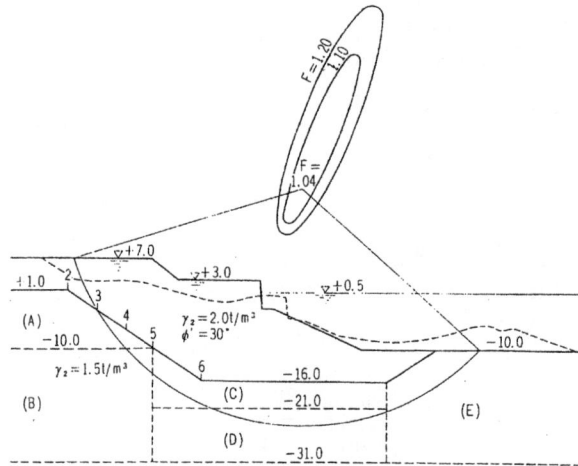

Fig. 16. Result of stability analysis—strength distribution (2) (case 3)

based on the results of borings made immediately before and after the failure. Fig. 15 shows the relationship between measured q_u and depth. In the figure the three straight lines (1), (2) and (3) are drawn to represent the distribution of the maximum, average and minimum strength with depth. In the determination of the strength distribution, the greatest weight was put on the measurement results at the end of the berth, where no failure occurred.

The practical computation was made for the circular slip surface passing through the six predetermined points Nos. 1 through 6, as shown in Fig. 14. The slip surface was assumed to be continuous through sand to clay. The Tschebotarioff's method was used in evaluating the part of restoring moment due to the frictional resistance of sand.

The result of analysis corresponding to the strength distribution (2) is shown in Fig. 16. In spite of an appreciable difference in the assumed strength distribution, positions of the critical circle were practically identical for the three cases of strength distribution. The minimum factor of safety was found to be 1.44, 1.04 and 0.74 corresponding to the strength distributions (1), (2) and (3) respectively.

In addition to the circular slip surface, the stability analysis was made for the composite slip surfaces. In this analysis a mass of sand was assumed to slide down along the inclined boundary plane between clay and replaced sand, and a circular slip surface developed in the foundation clay as a continuation of the straight slip surface. The results of the analysis were found to be practically the same to those in the case of a single circular slip surface.

5. DISCUSSIONS

It has been recognized that the accuracy of the stability analysis is governed by the accuracy in determining the strength of soil for the analysis. Results of analysis in the present paper show that this tendency is remarkable in the case where the unconfined compression strength is used for the analysis.

As the results of analysis of case records in the present paper, the occurrence of failure in cohesive soils seems to be explained reasonably by the $\phi_u=0$ analysis of stability, provided that the average distribution of measured unconfined compression strengths with depth is considered. The factors of safety obtained for three case records of failure were in the range of 0.93 to 1.04.

However the variation of factor of safety corresponding to the whole range of the measured strength was rather surprising. If the upper or lower limit of the measured strengths was considered, the deviation of the factor of safety from that corresponding to the average strength was found to be in the range of 13% to 38%.

In the present analysis the foundation soil was divided into several zones, in cross section, with respect to the strength distribution with depth. Each zone was defined by a set of vertical and horizontal lines. The above mentioned deviation in the factor of safety seems to be larger as the number of the zone division increases.

Change in the strength of foundation soil should essentially be continuous. Therefore the above mentioned division of foundation soil by a set of straight lines may result in some error in estimation of factor of safety. In order to obtain a more realistic distribution of undrained strength in the ground, the distribution of effective stresses in the ground must be known.

The accuracy in evaluating the effective stresses in the ground is said to be low, in particular, at the end of construction condition, to which the $\phi_u=0$ analysis is usually applied. The difficulty of this problem lies in an ambiguity of assessing the distribution of stress and pore pressure in the ground. This is just the reason why the stability analysis based on the concept of effective stresses, $c'-\phi'$ analysis, is not applied to the end of construction condition. Considering the above situation, an ambiguity of obtaining the distribution of undrained strength in the ground is not necessarily the characteristic defect of the $\phi_u=0$ analysis, but is one of the basic problems in soil mechanics.

Theoretical basis for a successful use of the average of measured strengths is not fully understood. Three factors, at least, appear to have some bearing on this problem. The first is the anisotropy of soil, i.e., the variation of undrained strength of clay with direction.[18],[19] The second is the reduction of strength due to creep, i.e., the decrease in strength of clay under sustained loads.[20] The third is the nonuniform mobilization of strength along a potential slip surface, i.e., an irrelevance of the use of conventional principle of limit design.

It is well known that the shape and position of the slip surface estimated in the conventional stability analysis is different from the actual one.[21] This may be due to the fact that conventional methods of stability analysis are simply derived from the equilibrium condition of forces acting on soil mass. Case records in the present paper have shown that the actual change in the foundation soil surface is observed far beyond the extent of the slip surface estimated by the stability analysis.

6. CONCLUSIONS

(1) As far as the unconfined compression strength is concerned, the mechanical disturbance in the whole process from sampling to testing has far greater influence on the accuracy in determining the undrained strength than the effect of stress release in sampling does.

(2) The $\bar{q}_u-z$ relationship is practically the same to the $q_{u\cdot1}-z$ and $q_{u\cdot1+2}-z$ relationships ; where $\bar{q}_u$ is the average of all the measured q_u values, $q_{u\cdot1}$ is the strength measured on the bottom sample in a sampler tube, $q_{u\cdot1+2}$ is the mean value of strengths measured on the samples at and next to the bottom of a sampler tube and z is a depth below the foundation soil surface.

(3) The occurrence of failure in cohesive soils can be explained reasonably by the $\phi_u=0$ analysis, provided that the q_u-z relationship is considered.

7. ACKNOWLEDGEMENTS

The Author is deeply indebted to staff of the District Port Construction Bureaus for offering valuable informations about the case records of failure.

The Author would like to express his gratitude to Messrs. T. Okumura, M. Sawaguchi and K. Kamiyama of the Port and Harbour Research Institute, who undertook a part of stability analysis of the case records. He also wishes to thank Mr. S. Yanase of the Port and Harbour Research Institute for making helpful remarks.

REFERENCES

1) Skempton, A. W. (1948), The $\phi=0$ analysis of stability and its theoretical basis, Proc. 2nd Int. Conf. S. M. F. E., Vol. 1, pp. 72–78

2) Bishop, A. W. and L. Bjerrum (1960), The relevance of the triaxial test to the solution of stability problems, Proc. Research Conf. Shear Strength of Cohesive Soils, ASCE, pp. 462–490

3) Skempton, A. W. and V. Sowa (1963), The behaviour of saturated clays during sampling and testing, Geotechnique, Vol. 13, No. 4, pp. 269–290

4) Noorany, I. and H. B. Seed (1965), In-situ strength characteristics of soft clays, Proc. ASCE, SM 2, pp. 49–80

5) Hvorslev, M. J. (1949), Subsurface exploration and sampling of soils for civil engineering purposes, WES, Vicksburg, p. 201

6) Fujishita, T., K. Matsumoto and H. Horie (1966), Study of boring and sampling of saturated clays, Report of Port and Harbour Research Institute, Vol. 5, No. 4 (in Japanese)

7) Skempton, A. W. and A. W. Bishop (1954), Soils, Chapter 10 of Building Materials, their Elasticity and Inelasticity, North-Holland Publ. Co., p. 460

8) Bishop, A. W. (1958), Test requirements for measuring the coefficient of earth pressure at rest, Proc. Brussel Conf. Earth Pressure Problems, Vol. 1, pp. 2–14

9) Nakase, A., S. Kishi and M. Katsuno (1965), Triaxial test of soils, Report PHRI, Vol. 4, No. 1 (in Japanese)

10) Skempton, A. W. (1957), Discussion of the planning and design of the New Hongkong Airport, Proc. ICE, 7, pp. 305–307

11) Nakase, A., S. Kurata and T. Okumura (1961), On the consolidation phenomena in the Kinkai levee, Report of Transportation Technical Research Institute, Vol. 11, No. 9 (in Japanese)

12) Soil investigations in Kasumigaura (1966), The 2nd District Port Construction Bureau and PHRI (in Japanese)

13) Taylor, D. W. (1948), Fundamentals of soil mechanics, John Wiley, pp. 432–441

14) Bishop, A. W. (1955), The use of the slip circle in the stability analysis of slopes, Geotechnique, Vol. 5, No. 1, pp. 7–17

15) Tschebotarioff, G. P. (1951), Soil mechanics, foundations and earth structures, McGraw-Hill, pp. 185–186

16) loc. cit. 13), pp. 200–204

17) Nakase, A. (1963), Method of computing the restoring moment in the $\phi=0$ analysis of stability, Soil mechanics and foundation engineering, Vol. 11, No, 4, pp. 34–36 (in Japanese)

18) Lo, K.Y. (1965), Stability of slopes in anisotropic soils, Proc. ASCE, Vol. 91, No. SM 4, pp. 85–106

19) Ladd, C. C. (1965), Stress-strain behaviour of anisotropically consolidated clays during undrained shear, Proc. 6th Int. Conf. S. M. F. E., Vol. 1, pp. 282–286

20) loc. cit. 7), pp. 450–451

21) Scott, R. F. (1963), Principles of soil mechanics, Addison-Wesley, p. 462

LIST OF SYMBOLS

A_f : Pore pressure coefficient at failure

c' : Apparent cohesion in terms of effective stress

c_u : Undrained strength of soil

I_p : Plasticity index

K : Coefficient of earth pressure at rest

p : Effective overburden pressure

q_u : Unconfined compression strength

$\bar{q}_u$: Average unconfined compression strength

$q_{u \cdot 1}$: Unconfined compression strength measured on the bottom sample in a sampler tube

$q_{u \cdot 1+2}$: Mean value of unconfined compression strengths measured on the samples at and next to the bottom of a sampler tube

z : Depth below the ground surface

γ_1 : Unit weight of partly saturated soil

γ_2 : Unit weight of saturated soil

γ_w : Unit weight of water

ϕ' : Angle of shear resistance in terms of effective stress

ϕ_u : Angle of shear resistance in undrained shear test in terms of total stress

(Received: January 10, 1967)

Soft Ground Engineering in Coastal Areas, Tsuchida et al. (eds)
© 2003 Swets & Zeitlinger, Lisse, ISBN 90 5809 613 0

Kansai International Airport-Construction of Man-Made Island **

A. Nakase   Professor of Geotechnical Engineering, Tokyo Institute of Technology, Tokyo, Japan

INTRODUCTION

In 1993, Japan will open her second major international airport in the Osaka Bay, five kilometers offshore and approximately 300 miles south-west of Tokyo. The new airport was proposed in order to solve serious operational limitations at the existing Osaka International Airport which has been caused by noise pollution problems. This new airport is expected to change the conventional air traffic patterns by allowing a smoother and more natural flow of the air traffic around Japan. It is also hoped by having the new international airport in the Osaka Bay, to enhance the industrial and cultural potentiality of the Kinki Region which covers cities of Osaka, Kyoto, Kobe, Nara and Wakayama. Locations of airports in Japan is shown in Fig.1.

The need for the new airport, the Kansai International Airport, was recognized in anticipation of environmental problems at the existing Osaka International Airport arising due to developing air traffic in the beginning of 1965. Actual study of this project was started in 1968. As the result of close examination of four proposed sites, the Council for Civil Aviation reported in 1974 that offshore of the "Senshu" area (south-eastern part of the Osaka Bay) was the most appropriate site for the new airport. The Council for Civil Aviation made the second report in 1980, that summarized a draft plan for the Kansai International Airport.

The Ministry of Transport submitted three reports of "The Kansai International Airport plan", "Environmental Impact Statement for the Kansai International Airport" and "Proposed Regional Development Concept" to three prefectures around the airport site, i.e. Osaka, Wakayama and Hyogo Prefectures, in May 1981. Five years were required for these three prefectures to approve these three reports in principle.

In June 1984, The Kansai International Airport Company Act was promulgated and the Company was founded and registered in October 1984. The Government approved the general policy for construction of facilities related to the airport in December 1985. Compensation agreement was concluded between the Kansai International Airport Company and the Fishermen's Association of Osaka Prefecture in April 1986.

Through the years from 1968 to 1986, continuous studies have been made on various aspects of engineering and environmental problems involved in the construction of the airport. Actual construction has started in January 1987, and its first phase is expected to be completed in 1993.

Construction of the Kansai International Airport is a challenging civil engineering project in view of its size as a marine airport and also its severe geotechnical conditions. In what follows, geotechnical aspects involved in the construction of the airport will be described.

Fig.1   Location of airports in Japan

** Reprinted from the Proceedings of ISSMFE VIII Asian Regional Conference, Vol.2,1987

OUTLINE OF THE AIRPORT

There are four basic ideas in planning the Kansai International Airport. They are : (i) to make a key airport handling either local, domestic and international air traffic, (ii) the airport is to be the round-the-clock one which is the first case in Japan, (iii) to pay full caution against any sort of environmental pollution, and (iv) to have a possibility of further expansion. Location of the Kansai International Airport is shown in Fig.2.

The Kansai International Airport is a marine airport constructed on a man-made island. The airport island is five kilometers off the shore line and the water depth at the site is 16 m to 19 m. Area of the airport is 511 ha (5,110,000 m^2) and a general plan of space allocation is shown in Table 1. The airport island is to be connected to the main land by a 3.8 kilometers long double deck bridge for road and railway, and the use of high speed boat is also being studied for access by the sea. Estimated access time to the airport is 30 min by railway from Osaka's southern gateway, 50 min by high speed boat from Kobe and 100 min by bullet train from Kyoto.

In anticipation of the new airport and the related high speed access routes that will bring more people into the airport area, the local municipalities have been making plans for the construction of new communities in their re-

Table 1   General plan of space allocation

Zone	Major Facilities	Space (ha)
Take-off & Landing	Runway, Taxiway, Navigation Aid	218
Apron	Loading, Nightstay, Maintenance	133
Passenger Terminal	Passenger Terminal	11
International Cargo Terminal	Warehouse, Government Offices	23
Domestic Cargo Terminal	Warehouse	5
Maintenance	Hanger	13
Supply & Disposal	Power, Gas, Water Aircraft Fuel, Sewerage	17
Administration	Government Offices, Airline and other offices	5
Access	Road, Railway, Station, Parking	60
Other	Seawall	26
Total		511

Fig.2   Location of the Kansai International Airport

spective areas. Among these, the Osaka Prefectural Government has decided to build a community with a shopping mall, a trade center, hotels, parks and other facilities for the convenience of visitors to and workers at the airport. 320 ha of coast land will be reclaimed around the access bridge junction.

Main reason of locating the new airport on the sea was to avoid any undesirable impact due to noise of air traffic. Noise exposure forecast contour map is shown in Fig.3, in which the Weighted Equivalent Continuous Perceived Noise Level (WECPNL) value is indicated. The WECPNL is a measure of noise level, which is estimated based on the noise of airplane and distribution with time of the number of take-offs and landings. It has been recommended in Japan that the WECPNL value should be less than 70 in residential areas. As seen in the figure, the contours of 70 of the WECPNL do not overlap the land area, therefore, assuring full protection of the land environment from noise pollution. The airport's 24 hour operations largely depend upon this assurance.

Plan for the first phase project of the Airport is shown in Fig.4. This airport of total area of 511 ha with a main runway of 3,500 m is expected to handle approximately 160,000 take-offs and landings per year.

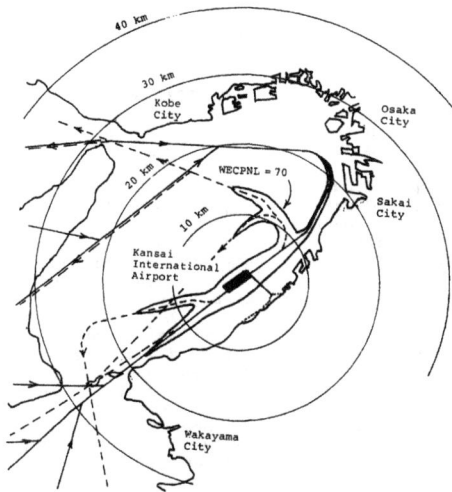

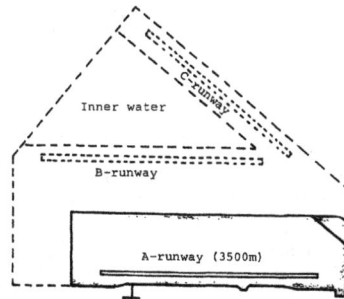

Fig. 5   Concept of final phase
(not authorized yet)

Fig. 3   Noise exposure forecast contour

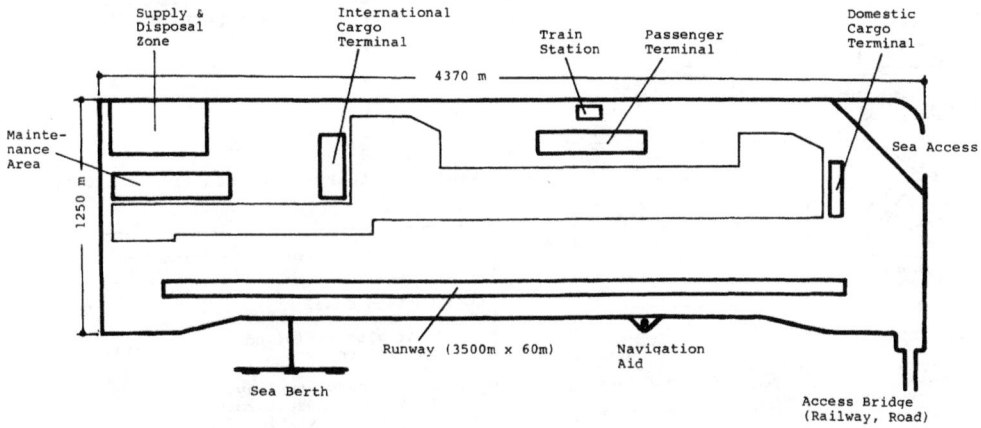

Fig. 4   The Kansai International Airport Plan (phase 1)

The final phase of the project has not been authorized yet, however, its concept is shown in Fig.5. After completion of the final phase, this airport is to have a total area of 1,200 ha with three runways at all. Fig.6 compares some of existing international airports in the world.

The estimated cost of the first phase is shown in Table 2, and the cost for airport island and related facilities is 800 billion yen, on the basis of costs in 1983. In addition to the construction of airport itself, the Government approved the general policy of constructing railways, expressways and ports, which will form a high speed link to the major cities in the Kansai region. The policy also includes the other facilities such as a residential area, parks, a water supply and a sewerage system. The cost for these development project is estimated 2,480 billion yen which is about 2.5 times the cost of constructing the airport itself.

SOIL CONDITIONS AT THE CONSTRUCTION SITE

Soil Exploration

During a period from 1977 to 1982, 65 test borings were carried out, which consisted of 24 borings down to C.D.L.-100 m, 34 borings to C.D.L.-150 m, 2 borings to C.D.L.-200 m and 2 borings down to C.D.L.-400 m. In the case of the deep borings down to C.D.L.-400 m, a newly developed wire-line system was employed in order to increase an efficiency of operation (2). In the wire-line system operation, either the Denison type samplers or hydraulic type stationary piston samplers were used for obtaining undisturbed soil samples.

In this soil exploration, a series of geological studies was carried out on all of the soil samples obtained. The geological studies included measurements of micro fossils of foraminifera, diatom, pollen, calcareous nannoplankton and others. Existence of the nannoplankton implies the sedimentation under marine condition, and its pattern of emergence with depth reflects the sedimentation environments. As the result of these geological studies, the sedimentation environment and age of the soil layers were traced.

The soil profiles along two axis lines of the airport island, parallel and perpendicular to the shore line, are shown in Fig.7. As shown in Fig.7, the sea bottom soil consists of alternating layers of cohesive soils and sandy or sand-gravel soils, and has an alternating patterns of marine sediments and non-marine sediments. The surface layer, denoted A_C in Fig.7, is an alluvial clay and its age is estimated as 6,000 years at the middle and 23,000 years at its bottom portion (4).

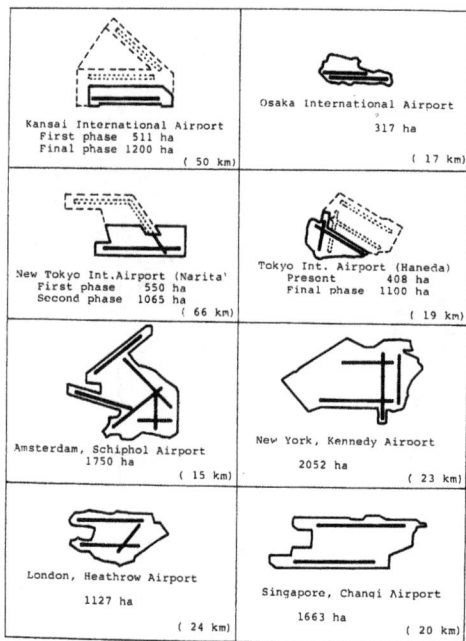

Kansai International Airport
First phase 511 ha
Final phase 1200 ha
(50 km)

Osaka International Airport
317 ha
(17 km)

New Tokyo Int.Airport (Narita'
First phase 550 ha
Second phase 1065 ha
(66 km)

Tokyo Int. Airport (Haneda)
Present 408 ha
Final phase 1100 ha
(19 km)

Amsterdam, Schiphol Airport
1750 ha
(15 km)

New York, Kennedy Airport
2052 ha
(23 km)

London, Heathrow Airport
1127 ha
(24 km)

Singapore, Changi Airport
1663 ha
(20 km)

( ) : Distance from city center

Fig.6   Existing international airports in the world

Table 2   Approximate construction cost

Project cost (billion Yen)	
Construction Cost	800
Cost of airport island reclamation	440
Cost of runway, taxiway, navigation aids, terminal facilities, aviation fuel supply facilities	240
Cost of the access bridge	120
Miscellaneous	200
Total	1000

Soil layers underlying the A_C layer are dilu- vial soils and their pattern of alternation of cohesive and sandy or sand-gravel soils continues down to C.D.L-400 m which is the limit of the depth of this particular soil exploration.

Geological structure of the sea bottom soil of the construction site is found to be a monocline type towards offshore direction, and its strike of bed is approximately paral- lel to the shore line.

Index Properties

Consistency properties of the alluvial clay A_C are shown in a plasticity chart in Fig.8. As seen in the figure, the A_C clay is mostly classified as an inorganic clay of high plas- ticity, CH. No appreciable difference was seen in the plots of the diluvial cohesive soils in the plasticity chart. Composition of soil particles of the cohesive soils is illustrated in a triangular soil classifica- tion chart in Fig.9. Particle size distribu- tion curve of the sandy soils is shown in Fig.10, where an extent of grading curves is illustrated.

(a)

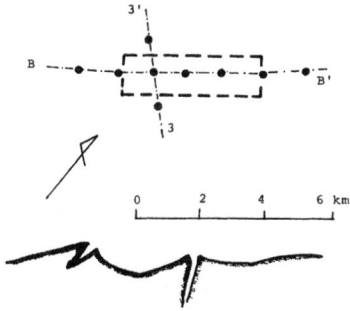

(b)

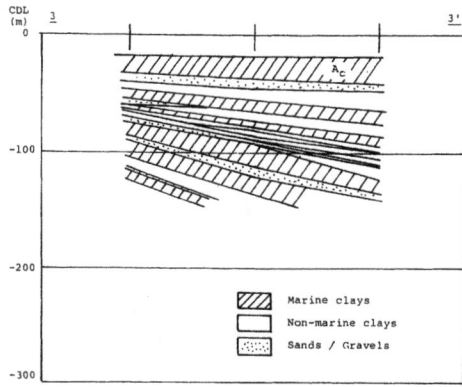

(c)

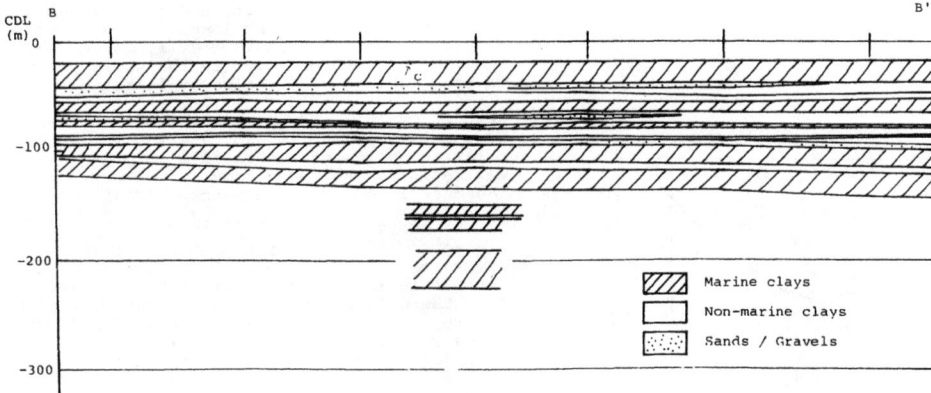

Fig.7   Soil profile at construction site

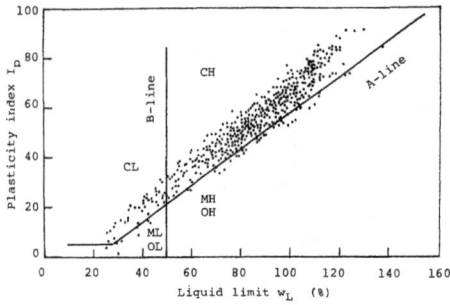

Fig. 8   Plasticity chart

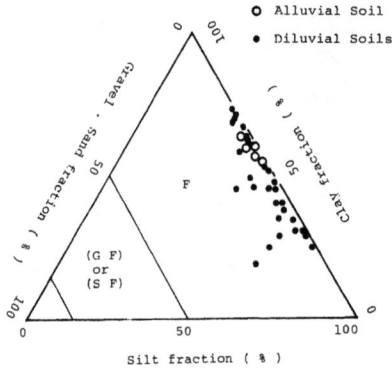

O   Alluvial Soil
•   Diluvial Soils

Fig. 9   Triangular soil classification chart

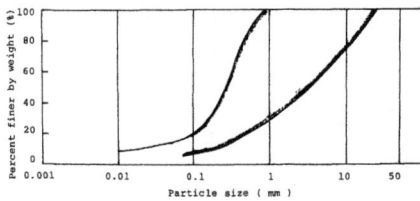

Fig. 10   Particle size distribution curve of sandy soils

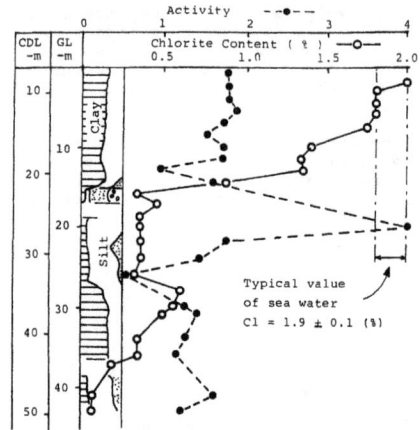

Fig. 11   Distribution with depth of chlorite content and activity

Looking over the index properties, the alluvial clay A_C of the construction site seems the typical soft marine clay in Japan. Experiences of harbour works in Japan indicate that there will be a considerable amount of compression of this A_C layer upon loading of weight of the airport island.

Fig.11 shows distribution with depth of the chlorite content and activity, which is the ratio of plasticity index to clay fraction of less than 2 micron particles.

Consolidation Characteristics

A number of consolidation tests were carried out on cohesive soils obtained by a series of the test borings and samplings, since an accuracy in predicting future settlement is considered the governing factor in the whole project of construction of the airport island.

Results of standard oedometer tests on the surface alluvial clay A_C indicate that, in a normally consolidated region, the compression index C_C is in the range from 0.5 to 1.7, and the coefficient of volume compressibility m_v

may be expressed in a form from $0.2\ p^{-1.1}$ to $0.4\ p^{-1.2}$ cm^2/kgf, where p is the consolidation pressure in kgf/cm^2 (3). Time settlement relationship in the oedometer tests indicate that there is a considerable amount of the secondary compression in each load step, and the primary consolidation ratio r is about 60 %. Coefficient of consolidation c_v is found to be in the range from 3.0 to 5.0 m^2/year, and the corrected value, by multiplying the primary consolidation ratio, $c_v' = r\ c_v$, is in the range from 1.5 to 3.0 m^2/year (5).

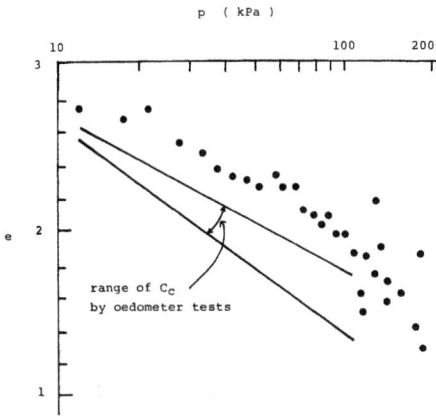

p (kPa)

range of C_C
by oedometer tests

Fig. 12   In situ effective overburden
stress and void ratio

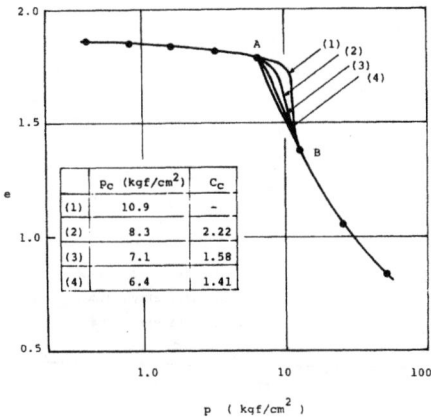

	p_C (kgf/cm²)	C_C
(1)	10.9	–
(2)	8.3	2.22
(3)	7.1	1.58
(4)	6.4	1.41

p (kgf/cm²)

Fig. 13   Dependence of precompression
stress and compression index on
shape of compression curve

As a whole the surface clay layer A_C appears to be in a normally consolidated state in view of the distribution of the precompression stress with depth.   Fig.12 shows the relationship between in-situ effective overburden stress and void ratio (6). As seen in the figure, the in-situ e-log p relationship is similar to that obtained by the standard consolidation tests.

There is some difficulty in interpreting the result of the standard consolidation tests on diluvial clays, which are much stiffer than the surface alluvial clay.   In the standard consolidation tests, the load increment ratio $\Delta p/p$ of unity is specified.   In the case of relatively stiff diluvial clays, the specified load step happens to straddle the probable precompression stress p_C , which causes some extent of error in evaluating the value of p_C as shown in Fig.13.   Then the special oedometer tests were carried out, in which the load increment ratio $\Delta p/p$ of either 0.5 and 0.25 was employed.   As is known, the value of p_C increases as the load increment ratio decreased.

Fig.14 compares the effective overburden pressure and the precompression stress obtained by consolidation tests, where three kinds of load increment ratio were employed.   As seen in the figure, the precompression stress obtained by consolidation tests with $\Delta p/p$ of 0.5 and 0.25 seems to give more realistic distribution of the precompression stress with depth.   According to this figure, the overconsolidation ratio OCR of the diluvial clays is about 1.3 and seems constant with depth.   This implies that the diluvial clays are in an overconsolidated state.

Permeability of the diluvial sands was found to be in the range from 10^{-6} to 10^{-3} cm/sec (3).   These values of permeability seem a little too small, however, this may be caused by the fact that the sampling of coarse sands from great depths is more difficult than the sand samples with some fraction of fine soil particles.

Strength Characteristics

Shear strength of clays was measured mainly by the unconfined compression tests and the unconsolidated undrained triaxial compression tests.   Field vane tests and static cone penetration tests were also made for the alluvial clay.   Soil samples taken out from greater depth are more susceptible to the effect of sampling disturbance due to a release of the in situ confining stress and also due to more severe mechanical disturbance through the severe sampling operations.   The unconsolidated undrained triaxial compression tests, therefore, were used mainly for the diluvial clays taken from greater depths.   Confining pressures in the triaxial tests were chosen in accordance with the total overburden stress at the depth of sampling.

Undrained shear strength c_u , in terms of half of the unconfined compression strength q_u, of the alluvial clay layer increases linearly with depth, as in an ordinary pattern in nor-

27

mally consolidated clays. When the undrained shear strength c_u is expressed in the form of $c_u = a + bz$, where a and b are constants and z is the depth below the ground surface, a is in the range from 0.009 to 0.014 kgf/cm^2, and b from 0.024 to 0.027 kgf/cm^3, z is in meter (4).

Undrained shear strength of the diluvial clays obtained by the triaxial tests was found to be larger than those obtained by the unconfined compression tests by 10 to 20 %.

Field vane tests and static cone penetration tests were carried out mainly for the alluvial clay layer. It was found that the undrained shear strength by the vane tests practically coincided to those by the unconfined

compression tests. Relationship between the cone resistance q_c and the unconfined compression strength q_u appeared to be expressed as $q_c = (8 \sim 10)q_u$.

c_u/p value obtained by the triaxial tests, both for alluvial clays and diluvial clays, are plotted against the plasticity index in Fig.15. Angle of shear resistance in terms of effective stress ϕ' was found 33° for alluvial clays and 30° for diluvial clays.

It has been pointed out that in the case of stiff diluvial clays at greater depths, the measured values of unconfined compression strength q_u show a considerable amount of scattering. This was the case of the shear tests at the airport island.

From the relationship between the measured unconfined compression strength and depth, a linear regression line is obtained. Working out the standard deviation from the mean value of q_u and the difference between the measured value and the value corresponding to the regression line, the coefficient of variation can be obtained. The coefficient of variation was found 0.267 for the alluvial clay and 0.369 for diluvial clay. If the test results of the specimens with appreciable amount of cracks or sand fraction are discarded, however, the coefficient of variation becomes 0.282 for the diluvial clay. It may be said that, though an actual amount of scattering in the q_u of diluvial clays appears enormous, the degree of scattering is not much different from that of soft alluvial clays.

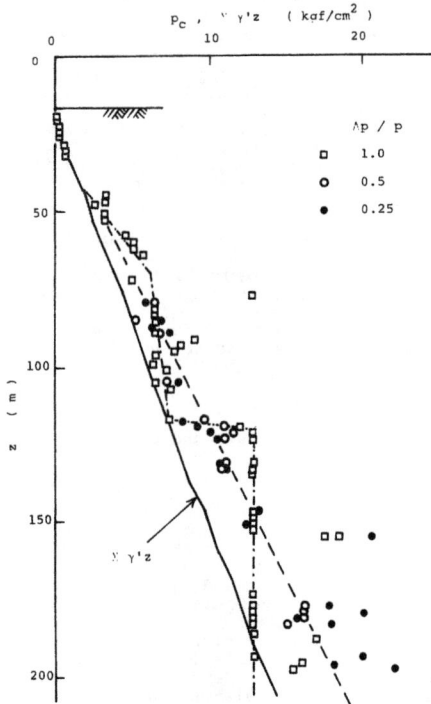

Fig. 14   Relationship between precompression stress and depth

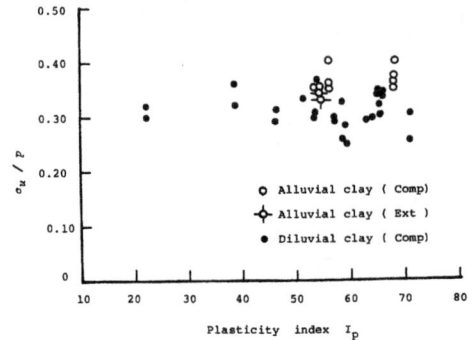

Fig. 15   Relationship between c_u/p and plasticity index

CONSTRUCTION OF AIRPORT ISLAND

Design Conditions

The airport island is located in the sea, so it is necessary to consider the forces acted by the surrounding sea, in addition to the force by earthquakes. As the result of study on the past record of sea conditions in the vicinity of the construction site, the storm surge of C.D.L.+3.2 m, which is the maximum in the past, and the wave of the return period of 50 years are considered in the designing. Considering this amount of storm surge, as well as pumping operation of inside water on the island, the surface elevation of the island is specified not lower than C.D.L.+4 m. As for earthquakes, the strong motion earthquake which is specified as the 5th degree on the Japan Meteorological Agency's scale is assumed. The Japanese seismic scale consists of seven degrees at all and the 5th degree corresponds to the earthquake with acceleration of 80 to 250 gals.

Construction Schedule

The airport island work is started with the construction of 11 km of seawalls surrounding the 511 ha of airport island. Since the surface alluvial clay A_C is not stiff enough to support a weight of the seawall, the clay beneath the seawall is to be improved. The whole construction work of the seawall, including the preceding ground improvement work, will take two years.

After completion of the seawall construction, the inside area of the seawall is to be reclaimed by the use of barges, which transport fill materials from nearby borrow pits.

This sequence of the seawall and reclamation work is decided from consideration to minimize an undesirable impact due to diffusion of muddy water upon dumping of fill materials. Total thickness of reclamation fill is estimated to be about 30 m. As described before, the surface alluvial clay of some 20 m thickness is not stiff enough and undesirable consolidation settlement is anticipated to continue for long time, therefore, the whole area inside the seawall is to be improved prior to the dumping of fill materials.

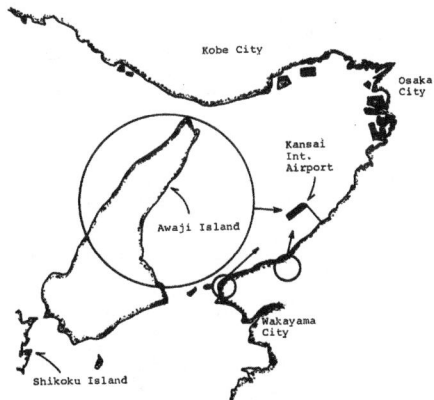

Fig. 16 Location of borrow pits

Table 3 Construction time table

Work Category	Year	1st	2nd	3rd	4th	5th	6th	7th
Seawall	Foundation improvement	━━	━					
Seawall	Seawall construction	━	━	━		━	━	
Reclamation	Foundation improvement	━	━	━				
Reclamation	Reclamation			━	━	━	━	━
Access bridge	Substructure	━	━	━	━			
Access bridge	Superstructure	━	━	━	━	━	━	━
Airport Facilities						━	━	━

Reclamation work is estimated to take five years including the ground improvement works, 166,000,000 m^3 of pit soils and 14,000,000 m^3 of beach sands are required for construction of the seawall and the reclamation. These materials are to be obtained from three borrow pits as shown in Fig.16.

Construction of airport facilities such as terminal building, hangars, railway station will start at the 4th year of the construction period. And these facilities will be completed at the same time of completion of the airport island itself.

Throughout the construction of seawalls and the inside reclamation, water quality and other environmental conditions are to be monitored in order to take an appropriate measure to minimize an undesirable impact due to these construction works. The construction time table for the Kansai International Airport is shown in Table 3.

Seawalls

Seawalls surrounding the airport island consist of four different types according to the wave actions and also to a possibility for the future expansion. General arrangement of the seawall and type of ground improvement are shown in Fig.17. At the shoreline side, three openings are to be provided for traffic of barges in the reclamation stage. After the reclamation work by barges is completed, these openings are to be closed as shown in Table 3. Cross sections of the seawalls are shown in Fig.18, together with the cross section of breakwater which is provided for maintaining a calm water area for access by sea.

Seawalls of types A and B are rubble mound embankments, however, armour blocks for wave dissipation are provided on outer face of the type A seawall, since this type of seawall is subjected to the most severe wave action. In the second phase work of this project, as shown in Fig.5, offshore side of the type A seawall is to be filled in the future expansion of the island.

The type C seawall is of steel plate cellular bulkhead, which is 23 m in diameter and 23 m in height. The type D seawall consists of concrete caisson, front face of which is provided with a series of slits in order to dissipate an energy of waves. Breakwater at the northern corner of the island is of vertical wall type consisted of continuous steel pipe piles. Sequence of events in construction of the seawalls is schematically illustrated in Fig.19.

As explained before, the surface layer of alluvial clay is soft. In order to obtain necessary bearing capacity of subsoil, therefore, the alluvial clay is to be improved prior to the construction of seawalls. Three kinds of ground improvement techniques are to be employed as shown in Fig.17. Geotechnical engineers in Japan, in particular those engaged in harbour works, have acquired an appreciable command of these three kinds of ground improvement techniques.

Some 430,000 vertical sand drains will be installed for accelerating the consolidation of the alluvial clay beneath the seawall of A and B types. The vertical drains are 40 cm in diameter and each drain is driven with a rectangular pattern of 2.5 m x 1.6 m.

Sand compaction piles of 2 m in diameter is to be pushed down to the alluvial clay beneath the seawalls of C and D types. Sand compaction pile method is a sort of forced replacement technique. In the case of subsoil of the seawall of types C and D, 70 % of soil mass is to be replaced by a volume of the sand compaction piles. When sand compaction piles are installed in cohesive soils with a close arrangement, surface of the soil is observed to heave up by as much as several meters. As a result of a series of full scale field tests, it has been decided to consolidate the heaved, remoulded clays as a part of the foundation soil, without removing it away. Some 36,000 sand compaction piles will be installed at all.

As shown in Fig.17, a chemical improvement method will be employed at two corners of the type A seawall, for increasing the alluvial clay's strength and rigidity. This technique is called the Deep Mixing Method, which has been developed by the Ministry of Transport for past fifteen years. In the Deep Mixing Method, cohesive soils are first remoulded by penetrating the rotating blades, then the soil is solidified by adding and mixing the chemical agent. A series of full scale tests have been carried out in the vicinity of the construction site. In the case of subsoil of the type A seawall, a cement milk is used as a solidifying agent.

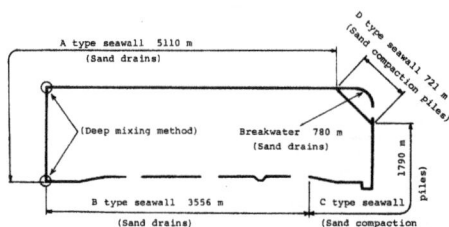

Fig. 17　General arrangement of seawall and type of ground improvement

Seawall (Type A)

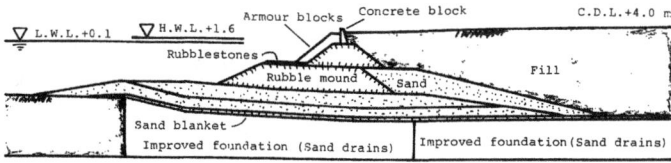

Armour blocks   Concrete block   C.D.L.+4.0 m
∇ L.W.L.+0.1   ∇ H.W.L.+1.6
        Rubblestones
                    Rubble mound   Sand   Fill
        Sand blanket
        Improved foundation (Sand drains) | Improved foundation (Sand drains)

Seawall (Type B)

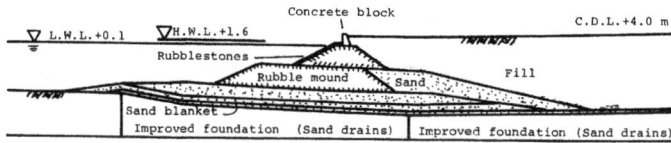

                              Concrete block   C.D.L.+4.0 m
    ∇ L.W.L.+0.1   ∇H.W.L.+1.6
         Rubblestones
                    Rubble mound   Sand   Fill
         Sand blanket
         Improved foundation (Sand drains) | Improved foundation (Sand drains)

Seawall (Type C)

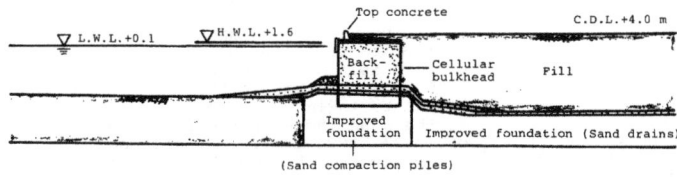

                        Top concrete   C.D.L.+4.0 m
    ∇ L.W.L.+0.1   ∇ H.W.L.+1.6
                              Back- ── Cellular
                              fill   bulkhead   Fill
                             Improved
                             foundation   Improved foundation (Sand drains)
                        (Sand compaction piles)

Seawall (Type D)

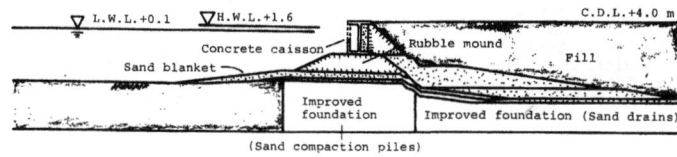

    ∇ L.W.L.+0.1   ∇H.W.L.+1.6   C.D.L.+4.0 m
                    Concrete caisson   Rubble mound
         Sand blanket   Fill
                             Improved
                             foundation   Improved foundation (Sand drains)
                        (Sand compaction piles)

Breakwater

                              Top concrete
    ∇ L.W.L.+0.1   ∇ H.W.L.+1.6
                 Rubble mound   ── Steel pipe piles

                    Improved foundation
                    (Sand drains)

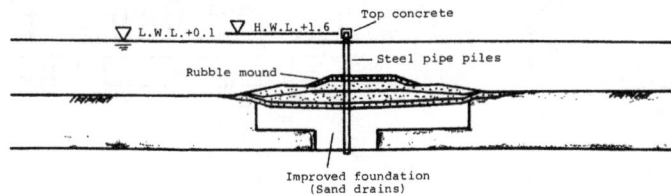

Fig. 18   Cross sections of seawalls and breakwater

31

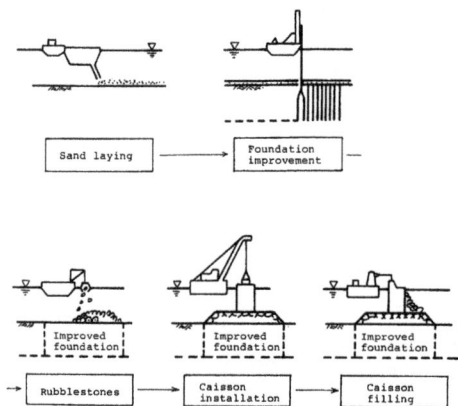

Fig. 19   Sequence of events in
           seawall construction

Reclamation Work

When the seawall is completed, the inside area
is to be reclaimed by the use of barges passing through the openings of the shore line
side seawall.   This type of dumping of soils
in the confined water area is taken because of
reducing possible environmental impact due to
diffusion of muddy water into open sea. Prior
to dumping soils, vertical sand drains are installed over the whole area inside of the seawall, in order to reduce the residual consolidation settlement in future.   The sand drain
of 40 cm in diameter is installed with a
square pattern of 2.5 m distance.   Total

number of the sand drain, including those installed beneath the seawall, is as much as one
million.

Tentative schedule of the reclamation work is
shown in Fig.20.   Offshore side of the airport island is to be completed earlier, because various kinds of airport facilities have
to be completed at the same time of the whole
work of the airport island, and also this
schedule depends on the location of the openings of the seawall.   When the inside water
depth decreases to an assigned value, the openings of seawall are closed as shown in Table 3.
Sequence of events in reclamation work is
schematically illustrated in Fig.21.

PREDICTION OF SETTLEMENTS

Accuracy in predicting future settlement is
the governing factor in geotechnical problems
involved in the Kansai International Airport
project.   Owing to its vast area, an error in
the settlement prediction will lead to a change
of huge amount of the fill material for maintaining the specified surface elevation of the
airport island.   Prediction of settlement consists of two items, the consolidation settlement with time and a probable unequal settlement, in particular, along the runway.

Extensive area of loading characterises the
prediction of consolidation settlement of the
airport island.   Size of the island is
1.25 km x 4.35 km, so the compression of soil
has to be taken into account for much greater
depths than those considered in ordinary size
of earth structures.   Prediction of a probable unequal settlement has been deemed difficult, since no reliable engineering procedures
have been established for evaluating the unequal settlements in the case such as this
airport island.

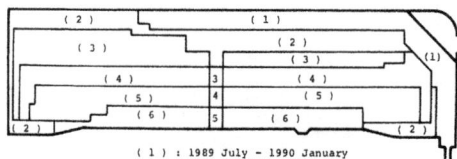

(1) : 1989 July – 1990 January
(2) : 1990 January – 1990 July
(3) : 1990 July – 1991 January
(4) : 1991 January – 1991 July
(5) : 1991 July – 1992 January
(6) : 1992 January – 1992 April

Fig. 20   Tentative schedule of reclamation

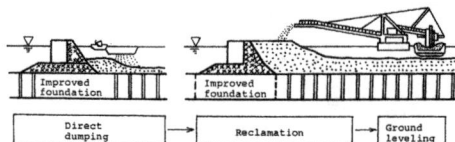

Fig. 21   Sequence of events
           in reclamation

Consolidation settlement

Settlement analysis has been made along three lines parallel to the runway. The A line is the nearshore side of the airport island, and the B and C lines pass through the center and the offshore side of the island respectively. In the past studies of the ground subsidence in the Kansai region, the diluvial soil layer with local name of Ma-6 has been considered practically incompressible (3). Elevation of the top of this stiff layer of the Ma-6 emerges at C.D.L.-139.5 m on the A line, -177.5 m on the B line and -206.5 m on the C line. Then the consolidation of cohesive soils was considered down to these particular depths.

In this particular case of settlement analysis the specification of drainage layers is a little complicated, because of an existence of a great many number of alternating layers of cohesive soils and sandy or gravely soils. As a rule, a soil layer with more than 70 % of sand fraction and more than one meter of thickness was specified the drainage layer. In some cases, however, this kind of judgement did not work satisfactorily, because of too many of ambiguous soil layers. In such a case, therefore, alternative sets of composition of soil layers were considered, where some number of ambiguous soil layers are assumed equivalent to single drainage layer.

The settlement analysis has been carried out mainly by the finite element method, and the analyses based on the compression curves and on the coefficient of volume compressibility are also added. In the analysis, the compression curves obtained by the consolidation tests were carefully examined, and those appeared unreliable to any extent were discarded beforehand.

Settlement analysis by the finite element method is that based on the Biot's consolidation theory (7). In this analysis the compression curve is idealized as shown in Fig.22. At first, the normal consolidation line and its slope λ are determined. The overconsolidation line is specified by a straight line passing through the state point on the compression

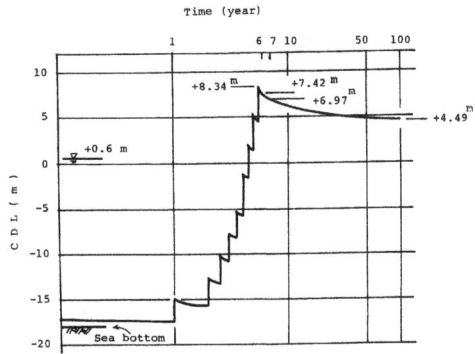

Fig. 23   Change in surface elevation of airport island with time

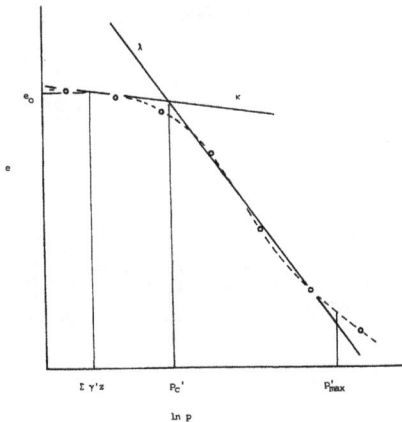

Fig. 22   Idealized compression curve

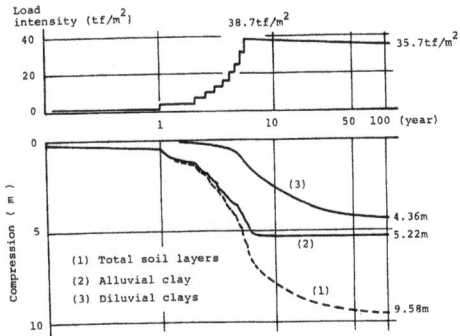

Fig. 24   Change in consolidation load and compression of soils with time

33

curve corresponding to the in situ effective overburden stress, and with the slope of $\kappa = \lambda/10$. Then the compression curve is to be represented by a set of two straight lines. Permeability is assumed to depend solely on the void ratio.

The surface alluvial clay layer A_C is improved by vertical sand drains, but deeper diluvial clays are consolidated without sand drains. For convenience of treating the whole problem as the one-dimensional consolidation type, consolidation of the surface A_C layer was converted to the one-dimensional type. In the analysis, therefore, a modified value of the coefficient of consolidation was considered so as to equalize its time of 50 % consolidation under one-dimensional consolidation to the time of 50 % consolidation by sand drains.

The reclamation work takes four years at all. In the analysis, therefore, the consolidation pressure was considered to be loaded in steps through the course of four years. In later steps of loading in which the top of the fill emerges above the sea level, the consolidation load intensity is to decrease as the airport island sinks down and the buoyancy increases. In the analysis, this kind of minor decrease in the consolidation load intensity with time was taken into account. Unit weight of fill material was assumed to be 2.0 tf/m^3, however, considering a possibility of mixing of large size rubbles, the cases of the unit weight of 2.1 tf/m^3 and 2.2 tf/m^3 were also considered.

As explained before, the surface elevation of the airport island is specified not lower than C.D.L.+4 m. A number of the settlement analyses, therefore, have been made under this requirement. Figs. 23 and 24 show a result of settlement analysis along the B line by the finite element method. Fig. 23 shows the change in the surface elevation of the airport island with time. Fig.24 illustrates the compression of soil layers with time, where the compression of the surface alluvial clay A_C and deeper diluvial layers are shown. It is noted that the subsoil consists of the top alluvial clay of 20 m thick and underlying alternating diluvial soil layers of total thickness of 140 m. The total settlement is 9.6 m,

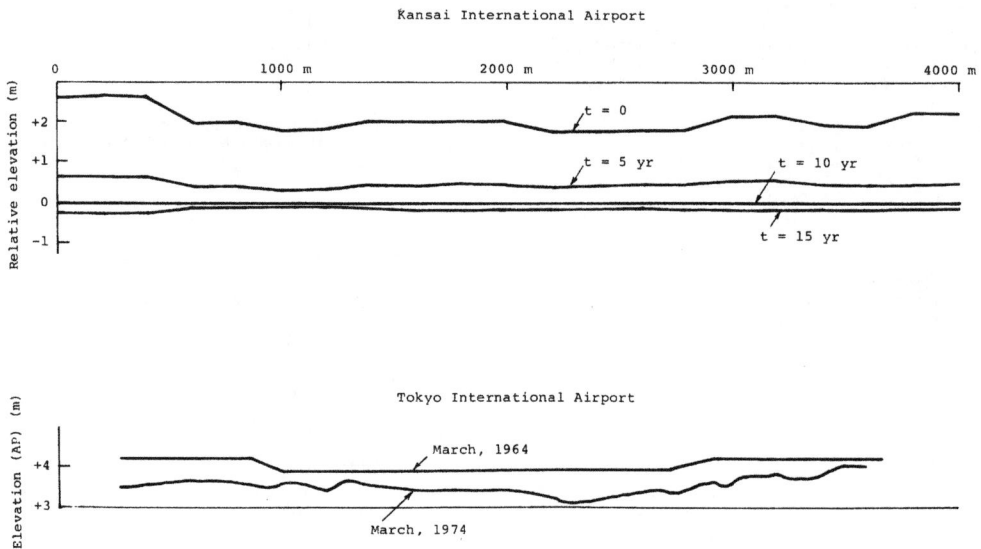

Kansai International Airport

Tokyo International Airport

Fig. 25   Predicted unequal settlement in the Kansai International Airport
and measured unequal settlement in the Tokyo International Airport

however, more than half of it is due to the compression of the thin soft clay layer A_c. Fig.23 indicates that if the elevation of airport island is made C.D.L.+8.3 m at the completion of reclamation, the final elevation will be C.D.L.+4.5 m.

Unequal Settlement

Unequal settlement will take place due to a number of factors, such as the spatial change in soil properties, change in the load intensity, change in the thickness of soil layer and change in the time elapsed after placement of the load from place to place. In the case of airports, in particular, an excessive unequal settlement will cause technical problems for take-off and landing of airplanes.

Engineering procedures for evaluating the unequal settlement do not seem to be established for such a case where the consolidation load intensity and soil condition are apparently uniform. A new computer program, therefore, has been compiled for coping with this kind of unequal settlement (8).

At first the airport island is divided into a number of square meshs of 200 m x 200 m. Then the reclamation schedule, i.e. loading steps, is specified for each mesh. Soil parameters required for the settlement analysis is allocated to each mesh by interpolating the results of soil test made on soil samples at nearby boring sites. In each mesh and in its various depths, type of the variation of coefficient of consolidation is assumed to be of the normal distribution, and the probability distribution for the variation of compression index, precompression stress and in situ void ratio. Simulation by the Monte Carlo method was carried out and some 50 iterations were made for variety of each parameter, and the difference in the consolidation settlement in adjacent meshes was examined. An example of this analysis is shown in Fig.25. This result shows that if the initial profile of runway is made so uneven as shown in the figure, the runway will be flat 10 years later. As a comparison, a change in the profile of runway of the Tokyo International Airport, Haneda Airport, with time is shown in Fig.25. The present profile of the runway in the Haneda Airport is well within the allowable range specified by the International Civil Aviation Organization.

CONCLUDING REMARKS

So far, a number of man-made islands of comparable size have been built in Japan. In view of great water depth and softness of subsoil, however, the construction of the Kansai International Airport may be said a challenging geotechnical work. Intense studies for construction of the airport island have been made for long time, and actual construction has just started. Results of geotechnical prediction for various problems are expected to be modified based on the results of close observations of actual behaviours of this earth structure.

REFERENCES

(1) The Kansai International Airport Company, (1986), "Introducing The Kansai International Airport"
(2) Horie,H., Zen,K., Ishii,I. and Matsumoto,K., (1984), Engineering properties of marine clays in Osaka Bay - (Part 1), Boring and sampling, Technical Note of Port and Harbour Research Institute, No. 498, 45 p,(in Japanese)
(3) The Third District Bureau for Port and Harbour Construction and the Kansai Branch of the Japanese Society of Soil Mechanics and Foundation Engineering, (1983), Report on geotechnical problems of the Kansai International Airport, 84 p, (in Japanese)
(4) Onodera, S.,(1984), Study on the engineering properties of soils in the Osaka Bay, Doctor Thesis, Tokyo Institute of Technology, (in Japanese)
(5) Japanese Society of Soil Mechanics and Foundation Engineering, (1979), Soil Testing Manual - Consolidation testing, pp. 372-422, (in Japanese)
(6) Nakase, A. and Kamei, T.,(1983), In situ void ratio, strength and overburden pressure anomalies in seabed clays, Proc. IUTAM Symposium on Seabed Mechanics, pp.9-15
(7) Kobayashi, M.,(1982), Numerical analysis of one-dimensional consolidation problems, Report of Port and Harbour Research Institute, Vol.21, No.1, pp.57-79 (in Japanese)
(8) Okumura, T. and Tsuchida, T.,(1981), Prediction of differential settlement with special reference to variability of soil parameters, Report of Port and Harbour Research Institute, Vol.29, No.3, pp.131-168, (in Japanese)

Keynote addresses

Soft Ground Engineering in Coastal Areas, Tsuchida et al. (eds)
© 2003 Swets & Zeitlinger, Lisse, ISBN 90 5809 613 0

The assessment of uncertainty in stability problems

N.R. Morgenstern
University of Alberta, Edmonton, Canada

ABSTRACT: The nature of uncertainty in geotechnical engineering is identified. For those issues not dominated by model uncertainty, probabilistic analysis provides a means of quantifying uncertainty. Recent developments in probabilistic slope stability analysis are summarized and case histories developed to illustrate its use in practice.

1 INTRODUCTION

Professor Nakase and I first met at Imperial College in 1962. At that time we both shared a keen interest in methods of slope stability analysis. In a sense we were building the tool-kit for our subsequent career, filled with the optimism that analytical methods in geotechnical engineering provided the most productive way forward. This may have been an appropriate perspective and at that time, and was certainly appropriate for us, in the early stages of our careers. Professor Nakase went on to make a number of important contributions to the stability of slopes and foundations in soft clays, and I still use and refer to them to-day. However, with the passage of the years, and with more varied experience, one becomes increasingly aware of the limitations of the analytical approach. This is not a novel recognition, as Terzaghi cautioned us in this regard even as he assembled the first synthesis of theoretical soil mechanics.

Experience teaches us that uncertainty is an intrinsic characteristic of many geotechnical designs, and perhaps nowhere more so than with stability problems, at least those dealing with natural materials.

The natural materials that the geotechnical engineer must deal with are complex and do not afford the luxury of replication. Geotechnical undertakings, either in-situ or associated with unit construction processes themselves, are performed under circumstances very different from the controlled environment of a manufacturing plant. The construction and testing of a prototype, prior to production, is a procedure rarely available to the geotechnical engineer. As a result, uncertainty is a perpetual component of geotechnical design and construction.

2 GEOTECHNICAL UNCERTAINTY

The literature on uncertainty and attempts to model it is extensive (e.g. National Research Council, 1994). For practical purposes Morgenstern (1995) adopted the following three sources of uncertainty:

i) Parameter uncertainty
ii) Model uncertainty
iii) Human uncertainty

Parameter uncertainty is readily understood and has received considerable attention in the geotechnical literature. It is concerned with input variables such as the spatial variations of parameters like strength or compressibility and the lack of data for key parameters. Many examples exist in the literature in which the statistical distribution, say, of strength is specified and the traditional factor of safety is replaced by a probability of failure. The model, an equation for factor of safety based on limit equilibrium assumptions, is taken as certain. However, if, for the real problem, the model itself is a major source of uncertainty, the seemingly sophisticated calculation is meaningless because the major source of uncertainty has not been addressed.

This is not to suggest that probabilistic analyses emphasizing parameter uncertainty are not useful. There are many instances where the opposite is true and their capacity to consider parameter uncertainty has become extremely powerful. Christian et al (1994) provided an example of the probabilistic design of dikes on soft clay illustrating how the components of uncertainty including data scatter, spatial variation, and systematic uncertainty of each soil parameter involved in the design can be considered. Developments in probabilistic analysis will be discussed in more detail.

Model uncertainty arises from gaps in the scientific theory that is required to make predictions on the basis of causal inference. Model uncertainty abounds in geotechnical practice. Vick (1994) has listed components of model uncertainty that affect the reliability of assessing failure of a particular dam. He emphasized that they encompass not just approximations in various methods of numerical analysis, but also uncertainties associated with conceptualization and interpretation of all of the various processes that could lead to the failure of a particular dam. Examples included seismic liquefaction triggering, post-earthquake behaviour, undrained versus effective strength characterization, and the progressive development of internal erosion.

While the objective of science is to provide explanations, that of engineering is to provide performance. Performance of engineering systems cannot be provided independent of human involvement and the functioning of social organizations. Human error can obviously overwhelm an otherwise effectively operating system and risk analysis that ignores or understates human involvement in geotechnical practice borders on naivety. Even corruption is not unknown.

3 CASE HISTORIES OF INSTABILITY ILLUSTRATING UNCERTAINTY

3.1 Failure of an embankment on soft fissured clays (Crooks et al., 1986)

This case history, while reflecting significant parameter uncertainty, is actually dominated by model uncertainty.

A highway relocation was required as part of a power project. The highway was to be carried over a mine haul road by means of a single span overpass structure with associated approach fills.

Following stripping of the surficial organics and topsoil, construction of the north fill began.

Fill consisted mainly of weathered shale placed and compacted in 150 mm lifts, generally achieving in excess of the specified 95% of Standard Proctor Density. The design fill height close to the overpass structure was about 12 m above the surrounding ground surface. Just as it was close to completion, with 10-12 m of fill in place, cracking developed on both sides of the embankment, with subsequent failure as shown in Figure 1. There was little evidence of deformation on the fill surface prior to failure.

The ground conditions, near Edmonton, Alberta, involve a highly plastic post-glacial lacustrine clay over glacial till. The lacustrine clay deposit can be subdivided into two units with the upper unit consisting of weathered soils, while the lower unit is unweathered. However, the lacustrine clay is fissured throughout its depth, with numerous slicken-sided surfaces. Original design was much influenced by a conservative interpretation of field vane tests. However, this proved to be excessively optimistic.

While laboratory UU tests of this material display considerable scatter, they reflect more clearly the influence of structure than the kinematically restrained vane test. Additional research into the clay showed that Ko varied within it as a result of having been deposited on an ablation till with subsequent melt-out. Moreover the stress-strain curve was strain weakening from a peak to a plateau mobilizing a frictional resistance of about 20°, before declining to a residual of 9°. Hence this clay was not amenable to simple normalization procedures for characterization and some element of progressive failure was involved in the embankment collapse. This clay has substantial variation of properties over a short distance, of the order of less than 100 m. The post-depositional melting of the ice in the underlying till had two effects on the clay, namely:

i) the formation of slickensides and fissures as a result of internal stressing due to deformation at the bottom of the clay layer,
ii) the spatial variation in the lateral stresses in the deposit which appear to be related to modest changes in topography.

Details are given in Chan and Morgenstern (1987).

The limited resistance offered by soft fissured clays had been identified before this failure (Rivard and Lu, 1978) but had evidently not been given enough weighting in the original design. Experience indicates that conventional undrained strength data is too variable to provide a reliable basis for design and that peak effective strength cannot be relied upon in effective stress analyses as a result of progressive failure. Hence design for this class of material remains essentially empirical, employing strengths based on back-analysis of past failures.

3.2 Kwun Lung Lau landslide (Morgenstern, 1994; Wong and Ho, 1997)

This case history contains elements of both parameter and model uncertainty. However it is dominated by human uncertainty.

At about 8:53 pm, on July 23, 1994 a landslide occurred below Block D at Kwun Lung Lau, Kennedy Town, Hong Kong. The landslide resulted in five fatalities and more injuries.

This landslide provoked considerable public concern in Hong Kong and resulted in technical detailed inquiries by the Geotechnical Engineering Office (GEO), Hong Kong Government (1994) and Morgenstern (1994). The case history is cited here because Hong Kong probably has the most advanced landslide reduction program of any major city in the world.

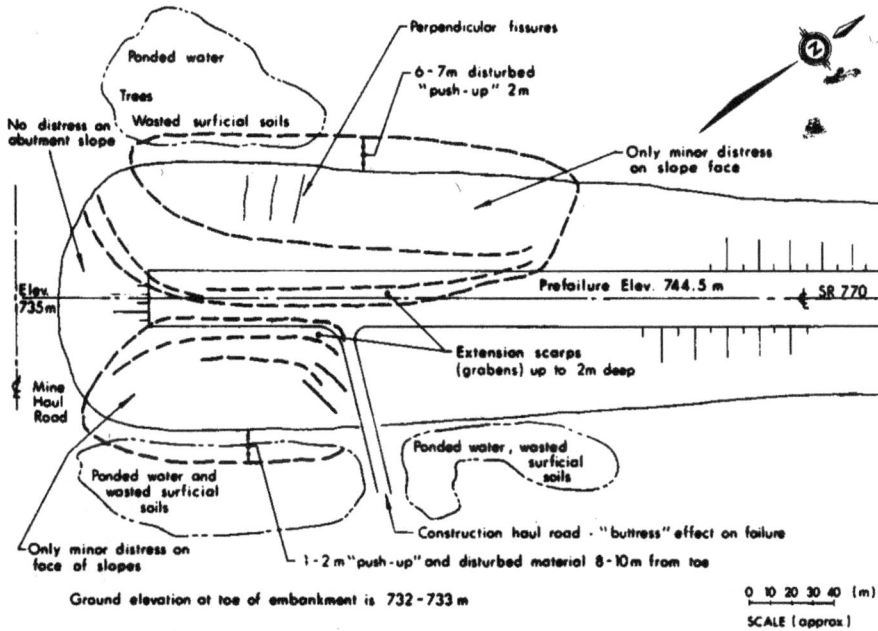

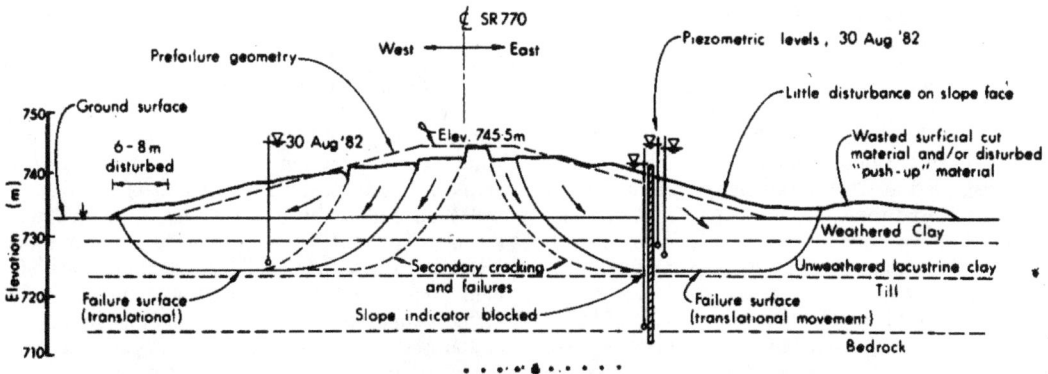

Figure 1. Failure of Embankment on Soft Fissured Clay (after Crooks et al., 1986).

Through the efforts of GEO, Hong Kong has developed a comprehensive catalogue of slopes, a risk-based ranking system and a phased approach to upgrade slopes, both in the public and private sectors. In addition, a landslide warning system has been developed based on correlations between landslide occurrence and rainfall. This warning system utilizes an extensive automated rain gauge system, operating in real time, and it has been demonstrated that few incidents precede that warning.

Together with these measures, the GEO has been instrumental in improving local geotechnical practice, highlighting responsibility for routine maintenance of slopes as well as instigating a number of other initiatives directed toward landslide prevention and risk reduction. There is little doubt that the work of the GEO has been very effective in reducing overall landslide risk in Hong Kong. Therefore it was particularly unsettling to discover that the Kwun Lung Lau landslide involved a slope and retaining wall that had been catalogued, a configuration that had been subjected to a preliminary study and assessed as adequate, a site that had been inspected periodically by qualified consultants, even shortly before the unfortunate occurrence, and had occurred when the landslide warning was in effect. These discoveries raised questions about the effectiveness of the whole risk management system and therefore extensive technical and public policy inquiries ensued.

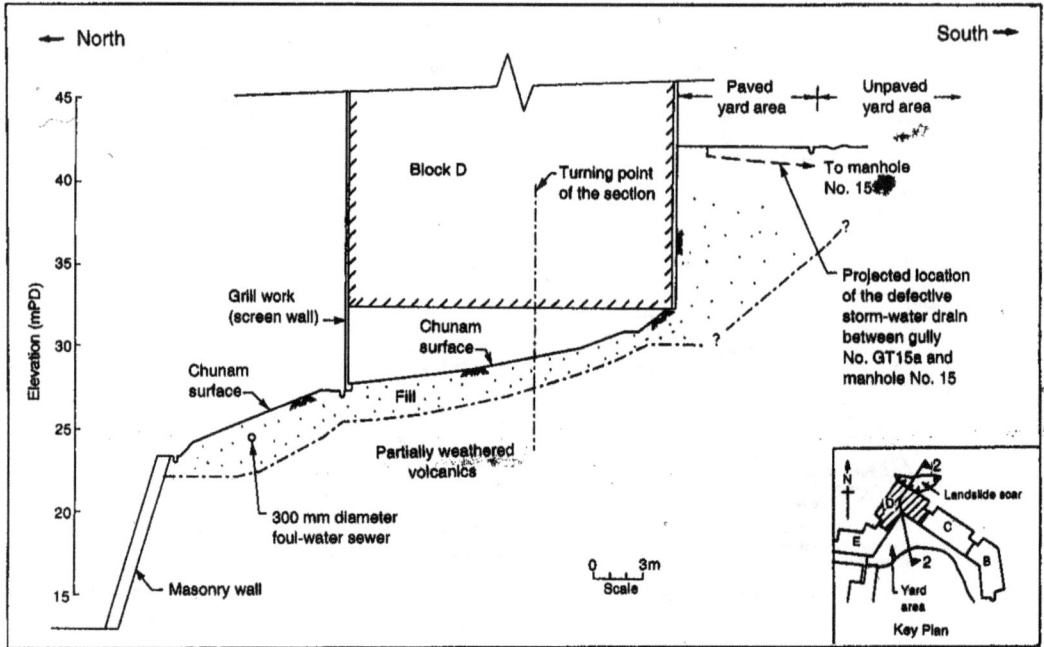

Figure 2. Kwun Lung Lau Landslide (After Wong and Ho, 1997).

A cross-section through the slide is shown in Figure 2. The slope was supported by a thin masonry wall that was in place by 1901. The soils consisted of loose fill and partially weathered volcanics and the slope was covered with chunam. The slope had been inspected at the time of heavy rainfall, about six hours before the slide, and no defects were detected. Rain was light at the time of the collapse although it had been extremely heavy some hours before. Failure was sudden, taking place over a very short period of time. The masonry wall burst out at about mid-height, followed by the instant collapse of the wall and slope. Had the failure mode been more ductile, it is likely that the disaster could have been avoided.

Detailed studies revealed that the landslide occurred as a result of sub-surface infiltration from defective, buried drainage systems. Flow from leaking storm water drains and a failed, foul-water sewer saturated the soil mass. Initially the wall, which was thin, provided support, but it finally buckled and collapsed in a brittle manner. The volume and mobility of the slide mass was actually increased by the presence of the wall. Had it been thicker, it likely would have deformed in a more ductile manner, which could have provided some warning of the impending danger. Further inquiry revealed that the drawings relied upon by GEO in their preliminary assessment of stability did not portray the wall correctly. The wall was shown to have a base width of 4 m on drawings approved in 1965 instead of the actual width of about 750 mm. Had the preliminary analysis been based on the actual width of the wall, the wall would likely have been found to be unsafe and the possibility of future instability may have been foreseen. While considerable attention was being paid to potential surface infiltration into slopes in Hong Kong, only limited attention was being given to sources of subsurface infiltration, particularly if a potential source was some distance away from the specific location of the slope under assessment. This has since been rectified.

The example of Kwun Lung Lau entailed parameter uncertainty, since geological and material characterization entered into the evaluation of stability. It also contained elements of model uncertainty since the study utilized limit equilibrium analyses, finite element analyses, distinct element analyses and analyses of flow through both saturated and unsaturated soils. However, these sources of uncertainty were overwhelmed by human uncertainty. In this case the filing and approval of inaccurate documents and the limited appreciation of potential sources of subsurface infiltration were dominant factors.

4 THE LIMITATIONS OF PREDICTION

4.1 *Prediction and performance*

Assuring performance through prediction is attractive. It presupposes sufficient knowledge and precisely enough models to allow quantitative forecasts of behaviour. Relying on prediction does not preclude intervention by application of the observational method to re-direct the performance assurance process.

Lambe (1973) has provided the strongest defense of prediction in geotechnical engineering. His classification of predictions; Class A - Before event, Class B - During event and Class C - After event, has entered the lexicon of geotechnical engineering. Lambe was an advocate of Type A predictions. He regarded them as more useful than Type B predictions even though he was fully cognizant of the limitations of the data available to the engineer and of the reality that mechanisms involved are rarely fully and correctly identified. Lambe was of the view that it would be desirable if Class A predictions permitted all the judgment decisions to be made at one stage and be clearly identified and discussed.

No one can doubt that quantitative forecasts of events to come are important in geotechnical engineering. For many assignments, such as those involving the assessment of settlement, they are the essence of the geotechnical undertaking. However, it is debatable whether an emphasis on Class A prediction in geotechnical practice is appropriate when compared with reliance on Class B prediction and other components of geotechnical risk management. At least it is of interest to evaluate the accuracy with which Class A predictions can be made before emphasizing their value.

Over the past twenty-five years or so a number of prediction competitions have been convened in geotechnical engineering. A synthesis of the results from these events provides some basis for assessing the accuracy of prediction in geotechnical engineering. However, it should be noted that even these competitions exaggerate the reliability of quantitative prediction in geotechnical practice. They usually present more comprehensive data than is normal,

sites tend to be homogeneous, there is a consensus that model uncertainty is not overwhelming and human uncertainty is essentially eliminated.

In order to assess the quality of the predictions the descriptors presented in Table 1 are proposed.

Table 1. Prediction quality classes

Accuracy of prediction (% actual)	Quality class
95 - 105% (within ± 5%)	Excellent
85 - 95% or 105 - 115% (within ± 15%)	Good
75 - 85% or 115 - 125% (within ± 25%)	Fair
50 - 75% or 125 - 150% (within ± 50%)	Poor
< 50° or > 150%	Bad

The first competition reviewed here is the MIT Prediction Symposium (MIT, 1974) involving the loading to failure of a large embankment on Boston Blue Clay. On the basis of comprehensive data provided before the embankment was loaded, ten teams or individuals were asked to predict the failure height of the embankment, as well as a number of pore pressures and displacements. Only failure height will be discussed in the following.

The ten predictions from the MIT competition discounting ranges of prediction are summarized in Figure 3 according to Quality Class. The range was 43 - 144% of the correct answer. Poor to Bad predictions embraced 70% of the attempts.

A more comprehensive prediction competition for an embankment on soft clay was undertaken in Malaysia (MHA, 1989) and it attracted 31 participants. A substantial amount of shear strength data was collected as well as index testing to characterize the site variability.

Details are provided in MAH (1989) and the results of the predictions are compiled by Poulos, Lee and Small (1990) and Brand (1995). Only predictions of the collapse height of the embankment are considered here.

Figure 4 summarizes the predictions according to Quality Class. The range was 52-170% of the correct

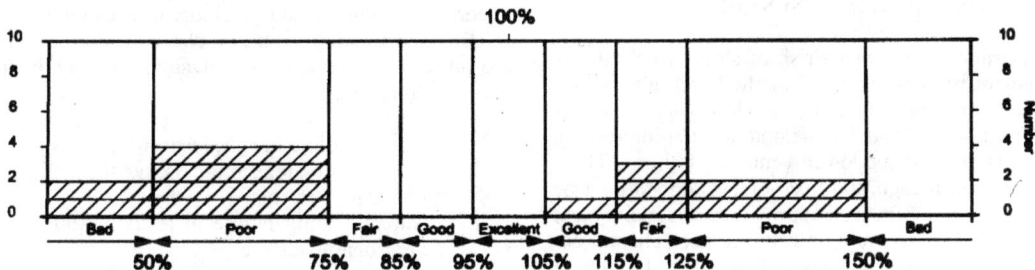

Figure 3. Prediction Quality Classification, MIT Embankment Prediction Competition (10 Predictors).

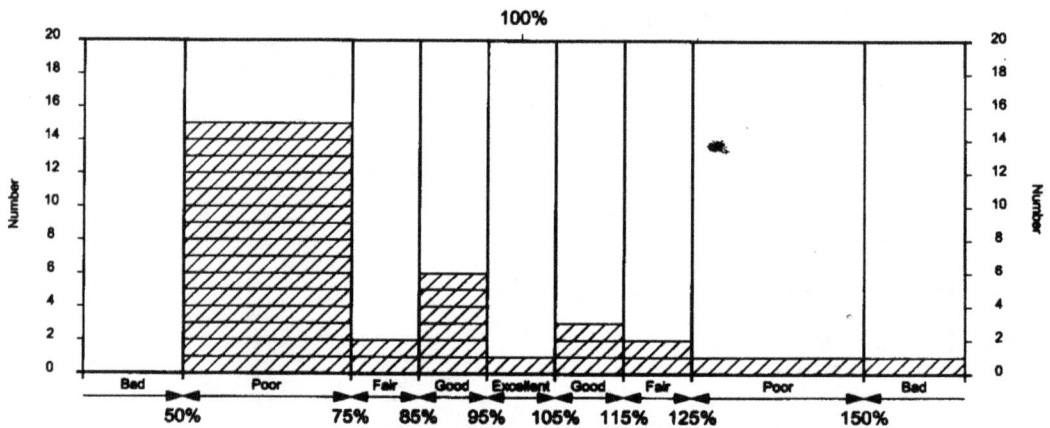

Figure 4. Prediction Quality Classification, Muar Embankment Prediction Competition (31 Predictors).

answer. Poor to Bad predictions accounted for 55% of the attempts.

Additional examples from competitions to predict foundation performance are given by Morgenstern (2000). They all illustrate that even under the near ideal conditions of a prediction competition, the accuracy of geotechnical prediction is poor.

In the face of the intrinsic uncertainty associated with geotechnical engineering, it is wise to remember Southwood's caution (1985):

"The things that we would like to know may be unknowable".

It is rare for the geotechnical engineers to rely only on quantitative prediction to meet his objectives. The practice of the geotechnical engineers is more modest. Risk must be managed to overcome the limitations of site characterization, knowledge of material properties, other unknowns and the vagaries of construction practice. In order to manage this risk effectively, it is essential that the geotechnical engineer maintain an on-going awareness of factors that contribute to unsuccessful performance and introduce this awareness into comprehensive risk management tools. This is discussed in more detail in Clayton (2001) and Morgenstern (2000).

5 LANDSLIDE RISK ASSESSMENT

It is common to manage risk of slope instability by means of the observational method and other forms of consequential risk analysis. However, success is far from perfect and the method is often impractical when applied to sudden fast-moving failures. There is increased recognition that some failures might be aided by a more focused approach to uncertainty and new questions are being put to the geotechnical engineer that cannot be answered in terms of factor of safety alone. This has led to an intensified interest in the potential of risk assessment procedures applied to the landslide/slope stability problem. There is much to be advanced at the interface between theory and practice in order to encourage change among practitioners. The status of landslide risk assessment, with emphasis on quantitative methods, is summarized in Cruden and Fell (1997) and Ho et al., (2000).

Where landslide statistics and characterization are abundant, the risk assessment procedures can be driven by data, and hence avoid excessive subjectivity. This is true for both site specific cases and societal risk assessments. Hong Kong is most advanced in applying this methodology in practice. A number of case histories have been published and an awareness of both the strengths and weaknesses has emerged (Ho and Wong, 2001).

Where empirical data is inadequate and a site specific risk assessment is being undertaken, one can proceed with subjective probability assessments, but the pitfalls associated with this direction are readily apparent. Probabilistic stability analyses offers an alternate way forward. Morgenstern (1997) cautioned against excessive optimism in this regard and urged that probabilistic analyses first be advanced for the classes of problems that are not dominated by model uncertainty. This would preclude, for example, soft or stiff fissured clays or brittle clays. The following are preferred candidates for advancing probabilistic methods in practice:

i) rotational failure in soft, insensitive clays
ii) sliding failure along pre-existing shear planes at residual strengths
iii) rotational/sliding failure in residual soils that are reasonable ductile.

6 PROBABILISTIC STABILITY ANALYSES

6.1 *Advances in probabilistic slope stability analyses*

Probabilistic slope stability analysis (PSSA) was first introduced into slope engineering in the 70's. In Japan, the early contribution of Matsuo and Kuroda (1974) is noteworthy. Tang, Yucemen and Ang (1976) and Vanmarcke (1977) provide additional examples of early contributions to PSSA.

The merits of probabilistic analyses have long been noted (Chowdhury, 1984; Whitman, 1984; Wolff, 1996; Christian, 1996). In spite of the uncertainties involved in slope problems and notwithstanding the benefits gained from a PSSA, the profession has been slow in adopting such techniques. The reluctance of practicing engineers to apply probabilistic methods is attributed to four factors. First, engineers often lack formal training in statistics and probability theory. Hence, they are less comfortable dealing with probabilities than they are with deterministic factors of safety. Second, there is a common misconception that probabilistic analyses require significantly more data, time and effort than deterministic analyses. Third, few published studies illustrate the implementation and benefits of probabilistic analyses. Finally, acceptable probabilities of unsatisfactory performance (a failure probability) are ill-defined and the link between a probabilistic and a conventional deterministic assessment is absent. For example, what is the probability of failure of a safe slope? All of these issues have been addressed in detail in El-Ramly (2001).

Published methods of PSSA vary in assumptions, limitations, capability to handle complex problems and mathematical complexity. Their limitations have been discussed by El-Ramly, Morgenstern and Cruden (2002) who advocate Monte Carlo simulation methods which have been made economic by recent developments in software. Recently developed PSSA based on Monte Carlo simulation has the advantage of being simple and not requiring a comprehensive statistical and mathematical background. The methodology is spreadsheet-based and makes use of the familiar and readily available Microsoft Excel and @Risk (Palisade, 1996) software.

The slope problem (geometry, stratigraphy, soil properties and slip surface) and the selected method of analysis (circular, various types of non-circular) are first modeled in an Excel spreadsheet. Available data are examined and uncertainties in input parameters are identified and described statistically by representative probability distributions. Only those parameters whose uncertainties are deemed significant to the analysis need to be treated as variables. A Monte Carlo simulation is then performed, selecting at random a value for each input variable within the defined probability distributions. The process is repeated sufficient times to estimate the statistical distribution of the factor of safety. Statistical analysis of this distribution allows estimating the mean and variance of the factor of safety and the probability of it being less than one. The procedure is applicable to any method of limit equilibrium analysis that can be represented in a spreadsheet.

Different approaches can be adopted to estimate the probability distribution of each input variable. Where there are adequate amounts of data, the cumulative distribution function (CDF) of the measurements can be used directly in the simulation process. Where observations are scarce or absent, parametric distributions can be assumed from the literature, or can be based on judgment alone. @Risk built-in functions allow great flexibility in modeling input variables.

It is not sufficient to characterize parameter uncertainty without considering the spatial variability of the input variables. This is not recognized in most methods of PSSA and, as emphasized by El-Ramly, Morgenstern and Cruden (2002a), this is a significant omission. Variance reduction due to spatial averaging of soil parameters over the length of the slip surface has to be considered. The reduction depends on the auto-correlation distance as well as the length of the slip surface which is not known beforehand. The new method takes the auto-correlation distance into consideration. Methods that ignore it are called simplified analyses (e.g. Nguyen and Chowdhury, 1984; Duncan, 2000).

The results of PSSA can be expressed in terms of P_u, the Probability of Unsatisfactory Performance (or failure), and the reliability index, β. @Risk also provides Spearman rank correlation coefficients between the factor of safety and the input variables. They can be used as a measure of the relative contributions of each input variable to the uncertainty in the factor of safety. This contribution comprises two elements: the degree of uncertainty of the input parameter and the sensitivity of the factor of safety to changes in that parameter. This is of considerable practical value.

6.2 *PSSA of & James Bay Dykes*

Application of the new procedures to the well-documented case of the James Bay dykes provides an opportunity to compare results with other probabilistic analyses. While the dykes were not built, the design was the subject of extensive studies quantifying the sources of uncertainty (Ladd et al., 1983; Soulie et al., 1990) and an analysis using the first-order second-movement (FOSM) procedure has been presented (Christian et al., 1994).

Figure 5 shows the geometry of the embankment and the underlying stratigraphy. Analysis was undrained circular slip with uncertainty considered for unit weights, the friction angle of the embank-

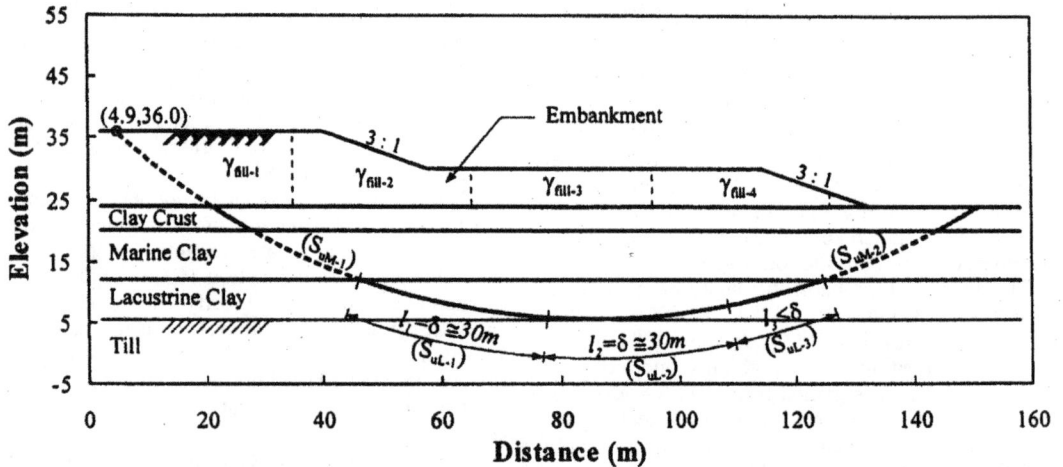

Figure 5. Cross-section and Stratigraphy of the James Bay dykes; showing the approach adopted to account for spatial variability of soil properties.

ment material, the thickness of the clay crust, the undrained strengths, the Bjerrum vane correction factor for the marine and lacustrine clays and the depth to the till layers. El-Ramly, Morgenstern and Cruden (2002a) summarize the input parameters as well as the autocorrelation length input.

The stability calculations are based on the Bishop method of slices. Figure 6 shows the computed histogram and probability distribution of the factor of safety. The mean factor of safety for this design is 1.46 with a standard deviation of 0.20. The probability of unsatisfactory performance is 4.7×10^{-3}. The reliability index is calculated to be 2.32.

A sensitivity analysis shows Spearman rank correlations for all input variables (Figure 7). It is interesting that many of the factors with major contributions to the uncertainty of the factor of safety are not related to soil property measurements. For example, Bjerrum's correction factor for the undrained shear strength of the lacustrine clay, the statistical uncertainty in the depth of the till, and the statistical uncertainty in the unit weight of the fill (which was evaluated judgmentally) are among the main factors affecting the reliability of the design. The results highlight the significance of the additional uncertainty that could be introduced by the designer through the use of empirical factors and subjective estimates of uncertainty. The results obtained from a FOSM analysis are similar to those quoted above. In order to estimate the P_u, a form of the probability density function of the factor of safety must be assumed. For a normal and a log normal probability distribution, P_u is estimated to be 8.4×10^{-3} and 2.5×10^{-3} respectively. The results are summarized in Table 2. In this case, the FOSM method is reasonable for estimating the means and variance of the factor of safety. How-

ever, the uncertainty about the shape of the probability density function of the factor of safety introduces uncertainties in estimating the probability of unsatisfactory performance.

Table 2 also includes the results of a simplified analysis which ignores the spatial variability of soil properties but assumes that all variables are normally distributed. The P_u is 2.37×10^{-2} and the reliability index is 1.84. This method overestimates the P_u by a factor of 5. This is attributed to a significant reduction in the uncertainty due to soil variability as a result of spatial averaging which is ignored in the simplified analysis.

6.3 *PSSA of Lodalen Slide*

Case histories are needed to benchmark the application of PSSA to practice. The Lodalen slide is a particularly valuable case because it is relatively simple and well-documented. It served, in the past, as an important case in benchmarking the application to engineering practice of the Bishop method of stability analysis in terms of effective stress. Details of the PSSA have been published by El-Ramly, Morgenstern and Cruden (2002b).

The slide occurred in 1954 in the area of the Lodalen marshalling yard near Oslo railway station, Norway. Failure occurred in a clay slope excavated 30 years earlier to expand the marshalling yard. Over this period the slope was steepened a few times. At the time of the failure, the cut was about 17 m high with a slope of about 26° (2H:1V). Failure was mainly rotational. Sevaldson (1956) provided a detailed description of the slide and the investigation into its causes. Modern effective stress analyses with the field observed pore pressures sensibly explain the event.

0.20

0.15

0.10

0.05

0.00

Frequency

P_u

1.0

0.86 1.10 1.34 1.59 1.83 2.08 2.32

Factor of Safety

Seed = 31069
E[FS] = 1.46
σ[FS] = 0.20
m = 32,000
$P_u = 4.7 \times 10^{-3}$

1.0

0.8

0.6

0.4

0.2

0.0

Cumulative Probability

0.6 0.9 1.2 1.5 1.8 2.1 2.4

Factor of Safety

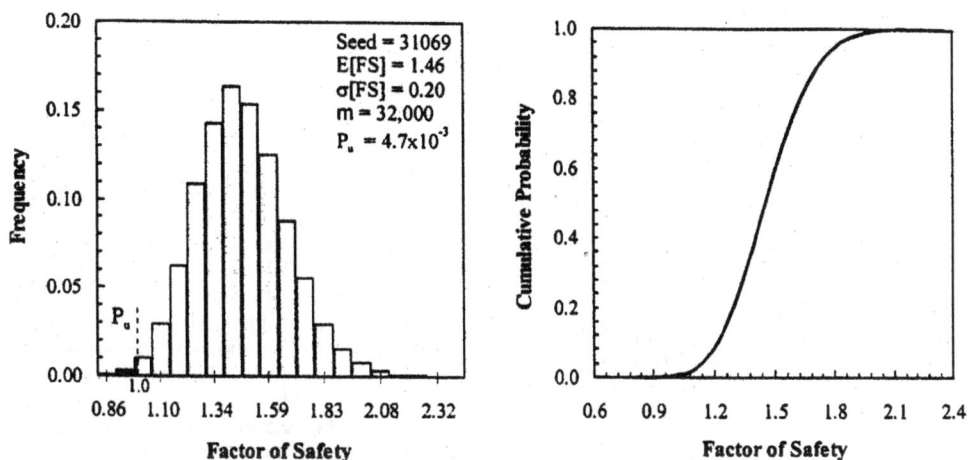

Figure 6. Histogram and probability distribution of the factor of safety.

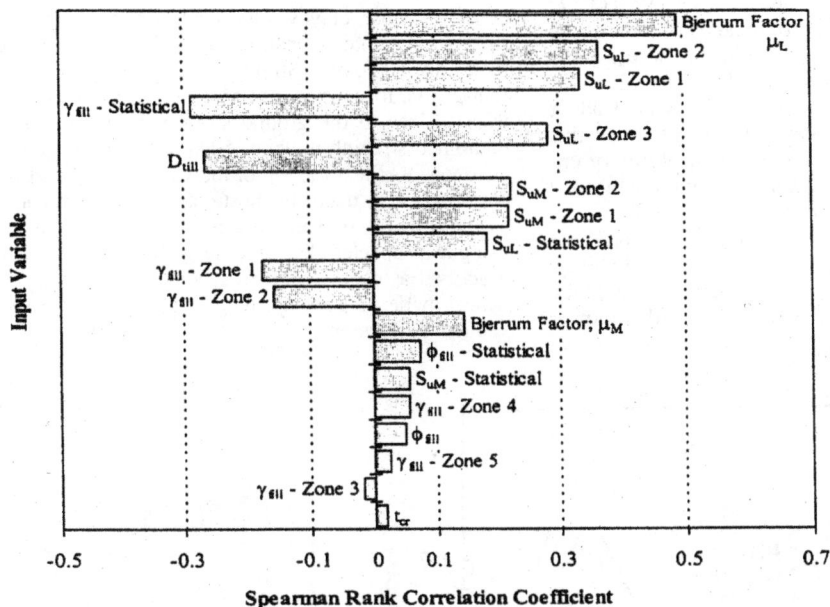

Input Variable

Bjerrum Factor μ_L
S_{uL} - Zone 2
S_{uL} - Zone 1
γ_{fill} - Statistical
S_{uL} - Zone 3
D_{till}
S_{uM} - Zone 2
S_{uM} - Zone 1
S_{uL} - Statistical
γ_{fill} - Zone 1
γ_{fill} - Zone 2
Bjerrum Factor; μ_M
ϕ_{fill} - Statistical
S_{uM} - Statistical
γ_{fill} - Zone 4
ϕ_{fill}
γ_{fill} - Zone 5
γ_{fill} - Zone 3
t_{cr}

-0.5   -0.3   -0.1   0   0.1   0.3   0.5   0.7

Spearman Rank Correlation Coefficient

Figure 7. Sensitivity analysis results; Spearman rank correlation coefficients for all input variables.

Table 2. Comparing the outputs of different analysis approaches

Method of analysis	E[FS]	σ[FS]	Skewness	P_u	β
Spread- sheet based probabilistic slope analysis	1.46	0.20	0.30	4.7×10^{-3}	2.32
FOSM	1.46	0.19	Not available	8.4×10^{-3a} 2.5×10^{-3b}	2.42
Simplified analysis	1.46	0.25	0.32	2.37×10^{-2}	1.84

[a] assuming the probability density function of the factor of safety is normal
[b] assuming the probability density function of the factor of safety is lognormal

The stratigraphy at the slide location comprised a clay crust approximately 1 m thick, overlying firm, homogeneous marine clay, of low to medium sensitivity. The clay had a moisture content of about 30%, a liquid limit of about 35% and a plastic limit of about 20%.

Input parameter considered the variability of the effective cohesion of the marine clay, of the effective friction angle of this clay and of the pore pressure, all based on field and laboratory observation. The auto correlation was also specified based on available data and varied over a likely range.

The calculated P_u was 0.77. Such a high probability of failure implies that failure was to be expected as the average factor of safety was 0.95. In order to bridge the gap between deterministic design and P_u, the slope was flattened in increments to (4H:1V) and a PSSA repeated from each case. Figure 8 shows the change of the P_u and the factor of safety with the slope angle. The plot indicates minimal increase in P_u as the slope increases from 14° (4H:1V) TO 17° (3.3H:1V). A slope of approximately 18° marks the beginning of a significant increase in the P_u. This corresponds to a factor of safety of about 1.12.

Conventional slope design practice would adopt an allowable factor of safety of 1.3 – 1.5. In this case, this would imply a near zero probability of unsatisfactory performance. Such a low probability is attributed to the small uncertainty in the input parameters. It should not be interpreted as a no failure condition, but rather as a high level of reliability. Armed with this assessment, a designer might be tempted to accept a factor of safety toward the lower limit of the allowable range.

6.4 *The Muar trial embankment*

As part of the effort to optimize the design of an express highway on a very soft marine clay, the Malaysia Highway Authority decided to construct a large-scale field trial embankment. They invited a large number of practitioners to predict the behaviour with some results of this competition summarized in Figure 4. Given the considerable data available for this case, if provided a good opportunity to calibrate PSSA. Details are available in El-Ramly (2001).

The stratigraphy of the site comprises a surface crest of weathered clay, about 2.0 m thick, overlying a very soft silty clay. The silty clay is about 6 m deep and is highly compressible. The strength of the clay increases almost linearly with depth. Below this, a layer of soft silty clay, 9.5 m deep, is encountered. Its shear strength is slightly higher than the overlying very soft clay and also increases with depth. A highly compressible peat, 0.7 m thick, underlies the soft clay. The peat, in turn, is underlain by a medium dense to dense clayey silty sand, which extends to substantial depth. Construction of the embankment took only 100 days. Strength characterization of the foundation, for purposes of analysis, relied upon vane tests. Close to the ground surface in the crust, the undrained strength is about 50 kPa and it reduced to as low as 8 kPa at the top of the very soft layer. Linear trends with depth fit most data, with some scatter around this trend. The embankment was constructed using a compacted decomposed granite that has been described as clayey sand to sandy clay. Triaxial test data provide undrained strength as well as effective strength parameters.

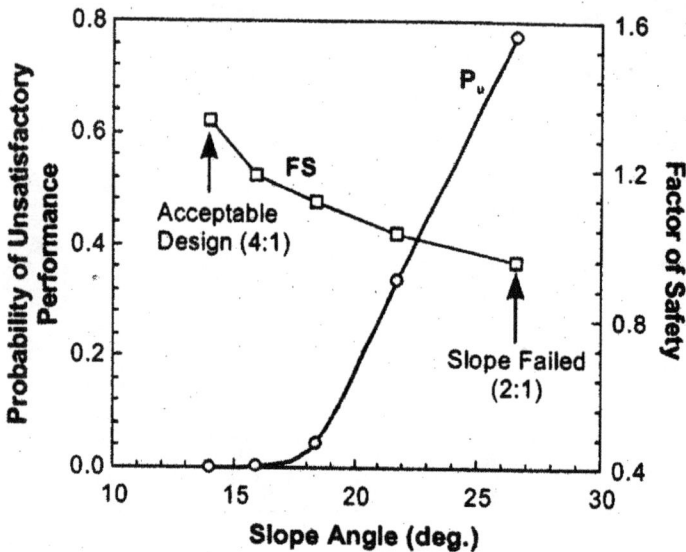

Figure 8. Variation of the probability of unsatisfactory performance, P_u and the factor of safety, FS, with the slope angle; Lodalen slide.

It is recognized that fill cracking may occur. However whether or not cracks occur is not embedded in stability analysis and is best handled as a matter of model uncertainty.

Stability analysis assumed circular failure. Statistical parameters are adopted for all layers, consistent with available data. The uncertainty of the Bjerrum correction factor is also considered. Autocorrelation distances are also specified.

Failure occurred at a height of 4.7 m. From the PSSA, the mean factor of safety was calculated to be 1.11 with a standard deviation of 0.15. The P_u was estimated to be 24%.

PSSA were repeated for lower heights and hence safer designs. Based on experience, conventional slope practice targets a design factor of safety in the order of 1.3-1.4 for the short term stability of embankments on soft soils. If the computed factor of safety at failure is 1.11, it is judged that an embankment height of 3.3 m, with a computed factor of safety of 1.42 would be an acceptable design. Figure 9 shows the variation of P_u with factor of safety and embankment height. The safe design has a P_u of 1.38%; which is significantly higher than the Lodalen case. The Spearman sensitivity analyses, Figure 10, indicates that the greatest uncertainty arises from the Bjerrum correction coefficient and the fill properties. This case has many characteristics in common with the James Bay dykes.

6.5 *Commentary*

The examples of PSSA presented here have been selected because they are soft clay problems, the major theme of this symposium. Other cases have been developed for cuts in medium clay and residual soils as well as an embankment on a clay shale foundation at residual strength. It is premature to propose probabilistic design criteria. However, it has been found that all failed slopes analyzed to date have reliability indices less than 0.8, where as the safe slopes have a minimum reliability index of 1.8 with most of the cases having indices higher than 2.0. A minimum reliability index of 2 corresponds to a probability of unsatisfactory performance on the order of 2×10^{-2}.

Conventional slope design practice addresses uncertainty only implicitly and in a subjective manner. Without proper consideration of uncertainty, the factor of safety alone can give a misleading sense of safety and is not a sufficient safety indicator. PSSA is a rational means to incorporate quantified uncertainties into the design process. PSSA can be applied in practice without any extensive effort over that needed in a conventional analysis. The stated obstacles impeding the adoption of such techniques into geotechnical practice are more apparent than real.

PSSA provides practicing engineers with valuable insights that cannot be reached otherwise. The level of uncertainty in the factor of safety is quantified through the variance of the factor of safety and the probability of unsatisfactory performance. This information could influence the decision of the design factor of safety. In addition, the practical value of quantifying the relative contributions of the various sorts of uncertainty to the overall uncertainty of the factor of safety through sensitivity analyses, using Spearman rank correlation coefficients, cannot be underestimated as a means for resource allocation.

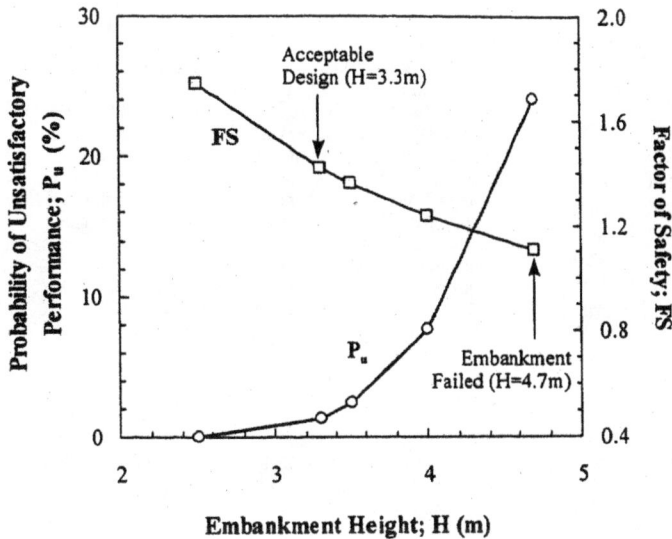

Figure 9. Variation of the probability of unsatisfactory performance and the factor of safety with embankment height, Muar Embankment – Proposed Methodology.

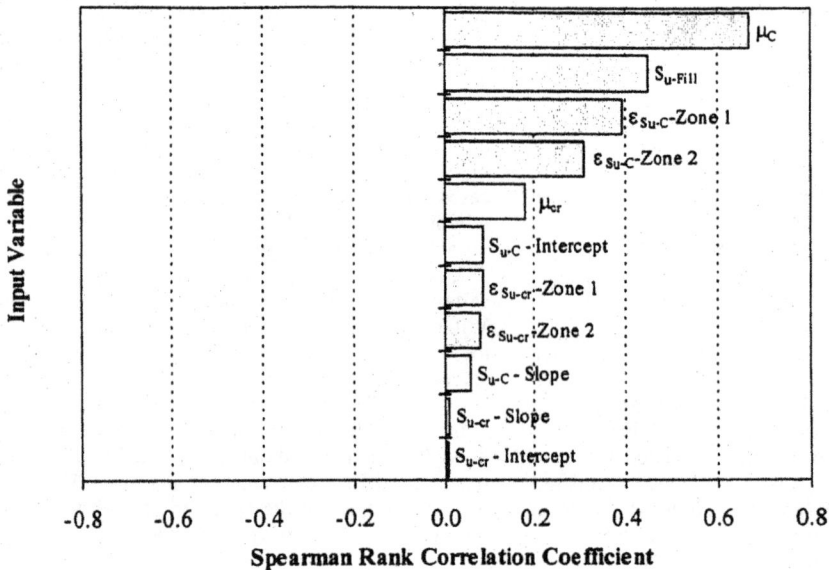

Figure 10. Sensitivity analysis results, Muar Embankment (H=3.3m) – Proposed Methodology.

7 CONCLUSIONS

Morgenstern (2000) concluded his Lumb lecture with the following:

"The assurance of geotechnical performance would be enhanced if geotechnical engineering shifted from the promise of certainty to the analysis of uncertainty".

In order to do so, it is essential to distinguish between the different types of uncertainty that arise in geotechnical practice. Examples have been provided.

Many slope stability problems are dominated by parameter uncertainty with minimal model uncertainty. In this case, the capacity to analyze uncertainty by means of probabilistic slope stability analysis is now practical and economic. Again, a number of examples have been provided.

It is anticipated that probabilistic slope stability analyses will be used increasingly in geotechnical practice and will add substantial to the judgment involved in assuring geotechnical performance.

REFERENCES

Brand, E.W., 1995. Stability of embankments on soft clay. Proceedings of the International Seminar on Stability of Cuts, Fills and Reclaimed Land, Osaka, p. 49-59.

Chan, A.C.Y. and Morgenstern, N.R., 1987. Influence of geological history on the properties of a clay. Proceedings International Symposium on Geotechnical Engineering of Soft Soils, Mexico, Vol. 1, p. 25-32.

Chowdhury, R.N., 1984. Recent developments in landslide studies: Probabilistic methods, State-of-the-Art Report. Proceedings of the 4th International Symposium on Landslides, Toronto, Vol. 1, p. 209-228.

Christian, J.T., 1996. Reliability methods for stability of existing slopes. In Uncertainty in the Geologic Environment: From Theory to Practice. Proceedings of Uncertainty '96. Geotechnical Special Publications No. 58, ASCE, Vol. 2, p. 409-419.

Christian, J.T., Ladd, C.C. and Baecher, G.B., 1994. Reliability and probability in stability analysis. Journal of Geotechnical Engineering Division, ASCE, Vol. 120, p. 1071-1111.

Clayton, C.R.L., 2001. Managing Geotechnical Risk. The Institution of Civil Engineers, London.

Crooks, J.H.A., Been, K., Mickleborough, B.W. and Dean, J.P., 1986. An embankment failure on soft fissured clay. Canadian Geotechnical Journal, Vol. 23, p. 528-540.

Cruden, D. and Fell, R., (eds.) 1997. Landslide Risk Assessment Proceedings of the International Workshop on Landslide Risk Assessment, Honolulu, USA, A.A. Balkema.

Duncan, J.M., 2000. Factors of safety and reliability in geotechnical engineering. Journal of Geotechnical and Geoenvironmental Engineering, ASCE, Vol. 126, p. 307-316.

El-Ramly, H., 2001. Probabilistic Analyses of Landslide Hazards and Risks: Bridging Theory and Practice. Ph.D. Thesis, University of Alberta, Canada.

El-Ramly, H., Morgenstern, N.R. and Cruden, D.M, 2002a. Probabilistic slope stability analysis for practice. Canadian Geotechnical Journal, Vol. 39, p. 665-683.

El-Ramly, H., Morgenstern, N.R., and Cruden, D.M., 2002b. Probabilistic stability analysis of Lodalen slide. Proceedings 55th Canadian Geotechnical Conference, Niagara Falls, Vol. 2, p. 1053-1060.

Geotechnical Engineering Office, 1994. Report on the Kwun Lung Lau Landslide of 23 July 1994, Geotechnical Engineering Office, Hong Kong.

Ho, K., Leroi, E. and Roberds, B., 2000. Quantitative risk assessment: Applications, myths and future direction. GeoEng 2000, Vol. 1, p. 269-312.

Ho, K.K. and Wong, H.N., 2001. Application of quantitative risk assessment in landslide risk management in Hong Kong. Proceedings 14th SE Asian Geotechnical Conference, Hong Kong, Vol. 1, p. 123-128.

Ladd, C.C., et al., 1983. Report of the subcommittee on embankment stability – annex II. Committee of Specialists on Sensitive Clays on the NBR Complex, Societé d' Energie de la Baie James, Montréal, Canada.

Lambe, T.W., 1973. Predictions in soil engineering. Geotechnique, Vol. 23, p. 149-202.

Malaysian Highway Authority, 1989. Proceedings of the International Symposium on Trial Embankments on Malaysia Marine Clays. Kuala Lumpur, 2 Volumes.

Matsuo, M. and Kurodo, K., 1974. Probabilistic approach to design of embankments. Soils and Foundations, Vol. 14, p. 1-17.

MIT, 1974. Proceedings of the Foundation Deformation Predictions Symposium, Massachusetts Institute of Technology, Cambridge, Massachusetts, 2 Volumes.

Morgenstern, N.R., 1994. Report on Kwun Lung Lau Landslide of 23 July, 1994; Causes of the Landslide and Adequacy of Slope Safety Practice in Hong Kong. Civil Engineering Department, Hong Kong.

Morgenstern, N.R., 1995. Managing risk in geotechnical engineering. The 3rd Casagrande Lecture. Proceedings 10th Pan-American Conference on Soil Mechanics and Foundation Engineering, Guadalajara, Vol. 4, p. 102-126.

Morgenstern, N.R., 1997. Toward landslide risk assessment in practice. In Landslide Risk Assessment, ed. by D. Cruden and R. Fell, A.A. Balkema, p. 15-23.

Morgenstern, N.R., 2000. Performance in geotechnical practice, the Lumb Lecture. Transactions of the Hong Kong Institution of Engineers, Vol. 7, p. 2-15.

National Research Council, 1994. Science and Judgment in Risk Assessment. National Academy Press, Washington.

Nguyen, V.U. and Chowdhury, R.N., 1984. Probabilistic study of spoil pile stability in ship coal mines – two techniques compared. International Journal of Rock Mechanics, Mining Science and Geomechanics, Vol. 21, p. 303-312.

Palisade Corporation, 1996. @ Risk: risk analysis and simulation add-in for Microsoft Excel or Lotus 1-2-3. Palisade Corporation, Newfield, NY.

Poulos, H.G., Lee, C.Y. and Small, J.C., 1990. Predicted and observed behaviour of a test embankment on Malaysian soft clays. Research Report No. R620, University of Sydney.

Rivard, P.J. and Lu, Y., 1978. Shear strength of soft fissured clays. Canadian Geotechnical Journal, Vol. 15, p. 382-390.

Sevaldson, R.A., 1956. The slide in Lodalen, October 6, 1954. Geotechnique, Vol. 6, p. 167-182.

Soulie, M., Montes, P. and Silvestri, V., 1990. Modeling spatial variability of soil parameters. Canadian Geotechnical Journal, Vol. 27, p. 617-630.

Southwood, T.R.E., 1985. The roles of proof and concern in the work of the Royal Commission on Environmental Pollution. Marine Pollution Bulletin, Vol. 16, p. 326-350.

Tang, W.H., Yucemen, M.S. and Ang, A.H.S., 1976. Probability-based short term design of slopes. Canadian Geotechnical Journal, Vol. 13, p. 201-215.

Vanmarcke, E.H., 1977. Reliability of earth slopes. Journal of the Geotechnical Engineering Division, ASCE, Vol. 103, p. 1247-1265.

Vick, S.G., 1995. Geotechnical risk and reliability - from theory to practice in dam safety. In The Earth, Engineers and Education - Whitman Symposium, Massachusetts Institute of Technology, p. 45-58.

Whitman, R.V., 1984. Evaluating calculated risk in geotechnical engineering. Journal of the Geotechnical Engineering Division, ASCE, Vol. 110, p. 145-188.

Wolff, T.F., 1996. Probabilistic slope stability in theory and practice. In Uncertainty in the Geologic Environment: From Theory to Practice. Proceedings of Uncertainty '96. Geotechnical Special Publication No. 58, ASCE, Vol. 2, p. 419-433.

Wong, H.N. and Ho, K.K.S., 1997. The 23 July 1994 landslide at Kwun Lung Lau, Hong Kong. Canadian Geotechnical Journal, Vol. 34, p. 825-840.

Soft Ground Engineering in Coastal Areas, Tsuchida et al. (eds)
© 2003 Swets & Zeitlinger, Lisse, ISBN 90 5809 613 0

The 1964 Niigata earthquake in retrospect

Y. Yoshimi
Tokyo Institute of Technology, Tokyo, Japan

ABSTRACT: The case history of the Niigata earthquake of 1964 is reviewed with emphasis on reinforced concrete buildings, many of which settled and/or tilted due to soil liquefaction with little damage to their superstructures. The Architectural Institute of Japan's design manual for building foundations before the earthquake was inadequate to cope with soil liquefaction. There had been a limited number of applications of the vibroflotaion method with mixed results. The simple soil profile consisting of clean, fine sand and the availability of reliable Standard Penetration Test data resulted in a critical N-value criterion for evaluating soil liquefaction potential. In an area where the ground surface moved more than a meter, a nine-story building remained stationary probably because the sand around the piles had been compacted by exceptionally hard pile driving.

1 INTRODUCTION

It is nearly four decades ago since a magnitude 7.5 earthquake occurred off the coast of the Sea of Japan near the City of Niigata. The extensive damage due to soil liquefaction took nearly everyone by surprise, except a farsighted few including Professors Tachu Naito (1886-1970) and Takeo Mogami (1911-1987). The geotechnical aspects of the earthquake were reported in two issues of *Soil and Foundation* (Japanese Society of Soil Mechanics and Foundation Engineering 1966). The earthquake was unique in that it gave us for the first time a wealth of case histories concerning the mechanism of and remedial measures for soil liquefaction and lateral displacements of nearly level ground (lateral ground spreading). The concept of a critical N-value for differentiating liquefiable and non-liquefiable conditions proposed by Dr. Y. Koizumi (1924-1983) has since been modified and become the most widely used practical method for evaluating the liquefaction potential of saturated sand (Koizumi, 1966).

2 MAIN FEATURES OF THE EARTHQUAKE

The main features of the earthquake may be summarized as follows:

(1) A relatively simple soil profile consisting of clean, fine sands covered a large area in the city as shown in Figure 1, in which the SPT N-values are less than five in the hatched zone.

(2) The groundwater table was quite shallow, often less than a meter.

(3) Because the earthquake occurred midday in summertime (1:01 p.m., June 16, 1964), many people were able to observe the liquefaction phenomena, resulting in numerous eye witness accounts, photographs and an 8-mm film of soil liquefaction in action. This was diametrically opposite to the situation during the Hyogoken-Nambu earthquake which occurred about 80 minutes before daybreak (5:46 a.m., January 17, 1995), for which there were few eye witnesses.

(4) The soil liquefaction caused significant damage to various urban facilities including bridges, quay walls, reinforced concrete (RC) buildings and buried structures such as septic tanks that had been designed without considering soil liquefaction.

(5) Various types of foundations were involved: spread footings, friction piles, end-bearing piles, and pneumatic caissons. A limited number of measures were used to reduce possible damage, e.g. compaction by the vibroflotation method, and confinement with sheet pile walls around basement walls.

(6) The relatively small number of deaths totaling 26, the tolerant mood of the society in which people

were not used to filing law suits over natural disasters, and the relatively limited urban area made it possible to do a thorough investigation of both damaged and undamaged facilities, which resulted in a number of well-documented reports. Many civil and building engineers, who were impressed by the dramatic damage, contributed a great deal of time and effort on the investigations.

(7) Thanks to the Japan Industrial Standard for the standard penetration test that was put into effect three years before the earthquake, a fair number of reliable SPT data using the *tombi* (a trigger that enables a free fall of the hammer) had been accumulated. The availability of young laborers was conducive to implementing the *tombi* method diligently.

3 REINFORCED CONCRETE BUILDINGS

3.1 *Settlement and tilt*

The earthquake caused damage to about 340 RC buildings that accounted for 22 percent of all the RC buildings in Niigata city. About two-thirds of the buildings settled and/or tilted with little damage to their superstructures. In Niigata city area where many buildings tilted more than 2.5 deg., the N-values to a depth of 10 m were less than 10 at most places, as shown in Figure 1.

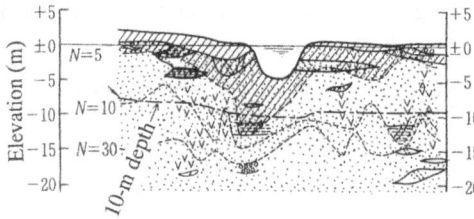

Figure 1. A cross section of the heavily damaged area in Niigata with contours of N-values.

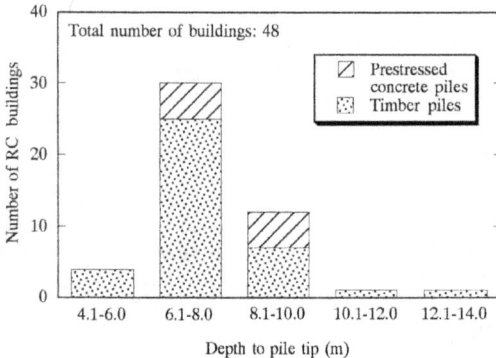

Figure 2. Histogram of pile tip depths where the N-values were less than 15 (data from Building Research Institute 1965).

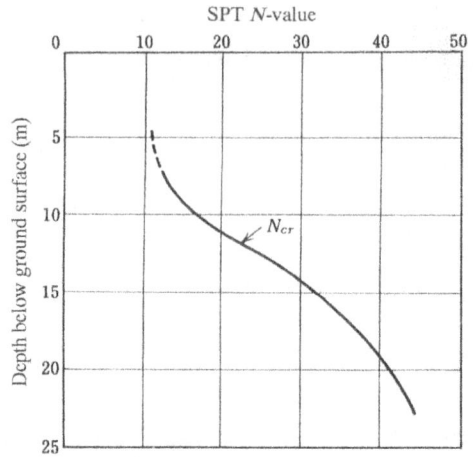

Figure 3. Critical N-value (Koizumi 1966).

Figure 2 shows a histogram of the depths of pile tips where the N-values were less than 15. As shown in the figure, 96 percent of the buildings were on piles whose tips were shallower than 10 m. Thus, the pile lengths of most buildings were inadequate according to the critical N-value, N_{cr}, for liquefaction as shown in Figure 3.

3.2 *Foundation design manual*

Prior to the Niigata earthquake, sandy soils had generally been considered superior to clayey soils for supporting building foundations. The idea was based on the fact that sands have a friction angle and low compressibility, and are free from time-dependent consolidation settlement. The design manual for building foundations published by the Architectural Institute of Japan[1] (1960) followed the general rule for the structural design of buildings in which the allowable stresses for "short-term (seismic and wind) loading" were set 50 to 100 percent higher than those for "long-term (static) loading." In other words, the factor of safety for short-term loading was set lower than that for long-term loading. For foundation design, the bearing capacity for seismic loading doubled that for static loading. This allowed many four-story RC buildings on loose sand where N-values were from 5 to 10 to be built on shallow foundations or short friction piles without any soil improvement.

However, the design manual did mention the possibility of soil liquefaction during strong earthquakes for loose sands whose N-values were less than five. The warning was apparently ignored by those who designed foundations for about 200 RC buildings

---

[1] This is a misnomer because the Institute deals with structural engineering and other branches of building technology in addition to architecture. A more appropriate name would be "Japanese Society for Architecture and Building Engineering."

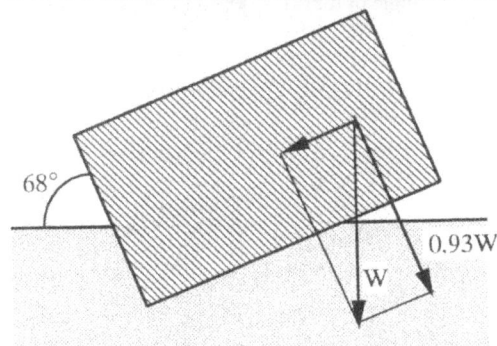

Figure 4. Apartment No. 4 at Kawagishi-Cho that tilted the most.

that were built between 1962 and 1964. In hindsight, the warning was not written with conviction, and the value of five for the critical blow count was too low, particularly for sands at some depths. The writer regrets that he did not emphasize the significance of the unusually loose sands in Niigata when he was sent to the city as one of the authors of the manual to explain its contents to practicing engineers. Such a continuing education type event has been standard practice at the Architectural Institute of Japan following the publication of a design manual's new version.

3.3 Little damage to superstructures

As shown in the photograph of Figure 4, the end wall of the RC apartment building was completely undamaged. This is remarkable when one realizes that the transverse load amounted to 93 percent of the weight of the upper part of the building that remained aboveground, equivalent to a seismic coefficient of 0.93, as shown in the line drawing of Figure 4. of the 72 RC buildings that settled more than 1.00 m or tilted more than 2.5 deg., 57 buildings (79%) sustained no damage to their superstructures, while only three (4%) required major repairs for their superstructures (Building Research Institute 1965).

The soundness of the superstructures may be attributed to (1) base-isolation effects of the liquefied

sand, (2) heavy RC grade beams[2] that were required to tie the column bases, and (3) extensive use of RC walls even for partitions. The last item is not the norm in many countries where hollow bricks are often used for walls, limiting the use of RC walls for fire protection purposes.

4 BANDAI BRIDGE AND SHOWA BRIDGE: A LESSON OF STRUCTURAL REDUNDANCY

The Bandai bridge built in 1929 was somewhat damaged but remained passable whereas the brand new Showa bridge collapsed as shown in Figure 5. The former consists of multi-span RC arches on pneumatic caissons while the latter is of simple-beam structure on single row steel piles. Although the excessive horizontal displacements of the piers of the Showa bridge due to soil liquefaction was the direct cause of the collapse, the statically determinate structure that lacks redundancy was also to blame.

The collapse of cantilevered expressway columns and simple-beam bridge sections during the Hyogoken-Nambu earthquake of 1995 were other examples of the vulnerability of statically determinate structures.

Figure 5. The Showa Bridge built in 1964.

5 REMEDIAL MEASURES FOR SOIL LIQUEFACTION

5.1 Intentional compaction

Professor Mogami, Dr. Nakase's mentor, played a leading role in promoting the vobroflotation method in the late 1950's and early 60's (Watanabe 1962). While the writer was teaching at Carnegie Institute

---

[2] The value of EI/L of the beam should be two to three times that of an inner column of the lowest story, where E is Young's modulus, I is the geometrical moment of inertia, and L is the length.

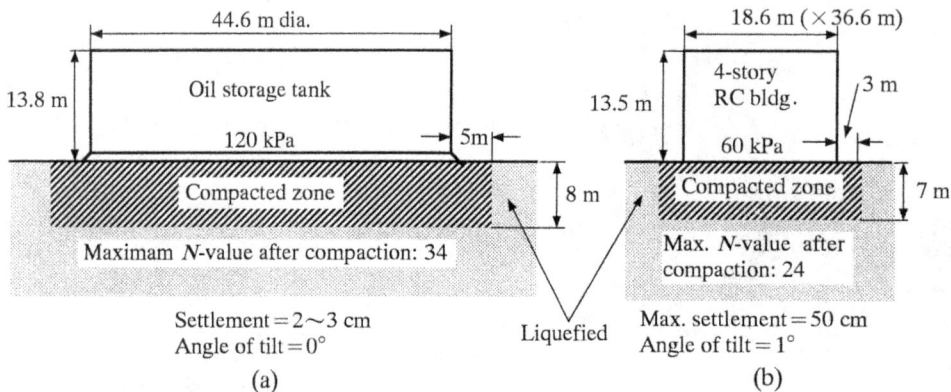

Figure 6. Performance of an oil storage tank and an RC building on ground compacted by vibroflotation (based on Watanabe 1966).

of Technology (currently Carnegie Mellon University), Pittsburgh, Pennsylvania, Professor Mogami visited the campus in 1961 on his return trip from Paris after attending the Sixth International Conference on Soil Mechanics and Foundation Engineering, and sought an interview with Dr.E. D'Appolonia who had written papers on vibroflotation.

The applications of the method in Niigata included some oil storage tanks and an RC building. The three oil storage tanks at Oze founded on compacted ground performed well during the earthquake as shown in Figure 6(a), while other tanks on unimproved ground suffered considerable damage.

On the other hand, a four-story RC building on compacted ground settled about 50 cm and tilted 1 deg. as shown in Figure 6(b). The less satisfactory performance of the building could be attributed to the facts that (1) the depth of compaction was limited to 7 m compared with 8 m for the tank, (2) the maximum N-value after compaction was 24 compared with 34 for the tank, and (4) the zone of compaction was 3 m beyond the perimeter of the structure compared with 5 m for the tank (Watanabe 1966). It is conceivable, however, that the tanks could also have suffered some damage if the earthquake ground motions had been stronger.

5.2 Unintentional compaction

The ground surface of the area north of the Niigata Railroad Station moved southeastward a meter or two as shown in Figure 7 (Hamada 1986). The permanent ground surface displacements were estimated by comparing aerial photographs taken before and after the earthquake. The study was initiated when large deformations of city gas lines were investigated in Noshiro, Akita Prefecture, following the Nihonkai-Chubu earthquake of 1983. The technique was later applied to Niigata city where horizontal displacements exceeding 8 m were observed near the Shinano River. The behavior of Building N and Building H in this area will now be reviewed.

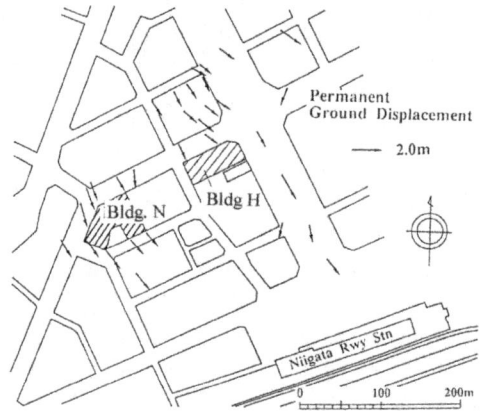

Figure 7. Lateral ground displacements around Buildings N and H (Hamada et al. 1986 modified to include Building H).

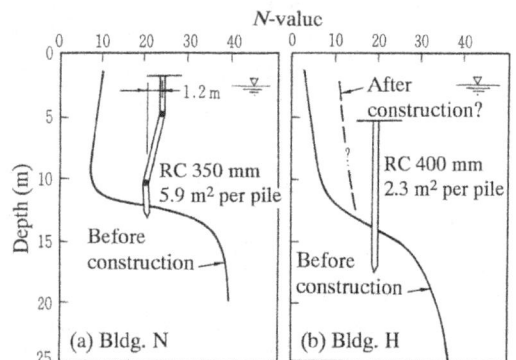

Figure 8. Buildings N and H compared with regard to N-values and pile positions (modified after Yoshimi 1997).

(1) Building N

This two-to-three-story RC building on 12-m RC piles tilted 0.76 deg., but was used until 1984 after leveling the floors. It was then demolished to make room for a new building with basement. The con-

56

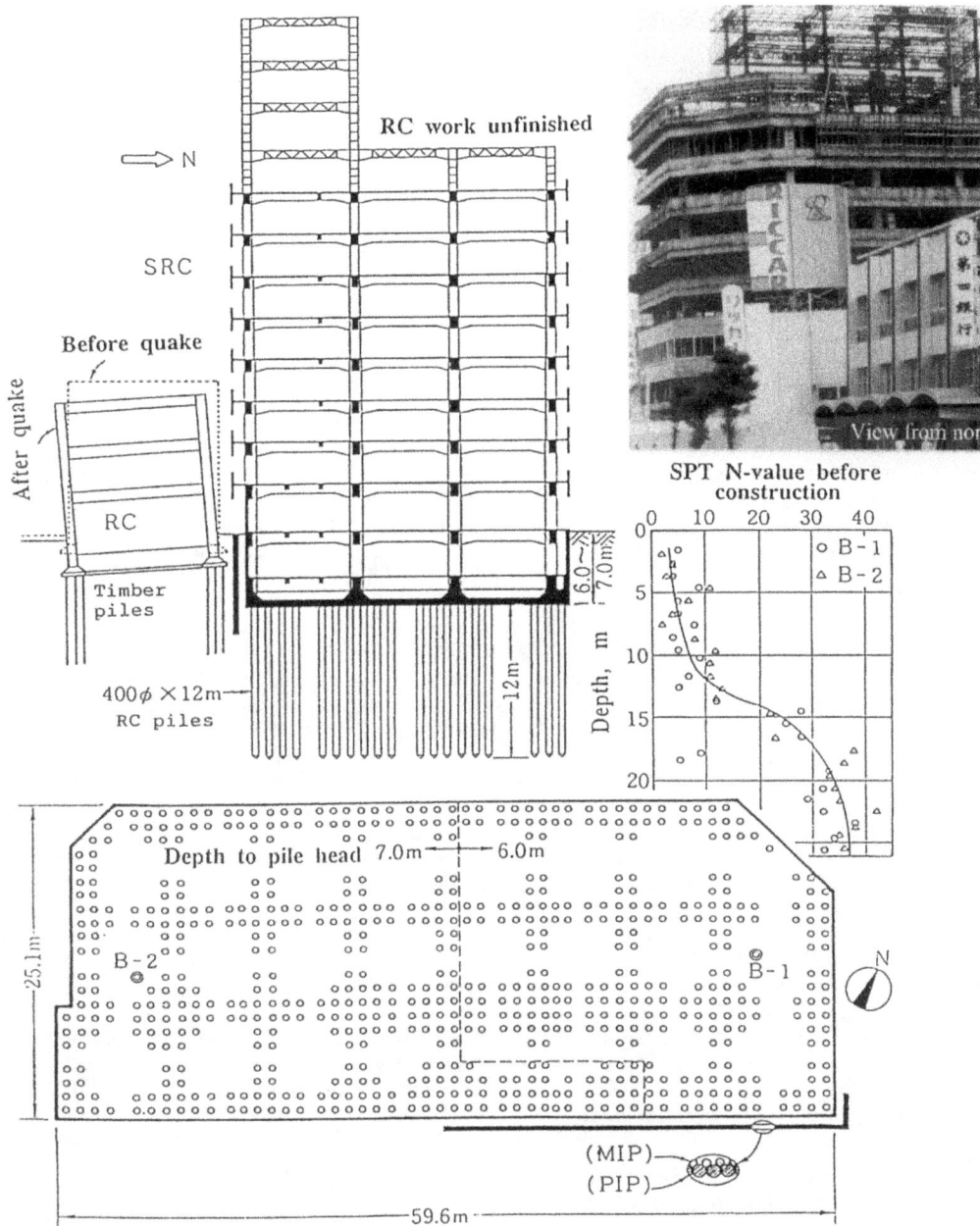

Figure 9. Cross section, pile layout and *N*-values before construction for Building H (Yoshimi 1991).

tractor, Taisei Corporation, made a commendable effort to recover the foundation piles intact by careful excavation. It was discovered that the RC piles were broken at two places as shown in Figure 8(a), the axes of the top and bottom parts being displaced laterally about 1.2 m which was compatible with the displacement of the ground surface around the building.

(2) Building H

This nine-story steel-skeleton RC building on end bearing piles remained stationary despite the fact that the surrounding ground moved more than a meter. (Nine-story buildings within a height limit of 31 m were the tallest allowed in those days.) As shown in Figure 9, the building was under construction with the concrete work above the eighth floor unfinished

(Yoshimi, 1991). The three-story RC building shown on the left on short timber piles settled 1.68 m maximum and tilted 3.7 deg., showing that the ground around Building H liquefied quite severely[3]. The successful performance of Building H despite the considerable ground movement can be attributed to the unintentional densification of the sand around the piles by vertical vibrations that were caused by driving the piles into the dense sand having N-values reaching 30[4], as shown in Figure 9. Finding it very difficult to drive the RC piles at a close spacing of 1.1 m to the prescribed depth of 12 m below the raft foundation, the contractor pleaded without success to the architect to let them stop driving short of the target depth. In fact, the pile driving caused some settlement of the street in front and cracked a marble countertop in the three-story building next door. Thus, the densified sand around the closely spaced piles is believed to have protected the piles from breakage as in the case of Building N. The broken line in Figure 8(b) shows qualitatively that the N-values could have increased after construction.

Incidentally, similar unintentional compaction that seemed to have mitigated liquefaction-induced ground settlement was observed at man-made Port Island, Kobe, during the Hyogoken-Nambu earthquake of 1995. The case involved the installation of sand drains for accelerating consolidation of the alluvial silty clay layer below the landfill consisting of sandy gravel, locally called *masa*. Normally, sand drains are not intended to compact the soil that they penetrate. In this case, however, the driving of the steel casings for sand drains met considerable resistance from a layer of slag dumped on the former seabed by a local steelmaker, the N-values of the slag reaching 40 at some places. It seems conceivable that the hard driving of the casings caused densification of the *masa* fill to about the same level of compaction achieved by rod compaction (Yasuda et al. 1996). According to Fukui (1995), the slag layer had to be punched with a solid steel rod before piles could be driven.

5.3 *Confinement with sheet pile walls around a building*

When Professor Naito of Waseda University, a leading figure in the field of earthquake engineering, was asked to give advice concerning the construction of the Niigata City Hall, he recognized the risk of soil liquefaction during a strong earthquake, and directed the municipal office to drive sheet piles that were longer than necessary for constructing the basement, and leave them buried so as to "prevent the sand below the building from running away" (Naito 1965). It was highly unusual in those days not to extract expensive steel sheet piles for repeated uses.

6 PROFESSOR SEED'S CONTRIBUTIONS

Professor H. B. Seed (1922-1989) of the University of California at Berkeley was keenly interested in the soil liquefaction phenomena in Niigata. He spent a week in Tokyo and Niigata with three American colleagues in July, 1965, to attend the Conference on Soil Dynamics under the Japan-U.S. Cooperative Science Program (Japanese Society of Soil Mechanics and Foundation Engineering 1966).

It appears that Professor Seed began his research program on soil liquefaction immediately following the Alaska earthquake of March 28, 1964, and the Niigata earthquake. Professor Seed and K. L. Lee (1931-1978), then a Ph.D. student, modified their cyclic triaxial apparatus originally designed to apply non-reversed cyclic stresses for road research to accommodate fully reversed cyclic shear stresses for simulating seismic loading on level ground. They cleverly noticed that they could keep the mean effective stress constant by applying fully reversed deviator stresses if the pore pressure coefficient B was kept 1 (Seed & Lee 1966). Thus, they were able to expedite their testing program by circumventing the need to vary the cell pressure simultaneously with the axial load. The writer clearly remembers that Professor Seed would not divulge anything about the ongoing research when asked by Professor Mogami during the Japan-U.S. Conference mentioned above.

A few years later, Professor Seed attempted to generalize the critical N-value criterion (Figure 3) by a procedure based on relative density of the sand (Seed & Idriss 1971). The choice of relative density to characterize the sand was a natural consequence of Lee's laboratory tests in which the specimens were prepared by pouring sand into a mold at various densities. The method has since been modified to replace the relative density with N-value because the relative density of sands in situ was difficult to assess, and because the relative density alone could not determine the liquefaction resistance of sands.

Professor Seed and his colleagues applied their analytical method to a typical soil profile in Niigata to estimate time histories of excess pore water pressures at some depths, as shown in Figure 10. Before publishing the paper in the ASCE Journal, Professor Seed came to Japan to orally present an abbreviated version of the paper at a domestic conference. The writer was surprised by the way in which the presentation shocked Professor Y. Ohsaki (1921-1999) of Tokyo University, who was perhaps writing a similar paper at the time.

---

[3] The wall labeled MIP and PIP were provided to minimize the effect of shock and vibration caused by driving sheet piles. The sheet piles for the rest of the perimeter were extracted after the construction of the basement.

[4] Based on good quality undisturbed samples of Niigata sand obtained by in situ freezing, the relative densities of sand of N=30 exceeded 82% (Yoshimi et al. 1984).

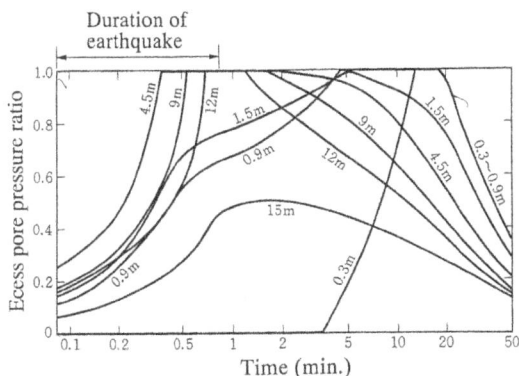

Figure 10. Time histories of excess pore water pressures during the Niigata earthquake (modified after Seed et al. 1976).

7 CONCLUSIONS

The Niigata earthquake of 1964 has been reviewed with emphasis on building foundations. The design manual for building foundations before the earthquake was inadequate to cope with soil liquefaction partly because the allowable bearing capacity for seismic loading doubled that for static loading. Unlike the majority of engineers who followed the manual mechanically, a few engineers had the foresight to recognize the danger of soil liquefaction and recommended remedial measures accordingly.

The simple soil profile and the availability of reliable SPT results using a free falling hammer were conducive to developing the critical N-value criterion to evaluate soil liquefaction potential.

A nine-story building remained stationary despite the fact that the surrounding ground moved more than a meter. This is believed to have resulted from a fortuitous, unintentional compaction of the sand around the piles due to exceptionally hard driving.

REFERENCES

Architectural Institute of Japan 1960. *Recommendations for Design of Building Foundations*

Building Research Institute, Ministry of Construction 1965. Damage to buildings due to the Niigata earthquake with particular emphasis on reinforced concrete buildings in Niigata city (in Japanese), *Bulletin of the Building Research Institute* No. 42: 72-97

Fukui, M. 1995. Private communication

Hamada, M. et al. 1986. *Study on Liquefaction Induced Permanent Ground Displacements*, Association for the Development of Earthquake Prediction: 30-31

Japanese Society of Soil Mechanics & Foundation Engineering 1966. *Soil & Foundation* 6 (1,2)

Kawamura, S. et al. 1985. Liquefaction-induced damage to piles discovered during excavation 20 years after the earthquake (in Japanese), *Nikkei Architecture* 244: 130-134

Koizumi, Y. 1966. Changes in density of sand subsoil caused by the Niigata earthquake, *Soil & Foundation* 6 (2): 38-44

Naito, T. 1965. *Nihon-no-Taishin-Kenchiku-to-Tomoni* (Japan's Earthquake-Resistant Buildings) (in Japanese): Sekka-Sha

Seed, H. B. et al. 1976. Pore-pressure changes during soil liquefaction, *Jour. Geotechnical Engineering Division, ASCE* 102 (GT4): 323-346

Seed, H. B. & Idriss, I. M. 1971. Simplified procedure for evaluating soil liquefaction potential, *Jour. Soil Mechanics & Foundation Division, ASCE* 97 (SM9): 1249-1273

Seed, H. B. & Lee, K. L. 1966. Liquefaction of saturated sands during cyclic loading, *Jour. Soil Mechanics & Foundation Division, ASCE* 92 (SM6): 105-134

Watanabe, T. 1962. *Research on the Vibroflotation Method* (in Japanese): Kajima Kensetsu Technical Laboratory

Watanabe, T. 1966. Damage to oil refinery plants and a building on compacted ground by the Niigata earthquake and their restoration, *Soil & Foundation* 6(2): 86-99

Yasuda, S. et al. 1996. Effects of soil improvement on ground subsidence due to liquefaction, *Soils & Foundations, Special Issue*: 104

Yoshimi, Y. 1991. A case history of a building that resisted liquefaction-induced lateral ground movement, *Proc. 9th Asian Regional Conf., International Society for Soil Mechanics & Foundation Engineering* 2: 493-495

Yoshimi, Y. 1997. A method to design piles that can accommodate lateral ground displacement (in Japanese), *Tsuchi-to-Kiso* 45(3): 9-12

Yoshimi, Y. et al. 1984. Undisturbed cyclic shear strength of a dense Niigata sand, *Soils & Foundations* 24(4): 131-145

Fundamental characteristics of clayey soils

Soft Ground Engineering in Coastal Areas, Tsuchida et al. (eds)
© 2003 Swets & Zeitlinger, Lisse, ISBN 90 5809 613 0

Compression behaviour of structurally complex marine clays

F. Cotecchia
Technical University of Bari, Italy

F. Santaloia
CNR-I.R.P.I., Bari, Italy

ABSTRACT: Due to the high degree of tectonic shearing occurred during Apennine orogenesis, the clay units outcropping in the chain areas of Southern Italy are structurally complex and have often a scaly fabric. This fabric is defined by an intricate network of fissures and shear planes which subdivide the clay into centimetric and millimetric fragments or lenses. The present paper deals with the compression behaviour and the microstructural changes of the Senerchia scaly clays, outcropping in a chain area in Southern Italy. The results of the oedometer tests outline a state boundary envelope in compression for the scaly which lies to the left of their intrinsic compression lines. These clays appear either to contract or to dilate with shear after consolidation respectively to low and high yield stress ratios. Their swelling behaviour is characterised by large unload-reload loops and a vertical shift of the swelling loops with compression post-gross-yield.

1 INTRODUCTION

The geological history of the Italian peninsula has resulted in the extensive outcropping of highly tectonized deposits, originally forming as flysch in pelagic basins. Due to the eastward tectonic movements of the southern Apennine orogenic system, which probably started in Upper Oligocene – Lower Miocene, these flysch deposits have been heavily deformed and displaced for hundreds of kilometers. The tectonic movements have often modified the original succession of the sediments, changing both the soil macro and micro-structures. In particular, for the clay strata the large soil strains associated to the tectonic movements have given rise to a highly consolidated and scaly structure. At the visual scale, the structure of these tectonized clay samples, i.e their "mesostructure" (Picarelli 1986), is characterized by a network of small slightly curved fissures, crossing each other at low angles and spaced from some millimeters to a few centimeters. These fissures define an arrangement of small clay fragments, called scales. The scales can be seen as the lenses produced by the system of minor shear planes developing within the shear bands crossing soil elements subjected to simple shear (Skempton 1966; Pini 1999). Sometimes, larger fissures, *second level fissures*, are superimposed on the network of small fissures separating the scales, and they may be spaced from centimeters up to meters. Due to these macro and meso-structural features, these tectonized materials are classified as "structurally complex formations" (Esu 1977). In par-

ticular, the scaly clays exhibit a mechanical behaviour different from that of either intact natural clays or mediumly fissured clays and are of particularly low strength, being as weak as soft clays, despite their very high consolidation indexes. Consequently, they are largely involved in instability processes.

The present paper presents some of the results of a research programme aiming at the definition of the framework of behaviour of scaly clays. The final objective of the research is to identify the relations between their structural features and some of their peculiar aspects of behaviour and thus to establish how their framework of behaviour differs from that expected according to critical state soil mechanics.

The materials discussed in the following are the Senerchia Clays, part of the slope below the town of Senerchia, where the Vadoncello landslide, discussed by Santaloia et al. (2001), took place. The clays are part of the Variegated Clay formation (Ogniben 1969), which largely outcrops in the epicentral area of the 1980 Irpinia earthquake. The data presented outline the compression behaviour of the scaly clays in relation to both their meso and microfabric, the latter of which has been investigated by means of the scanning electron microscope (SEM).

2 SAMPLE LOCATIONS AND MESOSTRUCTURES

Figure 1 presents the geological section of the Vadoncello slope and the changes in slope profile re-

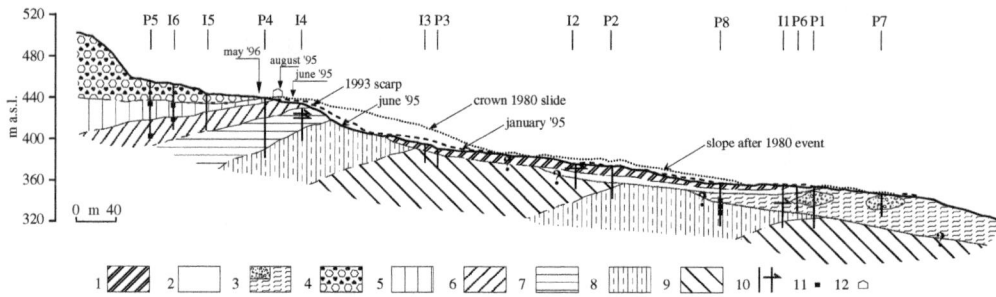

Figure 1. Cross section of the Vadoncello slope. Keys: 1) Complex G; 2) Complex H; 3) Complex I with calcareous blocks floating in it (a); 4) Complex A: calcareous mega-breccias; 5) Complex B: scaly marly clays with interbedding marls and marly limestone; 6) Complex C: scaly clays with marls and siltstone strata; 7) Complex D: scaly clays with calcareous marls; 8) Complex E: clays, marly clays, marls, marly limestones and silty sandstones; 9) Complex F: scaly clays with marly strata and siltstone layers; 10) piezometers (P1-P8) and inclinometers (I1-I6) with depth of shear; 11) undisturbed samples discussed in the paper; 12) house.

Table 1. Index properties of Senerchia samples

Sample	Complex	z	w	LL	PI	S	γ_s	γ	A	e_o
		m	%	%	%	%	kN/m³	kN/m³		
I6/1	B	61.0	14.36	57	33	93.7	27.40	22.08	0.541	0.420
P5/1	B	55.8	16.00	50	23	100	27.40	22.22	0.412	0.430
I6/3	C	53.0	16.00	66	41	100	27.40	22.16	0.774	0.437
P5/3	C	64.0	14.89	83	54	92.7	27.40	21.86	0.844	0.440
P8/5	I	53.1	19.50	57	36	92.4	27.18	20.75	0.681	0.578
P8/9	E	57.5	15.60	67	37	92.5	27.47	21.71	0.635	0.464
P8/12	E	58.8	17.53	nd	nd	87.8	27.56	20.90	nd	0.550

nd = not detected

sulting from the landslide activity from 1980 to 1996. According to the results of a thorough study of the landsliding process, Santaloia et al. (2001) report that this process is the combination of shallow roto-translational sliding and earthflowing, together with slow and long-lasting deep plastic deformations, which are active down to 30 m depth or more. Based upon the analysis of the borehole lithological profiles (Figure 1) and the soil index properties, Santaloia et al. (2001) divide the in-situ soils into different soil complexes: A to F, not involved in the landslide body, and G to I, within the landslide debris which are made up of marly, calcareous and sandy pebbles dispersed in a clayey or silty-clayey matrix. Although all the complexes, except for complex A (Figure 1), are basically formed of disarranged limestone strata, marls and scaly clays, they differ in the index properties and soil degradation degree. The clays in the complexes are generally scaly, although of different degree of second level fissuring.

Several undisturbed samples have been taken down the boreholes shown in Figure 1, which reports only the location of these samples subjected to mechanical testing and discussed in the following (Table 1). These samples belong to the in-situ complexes B, C and E and to the landslide debris I. They are homogeneously scaly, except for P8/5 and P5/1, which include some non-scaly matrix and some clasts. Some samples include some second level fis-

sures filled with calcite, which nonetheless do not seem to control the specimen behaviour. The sample mesostructures resemble that reported by Walker et al. (1979) as crossed by 5 to 10 mm spaced non-continuous fissures (Figure 2), or otherwise that classified as A2 by Esu (1977).

The samples have similar specific gravity G_s and total dry unit weight γ_d (on average 18.6 kN/m³).

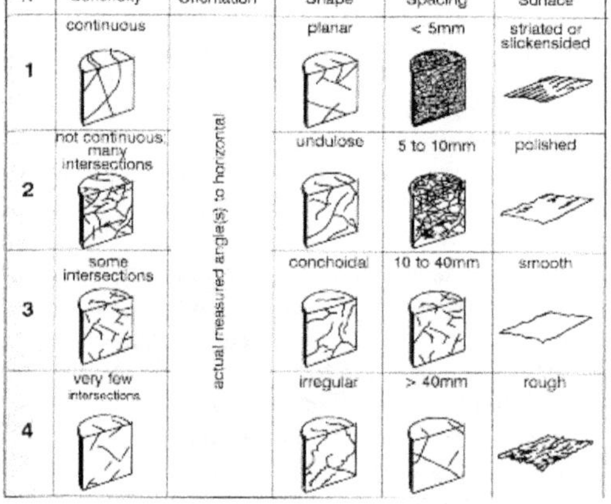

Figure 2. Classification criteria of fissured soils (Walker et al. 1987).

Both their mesostructural features and their void ratios do not vary significantly with depth; the void ratios seem to be slightly higher only for the landslide soils, as shown in Figure 3. This observation suggests that the mesostructural features of the samples influence their specific volumes more than their consolidation stress states and their different compositions. In fact, the void ratios of the in-situ scaly clays do not reflect the variations in their index properties.

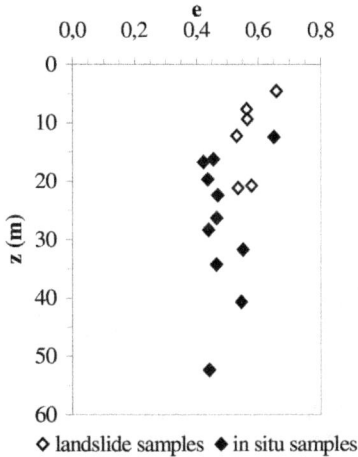

Figure 3. Void ratios of in situ and landslide samples.

3 MINERALOGY AND INDEX PROPERTIES OF THE SAMPLES

Figure 4 reports the grading curves of the samples subjected to mechanical testing, which have clay fraction 50%÷60%. The corresponding index properties are reported in Table 1, as measured according to standard testing (ASTM 1976), whereas the index properties of all the Senerchia samples characterized in the laboratory (Fearon 1998; Losurdo 2000)

are plotted in the plasticity chart in Figure 5, together with the properties of the scaly Bisaccia Clays (Pellegrino & Picarelli 1982; Picarelli & Urcioli 1993; Fenelli et al. 1992; Di Maio & Fenelli 1997; Olivares 1996) and those of Pappadai Clay, a non-scaly illitic blue marine clay (Cotecchia 1996).

In Figure 5, the Senerchia Clays appear to be of medium to high plasticity (CM-CH), less plastic than the Bisaccia scaly clays and as plastic as Pappadai Clay. However, several authors (Cotecchia 1971; Airo' Farulla & La Rosa, 1977; Rippa & Picarelli 1977; A.G.I 1979; Fearon & Coop 2000) have shown that the liquid limit of scaly clays can vary significantly with the energy used in the remoulding procedure, so that the scaly clay plasticity is generally underestimated by standard index testing. Fearon & Coop (2000) demonstrate that the plasticity index of the Senerchia Clays increases of more than 15% if measured on the clay after mincing in an industrial food mincer. Even the plasticity index of the clay after standard reconstitution in the laboratory (Burland 1990) may differ from that directly measured on the natural sample with standard index testing, although of only few percent. These observations highlight the existence of a strong bonding between the clay particles within the scales (intra-scale bonding), which is such that the index properties and activity values reported in Table 1 for the Senerchia samples do not reveal the real activity of the clay particles. Rather, this is revealed by the mineralogical analyses.

The results of X-ray diffractometry on the Senerchia Clays are reported in Table 2. The data show that the percentage of mixed-layer clay minerals (illite interlayered with smectite), which are highly swelling minerals, is quite high in all the samples, against small quantities of chlorite and kaolinite. Within the mixed-layers, the quantities of smectite

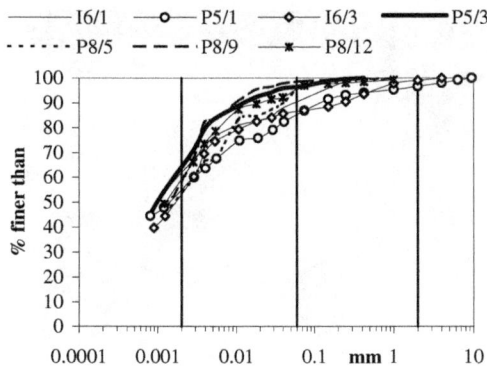

Figure 4. Grading curves for Senerchia samples subjected to mechanical testing.

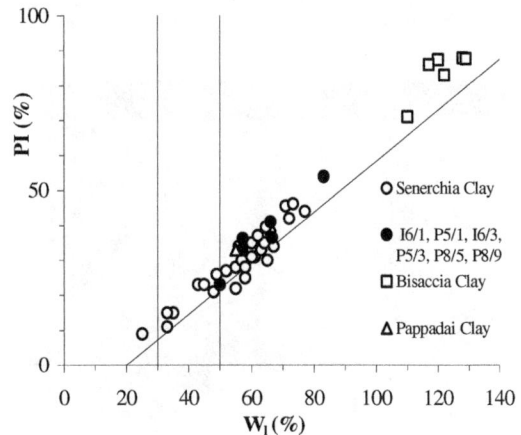

Figure 5. Plasticity chart for Senerchia, Bisaccia and Pappadai Clays.

mixed-layers, the quantities of smectite and illite are variable. The high percentage of swelling minerals recorded with X-ray diffraction is not reflected by the data in Figure 5 and Table 1. Thus, if the Senerchia Clays were less scaly, they would probably be more plastic than Pappadai Clay. The Bisaccia scaly clays, instead, are probably even richer than the Senerchia Clays in swelling minerals.

Table 2. Mineralogical composition (weight %) of the Senerchia samples

Sample	ML	I	K	C	Q	F	Cc	D	O	ML*
	%	%	%	%	%	%	%	%	%	%
I6/1	53	1	8	5	13	1	10	3	0-5	80(43)
P5/1	54	nd	nd	2	30	2	12	-	nd	82(66)
I6/3	51	2	5	5	16	2	15	2	0-2	76(60)
P5/3	50	1	6	3	12	3	17	3	0-6	83(45)
P8/5	60	1	6	5	15	2	7	1	0-2	78(48)
P8/9	54	2	5	3	10	1	18	2	0-2	67(50)
P8/12	60	tr	5	3	14	1	8	2	0-5	72(30)

Bulk sample: mixed-layers illite/smectite (ML), illite (I), kaolinite (K), chlorite (C), quartz (Q), feldspar (F), calcite (Cc), dolomite (D), salt minerals, siderite and hematite (O). Mixed layers illite/smectite in the clay fraction (ML*); the illite content (%) in the interlayers is specified in brackets. nd = not detected, tr = in trace

4 MICROFABRIC AND COMPRESSION BEHAVIOUR OF THE RECONSTITUTED SAMPLES

The reconstituted samples of scaly clays have plasticity index lower than that measured on the clay after mincing because they still include bonded aggregates of particles, which are relicts of the original scales. These aggregates are within a matrix formed of disaggregated clay particles (Fearon & Coop 2000). Thus, in the case of the scaly clays the differences in behaviour between the natural clay and the same reconstituted in the laboratory (which exhibits the soil intrinsic properties according to Burland

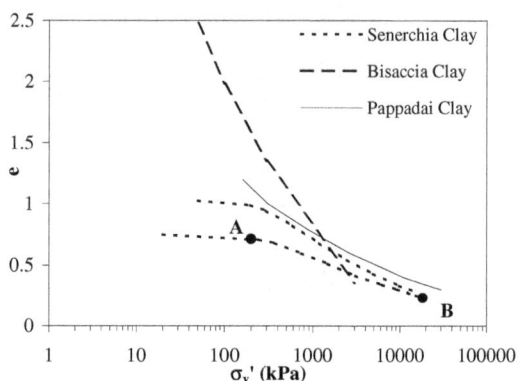

Figure 6. Intrinsic compression lines of Senerchia, Bisaccia and Pappadai Clays.

1990) are the result of differences between a clay homogeneously formed of scales and one formed of a matrix including only some scales.

Figure 6 shows the one-dimensional normal-consolidation lines (ICLs – Burland 1990) of the reconstituted clay from samples I6/1 and I6/3, together with the ICLs of Bisaccia Clay and Pappadai Clay. Although including a lower percentage of swelling minerals, Pappadai Clay plots to the right of the Senerchia ICLs, probably due to the presence in the reconstituted Senerchia Clay of relict scales, which reduce the clay plasticity index and compression index (C_c*). The high slope of the Bisaccia ICL, instead, reflects the decisively higher plasticity of this clay.

SEM analyses have been carried out on the Senerchia reconstituted clay samples at states A and B in Figure 6, after freeze-drying and gold-coating. The picture in Figure 7a shows that the reconstituted clay fabric on vertical fractures is aggregated in large flocculated arrangements, which are often chaotic. Stacks of densely packed particles can be seen in both Figures 7a and b. These oriented aggregates

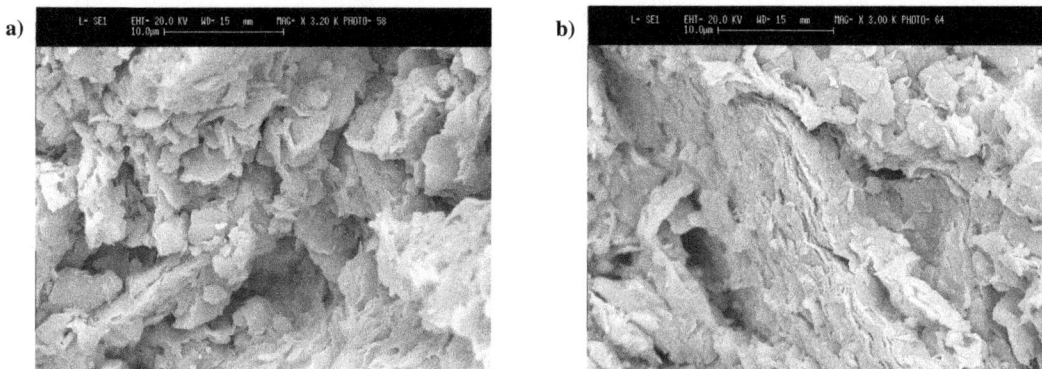

a)

b)

Figure 7. Fabric of the reconstituted clay on vertical fracture (point A in Fig. 6): edge-to-face and face-to face contact between the particles (a) and a relict of the original scales within the matrix fabric of the reconstituted sample (b).

Figure 8. Fabric of the reconstituted clay compressed to σ'_v=18 MPa (point B in Fig. 6).

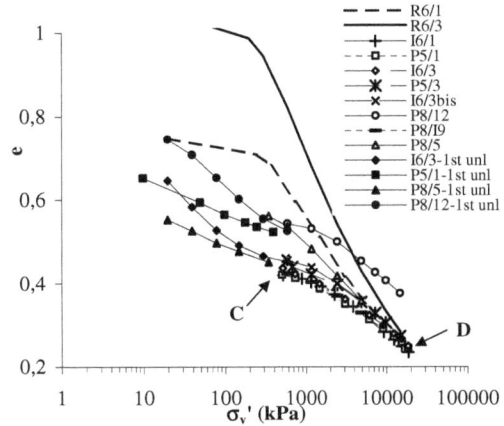

Figure 9. Compression behaviour of the natural and reconstituted samples.

are probably relicts of the natural scales. At high pressures (Figure 8; point B in Figure 6 – σ'_v=18 MPa) the fabric appears much more packed, but still quite chaotic. Many small particles, which are likely to be the most active and have the highest swelling capacity, are seen to cover the large domains.

The high quantity of swelling minerals recorded for the Senerchia Clays by X-ray diffraction is reflected in the very low values of the critical state friction angles, ϕ'_{cs}, measured on the reconstituted Senerchia Clay samples: $\phi'_{cs}\sim 14°÷18°$ (Fearon 1998), which are indicative of the poor intrinsic strength properties of these clays.

5 COMPRESSION BEHAVIOUR AND MICROFABRIC OF THE NATURAL SAMPLES

The results of eight compression and four swelling oedometer tests on natural scaly samples are reported in Figure 9, together with the Senerchia ICLs from Figure 6. Despite their variability in depth and index properties, the natural samples exhibit quite close compression curves, except for sample P8/12, which has a much higher smectite content (Table 2) within the intergrades (which may be responsible for its higher void ratio) and sample P8/5, which is the only one part of the landslide debris (Figure 1) and which results to be the most compressible. However, all the samples exhibit a compressibility quite lower than that of the reconstituted.

The ICLs in Figure 9 may be regarded as representative of the intrinsic compressibility of most of the samples in the figure, on the ground of their similarities in composition, except for sample P8/12, which must have a ICL located further to the right of those in the figure. Thus, it appears that the compression curves of the natural samples tend to converge with the ICLs only at very large pressures. In particular the compression curves of the natural samples I6/1 and I6/3 do not cross the corresponding

ICLs, R6/1 and R6/3 respectively, even at σ'_{vmax}=18 MPa.

The swelling curve of sample P5/1 is offset from the corresponding compression curve due to some variability in mesofabric within the sample. The other swelling curves, instead, appear to join the corresponding compression curves, as sketched in Figure 10, where three swelling curves and three compression curves are reported in order to outline a framework of response to compression which can be deduced for the scaly clays according to the test data. The interpretation in Figure 10 is indicative of a small compressibility of the natural samples for vertical pressures between 300÷500 kPa and 1200÷1700 kPa, where the interval 1200÷1700 kPa is a range of pressures where gross yield seems to take place. The term gross yield is here referred to a state beyond which the clay compressibility increases and the soil appears to follow a normal-consolidation line. When swelling from the initial state, the clay follows a rather flat curve initially. At pressures below σ'_v=300 kPa, the swelling curve becomes steeper for samples P8/12 and I6/3, even steeper than the compression curve at high pressures.

The occurrence of gross yield at σ'_y between 1.0 to 1.7 MPa appears to be confirmed by the logarithmic plot of the compression curves of the natural samples in Figure 11 (Butterfield 1980). Though, the yield process appears to progress in a large stress interval, so that the compression curves reach almost constant values of the compression index C_c only at pressures about σ'_v = 8 MPa, as shown in Figure 12. The reduction in oedometric stiffness, typical of pre-gross yield conditions (Janbu & Senneset 1979) applies only to the tests I6/1 and P8/12 (Figure 13). However, this reduction is often found to miss in the recompression of disturbed overconsolidated samples and the Senerchia Clay samples may be regarded as having been strongly disturbed by the tectonic processes.

67

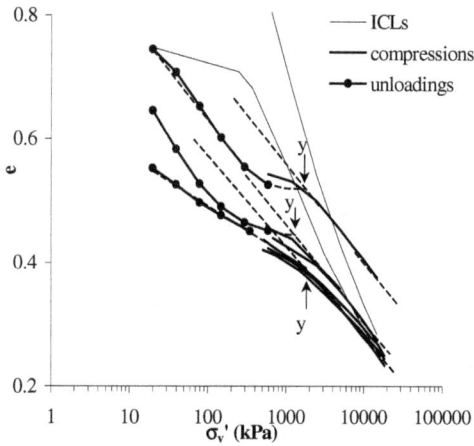

Figure 10. Sketch of the compression behaviour of the natural samples.

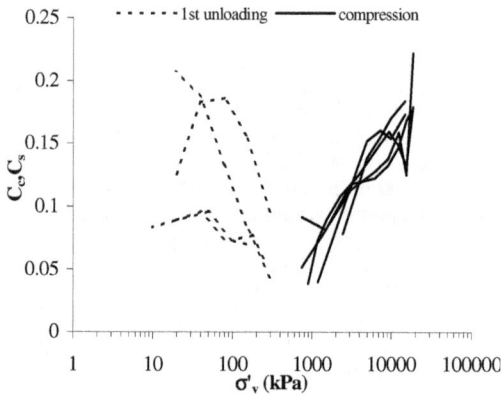

Figure 11. One-dimensional compression curves of the natural and reconstituted samples.

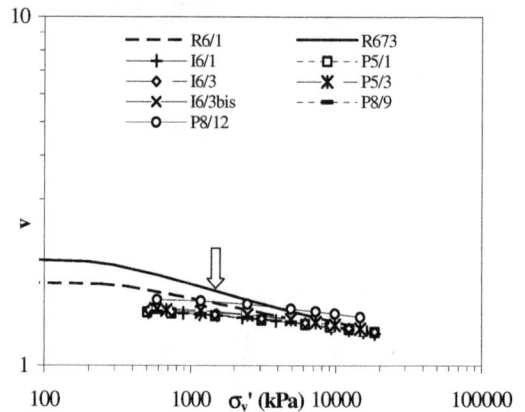

Figure 12. Compression and swelling indexes of the natural samples.

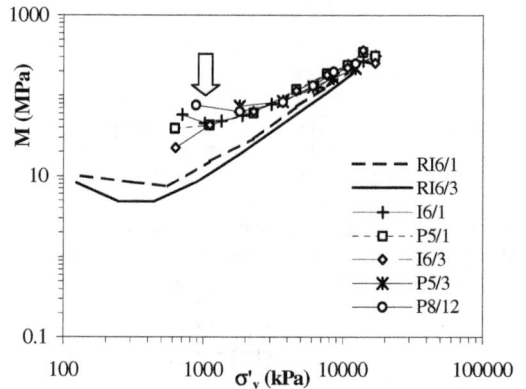

Figure 13. Stiffness in one-dimensional compression of the natural and reconstituted samples.

The observations reported above suggest the formulation of the framework of behaviour sketched in Figure 14 (sample S) to represent the response to compression of scaly clays. This is substantially different from that applying to non-scaly sensitive clays (sample I) also shown in the figure. According to this framework, the scaly clays have a state boundary in compression represented by a normal-consolidation line which lies to the left of the ICL. Therefore, their gross yield is located to the left of the ICL and their stress sensitivity ratio: $S_\sigma = \sigma'_y / \sigma'_e$ (Figure 14, Cotecchia & Chandler 2000) is lower than 1, against the $S_\sigma > 1$ of sensitive non-scaly clays. Thus $S_\sigma < 1$ characterizes scaly clays within the general framework of behaviour of clays (Cotecchia & Chandler 2000) and represents the mechanical effects of their structure, which Fearon & Coop (2002)

call a "negative" structure. This structure has adverse effects on the compression behaviour of the soil, since it makes it weaker than the reconstituted. Along with a $S_\sigma < 1$, scaly clays have a compression index C_c lower than that of the reconstituted clay C_c^*, against the $C_c > C_c^*$ applying to natural sensitive clays (Figure 14), which results from the important degradation of structure occurring in sensitive clays with post-gross yield compression. In particular, major degradation of the inter-particle bonding takes place with compression in clays with $S_\sigma > 1$, whereas in the scaly clays compression results in the reduction of the fissure openings and, eventually, in the increase of the interlocking of the scales, as observed in the analysis of SEM pictures discussed in the following.

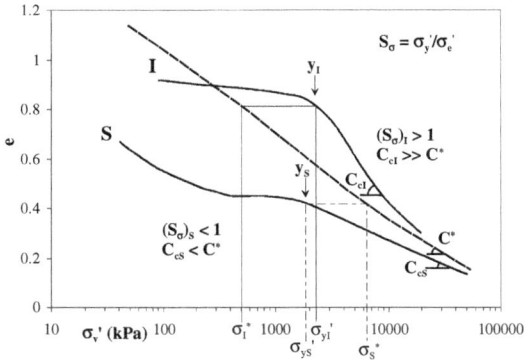

Figure 14. Compression behaviour of sensitive (I) and scaly clays (S).

SEM pictures of the fabric of sample I6/1 in states C and D (Figure 9) are shown in Figures 15 to 17. Sample I6/1 is a scaly clay with significant second level fissuring. Figure 15 shows that its fabric at state C is densely packed within aggregates separated by large elongated pores, which look like fissures. The aggregates in the figure are the magnification of portions of scales. Figure 16 shows the magnification of the clay fabric within two scales, which are separated by an open fissure. The domains within the scale are stacked in face to face contact. Diagenetic bonding within the scales is likely to be very strong, whereas no bonding agents between the scales can be detected by the SEM analysis. Figure 17 shows the fabric in sample I6/1 after oedometer compression to $\sigma'v$ = 18 MPa (state D in Fig. 9).

The fabric does not appear to have changed significantly, except for an increase in density. The particles within the aggregates are compressed in perfectly oriented stacks, but the aggregates are often bent and still separated by elongated pores which seem to be the relicts of the original fissures.

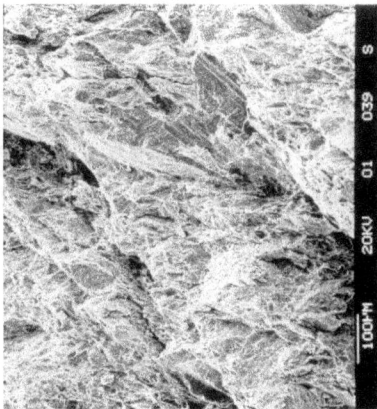

Figure 16. Fabric within the scales of the natural sample I6/1 on a vertical structure (point C in Fig. 9).

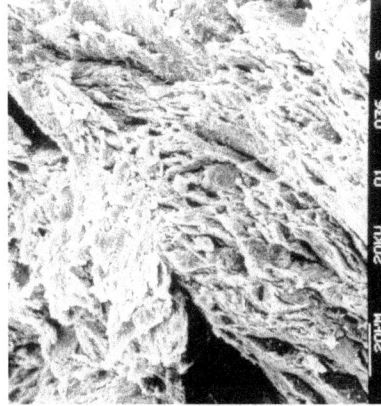

Figure 17. Fabric of the natural sample I6/1 on a vertical structure compressed to σ'_v=18 MPa (point D in Fig. 9).

Despite the fissuring, the coefficient of permeability of the clay is extremely low, reducing from 10^{-11} m/s to about $5*10^{-14}$ m/s with compression from σ'_v=500 kPa up to σ'_v=18 MPa, that is from void ratios about 0.41 to about 0.22.

The results of the SEM analysis highlight the role of scales as single elements separated by open fissures within the sample. The opening of the fissures is reduced by compression, but they still cross the material at high pressures. Large straining results in bending and deforming the scales and in increasing their packing, but the soil can still be seen as the packing of well-interlocked scales when at high pressures. The high consolidation level of the clays within the scales makes it little compressible, so that most of the sample volumetric strains result from the reduction in fissure opening. This is probably why the compression index of the scaly clays is quite low and the increase in compressibility associated to gross yield is very slow, occurring within large stress intervals and not being as sudden as for sensi-

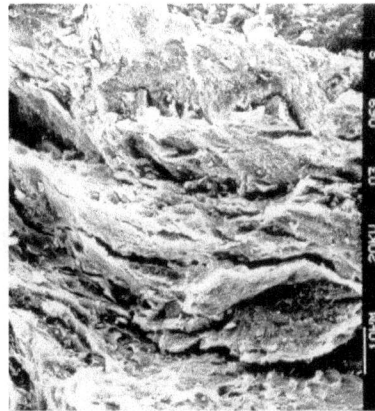

Figure 15. Fabric of the natural sample I6/1 on a vertical structure (point C in Fig. 9).

tive clays (Figure 14). With compression post-gross yield the bonding within the scales may degrade, but this process is likely to develop throughout compression post-gross yield. The SEM observations do not give evidence of significant modifications of the clay fabric with compression to high pressures, even when reaching void ratios as low as 0.25.

When shearing at high pressures, that is at mean effective stresses either about or above those here identified as relating to gross yield (at yield stress ratios YSR= σ'_y/σ'_v = 1; Burland 1990), the scaly samples exhibit a tendency to contraction, i.e. a wet behaviour, as would be expected for either reconstituted soils or natural sensitive clays at YSR<2. Figure 18 reports in the same plot the oedometer compression curves of undisturbed scaly samples, in the e-logσ'_v plane, and the consolidation e-p' states of samples subjected to shear testing; σ'_v or p' are plotted on the same abscissa. The consolidation states of the samples subject to shear are reported as open dots if exhibiting a dry behaviour and as full dots if exhibiting a wet behaviour. The data appear to define a left-side domain (high YSRs) of dry behaviour and a right-side domain (YSR<2.5) of wet behaviour, as expected for both reconstituted and natural sensitive clays. Therefore the shear data support the interpretation of the scaly clay behaviour represented by the framework in Figure 14, according to which the scaly clays become normally consolidated and exhibit a wet behaviour already to the left of the ICL. All the data presented above do not seem to support the hypothesis (Guerriero et al. 1995; Olivares 1996; Picarelli & Olivares 1998) that scaly clays reach gross yield at very large pressures, to the right of the ICL. according to which they would exhibit a wet behaviour even when at YSRs above 10-20. YSR is seen to be a parameter to which the scaly clay response can be related the same as for reconstituted clays.

It follows that scaly clays can be seen as packings of scales separated by closely spaced fissures, which on the whole either dilate or contract, depending on the yield stress ratio of the whole packing being lower or higher than about 2-3. Once compressed to extremely high pressures, higher than those generally reached in the experimental programmes developed so far, the material might exhibit a further increase in compressibility due to a major breakage of the intra-scale bonding, if this de-bonding has not progressed sufficiently before. However, within pressure intervals of engineering interest, scaly clays appear to exhibit a behaviour which might be interpreted, at least in first approximation, by the framework in Figure 14.

Results of both triaxial and direct shear tests on the Senerchia Clays (Fearon 1998; Santaloia et al. 2001) have shown that on the wet side the samples strain harden to maximum stress ratios lower than the critical state stress ratio of the corresponding reconstituted clay, whereas, on the dry side, dilation

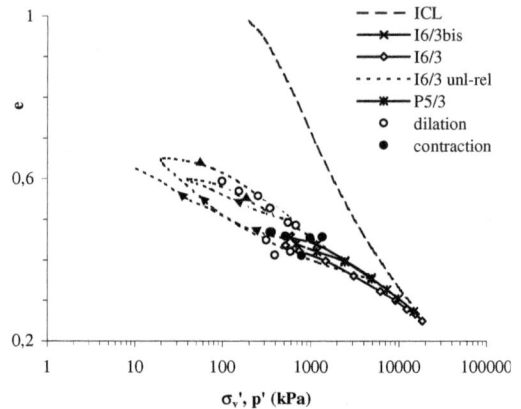

Figure 18. Compression behaviour of natural and reconstituted samples and state of the natural clays during shear (Fearon, 1998; Santaloia et al. 2001; Antonino, 1999 and new data).

allows for stress ratios only slightly higher than the maximum stress ratio reached on the wet side. The poor strength properties of these materials, poorer than for the corresponding reconstituted (Olivares 1996; Picarelli & Olivares 1998), result from the deformation mechanism taking place in the samples during shear, that is one of sliding along the network of fissures which cross the sample and that are characterized by very low friction properties. This sliding is accompanied by dilation on the dry side and by contraction of the wet side. For the Senerchia natural samples the maximum shear friction angle recorded on the wet side is about 15°-16° (Fearon 1998; Fearon & Coop 2002; Santaloia et al. 2001).

6 SWELLING BEHAVIOUR AND MICROFABRIC OF THE SWELLED SAMPLES

Figure 19 reports the results of three unloading-reloading oedometer tests and of two oedometer loading tests. The data show that the scaly clays have very high swelling indexes, even higher than the compression index when at high pressures (Figure 12), and that they follow very large hysteric loops in unloading-loading cycles. The loops of samples I6/3 and P8/12, that are of higher swelling capacity than the sample P8/5, appear to be larger than the loop followed by P8/5. After these big loops, with recompression the sample states go back to the state boundary curve reached by the samples subjected solely to compression from the initial consolidation state, i.e. the clay reaches always the same normal consolidation line (Figures 9, 10).

Samples P8/5 and P8/12 appear to follow a new swelling line during swelling after compression post-gross yield (σ'_y~1.2÷1.8 MPa; Figures 9, 10), which is parallel to but different from the initial one and displaced from this of an irrecoverable void ratio difference. This has been observed also for other samples, except for I6/3, which happens to exhibit the largest unload-reload loop. Thus, in general, the swelling loops appear to shift vertically with compression post-gross yield, as expected for reconstituted clays according to critical state soil mechanics. However, with scaly clays the unloading-reloading loops are so large as to make necessary to refer to kinematic hardening elasto-plastic models in order to represent the soil behaviour in unloading.

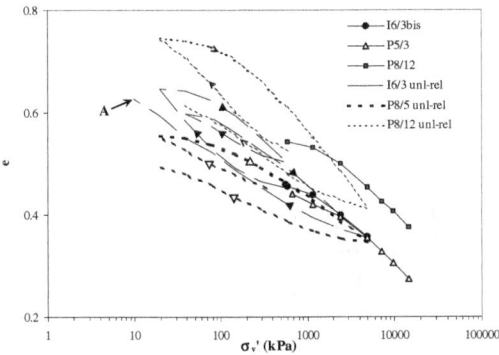

Figure 19. Loading and unloading-reloading oedometer tests of the Senerchia samples.

During reloading, the state path seems to join the normal consolidation line at void ratios higher than the gross-yield void ratio recorded in compression from the initial undisturbed state, at pressures lower than the initial gross yield pressure. Thus, swelling appears to cause a change in gross yield state, making it shift upwards and to the left. This effect differs from that observed for sensitive bonded clays (Cotecchia & Chandler 1998), for which destructuring during swelling causes a reduction in gross yield pressure associated to a reduction in void ratio.

The high swelling indexes observed for scaly clays (Cs = 0.06 ÷ 0.148) result from their high content in active swelling minerals (Table 2). These minerals tend to swell, so that the increase in void ratio with unloading is associated not only to an increase in both inter-scale and intra-scale pores, but also to an increase in volume of the swelling minerals themselves. Figure 20 shows a high magnification SEM picture of the fabric of the clay after swelling (state A in Figure 19). The picture shows the open fissures between the clay aggregates (presumably scales) and a mantle of small swelled clay particles covering the aggregates.

With compression post-gross yield, the swelling index of the natural Senerchia Clay does not increase. This observation would suggest that post-gross yield compression does not give rise to a significant weakening of the intra-scale bonding for Senerchia Clays.

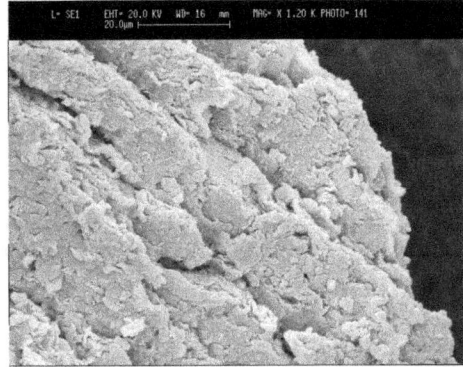

Figure 20. Fabric of the natural sample I6/1 on a vertical fracture after swelling (point A in Fig. 19).

7 CONCLUSIONS

The results of the investigation of the microfabrics of the Senerchia scaly clays and of their compression behaviour have outlined their type of structure and the relating response to compression to high pressures. The observed structural features and changes with compression, as well as the observed trends of mechanical response, are all seen to be consistent with the use of a critical state-type framework to represent the scaly clay behaviour. Accordingly, the yield stress ratio YSR is still seen as a parameter to which the soil response can be related, being gross yield located to the left of the ICL. For YSR values calculated with reference to this gross yield state, the soil appears to contract during shear after consolidation to low YSRs, and to dilate when consolidated to higher YSRs, as would be expected according to critical state soil mechanics. Thus, the stress sensitivity S_σ is lower than 1 for scaly clays, against the S_σ values above 1 of sensitive clays.

The presence of a network of fissures in the clay makes its strength properties particularly poor and makes the soil reach very low maximum stress ratios even when dilating, on the dry side. The presence of highly swelling minerals makes the unload-reload loops of scaly clays particularly large and causes the gross yield state to shift upwards along the normal consolidation line.

The research needs to be further developed for a deeper insight into the shear behaviour of the clay, but it seems possible to interpret the clay response as that of a material made of very packed scales, which are bent and deformed with loading and swell in unloading. The fissures separating the scales increase in opening with swelling and are progressively reduced in opening with compression. As a result, the clay compression curves are characterized by non-pronounced gross yield states.

ACKNOWLEDGMENTS

The authors thank the CNR-Institute of Clays Research (Tito, Potenza) for the mineralogical analysis and the CNR-Alpine and Quaternary Geodinamic Research Centre for scanning electron microscopy analysis.

The research has been carried out with the financial support of the CNR-GNDCI (Linea 2; 1999-2000) and of the Ministery for the University and the Scientific Research (MIUR, 2001).

REFERENCES

A.G.I. 1979. Some italian experiences on the mechanical characterization of structurally complex clay soils. International Society of Rock Mechanics (ed.), *Proc. 4th intern. congress of the International Society of Rock Mechanics*, 1: 827-846.

Antonino, D. 1999. Comportamento meccanico delle argille strutturalmente complesse dei pendii instabili di Serra dell'Acquara-Vadoncello (Senerchia, AV). *Tesi di Laurea*, Politecnico di Bari.

Airo' Farulla C. & La Rosa, G. 1977. An analysis of some factors influencing Atterberg limits determination of stiff fissured clay. *The Geotechnics of Structurally Complex Formations; Proc. intern. symp.*, Capri, 2: 23-29.

ASTM 1976. Annual Book of ASTM standard, Part 19. Philadelphia.

Bertolucci, P., Cotti, I. & Pagliara P. 1991. Caratteristiche di resistenza al taglio delle Argille Scagliose di S. Barbara. *Deformazioni in prossimità della rottura e resistenza dei terreni naturali e delle rocce*, Ravello, 27-28 February 1991.

Burland, J.B. 1990. On the compressibility and the shear strength of natural clays. *Géotechnique* 40 (3): 329-378.

Butterfield, R. 1980. A natural compression law for soils. *Géotechnique*, 24 (4): 469-480.

Cotecchia, F. 1996. The effects of structure on the properties of an Italian Pleistocene Clay. *PhD thesis*, Imperial college of Science, Technology and Medicine, London University, London.

Cotecchia, V. 1971. Su taluni problemi geotecnici in relazione alla natura dei terreni della regione pugliese. *Rivista Italiana di Geotecnica*, 1: 1-33.

Cotecchia, F. & Chandler, R. 1998. One-dimensional compression of a natural clay: Structural changes and mechanical effects. In A. Evangelista & L. Picarelli (eds), *The Geotechnics of Hard Soils-Soft Rock*; *Proc. intern. symp.*, Napoli, *12-14 October 1998*. Rotterdam: Balkema.

Cotecchia, F. & Chandler, R. 2000. A general framework for the mechanical behaviour of clays. *Géotechnique*, 50 (4): 431-447.

Di Maio, C. & Fenelli G. B. 1997. Clayey soil deformability: the influence of physicochemical interactions. *Rivista Italiana di Geotecnica*, 1: 695-707.

Esu, F. 1977. Behaviour of slopes in structurally complex formations. *The Geotechnics of Structurally Complex Formations*; *Proc. intern. symp.*, Capri, 2: 292-304.

Fearon, R. 1998. The behaviour of a structurally complex clay from an italian landslide. *PhD thesis*, City University, London.

Fearon, R. & Coop, M. R. 2000. Reconstitution - what makes an appropriate reference material? *Géotechnique*, 50 (4):471-477.

Fearon, R. & Coop, M. R. 2002. The influence of landsliding on the behaviour of a structurally complex clay. *Quarterly Journal of Engineering Geology*, 35: 25-32.

Fenelli, G. B., Picarelli, L. & Silvestri, F. 1992. Deformation process of a hill shaken by Irpinia earthquake in 1980. *Slope Stability in Seismic Areas*; *Proc. French-Italian conf.*, Borfighera.

Guerriero, G., Olivares, L. & Picarelli, L. 1995. Modelling the mechanical behaviour of clay shales: some experimental remarks. *Colloquium Mundanum "Chalk and Shales"*, Bruxelles, 2.1.20-2.1.30.

Janbu, N. & Senneset, K. 1979. Consolidation tests with continuous loading. *Proceedings of the 10th International Conference of Soil Mechanics and Foundation Engineering*, Stockholm, 1: 645-654.

Jappelli R., Liguori, V., Umiltà, G. & Valore, C. 1977. A survey of geotechnical properties of a stiff highly fissured clay. *The Geotechnics of Structurally Complex Formations*; *Proc. intern. symp.*, Capri, 2: 91-106.

Losurdo, V. 2000. Proprietà geotecniche delle Argille Varicolori a scaglie. Il caso di Senerchia (AV). *Tesi di Laurea*, Politecnico di Bari.

Ogniben, L. 1969. Schema introduttivo alla geologia del confine calabro-lucano. *Memorie della Società Geologica Italiana*, 8 (4): 453-763.

Olivares, L. 1996. Caratterizzazione dell'Argilla di Bisaccia in condizione monotone, cicliche e dinamiche e riflessi sul comportamento del "Colle" a seguito del terremoto del 1980. *PhD thesis*, Università degli Studi Federico II, Napoli.

Pellegrino, A. & Picarelli, L. 1982. Contributo alla caratterizzazione geotecnica di formazioni argillose intensamente tettonizzate. *Geologia Applicata ed Idrogeologia*, 20 (2): 155-192.

Picarelli, L. 1986. Caratterizzazione geotecnica dei terreni strutturalmente complessi nei problemi di stabilità dei pendii. Associazione Italiana di Geotecnica (ed.), *XVI Convegno Nazionale di Geotecnica, 14-16 Maggio, Napoli 1986*.

Picarelli, L. 1998. Properties and behaviour of tectonised clay shales in Italy. In A. Evangelista & L. Picarelli (eds), *The Geotechnics of Hard Soils-Soft Rock*; *Proc. intern. symp.*, Napoli, *12-14 October 1998*. Rotterdam: Balkema.

Picarelli, L. & Urcioli G. 1993. Effetti dell'erosione in argilliti ad alta plasticità. *Rivista Italiana di Geotecnica*, 1: 29-47.

Picarelli, L. & Olivares, L. 1998. Ingredients for modelling the mechanical behaviour of intensely fissured clay shales. In A. Evangelista & L. Picarelli (eds), *The Geotechnics of Hard Soils-Soft Rock*; *Proc. intern. symp.*, Napoli, *12-14 October 1998*. Rotterdam: Balkema.

Pini, G. A. 1999. Tectonosomes and Olisostromes in the Argille Scagliose of the Northern Apennines, Italy. The Geological Society of America (ed.), *Special Paper n. 335*, Colorado.

Rippa, F. & Picarelli, L. 1977. Some considerations on index properties of Southern Italy shales. *The Geotechnics of Structurally Complex Formations*; *Proc. intern. symp.*, Capri, 1: 401-406.

Santaloia F., Cotecchia F. & Polemio M. 2001. Mechanics of a tectonized soil slope: influence of boundary conditions and rainfalls. *Quarterly Journal of Engineering Geology*, 34: 165-185.

Skempton, A. W. 1966. Some observation of tectonic shear zones. International Society of Rock Mechanics (ed.), *Proc. 1st intern. congress of the International Society of Rock Mechanics*, 6: 329-335.

Walker, B. F., Blong, R.J. & McGregor, J. P. 1987. Landslide classification, geomorphology and site investigation. *Soil Slope Instability and Stabilisation*: 1-52, Rotterdam, Balkema.

Soft Ground Engineering in Coastal Areas, Tsuchida et al. (eds)
© *2003 Swets & Zeitlinger, Lisse, ISBN 90 5809 613 0*

Effect of drying and preparation method on microscopic analysis on Pleistocene clays using mercury intrusion porosimetry

M.S. Kang, Y. Watabe & T. Tsuchida
Independent Administrative Institution, Port and Airport Research Institute, Yokosuka, Japan

ABSTRACT: This research reviewed and investigated the effect of drying method for pore size distribution measurement according to various soils and test conditions, in utilizing Mercury Intrusion Porosimetry(MIP). For drying these soils, three types of drying methods such as air, oven and freeze drying were adopted for comparison. Through examining these results, it has been known that test results from MIP are largely affected by the drying process of specimen and that inappropriate adoption of drying method may lead to unrighteous understanding on the microstructure of fine-grained soils.

1 INTRODUCTION

Lately, microscopic instruments like Scanning Electron Microscope (SEM) for the observation of photographic images and Mercury Intrusion Porosimetry (MIP) for the evaluation of pore size distribution, have been considered as common procedures in the studies on the microscopic structure of fine-grained soils[1].

However, even though there are a number of results obtained by active researches and many of data have been accumulated so far, questions are still under discussion on the generalization of specimen preparation, testing procedures and test parameters when soils are used as the objective of studies.

For example, in both use of SEM and MIP for clays, the elimination of water in clays is mutually required while maintaining its inherent soil structure. The effect of the drying and preparation methods for specimen on test results is not well known and the specification on the selection of these methods considering the soil condition has not yet been established.

For these reasons, this study is to suggest the appropriate soil preparation method for the sample used to the microscopic studies, and this will be a help to the comprehensive and proper use of information and data provided by different researchers and institutes. In the fundamental process to this purpose, this research investigated the effect of drying method on the pore size distribution using mercury intrusion porosimetry according to various soil conditions and different drying methods.

2 TEST EQUIPMENTS AND METHODS

2.1 *Basic concept and principle of porosimeter*

Volumetric pore size distribution of soils can be determined using the mercury intrusion method. Measurement of intruded volume of mercury into pore spaces by staged-increase of external pressure results in the pore volumes equivalent to each pore entrance size. Mercury, a non-wetting fluid, due to its extremely large surface tension and high contact angle of fluid-to-solid (> 90°) will not enter into the pores without any application of external pressure. The mercury intrusion method uses this principle and in the basis of assumption that all the pores are cylindrical shape and are continuously connected, pore size of soils can be obtained by the following equation of relationship between the pore size (d_p) and external pressure (p) proposed by Washburn (1921).

$$d_p = -4\sigma \cos\theta / p \qquad (1)$$

where
σ : surface tension of mercury(N/m)
d_p : diameter of pore intruded(m)
p : applied external pressure(Pa)
θ : contact angle of mercury with solid (°)

The calculation of pore size diameter by the mercury intrusion type porosimeter is based on following assumptions and restrictions.
- All the pores are continuous in the cylindrical shape.
- The soil specimen tested should be fully dried beforehand.

- During the test, there is no change or compression of soil structure by the intrusion pressure. (Lawrence, 1978)
- There is the measurable range for pore sizes according to the minimum and maximum pressures by the capacity of equipment.
- Because the pore sizes are measured from large pores to small ones, there could be some errors of measurement in the case that a larger pore exists behind a smaller pore.

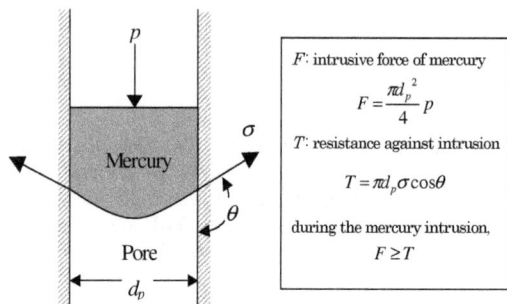

F: intrusive force of mercury

$$F = \frac{\pi d_p^2}{4} p$$

T: resistance against intrusion

$$T = \pi d_p \sigma \cos\theta$$

during the mercury intrusion,

$$F \geq T$$

Figure 1. Calculation of pore entrance size

2.2 Test method and equipment

The soil sample used in the porosimeter test should be dried causing the minimum change of volume and soil structure. The dried sample is carefully cut into the shape with an appropriate size.

The weight for a dried sample of ordinary clays is about 0.5-1g which is varied with the type and the water content of soil and with the capacity of the glass cell. In case of soil sample with the pore volume which is larger than 90% of the maximum test capacity of a glass cell, it is necessary to make the size of a specimen even smaller. After measuring the weight of soil specimen by the electrical decimal balance with the resolution of 1/1000g, the stem cell is assembled. The assembly of a cell is set into the main equipment.

As the test starts, the inside of the cell is vacuumed for eliminate the blocking air between the pores and then the mercury is intruded by the step increase of intrusion pressures. The pore size is calculated according to each stage of intrusion pressure.

Fig. 2 shows the picture of the porosimeter (Autopore III 9400, Micromeritics) which is equipped in Port and Airport Research Institute and used in this study. This porosimeter features the controlled mercury intrusion from sub-ambient pressure to 228MPa and the range of pore size distribution from 0.005μm to 369μm in diameter can be determined by the step increasing intrusion pressure.

It has been reported that surface tension of mercury which is dependent on the temperature, has little influence to the test results from porosimeter test and

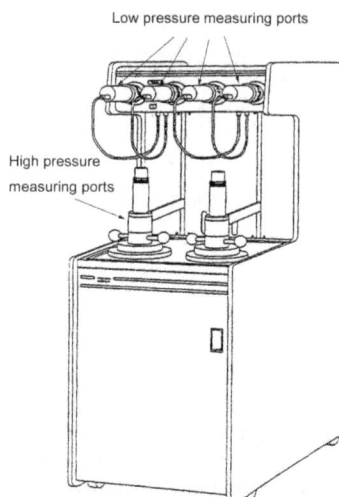

Figure 2. Porosimeter (Autopore III 9400)

generally the value of 0.484N/m at 25°C is adopted in the test. However for contact angle of mercury with soil, different values ranged from 130 to 160° by various researches Penumadu and Dean(2000) has reported that using 130° and 160° would result in the more than 30% difference between the two in calculating the pore diameter. Considering their results, in this study, the contact angle of mercury drop on the surface of clay sample was measured using the ultra-precision digital microscope as shown in Fig.3. As a result, 153° for freeze dried sample and 147% for oven and air dried sample in average, were obtained and used in the porosimeter test.

Considering that the increasing rate of intrusive pressure during the mercury intrusion into the pores by step pressure increase would affect the test result, preliminary tests with the varied equilibrium interval time for each step as 5, 10, 20 seconds, were performed. Fig.4 shows the results of this preliminary

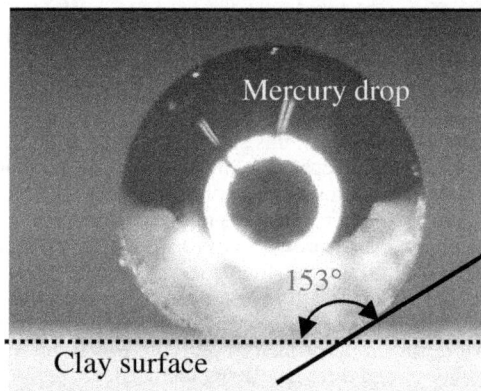

Figure 3. Measurement of contact angle

test. As it can be seen in the figure, the effect of equilibrium interval time is negligible. However, due to the fact that some instability of data was observed for the case of 5 seconds and that it may be difficult to sustain the pressure constantly for long time especially at higher pressure, equilibrium time for this study was determined as 10 seconds for each pressure step.

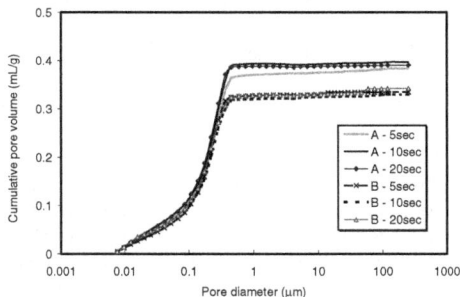

Figure 4. Effect of equilibrium time interval

2.3 Drying method

As the fundamental condition for the measurement of pore size of clays, it is necessary to remove the pore water completely without causing any structural changes to soil skeleton.

Oven drying and air drying are generally used drying methods for soil by their simplicities. Collins and McGown(1974) suggested that oven drying and air drying methods can be used to the clay specimens with large content of silt or sand. However no distinctive data were reported to support their proposition. Generally when fine grained soils are used, those methods are more likely to induce a large amount of shrinkage which eventually brings about the significant changes to soil structure specially of clays with higher water contents.

For this reason, the freeze drying method is the one often adopted in the researches on soil structure.

In freeze drying, the soil specimen is dipped in a liquefied nitrogen to be frozen instantaneously so that the water inside the soil is changed into noncrystallized ice avoiding the volumetric increase.

Then, the frozen specimen is put in a vacuumed cell of the freeze dryer in which drying by sublimation is sustained for about 24 hours.

However, by the freeze drying may also cause structural changes like shrinkage and cracks during the freezing and drying process for soils with higher water content. But the amount of shrinkage would be minimal compared to other drying methods. It is necessary to pay attention to determine the size of the specimen. The freeze drying method is only efficient with small specimens and instant freezing. Larger specimens and slow freezing may cause cracks in soils due to inconstant freezing.

To investigate the influence of drying method to the soil structure in terms of pore size distribution, three types of drying methods such as oven drying, air drying and freeze drying are compared. The soils used in this study are obtained from Holocene and Pleistocene deposit layers of Osaka Bay area in each depth of 20, 100 and 200m below the sea water level. Examining the soils from different depth could find out the effect of consolidation yielding pressure of soil.

For oven drying, soils are dried in the 110°C oven for 24 hours and for air drying, soils are dried at the atmospheric condition for 5 consecutive days and another 24 hours in 110°C oven to make sure the soil is fully dried. For freeze dried sample is prepared by the method explained above.

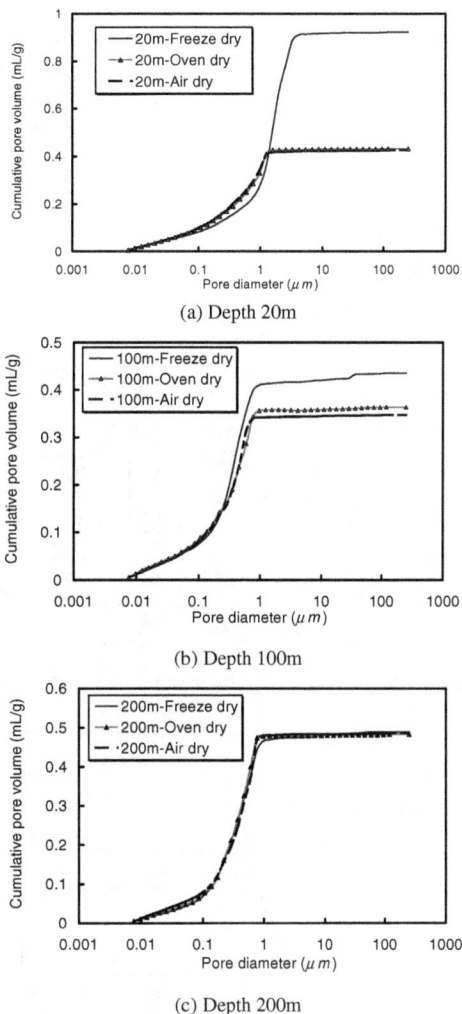

(a) Depth 20m

(b) Depth 100m

(c) Depth 200m

Figure 5. The cumulative intrusion volume with pore size distribution of soils from different depth

3 TEST RESULTS OF PORE SIZE DISTRIBUTION

3.1 *Test results and discussions*

In Fig.5, pore size distributions of soils from different depth as 20, 100 and 200m below sea level are compared. The soil sample of 20m which is the Holocene layer by air and oven drying reveals up to 50% of large shrinkage compared to the one by freeze drying. The difference between air drying and oven drying was minimal.

However, this amount of shrinkage by air and oven drying decreases with the increase of the soil depth, which eventually shows almost the same distributions by all three drying methods for the soil from 200m. Fig.6 rearranged these results into the classification of pore sizes defined by Matsuo and Kamon(1976), Miura et al. (1999).

From the soil from 20m dried by freeze drying, it can be said that almost 70% of pores is occupied with pores larger than 1μm, which are drastically decreased by the air and oven drying and that the pores larger than 1μm are dominantly involved with the shrinkage by drying. The results from 100m soil reveals that some of pores smaller than 1μm seem to take a part of role to the shrinkage.

Comparing the water content of 100m and 200m soils, it seems that water content has much less effect to the shrinkage than expected. Therefore, it can be estimated that the stress level of a soil may have a strong influence to the mechanism of shrinkage during drying process.

For this, a Pleistocene clay sample which was consolidated by the step loading according to standard consolidation test method, was tested by the porosimeter test to find out the effect of the stress level of soils. From the results of Fig.5 where the difference between the air drying and oven drying is negligible, only the oven dried sample is compared with the one freeze-dried. As it can be seen in the Fig.7, up to a certain level of consolidation stress, 1.26MPa for this test, the total intrusion volume of oven dried sample shows about 15% smaller value than the freeze dried one indicating that shrinkage is oc-

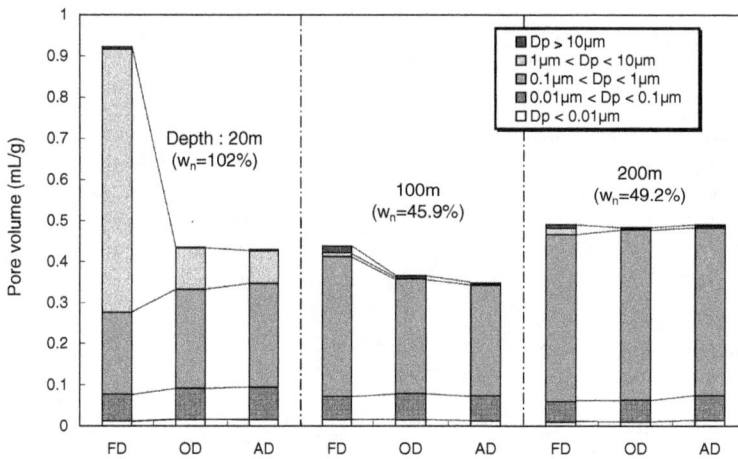

Figure 6. Comparison of pore size according to soils from different depth

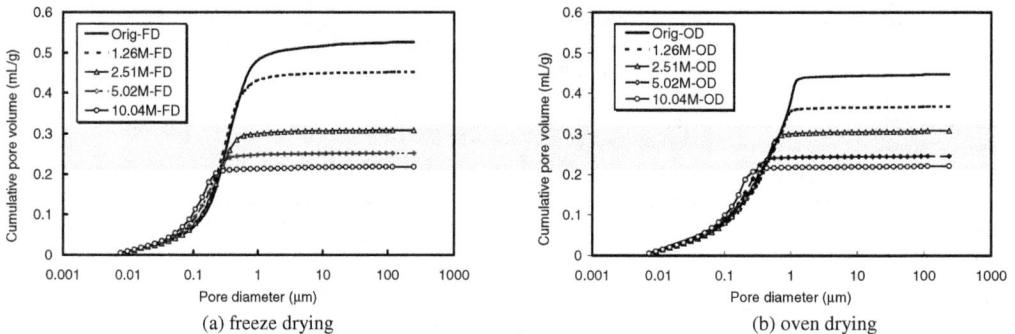

(a) freeze drying

(b) oven drying

Figure 7. Comparison of drying method for consolidation stress

curred. However after exceeding this stress level, larger stress than 1.26MPa, both samples by oven and freeze drying show almost the same value.

Shrinkage during the drying process is derived from the decrease of the distance between the adjacent soil particles due to the increase of meniscus curvature of pore water by evaporation dragging the soil particles closer. Therefore, it can be estimated that shrinkage depends on the resistance level of stress in the soil.

In this study, the amount of shrinkage, S after drying is expressed with the ratio of total intrusion volume (I_s) to total pore volume ($I_v=V_v/M_s$) as following;

$$S = (1 - I_s / I_v) \times 100 \qquad (2)$$

The total pore volume is obtained by the apparent density and water content of soil.

The relation of consolidation stress with shrinkage, S is represented in Fig.8 comparing the oven drying with freeze drying.

While the freeze dried sample showed almost constant value of shrinkage less than 5% against the consolidation stress, up to the stress level of about 2.5-3MPa, the shrinkage of oven dried sample showed radical decrease at first, then it came to a constant value less than 5% which is about the same amount of shrinkage by freeze drying.

Fig.9 compares all the specimens dried by three drying methods, in the relation of total intrusion volume with the pore volume calculated by water content. As it can be seen, freeze drying method is most likely to show well accordance with the pore volume by water content.

In Fig.10, e-log p curves by consolidation test results from standard consolidation test and constant rate of strain consolidation test (CRS) with mercury intrusion porosimetry results for freeze dried sample. These curves indicates that the curve by porosimeter test is a little smaller than the ones by consolidation test and that, however, it well represents the consolidation behavior of soil.

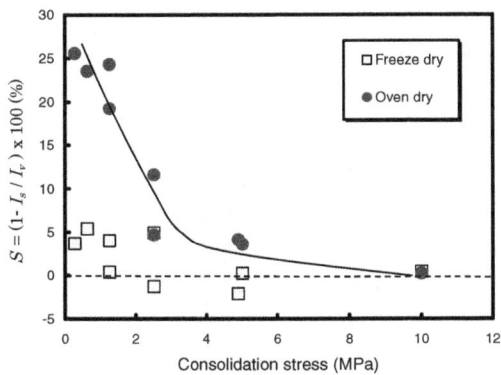

Figure 8. Consolidation stress and shrinkage

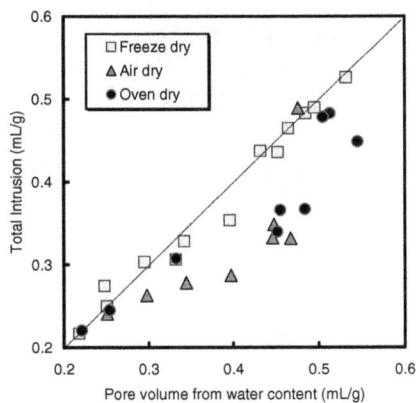

Figure 9. Comparison of all the drying methods used in this study

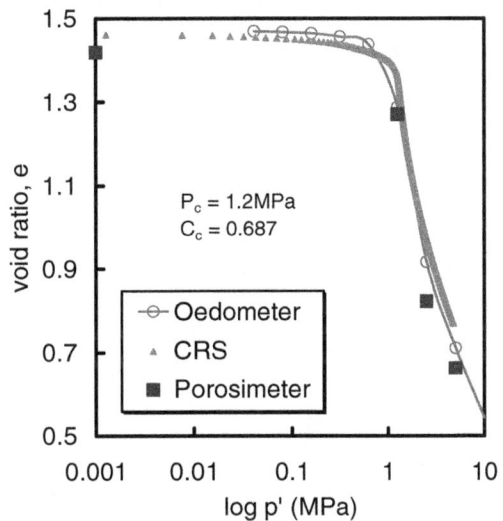

Figure 10. Comparison of e - log p' for MIP with consolidation test results

4 CONCLUSIONS

This study has investigated on the specimen preparation method, specially on the effect of drying method to the result of porosimeter test at the fundamental level. The results can be summarized as follows;

1) Results from different depths of soil layers revealed that the shrinkage by air and oven drying compared to freeze drying, increases with the

decrease of soil depth implying the shrinkage is is strongly dependent on the stiffness of soil. It was also found that pores larger than 1μm are dominantly involved with the shrinkage of soil. Soils with less amount of larger pores than 1μm, relatively showed less shrinkage. In either case, pores smaller than 0.1μm showed no change by drying.

2) The sample dried by oven and air drying method, showed the larger shrinkage than the one by freeze drying, while air drying and oven drying showed almost the same values through all the test results. The soil of higher water content or low yield stress resulted in larger shrinkage.

3) From the relation of consolidation stress of soils with the amount of shrinkage, freeze dried sample showed smaller shrinkage less than 5% which comprises that it is irrelevant to consolidation pressure. Meanwhile the shrinkage of oven dried sample showed radically decreasing tendency up to 2.5-3MPa of consolidation stress and after that, showed almost the same shrinkage as freeze dried sample bounded less than 5%.

REFERENCES

Collins, K. and McGown, A., The form and function of microfabric features in a variety of natural soils, *Géotechnique* 24, No.2, pp.223-254, 1974.

Delage, P. and Lefebvre, G., Study of the structure of a sensitive Champlain clay and of its evolution during consolidation, Can. Geotech. J. 21, pp.21-35, 1984.

Griffiths, F. J. and Joshi, R. C., Change in pore size distribution due to consolidation of clays, *Géotechnique* 39, Technical note, No.1, pp.159-167, 1989.

Lawrence, G. P., Stability of soil pores during mercury intrusion porosimetry, J. Soil Sci. 29, 299-304., 1978.

Matsuo, S. and Kamon, M., On the terms for structure of clays, Tsuchi To Kiso, Vol.24, No.1, pp.59-64, 1976.

Miura, N., Yamadera, A.and Hino, T., Consideration on compression properties of marine clay based on the pore size distribution measurement, Journal of JSCE, No.624, III-47, pp.203-215, 1999

Penumadu, D. and Dean, J., Compressibility effect in evaluating the pore-size distribution of kaolin clay using mercury intrusion porosimetry, Can. Geotech. J. 37, pp.393-405, 2000.

Washburn, E. W., A note on a method of determining the distribution of pore sizes in a porous material, Proc. National Academy of Science, Vol.7, pp.115, 1921.

Yamaguchi, H. and Ikenaga, H., Utilization of mercury intrusion porosimetry apparatus for evaluation of soil structure, Tsuchi To Kiso, Vol.41, No.4, pp.15-20, 1993.

Soft Ground Engineering in Coastal Areas, Tsuchida et al. (eds)
© 2003 Swets & Zeitlinger, Lisse, ISBN 90 5809 613 0

Estimation of the strain rate effect influenced to the consolidation characteristic of Osaka Bay Pleistocene clay

N. Ohmukai
OYO Corporation Core Laboratory Center, Saitama, Japan

F. Rito
OYO Corporation Tokyo Branch, Tokyo, Japan

H. Tanaka
Port and Airport Research Institute, Yokosuka, Japan

J. Mizukami
Kansai International Airport Co., Ltd., Osaka, Japan

ABSTRACT: Osaka Bay Pleistocene clay is called the quasi-over-consolidated clay. In the present study, a series of one-dimensional consolidation tests were carried out to examine effects of the strain rate on consolidation characteristics of Osaka Bay Pleistocene clay. Consequently, it has been verified that the ε_v – log σ_v curve shifts towards the left and the consolidation yield stress σ'_y becomes smaller, as the strain rate becomes slower. The consolidation yield stress is reduced by about 10% as the strain rate decreases to one tenth, assuming the strain rate $\dot{\varepsilon}_v = 3.3 \times 10^{-6} s^{-1}$ as the standard. Furthermore, this study proved that it is possible to estimate the strain rate effects on the consolidation yield stress of Osaka Bay Pleistocene clay by changing the strain rate during the consolidation course of CRS test, and that the strain rate effects are not influenced by the plasticity index I_P.

1 INTRODUCTION

Osaka Bay is located at the southwestern part of Japan and a number of reclamation development construction works in a large scale are taking place there in recent years. Along with these reclamations, problems on settlement are being enlarged not only on Holocene clay but also on Pleistocene clay. Pleistocene clay deposited on the sea bottom of Osaka Bay is aged normally consolidated clay that has never experienced any higher stress in the past than present effective overburden stress, which makes estimation of the settlement more difficult. In other words, over-consolidation of Pleistocene clay in Osaka Bay is quasi over-consolidation due to aging or a cementation effect formed by the elapse of time after deposition of clay, and it is important to fully understand the consolidation characteristics of such quasi-over consolidated clay.

The consolidation characteristics of Osaka Bay clay is usually determined by incremental loading consolidation test for 24 hours. However, continuous data cannot be obtained because the incremental loading consolidation test requires loading with incremental rate of 1, and determination of the accurate consolidation yield stress and estimation of the compressibility in the neighborhood is difficult as the loading interval is larger in the neighborhood of consolidation yield stress of Pleistocene clay sampled in great depth[1]. To solve this problem, the consolidation tests with constant rate of strain (CRS test) that enable us to obtain continuous data have come to be adopted. And yet, the CRS test generally requires a higher strain rate compared with incremental loading consolidation test, and as a consequence, higher consolidation yield stress is obtained as result. Therefore, it is more important for us to estimate effects of the strain rate on the consolidation yield stress.

Studies made by Larsson et al.[2] and Lerouil et al.[3] can be quoted as an example of the studies where a difference in the strain rate gives an effect on the consolidation yield stress. Larsson et al. showed a result of the experiment using both organic clay and ordinary clay where the consolidation yield stress would have been constant regardless of the strain rate, if a strain rate is smaller than $\dot{\varepsilon}_v=2.5\sim3.5\times10^{-6} \ s^{-1}$. In Swedish Geotechnical Institute(SGI), the standard strain rate for CRS test is fixed as $\dot{\varepsilon}_v =2.0\times10^{-6}s^{-1}$. On the other hand, Lerouil et al. reported results of the test using Canadian clay and Finnish clay that consolidation yield stress is reduced by about 10%, if the strain rate becomes smaller by one order. In this study, a series of CRS tests with different strain rates and creep tests were carried out to examine effects of the strain rate on the consolidation characteristics of Osaka Bay Pleistocene clay.

2 CHARACTERISTICS OF CLAY AND TEST PROCEDURES

In this study, using the Pleistocene clay layer Ma12 of Osaka Bay carried out three kinds of consolidation tests. Material and mechanical properties of the samples were; ρ_s=2.657, w_L=87.0%, w_P= 33.2%, w_n=65~72%, I_P=53.8, effective overburden stress σ'_{v0}=361kPa, consolidation yield stress σ'_y=530kPa, and OCR =1.47. Test procedures are as follows:
(a) CRS test

The CRS tests were carried out using 4 specimens from a same sampler for different strain rates for each specimen, namely $\dot{\varepsilon}_v$ =3.3×10^{-5}s^{-1}, 3.3×10^{-6} s^{-1}, 3.3×10^{-7}s^{-1}, 3.3×10^{-8}s^{-1}.
(b) Special CRS test (SCRS test)

The SCRS test is a test to grasp the strain rate effects using a single specimen, by changing the strain rate during the test within a range of $\dot{\varepsilon}_v$ =3.3× 10^{-5}~3.3×10^{-8}s^{-1}.

For your reference, CRS tests and SCRS tests were based on procedures prescribed in Japanese Industrial Standard (JIS). Backpressure was set at 98kPa in this test.
(c) Creep test

Creep tests were carried out using 8 specimens sampled from a same sampler. The specimens were, after preliminary loading, loaded with the stress equivalent to the effective overburden stress for 24 hours, and then loaded instantly with the effective stress set for respective specimen for 70 days to complete a creep test. The effective stress applied was 392, 431, 471, 510, 549, 667, 981 and 1608 in kPa respectively.

3 TEST RESULTS

3.1 CRS and SCRS tests

3.1.1 ε_v-log σ'_v curve

The ε_v-log σ'_v curve of CRS test results and SCRS test results are shown in Fig. 1 and Fig. 2. Strain values shown in these figures are adjusted as ε_{v0}=0 at σ'_{v0}=361kPa, where σ'_{v0} is the effective overburden stress.

It can be seen from Fig. 1 that the ε_v -log σ'_v curve moves to right side and the consolidation yields stress becomes larger, as the strain rate becomes higher. Furthermore, It can be found that ε_v - log σ'_v curves are almost parallel, except that the ε_v- log σ'_v curve for $\dot{\varepsilon}_v$ = 3.3×10^{-6}s^{-1} is projected at the neighborhood of consolidation yield stress. In Case 1 of Fig. 2, the strain rate is gradually made slower like ε_{v1}=3.3×10^{-5}s^{-1}→ ε_{v2}=3.3×10^{-6}s^{-1}→ ε_{v3}=3.3× 10^{-7}s^{-1} and then made faster like ε_{v3}=3.3×10^{-7}s^{-1} → ε_{v2}=3.3×10^{-6}s^{-1}→ ε_{v1}=3.3×10^{-5}s^{-1}. In Case 2, the test

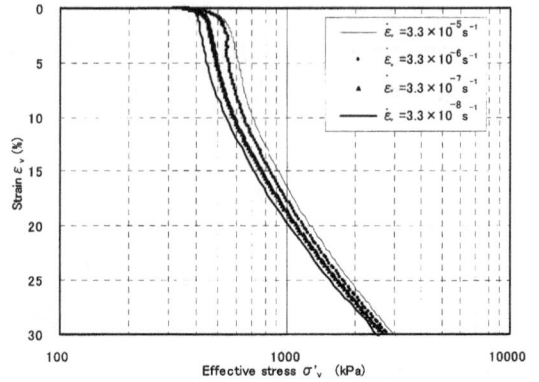

Figure 1. ε_v –log σ'_v curves from CRS tests

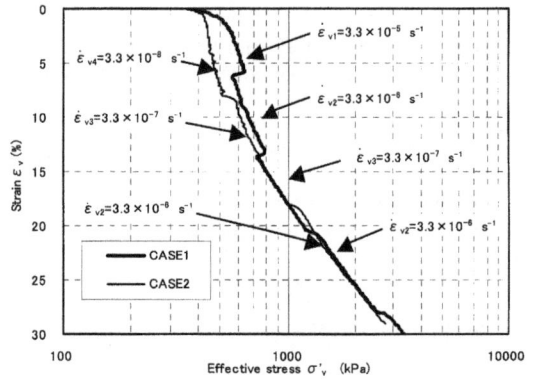

Figure 2. ε_v –logσ'_v curves from SCRS tests

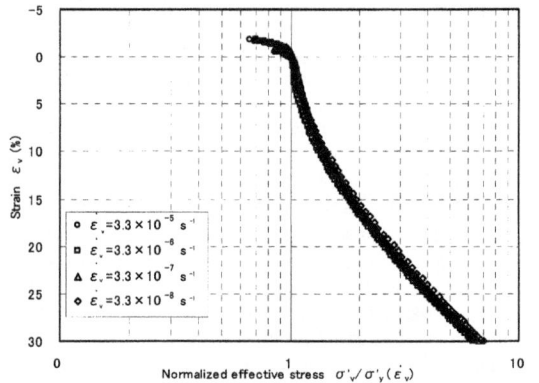

Figure 3. Relationship between strain and normalized effective stress

was carried out with a condition that the strain rate is further made faster in a manner like ε_{v4}=3.3×10^{-8}s^{-1} → ε_{v3}=3.3×10^{-7}s^{-1} → ε_{v2}=3.3×10^{-6}s^{-1}. All specimens used in Case 1 and Case 2 are prepared from the same sampler. When the strain rate is higher during the test using the same specimen, the ε_v-log σ'_v curve is shifted to the right, while the strain rate is slower, the ε_v-log σ'_v curve is shifted to the left. If the strain rate is kept constant in Case 1 and Case 2, the ε_v-log σ'_v curves will almost match to each other.

Fig. 3 shows the relationship of ε_v-log σ'_v/σ'_y rearranged in accordance with Fig. 1. The strain ε_v in Fig. 3 is rearranged by assuming the strain ε_y as zero at the time when it reached the consolidation yield stress in the CRS tests for respective strain rate, and that the normalized stress is a value obtained by normalizing the effective stress by the consolidation yield stress at respective strain rate. With this results, curves representing the relationship of ε_v-log σ'_v/σ'_y form nearly a same curve and the results were the same as those of tests conducted by Lerouil using clay from Canada and Finland.

3.1.2 Compression index

Using test results as shown in Fig. 1, relations between the compression index C_C and values (σ'_v/σ'_y) obtained by normalizing the effective stress with the consolidation yield stress for respective strain rate are shown in Fig. 4. Those results show that all C_C obtained have the same shape regardless of the intensity of strain rate except a few deviations in the neighborhood of the maximum C_C. The Compression index C_C of this clay shows the maximum value $(C_C \fallingdotseq 4.0)$ at a point slightly beyond σ'_y and rapidly turns smaller thereafter. In the neighborhood of $\sigma'_v/\sigma'_y = 2$, it shows $C_C = 1.0$, which is about one third of the maximum value. Thereafter, C_C reduces as σ'_v/σ'_y increases, and it becomes $C_C = 0.5$ at $\sigma'_v/\sigma'_y = 20$.

3.1.3 Excess pore water pressure

Fig. 5 shows behaviors of excess pore water pressure measured at the bottom of specimen in the CRS test and the SCRS test. The average effective stress within specimens in the CRS and SCRS tests is calculated from a formula $\sigma'_v = \sigma_v - (2/3) \times \Delta u_b$ according to Japanese Industrial Standard (JIS).

From the above-stated test results, it can be seen that the excess pore water pressure has remarkably increased due to an increase of the effective stress, as the strain rate becomes higher. Furthermore, the test result for the strain rate being $\varepsilon_v = 3.3 \times 10^{-7} s^{-1}$ shows that the excess pore water pressure slightly increases as the effective stress. Increases, but there will be no excess pore water pressure observed for the strain rate being $\varepsilon_v = 3.3 \times 10^{-8} s^{-1}$. Behaviors of the excess pore water pressure in the SCRS test where the strain rate is changed during the test showed a result that the excess pore water pressure changes in correspondence with changes of the strain rate. Compared with test results of the SCR test, it can be verified that the intensity of excess pore water pressure is almost equal, if the strain rate is kept equal.

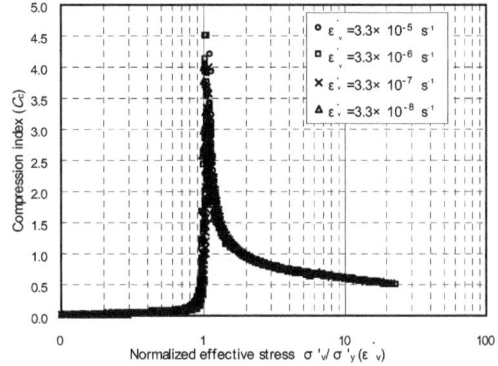

Figure 4. Relationship between compression index and normalized effective stress

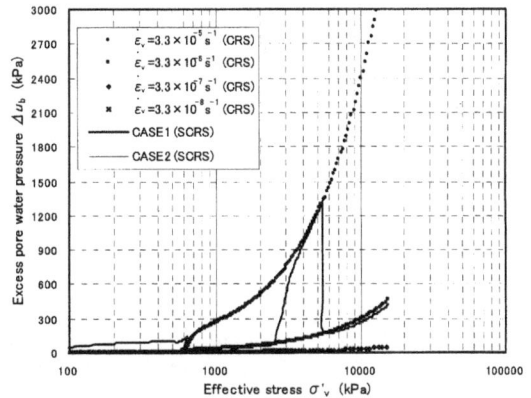

Figure 5. Excess pore water pressure from CRS tests and SCRS tests

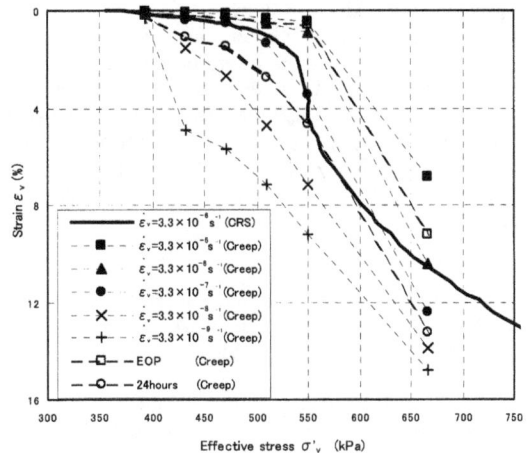

Figure 6. Constant ε_v curves from CRS tests and creep tests

3.2 Creep test

Fig. 6 shows results of a creep test with loading near a point of consolidation yield stress in a form of ε_v - σ'_v curve for respective strain rate. As a loading period was as long as 70 days, data for the strain rate $\dot{\varepsilon}_v$ = $3.3 \times 10^{-9} s^{-1}$ was obtained, which is one order slower than $\dot{\varepsilon}_v$ = $3.3 \times 10^{-8} s^{-1}$ in a case of a CRS test. Fig. 6 also shows test results on end of primary consolidation according to the $\sqrt{t}$ method and at the time of 24 hours elapsed, as well as results of CRS test for the strain rate $\dot{\varepsilon}_v$ = $3.3 \times 10^{-6} s^{-1}$.

According to the results, it can be seen that the ε_v - σ'_v curve shifts to the left and the consolidation yield stress becomes smaller, as the strain rate decreases. The strain rate on end of primary consolidation in the creep test is within a range of $\dot{\varepsilon}_v$ = $8 \times 10^{-6} \sim 2 \times 10^{-5} s^{-1}$ regardless of the loading and the ε_v - σ'_v curve is located to the right of the curve in the CRS test. And the strain rate after 24 hours remains within a range of $\dot{\varepsilon}_v$ = $7 \times 10^{-8} s^{-1} \sim 2 \times 10^{-7} s^{-1}$ except results of $\sigma'v$ = 392 kPa and the curve ε_v - σ'_v stays to the right of the curve in the CRS test. Over-shoot phenomenon of the compression curve in the neighborhood of consolidation yield stress is hardly observed when looking at a compression curve for $\dot{\varepsilon}_v$ = $3.3 \times 10^{-9} s^{-1}$.

4 DISCUSSION

It is shown in Fig. 1 and Fig. 6 that consolidation yield stress of Osaka Bay Pleistocene clay becomes smaller as the strain rate decreases. It is shown in Fig. 4 that the relation between the compression index and values (σ'_v/σ'_y) obtained by normalizing effective stress with consolidation yield stress for respective strain rate match to each other regardless of the strain rate. Also, as shown in Fig. 3, curves for ε_v-log (σ'_v/σ'_y) match to each other regardless of the strain rate. These results coordinate with those by Lerouil et al., and Lerouil et al. showed that the following formula is constituted from curves for two different strain rates $\dot{\varepsilon}_v = \dot{\varepsilon}_{v1}$ and $\dot{\varepsilon}_{v2}$ for which the consolidation yield stress are σ'_{y1} and σ'_{y2} respectively by giving effective stress at any desired strain[4].

$$\sigma'_{v1}/\sigma'_{y1} = \sigma'_{v2}/\sigma'_{y2} \tag{1}$$

It is possible from the above for us to assume the consolidation yield stress coping with changes of the strain rate in a SCRS test where the strain rate is changed.

Figs. 7 to 10 show test results conducted using 16 kinds of clays sampled from different depths than those adopted in the above-mentioned test. Those clays have a wide range of physical properties as understood by liquid limit w_L = 70 ~ 115%, plasticity index I_P = 45~75 and void ratio e = 1.1~2.0. Further,

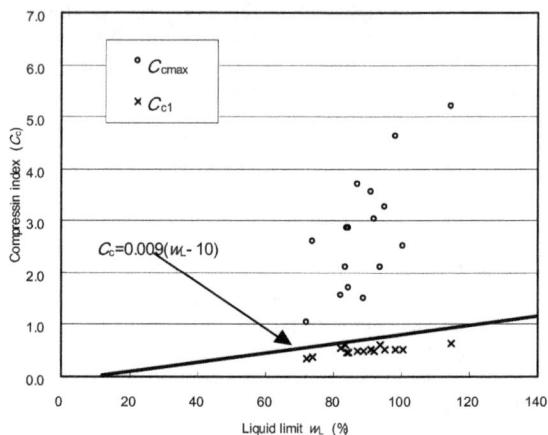

Figure 7. Relationship between compression index from CRS tests and liquid limit

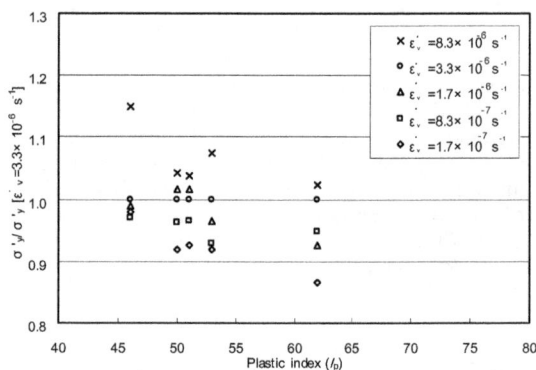

Figure 8. Relationship between normalized consolidation yield stress from CRS tests and plastic index

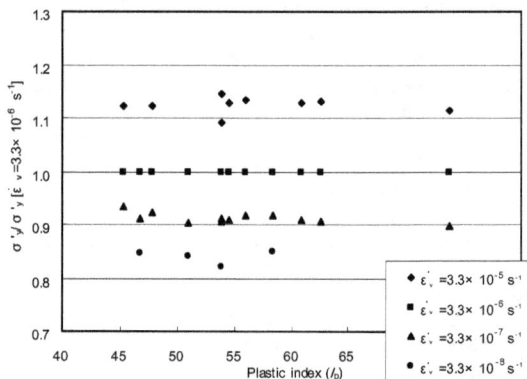

Figure 9. Relationship between normalized consolidation yield stress from SCRS tests and plastic index

82

the over-consolidation ratios possessed by those clays are all within a range of 1.2 ~ 1.6 in spite of the fact that they were collected from different depths.

Fig. 7 shows relationship between compression index and liquid limit of these clays. In this figure, C_{Cmax} represents the compression index in the neighborhood of consolidation yield stress of these clays and C_{C1} represents the compression index when the effective stress became large enough (in the neighborhood of $\sigma'_v = 10 \text{Mpa}$). Formulas in the figure are Skempton's equations[5]. Though C_{C1} indicates a smaller value than the value obtained from Skempton's equations, it can be seen that C_{Cmax} became a much larger value. This extremely high compressibility in the neighborhood of consolidation yield stress is a typical behavior of Osaka Bay Pleistocene clay.

Fig. 8 shows results of CRS tests conducted by using 5 clays with different liquid limits for the strain rate of $\dot{\varepsilon}_v = 8.3 \times 10^{-6} \text{s}^{-1}$, $3.3 \times 10^{-6} \text{s}^{-1}$, $1.7 \times 10^{-6} \text{s}^{-1}$, $8.3 \times 10^{-7} \text{s}^{-1}$ and $1.7 \times 10^{-7} \text{s}^{-1}$. This figure shows relationship between values obtained by normalizing the consolidation yield stress for respective strain rate with the strain rate of $\varepsilon_v = 3.3 \times 10^{-6} \text{s}^{-1}$ ($\sigma'_y / \sigma'_y [\varepsilon_v = 3.3 \times 10^{-6} \text{s}^{-1}]$) and the plasticity index I_P. Though the results are somewhat uneven, it seems that the consolidation yield stress slightly decreases as the plasticity index I_P increases. Secondly, Fig. 9 shows results of SCR tests conducted by using 11 clays with different liquid limits. The consolidation yield stress σ'_y in this figure is a value calculated from the aforementioned formula (1), and was rearranged like a case of Fig. 8. The results are hardly uneven as the same specimens are used. It can be said from those results that strain rate effects of the consolidation yield stress or the curve for ε_v-log $\sigma'v$ is hardly influenced by the plasticity index I_P within the range conducted in the present test.

Fig. 10 shows relationship between values obtained by normalizing the consolidation yield stress for respective strain rate with the consolidation yield stress for the strain rate $\varepsilon_v = 3.3 \times 10^{-6} \text{s}^{-1}$ and the strain rate. An approximate curve in Fig. 10 is summarized from results of the CRS test using 5 clays and results of the SCRS test using 11 clays. Forms obtained from approximate formulas for the CRS test and the SCRS test are similar to each other, and the consolidation yield stress is reduced approximately 10 %, as the strain rate becomes one order smaller, assuming the standard strain rate as $\dot{\varepsilon}_v = 3.3 \times 10^{-6} \text{s}^{-1}$. In the CRS test, where 4 to 5 different specimens are employed, results are slightly uneven and the relative coefficient is $R^2 = 0.72$. On the other hand, in the SCRS test where a single specimens is employed, uneven results are very small in number and the relative coefficient $R^2 = 0.98$. From the above, the strain rate effects of the consolidation yield stress of Osaka Bay Pleistocene clay, which is called quasi-over-

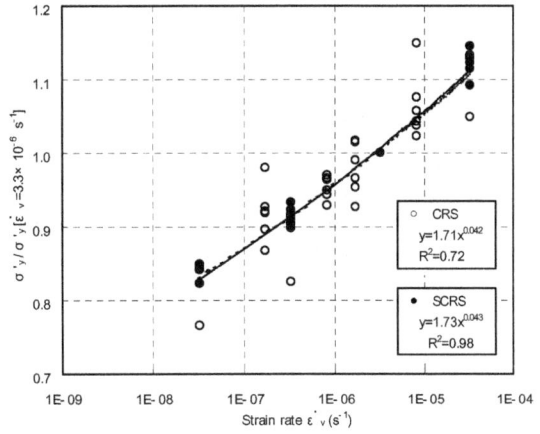

Figure 10. Relationship between normalized consolidation yield stress and strain rate

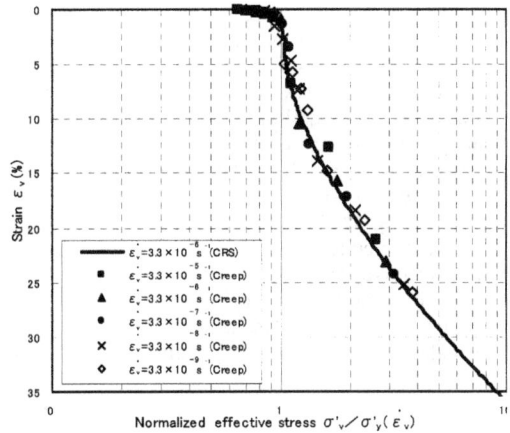

Figure 11. Relationship between strain and normalized effective stress from CRS tests and creep tests

consolidated clay, can be better estimated by the SCRS test where the strain rate is changed using a single specimen, and higher accuracy can be obtained with the use of a single specimen.

The ε_v-log (σ'_v / σ'_y) Curve of the CRS test and the creep test of Ma12 are shown in Fig. 11. The normalized consolidation stress σ'_v / σ'_y in the creep test is the one obtained by normalizing respective effective stress σ'_v with the consolidation yield stress σ'_y corresponding to the respective strain rate using an approximate expression in the SCRS test of Fig. 10. The approximate curve in Fig. 10 is prepared on results of the SCRS test for the strain rate up to $\varepsilon_v = 3.3 \times 10^{-8} \text{s}^{-1}$, but the consolidation yield stress for the strain rate $\varepsilon_v = 3.3 \times 10^{-9} \text{s}^{-1}$ is calculated by extrapolation. It can be confirmed from the above that results of the creep test can be expressed in an almost single ε_v-log (σ'_v / σ'_y) curve by normalizing the consolidation yield stress with the one corresponding to the strain rate.

From results of a series of consolidation test, it has been confirmed that, with regard to dependency of the consolidation yield stress of Osaka Bay on Pleistocene clay on the strain rate, the standard strain rate as shown in experiments by Larsson et al. cannot be proved when the strain rate becomes as small as $\varepsilon_v = 3.3 \times 10^{-9} s^{-1}$, and the consolidation yield stress becomes smaller as the strain rate becomes slower until it reaches to the limit of $\varepsilon_v = 3.3 \times 10^{-9} s^{-1}$.

5 CONCLUSION

Three kinds of consolidation test were carried out using Osaka Bay Pleistocene clay, and effects of the strain rate on the consolidation characteristics of Osaka Bay Pleistocene clay were estimated by results of these consolidation tests. Conclusion obtained is summarized as follows:

(1) The ε_v-log σ'_v curve of Osaka Bay Pleistocene clay is shifted to the left with the decrease of strain rate and the consolidation yield stress becomes small. Assuming the strain rate $\varepsilon_v = 3.3 \times 10^{-6} s^{-1}$ as the standard, the consolidation yield stress is reduced by about 10% as the strain rate decreases to one tenth. The ε_v-log σ'_v curve for respective strain rate are nearly parallel to each other, and they can be a single curve of ε_v-log (σ'_v/σ'_y) by normalizing the vertical stress with the consolidation yield stress corresponding to respective strain rate.

(2) With regard to strain rate effects of the consolidation yield stress of Osaka Bay Pleistocene clay, the standard strain rate as shown in the experiment by Larsson et al. cannot be recognized within a range of strain rate $\varepsilon_v = 3.3 \times 10^{-9} s^{-1}$. From such results, estimating the strain rate effects of consolidation yield stress or the ε_v-log σ'_v curve of Osaka Bay Pleistocene clay is available within a range of $\varepsilon_v = 3.3 \times 10^{-9} s^{-1}$ by conducting the SCRS test where the strain rate is changed during the test.

(3) In estimating the strain rate effects of Osaka Bay Pleistocene clay, the CRS tests for different strain rates using a plural number of specimens produces uneven results. However this can be improved by using a single specimen as in SCRS tests to be able to evaluate the strain rate effects without scattering data.

(4) Within a scope of experiment conducted at this time, the strain rate effects of the consolidation characteristics of Osaka Bay Pleistocene clay is not influenced by the plasticity index I_p.

REFERENCES

1) Ishii, I., Ogawa, T., Zen, K. 1980. Engineering property of sea bed deposits of Osaka senshu bay area.Port and harbor research institute engineering report, No. 498: pp. 47-86, (in Japanese).
2) Larsson, R. L. and Sallfors, G. 1986. Automatic Continuous Consolidation Testing in Sweden, Consolidation of Soils, *ASTM, STP*, No. 892: pp. 299-328.
3) Leroueil, S. 1996. Compressibility of Clays: Fundamental and Practical Aspects, *Proc. ASCE*, Vol. 122, No. 7: pp. 534-543.
4) Leroueil, S., Kabbaj, M., Tavenas, F. and Bouchard, R. 1985. Stress-strain-strain rate relation for the compressibility of sensitive natural clays, *Geotechnique*, Vol. 35, No. 2: pp. 159-180.
5) Skempton, A . W. 1970. The consolidation of clays by gravitational compaction, Q. J. Geological Society, Vol. 125: pp. 373-411.

Soft Ground Engineering in Coastal Areas, Tsuchida et al. (eds)
© 2003 Swets & Zeitlinger, Lisse, ISBN 90 5809 613 0

Engineering properties of intermediate soil consisting with grain size distribution

S. Suwa, M. Fukuda & T. Shimonodan
Geo-Research Institute, Osaka, Japan

ABSTRACT: Intermediate was defined for the soil that has the intermediate engineering properties varying between the both ends of characteristics of clay and sand. Nakase et al classified it based on the consistency properties of soil. However, there remains the undefined portion for the intermediate soil if insisting with this definition, because there is literally an intermediate soil with which the consistency testing isn't conducted.

In this paper, focusing on the relationship between grain size distribution and parameters of consolidation, the intermediate soil will be newly developed not consisting with the consistency properties, but covering this properties. Basic concept is the equivalent grain diameter method, however, non-logarithmic normal distribution is presupposed for the grain size distribution.

1 INTRODUCTION

The intermediate soil was defined as a soil that has the intermediate engineering properties localized between the both ends of the engineering properties of clay and sand. Representatively expressed, the intermediate soil has cohesion and internal friction angle. However, so far, this classification of intermediate soil was defined based on the consistency properties, especially on the amount of plasticity index by Nakase et al (Japanese Geotechnical Society, 1993). This definition has taken the important role the light of the inherent properties of soil having a variety of engineering characteristic, not consisting with the pure properties of clay and sand. However, this methodology of classification exclude some portion of literally intermediate soils, if wholly defined, because the coarser soil exists that is classified from the view point of grain size distribution, but the consistency test can't be conducted. This portion of intermediate soils has a finer fraction so that it isn't involved to figure the engineering properties of pure sand that hasn't the finer fraction of soil.

We have developed the equivalent grain diameter method to apply for the variety of engineering properties of soil from the viewpoint of the important governing factor of grain size distribution. The equivalent grain diameter method is to predict mainly the amount of hydraulic conductivity and internal friction angle in effective stress based on a grain size distribution and void ratio given. Therefore this method can cover the coarser soil different

from the pure sand but with the finer fraction that takes account of intermediate soil.

However, this methodology has the limitation of the assumption of logarithmic normal distribution imposed on the grain size distribution, even though actually grain size distributions of soils show the trend of non-logarithmic normal distribution. Hence, in this paper, the equivalent grain diameter method will be reviewed by the standpoint of non-logarithmic normal distribution. This revised assumption with the equivalent grain diameter method reinforced broader the definition engaged for the intermediate soil.

2 EXISTING DEFINITION OF INTERMEDIATE SOIL

Nakase et al proposed to derive the intermediate soil from the general compounding fractional soil, which has the both characteristics of clay and sand. They classified it by the amount of plasticity index as following (Japanese Geotechnical Society, 1993).

Pure clay $\geq$ Plasticity index of 20 to 30
Intermediate soil = NP to 30
Pure sand < NP

And they investigated the consolidation properties of intermediate soil consisting with hydraulic conductivity, coefficient of consolidation, compression index, coefficient of volume compression and the strength parameters with regard to the consistency index. However, according to the previously presented definition, soil containing the alternative volume of finer fraction, but labeled by NP, because that soil is impossible to carry out consistency test, therefore given by NP, exist away from sand.

3 EQUIVALENT GRAIN DIAMETER METHOD AND NON-LOGARITHMIC NORMAL DISTRIBUTION

Fundamental concept of the equivalent grain diameter method has been established based on the sphere shape of the particle for the purpose of simplification of the shape of natural soil particle, and logarithmic normal distribution. As the results of statistical approach, the following equations were presented (Fukuda et al. 1997).

$$d_c = \frac{0.3d_{50}}{\exp\left\{0.5(0.484 + 0.420\ell n U_c)^2\right\}} \qquad (1)$$

$$U_c = \frac{d_{60}}{d_{10}} \qquad (2)$$

$$h = d_c \cdot \frac{e}{G_s} \qquad (3)$$

$$log \, k_{sat} = 2.87(log \, k + 1) \qquad (4)$$

$$tan\phi' = 0.85 \cdot \frac{d^{0.09}}{h^{0.02}} \qquad (5)$$

Here, d_c : equivalent grain diameter (mm), d_{10}, d_{50}, d_{60} : grain diameter equivalent to percent finer than 10, 50 and 60% by weight (mm), U_c: coefficient of homogeneity, h : thickness of pore attached the surface of the particle of mean diameter (mm), k_{sat} : hydraulic conductivity of saturated soil (cm/s), φ': internal friction angle in effective stress (°).

Furthermore, Equations 6 and 7 are added to concern with the mean diameter.

$$\ln d_g = \ln \overline{d}_w - 3\ln^2 \sigma_w \qquad (6)$$

$$\ln \sigma_w = 0.484 + 0.420\ln U_c \qquad (7)$$

Here, d_g: means grain size, σ_w: standard deviation which is expressed experimentally by Equation 7. Specifically, this term is included in the denominator of Equation 1.

Important idea included in the equivalent grain diameter method is to change the grain size distribution in weight into a volumetric distribution of particles. This method can approximately predict the hydraulic conductivity and internal friction angle in effective using grain size distribution curve and void ratio.

However, there were found the discrepancy between the predicted values and the experimental data

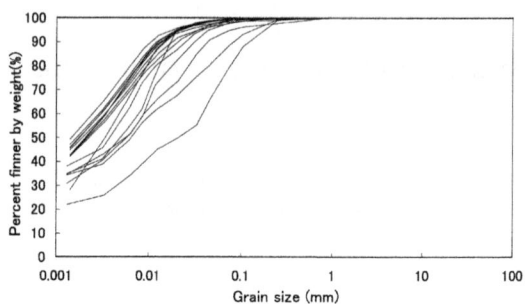

Figure 1. Representative examples of grain size distribution curves of clays.

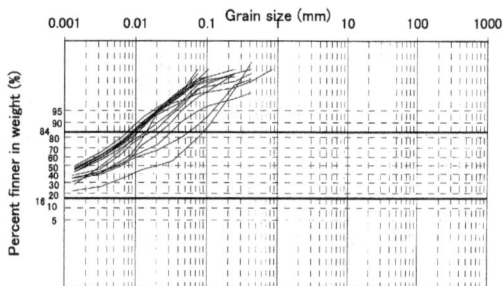

Figure 2. Logarithmic normal distributions of grain size distributions.

under some conditions. Especially the large discrepancies were to be found on the comparisons for clay and artificially mixed soil. As a result of this analysis, one of reasons was thought that for the discrepancy was that the application of logarithmic normal distribution on the grain size distribution of soil is restricted to emerge the discrepancy. Hence, in this paper, non-logarithmic normal distribution will be presupposed to develop the equivalent grain diameter method.

Figure 1 shows the burden of grain size distributions of clay concerned in this study. They are described in weight and transformed in the sheet of logarithmic normal distribution as shown in Figure 2. The later figure proves that the assumptions of logarithmic normal distributions on natural soils are insufficient with some grain size distributions to figure out by using the linear relationship. Therefore, to seek more general relationship, the non-logarithmic normal distribution was adopted.

Under the assumption of the unique specific gravity G_s , a number of particles ranged from diameter of d_i and d_j is assumed to correspond with Equations 8 and 9.

$$N_{ij} \frac{P_i - P_j}{\frac{\pi}{6}.d_{il}^{3} \cdot G_5} \qquad (8)$$

$$d_{ij} = \sqrt{d_i \cdot d_j} \qquad (9)$$

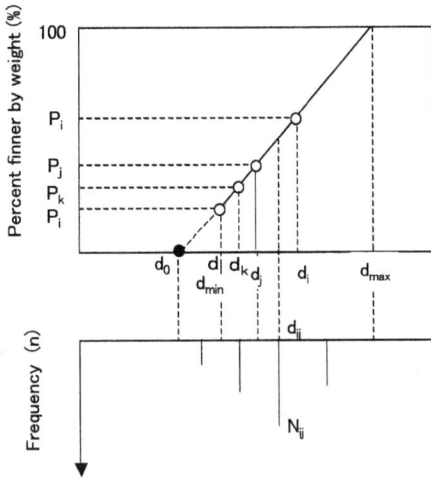

Figure 3. Mean diameter in every interval of grain diameter.

The Equation 8 means the number of mean particle ranged from P_i to P_j in percent finer in weight. The corresponding mean grain size is derived by the Equation 9. Taken the number of mean grain size in each divided sections of grain diameters, total number of particles can be calculated using the Equation 10. Furthermore, the existence of 0 percent finer in weight is assumed and the corresponding diameter d_0 is predicted to interpolate using P_k, P_ℓ, d_k and d_ℓ as follows.

$$N = \sum N_{ij} \tag{10}$$

P_e, P_k, d_e and d_k are experimental data as the smallest and second smallest diameter and the corresponding percent finer in weight similar with the previous definition. The mean grain diameter is calculated based on the moment equilibrium at the grain size d_0. This equation is shown as Equations 11 and 12.

$$\frac{P_e}{\log d_\ell - \log d_0} = \frac{P_k - P_\ell}{\log d_k - \log d_\ell} \tag{11}$$

$$N \cdot \left\{ \log d_g - \log d_0 \right\} = \sum N_{ij} \left\{ \log d_{ij} - \log d_0 \right\} \tag{12}$$

In this study, the mean diameter obtained from the Equation 11 is substituted the mean diameter in the Equation 6, and inversely the equivalent grain diameter d_c is given through the Equation 7. Here, even though the derived equivalent grain diameter is modified different from the original equivalent grain diameter subjected to the logarithmic normal distribution, however, the same word is used, because, later, the fundamental concept will be developed under the non-logarithmic normal distribution, but not with the logarithmic normal distribution. Hence, using same designation for avoiding the disturbance in terminology, the equivalent grain diameter d_c is calculated us-

ing the Equation 13. However, the hydraulic conductivity and the thickness of void h is assumed to predict using the same Equations 3 and 4.

$$d_c = 0.3 \, d_g^{1/6} \cdot d_{50}^{5/6} \tag{13}$$

4 PROPERTIES OF HYDRAULIC CONDUCTIVITY

Figures 4 to 6 are the examples comparing the predicted hydraulic conductivity based on the grain size distributions and void ratios to the consolidation test. Of these, Figure 4 shows the grain size distribution curves of samples used for analysis compatible with Figures 5 and 6. These figures show the corresponding predicted values. Test results for hydraulic conductivity are data not modified by the primary consolidation ratio in this study. Two figures show the relationship of the high suitability between the predicted values and test results ranged in the order of values. But this suitability is restricted in the normal consolidation pressure zone.

As shown in a later chapter, well suitability is proved with other samples of clay. Comparing with these results, although not revealed in this paper, the logarithmic normal distribution on the same grain size distribution lead to have larger discrepancy.

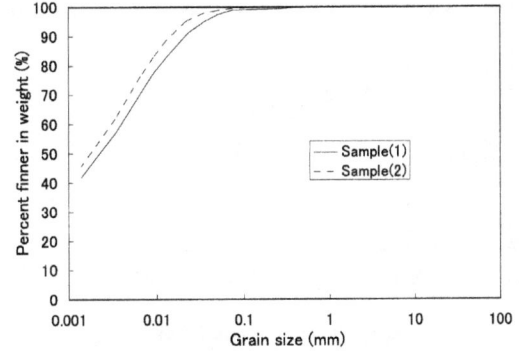

Figure 4. Grain size distributions samples of used for analysis.

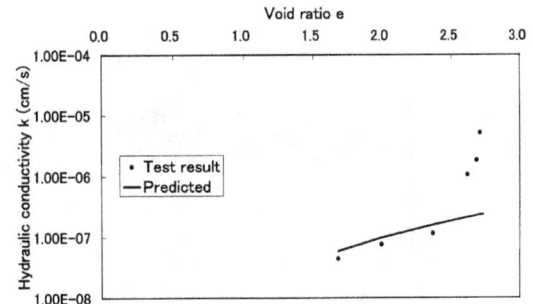

Figure 5. Relationship between experimental data and predicted value (sample 1).

87

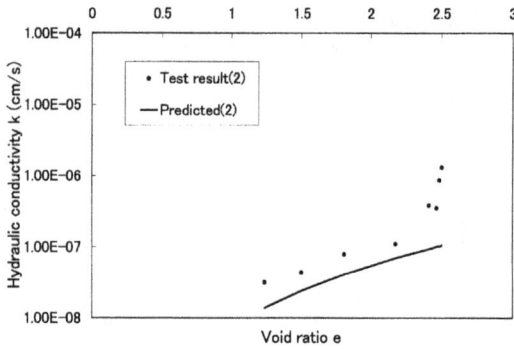

Figure 6. Relationship between experimental data and predicted value (sample 2).

5 RELATIONSHIP BETWEEN PARAMETERS OF CONSOLIDATION AND EQUIVALENT GRAIN DIAMETER METHOD

Nakase et al. summarized the relationship of coefficient of consolidation and coefficient of volume compression as the specific properties of intermediate soil. Focusing on these properties and investigating how well the equivalent grain diameter method can explain, Figures 7 to 11 show these properties relate to the equivalent grain diameter method. Figures 7 and 8 are the results of consolidation test and fundamental data for this investigation. Concerning grain size distribution of samples were already shown in Figure 1.

In Figure 9 vertically rising portion concern with in the over-consolidation pressure region, on the other hand the decreasing trend appear as the thickness h increases that attributed to the behavior in the normal consolidation pressure region. Figure 9 makes us image the existence of the concentrating narrow zone in the normal consolidation pressure, although they scatter in the over-consolidation pressure region.

Figure 10 shows the tendency on the same horizontal axis with the coefficient of volume compression, as data plotted for the thickness of void h. Even though there is found a largely scattering, however, if restricted in the normal consolidation pressure region, the linear tendency of coefficient of volume compression m_v decreasing as thickness h decreases with every each sample. But the wholly resemble slant of relationships is interesting to expect the existence of unique relationship any combination of factors.

However, there isn't the unique zone on the coefficient of volume compression even though in the normal consolidation region, and the relationship between the coefficient of volume compression and the thickness of void h is line on the logarithmic axis.

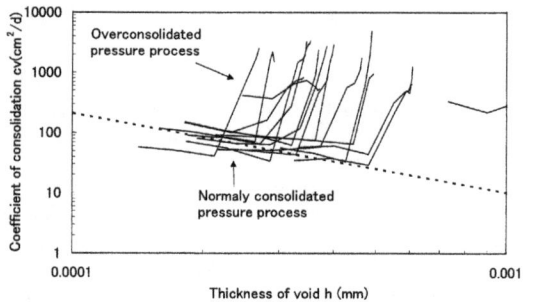

Figure 7. Relationship between coefficient of consolidation and consolidation pressure.

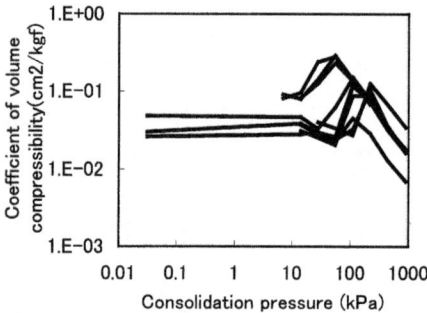

Figure 8. Relationship between coefficient of volume compressibility and consolidation pressure.

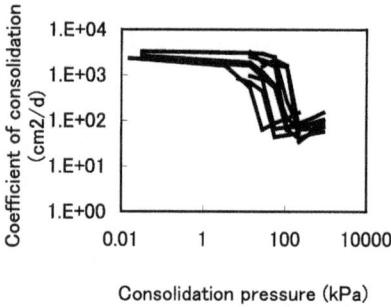

Figure 9. Relationship between coefficient of consolidation and mean thickness of void.

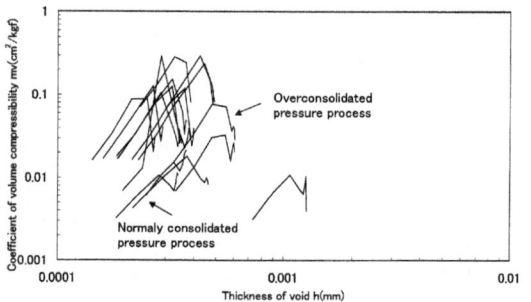

Figure 10. Relationship between coefficient of volume compressibility and mean thickness of void.

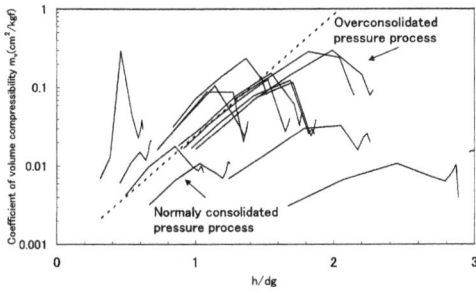

Figure 11. Relationship between coefficient of volume compressibility and function of h/dg.

Figure 11, although there scattered, investigates the uniqueness on the relationship between the coefficient of volume compression and the mean scale of void h/d_g that can be thought to correspond with the comparative resistance of a structure composed of particles. The tendency of concentrating in a unique band can be a little bit revealed.

6 RELATION OF LIQUID LIMIT WITH THE REPRESENTATIVE DIAMETER

Figure 12 shows the relationship between the liquid limit and the equivalent grain diameter d_c. The equivalent diameter d_c is described by the logarithmic axis. The linear relationship can be found between both factors and it is clear that there is a point where the consistency test can't be conducted, if the thickness of void h exceeds over. This point is the role of some value.

As dividing characteristics of clay and sand, it suggests that the boundary point can exist as possibility of engaging consistency test, and soil classified in the smaller region than this boundary is defined as soil with consistency. However, there isn't the minimum limit of boundary point dividing intermediate soil from clay. Hence, intermediate soil has properties continuously changing toward properties of pure clay. This means that the definition of intermediate soil based on the characteristics of consistency possibly contributes to make the existence range of intermediate soil smaller.

Same trend can be found on the relationship between plastic index and equivalent grain diameter as shown in Figure 13. Plastic index decreases as equivalent diameter increases, and the value of equivalent grain diameter where the plastic index disappears is approximately the same value.

Figure 14 shows the relationship between compression index and equivalent grain diameter. Although there are a few data, it is found that as equivalent grain diameter is lager, compression index decreases, because the larger equivalent grain diameter means to move to coarser side of soil.

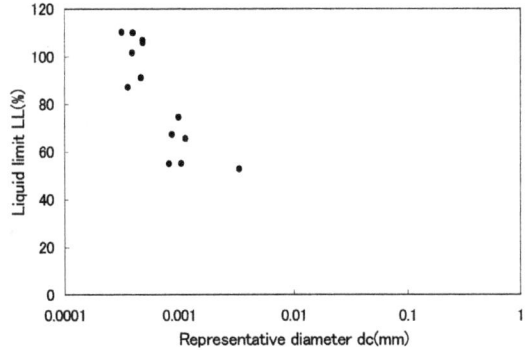

Figure 12. Relationship between equivalent grain diameter and liquid limit.

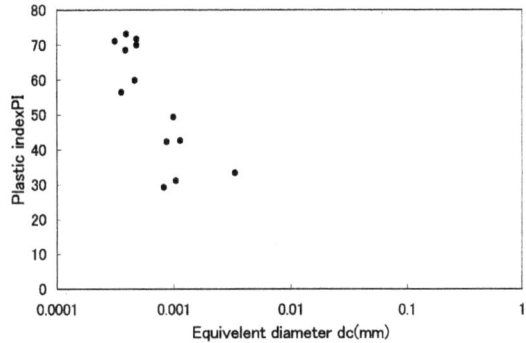

Figure 13. Relationship between equivalent grain diameter and plastic index.

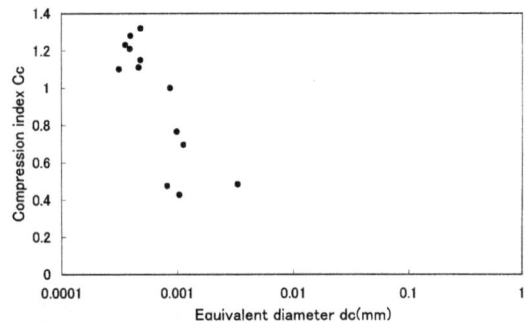

Figure 14. Relationship between equivalent grain diameter and compression index.

7 CONCLUSIONS

In this paper, the modified equivalent grain diameter method was developed under the assumption of non-logarithmic normal distribution of grain size distribution and the well suitability between the predicted hydraulic conductivity and the result of consolidation test.

Based on this finding, the intermediate soil was defined by the equivalent grain diameter method as follows.

1) Hydraulic conductivity in the normal consolidation pressure of clay obtained from the consolidation test can be approximately predicted within one order using the equivalent diameter method.

2) Both coefficient of consolidation c_v and coefficient of volume compression m_v in the normal consolidation pressure tend to be linear relationship with the thickness of void h.

3) The unique narrow zone exist in the coefficient of consolidation c_v and thickness of void h, where as, in the case of coefficient of volume compression, the unique linear relationship can be efficiently described with the logarithmic axis of h/d_g.

4) Using the equivalent grain diameter, the dividing point clearly appears that delineates sand and fine soil including intermediate soil and clay.

5) Intermediate soil and clay can be arranged without any boundary.

REFERENCES

Japanese Geotechnical Society, 1993. Geotech note, Intermediate soil –sand or clay-, Japanese Geotechinical Society, pp.1-6, in Japanese.

Japanese Geotechnical Society, 1993. Geotech note, Intermediate soil –sand or clay-, Japanese Geotechinical Society, pp.8, in Japanese.

Fukuda M. and Uno T., 1997. Roles of mean pore size index and grain size distribution in permeability phenomenon, *Journal of geotechinical engineering*, Japan Society of Civil Engineers, No.561/III-38, pp.193-204, in Japanese.

Fukuda M. and Uno T., 1997. Analysis of " Method of classification of soils" based on proposed " Diameter estimating grain-size distribution", *Journal of geotechnical engineering*, Japan Society of Civil Engineers, No.582/III-41, pp.125-136, in Japanese.

What does the q_u mean?

H. Tanaka
Port and Airport Research Institute, Yokosuka, Japan

ABSTRACT: The applicability of the q_u method, which is widely used in Japan to evaluate the undrained shear strength, is examined to marine clays recovered from various sites in the world. All samples studied in this paper were collected by the Japanese standard piston sampler. The residual effective stress, the rate effect and anisotropy were measured for each soil to examine as to whether or not these factors keep a balance and the q_u test provides a suitable strength for design. It is found from this study that the q_u method gives a reliable design value, except for Drammen clay, whose residual effective stress is considerably lost.

1 INTRODUCTION

Nakase (1967) successfully analyzed a failure of the embankment at Kinkai, using half of the unconfined compression strength ($q_u/2$) as the undrained shear strength (s_u), being combined the circle failure method (then we call it the q_u method). Since then, the unconfined compression test (q_u test) has been a standard test for evaluating s_u for cohesive soil in Japan, and the q_u method has been accepted by many geotechnical engineers to examine stability of clayey foundations. Several researchers in Japan have tried to understand the mean of $q_u/2$, and most of them have achieved a conclusion that the $q_u/2$ value is well balanced among factors over- or under-estimating the true strength.

The unconfined compression test is literally carried out under the atmosphere without in situ confining pressures so that the strength reduction may be anticipated. Even though the high quality sample sustains high residual effective stress in the specimen as negative pore water pressure, its magnitude is smaller than that corresponding pressures in situ. In this sense, the q_u test always provides lower strength than the true strength. On the other hand, the q_u test is performed at an axial strain rate of 1 %/min according to the standard of the Japanese Geotechnical Society (JGS), which is much faster than that at failure in the field. As the shear strength decreases due to decrease in the strain rate, especially for cohesive soil, such a large strain rate in the q_u test overestimates the strength. In addition, we have to consider anisotropy in the shear strength. Since the compression strength is usually the greatest in other strengths such as the extension strength,

the q_u test overestimates the strength in the stability analysis.

Tsuchida and Tanaka (1995) have studied on applicability of the q_u method to Japanese marine clays, and they have shown that as shown in Fig. 1, the $q_u/2$ value is in good agreement with the average value of compression and extension strengths measured by the recompression triaxial test where the specimen is consolidated under the in situ K_o condition and the strain rate is 0.01 %/min. If the sample quality is not suitable, however, the $q_u/2$ becomes smaller than the average value. Figure 1 implies that the factors underestimating the strength due to loss of the residual effective stress in the q_u method (we

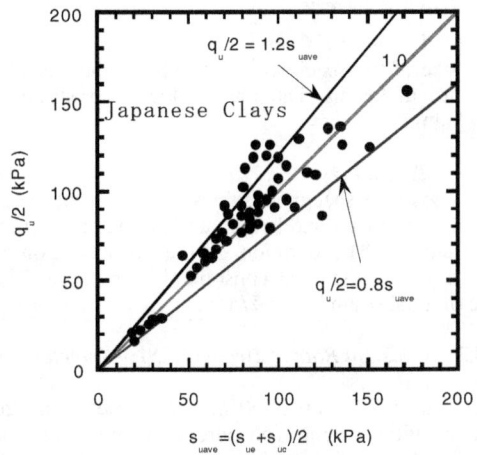

Figure 1. Relationships between $q_u/2$ and the mean strength from the recompression tests for Japanese clays.

may call it sample disturbance) compensates the overestimation caused by the rate effect and anisotropy.

The question is raised as to whether or not the q_u method can be applied to other clays than in Japan, where the environment for sedimentation as well as clay mineral seems much different in different regions. The author's geotechnical group has carried out site investigation in various sites all over the world, i.e., Drammen (Norway), Bothkennar (United Kingdom), Singapore, Bangkok (Thai land), Pusan (Korea) and Louiseville (Canada). In these investigations, the soil sample was retrieved by the stationary piston thin wall sampler, whose geometry and sampling method are strictly defined by the standard of the JGS. Soil samples were transported to the laboratory at Port and Airport Research Institute (PARI) and various tests were carried out. Using these data measured under the same conditions, the applicability of the q_u method is examined for various clays in the world to consider meaning of the q_u value.

2 USED SAMPLES AND TESTING METHODS

2.1 *Fundamental properties of used samples*

Fundamental properties of the samples used in this study, is available in literatures in the table in more detail.

2.2 *Sampling method*

All samples in this study were recovered using the Japanese standard stationary piston sampler. Geometry of the sampling tube is 75 mm in inside diameter, 1.5 mm in thickness and 1m long. The borehole was made until the sampling depth by a boring machine.

2.3 *Laboratory tests*

2.3.1 *Unconfined Compression (q_u) test*
The size of the specimen is 35 mm in diameter and 80 mm in height. The rate of the axial strain is 1 %/min.

2.3.2 *Triaxial test*
The size of the specimen is the same as that of the q_u test, i.e., diameter and height are 35mm and 80 mm, respectively. The duration of the consolidation is one day after the final consolidation pressure. The rate of axial strain is 0.1 %/min.

2.3.3 *Constant Rate of Strain (CRS) oedometer test*
The size of CRS test is 60 mm in diameter and 20 mm in initial height. The pore water pressure was measured at the bottom (the upper face is drainage). The consolidation pressure (p') was calculated by eq. (1).

$$p' = p - 2/3u \qquad (1)$$

where p and u are the applied force and the measured pore water pressure at the bottom, respectively.

2.3.4 *Residual effective stress*
The residual effective stress (p'_r) was measured before conducting the q_u test. The specimen trimmed for the q_u test was placed on a ceramic disc whose air entry value is 200 kPa.

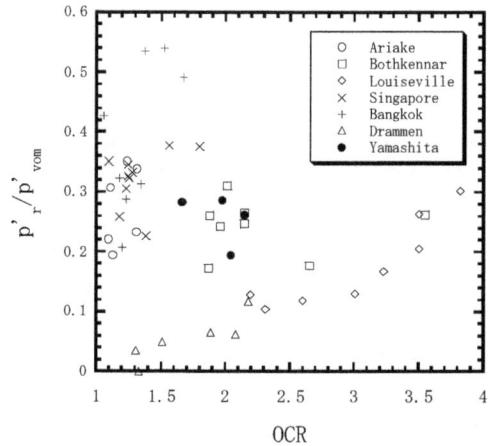

Figure 2. Residual effective stress for various soils.

3 TEST RESULTS

3.1 *Residual effective stress*

Figure 3 shows the residual effective stress (p'_r). The p'_r value is normalized by the in situ mean confining effective stresses (p'_{vom}), where the horizontal stress is estimated from K_0 calculated by eq. (2).

$$K_0 = (1 - \sin\phi')OCR^{\sin\phi'} \qquad (2)$$

where ϕ' is the effective internal friction angle and was measured from the compression triaxial test at the normal consolidation (see Fig. 4).

It is found in Fig. 3 that the order of p'_r / p'_{vom} is considerably different for each soil. As have already mentioned before, all samples were collected by the same sampling method, i.e., using the Japanese standard stationary piston sampler. The mechanical impact caused disturbance in the soil sample should be the same to every soil samples. The reason for difference in resultant p'_r for each soils is probably due to difference in physical properties such as grain size distribution, index properties as well as environmental conditions when they were deposited. There is a trend that the p'_r / p'_{vom} increases with increase in OCR, even taking account of increases in p'_{vom} due to large K_0 at large OCR. Especially, this trend is remarkably observed in Louiseville samples. How-

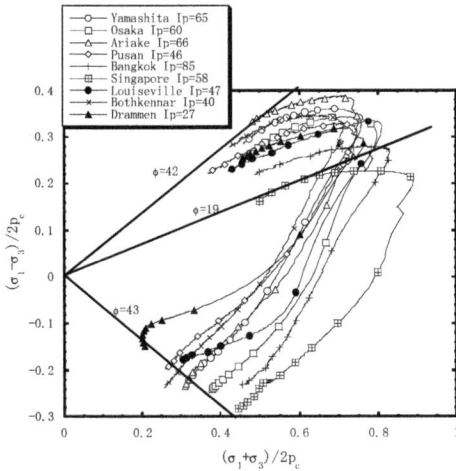

Figure 3. Stress path for various soils at the normally consolidated stage.

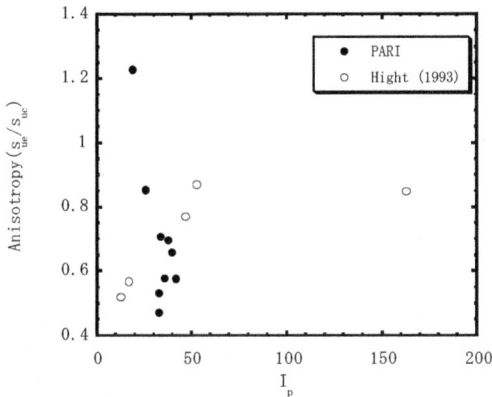

Figure 4. Relationship between anisotropy and I_p.

ever, even when the p/p is compared at the same OCR, the p/p is considerable different. For example, p'_r/p'_{vom} of Bangkok clay is more than 0.5, while the p'_r/p'_{vom} of Drammen clay is less than 0.1.

3.2 Strength anisotropy

Figure 4 shows stress path for various soils including in this study, where all specimens were consolidated under K_o conditions and consolidation pressures is large enough to exceed the consolidation yield pressure (p_y). It is observed from the figure that the stress path for each soils is different in terms of the shape as well as fundamental properties such as ϕ' in the compression side. It is interested in that the ϕ' in the extension side is nearly the same for each soils, i.e., 43 degree, although the maximum deviator stress in extension is different.

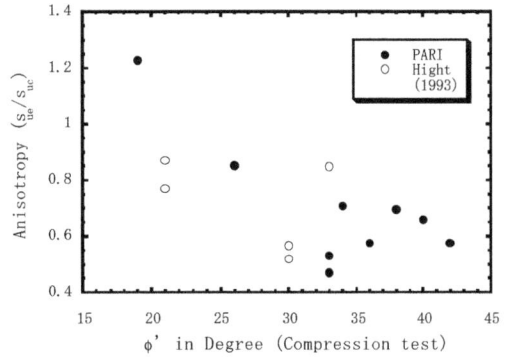

Figure 5. Relationship between anisotropy and ϕ'.

The undrained shear strengths for compression and extension (s_{uc} and s_{ue}, respectively) are defined as peak deviator strength. The ratio of s_{ue}/s_{uc} is plotted against plasticity index (I_p) in Fig. 5, including data by Hight (1993). Bjerrum (1973) has showed that s_{ue}/s_{uc} ratio can be correlated with I_p. However, recognizable relation between s_{ue}/s_{uc} and I_p can be found in this investigation. Instead, it is apparent that s_{ue}/s_{uc} ratio decreases with increase in ϕ' at the compression side. The reason for getting such a correlation is that from Fig. 4, the difference in the peak strength between compression and extension tests is nearly constant, regardless magnitude of ϕ'. The compression peak strength increases with increase in ϕ', in other words, the extension strength decreases with increase in ϕ'.

3.3 Strain rate

It may be considered that the undrained shear strength is strongly correlated with the p_y value. Therefore, the effect of the strain rate on p_y value was studied and its result is shown in Fig. 7. Bjerrum (1973) has also shown that the rate effect on s_u

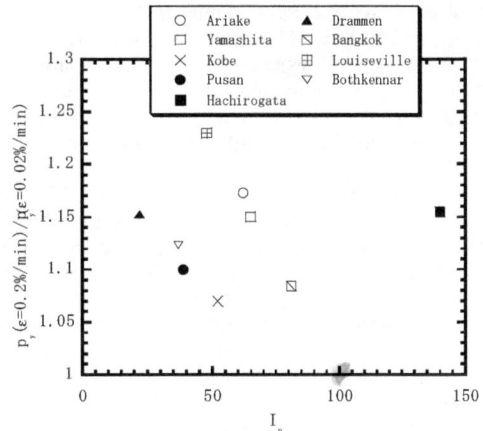

Figure 6. The rate effect on p_y.

93

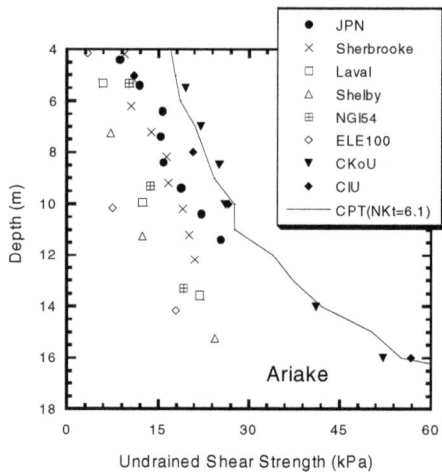

Figure 7. $q_u/2$ values for samples collected by various samplers at the Ariake site.

is much more prominent for soils with large I_p than that with small I_p. As seen in Fig. 7, however, there is no recognizable relation between the rate effect on p_y and I_p.

4 APPLICABLITY OF q_u TEST TO STUDIED SOILS

As have already shown, the p'_r/p'_{vom}, s_{ue}/s_{uc} and the rate effect are considerably different for studied soils and there is no systematic order among them. Is it true that these factors keep balance and the q_u test indeed provides a suitable strength for design?

4.1 q_u reduction due to lowering p'_r

In the following section, let consider how to make correlation between the reduction of p'_r and q_u value. At the site of Ariake, the geotechnical group at PARI used six different samplers to collect soil sample and examined sample quality retrieved by different samplers (Tanaka, 2000). Test results are shown in Fig. 8, indicating that the $q_u/2$ values are considerably affected by different samplers. Figure 9 shows corresponding p'_r to the $q_u/2$ values in Fig. 8. It is obvious that the $q_u/2$ value decreases in accordance with reduction of p'_r. Tanaka (2000) has shown that the q_u and p'_r have strong correlation provided that the soil structure are not destroyed by the sampling, that is, the shape of e-log$p$ curve is not changed and the recompression test can be applied.

Two strengths were also measured by recompression triaxial test at the Ariake site: consolidated under K_o in situ stress conditions (CK_oU) and mean in situ stress conditions, i.e., isotropically (CIU). Samples measuring these strengths were retrieved by the

Japanese sampler. These two strengths are nearly the same in spite of different consolidation methods. It may be said that these strengths are idealized one, in other words, corresponding to the strength for perfect sampling (s_{up}) because their specimens were consolidated under the in situ effective confining stresses. However, the rate effect should be considered because their tests were carried out at an axial strain rate of 0.1 %/min, which is 10 times slower than that of the q_u tests. Taking account into the rate effect, these strengths are multiplied by 1.16 times, which finger was obtained from the rate effect shown in Fig. 7. These perfect strengths can also be correlated with by the point resistance from cone penetration test (CPT) where the cone factor (N_{kt}) is adopted as 6.1, as shown in Fig. 8.

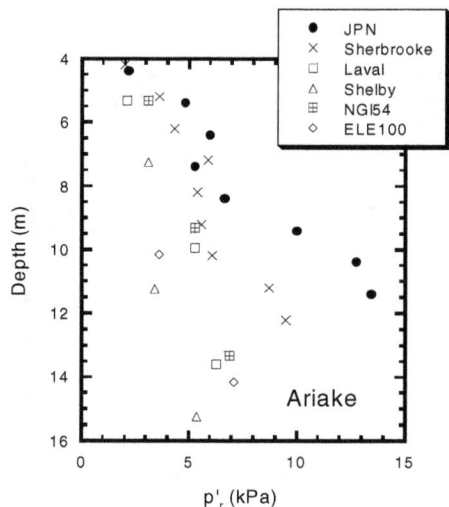

Figure 8. Residual effective stress for samples collected by various samplers at the site of the Ariake site.

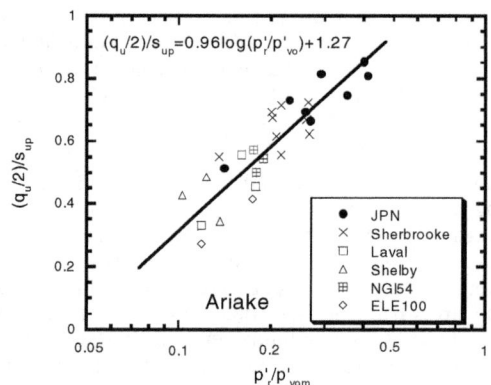

Figure 9. Reduction of q_u due to decrease in p'_r.

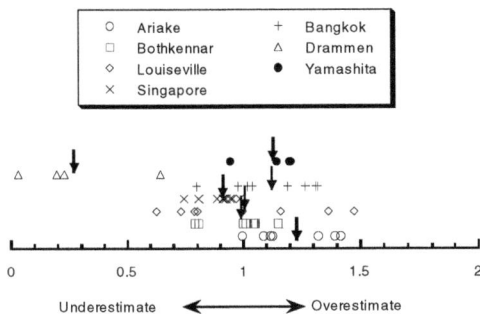

Figure 10. Comparison of $q_u/2$ with strength for design (s_{umob}).

The $q_u/2$ values are normalized by s_{up}, which was estimated by the CPT at the same depth. The p'_r value is also normalized by p'_{vom}. Then, the $q_u/2$ value may be correlated with p'_r as shown in Fig. 10 and by eq. (3).

$$(q_u/2)/s_{up} = 0.96\log(p'_r/p'_{vom})+1.27 \qquad (3)$$

Eq. (3) can present q_u and p'_r relation fairly well until p'_r/p'_{vom} is less than 0.4. When p'_r/p'_{vom} becomes more than 0.7, however, $q_u/2/s_{up}$ exceeds one. Therefore, strictly speaking, the relation between $(q_u/2)/s_{up}$ and $\log(p'_r/p'_{vom})$ is not liner, but when p'_r/p'_{vom} is over 0.5, $(q_u/2)/s_{up}$ is gradually approaching to 1.0 as p'_r/p'_{vom} p'_r/p'_{vom} increases.

4.2 Assessment of the $q_u/2$

Here, it is assumed that the reduction of q_u due to p'_r for other clays can be followed by eq.(3). Also, the rate effect of the undrained shear strength is the same as that of p_y, i.e, we can use Fig. 7. Hanzawa (1982) has indicated that the rate of strain at failure in the field is 0.01%/min, which is 100 times slower than that of the $q_u/2$ test. So, the rate effect on the $q_u/2$ strength should be two times as much as that indicated in Fig. 7. Anisotropy in the strength is obtained from Fig. 4, although this figure shows test results at the normally consolidated stage. Table 1 indicates factors of the rate effect (μ_r) and anisotropy ($\mu_a =0.5(1+s_{ue}/s_{uc})$).

Reduction of the q_u was calculated according to eq. (3) for each value indicated in Fig. 3. The ratio of $q_u/2$ to the strength for design (s_{umob}) may be given in eq. (4).

$$(q_u/2)/s_{umob}=\mu_r\mu_a(q_u/2)/s_{up} \qquad (4)$$

The final results for studied soils are shown in Fig. 11, where the arrow indicates the mean value for each soil. The figure shows that if $(q_u/2)/s_{umob}$ is over unity, $q_u/2$ overestimates the s_{umob}, in other words, $q_u/2$ gives a dangerous design parameter. It is a little fanny that the average of $(q_u/2)/s_{umob}$ for Ariake clay is 1.22, which is considerably large, even though Ariake clay is quite famous as a typical

Japanese clay. However, we have never heard that the q_u method cannot apply to Ariake clay. Yamashita is also Japanese marine clay, whose $(q_u/2)/s_{umob}$ is still 1.12. The reason for such a large $(q_u/2)/s_{umob}$ for Japanese clays, even though the q_u method can be extensively used in Japan, is due to overestimating the rate effect on the strength. Tsuchida and Tanaka (1995) have reported that the rate effect on the shear strength for Japanese clays is around 1.14 for two log cycles of the strain rate. This number is quite small compared with μ_r in Table 1.

Table 1. Rate and anisotropy for various soils

Site	Rate(μ_r)	Anisotropy(μ_a)
Ariake	1.34	1.27
Yamashita	1.3	1.24
Louiseville	1.46	1.31
Singapore	1.3	0.88
Drammen	1.3	1.33
Bothkennar	1.24	1.18
Bangkok	1.16	1.12

5 CONCLUSIONS

The applicability of the q_u method was examined for various soils in the world. Even though all soil samples were retrieved by the Japanese standard stationary piston sampler, the residual effective stress (p'_r) is considerably different for each sample. The rate and anisotropy effects (μ_r and μ_a, respectively) on the strength were also studied. As a result, the ratio of $(q_u/2)/s_{umob}$, where s_{umob} is the design value for stability analysis, is in the range of 0.9 and 1.22 for studied soils, except for Drammen clay. This range is rather small considering the variety of p'_r, μ_r and μ_a for each soil. The $(q_u/2)/s_{umob}$ for Drammen is considerably low, because p'_r is quite small due to low I_p.

REFERENCES

Bjerrum, L. 1973. Problems of soil mechanics in unstable soils. Proc. of 8th ICSMFE. 3:111-159. Moscow.

Hanzawa, H. 1982. Undrained strength characteristics of Alluvial clays and their application to short term stability problems, Doctor thesis of Tokyo University.

Hight, D. W. 1993. A review of sample effects in clays and sand, Offshore site investigation and foundation behaviou 28: 115-146.

Nakase, A. 1967. The φ=0 analysis of stability and confined compression strength, Soils and Foundations 7(2): 33-45.

Tanaka, H. 2000. Sample quality of cohesive soils: lessons from three sites, Ariake, Bothkenaar and Drammen. Soils and Foundations 40(4):57-74.

Tsuchida, T. & Tanaka, H. 1995. Evaluation of strength of soft clay deposit – a review of unconfined compression strength of clay. Report of the Port and Harbour Research Institute 34(1):1-37.

Soft Ground Engineering in Coastal Areas, Tsuchida et al. (eds)
© 2003 Swets & Zeitlinger, Lisse, ISBN 90 5809 613 0

Evaluation of in-situ lateral earth pressure at rest for marine clay by means of triaxial cell

Y. Watabe, M. Tanaka, H. Tanaka & T. Tsuchida
Port and Airport Research Institute, Yokosuka, Japan

ABSTRACT: In this study, a series of K_0-consolidation tests in a triaxial cell was carried out for marine clays collected from different areas in the world. From the test results, in the case using ϕ' corresponding to $(q/p)_{max}$, the equation $K_{0NC} = 1 - \sin \phi'$ proposed by Jáky (1944) underestimates K_{0NC} by 0.05; in contrast, in the case using ϕ' corresponding to $(q)_{max}$, the equation overestimates K_{0NC} by 0.05. For a representative undisturbed clay sample, K_0-consolidation and K_0-overconsolidation tests (i.e., SHANSEP test) were consecutively carried out to obtain $K - OCR$ relationship. The K-value corresponding to the OCR can be calculated as the in-situ K_0.

1 INTRODUCTION

In order to evaluate in-situ undrained shear strength by laboratory test, a high quality undisturbed sample should be examined by a reliable testing method that can reduce the influence of sample disturbance. One of the most useful testing methods is a recompression technique, in which a specimen sampled is consolidated under the in-situ effective stresses, i.e., $\sigma'_1 = \sigma'_{v0}$ and $\sigma'_3 = \sigma'_{h0}$, where σ'_{v0} is the in-situ overburden effective stress, σ'_{h0} is the in-situ lateral effective stress expressed as $\sigma'_{h0} = K_0\sigma'_{v0}$, and K_0 is the coefficient of lateral earth pressure at rest. This technique was proposed by Berre and Bjerrum (1973) and Bjerrum (1973).

K_0-consolidation test in the triaxial cell has been facilitated in an advanced geotechnical laboratory (e.g., Tsuchida and Kikuchi, 1991); however, it is not used in practical design procedure because of its complexity. Therefore, the method of anisotropic consolidation with a constant K_0-value is more preferable than the K_0-consolidation.

In order to carry out the anisotropic consolidation, the K_0-value must be estimated beforehand. It is ideal to measure it directly by an in-situ test, such as with self-boring type pressuremeter (Camkometer), hydraulic fracturing and flat dilatometer (DMT) tests as reported by Hamouche et al. (1995). However, these in-situ K_0 measurements are not often used in practice. Meanwhile, from the triaxial K_0-consolidation test, K_0-value in normal consolidation (K_{0NC}) can be obtained; however, the in-situ K_0 (K_{0IS}) that is in quasi-overconsolidation can barely be measured.

In addition, in finite element analysis (FEM) using elasto-plastic constitutive equations, a calculated result is influenced by the initial stress condition. Therefore, not only for stability analysis, in which undrained shear strength has been evaluated by the recompression method, but also for FEM analysis, in which the initial stress condition must be inputted, estimation of in-situ K_0-value is a very important geotechnical issue in practice.

Many previous researches on K_0-consolidation concerned the data for clays collected from a limited region (Díaz-Rodríguez et al., 1992; Mesri and Hayat, 1993). Some studies discussed the data for worldwide clays; however, the data were collected from other literatures (Mayne and Kulhawy, 1982). In the present study, a series of K_0-consolidation tests in the triaxial cell was carried out for marine clays with various characteristics collected from different areas in the world. In addition, a K_0-overconsolidation test from K_0-normal consolidation stage, i.e., SHANSEP test (Ladd and Foott, 1974), was also carried out for the clays. K_0-values obtained from the K_0-overconsolidation test corresponding to the overconsolidation ratio (K_{0OC}), and from the in-situ measurement (K_{0IS}), were compared.

2 CLAY SAMPLES

This study mainly dealt with eight marine clays from different areas in the world collected by Tanaka et al. (Tanaka, 2000; Tanaka et al., 2001) and the Osaka Bay Pleistocene clay for the Kansai International Airport project. Sampling sites for nine clays

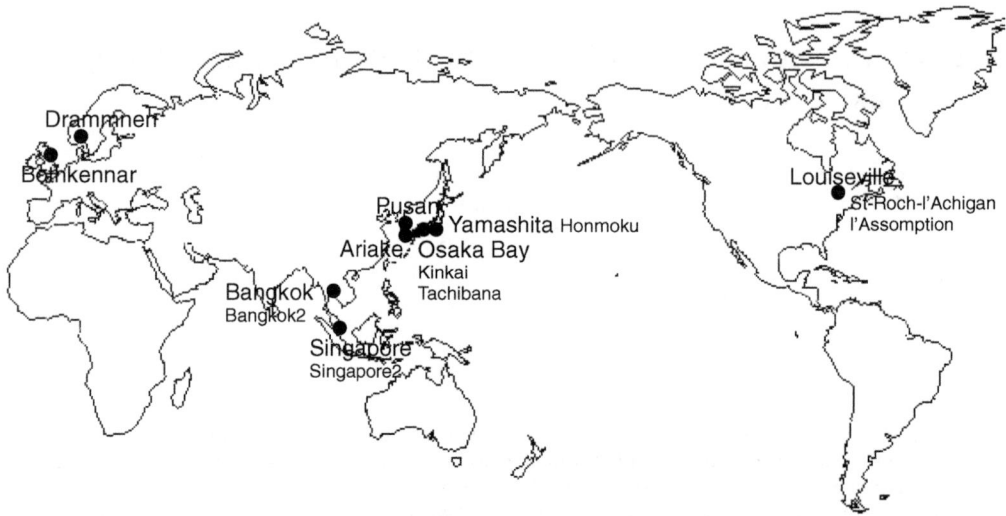

Figure 1. Sampling sites for clays dealt in this study.

are marked as solid circle in Figure 1: Drammen, Bothkennar, Louiseville, Singapore, Bangkok, Pusan, Ariake, Yamashita and Osaka Bay Pleistocene clays.

In addition to the nine clays above, K_0-consolidation tests were carried out for St-Roch-l'Achigan clay and l'Assomption clay sampled from the Champlain Sea, eastern Canada, Singapore2 clay from southwest of Singapore, Bangkok2 clay from Nong Ngoo Hao in Bangkok and Kinkai clay from Kinkai Bay, Japan. Remolded and reconstituted specimens of the St-Roch-l'Achigan clay (St-Roch(R)), the Osaka Bay Pleistocene clay (Osaka(R)), the Ariake clay (Ariake(R)), Honmoku clay (Honmoku(R)) dredged from Yokohama Port and Tachibana clay (Tachibana(R)) dredged from Tachibana Bay, Japan, were examined by the K_0-consolidation test. An industrial product of Kaolin clay (Kaolin(R)) was also examined.

3 TEST PROCEDURE

The specimen size is 35 mm in diameter and 85 mm in height. The specimen was surrounded by paper filter drain strips to promote the consolidation. Vertical strips were used for compression and spiral strips were used for extension. The water volume drained from the paper filter was corrected based on a calibrated relationship.

A specimen trimmed to size was placed on the rigid pedestal. The specimen was surrounded by drain filter paper strips. Between the specimen and both the top cap and the pedestal, a very thin latex membrane with silicone grease for lubrication was placed. The specimen was covered by a latex flexible membrane and was sealed by O-rings. A back

pressure (u_b) of 196 kPa was applied to the specimen and the preliminary consolidation was conducted by applying an isotropic stress equivalent to $u_b+\sigma'_{v0}/6$. The diameter of the specimen after terminating the preliminary consolidation was regarded as the initial diameter used to maintain the lateral strain ε_3 during K_0-consolidation. Vertical stress (σ_1) was increased linearly up to $\sigma'_{1max}= (3–4)\sigma'_{v0}$, which corresponds to $(1.5–3.0)\sigma'_y$, in 720–1440 min, while cell pressure was controlled to keep $\varepsilon_3 \approx 0$. Under this loading condition, the axial strain rate was at most 0.05%/min. After σ_1 reached its final value, 6–12 hours was required to dissipate the excess pore water pressure.

The apparatuses, in this study, can control as $|\varepsilon_3|$ < 0.01–0.05%, which satisfies the description ($|\varepsilon_3|$ < 0.05) in the JGS standard (JGS-0525). The performance of the control in K_0-consolidation depends not only on the apparatus itself but also on the clay, since it is improved if the compression index (C_c) is large.

In some cases, after terminating the K_0-consolidation test, a K_0-overconsolidation test, in which σ'_1 was decreased down to $\sigma'_{1max}/3$ (OCR = 3) while controlling σ_3 to keep $\varepsilon_3 \approx 0$, was consecutively carried out. This test is well known as the SHANSEP test proposed by Ladd and Foott (1974).

4 EVALUATION OF TEST RESULTS

4.1 K_0-value in normal consolidation

Relationships between K_{0NC} and I_p are plotted in Figure 2. No remarkable tendencies for these clays can be seen. After a K_0-consolidation test, the K_0-

98

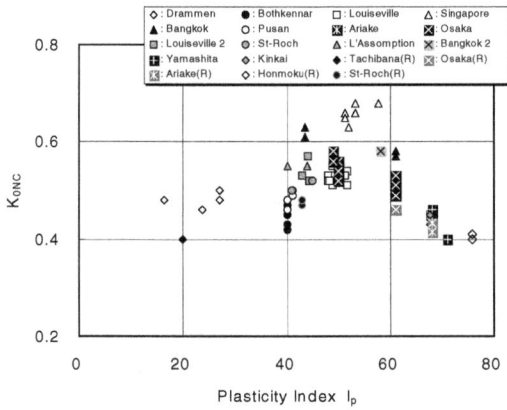

Figure 2. Relationship between K_{0NC} and I_p.

normally consolidated specimen was sheared with undrained condition for both compression and extension (CK_{0NC}UC and CK_{0NC}UE tests). Relationships between (a) deviator stress and axial strain ($q - \varepsilon$) and (b) deviator stress and effective mean stress ($q - p'$) normalized by the consolidation stress σ'_c(= σ'_{1max}) for K_0-normally consolidated clay from the undisturbed sample and the reconstituted sample are shown in Figure 3 and Figure 4, respectively. The relationships for CK_{0NC}UC, except for Kaolin(R), have a peak strength and show strain softening. The relationships for all CK_{0NC}UE show strain hardening without peak strength. Since shearing behaviors for both the undisturbed and the reconstituted samples

(see Ariake and Ariake(R), and Osaka and Osaka(R), respectively) are very similar to each other, it can be said that the shearing behavior from a normal consolidation stage for both undisturbed and reconstituted clays is the inherent one for the clay.

As a relationship between K_{0NC} and internal friction angle ϕ', Jáky (1944) proposed the following equation;

$$K_{0NC} = 1 - \sin \phi' \qquad (1)$$

The relationship between K_{0NC} and ϕ' evaluated by the K_0-consolidation test and undrained shear test respectively has been discussed in some literatures such as Mayne and Kulhawy (1982) and Mesri and Hayat (1993). However, it is often unclear whether ϕ' was defined as the critical state line (CSL) or the peak strength corresponding to the undrained shear strength. In addition, some other factors are often unclear as to whether ϕ' was evaluated by triaxial or direct shear tests, or whether ϕ' was investigated for normal consolidation stage from the undisturbed or the reconstituted specimen.

Singapore clay abundantly contains the kaolinite. Stress paths in comparison with Singapore and Kaolin(R) clays are shown in Figure 5. The values resemble each other well on the stress path, though the shapes are different. For Kaolin(R) clay, $(q)_{max}$ and $(q/p)_{max}$ coincide with each other, since strain hardening was observed for compression in Figure 4(a). Generally, the ϕ' is expressed with $M(= q/p')$ as the following equation;

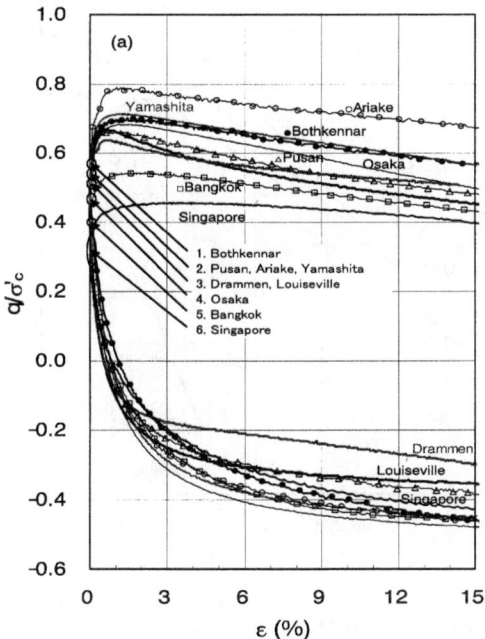

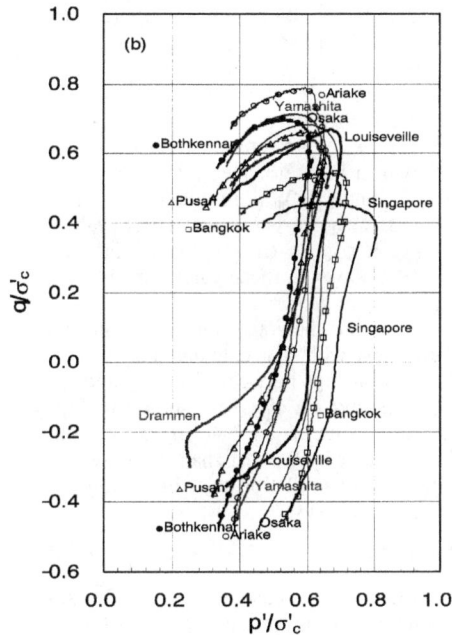

Figure 3. (a) Stress-strain curves and (b) stress paths observed in CK_{0NC}UC and CK_{0NC}UE tests for undisturbed clays.

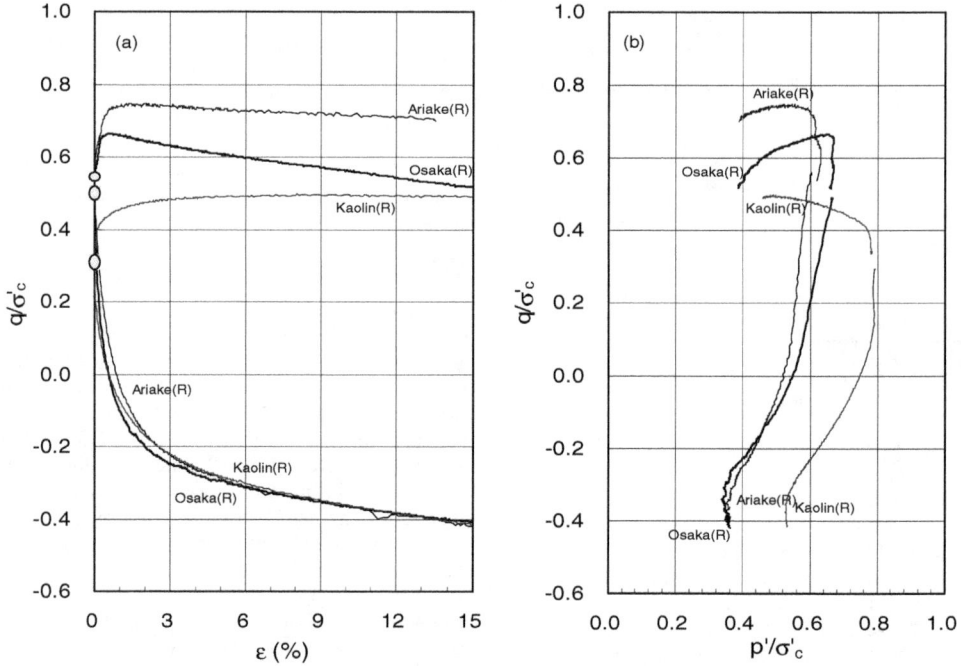

Figure 4. (a) Stress-strain curves and (b) stress paths observed in $CK_{0NC}UC$ and $CK_{0NC}UE$ tests for reconstituted clays.

$$\phi' = \sin^{-1} \frac{3M}{6+M} \qquad (2)$$

For Kaolin(R) clay, $(q)_{max}$ and $(q/p')_{max}$ coincide with each other with $M = 1.06$ which corresponds to $\phi' = 26.8°$ based on Equation (2). For Singapore clay, $M = 0.62$ & $\phi' = 16.3°$ and $M = 0.78$ & $\phi' = 20.2°$ are obtained for $(q)_{max}$ and $(q/p')_{max}$, respectively. In many textbooks on soil mechanics, $(q)_{max}$ and $(q/p')_{max}$ coincide with each other. Generally, however, M & ϕ' for $(q)_{max}$ are smaller than those for $(q/p')_{max}$ as shown in Figure 3. Here, M & ϕ' corresponding to $(q)_{max}$ and $(q/p')_{max}$ are noted as M_{peak} & ϕ'_{peak} and M_{CSL} & ϕ'_{CSL}, respectively. These values are $M_{peak} = M_{CSL}$ & $\phi'_{peak} = \phi'_{CSL}$ for a special clay such as Kaolin(R); however, these values are $M_{peak} < M_{CSL}$ & $\phi'_{peak} < \phi'_{CSL}$ in general.

In order to compare the K_{0NC} obtained from the K_0-consolidation test with that calculated by Jáky's Equation (1), relationships between (a) K_{0NC} and $\sin \phi'_{CSL}$ and (b) K_{0NC} and $\sin \phi'_{peak}$ are shown in Figure 6. In the case using ϕ'_{CSL}, $1 - \sin \phi'_{CSL}$ from Equation (1) is smaller by 0.0–0.1 than the K_{0NC} obtained from the K_0-consolidation test. In average, it can be expressed as follows;

$$K_{0NC} = 1.05 - \sin \phi'_{CSL} \qquad (3)$$

On the other hand, in the case using ϕ'_{peak}, $1 - \sin \phi'_{peak}$ from Equation (1) is greater by 0.0–0.1 than

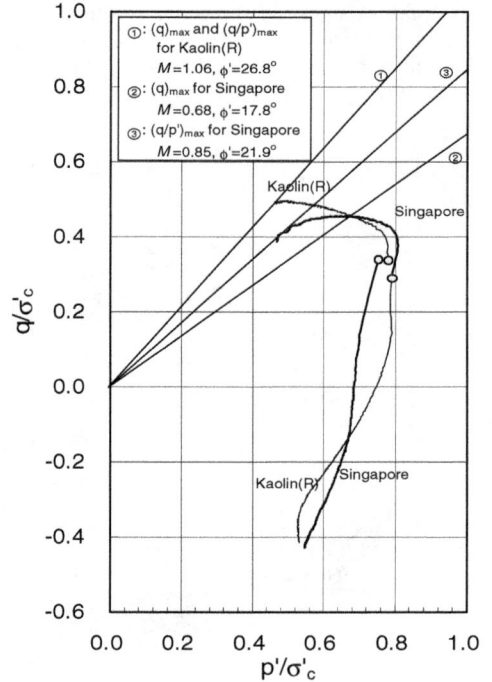

Figure 5. Stress paths observed in $CK_{0NC}UC$ and $CK_{0NC}UE$ tests for Singapore and Kaolin(R) clays.

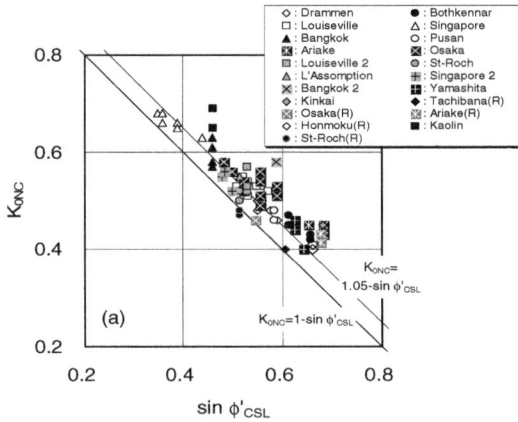

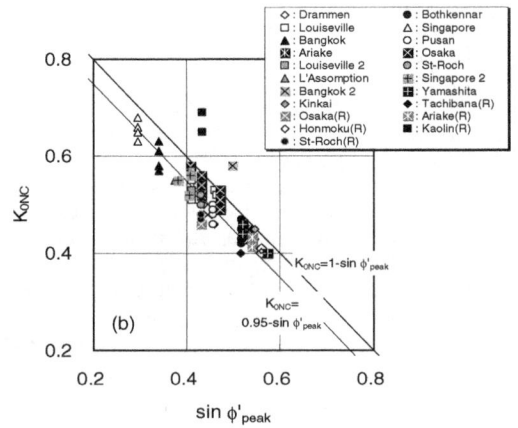

Figure 6. Relationship between (a) K_{0NC} and sin ϕ'_{CSL} and (b) K_{0NC} and sin ϕ'_{peak}.

the K_{0NC} obtained from the K_0-consolidation test. In average, it can be expressed as follows;

$$K_{0NC} = 0.95 - \sin \phi'_{peak} \qquad (4)$$

It can be said that Jáky's Equation (1) is useful for the relationship between ϕ' and K_{0NC} from the result of the triaxial K_0-consolidation test for marine clays collected from different areas examined in this study. However, it should be noted that Equation (1) overestimates the K_{0NC} if the ϕ' is defined based on $(q)_{max}$, and it underestimates the K_{0NC} if the ϕ' is defined based on $(q/p')_{max}$. Taking account of this fact, Equation (3) or (4) is more appropriate for the estimation of K_{0NC}.

4.2 In-situ K_0-value

In the site of (a) Drammen, (b) Bothkennar, (c) Louiseville, (d) Singapore, (e) Bangkok, (f) Ariake and (g) Yamashita clays, the in-situ K_0-value was measured by self-boring type pressuremeter, hydraulic fracturing and/or flat dilatometer test.

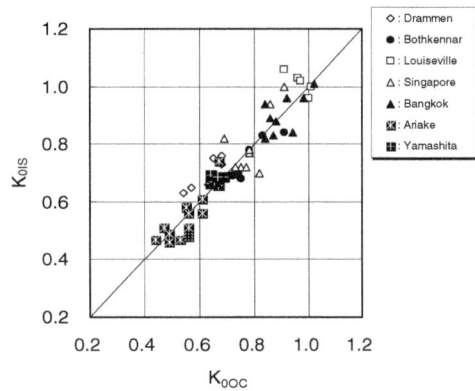

Figure 7. Relationship between K_{0IS} and K_{0OC}.

From the profiles of OCR and K–OCR relationships observed in SHANSEP K_0-overconsolidation tests, a K_0-value corresponding to the OCR (K_{0OC}) can be estimated. The values of K_{0IS} and K_{0OC} obtained by in-situ test and the triaxial K_0-overconsolidation test, respectively, are plotted in Figure 7, where the in-situ data sets of Drammen, Bothkennar and Louiseville clays are quoted from Lunne (1995), Nash et al. (1992) and Hamouche et al. (1995), respectively. It can be seen that the K_{0OC} compares well with the K_{0IS}.

As a relationship between K-value and OCR, Schmidt (1966) proposed the following equation;

$$K = K_{0NC}(OCR)^m \qquad (5)$$

where m is the slope in the relationship between logarithmic K-value and logarithmic OCR. In order to estimate the value of m from the internal friction angle (ϕ'), Mayne and Kulhawy (1982) proposed the following equation;

$$m = \sin \phi' \qquad (6)$$

Figure 8 shows the relationship between the observed m-value and sin ϕ'_{CSL}. Taking account of the difference between ϕ'_{CSL} and ϕ'_{peak}, the following equation can be derived from Figure 8:

$$\begin{aligned} m &= \sin \phi'_{CSL} - 0.05 \\ &= \sin \phi'_{peak} + 0.05 \end{aligned} \qquad (7)$$

From this study, in order to estimate a K_0-value, the following procedure can be proposed: For a representative clay sample, K_0-consolidation and K_0-overconsolidation tests in the triaxial cell (i.e., SHANSEP test) are consecutively carried out to obtain the K – OCR relationship. Using the profile of OCR obtained from an oedometer test, a K-value corresponding to the OCR can be calculated as the in-situ K_0. Generally, however, the K_0-overconsolidation test is not recommended as a practical design procedure, since conducting the K_0-consolidation test is

101

Table 1. Summary of test results.

Clay	Depth (m)	σ'_{v0} (kPa)	w_L (%)	w_P (%)	I_p (%)	w_n (%)	ρ_s (g/cm³)	p'_y (kPa)	OCR	K_{0NC}	ϕ'_{CSL} (deg)	ϕ'_{peak} (deg)	m	K_{0OC}	K_{0IS}
Drammen	8.5	83	48.1	21.1	27.0	39.9	2.813	140	1.69	0.49	33.4	25.9	0.49	0.63	0.76
	18.5	170	34.7	18.2	17.0	30.7	2.782	200	1.18	0.48	35.1	27.0	0.49	0.52	0.62
Bothkennar	8.1	60	71.8	31.7	40.1	66.3	2.694	127	2.12	0.44	39.5	31.2	0.55	0.67	0.70
Louiseville	12.1	74	70.4	21.6	48.8	64.9	2.777	201	2.72	0.53	32.7	24.3	0.55	0.92	0.99
Singapore	23.7	220	76.2	23.0	53.2	54.5	2.776	261	1.19	0.67	21.0	17.2	0.41	0.72	0.75
Bangkok	12.1	91	84.8	23.9	60.9	60.6	2.736	184	2.02	0.62	27.4	19.9	0.48	0.87	0.98
Ariake	12.0	57	112.0	44.2	67.8	105.5	2.611	77	1.35	0.45	40.7	32.0	0.53	0.53	0.53
Yamashita	31.5	240	121.8	53.7	68.1	95.0	2.664	484	2.02	0.45	38.7	31.4	0.55	0.66	0.66

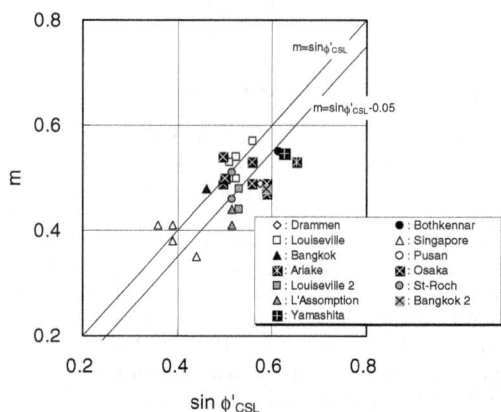

Figure 8. Relationship between observed m-value and sin ϕ'_{CSL}.

time consuming and costly. Therefore, the K_0-overconsolidation test is modeled by Equations (5), (6) and (7); however, the reliability of the estimated K_0-value is slightly low because the m value in Equation (5) might have some errors. If K_0-consolidation cannot be carried out, K_{0NC} is estimated by Equation (3) or (4); however, the reliability of the estimated K_0 becomes lower because K_{0NC} as well as m might have some errors.

Finally, some representative test results are summarized in Table 1.

5 CONCLUSIONS

In this study, a series of K_0-consolidation tests in the triaxial cell was carried out for marine clays collected from different areas in the world in order to evaluate the K_0-value. The relationships between the values of in-situ K_0 evaluated from the K_0-overconsolidation test corresponding to the OCR (K_{0OC}), and the measured value (K_{0IS}) were compared and discussed. The following conclusions are derived in this study:
1) It was confirmed that the famous equation K_{0NC} = 1 – sin ϕ' proposed by Jáky (1944) is useful for marine clays collected from different areas in the

world; however, the definition of the ϕ' should be carefully considered. In the case using ϕ'_{CSL} corresponding to the critical state line, Equation (3) is available. On the other hand, in the case using ϕ'_{peak} corresponding to the peak strength, Equation (4) is available.
2) K_0-values obtained from the K_0-overconsolidation test corresponding to the OCR (K_{0OC}) in laboratory and the in-situ measurement (K_{0IS}) in the field compare well.
3) In order to estimate the in-situ K_0, the following procedure is proposed from this study. For a representative clay sample, K_0-consolidation and K_0-overconsolidation tests (i.e., SHANSEP test) are consecutively carried out to obtain the K – OCR relationship. Using the profile of OCR obtained from an oedometer test, a K-value corresponding to the OCR can be calculated as the in-situ K_0.

REFERENCES

Beere, T. and Bjerrum, L. (1973): "Shear strength of normally consolidated clays," *Proceedings of 8th International Conference on Soil Mechanics and Foundation Engineering,* pp.39-49.

Bjerrum, L. (1973): "Problems of soil mechanics and construction on soft clays and structurally unstable soils," *Proceedings of the 8th International Conference on Soil Mechanics and Foundation Engineering,* State of the Art Report, pp.111-160.

Díaz-Rodríguez, J.A., Leroueil, S. and Alemán, J.D. (1992): "Yielding of Mexico city clay and other natural clays," *Journal of Geotechnical Engineering,* ASCE, Vol.118, No.GT7, pp.981-995.

Hamouche, K.K., Leroueil, S., Roy, M. and Lutenegger, A.J. (1995): "In situ evaluation of K_0 in eastern Canada clays," *Canadian Geotechnical Journal,* Vol.32, pp.677-688.

Jáky, J. (1944): "The coefficient of earth pressure at rest," *Journal of the Society of Hungarian Architects and Engineers,* pp.355-358.

Ladd, C.C. and Foott, R. (1974), "New design procedure for stability of soft clay," *Journal of the Geotechnical Engineering Division,* ASCE, Vol.100, No.GT7, pp.763-786.

Lunne, T. (1995): Personal communication.

Mayne, P.W. and Kulhawy, F.H. (1982): "K_0-OCR relationships in soil," *Journal of the Geotechnical Engineering Division,* ASCE, Vol.108, No.GT6, pp.851-872.

Mesri, G. and Hayat, T.M. (1993): "The coefficient of earth pressure at rest," *Canadian Geotechnical Journal*, Vol.30, No.4, pp.647-666.

Nash, D.F.T., Powell, J.J.M. and Lloyd, I.M. (1992): "Initial investigation of the soft clay test site at Bothkennar," *Géotechnique*, Vol.42, No.2, pp.163-181.

Schmidt, B. (1966): "Discussion of Earth pressures at rest related to stress history," Discussion. *Canadian Geotechnical Journal*, Vol.3, pp.239-242.

Tanaka, H. (2000): "Sample quality of cohesive soils: Lessons from three sites, Ariake, Bothkennar and Drammen," *Soils and Foundations*, Vol.40, No.4, pp.57-74.

Tanaka, H., Shiwakoti, D.R., Mishima, O., Watabe, Y. and Tanaka, M. (2001): "Comparison of mechanical behavior of two overconsolidated clays: Yamashita and Louiseville clays," *Soils and Foundations*, Vol.41, No.4, pp.73-87.

Tsuchida, T. and Kikuchi, Y. (1991): "K_0 consolidation of undisturbed clays by means of triaxial cell," *Soils and Foundations*, Vol.31, No.3, pp.127-137.

Estimation of consolidation settlement in the large-scale reclamation

Soft Ground Engineering in Coastal Areas, Tsuchida et al. (eds)
© 2003 Swets & Zeitlinger, Lisse, ISBN 90 5809 613 0

Some problems on settlement and stabilization of soft grounds

H. Aboshi
Hiroshima University, Japan

M. Mukai
Fukken Co., Ltd., Hiroshima, Japan

ABSTRACT: Two important problems on the settlement prediction and the soil stabilization of soft clayey grounds are discussed in this paper, in relation to the effect of secondary consolidation on the routine method of practice. Though the result published in the IS-Hiroshima 95 showed that the practical EOP (end of primary) obtained from long-term experiments and field measurements had become in between the A and the B hypotheses, the importance of gradual decrease of C_α is emphasized this time. A successful case record of the precompression technique carried out in the early 1980's is introduced, in order to show that it is possible to reduce the residual settlement including the secondary consolidation of a structure constructed on a very soft clayey ground to practically zero by utilizing this technique.

1 INTRODUCTION

When Dr.A.Nakase finished at the graduate course of education of the University of Tokyo in 1956 and engaged in the work of construction control of a seawall of the Kinkai Bay, Okayama Prefecture, the first author was working on a large-scaled experiment of three-dimensional consolidation of a clayey soil carried from the Kanaura Bay, the same prefecture, at a laboratory of testing materials of the Construction Ministry, Hiroshima. (Aboshi et al. 1961).

Dr. Nakase and the first author had experienced similar failure problems in their earlier career, which were a seawall at the Kinkai Bay, Okayama Prefecture in 1957 on Nakase, and a new highway embankment for Route No.2 across a bay in the same prefecture in 1951 on Aboshi, especially both cases being the earliest usage of sand drains in this country.

Similar experiences of failure for both of them, might have lead the research on soft foundation problems to their life works during the following half a century. In the occasion of the present Nakase Memorial Symposium, the first author would like to summarize their idea on the consolidation settlement of clayey soils, taking the secondary creep settlement into consideration. At the same time, they also wishes to emphasize the necessity of attaining no residual settlement in soil stabilization techniques, by showing a practical construction case record.

2 SECONDARY SETTLEMENT DURING LONG-TERM CONSOLIDATION

With his special lecture at IS-Hiroshima 95 (International Symposium on Compression and Consolidation of Clayey Soils), held in Hiroshima, Japan during May 10-12, 1995, the first author published a paper entitled "Case records of long-term measurement of consolidation settlement and their prediction." (Aboshi, 1995) In the lecture, he reported on three case records, one being a laboratory experiment and the others from two different fields, all of which being continuously measured during decades long period.

Essential data in the lecture necessary to the present paper are reproduced here.

2.1 *Large-scaled oedometer tests*

In order to clarify the size effect of oedometer tests on their secondary settlement characteristics, a series of five different oedmeters were used , the specimen size being shown in Table 1.(Aboshi,1973).

Table 1. Specimen size

Test No	1	2	3	4	5
Size Ratio	1	2.4	10	20	50
D cm	6	14.4	60	120	300
H cm	2	4.8	20	40	100

The settlement curves of each oedometers were compared under a same load from 20 to 78 kN/m², and the result obtained is shown in Fig.1.

A summary of conclusions obtained are as follows.

1 Settlement strain at EOP gradually increases with sample thickness.
2 Cv increases slightly with thickness.
3 Gradients of secondary tail are approximately equal to each other.
4 However, it is not at all anticipated that these parallel tails come on a single straight line in the future.
5 The gradient of the largest specimen shows a definite decrease after the passage of a decade time, the Cα being from its initial value of 0.84% to 0.34%.

These results indicate that the long-term consolidation settlement becomes in between the two hypotheses, the A(Terzaghi) and the B(Isotach). (The hypothesis C in Fig.1).

2.2 Shin-Ube power plant, Chugoku Electric Co., Ltd.

A reclamation fill of 15 ha was executed in the Bay of Ube on the coast of the Inland Sea in 1956 to construct a new power plant. Preliminary investigations had been carried out before the fill work and the settlement measurement had also been continued from the beginning of the work until 1994. Soil investigations had been performed at the site to check and assure these results even after the completion of the plant, twice in 1963 and in 1994. Especially the investigation performed in 1994 was a very precise and reliable one under the modern technology.

Geological conditions of the site seems to be an ideal case to testify the theory of one-dimensional

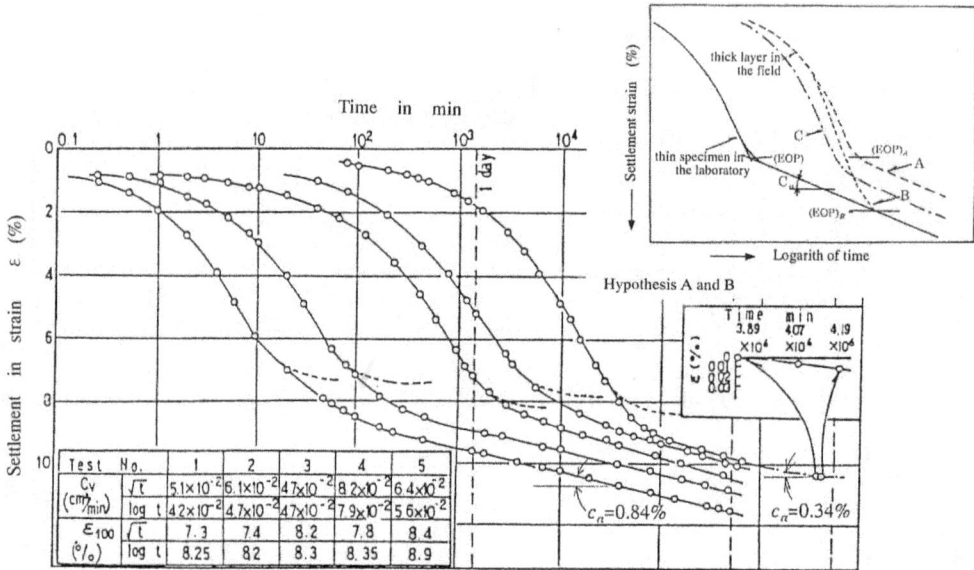

Figure 1. Comparison of settlement curves, laboratory experiment

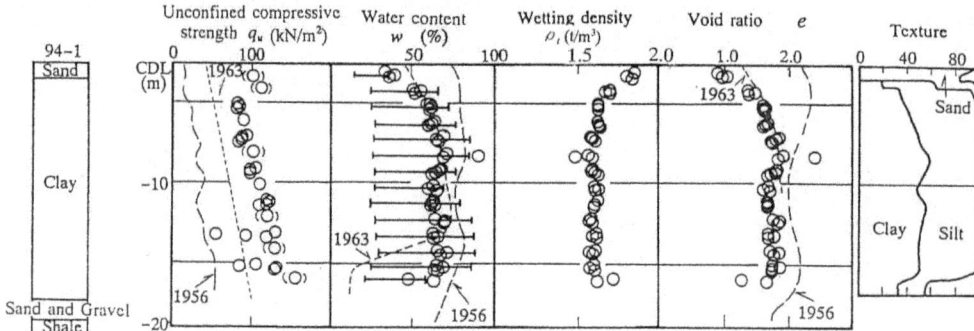

Figure 2. Soil properties in natural sand and clay, Shin Ube powerplant, 1994

consolidation, an alluvial clay layer of 18m thick being sandwiched with natural sand layers at its top

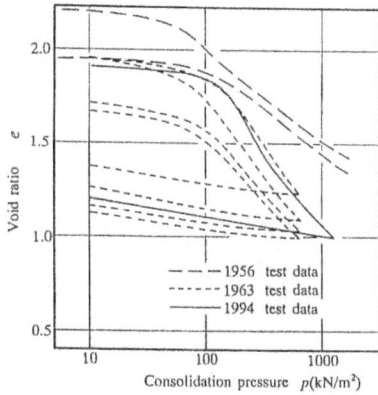

Figure 3. Comparison of e~p relations, '56, '63 and '94

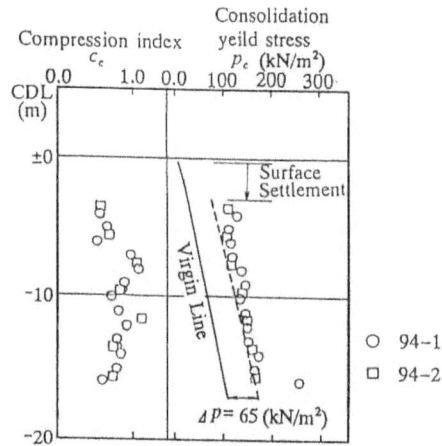

Figure 4. c_c and p_c at 94-1, 94-2

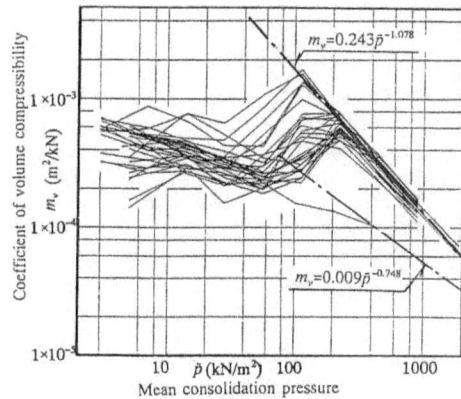

Figure 5. Consolidation characteristics by routine tests m_v ~ $\bar{p}$, 1994

and the bottom, and almost homogeneous without any sand seams or any sand lens, as shown in the summary figure of the layer in Fig. 2.

Consolidation characteristics obtained mainly from the 1994 investigation are shown in Fig. 3, 4, 5, 6, 7 and 8.

In Fig. 3, the gradient of e~log p curves increases with the year of the test, indicating the progress of sampling techniques without disturbance of the soil structure. It also caused the earlier prediction of the settlement at their smaller deviations. Increase and parallel movement of pc values shown in Fig.4 indicates that the primary consolidation of the clay layer concerned has already finished.

The measured settlements during 40 years are shown in Fig. 9 and Fig. 10 in natural and semi-logarithmic scales respectively.

The settlement curve, which is expressed in the form of time – settlement strain same as in Fig. 1, is

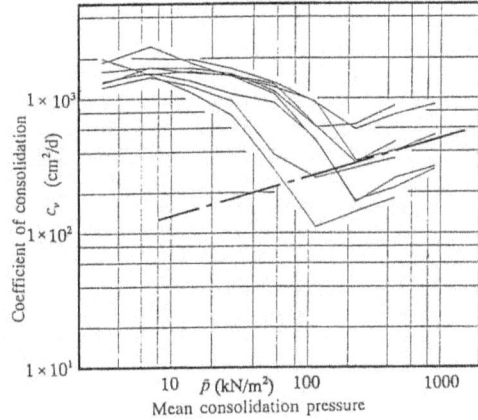

Figure 6. Consolidation characteristics by routine tests c_v ~ $\bar{p}$ (-0.2m ~ -3.0m), 1994

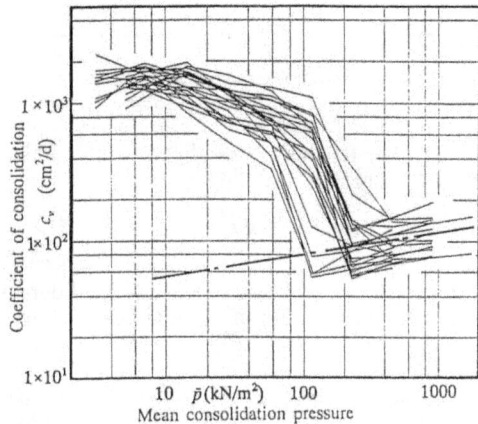

Figure 7. Consolidation characteristics by routine teests c_v ~ $\bar{p}$ (-3.0m ~ -17.9m), 1994

109

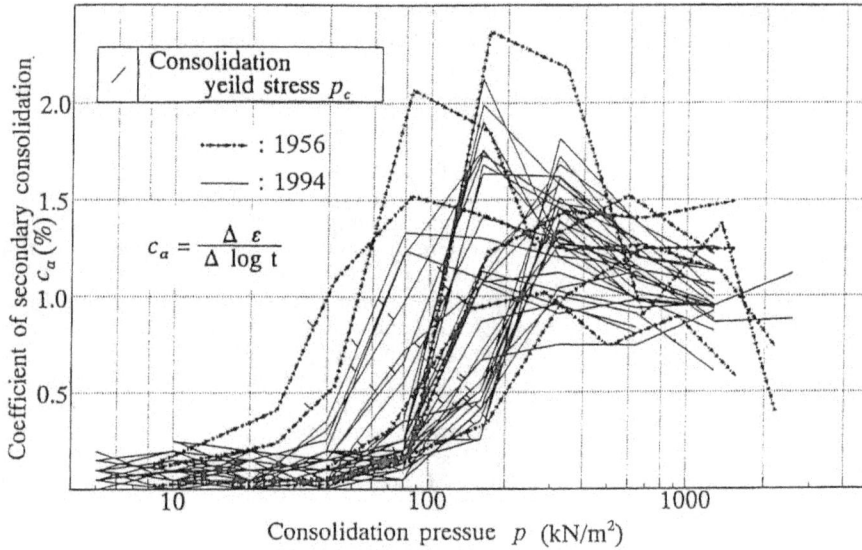

Figure 8. Consolidation characteristics by routine tests $c_a \sim p$

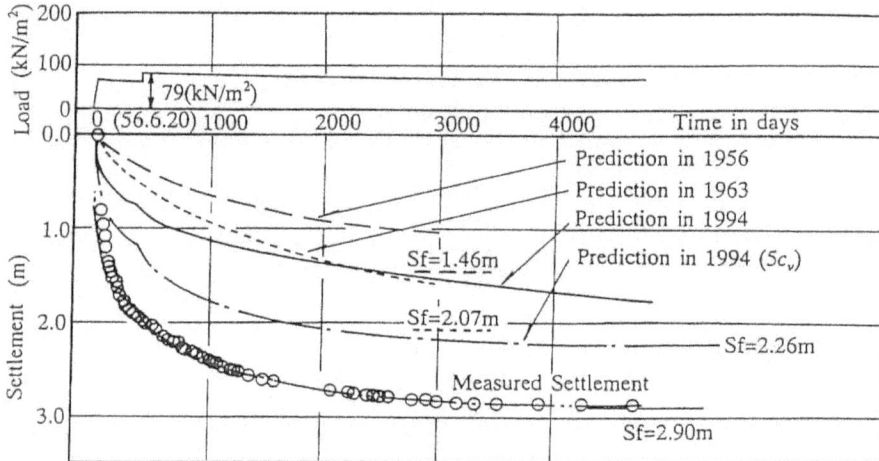

Figure 9. Prediction and performance in natural scale

shown in the following figure, Fig.11. This figure also shows that the practical settlement comes in between the A and the B hypotheses.

2.3 *West Hiroshima Development Project 1971 – 1981*

A reclamation of 328ha had been performed during 1971 to 1981, on the coast of the Inland Sea, western suburb of Hiroshima City. This is the first case of a new reclamation technique, in which the whole area of the reclaimed land had been stabilized with sand drains, executed from floating barges before the reclamation fill was started. The spacing of the drains had been altered by considering the land usage of the

region from 2.1 m to 4.5 m and a part of the land was filled up without drains, by the reason that the floating barge could not enter into a shallower area near the old coast line.

The whole project had become a model test to compare the effectiveness of the sand drain technique to hasten consolidation settlement of clayey grounds. Several basic soil properties are shown in Fig.12, 13, 14, 15 and 16.

A representative example of the measured settlement is shown in Fig.17 and Fig.18, plotted in natural and semi-logarithmic scales. Even though the EOP in the case of sand drains had been reached in several months or a few years period, a slight in-

110

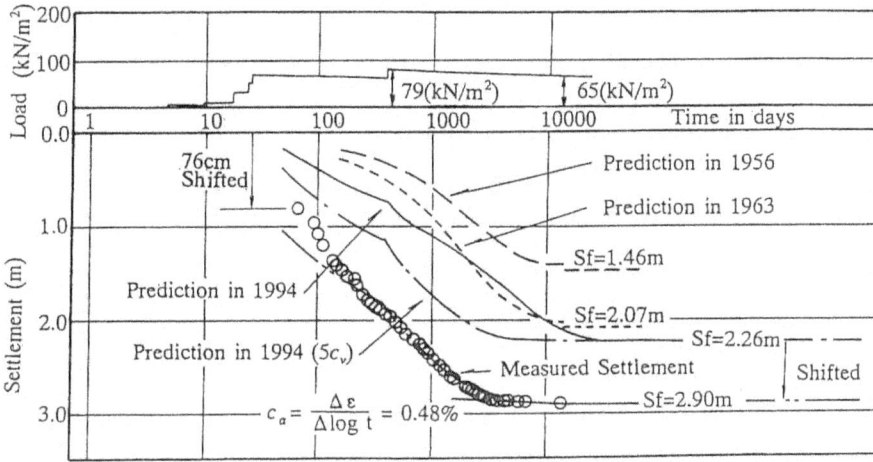

Figure 10. Prediction and performance in semi-logarithmic scale

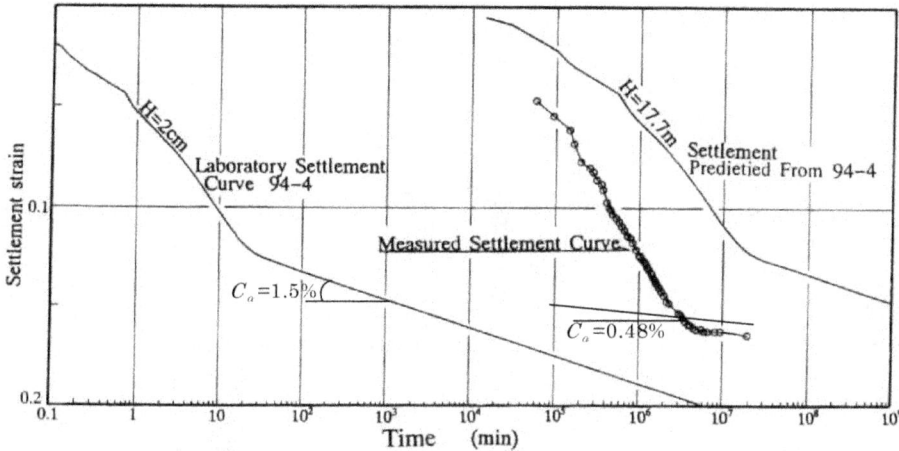

Figure 11. Prediction and performance in strain at settlement station

crease of EOP is shown, compared with the theoretical prediction, with a slight increase of effective C_v value at the same time. The most important point of interest in this case, obtained during the last decade after the former lecture, is the recent tendency of decrease of the gradient of secondary consolidation as shown in Fig.18.

2.4 Characteristics of long-term creep settlement

In the former lecture (Aboshi 1995), it was emphasized that the long-term consolidation settlement measured both in the laboratory and in the field, had become in between the A hypothesis (Terzaghi's theory) and the B hypothesis (Isotach). The case of the West Hiroshima Development Project, the measurement of which being continued even now, might seem to indicate the importance of gradual decrease of C_α in decades time scale, as shown in Fig.18. Whether this phenomenon means that the final settlement converges to the same target, as is suggested by a modified hypothesis B with decreased C_α, the authors could not assure through practical measurement while their lives. However, it seems to be significant to plot the ratio of the decreased C_α values of each case to their initial test data, with the passage of decades time. Fig.19 shows the relation of the ratio of decreased C_α at a certain time to its initial value measured in the laboratory, against the elapsed time, using the above data of three cases .

From the extrapolation of the line in Fig.19, the secondary creep settlement seems to reach at its final stage of zero in a hundred year time.

111

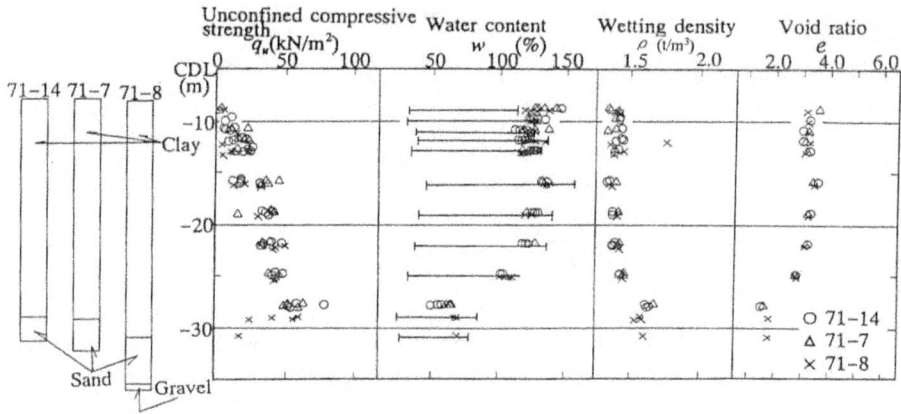

Figure 12. Soil properties from borings No.7,8,14('71) located perpendicular to the coast line

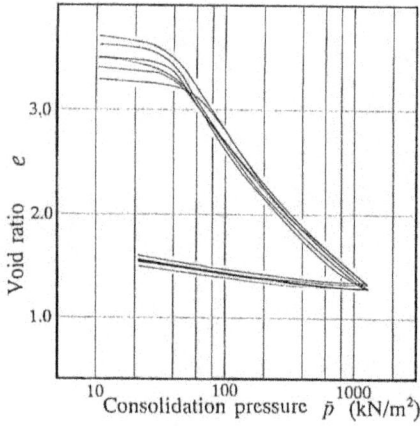

Figure 13. $e \sim \bar{p}$

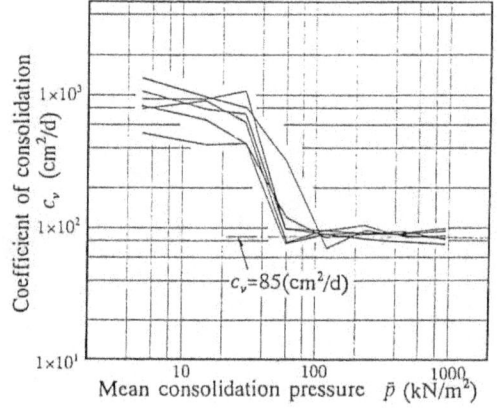

Figure 14. $m_v \sim \bar{p}$

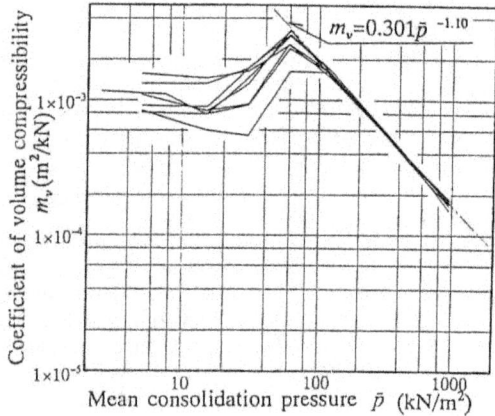

Figure 15. $c_v \sim \bar{p}$

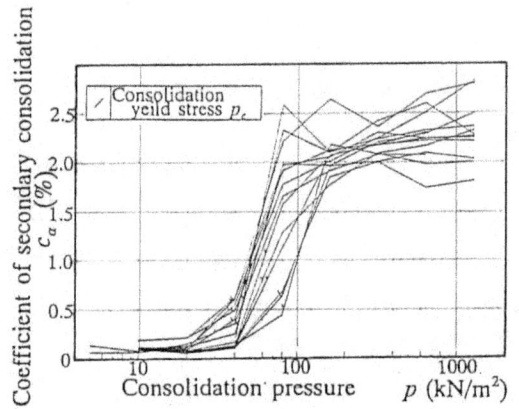

Figure 16. $c_a \sim p$

112

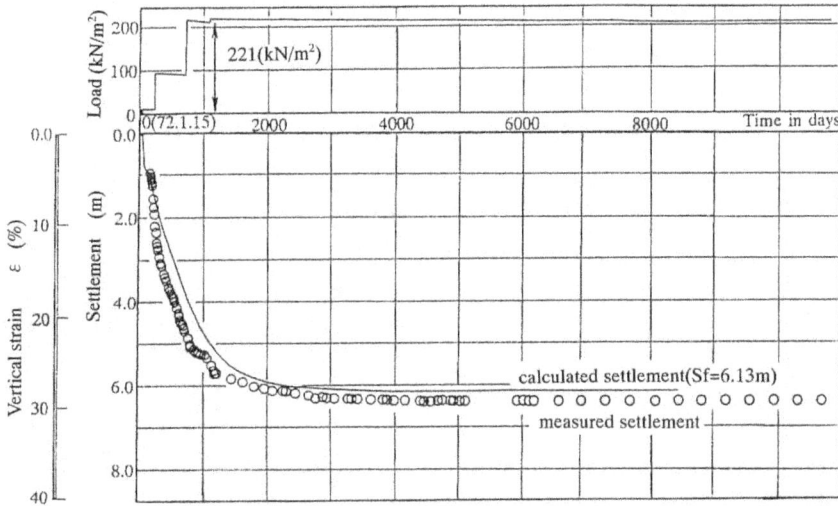

Figure 17. Measured and calculated settlement at S-7 in natural scale

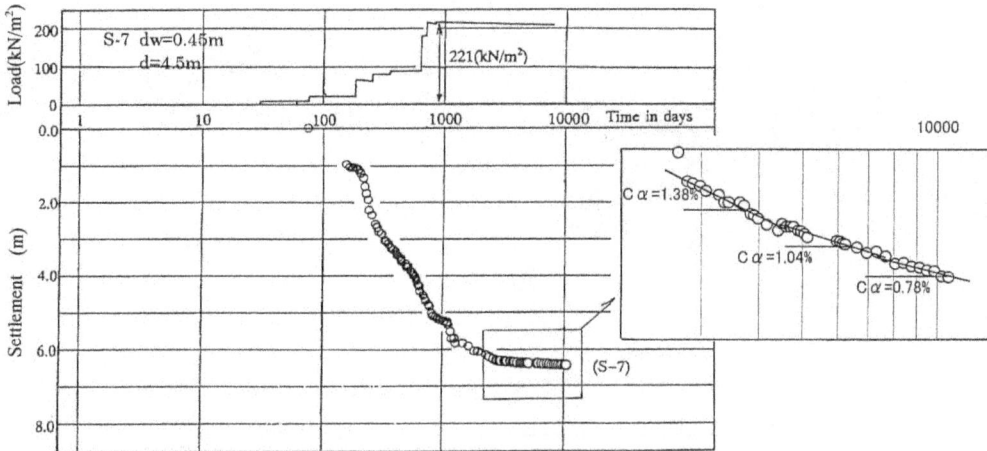

Figure 18. Measured and calculated settlement at S-7 in semi-logarithmic scale

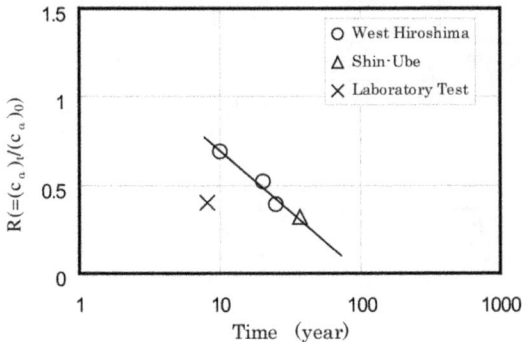

Figure 19. $R(=(c_\alpha)_t/(c_\alpha)_0) \sim t$

3 A CASE OF PRECOMPRESSION TECHNIQUE WITH NO RESIDUAL SETTLEMENT

In almost every cases in soft foundation problems, where soil stabilization techniques were executed, there remained a certain quantity of residual settlement. Civil engineers including specialists on foundation engineering, usually accept it as a matter of course. However, ordinary people do not think the matter like that. On the contrary, they recognize the land as defective, when it settles after their land usage, following the Civil Law Act. A case record of precompression technique where there was no residual settlement after the completion of a structure on it, is introduced here, in order to emphasis that it is possible to reduce residual settlement to an infinitesimally small quantity by utilizing this technique.

113

3.1 *Basic experiments on precompression and C_α*

In order to clarify the effect of precompre-ssion on C_α, a series of consolidation tests had been performed using a separate-type consolidometer (Aboshi et al. 1981). The outline of the consolidometer is shown in Fig. 20. It consists of five oedometers of standard size which are connected in series and loaded hydraulically under a constant pressure. It was developed for the purpose of measuring precisely strain and pore pressure distributions in a consolidating clay specimen. Fig.21 is a schematic diagram of the apparatus.

An example of the settlement strain–time relation is shown in Fig.22. This figure is obtained when the load is once released to zero at a certain consolidation degree, and again surcharged with a smaller load as in Fig.23. The dotted line shows the effective stress path. This is a model of loading in precom-pression techniques, and each curve in Fig.22 shows the settlement of the five separate portions in Fig.20.

The gradient of each tail C_α is considerably different with the progress of consolidation of each portion. As the average effective stress at each portion can be determined by pore pressure measured on both surfaces, the relation between C_α and effective stress is plotted.

Fig.24 shows the distribution of $C_\alpha (= d\varepsilon/d\log t)$ and σ/σ_f, where σ is an attained effective stress by precompression and σ_f is a stress by a structure constructed on the preconsolidated clay layer.

This basic research shows that the ordinary concept of precompression technique is insufficient to reduce the residual settlement including secondary creep to zero, that is, C_α being about 1% when $\sigma/\sigma_f = 1.0$ as in Fig.24. The ratio must be about 2.0 or more to minimize the residual settlement in precompression technique. (Aboshi et al. 1983)

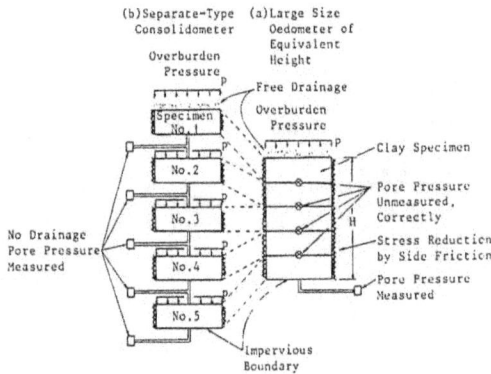

Figure 20. Comparison between separate-type consolidometer and large size oedometer of equivalent height

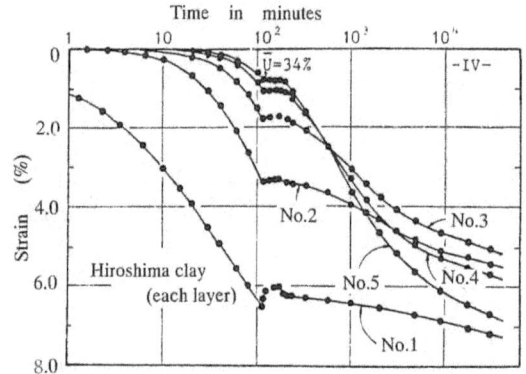

Figure 22. Settlement in rebound-recompression test, separate layers

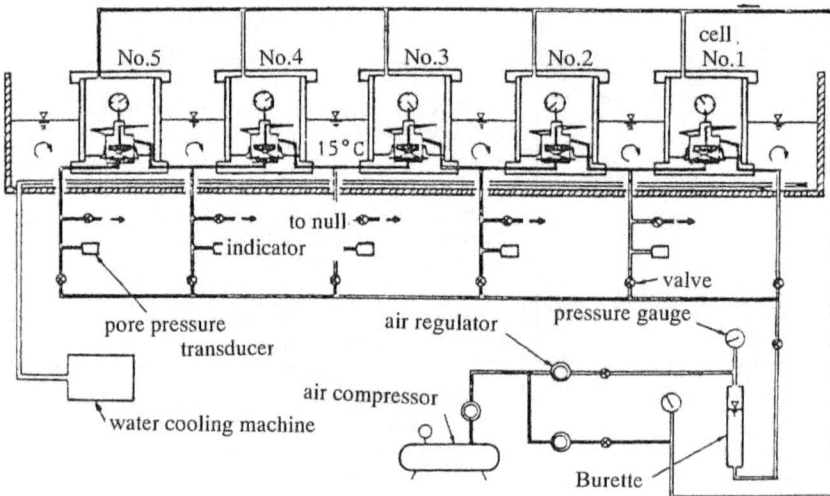

Figure 21. Separate-type consolidometer

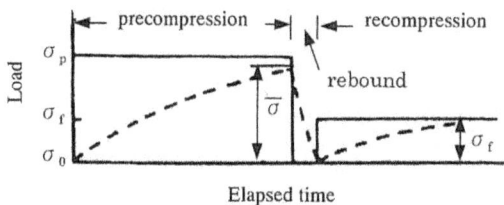

Figure 23. Typical loading path in rebound-recompression test

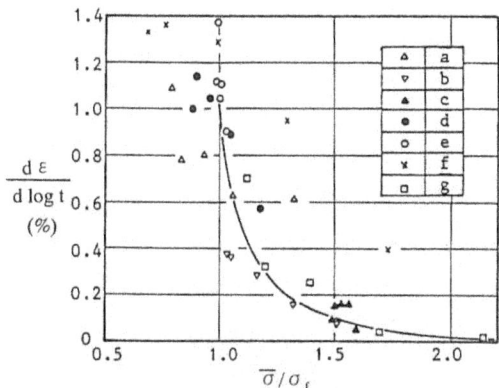

Figure 24. Relationships between the gradient of creep settlement and effective overconsolidation ratio

3.2 *A successful example of precompression technique*

A huge sewage treatment plant, a water-pool-like structure of 220 m in length and 80 m in width, was constructed on a soft alluvial clayey ground, on a coast of the Inland Sea, western suburb of Hiroshima City from 1977 until 1980. (Aboshi et al. 1982) It was a terminal treatment plant of sewage water from the western half of the city area, and the first experience for the engineers concerned to construct such a

huge pool-like structure on a very soft grounds. Originally it was planned to be supported totally by steel pipe piles of 80 cm in diameter and 35 m in length, with the spacing of 2.5 m.

As the site was a newly reclaimed fill, and the total consolidation settlement was estimated to be about 4 m, so the usage of pile foundation was not recommended by the reason that there might be anticipated a large difference of settlement between the structure and the surrounding ground, causing the inverse inclination or damage on the connection of pipe lines. Instead of the deep foundation, a plan to utilize soil stabilization of higher quality was accepted.

Outline of the soil condition is shown in Fig.25 and the surcharge was executed by filling up an embankment up to the limit of the bearing capacity and in addition to the embankment load, −9 m water table lowering was performed by deep well pumping.

Total diagram of soil stabilization plan is shown in Fig.26. (Aboshi et al. 1984).

About one and half years after the surcharging, it was confirmed that 96% consolidation had been attained through the measurement of settlement and pore pressure.

The surcharge was removed and the construction of structures was carried on.

Total settlement was 2.3m, excluding about 2m of settlement by reclamation before the execution of sand drains.

After the preloading stage, which seemed to reach $\sigma/\sigma f \fallingdotseq 2.0$, the ground was somewhat heaved by foundation excavation of the structure. The movement of ground surface is shown in Fig.27 in a different scale. The rebound was about 5mm, and the settlement of structure after its completion was almost negligible. The precompression technique had been perfectly carried through under observational procedures. (Aboshi et al. 1982).

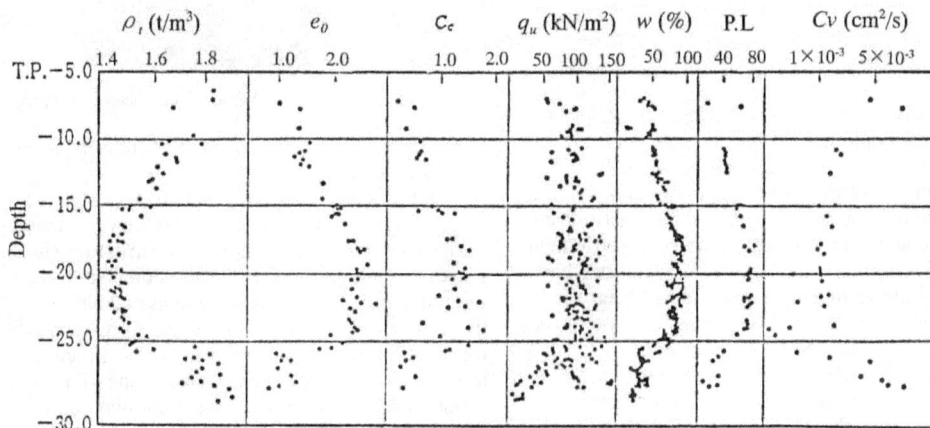

Figure 25. Soil condition of the site

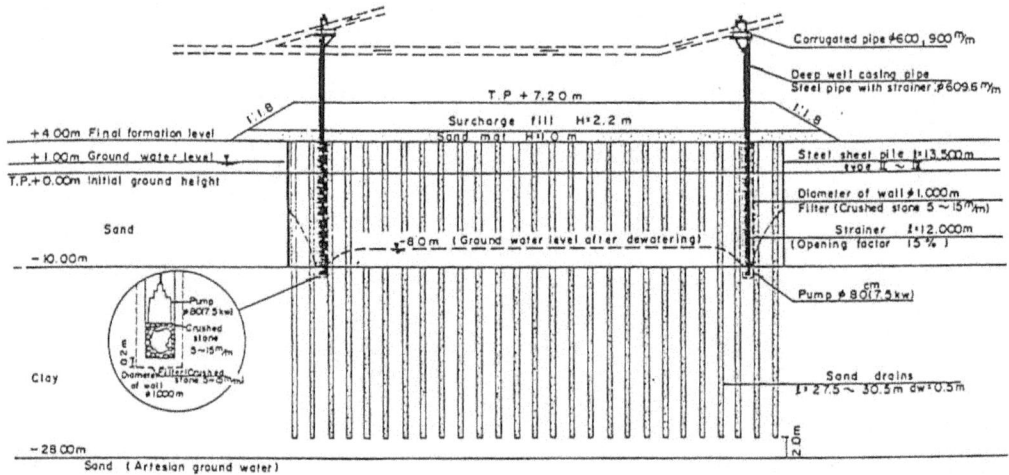

Figure 26. Illustration of soil stabilization works

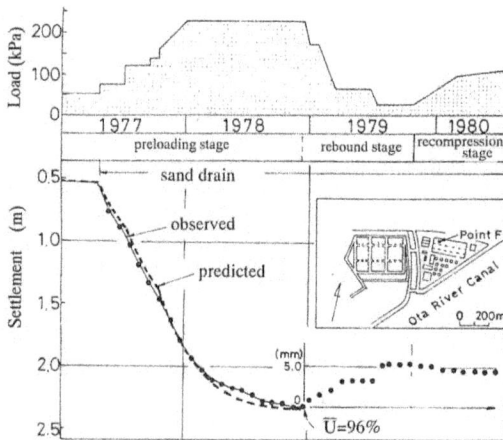

Figure 27. Settlement by preloading, west Hiroshima sewage treatment plant

Photo 1. Photograph of 3m diameter oedometer test, 1985

4 CONCLUSION

Two interesting topics on consolidation and pre-compression technique of soft clayey grounds are referred in the present paper. The first topic is on the secondary creep settlement, which continues for a long period after the primary consolidation finished. Through decades long measurement of settlement both in the laboratory and fields, the authors have shown that long-term EOP comes in between the A and the B hypotheses (nominated as C hypothesis) and they treat mainly the change of gradient C_α this time. C_α gradually decreases in decades time scale, and is conjectured to become zero in a hundred year. However, whether the final creep settlement comes on a single line as the B hypothesis insists, or not,

the authors reserve the opinion by reason of having no practical data.

Another case referred in this paper is an example of successful soil stabilization technique carried out in the late 1970's. In practical construction sites, engineers usually accept that the residual settlement after soil stabilization might be unavoidable, however, there was a successful case with no residual settlement and the cost of construction was about two-third of the case of deep foundations. The authors insist that the more rigorous treatment of residual settlement problems is needed before the land treated is handed over to no-professionals.

Photo 2. Aerial photograph of Shin-Ube powerplant, 1990

Photo 3. Soil stabilization work,west Hiroshima sewageplant, 1979

117

REFERENCES

Aboshi, H. & H. Monden, 1961. Three- dimensional consolidation of saturated clay. *Proc. 5th ICSMFE (Paris)* Vol.1, pp.559-562

Aboshi, H., 1973. An experimental investigation on the similitude in the consolidation of a soft clay, including secondary creep settlement, *Proc 8th ICSMFE(Moscow)* 4, 3, 88

Aboshi, H., H. Matsuda & M. Okuda, 1981. Precompression by separate-type consolidometer. *Proc. 10th ICSMFE (Stockholm)* pp. 577-580

Aboshi, H., K.Ishii & T.Inoue, 1982. Observational procedure used in the soil stabilization work, west Hiroshima sewage treatment station. *Soils and Foundations* Vol. 30, No.7, pp. 37-44

Aboshi, H & H. Matsuda, 1983. A study on precompression technique by separate-type consolidometer. *Jour. Geotech Eng. JSCE* No.340, pp. 139-144

Aboshi, H., K.Ishii & T.Inoue, 1984. Construction of west Hiroshima sewage treatment station by preloading. *Soils and Foundations,* Vol.33, No.5, pp. 29-34

Aboshi, H., 1995. Case record of long-term measurement of consolidation settlement and their prediction. *Special lecture IS-Hiroshima 95, Compression and Consolidation of Clayey Soils, Balkema*, Vol.2, pp. 847-872

Soft Ground Engineering in Coastal Areas, Tsuchida et al. (eds)
© 2003 Swets & Zeitlinger, Lisse, ISBN 90 5809 613 0

Settlement analysis and observational method for the reclamation with dredged clay – A case record at New-Kitakyushu Airport

K. Egashira
Coastal Development Institute of Technology, Tokyo, Japan

N. Yamagata & T. Takada
Ministry of Land, Infrastructure and Transport, Shimonoseki, Japan

M. Katagiri & M. Terashi
Nikken Sekkei Nakase Geotechnical Institute, Kawasaki, Japan

T. Yoshifuku & S. Murakawa
Nikken Sekkei Civil Engineering, Tokyo, Japan

ABSTRACT: A New-Kitakyushu airport is under construction on a reclaimed land by dredged clay. Along with the progress of reclamation by pump-dredged clay, the detailed soil investigation and monitoring had been performed together with the numerical simulation by "CONAN", for predicting the consolidation behavior of dredged clay layer and for modeling the soil profiles prior to ground improvement. The accuracy of prediction using CONAN was improved by the parameters identified periodically from the soil investigation and monitoring. The design of ground improvement using vertical drain was performed for ground model predicted by CONAN analysis with the identified parameters. After the placement of fill on top of dredged clay, the settlement behavior that is predicted based on the ground model is confirmed consistent with the measured data.

1 INTRODUCTION

Two projects are going on simultaneously on the sea northeast of Kyushu Island as shown in Figure1.

One is the maintenance dredging of sea bottom sediment to keep the required depth of navigation channels and anchorage areas. For the protection of marine environment, dredged materials, mostly clays have been and will have to be discharged into a pond surrounded by containment dikes on the sea. This in turn continuously creates a new artificial island of extremely soft soil condition.

The other is a New Kita-kyushu Airport construction project that is a relocation of existing one inland, thereby reducing the air pollution and noise to the residential area and increasing the capacity of airport. The site for the airport was selected on an artificial island created by the maintenance dredging. The merge of two projects had a merit of reducing the amount of hill-cut materials substantially that might otherwise be enormous and cause a destruction of the environment at borrow area.

When pump-dredged clay slurry with high water content like a liquid is discharged into the pond, suspended soil particles settles loosely with the water content between 200 to 300 %. The dredged clay layer thus created is in the unconsolidated condition and subsequently consolidates largely due to its own weight in the long term. To create a reliable foundation ground for the airport at a specified elevation, the placement of good quality fill material over the dredged clay layer is necessary.

The airport construction required strict time schedule until its opening. It was necessary to accelerate the consolidation of soft reclaimed clay layer by vertical drain. In the ordinary case, the design of ground improvement using vertical drain, that is the determination of drain pitch, is performed on the ground model determined by the soil investigation. In this project, however, the time from the completion of reclamation with dredged clay to surface soil

Figure 1. Location of New Kitakyushu airport

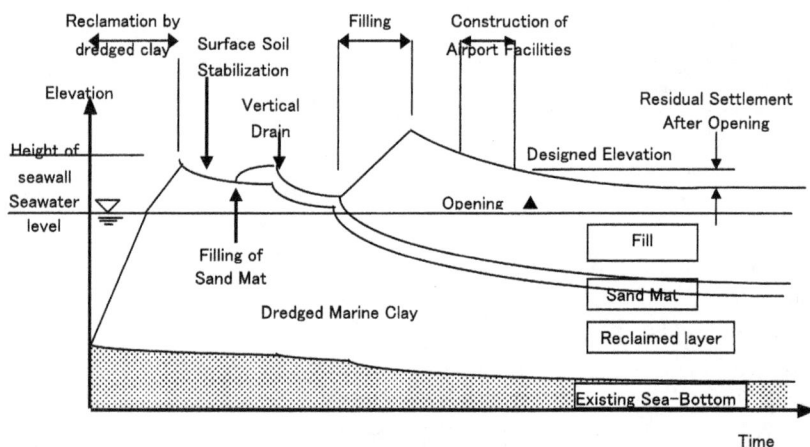

Figure 2. Change of ground elevation from reclamation with dredged clay to opening of airport

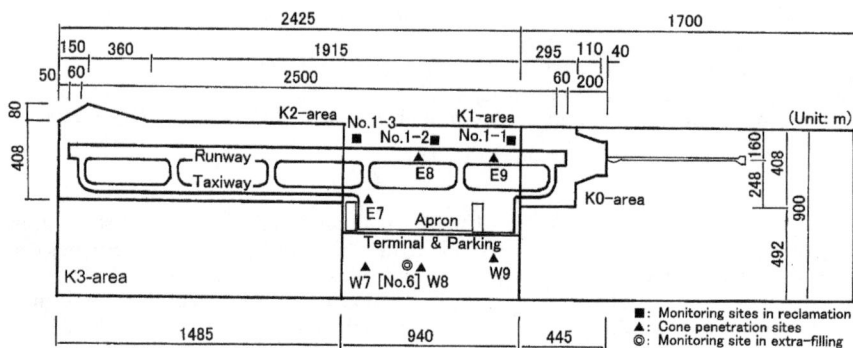

Figure 3. Plan view of New Kita-Kyushu airport and Monitoring sites in K-1 area

stabilization was too short to perform detailed soil investigation. Therefore, the design of ground improvement had to be performed during reclamation without the knowledge of completed soil profile. Whereas the accurate prediction of the time dependent settlement after the filling is important in order to estimate the amount of fill material, and to keep the final elevation to a required level. The problems in design here are the accurate assumption of ground model and determination of consolidation parameters.

In this paper, an outline of the new Kita-kyushu airport construction and the results of field observation and settlement analysis will be described. The validity of the ground model determined by numerical analysis will be discussed using settlement behavior measured during and after filling.

2 OUTLINE OF NEW KITA-KYUSHU AIRPORT CONSTRUCTION

The time history of ground surface level from the start of reclamation with dredged clay to the opening of airport is schematically shown in Fig. 2. The horizontal axis is time, and the vertical one shows the elevation. Term of each construction process and the change of each layer are also illustrated. The allowable residual settlement in this project is 35 cm after the opening of airport. The ground improvement was designed in consideration of this allowable value.

The project site consists of four areas (four separate disposal ponds) as shown in Fig. 3. Airport facilities such as a runway with 2,500 m long, taxiway and apron will be constructed across three areas K0, K1 and K2. In the three areas, the construction of airport is carried out as shown in Fig. 2, though the origin of time axis is different for each area. The K0 area had been reclaimed from the sea by dredged materials for about 18 years from July 1979 to March 1997. The K1 area had been reclaimed for 32 months from October 1996 to June 1999, and the K2 area started in December 1998 and ended in March 2002.

The airport should be in operation no later than October 2005. From this tight time plan, immediately after the reclamation with dredged clay, the surface soil stabilization, placement of sand mat, in-

120

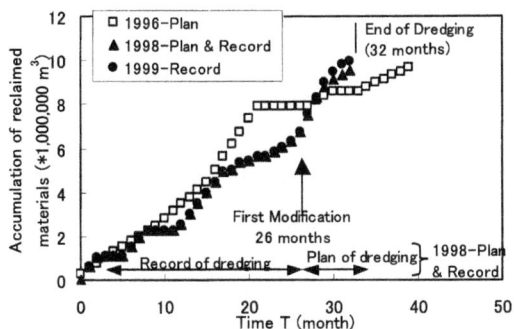

Figure 4. Reclamation history of K1-area

stallation of drains, and filling had to be performed continuously without any intermission. As described before, the design of ground improvement had to be performed during the reclamation without the precise knowledge on completed clay layer.

In the following sections, the design of ground improvement based on reclamation analysis will be discussed on the K1-area.

3 CONDITIONS OF K1-AREA

The K1-area is 900 x 940 m in plan and the average depth of seabed and the thickness of Holocene clay layer are DL-7.7m and 5 m before the reclamation, respectively. The sandy layer is lying beneath the Holocene clay layer.

Reclamation by dredged materials is planned up to DL+7 m, which corresponds to the height of seawall. The dredged materials are mainly marine clays from three navigation channels located near the con-

struction site as shown in Fig. 1. The physical properties of the materials are as follows; density of soil particle ranges from 2.55 to 2.71 g/cm^3, with the average value of 2.64 g/cm^3, natural water content 73 to 145 %, with the average value of 101 %, Liquid limit: 47 ~ 99 %, Plasticity index: 26 ~ 69.

For the numerical analysis, true mass of solid part of the reclaimed materials is needed. The mean water content of 101 % obtained from the soil investigation was used for the calculation of true mass.

The reclamation history of the K1-area has been changed during the project as shown in Fig. 4. Total amount of dredged soil at their borrow area has not changed and was about 10 million cubic meters. But the discharge term has changed. The project started with the 1996-Plan (□), in which the reclamation term was 40 months. The actual reclamation (●) was carried out in 32 months (1999-result). In the intermediate analysis at 26 months, the actual record of reclamation up to 26 months and the plan at that time (▲) were used as the basis for calculation.

Before reclamation, the preliminary prediction for reclamation with dredged clay was performed using the 1996-Plan. For the intermediate analyses at 26 months, the 1998-Plan & Record mentioned above was used. Then, for the back analyses performed after the reclamation with dredged clay, the 1999-Record was used. The numerical results will be described in section 5.

4 CONSOLIDATION ANALYSES

In this study, 2 types of consolidation analyses were used according to the construction process. As shown in Fig. 5, the CONAN analysis in one-

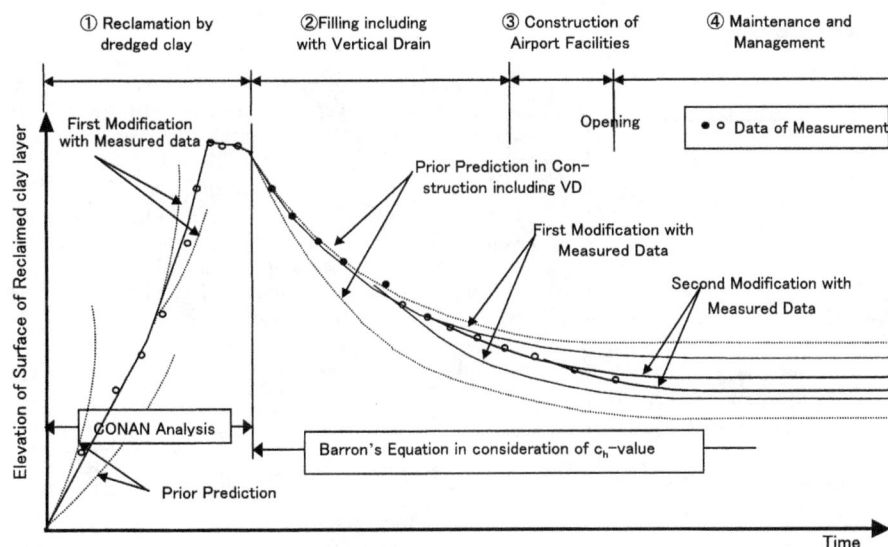

Figure 5. Schematic settlement behavior of reclaimed clay layer and applied range of two analysis methods

121

dimensional consolidation was applied in the reclamation stage with dredged clay until the installation of vertical drains. The Barron's Equation together with Cc-value was used after that.

4.1 CONAN in one-dimensional consolidation

The numerical method used in the reclamation with dredged clay was 'CONAN', which was developed based on a generalized consolidation theory (Imai, 1995). The numerical code was developed by considering the accumulation of reclaimed layers based on the technique proposed by Yamauchi et al.(1991). The detailed procedure of this numerical method was described by Katagiri et al. (2000).

As described earlier, the consolidation analysis of dredged clay by CONAN was repeated during the reclamation process; preliminary, intermediate and back analyses to improve the accuracy of prediction by incorporating the results of monitoring and periodical soil tests on samples taken from the reclaimed area. For the consolidation analysis of extremely soft dredged clay, the consolidation parameters spanning

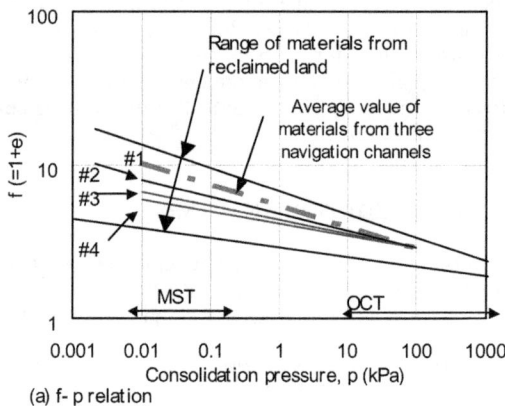

(a) f- p relation

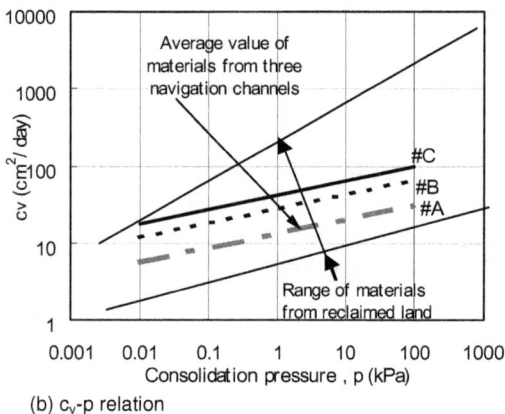

(b) c_v-p relation

Figure 6. Range of consolidation parameters of samples obtained from K1-area and assumed consolidation parameters.

wide stress range as shown in Fig.6 are necessary. To determine the consolidation parameters over such a wide stress range, the multi-sedimentation tests (MST) proposed by Yamauchi et al. (1990) and the ordinary consolidation tests (OCT) were carried out.

For the preliminary prediction, only the relations #1 and #A (herein after called "average relations") were available that were determined by averaging the tests results on three samples obtained from the borrow area (nearby navigation channels).

Along with the progress of reclamation, the soil samples were taken from the K1-area after the grain size sorting due to sedimentation. The consolidation parameters thus obtained had a wider range as shown in Fig.6 (Sato et al., 2000). In the intermediate prediction and back analyses, the consolidation parameters are modified based on the "average relations" as shown in Fig. 6. For the compressibility, the inclination of $\log f - \log p$ relation was changed by fixing a point at $p = 1,000$ kPa on the average relation, as case #2 to #4. The permeability was changed by shifting the $\log c_v - \log p$ relation in parallel to the average one (#A), as case #B and #C. These assumed relations are within the range obtained by the samples from the K1-area.

The conditions of the existing seabed are as follows. The Holocene clay layer before the reclamation is 3 to 5 m thick, and is in a normally consolidated sate. The consolidation parameters are determined by the test results of undisturbed samples from the seabed, and are $Cc = 1.05$ and $c_v = 50$ cm^2/day. Drainage at the bottom of Holocene clay layer is permitted. The boundary condition at the top of the reclaimed layer is fixed at 9.8 Pa as proposed by Yamauchi et al. (1990), and is in the drained condition.

4.2 Yoshikuni's modified Equation in multi-dimensional consolidation

After installation of vertical drains, excess pore water in the clay layer dissipates radially toward the vertical drains by the horizontal flow. The settlement of ground occurs vertically. To calculate the consolidation process after filling, the Yoshikuni's Equation modified from Barron's Equation (Yoshikuni, 1979) was used. The magnitude of settlement was estimated by the compression index, Cc.

Eq. (1) shows Yoshikuni's Equation taking the well resistance into account.

$$U(T_h) = 1 - \exp\left[-\frac{8}{F(n) + 0.8 \cdot L_{cwR}} \cdot T_h\right] \quad (1)$$

Where,

$$F(n) = \frac{n^2}{n^2 - 1} \cdot \ln n - \frac{3n^2 - 1}{4n^2}, \quad L_{cwr} = \frac{32}{n^2 - 1} \cdot \frac{k_c}{k_w} \cdot \left(\frac{H}{d_w}\right)^2$$

$U(T_h)$ is the degree of consolidation. T_h is the time factor concerning settlement with horizontal drainage as follows;

122

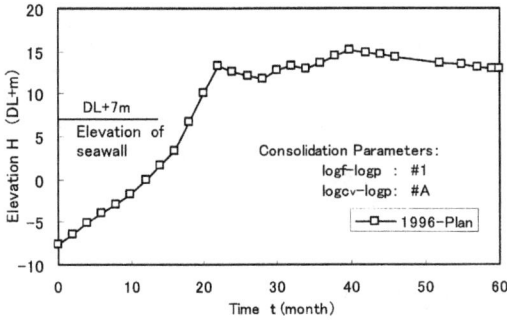

Figure 7. Preliminary prediction of elevation of top surface of dredged clay

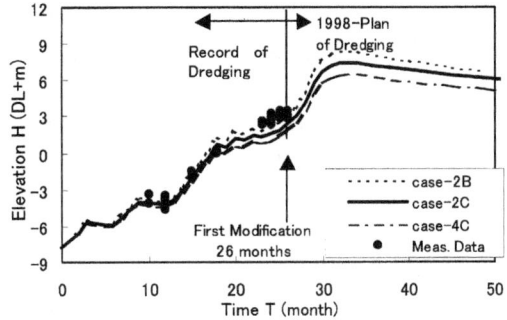

Figure 8. Intermediate prediction of elevation of top surface of dredged clay

$$T_h = \frac{c_h \cdot t}{d_e^2}$$

where, c_h is a coefficient of consolidation in horizontal direction, and t is real time. d_e is the effective radius as function of drain pitch and arrangement of drain. n is the ratio of d_e to equivalent radius of vertical wicked drain, d_w. k_c and k_w are horizontal permeability of undisturbed clay and permeability of drain material, respectively. H is a length of drain.

5 EVALUATION OF CONAN ANALYSIS BY MONITORING RESULTS

Along with the progress of reclamation, back analyses were carried out in several stages and parameters were identified for the prediction of next stage. The following is the history of predictions of the elevation in a couple of stages together with the measured data.

The consolidation of clay layer is governed by two parameters, compressibility and rate of consolidation. Therefore, the numerical analysis of consolidation behavior must fit not only with the time-dependent elevation of the clay layer but also with the soil profile represented, for example, by water content distribution and/or excess pore water pressure distribution.

5.1 Preliminary prediction

Before the reclamation, the preliminary prediction of elevation of top surface of dredged clay layer was performed (Fig. 7) under the 1996-Plan and average parameters of sampled materials (#1 and #A). By this prediction, the elevation of dredged clay surface exceeded the designed height of seawall at 20 months, which meant that the total volume of dredged clay in 1996-Plan could not be discharged into the reclamation area. However, the accuracy of prediction in this stage was low because the parameters were the average value of only three samples taken from the borrow area and because those were not confirmed to represent the reclaimed materials.

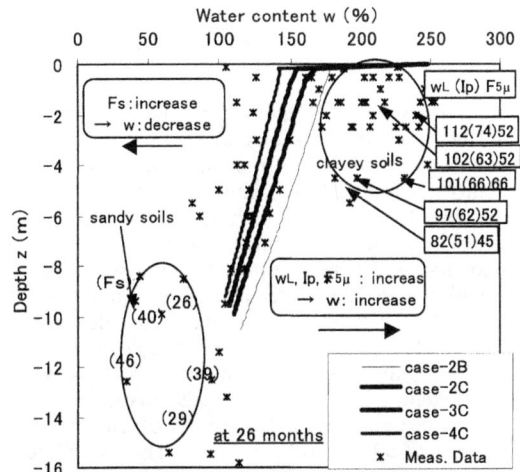

Figure 9. Comparison between measured and analyzed water content distribution of reclaimed layer at 26 months after reclamation

5.2 Intermediate prediction

At 26 months after the start of reclamation, intermediate prediction was performed. Intermediate prediction is actually a back analysis up to that period of time and the best-fit solution will be utilized for the prediction in the following reclamation. Figure 8 shows the measured elevation of the reclaimed land monitored from the beginning of reclamation to 26 months together with some of the intermediate prediction, with different combinations of consolidation parameters as shown in Fig. 6. The consolidation parameters used were denoted in such a way as the case-2C, for example, which means a set of #2 log f – log p relation and #C log c_v – log p. In these analyses, the actual record of discharge up to 26 months from the beginning and subsequent plan (1998-Plan & Record) as shown earlier in Fig. 4 was used. The prediction by case-2B, 2C, and 4C equally simulated the actual measurement as far as the surface elevation is concerned. The prediction by case-3C is lo-

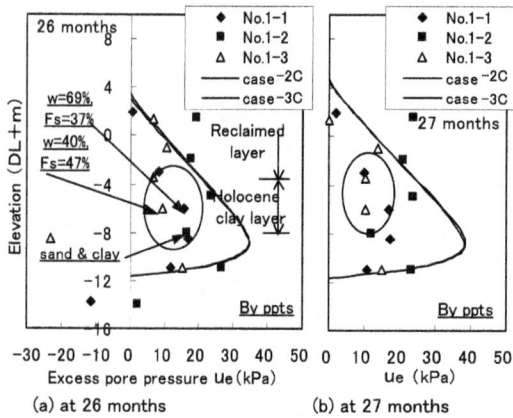

Figure 10. Excess pore water pressure distributions in the middle of reclamation with dredged clay

cated between case-2C and 4C. By the solution case-2C, the predicted peak of elevation would become about 8.0 m at the end of reclamation, and would be slightly larger than the designed elevation of 7.0 m.

Figure 9 shows the water content distribution obtained on a number of samples taken at the K1 area at 26 months. The predictions with some sets of consolidation parameters are also shown in the same figure. As the grain size sorting occurs during the sedimentation process, the grain size distribution and Atterberg limits are also investigated on all the samples taken. The original characteristics of the dredged materials at the borrow area were $Ip = 59$-69, $Fc = 43$-50 % and $Fs = 4$-8 %. The data plotted on the upper right of the figure with water content above 200 % and enclosed with a circle are obtained from the samples with higher liquid limit and with larger clay fraction content (< 5 µm) in comparison with original soils. The data with lower water content on the lower left enclosed with another circle are from the samples which contains sand fraction more than 25 %. When ignoring the data enclosed with two circles, the predicted water content distributions by the case-2C, 3C and 4C agree well with the measured ones.

Figure 10 compares the measured and calculated distributions of excess pore pressure at 26 months after reclamation (Fig. 10(a)) and 27 months (Fig. 10(b)), when the water level in the reclaimed land was almost the same as the seawater level outside the reclaimed land. Pore water pressures were measured at three points, No. 1-1, 1-2 and 1-3 that are indicated as black squares in Fig. 3. Here, excess pore pressure is defined as the measured pore pressure minus hydrostatic pressure. The results of numerical simulation with two sets of consolidation parameters, case-2C and 3C, almost coincide each other. The grain size distribution and the water content were investigated on the samples taken at the location of transducers. The data enclosed with a circle

in this figure are measured by the transducers embedded in the soil with a large sand fraction. When these data are ignored, the calculations by case-2C and 3C agree well with measured ones.

At this intermediate stage, the suitable set of consolidation parameters was determined as the case-2C because the predicted relation for that set simulated the overall behavior until 26 months.

5.3 Back analysis just before ground improvement

After the completion of discharging the dredged clay into the K1-area, a back analysis of reclamation was performed to estimate accurately the soil profile of reclaimed land prior to ground improvement.

Figure 11 shows the measured elevations against time and several numerical results. It is confirmed that the numerical result of case-3C is closer to the measured data than that of case-2C.

Figure 12 shows the comparison between the measured and calculated distributions of excess pore pressure at 36 months after the beginning of reclamation. The measured points are expressed as solid squares in Fig. 3. The pore water pressures in the reclaimed land were measured by the pore pressure dissipation tests using a piezometer cone, and were obtained from the assumption that the ground water table was consistent with the ground surface of reclaimed land. The thickness of the Holocene clay underlying the dredged soil layer is different within the K1-area. In the numerical analyses, the results of case-3C for 3 and 5 meters of thickness of Holocene clay layer were drawn.

In the reclaimed layer, every measured data located relatively in a narrow range, and its distribution has the maximum value at the bottom of reclaimed layer. The numerical results also show the same tendency. The calculated distributions are located higher than the average measured data but are consistent with the measured ones.

Considering the time record of elevation of reclaimed land in Fig. 11, especially after the complete of reclamation, the best-fit set of consolidation parameters were identified as case-3C.

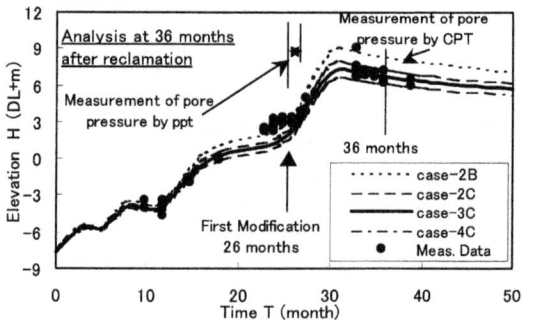

Figure 11. Comparison between measured and numerical results of time record of top surface of reclaimed land

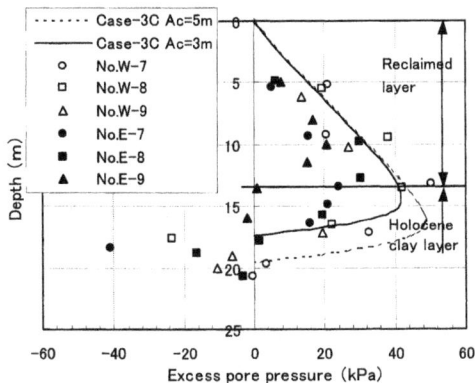

Figure 12. Excess pore water pressure distributions after complete reclamation with dredged clay

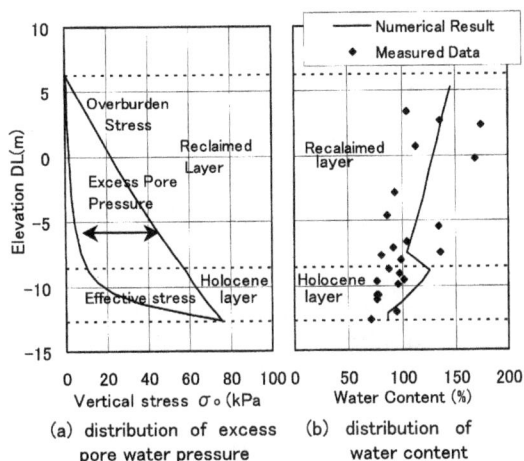

(a) distribution of excess pore water pressure

(b) distribution of water content

Figure 13. Soil profile for ground improvement that calculated by CONAN with case-2C

6 GROUND IMPROVEMENT DESIGN FOR FILLING

6.1 Estimation of soil profiles for ground improvement

Although the case-3C was found to be the best-fit for the CONAN analysis after the completion of the reclamation by dredged clay, the design of the ground improvement had to be done earlier at 26 months due to the tight time schedule of the project. At 26 months, the design of ground improvement was performed using the ground model assumed by the best-fit result of CONAN analysis then with case-2C.

Figure 13 shows the estimated soil profile prior to ground improvement that was obtained from the CONAN analysis using the identified consolidation parameters (case-2C). Figure 13 (a) shows the excess pore pressure distribution that is defined as total overburden stress minus effective stress calculated by CONAN. The settlement of reclaimed land under the fill was calculated in consideration of this initial excess pore pressure distribution and load of fill, using compressibility Cc. The rate of consolidation was evaluated by Eq. (1).

The initial condition for the design was also obtained from CONAN analysis as shown in Fig. 13 (b). In this figure, the water contents measured at 39 months are also plotted. In the Holocene clay layer, the prediction is slightly larger than the measured ones, but in the reclaimed layer, the prediction is located in the middle within the distribution of measured data. The numerical result is confirmed to agree with the measured data when the appropriate consolidation parameters are identified through back calculations.

6.2 Estimation of consolidation parameters of reclaimed land for ground improvement

For the design of ground improvement using vertical drain, the consolidation parameters such as com-

pressibility and rate of consolidation of reclaimed layer are necessary.

In this project, the assumption of consolidation parameters for ground improvement was estimated by extrapolating the parameters identified by the CONAN analysis during the reclamation with dredged clay. The stress level in the reclaimed land during the filling was thought to be from 1 to 100 kPa, and that in the reclamation with dredged clay was less than 3 kPa.

For the compressibility, extrapolation of the $\log f$ – $\log p$ relation of reclaimed clay layer was used as shown in Fig. 14. The solid curve in the range from 0.01 to 1 kPa is the identified one, and the line over 1 kPa is linear extrapolation with Cc = 1.05. The plots in the same figure show the test results of undisturbed samples taken from the deep part of reclaimed layer in the K1-area during the reclamation. The assumed relation is located in the middle of the test results, and is thought to be appropriate.

For the rate of consolidation, 50 cm²/day that is relevant to the lower value of case-#C in Fig. 6(b) in the range from 1 to 100 kPa was selected. The test results of undisturbed samples were distributed from 20 to 120 cm²/day in normally consolidated state. The coefficient of consolidation is also adequate.

For the design of ground improvement using vertical drain, the horizontal coefficient of consolidation is necessary. In this project, the result of field experiment at the K0-area (Egashira et al., 2001) and Kashii Park-port in Fukuoka (Matsuoka et al., 1998) were used. The field experiments were carried out to evaluate the horizontal coefficient of consolidation by back analyses. The results indicate that the horizontal coefficient of consolidation is influenced by the pitch of drain. The relationship between pitch of drain, d and ratio of horizontal coefficient of consolidation, c_h to vertical coefficient of consolidation,

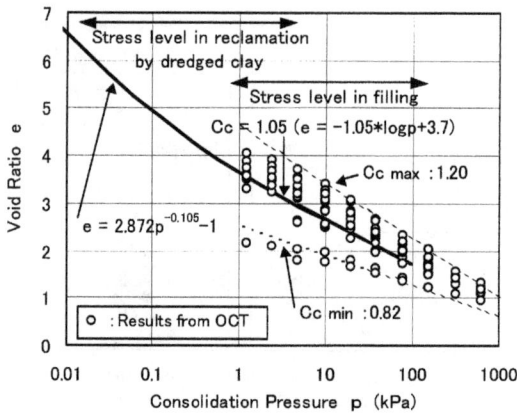

Figure 14. Identified and measured compressibility of reclaimed clay layer

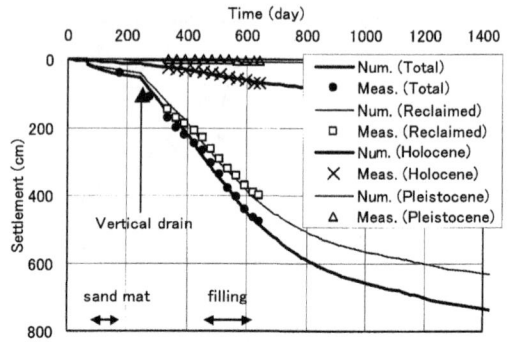

Figure 15. Predicted and measured settlements of reclaimed land with vertical drain during filling

c_v obtained from the consolidation tests on the sampled specimens is liner. The c_h value for the present calculation was estimated by the linear $d - c_h/c_v$ relation. As results, the pitch of vertical drain was determined 1.4 m.

6.3 Monitoring of settlement and evaluation of design of ground improvement

Figure 15 shows the monitored settlement of reclaimed land under the fill pressure. The predicted results are also drawn in the same figure. The measured point is No. 6 in Fig. 3 (marked as ◎). Time in the horizontal axis starts at the end of reclamation with dredged clay.

At No.6 location, the measured settlement data for individual layer is consistent with each prediction. At the latest measurement (660 days), the total settlements measured at other points were varied form 450 to 540 cm. The corresponding prediction was approximately 490 cm and was equivalent to the mean value. Therefore, the prediction method used in this project is practically useful.

This variation is caused by the inhomogeneity of reclaimed layer, and will cause differential settlement in a future. Considering the function of airport, the countermeasure for the differential settlement will be taken in due course.

7 CONCLUSIONS

From this study, the following conclusions can be drawn:

1 The accuracy of predicted reclamation behavior is improved by the parameters identified through the detailed soil investigation and monitoring.
2 Using the identified parameters, the consolidation behavior of dredged clay layer under fill pressure can be estimated with high accuracy.

ACKNOWLEDGMENT

The authors would like to thank Prof. H. Ochiai of Kyushu University and Prof. G. Imai of Yokohama National University for their valuable advice.

REFERENCES

Egashira, K., et al., 2001. Horizontal coefficient of consolidation in reclaimed ground by dredged clay (in Japanese), *Proc. of 36th Annual meeting of JGS*: 1003-1004.
Imai, G., 1995. Analytical examinations of the foundations to formulate consolidation phenomena with inherent time-dependence. *Proc. of IS-Hiroshima '95*, 2: 891-935.
Katagiri, M., et al., 2000. Change of consolidation characteristics of clay from dredging to reclamation, *Proc. of IS-Yokohama 2000*, 307-313.
Matsuoka, T., 1998. Evaluation of coefficient of consolidation on the plastic-board drain method. (in Japanese), *Proc. of 33rd Annual meeting of JGS*: 2129-2130.
Sato, T., et al., 2000. Reclamation control of pump-dredged clay by CONAN, *Proc. of IS-Yokohama 2000*, 507-513.
Yamauchi, H., et al., 1990. Effect of the coefficient of consolidation on the sedimentation consolidation analysis for a very soft clayey soil (in Japanese), *Proc. of 25th Annual meeting of JSSMFE*: 359-362.
Yamauchi, H., et al., 1991. Sedimentation-consolidation analysis of pump-dredged cohesive soils, *Proc. of Geo-coast '91*: 129-134.
Yoshikuni, H., 1979. Design and execution management of vertical drain method (in Japanese), Gihodo-shuppan.

Soft Ground Engineering in Coastal Areas, Tsuchida et al. (eds)
© 2003 Swets & Zeitlinger, Lisse, ISBN 90 5809 613 0

An evaluation of the settlement of offshore man-made island improved with large-scale preloading

N. Koushige, D. Karube, S. Honda & T. Shigeno
Nikken Soil Research Co., Ltd., Osaka, Japan

ABSTRACT: When a deep water sea area is reclaimed on a large scale, the effect of the fill intrudes deep into the sea bottom soil. It is known from recent researches that settlement estimation by the Cc Method and the Hyperbolic Curve Approximation Method suffers difficulties in making calculated settlement agree with actually measured value, and that the secondary consolidation, in addition to the primary consolidation of the diluvial formation should be taken into consideration. In this paper, the settlement analysis by the 2-dimentional viscoelastplasticity finite element methods will be performed so that the degrees of effectiveness of the settlement analysis methods will be discussed. Kobe Port Island was taken up as the subject of this study. Attention will be given to areas within this island for which reclamation history, soil information, soil improvement elements and long-term settlement observation results have already been obtained.

1 INTRODUCTION

In the recent days, many sea area reclamations are located in deep sea areas and are intended to create large area man-made islands. These construction conditions result in placing a large load increase down to a considerable depth in the sea bottom soil. In other words, a territory of soil engineering devoid of accumulated experience cannot but be entered. Specially, in the diluvial formation, the effect of secondary consolidation that cannot be dealt with by the primary consolidation theory is so significant that the resultant settlement is reaching such level that cannot be disregarded as a realistically problem from the view point of engineering. The theory to estimate settlement behaviors has already been established. The settlement curve for any location in any reclaimed ground can be sought when information is known on the 3-dimensional distribution of soil layers, physical and mechanical properties of soil. On the other hand, the current soil investigation techniques do not always give the correct physical and mechanical properties of soil down to the depth contemplated. These problems may be improved one after another as the investigation techniques are improved. However, if they are to be used for settlement behavior estimation at the present time, verification and study from different angles may be necessary and indispensable.

In this paper, attention will be given to the settlement records of the soil and structures thereupon at Kobe Port Island, one of the large-scale deep sea reclamation projects performed earliest in this country. The subjects of attention include 3 cases in 2 sites with different soil improvement histories. The degrees of effectiveness of the settlement analysis by the 2-dimentional viscoelastplasticity finite element methods will be discussed by comparison between the actual settlement measurement results and the current settlement estimation theory.

2 GENERAL DESCRIPTION OF KOBE PORT ISLAND

The reclamation of Kobe Port Island started in 1966 and was fully completed in 1981. It is a man-made island of 436 ha in area (Phase I Island only). The soil filled was approximately 20m deep requiring approximately 80 million cubic meters of borrow soil, which mainly consisted of "Masa-do" of weathered granite soil. The fill soil was transported by the pusher barge and was directly dumped by opening the bottom of the boat. As the reclamation pro-

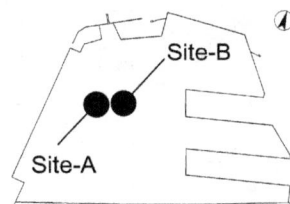

Figure 1. Plan of Kobe Port Island

gressed, it has become difficult for barges to penetrate so that thereafter the soil was unloaded onto the reclaimed land by bucket-type unloaders and was pushed out of the reclaimed land to the surrounding sea areas. Fig-1 shows the plan view of Kobe Port Island Phase I Island, and Fig-2 shows the common Quaternary period stratiform order in the Hanshin District. The Holocene epoch Ma13 is the "alluvial clay formation" that forms the bottom of the sea, and this was the subject of the soil improvement in the reclamation areas including this island. Ma12 is the newest "diluvial clay formation". Below those layers, Ma11 and the sea clay of the Osaka Group continue. Fig-3 shows the results of the soil boring log performed in 1975 down to a depth of 200m at a point slightly west of the center of the island. These are the boring data obtained when the ground elevation reached approximately KP+5.4m. The reclamation continued thereafter up to KP-13.5m until the fill thickness reached 19m. The fill material mainly consisted of "Masa-do" and was partially mixed with waste soil from within the city as well as slag, etc. Under the reclamation deposit soil, the alluvial clay (Ma13) that used to form the sea bottom is uniformly distributed to approximately KP-25m. The layer thickness is about 12m. At this time, the consolidation yield stress pc of the alluvial clay layer was 88 to 206kPa, which was less than the effective overburden pressure of $\sigma v'$=206 to 284kPa so that the consolidation is not complete. The soil below the alluvial clay layer down to a depth of KP-152m is classified as upper diluvial formation and there under the Osaka Group occurs. The upper diluvial formation is composed of distinct sea clay layers Ma12 and Ma11 (3 layers recognized), sand, gravel and clay deposit occurring alternately.

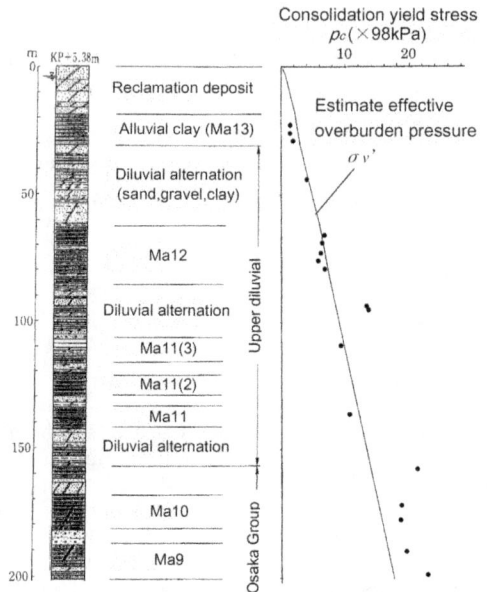

Figure 3. 200m Boring log on Kobe Port Island

Especially, the diluvial alternate layer continuing from KP-25m to KP-57m directly below the alluvial clay is serving as the bearing layer of the foundation piles for highrise buildings. The consolidation yield stresses of Ma12 and Ma11 are 451 to 794kPa and 902 to 1029kPa respectively so that the consolidation against the fill load of KP+5.4m is not complete yet. As to the Osaka Group distributed from KP-150m, 2 sea clay layers Ma10 and Ma9 have been confirmed. They have consolidation yield stresses of 1800 to 2200kPa, both of which are larger compared with the effective overburden pressure so that they are in the state of over-consolidated clay.

3 RELATION BETWEEN SOIL IMPROVEMENT SPECIFICATIONS AND SETTLEMENT

Ground settlement measurement is performed at many points in Kobe Port Island. Out of them, the long-term settlement records of diluvial formation at 2 sites whose preloading values are different from each other may be discussed.

Site A is situated slightly to the west of the center of Kobe Port Island. For soil improvement there, a large-scale preloading of 10m in height (flat size 272m x 167m) was applied for about 2 years from 1976 to 1978 to accelerate the consolidation of the alluvial clay layer. Site B adjoins to Site A on the east of the latter and is situated in the center of Kobe Port Island. This site was treated with sand drain to promote the consolidation of the alluvial clay layer and preloading of 6m in height for about 6 months.

Figure 2. Underground layer on Hanshin district

128

The layout of the settlement measurement points in Site A as overlapped by preloading points will be shown in Fig-4. The highrise building built in Site A was provided with 40 settlement observation points, and the differential settlement gauge was set at 2 points within the building (Lay-1, Lay-2) and another set at a point away from the building (Lay-4). The highrise building is supported by piles reaching the diluvial alternate formation through the alluvial clay layer. In order to keep balance between the building weight and the displaced soil weight, basements were provided after excavation down to a depth of 10m. At present, the underground water table is found at a depth in the vicinity of GL-4m so that the displaced soil weight is approximately 60kPa larger than the building weight in terms of loading intensity. Therefore, the loading condition for the highrise building, compared with areas without excavation, at the present time is such that the fill load is approximately 60kPa less or, in terms of fill soil thickness, approximately 3m less.

Fig-5 shows one example of settlement measurement results. The initial value of the figure corresponds to the initial day of February 1981 when the highrise building weight was stabilized. The average subsidence of the highrise building in about 17 years is 292mm. The minimum value of 258mm appeared

on the northwest side which was equal to the settlement of the diluvial formation measured by the differential settlement gauge Lay-2 at the same position. The settlement of the diluvial formation referred to hereinabove corresponds to the settlement of the element placed at the lower end of the alluvial clay layer or at the top of the diluvial alternate formation. The differential settlement gauge Lay-1 was located close to the center of the building, and its value was almost equal to the average subsidence of the building. On the other hand, compared with other locations, the settlement of the diluvial formation was apparently larger at Lay-4, where excavation soil displacement was not performed.

In Site B, sand drain was placed in square layouts of 2.0m to 3.4m according to the scale and layout of the buildings, and settlement was measured at many points. Fig-6 shows the layout of the measuring instruments, and Fig-7 shows the settlement record of Building No. 1 in southwest section of the site where the standard preloading (height 6m, loading period 6 months) was applied. This figure graphically shows the settlement of the 8-stories apartment building supported by piles on the diluvial alternate formation as well as the settlement of the diluvial formation measured by the differential settlement gauges. The settlement of the diluvial formation means the data

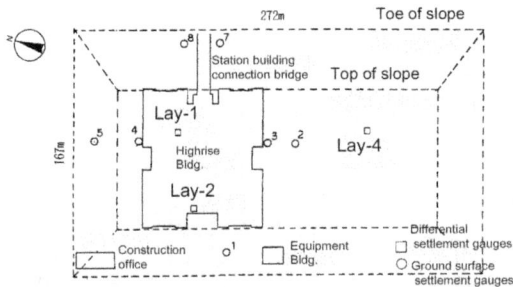

Figure 4. Preload, building and measuring point position (SITE-A)

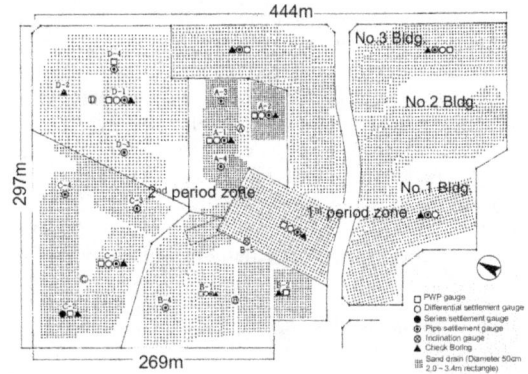

Figure 6. Measurement apparatus position (SITE-B)

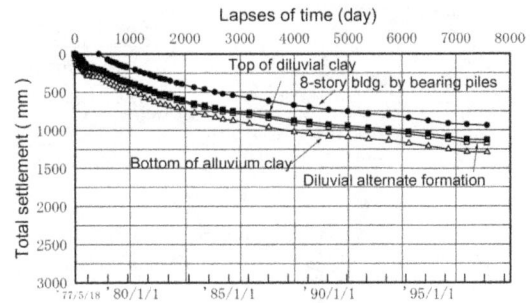

Figure 5. Settlement records of highrise building and diluvial layer (SITE-A)

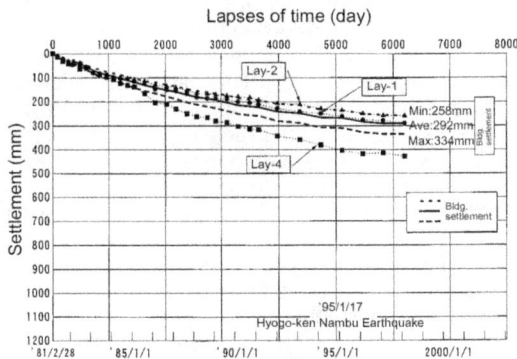

Figure 7. Settlement records of high-rise building and diluvial layer (SITE-B)

of the settlement element placed at the depth corresponding to the depth of the tips of the foundation piles in the diluvial alternate formation and the settlement element placed at the top of the diluvial clay layer Ma12. The building subsidence measurement was started about a year later, but it could be understood that the subsidence of the building supported by piles on the diluvial alternate formation is nearly equal to the settlement of the diluvial formation.

The reclamation history reveals that the reclamation of Site B was executed about 2 years later than Site A. In order to compare Fig-5 with Fig-7, those sections whose lapses of time since preloading removal and building load stabilization were equal were taken out and their settlement values were compared. The results thereof as summarized are shown in Table-1. The extraction period is from February 1981 to February 1996 in the case of Site A and from April 1983 to February 1998 in the case of Site B. Table-1 summarizes the settlement of the building and the settlement of diluvial formation obtained by the differential settlement gauges classifying the loading condition to 3 cases. CASE-1 deals with the settlement of the highrise building area where 10m deep excavation soil displacement was performed within Site A having received 10m high preloading for about 2 years. CASE-2 deals with the settlement measured by the differential settlement gauge (Lay-4) where underground excavation was not performed within Site A. CASE-3 deals with the subsidence of the buildings and the settlement of the diluvial formation obtained by the differential settlement gauges in Site B where 6m high preloading was applied for 6 months. Table-1 shows that in CASE-2 wherein a large-scale preloading was applied, the settlement of the diluvial formation has decreased to about 80% compared with CASE-3. Further in CASE-1, wherein underground excavation was performed, the settlement was about half of that in CASE-3.

Table 1. Settlements comparison of the diluvial of SITE-A and SITE-B

	SITE A Measurement term 1981/2~1996/2		SITE B Measurement term 1983/4~1998/2
BUILDING SETTLEMENT	High Bldg. (12F) 28.3cm	—	A-tower (8F) 50.2cm
LAYER SETTLEMENT	Lay-1 27.6cm Lay-2 24.7cm	Lay-4 41.8cm	Lay01 51.9cm
LOADING CONDITION	Preload hight 10m Loading term 2 years		Preload hight 6m Loading term 0.5 year No-excavate
	Excavate to 10m	No-excavate	
KIND	CASE-1	CASE-2	CASE-3

4 SETTLEMENT BEHAVIORS UNDER DIFFERENT LOADING CONDITIONS

Consolidation settlement analysis was made in order to quantitatively evaluate the effects which the differences in loading conditions as shown in the three cases as in the preceding section place upon the settlement of diluvial formation. The superimposed load was stress-dispersed by Boussinesq so that the settlement values were sought by the Cc Method. Fig-8 shows the foundation soil models and calculation parameters used in the analysis. The foundation soil was assumed based on the results of the 200m deep soil boring test performed in Site A, and the foundation soil property was based on the physical property values obtained from the consolidation test on the clay sampled by the 250m deep soil boring test performed prior to building construction on Phase II Island. Further, considering the characteristics of the consolidation test, calculations were made also using models for which smaller consolidation yield stresses were set. The results of the analysis will be shown in Fig-9 and Table-2. Table-2 shows that in all cases the settlement of the diluvial formation obtained by actual measurement exceeds the analysis value so that the analyses did not sufficiently simulate the actual measurement values.

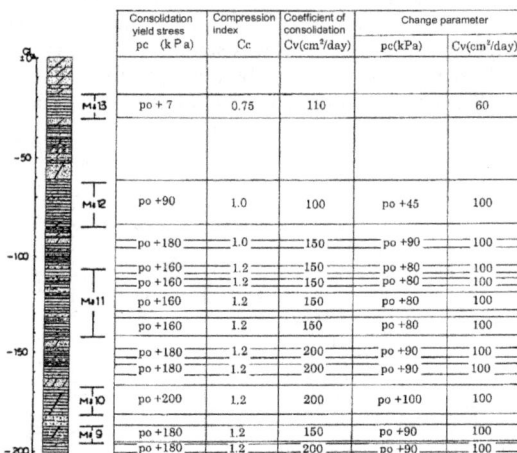

Figure 8. Parameter for calculation of consolidation settlement

On the other hand, when the CASE-3 settlement is assumed to be 100, the settlement percentages of CASE-1 and CASE-2 comparatively conform to their actually measured values, and this seems to qualitatively endorse the assumption that the difference of loading condition such as preloading places effects upon the long-term diluvial settlement thereafter. The reason why the analysis results showed values less than those of the actual measurement could be that the analysis by the Cc Method deals only with the primary consolidation and that the model depth range of 200m is small for the loading area. Especially, it suggests that when the long-term settlement is considered, the weight of the secondary consolidation included in the actual settlement is great.

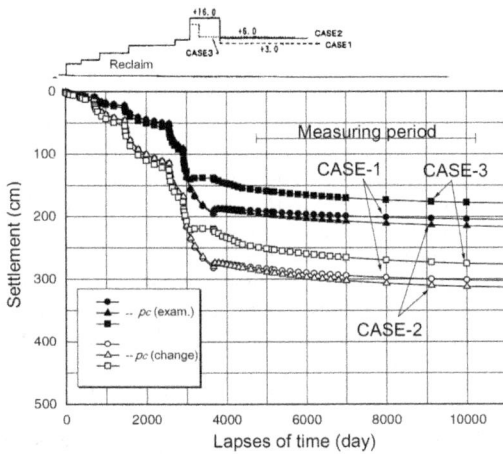

Figure 9. Settlement of diluvial formation by loading conditions

Table 2. Comparison of the amount of survey settlements of a diluvium, and an analysis value

		CASE-1 Preload 10m + Excavate	CASE-2 Preload 10m	CASE-3 Preload 6m
Analysis Value	①	13. 0cm (64%)	19. 6cm (97%)	20. 3cm (100%)
	②	19. 4cm (74%)	25. 9cm (99%)	26. 2cm (100%)
Measurement		26. 9cm (53%)	41. 8cm (82%)	51. 1cm (100%)

Note-1; The analysis value ① set up pc based on the soil-examination. And the analysis value ② set up pc small (refer to Fig-8).

Note-2; Figure of () shows the rate when setting the amount of settlements of CASE-3 to 100.

5 ESTIMATION OF LONG-TERM SETTLEMENT BY THE HYPERBOLIC CURVE METHOD

As a common settlement estimation technique, the settlement-time relation was evaluated by the Cc Method in the preceding section. However, this work requires much labor and is apt to leave many ambiguous areas unqualified including initial conditions such as layer thickness as well as establishment of soil constants. Further, although it showed that the qualitative settlement behaviors are possible of simulation, it clarified that results cannot be obtained with such accuracy that enables long-term settlement estimation.

On the other hand, as settlement measurement in a reclaimed ground is generally performed on actual structures thereupon since their construction completion, the health of such structures may be effectively evaluated if the settlement measurement is continued thereafter for a certain period of time so that their future settlement may be estimated easily with a certain accuracy based upon the data of such actual measurement.

In this section, a long-term settlement analysis will be performed using the actually measured values of settlement as mentioned in the above. Especially, the Hyperbolic Curve Method was taken up as a settlement estimation technique that could be readily used, and the approximation degree of the Hyperbolic Curve Method against the actually measured values was studied.

The study was made using the results of actual settlement measurement on structures in Site A and Site B. In this analysis, the (time t/settlement s) ~ (time t) relations were expressed using the data after removal of preloading, and those sections in the figure that could be regarded as straight lines were extracted. Then, the actual measurement value for about 1 year from the beginning of each straight line section was approximated to a hyperbolic curve to draw a settlement forecast extrapolation line.

Fig-10 shows the typical drawing from the analysis results, and Fig-11 shows the hyperbolic line approximation results. Where the Site A structure data are used, the section that can be regarded as a line sufficiently straight from the typical drawing, is the period of about 1 year beginning 962 days after the building load was considered to have been substantially stabilized in February 1981 and ending 1,325 days after such building load substantial stabilization. The settlement as of February 1998 as estimated using this data is 270mm against the actually measured value of 291mm so that the accuracy is 93%. However, the beginning of the straight line that could be approximated was 962 days after the building load substantial stabilization and it was four and half years after the measurement started in April 1979 so that this lapse of time is too long to enable realistic estimation.

In the next analysis, the data of the ground surface settlement plate was used as the measurement results in Site B. A hyperbolic curve was approximated from the typical drawing (Fig-12) using the data covering a period of about 1-plus year beginning 794 days after the preloading removal in July 1980 and ending 1,195 days after the preloading removal. Against the estimated value of 463mm after 4,966 days as shown in Fig-13, the actual measurement value was 564mm resulting in an accuracy of about 82%.

As far as the data at the 2 sites mentioned in the above are concerned, the approximated estimation values by the hyperbolic curve were less than the actually measured values. Assumedly, it has something to do with the fact that according to the conventional secondary consolidation theory, the secondary consolidation progresses proportional to the time logarithm and does not have a convergent value while a hyperbolic curve has a convergent value.

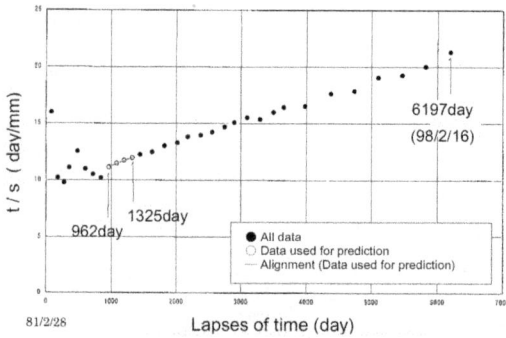

Figure 10. Typical drawing (SITE-A structure)

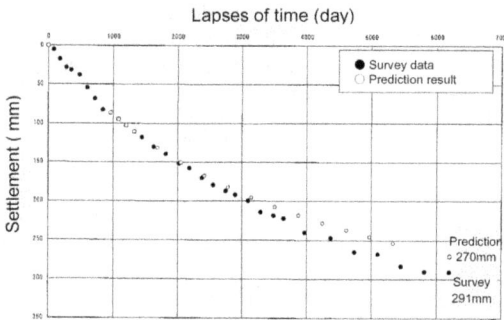

Figure 11. Records of settlement (SITE-A structure)

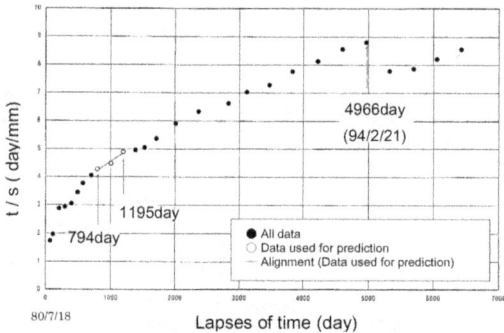

Figure 12. Typical drawing (SITE-B surface)

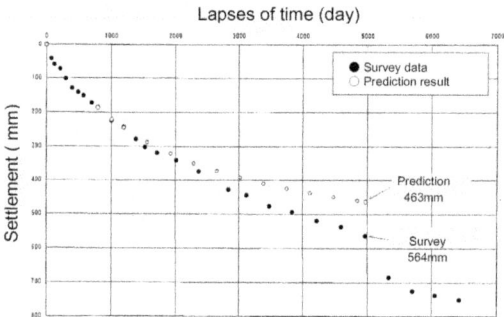

Figure 13. Records of settlement (SITE-B surface)

6 SETTLEMENT ANALYSIS BY VISCOELASTPLASTICITY FINITE ELEMENT ANALYSIS

The importance of secondary consolidation in long-term settlement was re-recognized in settlement calculation by the Cc Method and in the study of settlement analysis by the Hyperbolic Curve Method in and before the preceding section. It is thought that long-term settlement estimation should be performed by techniques using a theory including secondary consolidation. In this section, therefore, the analysis of consolidation settlement was made using the viscoelastoplasticity finite element analysis tool "DACSAR" which utilizes "Sekiguchi-Ohta Model".

While the finite element analysis technique is an effective tool which enables expression of various phenomena at the same time including shear and consolidation, the setting of initial conditions to be inputted to seek high accuracy calculation results will become important.

In this analysis, an attempt was made to grip loading conditions as extensively as possible based on past reclamation data and to reproduce them faithfully in the calculations.

Fig-14 shows the drawings for the analysis models. The soil layer composition was based upon the results of the 200m soil boring log performed in Site A (See Fig-2). The analysis domain, as shown in Fig-14, was the half section in the shorter side direction of Site A, portion of which about 500m long horizontally and 200m deep under the sea surface was modeled. Further, characteristics of clay were determined with reference to the results of the in-room soil test on the undisturbed clay sampled by boring before the reclamation of Phase II Island. Other physical properties required by the analysis were calculated by estimation equations. At this point, the secondary consolidation coefficient $C\alpha$ was assumed to be 15% of Cc. Table-3 shows a list of the representative physical properties of clay.

The analysis was performed under plane strain condition. The drainage boundary conditions were that the model bottom was undrained; the right end, being the axis of symmetry, was undrained; and the left end was drained. However, in actuality, wherein fill soil is loaded all over the analytical domain, the model left end will not be under static hydraulic pressure condition. Therefore, in consideration of the distance to the end of Port Island, the water permeability coefficient of the model left end element in horizontal direction was reduced to 1/100 to decrease the lateral direction water permeation velocity.

The analysis simulated the reclamation histories with calculations covering an approximate period of 30 years from January 1966 to 1998. The calculations dealt with 3 cases including CASE-1 (3m excavation after 10m preloading) and CASE-2 (10m preloading only) both in Site A and CASE-3 (6m preloading) in Site B. Fig-15 shows the reclamation histories. The unit volume weight of the fill element was assumed to be 1.8t/m³ (submerged weight 1.0t/m³). Further, the buoyancy to be generated by settlement due to filling was expressed by adding to the following step calculations the buoyancy, as upward force, obtained by inverse calculation from the volume of the fill material submerged.

The aforementioned 200m soil boring layer thickness, used in the analysis, resulted from the investigation performed at the time when the ground elevation reached KP+5.4m so that the settlement due to filling had already been reflected. Therefore, in order to estimate the soil thickness before reclamation, the analysis of fill layers was made up to KP+5.4m and the compaction values of respective layers were added up to establish the thickness before reclamation.

The sand drains installed in CASE-3 were considered through establishment of equivalent coefficients of permeability. Table-4 shows those coefficients of permeability established.

The results of analyses are shown in Fig-16. The figure shows the settlement time history of the diluvial formation top (bottom of alluvial clay layer Ma13) from the initial day of the analysis (June 1966) up to January 1997 in Site A and the same up to February 1998 in Site B. The results in Site A show that in about 30 years from reclamation initiation up to 1997, a ground settlement of 2.1m occurred in CASE-1 and that of 2.3m in CASE-2. For the same period, a ground settlement of about 2.4m was calculated in CASE-3 in Site B.

Further, Fig-17 and Fig-18 show the actually measured settlement time history of diluvial formation (settlement element at the bottom of alluvial clay layer) in Site A and Site B respectively.

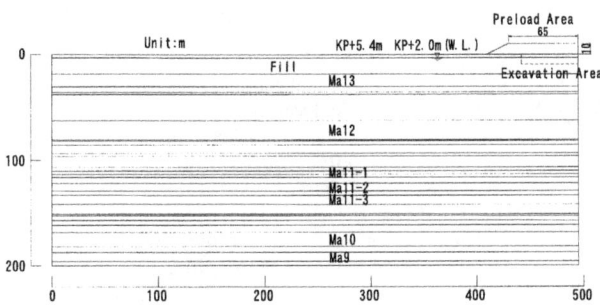

Figure 14. Analysis model

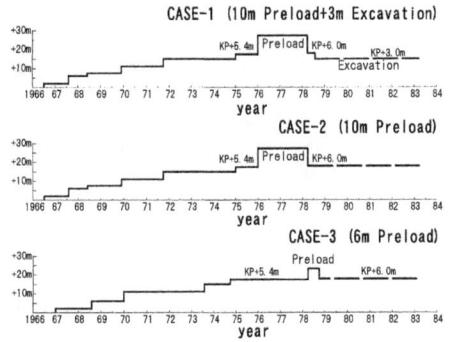

Figure 15. Reclamation histories

Table 3. Input parameter

Layer Name	Cc	Cs	Cv (m²/day)	OCR	φ′	D	Λ	M	ν′	k (m/day)	Ko	Ki	α	dvo/dt
Ma13	0.75	0.075	0.007	1.0	35.00	0.062	0.810	1.418	0.299	1.63E-03	0.426	0.426	1.63E-02	9.34E-07
Ma12	1.00	0.100	0.010	1.2	38.00	0.092	0.885	1.549	0.278	6.00E-05	0.384	0.406	2.41E-02	8.45E-07
Ma11	1.00	0.100	0.015	1.2	38.00	0.099	0.885	1.549	0.278	4.94E-05	0.384	0.406	2.61E-02	4.23E-05
Ma10	1.20	0.120	0.020	1.2	38.00	0.122	0.885	1.549	0.278	4.70E-05	0.384	0.406	3.19E-02	4.19E-06
Ma9	1.20	0.120	0.015	1.2	38.00	0.122	0.885	1.549	0.278	3.20E-05	0.384	0.406	3.19E-02	7.81E-06

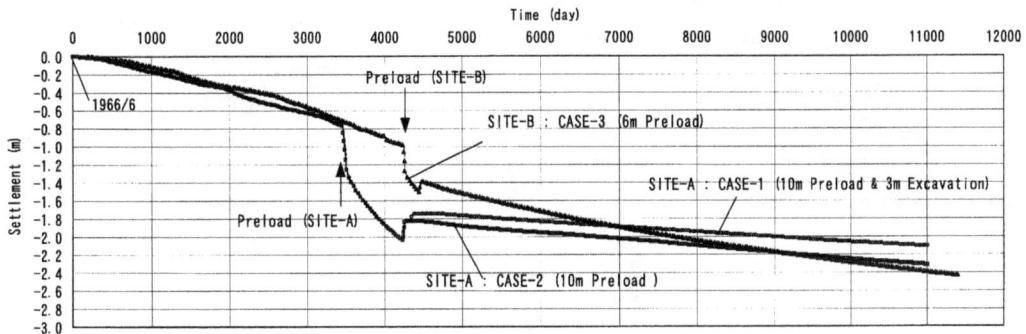

Figure 16. Relationship between settlement and lapses of time (the top of diluvial formation)

133

Table 4. Equivalent coefficient of permeability

Layer Name	Changed kx (m/day)	Changed ky (m/day)	kx, ky original (m/day)
Ma13	1.954×10^{-3} ~ 2.959×10^{-4}	1.44×10^{0}	1.628×10^{-3} ~ 2.466×10^{-4}

The initiation date of these data is February 1981, and the analytical values for the corresponding periods are additionally noted in the figures.

In Site A, as shown in Fig-17, the analytical values form a gentle convex upward from the initial date (February 1981) up to the vicinity of 3,000th day showing no difference in settlement trend between CASE-1 and CASE-2. In both cases of analysis, the settlement obtained is less than the actually measured value. Thereafter, the settlement curve forms a gentle convex downward, with the analytical settlement tending to exceed the actually measured value. In the case of Site B, as shown in Fig-18, the calculated settlement curve forms a gentle convex downward showing the same trend as the actually measured values, except that the calculated value begins to exceed the actually measured value from the vicinity of 2,500th day in such a direction as to depart away from the actually measured values. In Site A, the reason why the analytical results in the early period (up to about 3,000th day) are generally less than the actually measured values and do not conform to the latter could be considered to be partially attributable to the fact that the 2-year application of 10m preloading performed in Site A was not simulated well making the residual primary consolidation became smaller than actual.

Further, Table-5 shows the records, in summary as combined with the corresponding analytical results, of the settlement in Site A from February 1981 to February 1996 and the settlement in Site B from April 1983 to February 1998 in order to compare the settlement of diluvial formation in Site A with the same in Site B to find differences between them due to different soil improvement treatments given to Site A and Site B.

According to this table, the actually measured settlement in CASE-3, wherein 6m preloading was applied for 6 months, is about 50cm, which is the largest, while the corresponding analytical value is about 70cm resulting in some over-evaluation. In CASE-2, wherein 10m preloading was applied for 2 years in Site A, the actual settlement was about 42cm as against the analytical value of 38cm.

In CASE-1, wherein 3m excavations were made after 10m preloading for 2 years, the actual settlement was about 26cm as against the analytical value of about 30cm.

The amount of layer compression of Ma9 and Ma10 with the passage of time in CASE-2 is put on Fig-19. Ma9 is located in the lowest layer of this analysis model. As for Ma9, about 7cm is accepted

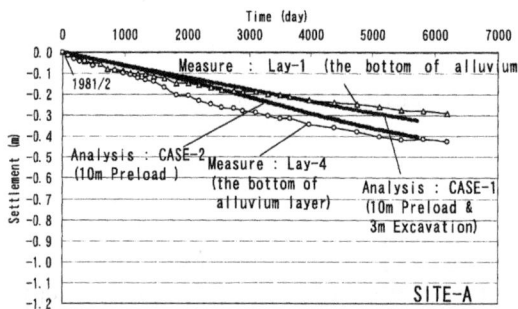

Figure 17. Comparison of the amount of survey settlements of a diluvium and FEM results (SITE-A)

Table 5. Comparison of the amount of survey settlements of a diluvium, and FEM results

	CASE-1 Preload 10m + 3m Excavate	CASE-2 Preload 10m	CASE-3 Preload 6m
Measurement	26.9cm	41.8cm	51.1cm
FEM results	30.9cm	38.6cm	70.4cm

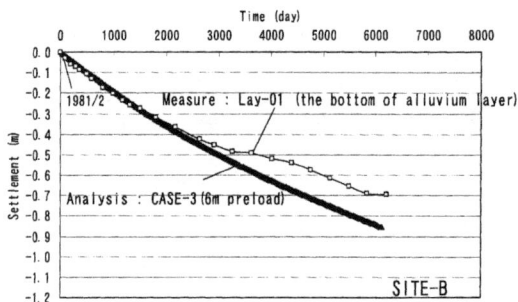

Figure 18. Comparison of the amount of survey settlements of a diluvium and FEM results (SITE-B)

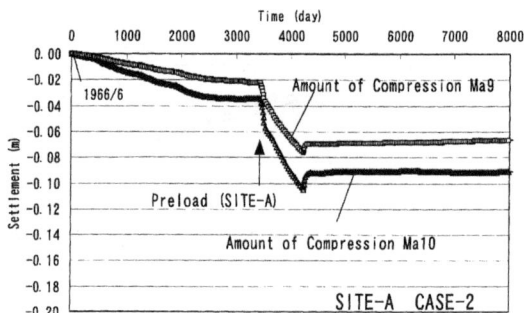

Figure 19. Amount of Compression Ma10 and Ma9 from FEM results (CASE-2 10m preload SITE-A)

and, as for Ma10, about 9cm layer compression is accepted by reclamation and preload. It is 4% or less of all settlements (about 2.3m) of a diluvium. Especially, settlement hardly accepts in the neglect period after the preload end.

7 CONCLUSIONS

The discussions made in the above may lead to the following conclusion concerning the applicability of the viscoelastoplasticity finite element analysis on the long-term settlement estimation for Kobe Port Island:

1 An attempt was made to quantitatively express by the viscoelastoplasticity finite element analysis the differences in settlement of grounds having received soil improvement treatments of different specifications. There were some periods where the settlement curve pattern did not fit the actually measured curve pattern well, but in some cases, the calculated settlement values exceeded the actual settlement values. Thus, the possibility was revealed that the introduction of secondary consolidation would enable reproduction of actual behaviors with a high accuracy.

2 As to the depth in which modeling (for settlement calculation) should cover, it is considered that the actual measurement values can be simulated identifying the physical property values within the practical range contemplating a soil composition to a depth of about 200m as adopted this time.

3 In case a several decade long-term settlement is to be estimated in the future, it is necessary and indispensable, for acquisition of accurate analytical solutions, to collect information as detailed as possible concerning initial loading conditions such as filling and soil improvement that would-determine the behaviors during the long untouched period and to establish input conditions very carefully, not to mention the importance of the physical property values of soil and analytical models.

4 Actual soil conditions behave 3-dimensionally. In the preloaded and excavated areas in the sites contemplated this time, it is thought that essentially the effect in the depth direction cannot be disregarded. In order to enable still higher accuracy analyses, studies taking 3-dimensional analyses into consideration are hoped for as future prospect.

REFERENCES

A.Iizuka and H.Ohta., Sept. 1987.. Soils & Foundations, Vol.27, No.3, "A determination procedure of input parameters in elasto-viscoplastic finite element analysis.", pp.71-87.

D.Karube, Y.Fukagawa, S.Honda and K.Kawai, 200.12 Journal of Geotechnical Engineering(III) No.665, "An evaluation of the long-term settlement of offshore man-made islands", VI-49,1-18 (in Japanese)

M.Fukui, A.Matsui and R.Tamaki, 1982.4 , Nikken Sekkei Technical Report, No.73, "Geotechnical and Foundation of Kobe Port Island" (in Japanese)

M.Fukui, R.Kobori, M,Tanaka, I.Ito and S.Honda 1998.9 Summaries of Technical Papers of Annual Meeting Architectural Institute of Japan, "Long term Settlement Behavior of Buildings on Kobe Port Island (Report No.2 : Settlement Measurement Results), pp813-814. (in Japanese)

N.Koushige, R.Kobori, H,Yoshida, M,Okamoto and S.Honda 1998.9 Summaries of Technical Papers of Annual Meeting Architectural Institute of Japan, "Long term Settlement Behavior of Buildings on Kobe Port Island (Report No.3 : Settlement of the Diluvium Soil Deposits), pp815-816. (in Japanese)

Y.Gyoten, K.Tanimoto, M.Fukui and S.Honda, 1997.9 34th Soil Engineering Research Paper Reading (Tokyo), "Settlement Observation of Highrise Buildings with Bearing Pile Foundations Constructed on Reclaimed Land (Section 2)" pp.1395-1396.

Soft Ground Engineering in Coastal Areas, Tsuchida et al. (eds)
© 2003 Swets & Zeitlinger, Lisse, ISBN 90 5809 613 0

Interaction of two adjacent man-made islands on soft ground

M. Kitazume & S. Miyajima
Port and Airport Reserch Institute, Yokosuka, Japan

M. Hirose
Kansai International Airport Land Development Co., Ltd., Izumisano, Japan

S. Suzuki
Kansai International Airport Co., Ltd., Izumisano, Japan

M. Taki
Fukken Co., Ltd., Takamatsu, Japan

H. Hashizume
Geodesign Co., Ltd., Minato-ku, Japan

ABSTRACT: When two man-made islands are constructed close to each other in comparison of their plane dimension, an interaction of these two islands might cause any influence to the deformation of the islands. FEM analysis has been frequently applied to investigate the behavior of ground deformation. However, this kind of problem is relatively new problem. Further research efforts are required to investigate the interaction of these islands and to evaluate the deformation with relatively high accuracy. In this study, a series of centrifuge model tests and FEM analyses was carried out to investigate the influence of the islands to the deformation of the ground.

1 INTRODUCTION

Kansai International Airport was opened in September 1994 on a man-made island (hereafter cited as 1st Phase) on very soft clay ground under the sea with 18m in average depth. Another man-made island (hereafter cited as 2nd Phase) is now being constructed to expand the airport with 200m offshore of the existing island. Since these islands are close to each other in comparison of their plane dimension, an interaction of these two islands might cause any influence to the deformation of the ground. FEM analysis has been frequently applied to investigate the behavior of ground deformation. However, the above-mentioned problem is relatively new problem, therefore further research efforts are required to investigate the interaction of the islands and to evaluate the deformation with relatively high accuracy.

In the present study, a series of model tests and FEM analyses was carried out to investigate the deformation of the ground with sufficient accuracy. The model tests were performed in the geotechnical centrifuge at Port and Airport Research Institute (PARI) (Kitazume and Miyajima, 1995). In the tests a normally consolidated clay ground was subjected to loads of two embankments as a simulation of the islands, which were constructed with different height and construction speed, to examine the interaction of the embankments. In the analysis, a FEM code developed by PARI (Kobayashi, 1984) was used to simulate the test results under a plain strain condition and to evaluate sensitivity of each parameter to the analyzed results.

2 CENTRIFUGE MODEL TESTS

2.1 Model ground

The schematic view of the model ground is illustrated in Fig. 1. The model ground consists of a drainage layer, a clay layer and two embankments to be constructed later. The left hand side embankment will be constructed first and the right hand side embankment will be constructed secondly. Kaolin clay (w_l=59.0%, w_p=16.8%, I_p=42.2) was used for the clay layer. Toyoura sand, which has a uniformity coefficient, U_c=1.38 and an effective grain size, D_{10}=0.13mm, was used for the drainage layer at the bottom and the two embankments.

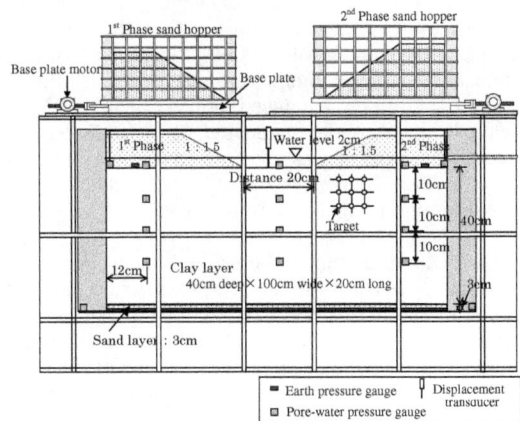

Figure 1. Schematic view of model ground

2.2 Model ground preparation

The model ground was constructed by the following procedure.

The drainage layer of 3cm in thickness was formed at the bottom of a strong specimen box, whose dimensions were 60cm in depth, 120cm in width and 20cm in length. This layer was compacted to a relative density of about 90%, and then was saturated by percolation of water from the bottom inlet of the specimen box. Saturated clay slurry with a water content of 120%, which was sufficiently higher than the liquid limit, was poured on the sand layer, and it was pre-consolidated with the vertical pressure of $9.8kN/m^2$ on a laboratory floor for about two weeks. After the completion of the pre-consolidation of the clay layer, the front window of the specimen box was removed to place target markers on the front side of the clay with a 2cm grid pattern in order to trace the ground deformation. After reassembling the specimen box, some pore water pressure gauges were installed into the clay layer. The other water pressure gauges, earth pressure gauges and displacement gauges were also placed on the clay ground surface to measure the overburden pressure of the embankment and the vertical displacement of the clay ground. After preparing the ground, two sand hoppers were installed on the specimen box for embankment construction, which were filled with dry sand. The model ground thus prepared was allowed to consolidate under a centrifugal acceleration of 50g for about 15 hours. This procedure provides the normally consolidated clay ground having a strength increasing linearly with depth.

2.3 Test procedure

After the consolidation in the centrifuge, the sand embankment for the 1st Phase island was constructed at the left hand side of the ground by falling down the sand from the sand hopper, which was a simulation of the first construction phase in the field. To simulate the interval of the field construction of two embankments, the model ground was allowed to consolidate by the enhanced embankment weight to achieve a consolidation degree of 70%. Then another embankment was constructed at the right hand side of the ground for the 2nd Phase island. The model ground was allowed to consolidate again by the enhanced embankment load for about three hours. During and after the embankments' construction, the pore water pressures and the earth pressures were measured, as well as taking photographs to analyze the ground deformation. For ease of referring the loading phases, these test processes were named in this text as 1st Phase construction, 1st Phase consolidation, 2nd Phase construction and 2nd Phase consolidation, respectively.

2.4 Test cases

Six test cases were carried out as listed in Table 1. The series of model tests focused on the effect of the dimensions and the construction speed of the 2nd Phase embankment on the ground deformation. The distance between two embankments was 20cm at the toes of the embankments throughout the test series. The slopes of both embankments were kept constant about 1 to 1.5.

Table 1. Test cases

case	Height (cm)		Construction Speed (cm / 10min)	
	1st Phase	2nd Phase	1st Phase	2nd Phase
case1	11.5	17.1	1.4	1.0
case2	11.1	17.2	1.4	5.1
case3	7.4	10.9	1.9	5.0
case4	6.5	10.5	1.6	1.0
case5	6.6	6.1	1.6	4.4
case6	5.3	6.6	1.3	1.0

2.5 Test results

Figure 2 shows the relations between the earth pressure beneath the 1st Phase embankment and the vertical displacement of the ground during the 1st Phase construction. Results of all the test cases are plotted in this figure. Although there is relatively large scatter in the measured data, it is confirmed that there is a tendency that the vertical displacement of the ground increases significantly with the increase of the earth pressure by the 1st Phase. These measured data were approximated by a hyperbola as shown in the figure. From this approximation, the ultimate earth pressure by the embankment loading (P_u), which means the yield pressure of the ground, was estimated as $85kN/m^2$. The ratios of P_u to the earth pressure beneath the embankment in case1 and 2 were 1.18 and 1.25 respectively, which means the

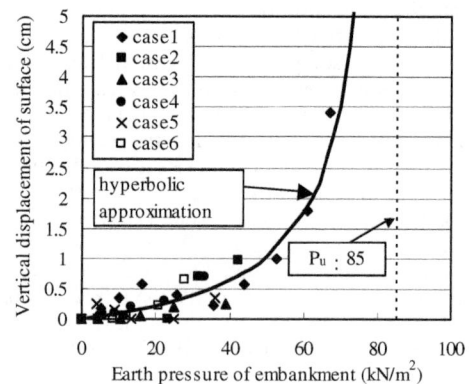

Figure 2. Relations between the earth pressure and the vertical displacement beneath 1st Phase embankment during 1st Phase construction

ground was subjected to comparatively large embankment load. In the test cases from case3 to case6, in which the ground was subjected to comparatively small embankment load, the ratio were 2.14, 2.53, 2.32 and 2.98 in case3, 4, 5 and 6, respectively.

Relations between the earth pressure at the bottom of the 2nd Phase embankment and the vertical displacement of the ground during the 2nd Phase construction is shown in Fig. 3. The approximation curve of hyperbola for the 1st Phase, which was the same as shown in Fig. 2, was also shown in this figure to compare the deformations by the construction of the embankments. It is found in case1 that the vertical displacement of the ground surface during the 2nd Phase construction is smaller than that at the 1st Phase construction. Also, the vertical displacement shows gradual increase even if the embankment pressure exceeds P_u. This tendency is much different from that of the relations during the 1st Phase construction. On the other hand, in case4 and case6, the relations between the earth pressure and the vertical displacement shows similar tendency to the hyperbolic approximation of the 1st Phase. From these results, it is considered that the 1st Phase embankment in case 1 functioned as a counterweight.

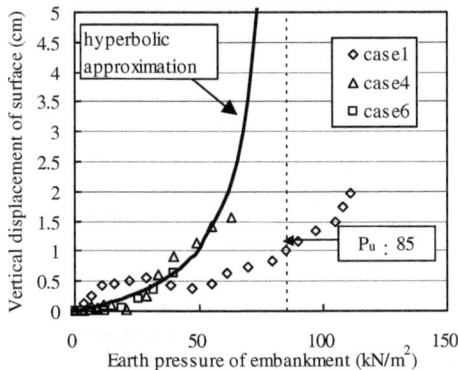

Figure 3. Relations between the earth pressure and the vertical displacement beneath 2nd Phase embankment during 2nd Phase construction

Figure 4 shows the vertical displacement distributions at the clay ground surface analyzed with photographs at four loading steps in case6. The test data plotted in this figure are increment value from the values just before the 1st Phase construction. From the vertical displacement after the 1st Phase construction, it is observed that the 1st Phase embankment construction caused large vertical displacement in the clay ground under the embankment; at the same time upheaval deformation of the ground was observed outside the embankment. The upheaval deformation due to the 1st Phase construction occurred at the area about 10cm from the toe of the 1st Phase embankment, and it didn't reach the area where the

2nd Phase is planned. During the 1st Phase consolidation, further vertical displacement was observed in the ground beneath the 1st Phase embankment. In the figure, the vertical displacement from the 1st Phase consolidation to the 2nd Phase consolidation was shifted to remove the influence of self-weight consolidation during the 1st Phase consolidation. During the 2nd Phase construction, the vertical displacement of the clay ground occurred beneath the 2nd Phase embankment, which is similar phenomenon to that of the 1st Phase construction. However, negligible vertical displacement took place at the clay ground beneath the 1st Phase even if the 2nd Phase embankment was constructed.

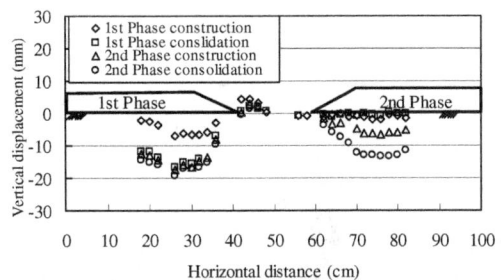

Figure 4. Distributions of vertical displacement of the clay ground surface during each process (case6)

Figures 5-(a) and (b) show the horizontal displacement distributions in the clay ground obtained at the toe of the 1st Phase embankment and the 2nd Phase embankment respectively. In these figures, the x-axis shows the measured horizontal displacement, where positive value of the x-axis means a displacement toward the 2nd phase embankment, and the y-axis shows the depth. The test data in these figures were initialized with the values just before the 1st Phase construction. From the test data, after the 1st Phase construction, it is observed that the clay layer at the toe of the 1st Phase embankment displaced horizontally toward the 2nd Phase embankment. However, significant horizontal displacement was not observed at the toe of the 2nd Phase embankment during the 1st Phase construction, as shown in Fig. 5(b). In the process of the 1st Phase consolidation, further horizontal displacement was observed at the toe of the 1st Phase embankment as well as at the toe of the 2nd Phase embankment. On the other hand, in the processes of the 2nd Phase construction and the consolidation, it is found the clay layer at both toes of the 1st and the 2nd Phases moved toward the 1st phase embankment. Especially, it is observed that the horizontal displacement at the toe of the 2nd Phase embankment turned over across the initial point.

Figure 6 shows the vertical displacement distributions together with the horizontal displacement distributions measured during the 1st Phase construction

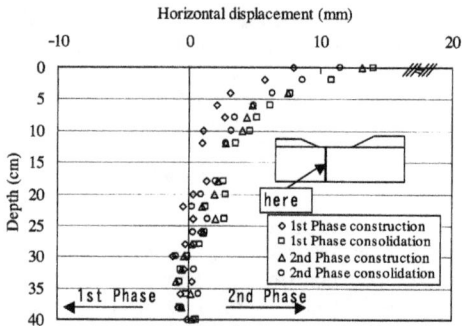

Horizontal displacement (mm)

Figure 5(a). Horizontal displacement in the clay layer at around of the toe of 1st Phase embankment (case6)

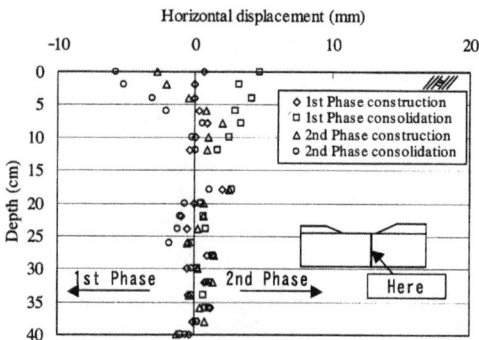

Horizontal displacement (mm)

Figure 5(b). Horizontal displacement in the clay layer at around of the toe of 2nd Phase embankment (case6)

in case1 and case4. The x-axis shows the distance measured from the toe of the 1st Phase embankment, and the y-axis shows the displacement at the clay surface in vertical and horizontal directions. Here, the positive value of the vertical displacement means upheaval deformation, and the positive value of the horizontal displacement means the movement toward the 2nd Phase embankment. As shown in Table 1, the height of the embankment in case1 is larger than that in case4, accordingly the displacements in both of vertical and horizontal directions in case1 are larger than that in case4. Additionally, the areas where large displacement took place are wider in case1 than those in case4. In comparison of the vertical displacement to the horizontal displacement, the horizontal displacement toward the 2nd Phase embankment was wider than vertical displacement irrespective of the height of the embankment.

To illustrate the influence of the height of the 2nd Phase embankment on the displacement at the ground surface, Figure 7 shows the vertical and the horizontal displacement distributions during the 2nd Phase construction in case3 and case5. The x- and the y-axes in the figure show the distance and the displacements in the similar manner to that in Fig. 6, except that the displacement is shown turned over right-side left for the ease of comparison between

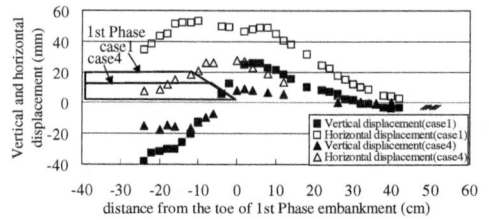

Figure 6. Distributions of vertical and horizontal displacement at the clay surface during 1st Phase construction (case1, case4)

Figs. 6 and 7. The displacements in both of the vertical and the horizontal directions in case3 are larger than those in case5, since the height of the embankment in case3 is larger than that in case5. This tendency is same as that measured during the 1st Phase construction. However, no significant difference regarding to the height of the embankment was seen in the plane extensions where large displacement was observed in case3 and in case5. Also, the plane extension where large vertical displacement was observed was similar to that for the horizontal displacement. In comparison between the vertical and the horizontal displacements, the horizontal displacement toward the 2nd Phase embankment was wider than the vertical displacement irrespective of the embankment height, which is similar tendency to that during the 1st Phase construction.

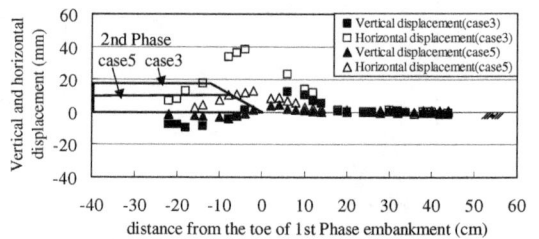

Figure 7. Distribution of vertical and horizontal displacement at the clay surface during 2nd Phase construction (case3, case5)

To investigate the influence of the construction speed of the 2nd Phase on the displacement at the ground surface, Figure 8 shows the vertical and the horizontal displacement distributions during the 2nd Phase construction in case1 and case2. This figure is drawn in the same manner as Fig. 7. As shown in Table 1, the test conditions of case1 and case2 are the same, except the construction speed of the 2nd phase embankment is faster than that of the 1st phase embankment. The displacements in both of the vertical and the horizontal directions in case2 are larger than that in case1, since the construction speed in case2 is faster than that in case1. However, no significant difference regarding to the construction speed was seen in the plane extension where large displacement was observed in case1 and case2. Also,

140

the plane extension where large vertical displacement was observed was similar to that for the horizontal displacement. In comparison between the vertical and the horizontal displacements in the area outside of the embankment, the horizontal displacements show the similar tendency to that of the vertical displacement irrespective of the construction speed.

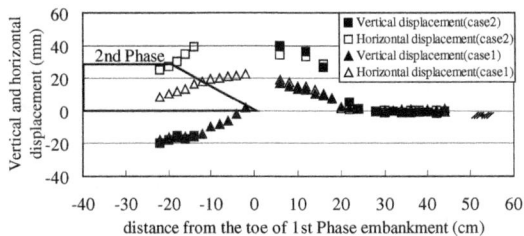

Figure 8. Distribution of vertical and horizontal displacement at the clay surface during 2nd Phase construction (case1, case2)

3 FEM ANALYSIS AND DISCUSSION

3.1 FEM analysis procedure

A series of FEM analyses was carried out to compare with the model test results. In the analyses, a sort of sensitivity calculations was also carried out to investigate the applicability of the FEM to this type of problem, where some soil parameters were changed. The FEM code named as GeoFem was used for this analyses, which has been frequently applied to practical constructions. In the calculations, the Sekiguchi-Ohta's model, an elasto-visco-plastic model, was applied to the clay layer, and an elastic model was applied to the embankments and sand drainage layer. Material properties of the clay layer are summarized in Table 2.

Table 2. Material properties of the clay layer

Input data		test	estimate
ϕ'	36.86°	(25.0°)	○
K_0	0.500		○
M	1.500	(0.984)	○
ν	0.333		○
η_0	0.250		○
e_0	each layer	○	
λ	0.252	○	
κ	0.0252		○
c_v	120cm²/d	○	
α	0.0027	○	
v_0	0.0051	○	

The values of e_0, void ratio at pressure of unity, and λ, compression index, were determined from the e~log p curve measured in the oedometer test, then the value of κ, expansion index, was estimated as

$\lambda/10$. The values of α, coefficient of secondary consolidation, and v_0, strain rate, were determined from the results of triaxial compression tests. The value of c_v was obtained by back-calculation from the time-displacement relations at the self-weight consolidation process in the centrifuge. The value of K_0 was estimated as a typical value of normally consolidated clay, then the values of ν and η_0 were estimated from K_0. As for M, critical state parameter, the value was easily obtained from the triaxial compression test results. However, thus obtained M gave diverged solution in these analyses. Therefore the minimum value of M that gave converged solution at the calculations for all the test cases was used in the FEM analyses.

For the model test case6, two FEM calculations were carried out with various combinations of soil parameters. In the first calculation, named as Type1, the soil parameters of the clay layer summarized in the Table 2 were used. In another calculation, named as Type2, the value of M of the clay layer at the depth smaller than 8cm was slightly decreased, because the horizontal displacement calculated in the Type1 analysis was found to be quite small compared with the model test results. Then the M was determined to 0.98 at the ground between the ground level and 4cm in depth, and was 1.2 between 4cm and 8cm in depth. The calculation results of the FEM analyses are shown in Figs. 9 and 10 for the model test case 6.

3.2 Result of FEM analyses

The calculated vertical displacement distributions at the clay ground surface are shown in Figs. 9(1-a) to 9(4-a), together with the model test results. It is found in the figures the calculated vertical displacements almost coincide with the model test results although their magnitudes are slightly smaller than the model test results. It can be observed in the figure that the upheaval displacements take place at the toe of the embankment at the end of embankment construction but the downward displacements take place at another surface. It is found that the Type1 calculation cannot simulate this upheaval displacement at the ground surface. This is because the shear strength at the ground surface is overestimated in the analysis. However, Type2 calculation, with slightly smaller shear strength in the shallow depth, can simulate the upheaval displacement reasonably.

The calculated horizontal displacement distributions under the toe of the 1st Phase embankment are shown in Figs. 9(1-b) to 9(4-b), together with the model test results. In these figures, the positive value of the x-axis shows a horizontal displacement of the ground toward the 2nd phase embankment. It is found in the figures that the model test results for the horizontal displacement increase from GL-28cm to the clay surface. Similar tendency is observed in Figs. 9(1-c) to 9(4-c), in which the horizontal displacement distributions along the depth at the toe

placement distributions along the depth at the toe of the 2nd Phase embankment are illustrated. On the other hand, Type1 calculations show large displacement at around GL-35cm and shallower, from the process of the 1st Phase construction. It is found that Type2 calculations can predict the test results better than the Type1, however there is still relatively large difference between the analytical and the test results in this study.

Calculated vertical displacements at the clay surface beneath the two embankments are shown in Fig. 10, together with the model test results. It is found in

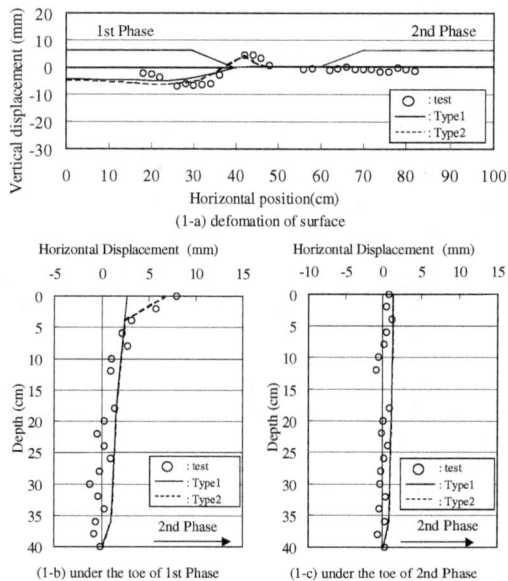

Figure 9(1). Deformation of the model test and of the calculations on 1st Phase construction

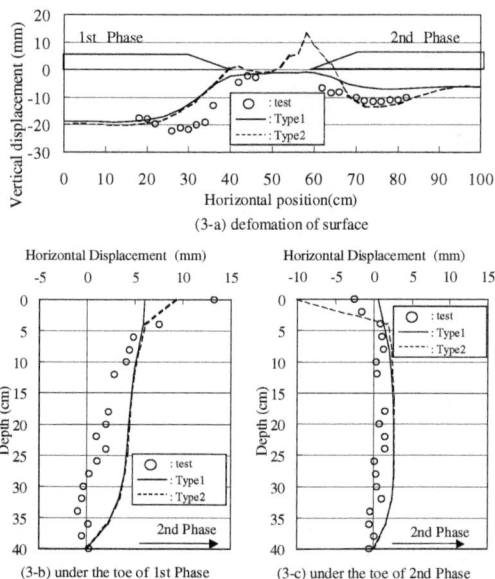

Figure 9(3). Deformation of the model test and of the calculations on 2nd Phase construction

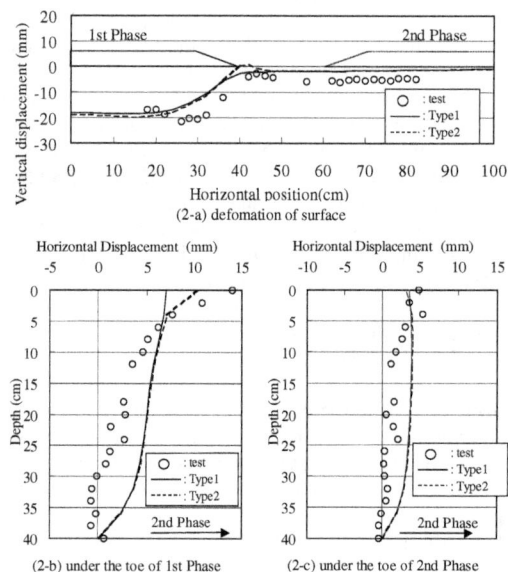

Figure 9(2). Deformation of the model test and of the calculations on 1st Phase consolidation

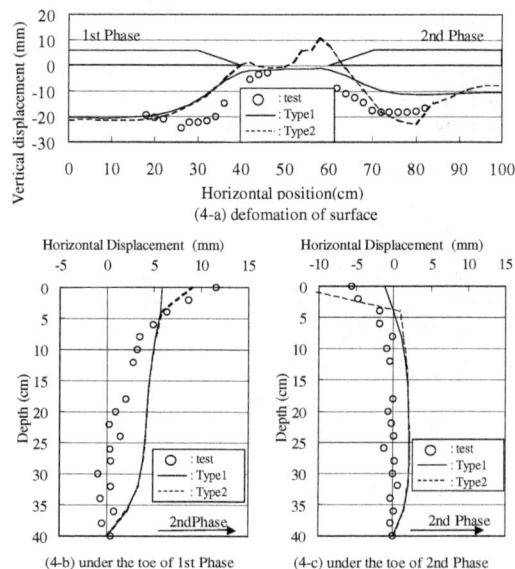

Figure 9(4). Deformation of the model test and of the calculations on 2nd Phase consolidation

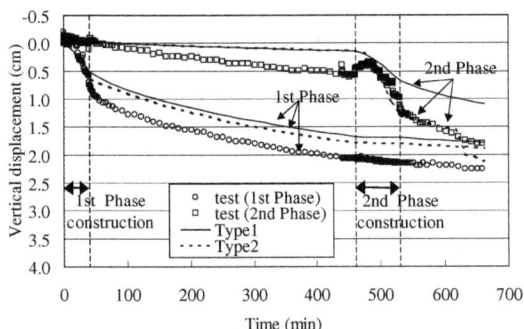

Figure 10. Time course of the vertical displacement of the clay surface beneath two embankments (case6)

this figure that the calculated vertical displacement in Type1 calculation is smaller than that of the model tests and the difference is larger after the 2nd Phase construction than before then. The Type2 calculation gives the results slightly close to the model test results. Especially, the calculated vertical displacements during and after the 2nd Phase construction coincide well with the tests results. It is found that the calculations with slightly smaller shear strength at the shallow area can simulate the vertical displacement at the clay surface beneath the two embankments with high accuracy.

From the above-mentioned comparison between the analytical results and the model tests results, it can be concluded that the FEM analysis can well predict the vertical displacement but still have some subjects for accuracy to improve. However, it is not easy for the FEM analysis to predict the distribution and the amount of the horizontal displacement qualitatively and quantitatively. In the next section, the sensitivity analyses were discussed to improve the accuracy of the analysis.

3.3 Sensitivity analyses of input data

The sensitivity analysis, which investigate the influence of input data on the results of FEM, was carried out by changing the value of parameters, λ, κ, ν_0 and c_v, separately. In this section, the results of sensitivity analysis for λ is described. As for the results for the other parameters, there was negligible influence by changing the value of κ and ν_0, and comparatively large influence by changing the value of c_v (Kitazume, et al., 2001).

In the analyses, the value of λ was changed from 0.5 times to 2.0 times the value of λ for the basic calculation shown in Table 2 (hereafter cited as λ_0). The vertical and the horizontal displacements obtained by the sensitivity calculations are compared with the model test results in Figs. 11(1) and 11(2), where the vertical displacements at the base and the center of two embankments are focused. In the Figs. 11(1) and 11(2), S and δ in the y-axis are the vertical

and the horizontal displacements, and the subscript as m and c mean the model test results and the calculation results, respectively. It is found in Fig. 11(1) that the plotted data for the vertical displacement are considerably close to the $S_c = S_m$ line if the λ/λ_0 is about 1.2. It is found in Fig. 11(2), on the other hand, that the plotted data for the horizontal displacement at the middle depth of the clay layer is close δ_c/δ_m line if the λ/λ_0 is about 0.5, and those at the surface of the clay layer is close to the δ_c/δ_m line if the λ/λ_0 is about 2.0. These phenomena show that the calculated horizontal displacements at the ground surface are not easily coincided with the model test results even if the value of λ is changed. In addition, it is found the value of λ gives a large influence to the calculated horizontal displacement.

4 CONCLUSION

Following conclusions were obtained from this research.
(1) From the centrifuge test, the construction of the embankments causes the vertical displacement and

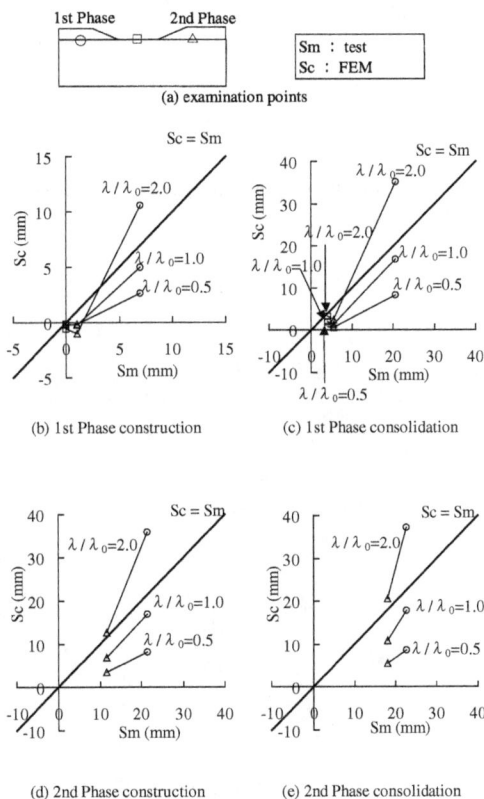

Figure 11(1). Results of the sensitivity analyses (vertical displacement)

(a) examination points

| δ m | : test |
| δ c | : FEM |

(b) 1st Phase construction

(c) 1st Phase consolidation

(d) 2nd Phase construction

(e) 2nd Phase consolidation

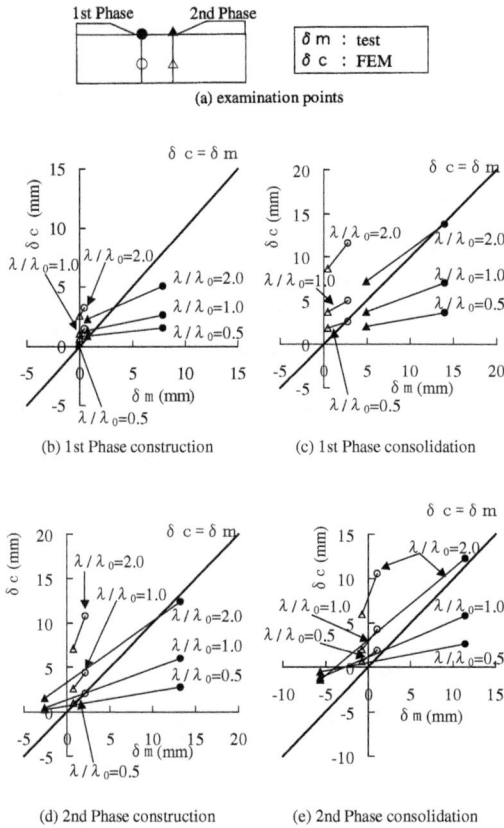

Figure 11(2). Results of the sensitivity analyses (horizontal displacement)

the upheaval deformation of the clay surface around the embankment. When one embankment is constructed, large displacement occurred by larger embankment load, and the amount of the horizontal displacement is larger than the vertical displacement at the outside the embankment. When the second embankments are constructed later, the deformation of clay by the second embankment is smaller than that by the first embankment. As for the range where comparatively large deformation is observed, it is narrower at the second embankment than that at the first embankment. The deformation of the clay surface becomes large when the embankment load is large or construction speed is fast.

(2) FEM analyses can simulate comparatively well the vertical displacement of the clay ground during the construction and the consolidation of the embankment. However, as far as this study's condition, the FEM analysis cannot succeed to estimate the horizontal displacement. Further analytical research is required to improve the horizontal displacement estimation.

REFERENCES

Kitazume, M. and Miyajima, S. 1995. Development of PHRI Mark II Geotechnical Centrifuge. *Technical Note of the Port and Harbour Research Institute*, 817: 33p.

Kitazume, M., Miyajima, S., Hirose, M., Suzuki, S., Taki, M. and Hashizume, H. 2001. Centrifuge Model Tests and FEM Analyses on Behavior of Soft Clay Ground. *Proceedings of the 46th Symposium on Geotechnical Engineering*, Japanese Geotechnical Society, 103-108, (in Japanese).

Kobayashi, M. 1984. Stability Analysis of Geotechnical Structures by Finite Elements. *Report of the Port and Harbour Research Institute*, 23(1): 83-101, (in Japanese).

Soft Ground Engineering in Coastal Areas, Tsuchida et al. (eds)
© 2003 Swets & Zeitlinger, Lisse, ISBN 90 5809 613 0

Evaluation of long-term settlement of Pleistocene deposits in Osaka Bay

M. Mimura
Disaster prevention Research Institute of Kyoto University, Uji, Japan

W.Y. Jang
Kyoto University, Kyoto, Japan

ABSTRACT: A newly developed elasto-viscoplastic finite element procedure is introduced to assess the long-term settlement taken place in the reclaimed islands in Osaka Bay. In order to describe the serious time dependent behavior of the Pleistocene clays in the stress range less than p_c, an elasto-viscoplastic behavior is assumed to occur not only in the normally consolidated region but also in the overconsolidated region, on the basis of the achievement from laboratory experimental data as well as in-situ measurement. The calculated performance with the proposed procedure is validated by comparing with the monitored results of the total and differential settlement at Maishima Reclaimed Islands.

1 INTRODUCTION

In Osaka Bay, the development of the metropolises has been associated with the expansion of coastal areas as well as the construction of the offshore reclaimed islands. The reclaimed island called Sakishima was constructed in 1980 followed by Maishima and Yumeshima Islands in Osaka Port. Kansai International Airport is one of the most significant representatives of those offshore reclaimed islands. However, it is also well known that those man-made islands have suffered from serious long-term settlement mainly due to the delayed compression of the Pleistocene clays.

Figure 1 shows the oval shape of Osaka Bay (60, 30 km in size). This beautiful creation was shaped by the couple of rising of the surrounding mountains and precipitation of the base ground in the center of

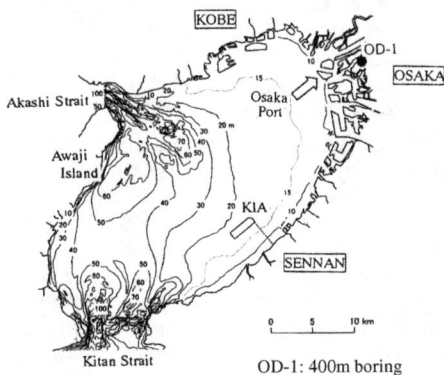

Figure 1. Plan view of Osaka Bay and seabed level.

the bay. The seabed deposits of the Osaka Bay have been formed due to the soil supply from the rivers and sedimentation on the base that has been shaped by the tectonic movement mentioned above. This stratification was confirmed in 1960s by the boring log, called OD-1, and fourteen marine clay deposits are called with a name of "Ma" as shown in Figure 2. The superficial marine clay, Ma 13, is the soft Holocene deposit about 10,000 years ago, while Ma12~Ma0 are the stiff Pleistocene deposits for the several thousands to one million years ago.

Although it is common that the clay deposits formed under this environment should be normally consolidated, the Pleistocene clays in Osaka Bay exhibit slight overconsolidation with OCR of 1.2 to 1.5 in average (see Figure 2). This apparent overconsolidation is thought not to arise from the mechanical reason but to be subjected to the effect of diagenesis, such as aging effect and/or development of cementation among clay particles. In the sense, the Pleistocene clay deposited in Osaka Bay is so-called "quasi-overconsolidated clays" without definite mechanical overconsolidation history as seen for the post-glacial clays in North America and Europe. It is true that long-term and large deformation has been a serious issue because the loading increment, $\Delta\sigma$, has become large for the overconsolidated Pleistocene clay enough to undergo plastic yielding due to large-scale reclamation in a deep-sea area, but it is much more important to note that the long-term settlement has also been monitored where the total stress, $\Delta\sigma$ is less than the consolidation yielding stress, p_c. The conventional framework to assess the settlement, such as Terzaghi's one-dimensional calculation or elasto-plastic approach can not function for the time-dependent behavior in the overconsolidated region.

In the present paper, the refined elasto-viscoplastic finite element procedure is introduced to assess the overall viscoplastic behavior of those quai-overcolidated Pleistocene clay deposits in Osaka Port. In the present procedure, elasto-viscoplastic behavior is assumed to occur even in the overconsolidated region because the Pleistocene clays in Osaka Bay can be considered normally consolidated aged clays with the apparent overconsolidation on the basis of the sedimentation environment of this area stated above. The monitored achievement of settlement at Maishima Reclaimed Island is used to validate the calculated performance by the present procedure.

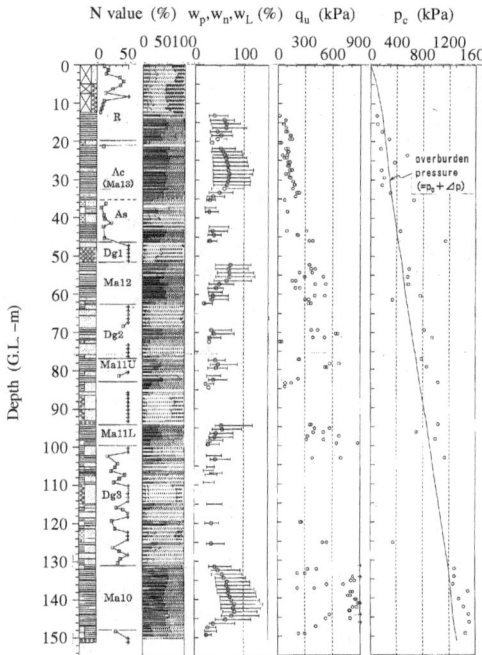

Figure 2. Representative subsoil conditions of Osaka Port and physical properties.

2 ELASTO-VISCOPLASTIC FINITE ELEMENT ANALYSIS

2.1 Constitutive model and formulation

The elasto-viscoplastic constitutive model used in this paper was proposed by Sekiguchi (1977). Sekiguchi et al. (1982) modified the model to a plane-strain version. The viscoplastic flow rule for the model is generally expressed as follows:

$$\dot{\varepsilon}_{ij}^p = \Lambda \frac{\partial F}{\partial \sigma_{ij}'} \qquad (1)$$

in which F is the viscoplastic potential and Λ is the proportional constant. Viscoplastic potential F is defined as follows:

$$F = \alpha \cdot \ln\left[1 + \frac{\dot{v}_0 \cdot t}{\alpha} \exp\left(\frac{f}{\alpha}\right)\right] = v^p \qquad (2)$$

in which α is a secondary compression index, v_0 is the reference volumetric strain rate, f is the function in terms of the effective stress and v^p is the viscoplastic volumetric strain. The concrete form of the model is shown in the reference (Mimura and Sekiguchi, 1986). The resulting constitutive relations are implemented into the finite element analysis procedure through the following incremental form:

$$\{\Delta\sigma'\} = [C^{ep}]\{\Delta\varepsilon\} - \{\Delta\sigma^R\} \qquad (3)$$

Where $\{\Delta\sigma'\}$ and $\{\Delta\varepsilon\}$ are the associated sets of the effective stress increments and the strain increments respectively, and $[C^{ep}]$ stands for the elasto-viscoplastic coefficient matrix. The term $\{\sigma^R\}$ represents a set of 'relaxation stress', which increases with time when the strain is held constant. The pore water flow is assumed to obey isotropic Darcy's law. In relation to this, it is further assumed that the coefficient of permeability, k, depends on the void ratio, e, in the following form:

$$k = k_0 \cdot \exp\left(\frac{e - e_0}{\lambda_k}\right) \qquad (4)$$

in which k_0 is the initial value of k at $e=e_0$ and λ_k is a material constant governing the rate of change in permeability subjected to a change in the void ratio.

Note that each quadrilateral element consists of four constant strain triangles and the nodal displacement increments and the element pore water pressure is taken as the primary unknowns of the problem. The finite element equations governing those unknowns are established on the basis of Biot's formulation (Christian, 1968, Akai and Tamura, 1976), and are solved numerically by using the semi-band method of Gaussian elimination.

2.2 Maishima Reclaimed Island

In Osaka Port, Maishima Reclaimed Island has started to be constructed in 1969's by dumping the dredged disposals and at after 7 years, sand drains were driven following the set of sand mat. Then the large-scale filling was started with dredged materials. Drains were also driven in the dredged materials to promote dewatering from the dredged materials followed by the final embankment construction. The construction sequence is shown in Figure 3 together with the monitored data. During the 1st and 2nd fill-

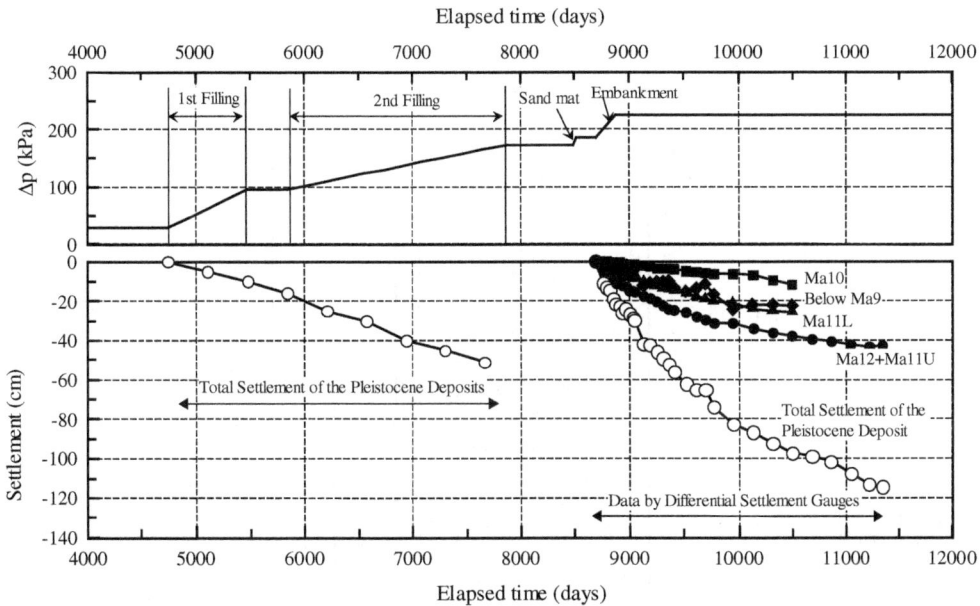

Figure 3. Construction sequence and the monitored settlements at Maishima Reclaimed Island.

ing with dredged materials, the total settlement of the Pleistocene deposits was monitored by double tube settlement gauges. Again during the final sand mat setting prior to embankment, a set of differential settlement gauges was installed to measure the compression of each Pleistocene clay layers, such as Ma12 and Ma11U, Ma11L, Ma10 and the below Ma9. As shown in Figure 3, as it is a pity that the continuous measurement could not be done in this site, we can get only the total settlement of the Pleistocene deposits during the 1st and 2nd filling and the differential settlements together with those summation during embankment construction. It should be noted that a large amount of settlement continues for the Pleistocene deposits, particularly, even below Ma9, non-negligible long-term settlement takes place with the elapsed time.

2.3 Problems setup and parameters

In the present study, a new assumption is introduced to the conventional procedure. Based on the in-situ measured data, the long-term settlement has been found to take place in the reclaimed islands of Osaka Port even in the Pleistocene clay layers that do not undergo the plastic yielding ($\sigma_v \leq p_c$). As is also stated that the seabed deposits of the Osaka Bay have been formed due to the soil supply from the rivers there are no further authors place the cursor one space behind the word ABSTRACT: and type your abstract of not more than 150 words. The top of rivers on the sinking base. Therefore, those sediments such as the Pleistocene clays should be normally consolidated. However, they exhibit the ap-

parent overconsolidation with OCR of 1.3 to 1.5. The authors consider those Pleistocene clays as "normally consolidated aged clay" with apparent overconsolidation due to diagenesis effect. In the sense, the following assumption is introduced:

The Pleistocene clays in Osaka Bay is "normally consolidated aged clays" that exhibit the elasto-viscoplastic behavior even in the region $\sigma_v \leq p_c$ as schematically shown in Figure 4.

Here, the Pleistocene clays are assumed to exhibit a time-dependent behavior not only in the normally consolidated but also in the overconsolidated region. But once those clays undergo additional loading, the corresponding stress should be p_c and unloading - reloading region clays exhibit elastic behavior as is conventionally assumed. In short, only the virgin loading for the Pleistocene clays can provide elasto-

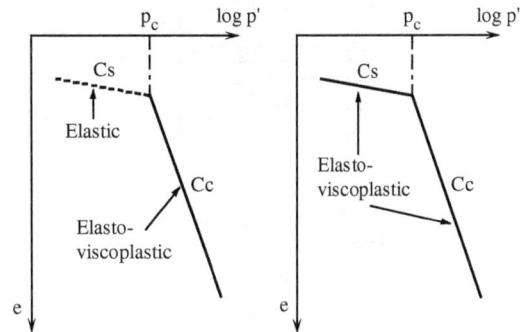

Figure 4. Assumption for the visco-plastic behavior for the Pleistocene clays.

Table 1. Material parameters of soils.

MTYP	OC region				NC region				M	v'	σ_{vo}' (kPa)	σ_{vc}' (kPa)	e_o	k_o (m/day)	λ_k	
	λ_{oc}	κ_{oc}	α_{oc}	$v_{o\,oc}$ (day^{-1})	λ_{nc}	κ_{nc}	α_{nc}	$v_{o\,nc}$ (day^{-1})								
23	0.077	0.008	1.15×10^{-3}	2.49×10^{-6}	0.768	0.077	1.15×10^{-2}	2.49×10^{-6}	1.3	0.36	285	385	2.34	8.64×10^{-5}	0.768	Ma12
22	0.109	0.011	1.64×10^{-3}	3.55×10^{-6}	1.089	0.109	1.64×10^{-2}	3.55×10^{-6}	1.3	0.36	305	412	2.32	8.64×10^{-5}	1.089	Ma12
21	0.084	0.008	1.60×10^{-3}	3.46×10^{-6}	0.842	0.084	1.60×10^{-2}	3.46×10^{-6}	1.3	0.36	325	439	1.63	8.64×10^{-5}	0.842	Ma12
20	0.02	0.002	4.81×10^{-4}	1.04×10^{-6}	0.204	0.020	4.81×10^{-3}	1.04×10^{-6}	1.3	0.36	346	467	1.12	8.64×10^{-5}	0.204	Ma12
19	—	—	—	—	—	—	—	—	—	0.33	429	558	0.93	2.16×10^{-2}	—	Sand
18	0.031	0.003	6.82×10^{-4}	1.03×10^{-5}	0.308	0.031	6.82×10^{-3}	1.03×10^{-5}	1.3	0.36	513	667	1.26	6.05×10^{-5}	0.308	Ma11U
17	0.024	0.002	5.96×10^{-4}	8.97×10^{-6}	0.243	0.024	5.96×10^{-3}	8.97×10^{-6}	1.3	0.36	537	698	1.04	6.05×10^{-5}	0.243	Ma11U
16	—	—	—	—	—	—	—	—	—	0.33	582	728	0.93	2.16×10^{-2}	—	Sand
15	0.077	0.008	1.40×10^{-3}	1.34×10^{-5}	0.774	0.077	1.40×10^{-2}	1.34×10^{-5}	1.3	0.36	625	750	1.76	5.18×10^{-5}	0.774	Ma11L
14	0.049	0.005	9.77×10^{-4}	9.36×10^{-6}	0.490	0.049	9.77×10^{-3}	9.36×10^{-6}	1.3	0.36	649	779	1.51	5.18×10^{-5}	0.490	Ma11L
13	0.024	0.002	6.36×10^{-4}	6.09×10^{-6}	0.243	0.024	6.36×10^{-3}	6.09×10^{-6}	1.3	0.36	676	811	0.91	5.18×10^{-5}	0.243	Ma11L
12	—	—	—	—	—	—	—	—	—	0.33	865	1038	0.93	2.16×10^{-2}	—	Sand
11	0.063	0.006	1.33×10^{-3}	2.15×10^{-6}	0.629	0.063	1.33×10^{-2}	2.15×10^{-6}	1.3	0.36	1051	1261	1.45	2.59×10^{-5}	0.629	Ma10
10	0.081	0.008	1.47×10^{-3}	2.37×10^{-6}	0.812	0.081	1.47×10^{-2}	2.37×10^{-6}	1.3	0.36	1075	1290	1.87	2.59×10^{-5}	0.812	Ma10
9	0.092	0.009	1.43×10^{-3}	2.30×10^{-6}	0.916	0.092	1.43×10^{-2}	2.30×10^{-6}	1.3	0.36	1102	1322	2.21	2.59×10^{-5}	0.916	Ma10
8	0.056	0.006	1.25×10^{-3}	2.01×10^{-6}	0.560	0.056	1.25×10^{-2}	2.01×10^{-6}	1.3	0.36	1126	1351	1.24	2.59×10^{-5}	0.560	Ma10
7	—	—	—	—	—	—	—	—	—	0.33	1226	1471	0.93	2.16×10^{-2}	—	Sand
6	0.044	0.004	1.12×10^{-3}	6.01×10^{-7}	0.438	0.044	1.12×10^{-2}	6.01×10^{-7}	1.3	0.36	1330	1596	0.95	3.46×10^{-5}	0.438	Ma9
5	0.072	0.007	1.51×10^{-3}	8.08×10^{-7}	0.716	0.072	1.51×10^{-2}	8.08×10^{-7}	1.3	0.36	1352	1622	1.37	3.46×10^{-5}	0.716	Ma9
4	0.072	0.007	1.51×10^{-3}	8.08×10^{-7}	0.716	0.072	1.51×10^{-2}	8.08×10^{-7}	1.3	0.36	1389	1667	1.37	3.46×10^{-5}	0.716	Ma9
3	0.064	0.006	1.22×10^{-3}	6.54×10^{-7}	0.638	0.064	1.22×10^{-2}	6.54×10^{-7}	1.3	0.36	1420	1704	1.61	3.46×10^{-5}	0.638	Ma9
2	0.064	0.006	1.22×10^{-3}	6.54×10^{-7}	0.638	0.064	1.22×10^{-2}	6.54×10^{-7}	1.3	0.36	1446	1735	1.61	3.46×10^{-5}	0.638	Ma9
1	0.095	0.010	2.21×10^{-3}	1.18×10^{-7}	0.950	0.095	2.21×10^{-2}	1.18×10^{-7}	1.3	0.36	1464	1757	1.15	3.46×10^{-5}	0.950	Ma9

viscoplastic behavior and once clays experience additional load increment $\Delta\sigma$, they exhibit the conventional OC behavior in the range of σ_0 to $\sigma_0 +\Delta\sigma$.

One-dimensional finite element analysis is adopted in the present study to assess the long-term settlement of Maishima Reclamied Island. The adopted subsurface condition is shown in Figure 5. As is clearly seen in Figure 5 that the sand layers in Osaka Port is thick and their continuity is guaranteed (Kaitei Jiban, 1995).

Then, one-dimensional approach can be accepted because the effect of the permeability boundary and mass permeability is unnecessary to be taken into account. The soil parameters required for the input to the finite element code were determined rationally

based on the prescribed procedure (Mimura et al., 1990). The set of parameters are shown in Table 1. In the present analysis, the compression line in the OC region with the inclination of $\kappa =\lambda/10=\lambda_{OC}$ is assumed to have viscoplastic component as same as in the NC region. It should be noted that e - logp relation for the corresponding Pleistocene clays is unchanged at all except the virgin compression contains irreversible time-dependent behavior. Secondary compression index, α is determined as relating to the compressibility in terms of λ (Tsuchida et al., 1983). It is also consistent that the values of α in the OC region (α_{OC}) keep the same relations with λ_{OC} as in the NC region in this particular study. The rate of consolidation is another important factor to control the process of deformation. From the Geodatabase for Osaka Pleistocene clays, consolidation coefficient, c_v in the OC region is about 10 times as that in the NC region. The characteristic time, t_c in the present constitutive model is closely related to c_v. Based on the above data of c_v for the Osaka Pleistocene clays, the values of t_c is determined as 1/10 in the OC region compared to those in the NC region. Then, the initial volumetric strain rate, in the OC region is derived as the same as that in the NC region.

Elevation (O.P.)

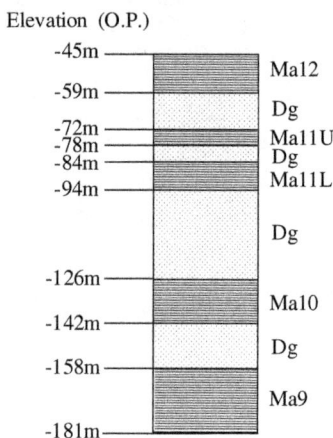

Figure 5. Subsurface model for Maishima Reclaimed Island.

3 RESULTS AND DISCUSSIONS

Figures 6 show the calculated e - logp relations during reclamation at Maishima. Ma 12 in Figure 6(a) is a representative of the Pleistocene clay layers that undergoes plastic yielding. Ma10 in Figure 6 (b) is a representative of those layers that the final stress be-

148

Figure 7. Comparison of advance in settlement for each Pleistocene clay layer.

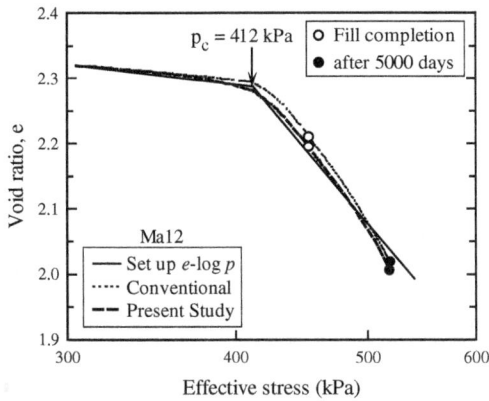

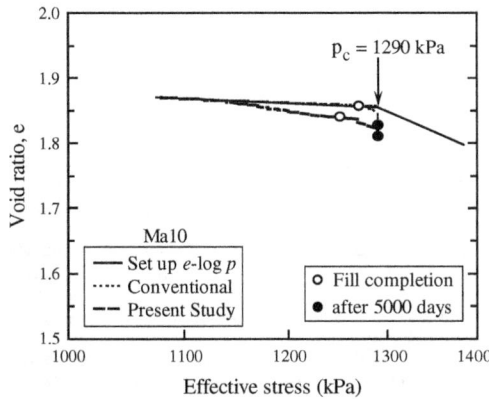

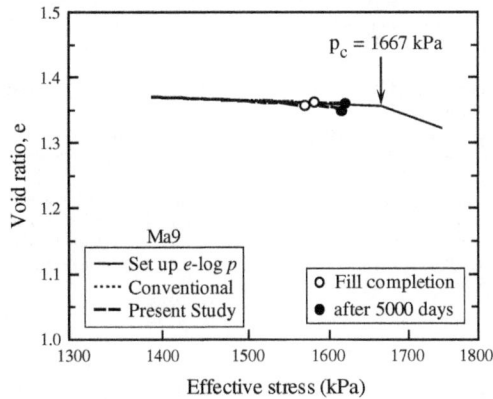

Figure 6. Calculated compression curves, e - logp relations for Ma 12, Ma 10 and Ma 9 at Maishima Reclaimed Island.

comes close to pc and Ma 9 in Figure 6 (c) is a representative of those layers that the final stress remains less than pc respectively. In all Figures, the setup compression curves are illustrated in the hatched lines and the results by the conventional elasto-viscoplastic FEM are shown by the dotted lines. In all cases, the calculated compression lines in terms of e - logp relations move a slightly lower

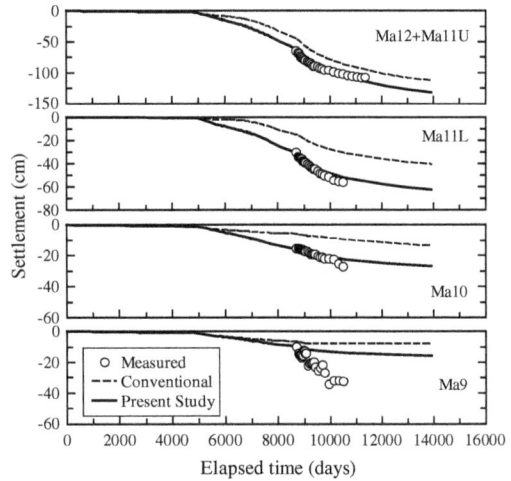

than the setup e - logp curves whereas those by the conventional FEM are almost move on the setup curves. It is natural that the time-dependent behavior such as shown in Figures 6 takes place because the present procedure assumes that the time-dependent behavior occurs even in the OC region. Due to the assumption introduced in the present study, the advance in compression of the Pleistocene clays in the region less than p_c is found to be described.

The calculated performance is compared with the measured settlement for each layer in Figure 7. Here, the sold lines denote the results of the present study while the hated lines are those by the conventional FE analysis. For Ma 12 and Ma11U, the calculated and measured results show a good match. As is seen from Figure 6(a), the upper Pleistocene clay layers such as Ma 12 and 11 undergo the plastic yielding due to reclamation work. So, it is very natural that the calculated performance by both the conventional and the proposed procedures can well describe the stress - deformation characteristics of those clay layers Ma 12 and 11U that become normally consolidated. On the other hand, the different behavior is expected in Ma 11L, Ma10 and Ma9 the stress levels of which are close to p_c and less than p_c. For Ma 11 and 10, the calculated performance with the proposed procedure shows much better coincidence with the measured settlement, whereas the conventional approach shows a remarkable underestimation. As for the settlement of Ma9, the calculated performance with the proposed procedure shows larger settlement than that by that with the conventional elasto-viscoplastic FEM because of the introduction of the time-dependent behavior in OC region. But here, as it is impossible to separate the contribution of Ma9 itself from the measured data, the comparison of the calculated performance for

149

Ma9 to the measured data is not appropriate. However, the measured amount of settlement taken place below Ma9 is more than 20cm in this limited monitoring period. Although it is not so clear that which layers in the lower Pleistocene clay deposits are affected by the reclamation load, it is very serious that such deeply sedimented Pleistocene clays can contribute the advance in settlement.

Figure 8 shows the advance in the settlement of the Pleistocene deposits. Attention should again be paid to the fact that the measured data contains the contribution by the layers below Ma9. But, the calculated performance with the proposed procedure can describe the settlement of the Pleistocene deposits much better than that with the conventional FEM.

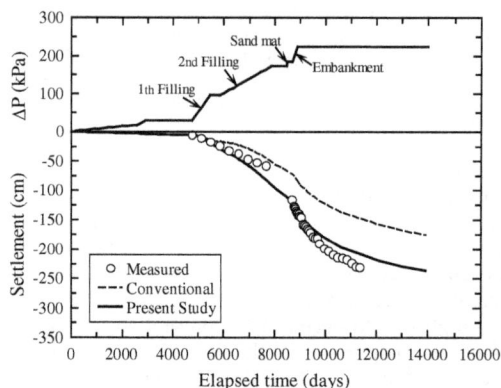

Figure 8. Comparison of advance in total settlement for Pleistocene clay deposits.

4 CONCLUSIONS

Long-term settlement has taken place in the reclaimed coast and the reclaimed islands in Osaka Bay. The compression of the quasi-overconsolidated Pleistocene clays (normally consolidated aged Pleistocene clays) has been confirmed by the in-situ measurement. The difficulty of evaluating those phenomena consists in that the serious time-dependent behavior of the Pleistocene clays occurs even in the overconsolidation region. The conventional procedure, such as elastic - elastoviscoplastic approach is found to hold serious limitation to assess the long-term settlement taken place in Osaka Bay. The revised procedure in terms of the elasto-viscoplastic finite element analysis was proposed in this paper with the assumption that an elasto-viscoplastic behavior takes place even in the overconsolidated region as well as the normally consolidated region. The calculated performance with the proposed procedure was found to well function to describe the time-dependent behavior of the Pleistocene clays in both pre- and post yielding region as well. The measured settlements at Maishima Reclaimed Island were better evaluated with the proposed procedure compared to the conventional analysis.

REFERENCES

Akai, K. & T. Tamura 1976. An application of nonlinear stress-strain relations to multi-dimensional consolidation problems. *Annuals DPRI, Kyoto University*, 21(B-2) : 19-35 (in Japanese).

Christian, J.T. 1968. Undrained stress distribution by numerical method. *Journal of Soil Mech. and Foundation Div., ASCE*, 94 (SM6) : 1333-1345.

Kaitei Jiban −Osaka Bay- 1995. Kansai Branch, JGS, (in Japanese).

Mimura, M. & H. Sekiguchi 1986. Bearing capacity and plastic flow of a rate-sensitive clay under strip loading. *Bulletin of DPRI, Kyoto University*, 36(2) : 99-111.

Mimura, M., T. Shibata, M. Nozu & M. Kitazawa 1990. Deformation analysis of a reclaimed marine foundation subjected to land construction. *Soils and Foundations*, 30(4) : 119-133.

Sekiguchi, H. 1977. Rheological characteristics of clays. *Proc. 9th ICSMFE*, 1 : 289-292.

Sekiguchi, H., Y. Nishida & F. Kanai 1982. A plane-strain viscoplastic constitutive model for clay. *Proc. 37th Natl. Conf., JSCE* : 181-182 (in Japanese).

Tsuchida, T., Y. Kikuchi, K. Nakashima and M. Kobayashi 1984. Engineering properties of marine clays in Osaka Bay (Part 3), Static characteristics of shear. *Technical Note of the Port and Harbour Research Institut , Ministry of Transport*, 498 : 87-114 (in Japanese).

Soft Ground Engineering in Coastal Areas, Tsuchida et al. (eds)
© 2003 Swets & Zeitlinger, Lisse, ISBN 90 5809 613 0

Evaluation of consolidation process of soft clay ground by cone penetration test

S. Murakawa & T. Yoshifuku
Design Department, Nikken Sekkei Civil Engineering Ltd., Tokyo, Japan

M. Katagiri & M. Terashi
Nakase Geotechnical Institute, Nikken Sekkei Ltd., Kawasaki, Japan

ABSTRACT: It is necessary to monitor the change of ground properties accurately and appropriately in case of rapid construction on soft clay ground such by preloading with prefabricated drains. Monitoring and management of the consolidation process involve measurements of settlement, pore water pressure, shear strength, and consolidation yield stress of the improved ground. Ordinary practice for the latter two is based on the laboratory tests on undisturbed samples taken through boring. Boring and sampling need, however, considerable cost and time. The authors propose the use of cone penetration test (CPT) for the evaluation of consolidation process, which is rapid and economical. Furthermore, it enables monitoring of pore water pressure as well. In this paper, the applicability and advantage of the method is described based on the application at the Island City project in Fukuoka.

1 INTRODUCTION

In the case of rapid construction on soft ground such by preloading with prefabricated drains, it is necessary to monitor the change of ground properties accurately and appropriately for determining the time of removing preload embankment. Monitoring and management in the consolidation process involve measurements of settlement, pore water pressure, shear strength and consolidation yield stress of the improved clay ground. Ordinary practice for the latter two is based on the laboratory tests on undisturbed samples taken through boring. Boring and sampling need considerable cost and time and even more it is difficult to take undisturbed samples in the case of extremely soft soil. Ordinary laboratory tests such as the shear and oedometer tests require further time and cost. The authors propose the use of cone penetration test (CPT) for evaluation of consolidation process, which is rapid and economical. Furthermore it enables monitoring of pore water pressure as well. In this paper, the applicability and advantage of the proposed method is described based on the application at the Island City project in Fukuoka.

2 OUTLINE OF ISLAND CITY PROJECT

The Island City shown in Fig. 1 is an urban space developed in a man-made island reclaimed from the sea in Hakata Bay, Fukuoka. An average depth of the

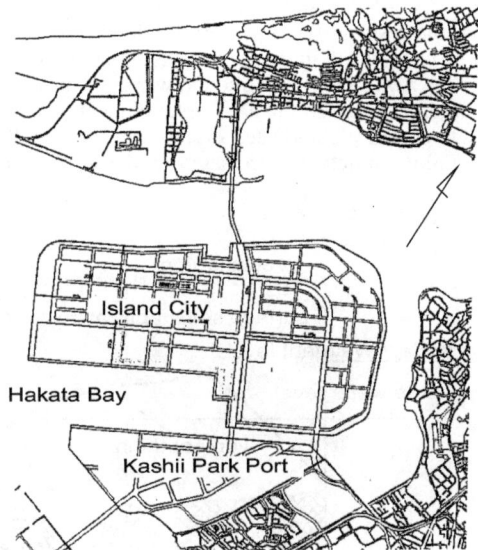

Figure 1. Location of Island City in Fukuoka

existing seabed before the reclamation was DL-3.5 m, and the seabed was Holocene clay layer of 8 m thick. A main material for the reclamation is marine clay dredged from nearby navigation channels, anchorage area, etc., produced in port construction (see Fig. 2). The reuse of dredged clay is friendly for the environment and economy. However, because the ground reclaimed with the dredged clay is under

151

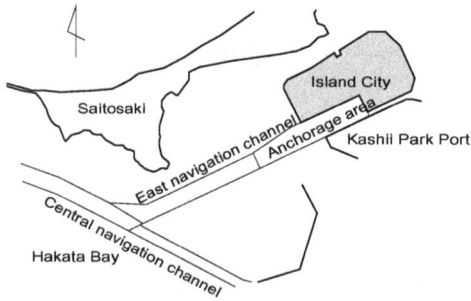

Figure 2. Location of dredged area for Island City.

consolidation and the water content of reclaimed layer exceeds 200%, very large settlement would occur due to its self-weight. The reclamation will complete by placing relatively good quality fill material on top of the dredged clay. The weight of the fill material will further increase the settlement. Prior to building urban infrastructures, it is necessary to create sound ground by accelerating consolidation by preloading with vertical drainage.

In this project, the maximum height of the dredged clay during reclamation was limited to DL + 6.5 m which is the height of revetment surrounding the artificial island. The elevation of the land was planned to be DL + 5.0 m at the start of the servicing. The allowable residual settlement by the load of facilities to be constructed in the future was set at 30 cm (see Fig. 3). And a rapid construction was demanded.

In order to attain the above requirements, the preloading with prefabricated drains was adopted. The overall construction processes from the sea reclama-

tion to the construction of facilities are shown in Fig. 3; i) reclamation with dredged clay, ii) surface soil stabilization, iii) installation of vertical drains, iv) overburdening by fill material including preload, v) removal of preload fill and vi) building facilities. The accuracy of settlement prediction influences the volume of fill and preload material that in turn influences the estimated construction cost.

The Island City project is divided into several sections and the reclamation progresses from one section to another in a series according to the construction schedule that meets the predicted demand for the land. In this process, the most efficient and economical construction method is to reuse the preload material in the previous section for the fill material in the next section. In order to complete the reclamation within a limited construction term and to keep the residual settlement within the allowable value, it is necessary to perform an accurate initial settlement prediction in the ground improvement design and construction management associated with detailed monitoring.

3 SETLEMENT PREDICTION AND CONSRUCTION CONTROL

3.1 *Settlement prediction*

The purposes of the initial settlement prediction are i) to determine the ground improvement specification (length and spacing of drain) so as to satisfy the planned construction term, and ii) to determine the quantity of the overburdening soil including preload

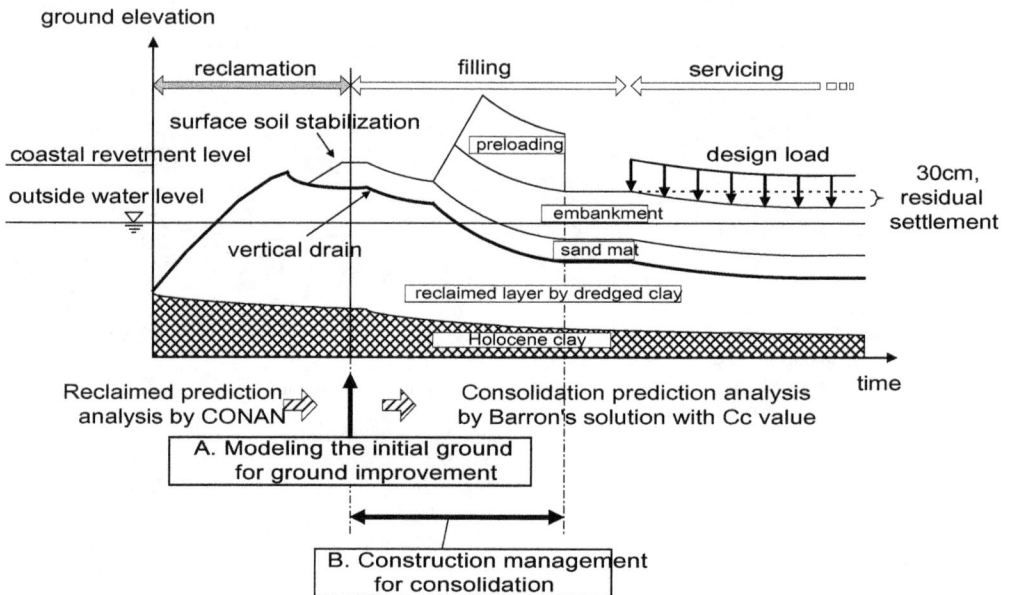

Figure 3. Outline of construction for the Island City.

that satisfies both the planned final elevation and allowable residual settlement. In general, the initial ground condition for the ground improvement design is determined by the soil investigation of the reclaimed ground. Due to the tight time schedule of Island City project, however, the ground improvement design should be carried out when the reclamation was still in progress. The initial ground condition for the ground improvement design was estimated by the analysis code by Katagiri et al. (2000), which is developed based on the one-dimensional consolidation concept proposed by Imai (1995) and the consideration of reclamation (increasing thickness) proposed by Yamauchi et al. (1990).

The outline of modeling the initial ground is as follows: i) Consolidation parameters are determined by the multi-layer sedimentation test using samples from dredging site. ii) Using the parameters, the prediction analysis is performed. iii) In order to improve the accuracy of the prediction, the consolidation parameters are modified by comparing the analyzed and measured values of settlement, the water content and the excess pore water pressure distributions along with reclamation process. iv) Subsequent prediction analysis is performed using the modified parameters.

For the ground improvement design, Barron's solution considering the well resistance is applied to the estimated soil profile of the reclaimed ground just before improvement (just after surface soil stabilization).

3.2 Construction management with the aid of CPT

The construction management based on the observational method was employed to supplement the initial ground improvement design. The purposes of the construction management are i) to determine appropriate time of removing preload and ii) to take countermeasures such as additional preloading where the consolidation occurs slower than expected.

Re-calculation was performed correcting consolidation parameters set up in the initial prediction, based on the results of field observation using differential settlement gauges, settlement plates and pore pressure gauges and soil investigation, then the time of removing preload was finally determined.

The authors proposed a method to know the spatial distribution of ground properties rapidly, using the cone penetration tests in addition to the conventional boring investigations, and applied to Island City project.

3.3 Outline of CPT

The cone penetration test (CPT) is an in-situ test that can continuously measure the vertical distributions of penetration resistance, skin friction and pore water pressure, by penetrating the cone as shown in Fig. 4 statically into the ground.

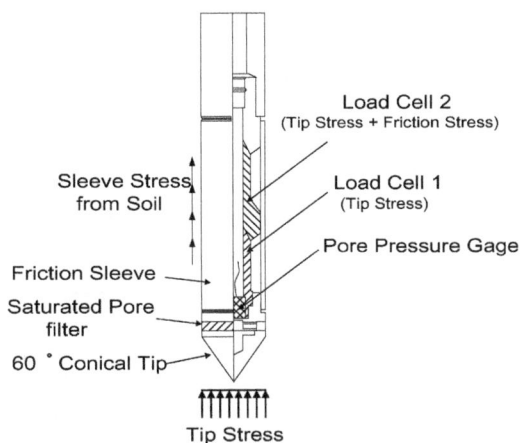

Figure 4. Structure of cone penetrometer.

The measurements of excess pore water pressure can be conducted by CPT with water pressure dissipation test. Water pressure dissipation test measures the change of pore water pressure at a point in the ground where cone penetration is stopped and left for a certain period.

Cone penetration resistance (cone index) q_t has a high correlation with shear strength of the ground. For calculating undrained shear strength, c_u of clay from the cone penetration test, Eq.(1) is widely accepted.

$$c_u = (q_t - \sigma_{v0}) / N_{kt} \qquad (1)$$

where q_t is cone index, σ_{v0} is total overburden pressure and N_{kt} is cone coefficient. Currently this N_{kt} is given experimentally because it is influenced by many factors, in spite of several theoretical approaches. It is shown that cone indices N_{kt} of Japanese clays distribute approximately in a range between 8 and 15 with a large variance (Tanaka, 1996). However, the variance for a particular soil is small, then it is possible to determine appropriate N_{kt} in the ground of interest, when calibrations with other strength tests such as the unconfined compression and vane shear tests are performed. If N_{kt} is constant not depending on the increase in strength of clay during consolidation period, c_u and q_t would have a one-to-one correspondence, which is convenient for the settlement management.

4 APPLICATION OF PROPOSED METHOD TO ISLAND CITY PROJECT

4.1 Outline of ground investigation and field observation

Immediately before surface soil improvement, sampling using boring hole (B-1 ~ B-3) and cone penetration tests (C-1 ~ C-9) were performed in the

153

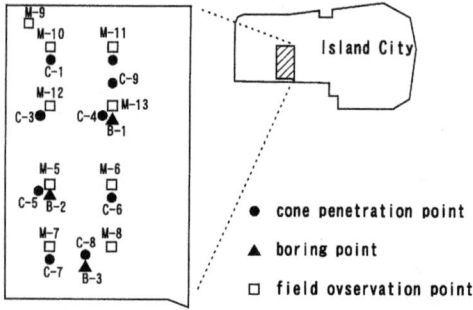

Figure 5. Location of initial investigation and field observation points.

hatched area (580 * 350 m) in Fig. 5, to confirm the validity of initially estimated ground profile.

After the ground improvement work, differential settlements and pore pressures were monitored at the field observation points (M-5 ~ M-13) in the area. While installing the settlement gauges and pore pressure gauges, undisturbed soil samples were taken at these observation points.

In addition to the above, the cone penetration tests and undisturbed soil sampling are carried out at the intermediate phase and final phase as shown later in Fig. 11.

To determine soil parameters, the laboratory tests such as unconfined compression and oedometer tests were performed using undisturbed samples.

4.2 Evaluation of initial ground properties predicted by CONAN

To evaluate the ground condition prior to ground improvement that was predicted by CONAN, the soil profiles obtained from boring and laboratory tests on undisturbed samples were compared with the prediction. The comparisons are shown in Fig. 6.

From the soil investigation, it is confirmed that the part above DL -4.0 m is the new reclaimed soil layer, DL -4.0 m ~ DL -11.0 m is the original Holocene clay layer and below DL -11.0 m is the Pleistocene layers. The reclaimed clay layer is very soft so that the unit weight of soil γ_t (Fig. 6(a)) and the void ratio e (Fig. 6(b)) at its upper part are 13 ~ 16 kN/m^3 and 3 ~ 4, respectively. In the Holocene clay layer, the unit weight of soil increases with depth, and the void ratio decreases. The predictions by CONAN in both the relations are located in the mean of variance of soil investigation results.

Figure 6(c) is the distribution of the consolidation yield stresses, p_c obtained from the oedometer tests. The effective stresses predicted by CONAN analysis are also plotted in the figure. The consolidation yield stress and the effective stress are equivalent for the reclaimed and Holocene layers because of the reclamation history. It is also confirmed that the CONAN analysis can simulate the effective stress with high accuracy.

Figure 6(d) shows the distribution of undrained strength obtained from the unconfined compression tests on undisturbed samples. The reclaimed layer is very weak and the strengths are less than 5 kPa. In the Holocene clay layer, the strength increases with depth, and the strength at DL -10 m is over 20 kPa.

Figure 6(e) shows all the distributions of cone resistance, q_t measured at C-1 to C-9. The $z - q_t$ relations are consistent with each other, except for two cases at two depths. At two depths, the q_t values are larger than those of the others, and it is recognized

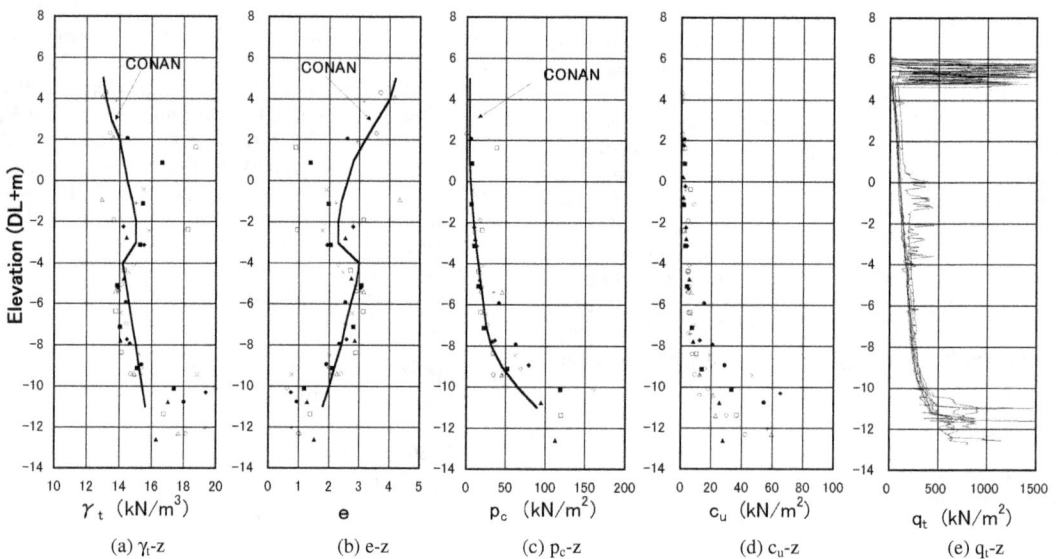

Figure 6. Profile of the reclaimed land from initial soil investigation.

that sand seams exist locally in the reclaimed clay layer.

About 1.5 m from the ground surface is the sand mat and high q_t values are measured. In the reclaimed and Holocene clay layers, the q_t increases with depth. It is difficult to recognize the boundary between reclaimed and Holocene layers from the measured distribution.

4.3 Calibration of cone coefficient by initial soil investigation

At the initial soil investigation, the cone penetration tests and sampling were carried out at the same points of C-4 (B-1), C-5 (B-2) and C-8 (B-3) in Fig. 5. To obtain the cone coefficient N_{kt}, the q_t values were compared with the undrained shear strengths from the unconfined compression tests on the undisturbed samples. From these cone coefficients N_{kt}, shear strength c_u is correlated to q_t, using Eq. (2), which is a developed form about q_t of Eq. (1).

$$q_t = c_u * N_{kt} + \sigma_{v0} \tag{2}$$

Figure 7 shows the relationship between $q_t - \sigma_{vo}$ and c_u at the same depth in the same point. Although there is some scattering, the average slope of $q_t - \sigma_{vo}$ and c_u is determined as 15 irrespective to the reclaimed and Holocene layers.

In addition, using the strength increase ratio of clay, effective overburden stress or consolidation yield stress is estimated by Eq. (3).

$$q_t = m * p_c * N_{kt} + \sigma_{v0} \tag{3}$$

where m is the strength increase ratio of clay. In this project, the m-value was assumed as 0.3.

Figure 8 shows the distributions of measured and converted q_t-values. The open circles in Fig. 8(a) are the q_t values converted from the shear strength c_u obtained by unconfined compression tests at B-1 to B-3. The open circles in Fig. 8(b) are the converted q_t from the consolidation yield stress p_c obtained by oedometer tests also at B-1 to B-3. All the measured and converted q_t distributions exhibit very good coincidence.

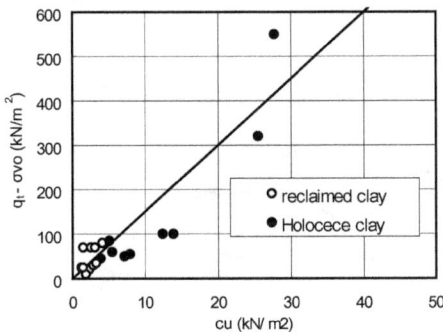

Figure 7. $q_t - \sigma_{v0}$ vs c_u relation.

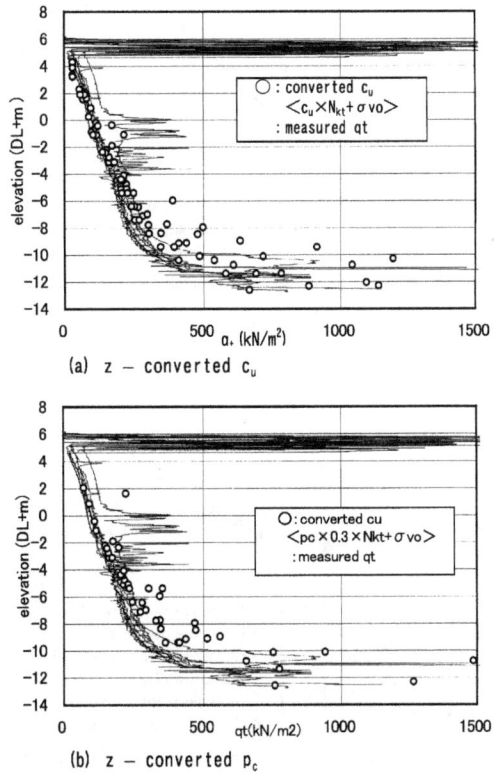

(a) z – converted c_u

(b) z – converted p_c

Figure 8. Distributions of z and converted c_u (a) and p_c (b).

From these results, it is confirmed that the cone penetration test provides not only shear strength distribution but also consolidation yield stress distribution.

4.4 Confirmation of cone coefficient

Intermediate soil investigation by cone penetration, undisturbed soil sampling and testing were conducted 4 months after the completion of the second filling including the preloading fill. The purpose of the investigation was twofold. One is to improve the settlement prediction and the other is to investigate the time-dependency of cone coefficient.

The q_t values and shear strengths obtained at the intermediate investigation are compared in Fig. 9. In this figure, the results at the initial soil investigation mentioned earlier are also plotted. Both the magnitudes of q_t and c_u at the intermediate investigation are larger than those at the initial one due to the progress of consolidation. The relationship between $q_t - \sigma_{vo}$ and c_u is almost linear, and the slope of relation is regarded as 15, which is the same as that at the initial investigation shown in Fig.7. It is confirmed that the cone coefficient is independent of consolidation process.

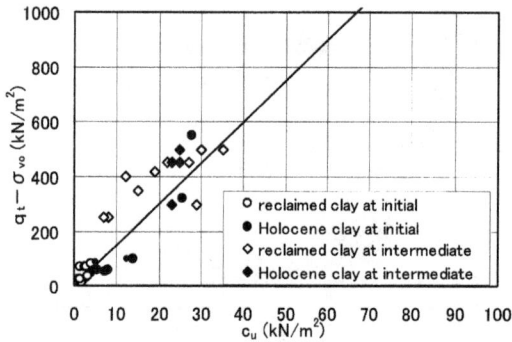

Figure 9. c_u vs $q_t - \sigma_{vo}$ relation at intermediate soil investigation.

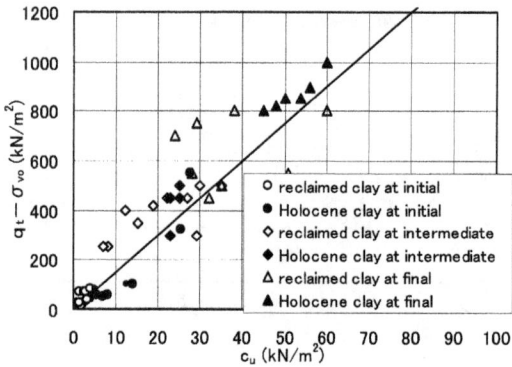

Figure 10. c_u vs $q_t - \sigma_{vo}$ relation at final soil investigation.

Before the removal of extra-fill for preload, the final soil investigation was performed. Figure 10 shows all the data on $q_t - \sigma_{vo}$ and c_u relation from the initial to the final investigations. Along with the process of consolidation, the q_t and c_u increase, and

the ratio of $q_t - \sigma_{vo}$ to c_u is independent of consolidation process.

4.5 Evaluation of consolidation process by CPT

The field observations are performed at M-5 ~ M-13 in Fig. 5. Settlements of the reclaimed and Holocene layers were measured using differential settlement gauges after drains were installed.

Figure 11 shows the time history of construction and settlement of reclaimed and Holocene layers. The upper figure indicates the loading history, the times of drain installation, instrumentation and soil investigations. The second filling includes the pre-loading fill. The magnitude of preloading in this project was nearly equal to the weight of structures to be built in the future. The lower shows the measured and predicted settlements of the reclaimed, Holocene and Pleistocene clay layers. Up to the installation of differential gauges, only the settlement of surface of the reclaimed land had been measured by optical instrument. After the installation of differential gauges, the settlements of each layer and total layers are measured.

The solid curves in the lower figure are the measured settlement. The dotted curves are the predicted settlements that are recalculated after the intermediate investigation in order to fit the settlement record up to that time by incorporating the modified consolidation parameters. The horizontal solid line is the predicted final settlement. In the present project the residual settlement of 30 cm is allowed. The dotted horizontal line shows the settlement 30 cm smaller than the final settlement and this becomes the target settlement by preloading.

Using total settlement predicted, the condition of ground at the intermediate soil investigation was calculated as the degree of consolidation of 85 % in

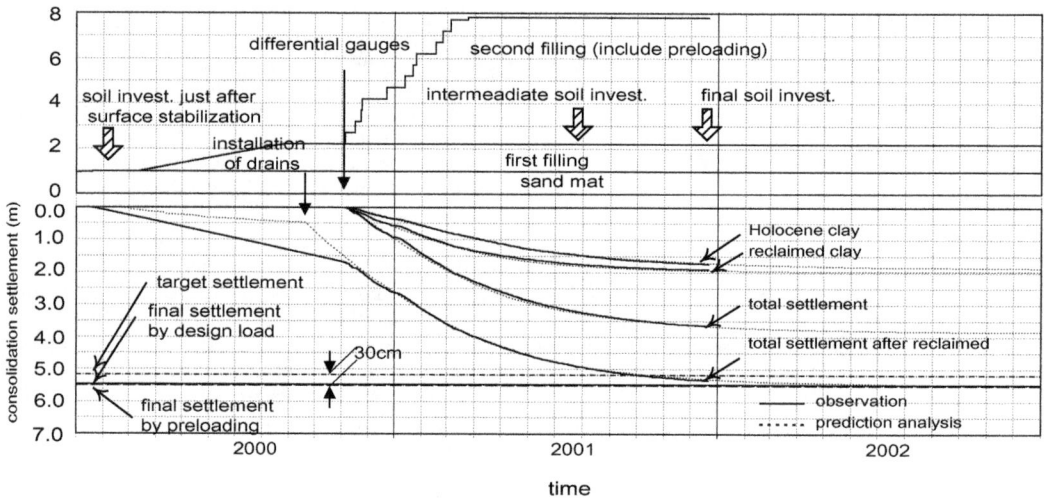

Figure 11. Results of field observation at M-7 point and prediction analysis.

156

terms of settlement (Us = 85 %) and as 65 % in terms of effective stress (U_σ = 65 %). The degree of consolidation at which the allowable residual settlement of 30 cm was satisfied, was estimated as Us = 95 % and as U_σ = 85 %. Here, the degree of consolidation in terms of effective stress was determined as the overburden stress minus excess pore pressure that is concerning with subsequent settlement and that is calculated with compression index, Cc.

The time of removing preload fill has to be determined by the settlement of ground and soil profile, because the settlement of ground shows the total magnitude, but cannot assess the local behavior. Therefore, it is necessary to evaluate the condition of clay layer. In this project, it is decided that all the effective stresses should exceed 85 % of designed load before the removal of preload fill. The laboratory tests such as unconfined compression and oedometer tests on the undisturbed samples are generally required to confirm the degree of consolidation by spending time and cost.

In the final soil investigation to determine the removal of preload fill, some of the laboratory tests on undisturbed samples were replaced by cone penetration tests. As a new trial, the cone penetration tests with measurement of pore water pressure at 1 day after penetration was performed. Figure 12 shows the distributions of consolidation yield stress, shear strength, cone index, and pore water pressure obtained from the cone penetration tests and laboratory tests.

The solid line in Figure 12(a) is the final effective stress under the preload. The broken line in the same figure shows the distribution of target vertical effective stress at which the allowable residual settlement of 30 cm was satisfied (U_σ = 85 %). The open circles show the consolidation yield stress of samples taken from the improved ground at the final soil investigation. The measured data are almost consistent with the target distribution.

Figure 12(b) shows the distributions of measured shear strength at the initial and final soil investigations. The predicted relationships between shear strength and elevation at the target (U_σ = 85 %) and the designed load (solid curve) are also drawn in the same figure. Here, the target and designed shear strengths were determined from the target and designed effective stresses shown in Fig. 12(a), using $c_u = m * \sigma_v'$ with $m = 0.3$. The initial shear strength distribution was calculated from the results of CONAN analysis using the strength increase ratio, $m = 0.3$.

At the initial soil investigation, shear strengths from the unconfined compression tests are less than 10 kN/m^2 in the reclaimed layer. At the final soil investigation, the shear strengths of reclaimed layer become larger, and are from 30 to 60 kN/m^2. Each predicted shear strength distribution is located within the range of scattering data. The method for prediction of shear strength and assumed strength increase ratio are found acceptable.

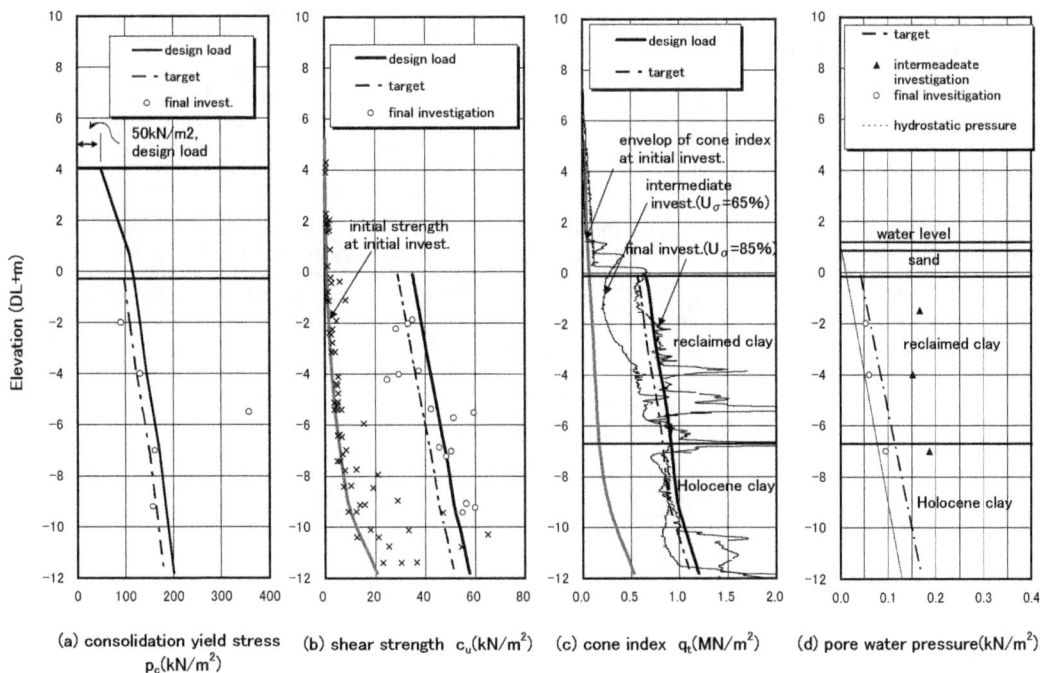

Figure 12. Results of intermediate and final investigations.

(a) consolidation yield stress p_c(kN/m^2)

(b) shear strength c_u(kN/m^2)

(c) cone index q_t(MN/m^2)

(d) pore water pressure(kN/m^2)

Figure 12(c) shows the distributions of cone indices at the initial, intermediate and final soil investigations. The initial and target curves between elevation and cone index are indicated in the same figure. The target curve indicated by a broken line corresponds to the condition of allowable residual settlement of 30 cm, and is calculated from the curve in Fig. 12(b) and Eq.(3). At the time of removing preload fill, the q_t value at each point must exceed the magnitude given by the target curve.

The result at the intermediate soil investigation, the magnitude of q_t at almost all parts did not reach the target curve. Along with the progress of consolidation, the q_t values at the final soil investigation exceeded the target value. At the time of final investigation, the settlement also reached the target magnitude. Therefore, the judgment of removing the preload fill was delivered.

Figure 12(d) shows the measured pore water pressure distributions at the intermediate and final soil investigations. The figure also shows the hydrostatic pressures and target value of pore pressure at the time of removing preload fill. By the intermediate investigation, the measured values (▲) are larger than the target value. In the final investigation, the measured data are smaller than the target value.

By the results explained above, the measured shear strength c_u and consolidation yield stress p_c are consistent with the calculation based cone index q_t. Therefore, spatial distribution of ground properties can be known efficiently by the cone penetration tests as far as the appropriate calibrations with unconfined compression tests and oedometer tests on undisturbed samples are performed.

5 CONCLUSIONS

The following conclusions can be drawn from the present study;
1 The initial ground condition immediately after the reclamation was successfully predicted by the CONAN analysis.

2 The cone coefficient N_{kt} is constant during consolidation process. Therefore, increase in strength of clay due to consolidation can be easily known by the cone penetration tests.
3 The applicability of the consolidation management method with the aid of cone penetration tests was confirmed at the Island City Project.
4 The ground properties can be known efficiently by the cone penetration tests when the appropriate calibrations are performed at the site of concern.

In the conventional method, 16 boring investigations would have been necessary for the settlement management on the hatched area shown in Fig. 5 (about 150,000 m^2), while by the proposed method using CPT, 8 boring investigations and 8 cone penetration tests are satisfactory, that takes no more than 70 % cost of the former.

ACKNOWLEDGEMENT

This study was carried out under the contract with the Port & Harbor Bureau, Fukuoka City. The authors thank Mr. Satomi Hirakawa, Port & Harbor Bureau, Fukuoka City for allowing the authors to publicize the test data.

REFERENCES

Henmi, K., Katagiri, M., Terashi, M. & Fukuda, K. (2000): Case history of the reclamation at Island City in Fukuoka. *IS-Yokohama 2000*, 299-305.
Imai, G. (1995): Analytical examinations of the foundations to formulate consolidation phenomena with inherent time-dependence, *IS-Hiroshima 1995*, 2: 891-935.
Katagiri, M., Terashi, M., Henmi, K. & Fukuda, K. (2000): Change of consolidation characteristics of clay from dredging to reclamation, *IS-Yokohama 2000*, 307-313.
Tanaka, H. (1996): National Report - The current state of CPT in Japan, *Proceedings of the International Symposium on Cone Penetration Testing*, 1: 115-124.
Yamauchi, H., Imai, G. & Yono, K., 1990. Effect of the coefficient of consolidation on the sedimentation consolidation analysis for a very soft clayey soil (in Japanese), *Proc. of 25th Annual meeting of JSSMFE*, 359-362.

Soft Ground Engineering in Coastal Areas, Tsuchida et al. (eds)
© 2003 Swets & Zeitlinger, Lisse, ISBN 90 5809 613 0

Settlement analysis and intelligent site management of the second-phase land reclamation works for Kansai International Airport

M. Shinohara
Kansai International Airport Co., Ltd., Osaka, Japan
Project Team for the 2nd-phase Land Reclamation, Kansai International Airport Land Development Co., Ltd.

ABSTRACT: The 2nd-phase airport construction project entails land reclamation of approximately 40m-thick layers of sand and gravel on 20m-deep soft seabed of Holocene clay layer with 25m thickness. Underneath Holocene clay lie Pleistocene layers of alternate clay and sand with more than 300m thickness. For Holocene clay layer, sand drain works have been applied as a method of accelerating consolidation process. Settlement analysis of Holocene clay layer is efficiently conducted with the application of the intelligent site management utilizing a state-of-the-art settlement monitoring technology and a precise bathymetric surveying method. For settlement prediction of Pleistocene clay layers, a finite-element elasto-viscoplastic analysis is carried out considering the actual settlement data of the 1st-phase airport island. In order to accurately monitor the consolidation process of each Pleistocene clay layer, two platforms of 350m-deep measuring rigs with magnetized devices placed in each layer have been set up to monitor the settlement of Pleistocene clay layers as well as pore water pressure of sand and clay layers. The total amount of settlement of Holocene and Pleistocene layers combined is predicted to reach 18m in 50 years.

1 INTRODUCTION

1.1 *Overview of the project*

Kansai International Airport opened in 1994 as the first around-the-clock offshore airport in Japan. It is expected to meet the steady future growth of demands on air transport right after the airport opening and to contribute to further development of the economy and the society of not only Kansai region but also Japan as a nation. To fully function as international and domestic hub airport, it is indispensable to provide adequate and necessary facilities and to expand the airport as needed.

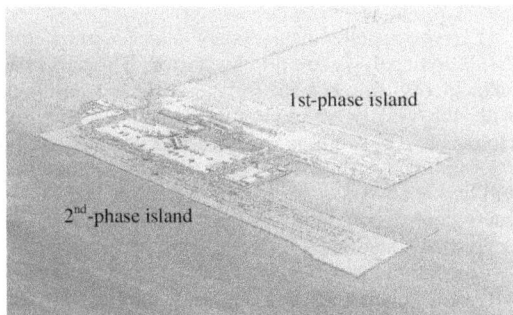

Figure 1. Artist's rendering of Kansai International Airport at the completion of the 2nd-phase construction

From such a background, in 1999, Kansai International Airport Company (KIAC) and Kansai International Airport Land Development Company (KALD) jointly commenced the 2nd-phase airport development project, aiming towards operation of the 2nd-ruway in 2007. This paper describes settlement analysis and intelligent site management that are applied in the 2nd-phase land reclamation works.

1.2 *Comparison between the 1st and the 2nd phase construction*

The 2nd-phase island is expected to settle more than the 1st phase because ground conditions underneath the seabed of the reclamation site feature thicker layers of compressible clay in addition to a greater water depth. Despite under such severe conditions, extraordinary large-scale construction must be executed in a limited period of time. KALD makes every effort to satisfy necessary conditions as an airport island with more speed, reliability, and environmental harmony. A comparison between the 1st phase and the 2nd phase is shown in Table 1.

1.3 *Construction schedule*

2nd-phase airport construction commenced in July of 1999. First, ground improvement works for Holocene clay layer by a sand drain method were carried out throughout the entire seawall area and then rec-

Table 1. Comparison of scale and natural conditions

	Scale			Natural conditions	
	Reclamation Area	Volume	Seawall length	Water depth	Thickness of compressible clay
1st phase	510ha	180 Mm³	11km	18m	150 - 200m
2nd phase	545ha	250 Mm³	13km	20m	250 - 300m

(M = million)

lamation area. In December of 1999, the seawall construction started.

In November of 2001, outer parts of the seawall have been nearly completed, thus enabling direct dumping of sand and gravel into the reclamation area. Earth-heaping by a reclaimer barge started in May of 2002, and the reclamation for the main area is scheduled to be completed by the middle of 2005. A bar chart for the construction schedule of the 2nd-phase airport development project is shown in Table 2.

Table 2. Construction schedule of the 2nd-phase airport development

Year	1999 2000 2001 2002 2003 2004 2005 2006 2007
Sand drain	▬▬▬▬
Seawall	▬▬▬▬▬
Reclamation	▬▬▬▬▬
Airport facilities	▬▬▬

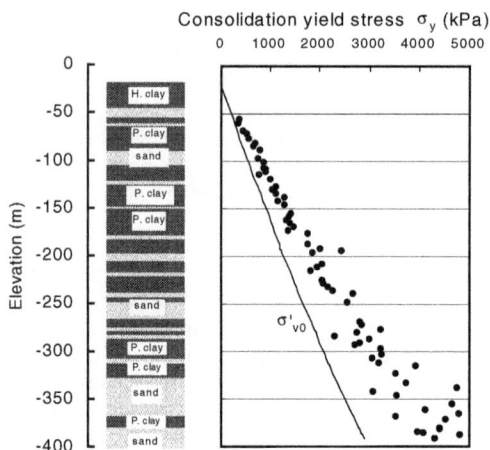

Figure 2. Representative soil profile at the 2nd-phase site (Tsuchida et al., 2002)

Table 3. Soil properties of Pleistocene clay (Tsuchida et al., 2002)

Natural water Content, (%)	Liquid limit, (%)	Plastic limit, (%)	Specific gravity, G_s
53.8~57.8	73.6~108	31.1~36.6	2.69~2.72

2 GEOTECHNICAL CONDITIONS AND CONSTRUCTION WORKS

2.1 Geotechnical conditions

The ground under the seabed of the 2nd-phase airport island consists of a soft layer of Holocene clay and alternate layers of Pleistocene clay and sand. Average thickness of Holocene clay layer is 25m, while total thickness of compressible Pleistocene clay and sand layers amounts to 250 to 300m. Pleistocene clay deposited from 2 million to 10 thousand years ago during repeated glacial and interglacial periods.

Prior to commencing the 1st-phase island construction, a total of 65 drilling surveys including two of 400m depths were conducted. In addition, four drilling surveys of 400m depths were done before the 2nd-phase construction. A representative soil profile with consolidation yield stress is shown in Figure 2, and soil properties of Pleistocene clay of 150-200 m depth below sea level are shown in Table 3 (Tsuchida et al., 2002).

2.2 Ground improvement

In the 2nd-phase construction as well as in the 1st-phase, a sand drain method is applied for most parts of the seawall construction area and all of the reclamation area. More than 1.2 million sand piles of 40-cm diameter are driven into Holocene clay with the average thickness of 25m. Spacing of the sand piles is set at 2.5m × 2.5m square grid, excluding the seawall construction area with spacing of 1.6m × 2.5m rectangular grid as is shown in Figure 3.

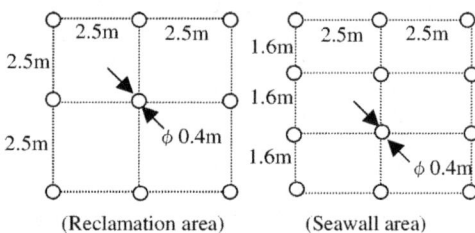

Figure 3. Spacing of vertical drain sand piles

The two spacing distances are determined by consolidation analysis solution by Barron. The narrower spacing of 1.6m × 2.5m for the seawall area is set with the aim of achieving early strengthening of Holocene clay with a more accelerated degree of consolidation. According to a finite-element consolidation analysis, the spacing of 1.6m × 2.5m is equivalent to a spacing of 2.0m × 2.0m in terms of degree of consolidation (Kobayashi, 1976). Considering continuous and seamless placing of sand piles between reclamation area and seawall area, the rectangular spacing is applied for the seawall area (Maeda, 1989).

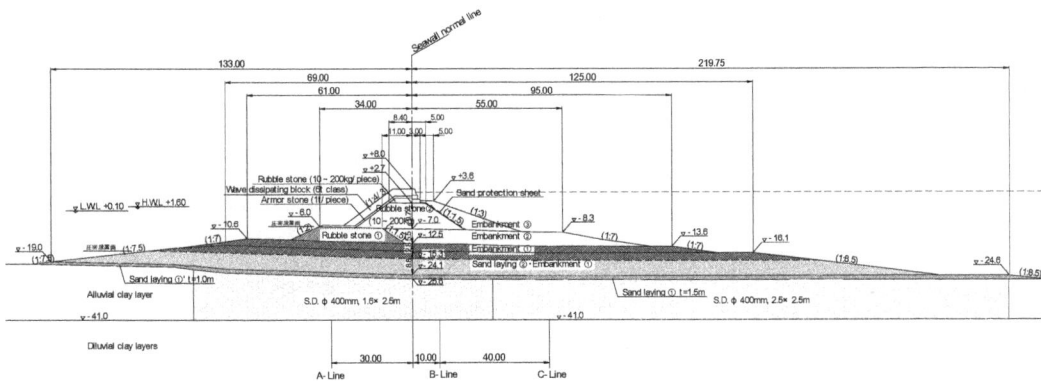

Figure 4. Typical cross section of rubble-mound type seawall

2.3 Seawall construction

In the 2nd-phase construction as well as the 1st-phase, more than 90% of 13-km-long seawall has rubble-mound type. A typical cross section of the rubble-mound type seawall is shown in Figure 4. Other parts of seawall have different types such as upright wave-dissipating caisson and steel cellular bulkhead, depending on usage of the seawalls.

The gently sloping rubble-mound type is adopted because of its structural flexibility to uneven settlement of the soft ground, relative low-cost advantages, and environmental friendliness to surrounding marine habitats.

2.4 Reclamation works

Reclamation works start in full swing after most parts of seawall construction has been completed. Reclamation process, as is shown in Figure 5, starts with sand and gravel dumping by a hopper-barge on a sea-sand blanket of 1.5m thickness, which covers and protects the sand piles. After the soil dumping piles up to 3m under the sea level, placement of sand and gravel above water is executed by a reclaimer-barge with a belt-conveyer, heaping earth up to ground level of 10m above sea level. The final earth-moving work involves reclaimer barges, large-scale dump trucks, bulldozers and vibratory rollers, which apply roller compaction to heaped soil.

The total thickness of layers of soil from the seabed up to the final ground surface amounts to 40 to 43m, including the sand blanket.

3 SETTLEMENT ANALYSIS OF HOLOCENE CLAY

3.1 Settlement measurement

It is imperative for the large-scale, rapid reclamation works on the soft grounds to precisely monitor the

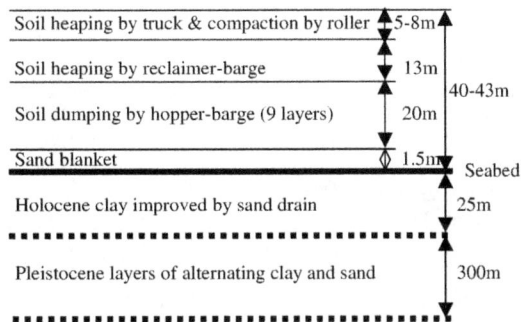

Figure 5. Profile of reclamation layers and seabed grounds

settlement of Holocene clay layer during construction in a consistent way. In order to efficiently measure settlement data for seawall construction and reclamation, two types of settlement measuring device are placed in the 2nd-phase construction site: a settlement plate and a hydraulic pressure gauge with a magnetic transmitter.

Settlement measuring by a settlement plate is one of the most basic settlement management methods. A typical settlement plate made of steel is shaped like upside-down T figure. It consists of a bottom plate and plural upright cylinders used for testing soil strengths of improved grounds by boring. A settlement plate is placed on the surface of sand blanket (see Figure 5) right after driving sand drain piles. Ground settlement caused by weight of embankment and reclamation of soil is measured by gauging water depth of the top of the cylinder. This method has many previous experiences, and it enables accurate measurement, but has some disadvantages: A bad weather sometimes prevents weekly measurement of settlement because the settlement-measuring work depends on a diver who puts down a handy water pressure gauge on the top of the cylinder. Also the settlement plate itself may become an obstacle against navigation of construction work vessels such as a hopper barge.

The other type of settlement-measuring device is a hydraulic pressure gauge with a magnetic transmitter (hereinafter referred to as magnetic settlement device), which is placed on the seabed mainly in the reclamation area. This settlement-monitoring method is also based on measuring the increase of water pressure due to the settlement of grounds. This device detects water pressure every two hours and transforms it to settlement data automatically. Then the device transmits data via a magnetic transmitter to a boat equipped with a magnetic data receiver as is shown in Figure 6. Significant merits of this method are: the device can be placed anywhere; it is not an obstacle to construction work vessels; and it makes possible continuous, unhindered settlement measurement.

The seawall construction area has 34 settlement plates alongside the 13km-long seawall. The construction area is divided into 34 site management blocks, each one having one settlement plate. A settlement plate equipped with borehole-cylinders is installed at 19 blocks, and a settlement plate without a borehole-cylinder is installed at another 15 blocks. In the reclamation area there are 37 magnetic settlement devices and 17 settlement plates. Magnetic settlement devices are placed at a rectangular grid of 350m × 250m.

These two settlement-measurement devices are to monitor the combined total settlement amount of Holocene clay and Pleistocene clay layers. In order to distinguish the settlement amount of Holocene clay from that of Pleistocene clay layers, there exists another type of measuring device: an anchor-rod type settlement gauge made of steel rod having a rectangular cross section. The bottom end of a rod is placed upon top of the uppermost Pleistocene clay layer so that the total settlement amount of Pleistocene clay layers can be separately measured. Two anchor-rod type settlement gauges are placed in the seawall construction area, and 7 gauges in the reclamation area.

Measuring settlement data in such a multiplex way enables data verification by comparing those data with each other. In addition, this multiplex system of measurement redundancy prevents lack of measurement due to an unexpected accident or a device failure (Shinohara et al., 2002).

3.2 Settlement prediction

Settlement prediction of Holocene clay is based on two methods: One is "M_v method" for the site management purpose, comparing actual settlement data monitored during construction with predicted settlement calculation; And the other is elasto-visco plastic analysis by a finite-element method. In this section, M_v method is to be explained, and the finite-element analysis will be mentioned later in Section 4. M_v method is adopted mainly for site management purposes because it is relatively ease to use and it is built into a set of computational programs, "Settlement-stability analysis system of Holocene clay", developed by KIAC for consultants and contractors to utilize. In M_v method, S_t, settlement amount at time t, is calculated by the following formula:

$$S_t = M_v \, p \, H \, U_t \qquad (1)$$

where, M_v is a coefficient of volume compressibility, p is a consolidation pressure, H is thickness of Holocene clay layer, *and* U_t is a degree of consolidation at time t, which is calculated by Barron's formula, using C_v, a coefficient of consolidation determined by consolidation tests. It is practically assumed that C_v (vertical coefficient) of the Holocene clay at the construction site is equal to C_h (horizontal coefficient) (Maeda, 1989). M_v is also determined by consolidation tests. These test results for M_v and C_v are shown in Figure 7 and 8, respectively.

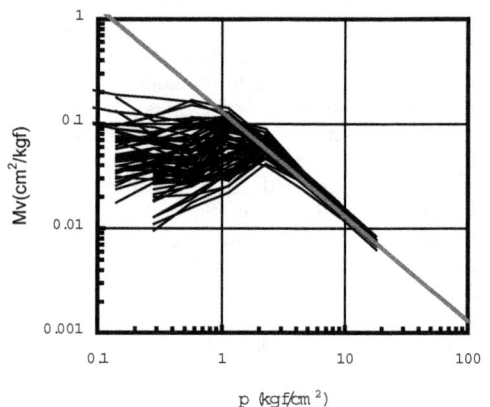

Figure 7. Consolidation tests for M_v and p

From these test results, M_v and C_v are set as the following:

$$M_v = 0.13 \, p^{-1.0} \ \text{cm}^2/\text{kgf} \qquad (2)$$

$$C_v = 90 \ \text{cm}^2/\text{day} \qquad (3)$$

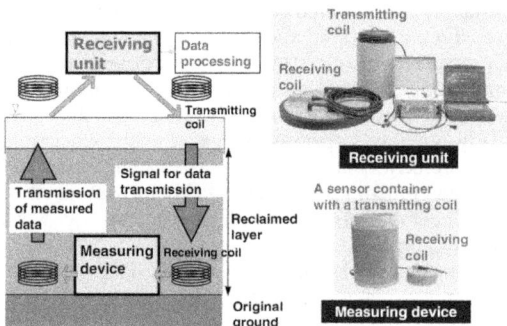

Figure 6. Hidraulic pressure gauge with a magnetic transmitter

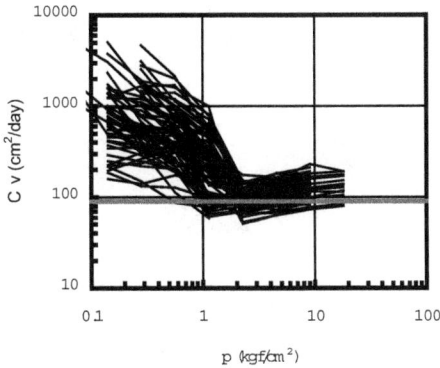

Figure 8. Consolidation tests for C_v and p

In order to determine input data of an incremental consolidation pressure, p, by each loading step of seawall construction for the Mv-method computational program, a cross-sectional graph of construction schedules and a crest height sequential graph are drawn as in Figure 9. The left figure shows cross sectional sequential data of seawall construction by bathymetric sounding of a seawall construction site. The right figure shows a crest height sequential graph plotted with elapsed time from the start of construction.

On the crest height sequential graph the day of each loading step for the settlement calculation is set at the middle point of each step. We calculate settlement curve from the day of the loading step for management of reclamation thickness, settlement, and stability.

Figure 10 shows observed settlement data plots compared with calculated settlement curves by Mv method for a seawall site management block. One seawall site management block has one set of three settlement plates: A, B, C - line in the legend represents the particular location of the three settlement plates as shown in Figure 4. B - line is placed near the seawall normal line, which has heavier loadings than A and C - line. Around 350 days after starting seawall construction, the three settlement measuring points show approximately 3m settlement. For the

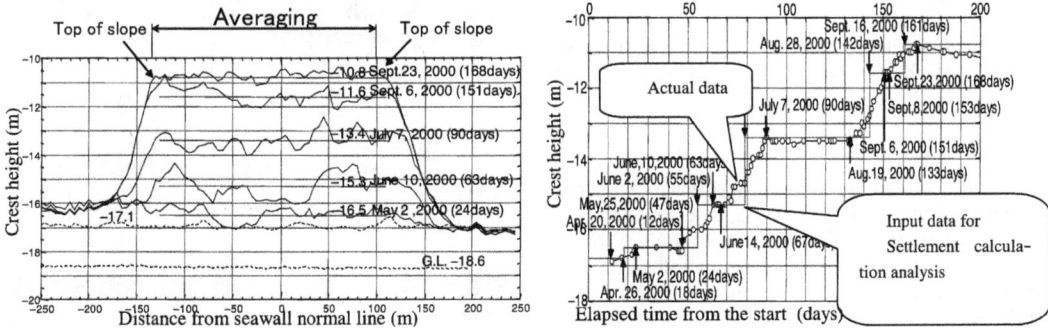

Figure 9. A cross-sectional view graph by bathymetric survey and a crest height sequential graph of seawall construction

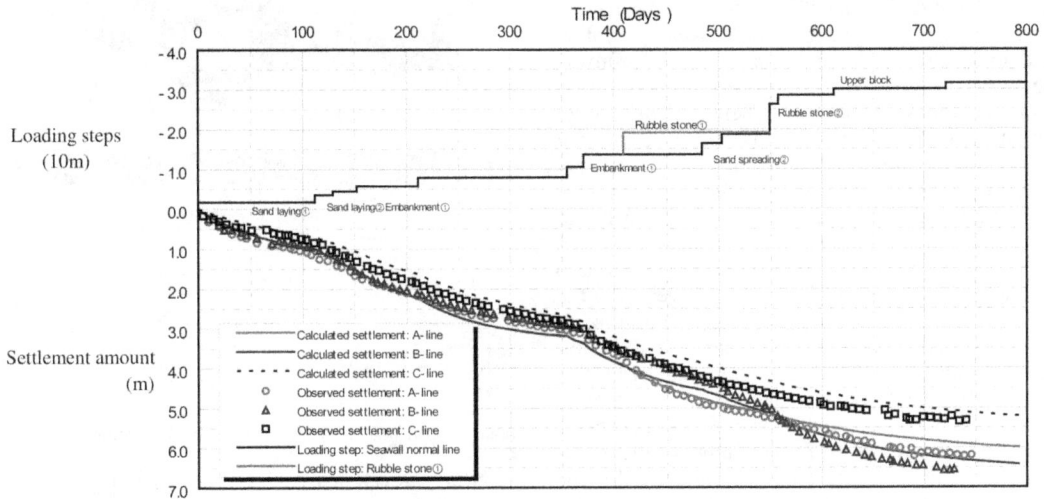

Figure 10. Observed settlement data plots compared with calculated settlement curves

163

loading step of rubble stone①, it shows more than 4m settlement amount. It is predicted that the final amount of settlement of Holocene clay for the B-line will reach 7m on average for 25m thickness of the clay layer after 100% degree of consolidation.

As an indicator of evaluating and analyzing observed settlement measurement data, we apply "settlement ratio": That is the ratio of the calculated settlement amount divided by the observed settlement amount in a specific time period. When noticeable discrepancies appear between observed settlement and calculated settlement, we may slightly modify the M_v and C_v so that the calculated settlement curve better fit in with the observed settlement data. If abnormal settlement behavior is detected, we look into the cause of the abnormal ground behavior by conducting in-situ tests such as cone penetration, and take appropriate measures immediately.

In most cases, the observed settlement data are plotted within a predicted range of calculated settlement data because of high accuracy of prediction supported by intelligent site management, to be described in the next section. In some cases, though, calculation results do not seem to fit in well with observed data. Therefore, it is important that settlement management should be carried out with careful consideration to varied conditions and circumstances of soil properties, loading steps, etc.

3.3 Intelligent site management

Considering construction planning of various airport facilities on the reclaimed land, the reclamation work should be executed, giving special attention to predicting residual settlement amount after completion of reclamation, which may cause damage to airport facilities such as aprons and terminal buildings. To cope with this challenge, KALD maps out plans for reclamation development processes with deliberate sequential data of landfill works in order to predict the residual settlement amount, which is calculated to reach more than 5m after the 2nd runway opening.

In this context, acquiring sequential data of fluctuation on settlement trend and exact location and volume of soil dumping and heaping should be very important information for the site management. One of the technical challenges for the 2nd-phase island construction is to conduct the land reclamation process as evenly and uniformly as possible, taking into consideration spatial distribution of the weight of the dumped soil on the seabed, and thereby minimizing the uneven settlement that will occur with the completed airport island. While the similar care was also taken during the 1st-phase reclamation, recent rapid advances in measuring and information technologies have made it possible to further improve the evenness and uniformity of the land reclamation for the 2nd-phase construction (Hirose, 2001).

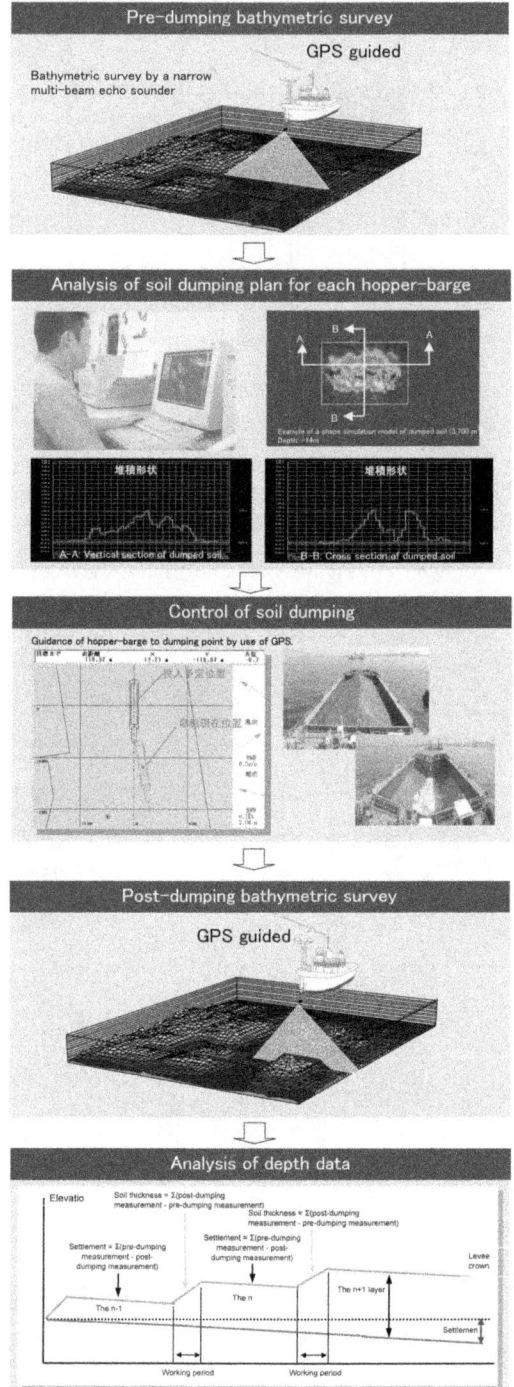

Figure 11. Site management system for soil dumping

Newly developed "VS10" system is installed on all hopper-barges for land reclamation operation in the 2nd-phase construction. The VS10 system manages all data, including each barge's name, soil dumping position, soil dumping volume and soil source location data, in a single database.

In addition, a narrow multi-beam echo sounder is used to monitor gradual buildup of dumped soil layers on the seabed. Data obtained from this surveying system is combined with those generated by the VS10 to provide all information resources necessary for effectively managing the land reclamation tasks for each load carried by a hopper-barge.

This information database has made it possible to estimate how the buildup of reclaimed land will take shape according to each hopper-barge through preliminary dumping simulations. This makes the effect on uniformity to be confirmed in advance and allows adjustments in soil dumping position to be made accordingly. The overall thickness distribution based on the accumulation of dumped soil is analyzed by measuring depth contours using the narrow multi-beam echo sounder before and after soil is dumped from a barge.

For thickness analysis, differences obtained from these depth measurements before and after soil dumping can be calibrated using values from constant monitoring of settlement obtained from magnetic settlement devices. Accuracy of this layer-thickness analyzing system has been verified through results of preceding construction of the seawall and initial land reclamation works by soil dumping with hopper-barges, thus enabling us to create a standardized site management system (see Figure 11).

4 SETTLEMENT ANALYSIS OF PLEISTOCENE CLAY

4.1 Settlement measurement

It is imperative to monitor settlement amounts of Pleistocene clay layers and pore water pressures of Pleistocene sand and clay layers in order to predict the final settlement amount as accurately as possible. Two "offshore oil-rig" type platforms are placed in the 2nd-phase reclamation area in order to install various measurement devices into Pleistocene layers deep down to 350m. The measurement devices installed in Pleistocene layers in the 2nd-phase construction site include: an anchor-rod type settlement gauge, a differential settlement gauge, and a pore water pressure gauge (see Figure 12).

An anchor-rod type settlement gauge made of long steel pipe is installed in Ma2 (clay) and Ds10 (sand) stratum in order to measure settlement amount of deepest compressible clay layers. Settlement is measured by monitoring height of the upper end of the steel pipe. An inclinometer is used to compensate for inclination of the pipe.

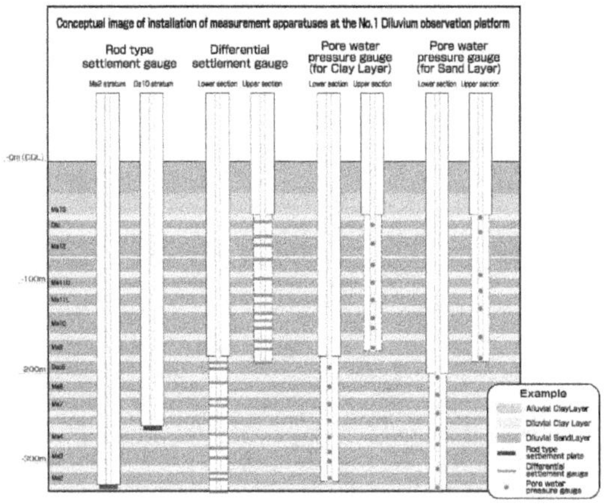

Figure 12. Measurement devices for Pleistocene layers

A differential settlement gauge is installed in every clay layers from the lowest Ma2 stratum to Dtc stratum, the uppermost Pleistocene clay layer. In a thicker clay stratum such as Ma 10, there are two or more gauges installed at the different level of the same stratum in order to measure the difference of degree of consolidation within the same clay stratum. A mechanism of the differential settlement gauge is shown in Figure 13. Data obtained from this device is cross-checked with settlement data obtained by the anchor-rod type settlement gauge.

A pore water pressure gauge is installed in every clay and sand layers. This gauge measures a pore water pressure by finding the particular pressure value of gas in a pressure line that balances out with the water pressure around a measuring device (see Figure 14). Excess pore water pressure can be known from by comparing hydrostatic pressure with the pore water pressure.

The above-said installation methods and measuring systems of these devices have been greatly improved based on the past experiences of the settlement monitoring of the 1st-phase island.

4.2 Settlement prediction

Prediction of final settlement amount of clay layers of both Pleistocene and Holocene at the 2nd-phase reclamation site is based on "FCAP" program. FCAP (Finite-element Consolidation Analysis Program) is composed of a one-dimensional finite-element consolidation analysis model and a separate two-dimensional seepage analysis. The computational model was developed by KIAC and geotechnical consultants in close collaboration with a research group at Port and Harbour Research Institute and later it has been revised and refined with academic cooperation from distinguished scholars of an investigation committee for settlement prediction of the 2nd-phase island, considering observed data of actual settlement and pore water pressure at the 1st-

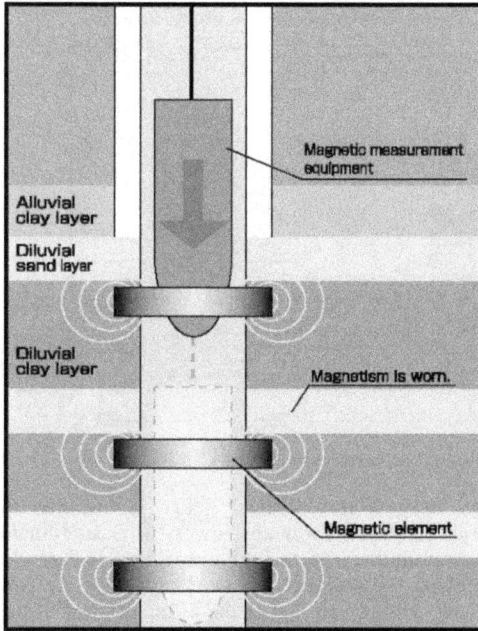

Figure 13. Mechanism of a differential settlement gauge

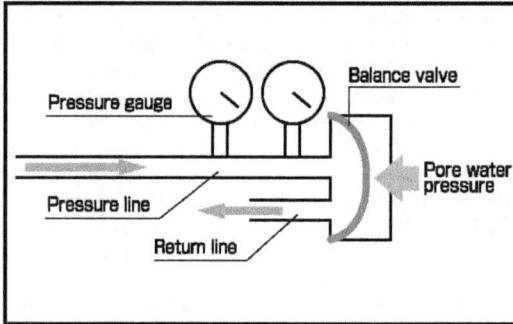

Figure 14. Mechanism of a pore water pressure gauge

phase airport island. The fundamental consolidation equations of FCAP program are based on the following:

$$\varepsilon = \varepsilon_e + \varepsilon_p \qquad (4)$$

$$\varepsilon_e = \kappa/(1 + e_0) \ln(p/p_0) \qquad (5)$$

$$\varepsilon_p = \lambda/(1 + e_0) \ln(p/p_c) \qquad (6)$$

where ε is total strain, ε_e is elastic strain, ε_p is plastic strain, κ is swelling index, λ is compression index, e_0 is initial void ratio, p_0 is overburden pressure, and p_c is consolidation yield stress. κ and λ are determined by e-log p curves (see Figure 15) for each clay layers based on various types of consolidation tests. The equations are solved by a one-dimensional finite-element analysis with the boundary conditions

of pore water pressures of each Pleistocene sand layers that are calculated by a separate two-dimensional underground water seepage analysis.

In order to account for secondary consolidation due to apparent viscous behaviors (time-dependency) of deeper Pleistocene clays, long-term settlement amount of deep clay layers such as Ma2 and Ma3 are calculated in the following equation:

$$\varepsilon = \kappa/(1 + e_0) \ln(p/p_0) + \alpha \ln(t/t_0) \qquad (7)$$

where α is secondary consolidation coefficient, and t_0 is initial loading date set by an inverse analysis, considering the measured consolidation data of deep clay layers at the 1st-phase island. In this sense, FCAP method is not a genuinely elasto-viscoplastic analysis, but an elasto-plastic analysis complemented with a simplified viscous model.

Soil parameters of FCAP model are set in accordance with various types of soil testing results of numerous undisturbed samples, considering the actual settlement amount and measured pore water pressure data of Pleistocene layers under the 1st-phase airport island. Figure 15 and 16 shows consolidation test results for Ma12 and Ma7 clays, respectively. FCAP predicts 8m settlement amount for Holocene clay and 10m for Pleistocene clay layers on average for the 2nd-phase airport island 50 years after the 2nd-runway opening. Figure 17 shows observed settlement data of a magnetic settlement device compared with a predicted settlement curve. This settlement device is placed within connecting taxiway area in the 2nd-phase airport island. At this location total settlement as of September 2002 amounts to 7m, 1.5m of which are caused by Pleistocene clay consolidation.

As is earlier explained, FCAP model may deserve some refinement regarding theoretical treatment of secondary consolidation behavior and pore water seepage analysis of sand layers, which is conducted independently from consolidation calculation of clay layers. In this context, a new program called KCAP (KALD-Kobayashi Consolidation Analysis Program) is now being developed by KALD and Kobayashi (Kobayashi, 1982) in order to establish theoretical authenticity while pursuing a user-friendlier interface than FCAP. In KCAP, which is a one-dimensional finite-element elasto-viscoplastic soil-water coupled analysis program, will be incorporated peculiar and complex characteristics of Pleistocene clays such as high compressibility around a consolidation yield stress and long-term secondary consolidation behaviour.

As a joint-research project with KALD, the research group in Port and Airport Research Institute has been conducting intensive studies on mechanical properties of aged clays having structures formed during the process of long-term sedimentation and consolidation (Tsuchida et al., 2002).

166

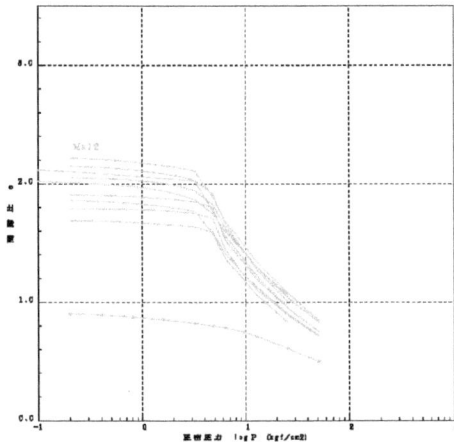

Figure 15. *e*-log *p* curve for Ma12 clay

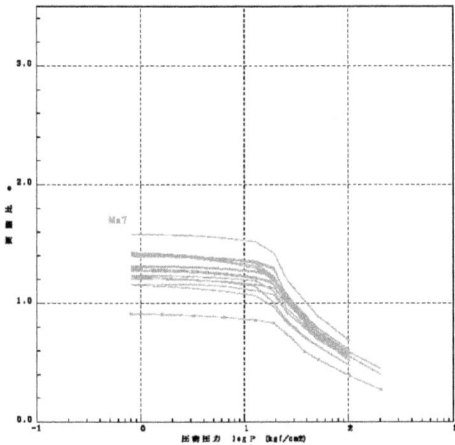

Figure 16. *e*-log *p* curve for Ma7 clay

STWE

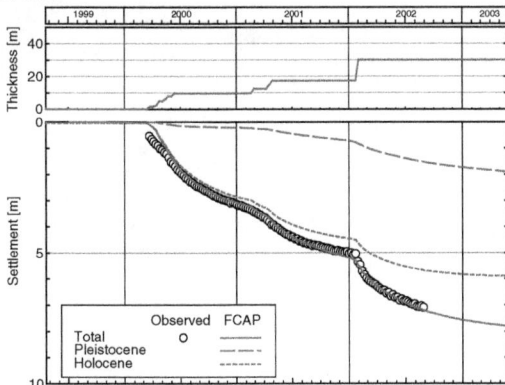

Figure 17. Observed vs. FCAP-calculated settlement data

5 CONCLUSION

In this large-scale reclamation project with extraordinary amount of ground settlement, it is an absolute necessity to establish an intelligent site management system that entails efficient reclamation planning, agile and flexible modifications of construction schedules, and information-based surveying and monitoring technology. It is also imperative to constantly re-examine prediction of ground settlement based on quasi-real-time feedbacks from observed data during reclamation processes.

The prediction of settlement in the 2nd-phase island has been carried out based on the experiences of the 1st- phase, and the average predicted total settlement in the reclaimed island amounts to18m, which is probably the largest consolidation settlement geotechnical engineers in the world have ever experienced. Accurate prediction of long-term settlement amount for Pleistocene clay layers needs a one-dimensional finite-element analysis of consolidation processes in conjunction with two-dimensional seepage analysis of pore water pressure in Pleistocene sand layers. KIAC has already established FCAP model as a reliable prediction method for consolidation of Pleistocene clay layers. We now endeavor to pursue a more theoretically refined computational model called KCAP.

REFERENCES

Hirose, M. (2001):"Second phase airport island construction of Kansai International Airport", *Proceedings of International Airport Symposium 2001*

Kobayashi, M. (1976):"Analysis of consolidation problems by a finite-element method", *Technical Note of Port and Harbour Research Institute*, No.247. (in Japanese)

Kobayashi, M. (1982):"Numerical analysis of one-dimensional consolidation problems", *Report of the Port and Harbour Research Institute*, Vol.21, No.1. (in Japanese)

Maeda, S. (1989):"Research on settlement and stability management system of soft grounds improved by sand drain for a large-scale offshore artificial island", Thesis of Dr. Eng. (in Japanese)

Shinohara, M. and Oyaizu, D. (2002):"Prediction of ground settlement and site management of ground improvement works for the 2nd-phase project of Kansai International Airport", *Proceedings of the 4th International Conference on Ground Improvement Techniques*, Vol.2.

Tsuchida, T., Watabe, Y., and Kang, M. (2002):"Evaluation of structure and mechanical properties of Pleistocene clay in Osaka Bay", *Report of Port and Airport Research Institute*, Vol.41, No.2.

Soft Ground Engineering in Coastal Areas, Tsuchida et al. (eds)
© 2003 Swets & Zeitlinger, Lisse, ISBN 90 5809 613 0

Selection of soil improvement methods for reclaimed land and actual examples

T. Shigeno, S. Honda & N. Koushige
Nikken Soil Research Co. Ltd., Osaka, Japan

M. Fukui
Otemae University, Itami, Japan

ABSTRACT: In this paper, soil improvement work examples will be taken up including such large-scale pre-loading and sand drain as performed in the Kobe Port Island and Rokko Island for the purpose of accelerating consolidation settlement of alluvial clay layers in the construction of highrise buildings, large-scale apartment buildings, detached houses, warehouses and civil work structures (elevated railroad) and, our approaches to the engineering and installation work in such soil improvement efforts will be discussed. Further, discussions will be made concerning ground surface settlement and alluvial clay settlement behaviors, etc. as measured over a long period of time since the latter half of the year 1970. Furthermore, the recognized building damages inflicted by the Hyogoken-Nambu Earthquake, which took place in 1995, and the effects of soil improvement installed in the affected areas will be described and introduced as exemplary case histories of foundation structural engineering and soil improvement in reclaimed land.

1 INTRODUCTION

There are a number of large-scale reclaimed land in the Osaka Bay coastal areas represented by Kobe Port Island, Rokko Island, Osaka Nanko and Kansai International Airport, etc. At the present time, there are many buildings constructed within the sites functioning as port and harbor or urban facilities on such reclaimed land, and they are connected to each other by the Osaka Bay Coastal Road.

Those structures suffered heavy and unprecedented damages when the Hyogoken-Nambu Earthquake took place on 17 January 1995. Especially in the coastal reclamation land composed of soft foundation soil, sand boil due to soil liquefaction was recognized at various points which badly damaged roads, buildings and port facilities. The investigations conducted after the earthquake revealed that such damages occurred mostly in areas with no soil improvement treatment applied. It deserves a special attention that damage by the earthquake was alleviated not only in areas where compaction type soil improvement had been applied for foundation soil compaction but also in areas where soil improvement had been applied for acceleration of alluvial clay consolidation. The conditions of damage, the measures taken thereafter, etc. have been already described in detail in the reports by related academic societies and research institutes.

In this paper, the problem of soft soil stabilization treatment which is one of important problems at off-

shore reclamation land will be taken up, and the author's perspectives on soil improvement cherished at the time they were involved in those representative buildings on Kobe Port Island and Rokko Island as well as the history of ground settlement there to date and the conditions after the earthquake will be described.

2 SELECTION OF SOIL IMPROVEMENT METHODS FOR OFF-SHORE RECLAMED LAND

2.1 *Building plan and soil investigation plan*

Technical problems in a reclaimed land can be generally classified into liquefaction of the fill material, consolidation settlement of the alluvial and diluvial clay layers and evaluation of the bearing capacities of load bearing soil layers, added, depending on the case, by dynamic problems or technical problems related to construction work such as excavation.

Fig. 1 summarizes technical problems and investigation subjects by foundation types. The building planning and soil investigation planning are performed simultaneously, but the soil investigation plan must be based upon the building foundation type. In the case of spread foundation and friction pile foundation, the bearing capacity and settlement of the fill layer must be studied, and in the case of bearing pile foundation, comprehensive studies must be made involving estimation of the settlement above the load bearing soil layer, forecast of negative friction generation/non-generation and estimation of the settlement of the load bearing soil layer and layers below, etc. and etc. Since 1965, damages to buildings in areas suffering large ground settlement have been reported one after another. They were due to differential settlement by negative friction acting upon bearing piles and relative draw-up of buildings. Thereafter, researches and experiments were repeated and as the result, in many cases in Kobe Port Island and Rokko Island, Belled out piles are used and an asphalt compound is applied to the upper portions of piles.

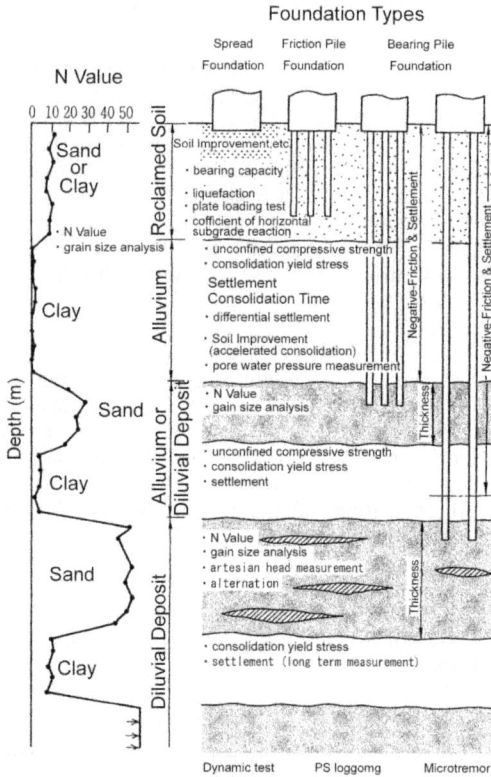

Figure 1. Foundation types and soil investigation study subjects

2.2 Foundation construction methods and soil improvement methods for reclaimed land

Building foundations on reclaimed land can be roughly classified into the following 3 types:
1. Bearing pile foundation
2. Friction pile foundation
3. Spread foundation

In the case of any foundation type, it is desirable to decrease the absolute settlement value by providing basements, etc. to keep balance between the building weight and the displaced soil weight so that the building weight does not result in an additional

load. In the case of bearing pile foundation, it is necessary to study negative friction. Tab. 1 shows foundation construction methods for reclaimed land areas and countermeasures. Fig. 2 shows outline drawings of foundation construction methods.

Generally speaking, reclaimed land is a newly-created soil foundation so that the soil compaction is loose and the bearing capacity is supposed to lack consistency. Therefore, as countermeasures against ground settlement and soil liquefaction, soil improvement for the purpose of increasing the N-value of the reclamation soil layer and soil improvement for accelerating the consolidation settlement of the alluvial clay layer are conceivable. Tab. 2 and Fig. 3 show the outlines of representative soil improvement methods.

Table 1. Foundation construction methods for reclaimed land

Method	Pile Type	Negative-Friction Countermeasures	
Bearing Pile Foundation	Precast Pile	Asphalt Slip Layer	Pile Group
	Cast-in-place Concrete Pile	Bell-pile(with Asphalt Slip Layer)	
Spread Foundation	Precast Pile	·Improved Reclaimed Soil ·Buiding Jack-up	
Friction Pile Foundation		·Improved Reclaimed Soil ·Buiding Jack-up	

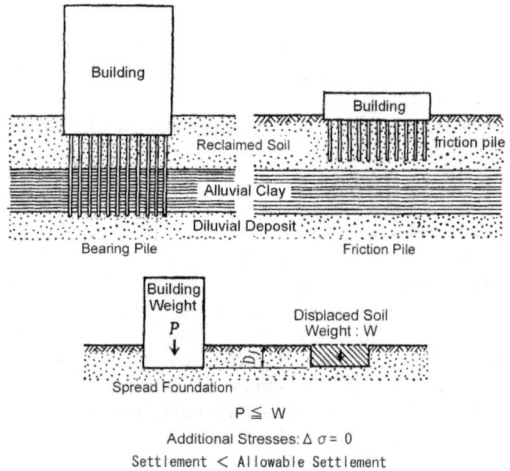

$$P \leq W$$
Additional Stresses: $\Delta \sigma = 0$
Settlement < Allowable Settlement

Figure 2. Outlines of foundation construction methods

Table 2. Soil improvement methods

Soil Improvement Methods	Reclaimed Soil	Alluvial Clay	Diluvial Clay
Vibration Compaction Method (Sand Compaction Pile Method)	○		
Sand Drain Method (with Preloading)	○	○	
Preloading Method	○	○	○

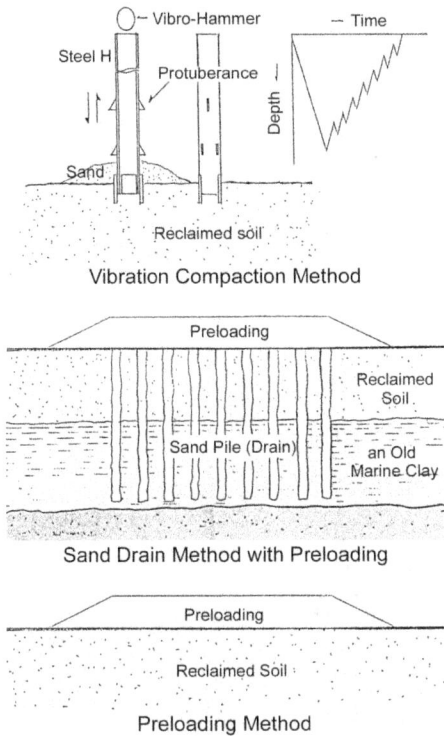

Vibration Compaction Method

Sand Drain Method with Preloading

Preloading Method

Figure 3. Outlines of soil improvement methods

3 KOBE PORT ISLAND

3.1 *Soil characteristics and foundation construction methods*

Kobe Port Island is a man-made island of 436ha in area whose reclamation work started in 1966 and ended in 1978. The fill material was mostly Masa-do which is soil of good quality from the Rokko mountain system placed to a thickness of approximately 20m underlain by an alluvial clay layer of 10m to 15m in thickness and a sandy soil and viscous soil alternate layer about 30m thick and thereunder by a diluvial clay soil formation about 20m thick. The perimeter sections filled earlier accommodate the port and harbor facility buildings for which steel pipe bearing piles were used. Within the inner area, houses, hospitals, hotels, schools, etc. were built. Pre-cast reinforced concrete bearing piles were used to support commercial buildings. As more buildings were built, the noise and vibration from pile driving operation created problems, and thereafter cast-in-place concrete piles were adopted.

In the case of bearing pile foundation, countermeasures were indispensable to prevent relative draw-up of piles by soil settlement as well as differential settlement by negative friction of piles. The load bearing soil layer in this case was an alternate layer so that securing the pile-driving penetration refusal and the load bearing capacity presented special problems. In some cases, friction pile foundation or spread foundation was suitable rather than bearing piles depending on the scale of the building, and such type foundations were used more in later days.

3.2 *Damages by the earthquake*

Fig. 4 shows the sand boil by soil liquefaction caused by the Hyogoken-Nambu Earthquake and the soil improvement areas on Kobe Port Island. Sand boil by soil liquefaction could hardly be seen in the areas where soil improvement had been given to prove the effect of soil liquefaction restriction by soil improvement. Especially, the liquefaction restriction effect of compaction type soil improvement was apparent. In the following section onward, the representative structures with which the Authors of this paper were concerned, as shown in Fig. 4, will be taken up as examples, and the outlines of soil improvement and the behaviors of foundation soil relating to said structures before and after the earthquake will be described.

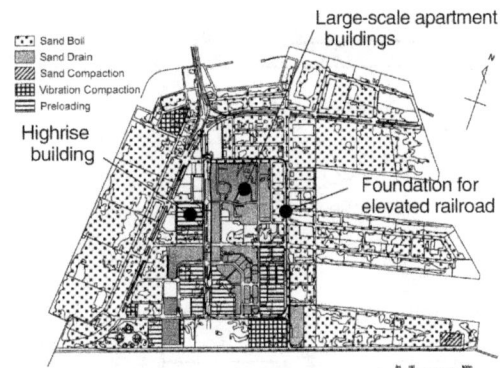

Figure 4. Sand boil and soil improvement areas in Kobe Port Island

3.3 *Large-scale apartment buildings*

In subject area, there are a few apartment buildings 7 to 14 stories high built on load bearing pile foundations and 2 other buildings supported on friction pile foundations. In order to accelerate the residual consolidation settlement of about 40cm of the alluvial clay layer distributed below the fill soil layer, soil improvement was performed by sand drains and preloading of approximately 6m in height. Fig. 5 shows the measurement results of the alluvial clay consolidation settlement by preloading. Sand pile of 50cm in diameter and 30m long were installed in square layout (2.00m×2.00m to 3.20m×3.20m) (Fig. 6 and Fig. 7). Sand drain, originally, is not a soil improvement technique to obtain a compaction effect, but the N-value of 10 to 40 before improvement increased to 14 to 50 after improvement (See Fig. 8).

171

It is known that slag had been dumped locally on the former sea bottom in the vicinity. It was therefore supposed that in order for a pile to penetrate the slag layer, a considerably large number of percussion was required (3,000 to 11,000 in fill layer in some extreme cases), and this fact could have contributed to the vibration compaction effect on the fill soil layer. The measurement of settlement was performed continuously for more than 19 years from 1977 on 2 pile-supported buildings, 2 friction pile supported buildings as well as foundation soil layer by layer. Fig. 9 shows an example of the time-serial changes of ground surface settlement and building subsidence. The total subsidence of pile-supported building was 81.2 to 93.9cm.

In addition to the above-mentioned measurement results, site investigations were made after the Hyogoken-Nambu Earthquake. Radical settlement by the earthquake could not be recognized involving pile-supported foundation buildings, but buildings on friction pile foundations, as well as the ground surface, subsided 5cm to 10cm due to the earthquake.

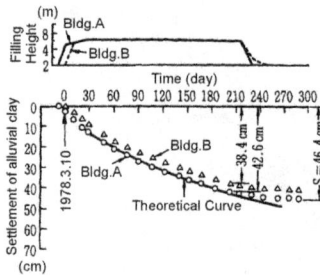

Figure 7. Outline of subject soil foundation and elevation of sand drain

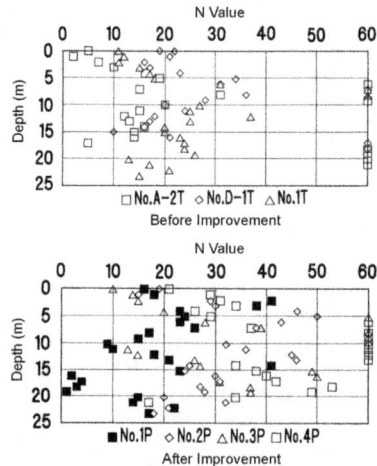

Figure 5. Settlement of alluvial clay by preloading

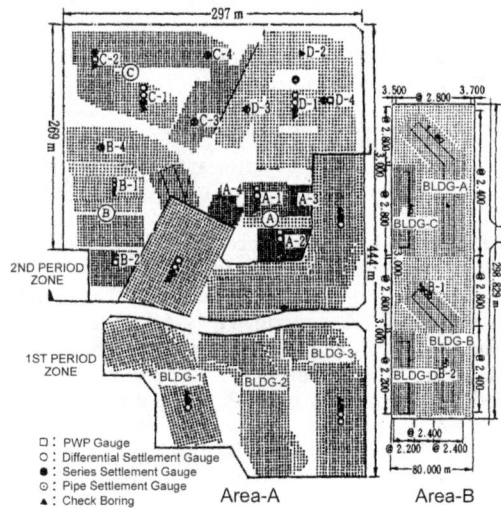

Figure 8. N-value changes before and after soil improvement

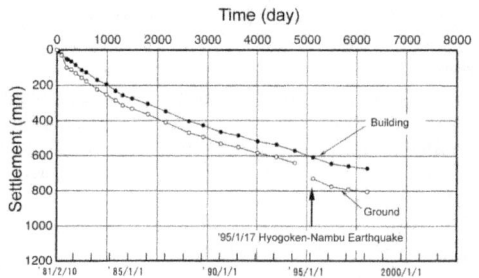

Figure 6. Sand drain layout drawing

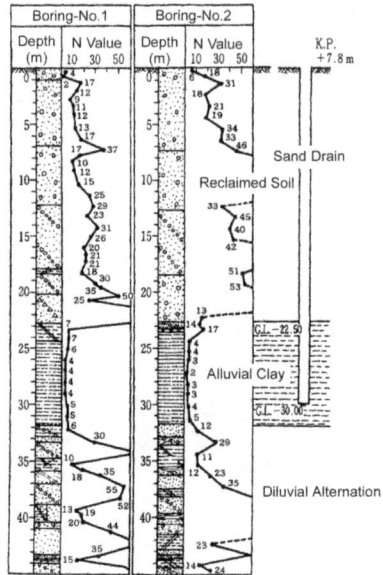

Figure 9. Subsidence of pile-supported buildings and ground surface

172

There were no large-scale sand boil caused by the earthquake, and the buildings did not seem to have been seriously damaged.

3.4 Highrise building

In subject area, the fill soil of approximately 20m in thickness is underlain by an alluvial clay layer of approximately 12m in thickness. In order to accelerate the consolidation of the alluvial clay deposit, a large-scale preloading was applied. Soil fill-up 10m high was placed on the site, size of 272m×167m, for about 2 years from October 1975 to March 1978 (Fig. 10, Pic. 1). Fig. 11 shows subject building and the soil layer section, and Fig. 12 shows the results of the consolidation settlement measurement of the preloading.

A highrise building, 1 story underground and 12 stories above-ground, was constructed after preloading removal. Its footprint area is approximately 80m×90m, building weight approximately 96,000 tons and soil displacement weight approximately 101,000 tons. The building is supported by load bearing pile foundations (Fig. 13).

Measurement of ground surface settlement and layer-by-layer settlement continued over the period of 17 years from February 1981. The measurement results are shown in Fig. 14. The building subsidence during said period was approximately 29.2cm. 14 years has passed since construction of this building, and the building encountered the Hyogoken-Nambu Earthquake. However, the building itself did not suffer rapid settlement or declination, and no damages were recognized. The ground surface settlement around the building was 22.1cm to 27.6cm

Picture 1. Large-scale preloading overall view

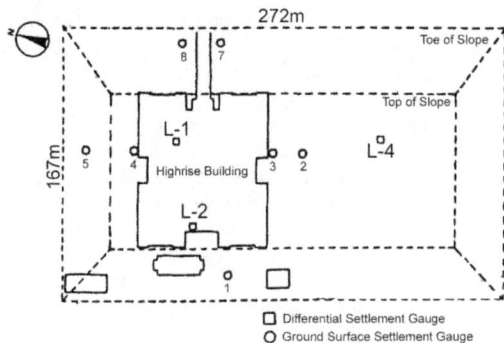

Figure 10. Preloading and settlement measurement points layout

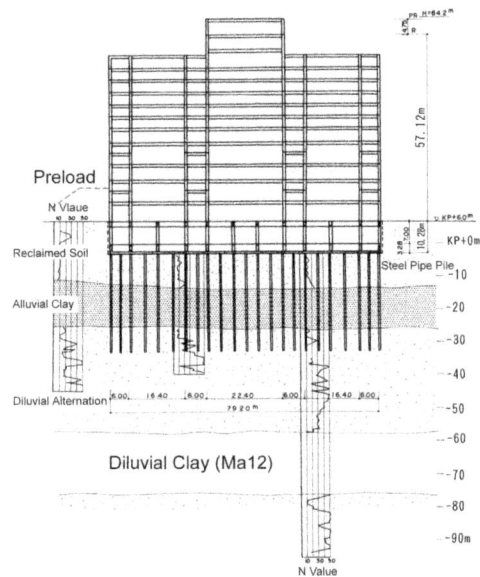

Figure 11. Building and soil layer sections

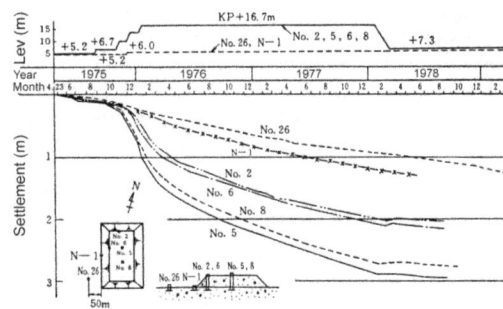

Figure 12. Settlement of alluvial clay by preloading

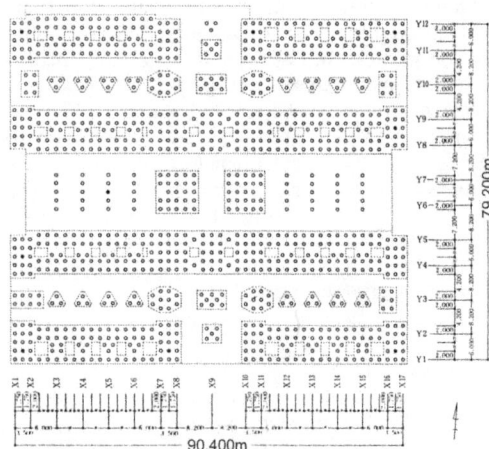

Figure 13. Pile layout for subject building

a) ground surface & Building Settlement

b) Differential Settlement

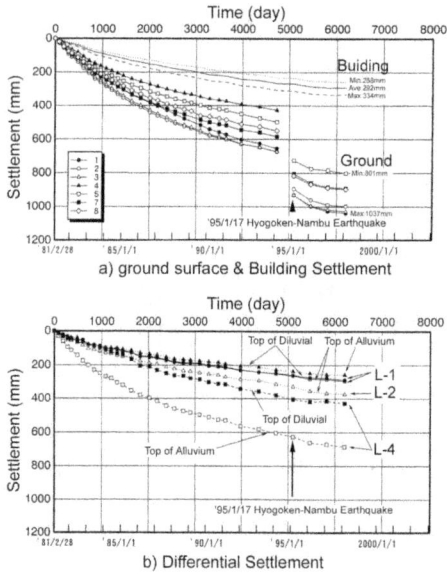

Figure 14. Actual settlement measurement at respective points

(average 24.5cm) and the ground settlement close to the building was 38cm. The reason why the ground settlement close to the building was larger is considered to lie in the underground backfill after excavation by the open cut method. The conditions around the building before and after the earthquake are shown in Pics. 2 to 5.

3.5 Foundation for elevated railroad

Subject structure is supported by load bearing pile foundations (φ900 steel pile with open tip, asphalt-treated) resting on the diluvial sand-gravel layer distributed about 30m deep and deeper. In consideration of the nature of the structure, level differences at ground surface would hardly cause functional problems so that no soil improvement to the alluvial clay was performed. Fig. 15 shows the settlement of the foundation footings of subject structure and the settlement curve of the ground surface close by. During the period of about 7.3 years from February 1981 to May 1988, the settlement of the ground surface was 86.3cm, but during the period of about 11.2 years from January 1980 to March 1991, the settlement of the foundation was 54.0cm. This difference of settlement may be attributable to the consolidation settlement of the alluvial clay.

4 ROKKO ISLAND

4.1 Soil characteristics and foundation construction method

Rokko Island is a man-made island of 580ha in total area. Its reclamation started in 1972 and was sub-

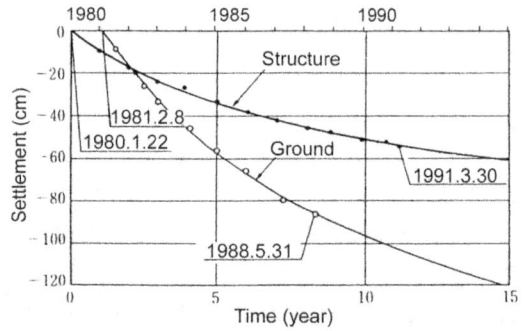

Figure 15. Settlement of ground surface and foundation for structure

Picture 2. to 5. Conditions around building before and after the earthquake

stantially completed in 1990. The stratiform structure of its foundation soil is almost identical to that of Kobe Port Island, except that the fill soil mainly consisted of fine quality mountain soil from the Rokko mountain system as mixed with part of the Kobe Group (tuff, sandstone, shale, etc.). The diluvial formation below the alluvial clay layer is composed of alternate layers, approximately 30m thick in all, of sandy clay and viscous soil and is a formation in which the load bearing pile refusal position is difficult to locate.

The port and harbor facilities in the island perimeter sections where reclamation was performed earlier than other sections were treated with a combination of soil improvement and friction pile foundations considering the experience at Kobe Port Island that both bearing pile foundation draw-up and differential settlement occurred there and the residual settlement of the alluvial clay layer was large there resulting in a general recognition that the bearing pile foundation was unsuitable. In the central area for urban function facilities, office buildings,

174

highrise apartment buildings, schools, detached houses, etc. were constructed. Various foundation construction methods were adopted depending on building scale, importance degree, economy, etc. As this man-made island was urbanized, noise and vibration created more problems so that more cast-in-place concrete piles were used. For the foundation soil at this location, the negative friction countermeasure method was adopted for the first time on cast-in-place concrete piles.

4.2 *Damages by the earthquake*

Fig. 16 shows the sand boil by soil liquefaction caused by the Hyogoken-Nambu Earthquake and soil improvement areas on Rokko Island. As in Kobe Port Island, the effect of the soil liquefaction restriction is recognized in the soil improvement areas of this island. In the following section onward, the representative structures with which the Authors of this paper were concerned, as in Fig. 16, will be taken up as examples, and the outlines of soil improvement treatments and the behaviors of their foundation soil before and after the earthquake will be described.

4.3 *Highrise apartment building*

Several 14-story apartment buildings were constructed in the central area of Rokko Island (Pic. 6). In this area, alluvial clay layers 11.5m thick were distributed below the fill layer, and their residual settlement after construction was estimated at approximately 90cm. In order to accelerate the consolidation settlement of the alluvial clay layers, sand drains of 500mm in diameter were installed in square layout of 3.5m ×3.5m.

After the earthquake, neither level difference between the buildings and ground surface nor any trouble could be seen.

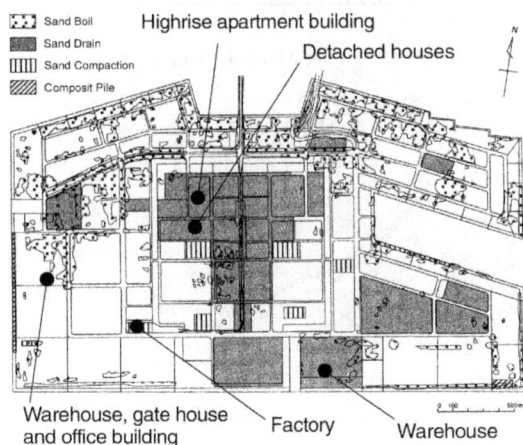

Figure 16. Sand boil and soil improvement areas in Rokko Island

4.4 *Detached houses*

A residential site, size of 134.4m×318.8m, (Pic. 7) is situated on the south side of the above-mentioned site for the highrise buildings. The soil columns sampled in this area as shown in Fig. 17 indicate that the thicknesses of the alluvial clay layers in this area change between 13m and 18m and their residual settlement was estimated at approximately 200cm. As acceleration of consolidation was indispensable, sand drains (sand pile of 500mm in diameter) were installed in square layout of 2.2m×2.2m (preloading fill-up 6.0m high) and 3.3m×3.3m (preloading fill-up 4.0m high). As shown in Fig. 18 and Fig. 19, a settlement of about 190cm was accelerated during the measurement period. In this residential site, the earthquake caused no ground liquefaction or damage by differential settlement to houses with spread foundations.

Picture 6. Highrise apartment building area

Picture 7. Residential area for detached houses

4.5 *Warehouse, gate house and office building*

For the warehouse, office building and gate house constructed in the center of the west side of Rokko Island, sand drains (sand pile of 400mm in diameter) were installed, working from the sea surface, into

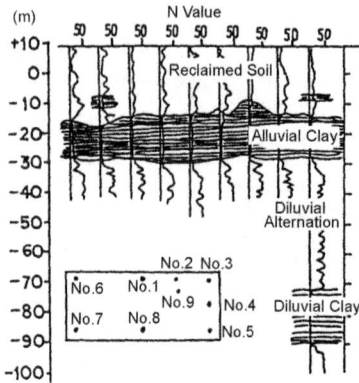

Figure 17. Foundation soil conditions of subject area

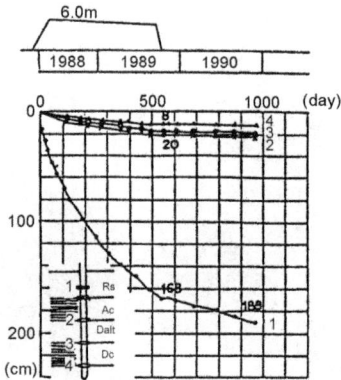

Figure 18. Conditions of settlement of respective layers by preloading (Highrise building area)

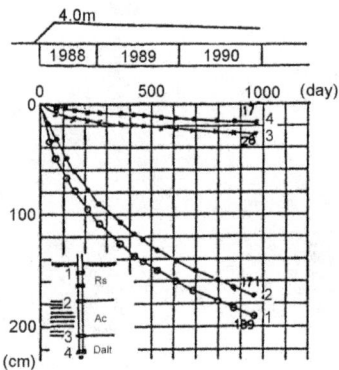

Figure 19. Conditions of settlement of respective layers by preloading (Detached house area)

the alluvial clay layer in square layout of 4.0m×4.0m in February 1981 (before reclamation). The alluvial clay layer was 14m thick, and the residual settlement after construction was estimated at 210cm. Bearing

pile foundations were determined to be unsuitable, and reinforced concrete nodular piles were adopted. Pic. 8 shows the overall view of the buildings, and Fig. 20 shows the soil improvement areas and the building layout. Settlement measurement is still being made at said buildings, and the results thereof are shown in Fig. 21. This figure does not include data at time of the earthquake, but no differential settlement is seen after the earthquake so that the buildings are functioning normally.

Picture 8. Overall view of buildings

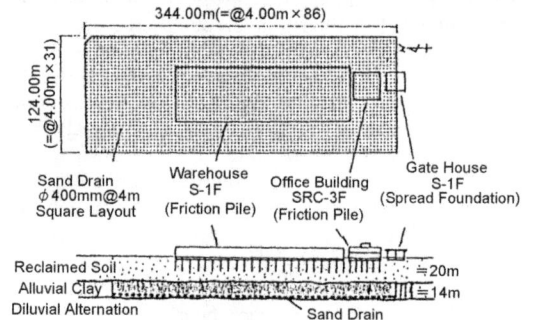

Figure 20. Soil improvement areas and building layout

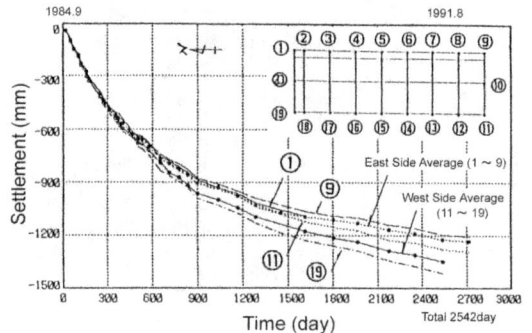

Figure 21. Building settlement measurement results

4.6 Warehouse

The site for the warehouse is situated along the quay at the south end of the central section of Rokko Island. A 1-story warehouse was built along the quay and a 4-story office building was built slightly away from the quay on friction pile foundations. The thickness of the alluvial clay layer in said area was 14m to 20m, and the residual settlement was estimated at about 180cm. In order to accelerate consolidation of the alluvial clay layer, sand drains (sand pile of 500mm diameter) were installed in square layout of 2.5m×2.5m. In anticipation of fast settlement, an extra fill of 1.5m was given. Fig. 22 shows the relation between the quay and the buildings, and Pic. 9 shows the appearances of subject buildings.

The earthquake deformed the quay toward the sea and cracked the ground back of the quay with a certain amount of settlement (Pic. 10), but did not cause any damage to the buildings. Settlement of about 20cm occurred all over the site but no soil liquefaction occurred.

Picture 9. Overall view of the building

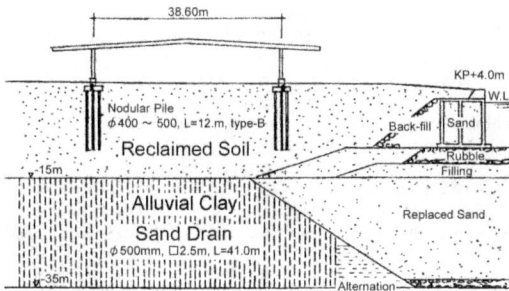

Figure 22. Buildings and foundation soil sections

Picture 10. Cracks in ground back of the quay by the earthquake

4.7 Factory

A 2-story steel frame factory was constructed in the southwest section of Rokko Island. The alluvial clay layer was 15m thick, and the residual settlement was estimated at about 300cm, but the sand compaction method was adopted expecting compaction of the fill material. Sand piles were 700mm in diameter and 15.0m in length and were installed in square layout (2.0m×2.0m to 3.0m×3.0m). The targeted N-value after the soil improvement was 14 in the central area of the buildings and 10 in the perimeter sections. After installation of the sand piles, N-values were carefully measured by soil survey boring. The buildings rested upon strip footing (bearing capacity of soil 150kN/m^2) and partially reinforced concrete nodular piles were used. As the result, no soil liquefaction, level difference between buildings and ground or other damages were inflicted by the Hyogoken-Nambu Earthquake. Pic. 11 shows the overall view of the building. Fig. 23 shows the building and soil layer section and Fig. 24 shows the sand compaction pile layout.

Picture 11. Overall view of the building

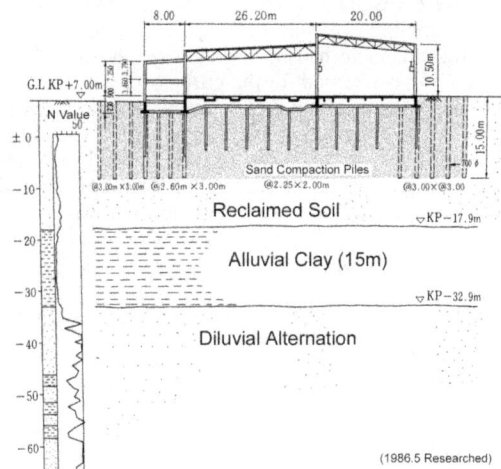

Figure 23. Building and soil layer section

177

Sand Pile : φ700mm, L=15m, Num=1030

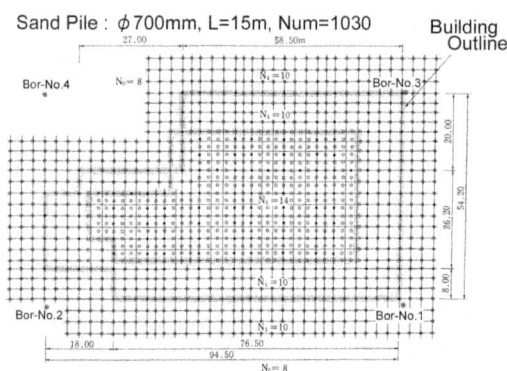

Figure 24. Sand compaction pile layout

quantitative evaluation has not been attained yet due to differences in quality of fill materials, etc. It is a problem that must be solved in the future. In the recent days, researches concerning the vibration isolation effect of liquefied soil layer are in progress so that the expectation for soil improvement in soft soil ground is growing.

It cannot be denied that there still is a trend to depend on conventional load bearing pile foundations when comparatively large structures are planned on soft soil reclaimed ground. Appropriate foundation construction methods and soil improvement methods must be selected depending on the building scale and importance. We shall be very pleased if this paper will be a help in selecting the suitable foundation type in such a case.

5 CONCLUSION

In about 1965, the problem of damage to buildings by ground settlement began to draw general attention in reclaimed land areas where foundation soil settlement was large so that many researches and experiments have been repeated since around 1975. It goes without saying that the results of such efforts have led to today's negative friction technique contributing to the development of bearing pile foundations. On the other hand, the use of bearing piles has demonstrated functional problems such as building float-up so that friction pile foundations and spread foundations have been re-recognized as foundation techniques not dependent on load bearing piles. The soil improvement is the very technology that has been developed for the purpose of supporting the functions of the foundations for structures. Its usefulness was sufficiently proven also by the Hyogoken-Nambu Earthquake.

It deserves special attention that no major damages were seen by soil liquefaction in areas where soil improvement had been given by sand drain and preloading performed for the purpose of accelerating consolidation of alluvial clay layer. Assumedly, it is because the compaction effect by vibration caused at time of driving casings for sand drains and the relative density increase by preloading have contributed to the restriction of soil liquefaction. However, the

REFERENCES

1995. "Report on Investigation of Reclaimed Land Deformation by Hyogoken-Nambu Earthquake", Kobe Municipal Development Bureau, pp.77-78.

Report on Investigation of Damages by Hanshin-Awaji Great Earthquake: Architectural Institute of Japan

Yoshio Suzuki, Minoru Fukui and Fumiya Osugi, February. 1998, "Selection of Soil Improvement Methods as Seen in Actual Examples Involving Various Type Structures", Engineering, Soil and Foundation, pp.57-60.

Kohori, Tanaka, Ito, Fukui, Honda, Okamoto and Koushige, September 1998, "Long-term Building Settlement Measurement at Kobe Port Island (Section 1 to Section 4)", Architectural Institute of Japan, General Meeting Scientific Lectures Synopses (Kyushu), pp.811-818.

Yoshihisa Gyoten, Kiichi Tanimoto, Minoru Fukui, Shuji Honda, July 1997, "Settlement Observation of Highrise Buildings with Bearing Pile Foundations Constructed on Reclaimed Land (Section 2)", 34th Soil Engineering Research Paper Reading (Tokyo), pp.1395-1396.

Minoru Fukui, April 2000, "Building Foundation Structures on Reclaimed Land - An Experience in Hyogoken-Nambu Earthquake", Soil Engineering Institute Kansai Branch 2000 General Meeting Lecture.

Shuji Honda, Minoru Fukui, Kiichi Tanimoto, December 1992, "Settlement of Soil and Structures at Coastal Reclamation Land", Foundation Engineering, pp.99-105.

Minoru Fukui, November 2001 & January 2002, "Selection of Soil Improvement Methods as in Actual Examples and Engineering Lecture Class Materials", Soil Engineering Institute.

Soft Ground Engineering in Coastal Areas, Tsuchida et al. (eds)
© 2003 Swets & Zeitlinger, Lisse, ISBN 90 5809 613 0

A non-linear equation for the consolidation with vertical drain and its applications

K. Suzuki
Design Department, Civil Engineering Headquarters, TOA Corporation, Tokyo, Japan

K. Yasuhara
Department of Urban and Civil Engineering, Ibaraki University, Japan

T. Fukasawa
Technical Research Institute, TOA Corporation, Yokohama, Japan

ABSTRACT: A non-linear equation for the consolidation with vertical drain is derived in order to propose the way to evaluate consolidation in terms of stress and strain separately, which the conventional equation cannot provide with. The usefulness of the derived equation is verified by observing the settlement of the clay specimen with 200 mm diameter. The derivation of the equation is firstly described in this paper, followed by the laboratory test results and analysis for the test based on the equation. Then, application of this equation to the actual consolidation problem taken place in the construction of a seawall for the second island of the Kansai International Airport is also presented with regard to not only settlement behavior but also shear strength increment measured by a series of cone penetration tests. It can be concluded through the analysis and comparison with the actual behavior that the derived equation is useful to evaluate settlement and shear strength increment of clays due to consolidation.

1 INTRODUCTION

In coastal areas of Japanese islands, soft clay deposits are widely found on the seabed. Construction works in these areas, therefore, start with improvement of the soft clay deposits and vertical drain method frequently used to improve them. To analyze the consolidation with vertical drain, Eq. (1) has been commonly employed.

$$\frac{\partial u}{\partial t} = c_v \frac{\partial^2 u}{\partial z^2} + c_h \frac{\partial^2 u}{\partial r^2} + \frac{c_h}{r} \frac{\partial u}{\partial r} \qquad (1)$$

Mikasa (1963) pointed out the importance of distinguishing the progress of consolidation in terms of stress and in terms of strain. However, Eq. (1) cannot evaluate the difference between consolidation in terms of stress and that of strain, because Eq. (1) requires a linear stress-strain relation to be derived.

Vertical drain method is usually coupled with multi-stage loading technique, and design of the vertical drain method frequently requires engineers to predict increment of shear strength due to consolidation. Since the shear strength directly relates to effective stress, consolidation equation that can separate stress from strain is desirable.

In order to separate stress from strain, a non-linear consolidation equation is derived in this paper. Verification of the equation by laboratory test is then described. The equation is also applied to evaluate the consolidation problem encountered in the construction of a seawall for the second island of the Kansai International Airport. Successful results of the analysis are also reported lastly in this paper.

2 CONSOLIDATION EQUATION

We will consider a soil element as indicated in Fig. 1. We assume that compressive strain ε in the soil element is limited to work only in z-direction (vertical direction). When $d\varepsilon$ denotes compressive strain that takes place within a small time increment dt, volume change of the element ΔV during dt can be expressed by Eq. (2).

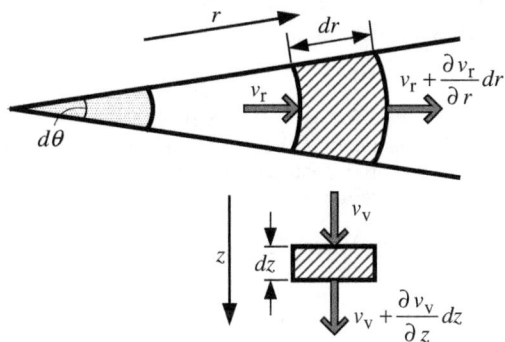

Figure 1. Soil element and water flow.

$$\Delta V = r d\theta dr dz d\varepsilon = r d\theta dr dz \frac{\partial \varepsilon}{\partial t} dt \qquad (2)$$

The amount of pore water flow ΔV_w out of the element during dt can be given by Eq. (3-1) for z-direction and by Eq. (3-2) for r-direction (radial direction), ignoring terms of second order. The values of v_v and v_r in the equations are the apparent speed of pore water flow in z- and r-directions, respectively.

$$\Delta V_{w(v)} = \left\{ r d\theta dr \left(v_v + \frac{\partial v_v}{\partial z} dz \right) - r d\theta dr v_v \right\} dt$$
$$= r d\theta dr dz \frac{\partial v_v}{\partial z} dt \qquad (3\text{-}1)$$

$$\Delta V_{w(r)} = \left\{ (r + dr) d\theta dz \left(v_r + \frac{\partial v_r}{\partial r} dr \right) - r d\theta dz v_r \right\} dt$$
$$= r d\theta dr dz \left(\frac{\partial v_r}{\partial r} + \frac{v_r}{r} \right) dt \qquad (3\text{-}2)$$

Assuming that soil particles and pore water are incompressible, we get

$$\Delta V = \Delta V_{w(v)} + \Delta V_{w(r)} \qquad (4)$$

Substituting Eqs. (2) and (3) into Eq. (4), we obtain

$$\frac{\partial \varepsilon}{\partial t} = \frac{\partial v_v}{\partial z} + \frac{\partial v_r}{\partial r} + \frac{v_r}{r} \qquad (5)$$

We can express the apparent speed of pore water flow by using Darcy's law as

$$v_v = -\frac{k_v}{\gamma_w} \frac{\partial u}{\partial z}, \quad v_r = -\frac{k_h}{\gamma_w} \frac{\partial u}{\partial r} \qquad (6)$$

where u is excess pore water pressure, γ_w is unit weight of the pore water, and k_v and k_h are coefficients of permeability in vertical and horizontal directions, respectively.

Effective vertical stress σ'_v inside the ground under consolidation can be given as

$$\sigma'_v = \sigma'_0 + \sigma_c - u \qquad (7)$$

where σ'_0 is effective overburden stress before the start of consolidation and σ_c is consolidation pressure in total stress. Partial differentiation of Eq. (7) in z- and r-directions are

$$\frac{\partial \sigma'_v}{\partial z} = \frac{\partial \sigma'_0}{\partial z} + \frac{\partial \sigma_c}{\partial z} - \frac{\partial u}{\partial z} = \gamma' - \frac{\partial u}{\partial z} \qquad (8\text{-}1)$$

$$\frac{\partial \sigma'_v}{\partial r} = \frac{\partial \sigma'_0}{\partial r} + \frac{\partial \sigma_c}{\partial r} - \frac{\partial u}{\partial r} = -\frac{\partial u}{\partial r} \qquad (8\text{-}2)$$

respectively, from which we get

$$\frac{\partial u}{\partial z} = \gamma' - \frac{\partial \sigma'_v}{\partial z}, \quad \frac{\partial u}{\partial r} = -\frac{\partial \sigma'_v}{\partial r} \qquad (9)$$

where γ' is unit weight of the soil. Therefore, substituting Eq. (9) into Eq. (6), we obtain Darcy's law in terms of the effective stress as

$$v_v = \frac{k_v}{\gamma_w} \left(\frac{\partial \sigma'_v}{\partial z} - \gamma' \right), \quad v_r = \frac{k_h}{\gamma_w} \frac{\partial \sigma'_v}{\partial r} \qquad (10\text{-}1)$$

or

$$v_v = \frac{k_v}{\gamma_w} \left(\frac{\partial \sigma'_v}{\partial \varepsilon} \frac{\partial \varepsilon}{\partial z} - \gamma' \right), \quad v_r = \frac{k_h}{\gamma_w} \frac{\partial \sigma'_v}{\partial \varepsilon} \frac{\partial \varepsilon}{\partial r} \qquad (10\text{-}2)$$

By using Eq. (10-1), Eq. (5) can be rewritten as

$$\frac{\partial \varepsilon}{\partial t} = \frac{\partial}{\partial z} \left\{ \frac{k_v}{\gamma_w} \left(\frac{\partial \sigma'_v}{\partial z} - \gamma' \right) \right\}$$
$$+ \frac{\partial}{\partial r} \left(\frac{k_h}{\gamma_w} \frac{\partial \sigma'_v}{\partial r} \right) + \frac{1}{r} \frac{k_h}{\gamma_w} \frac{\partial \sigma'_v}{\partial r} \qquad (11)$$

From Eq. (11), we get

$$\frac{\partial \varepsilon}{\partial t} = \frac{\partial}{\partial z} \left(c_v \frac{\partial \varepsilon}{\partial z} \right) - \frac{\partial}{\partial z} \left(\frac{k_v}{\gamma_w} \gamma' \right)$$
$$+ \frac{\partial}{\partial r} \left(c_h \frac{\partial \varepsilon}{\partial r} \right) + \frac{c_h}{r} \frac{\partial \varepsilon}{\partial r} \qquad (12)$$

where

$$c_v = \frac{k_v}{\gamma_w} \frac{\partial \sigma'_v}{\partial \varepsilon}, \quad c_h = \frac{k_h}{\gamma_w} \frac{\partial \sigma'_v}{\partial \varepsilon} \qquad (13)$$

Assuming that the values of c_v and c_h are constant during consolidation, we get

$$\frac{\partial \varepsilon}{\partial t} = c_v \frac{\partial^2 \varepsilon}{\partial z^2} - \frac{\partial}{\partial z} \left(\frac{k_v}{\gamma_w} \gamma' \right) + c_h \frac{\partial^2 \varepsilon}{\partial r^2} + \frac{c_h}{r} \frac{\partial \varepsilon}{\partial r} \qquad (14)$$

In order to simplify Eq. (14), we consider erasing the second term on the right hand side of Eq. (14). When we assume that γ' is larger than zero and constant, k_v should be constant to erase the term. Since we already assume c_v and c_h being constant, constant value of k_v means that $\partial \sigma'_v / \partial \varepsilon (= 1/m_v)$ should be constant according to their definition expressed by Eq. (13). Consequently, non-linearity of the stress-strain relation is lost.

Assumption that γ' is equal to zero can also delete the second term on the right hand side of Eq. (14), but this assumption distort Darcy's law expressed by Eq. (10). However, when the radial flow is predominant, the effect of distortion is expected to be negligible.

Here, in order to keep non-linearity in the stress-strain relation, we assume $\gamma' = 0$. Therefore, Eq. (14) can be simplified as

180

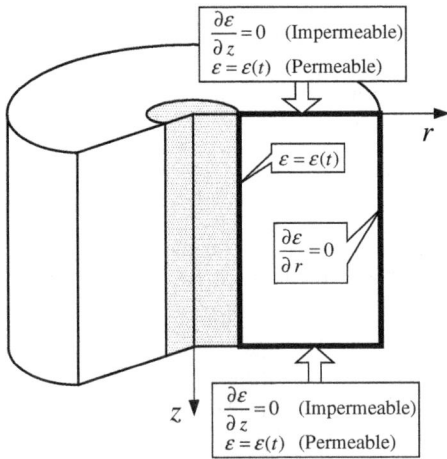

Figure 2. Boundary conditions.

$$\frac{\partial \varepsilon}{\partial t} = c_v \frac{\partial^2 \varepsilon}{\partial z^2} + c_h \frac{\partial^2 \varepsilon}{\partial r^2} + \frac{c_h}{r} \frac{\partial \varepsilon}{\partial r} \quad (15)$$

This equation has the same form as Eq. (1), but uses ε instead of u.

Fig. 2 shows boundary conditions for a cylindrical mass of soil with a drain well at its center. On the permeable boundary including surface of the drain well, consolidation finishes instantaneously when a load is applied. Compression strain, therefore, can be determined from the given stress-strain relation. On the impermeable boundary and on the outer surface of the cylinder, speed of the water flow is zero. Consequently, from the assumption that γ' is equal to zero and Eq. (10-2), we obtain

$$\frac{\partial \varepsilon}{\partial z} = 0, \ \frac{\partial \varepsilon}{\partial r} = 0 \quad (16)$$

3 LABORATORY EXAMINATION

In order to verify Eq. (15), laboratory test was conducted on a reconstituted clay specimen with the diameter of 200 mm. Fig. 3 illustrates drainage conditions of the test. Drainage from the loading plate is limited to its center area (the diameter is 6 cm) so as to generate radial flow of the pore water.

The clay specimen was prepared from slurry (water content = 130 %) by consolidation under $\sigma'_v = 10$ kPa inside the container. Then the specimen was consolidated under Type-1 drainage condition from $\sigma'_v = 10$ kPa to 40 kPa. The value of c_v can be determined from this stage of the test. Type-2 drainage was adopted to the stage of $\sigma'_v = 40$ kPa to 70 kPa and $\sigma'_v = 70$ kPa to 100 kPa. After Type-2 drainage, Type-3 drainage was started as the final loading stage from $\sigma'_v = 100$ kPa to 130 kPa.

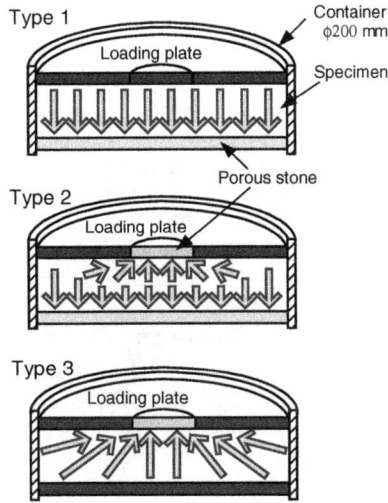

Figure 3. Types of drainage in laboratory test.

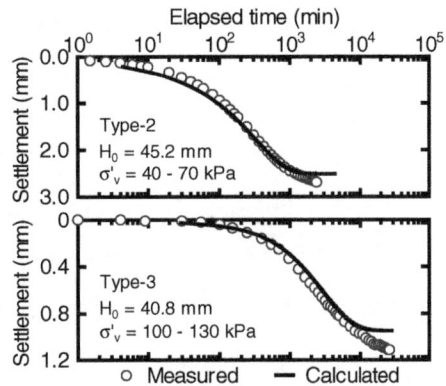

Figure 4. Comparison of measured and calculated settlement for Type-2 and Type-3 drainage.

Adopting the square root t method to Type-1 drainage, 30 cm^2/d was determined as the value of c_v. Consolidation analysis for Type-2 and Type-3 was carried out with this value. According to the report made by Pradhan et al. (1993), which implies that c_h of reconstituted sample does not differ from c_v, the analysis was performed with assuming $c_h = c_v$.

Fig. 4 presents the comparison between the measured settlement of the specimen and the results of the analysis with finite difference method based on Eq. (15). The settlement curves obtained from the analysis well agree with the observed values both for Type-2 and Type-3 drainage as indicated in the figure. It can be concluded from the laboratory test that Eq. (15) gives good prediction of consolidation process when appropriate values of c_v and c_h are found.

181

4 APPLICATION TO FIELD BEHAVIOR

Here in this section, analysis with Eq. (15) is applied to the actual consolidation behavior observed in the construction site of a seawall for the reclamation of the second island for the Kansai International Airport. The locations of pre-construction investigation and settlement monitoring are shown in Fig. 5. The reclamation area is located more than 5 km off the coast.

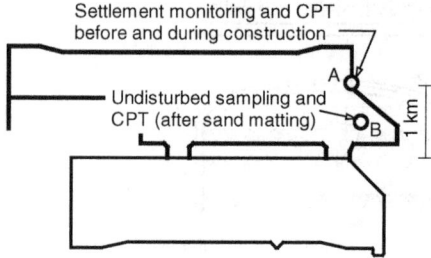

Figure 5. Locations of soil investigation and field monitoring.

Fig. 6 demonstrates the cross section of the seawall at location A in Fig. 5. As indicated in the figure, the seawall was constructed on a Holocene clay. The clay deposit was improved by sand drain method after it was covered with 1.7 m thick sand mat. The diameter of the drain wells was 40 cm, and they were installed down to the bottom of the clay deposit with the arrangement of 2.5 m by 1.6 m right under the seawall and 2.5 m by 2.5 m in the reclamation side. In order to monitor the settlement, three settlement plates (SP1, SP2, and SP3) were placed on the sand mat.

After the completion of fill No. 1, no filling materials were spread for a period of four months so that the clay gains enough shear strength to support the weight of further filling. During this consolidation period, CPT was penetrated twice (CPT-1 and CPT-2) into the clay to confirm the actual increment of the shear strength due to consolidation. CPT was again conducted after the completion of rubble mound No. 1 and fill No. 2 (CPT-3 and CPT-4). These CPTs were carried out in the vicinity of SP2 and in the center of the grid formed by the drain wells.

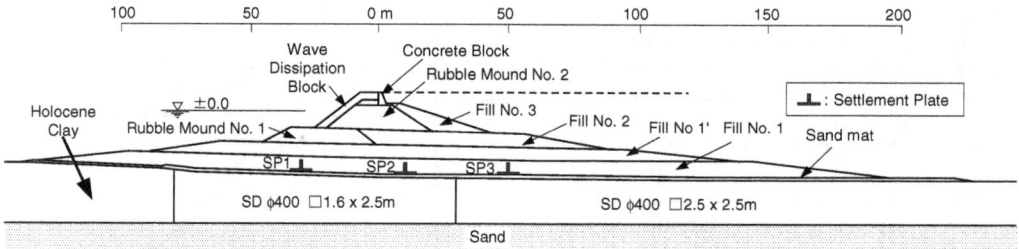

Figure 6. Typical cross section of the seawall.

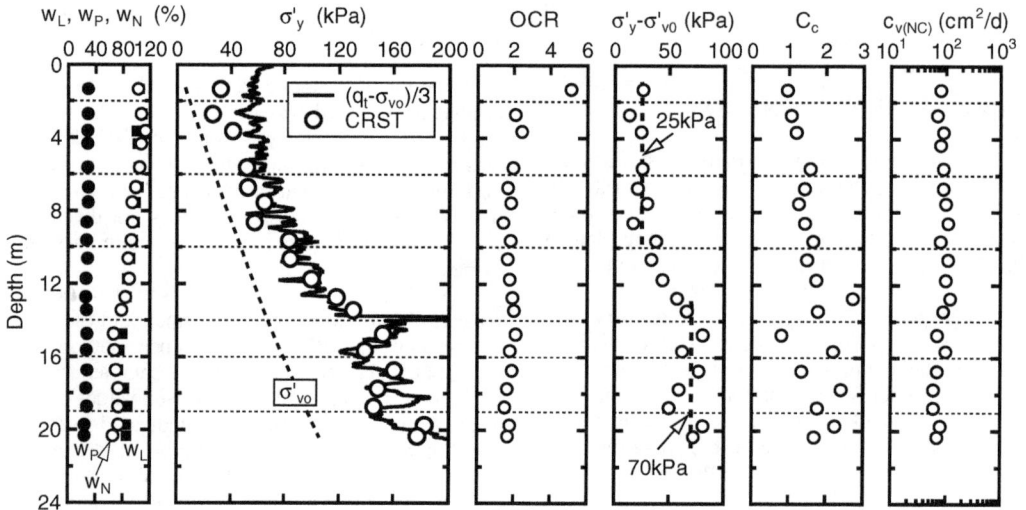

Figure 7. Consolidation characteristics of the Holocene clay in the site.

4.1 Consolidation characteristics of the Holocene clay

Pre-construction investigation including undisturbed sampling and CPT were carried out at location B in Fig. 5 after the spreading of sand mat in order to study the engineering properties of the clay. Fixed piston thin wall sampler was used for undisturbed sampling. This sampler, very common in Japan, has been proved to produce high quality undisturbed samples (Tanaka, 2000).

Consolidation characteristics obtained from constant rate of strain consolidation test CRST are summarized in Fig. 7 together with Atterberg limits. The size of the specimens for CRST was 6 cm in diameter and 2 cm in height, and they were compressed at a strain rate of 0.02 %/min. This strain rate is the same as that chosen by Hanzawa et al. (1990) to study consolidation parameters of Ariake clay. As indicated in the figure, consolidation yielding stress σ'_y is larger than effective overburden stress σ'_{v0}, revealing that the clay is in overconsolidated state. The value of overconsolidated ratio OCR decreases with increasing depth down to GL-10 m, and constant with $OCR = 1.8$ below GL-10 m. The amount of overconsolidation in terms of σ'_y-σ'_{v0} is about 25 kPa from the surface to the depth of 10 m, and about 75 kPa beneath the depth of 14 m. Compression index C_c increases with increasing depth from 1.0 to 2.5, while coefficient of consolidation in normally consolidated state $c_{v(NC)}$ is almost constant through the depth.

Hanzawa (2000) reported that the relation between σ'_y from CRST conducted with 0.02 %/min and cone resistance obtained from clays in Japan and Southeast Asia can be averaged out to Eq. (17), where q_T is the point resistance corrected by pore water pressure measured at the shoulder of cone tip and σ_{v0} is total overburden stress. The value of σ'_y from CPT in Fig. 7 was derived from this equation.

$$\sigma'_{y(CRST)} = (q_T - \sigma_{v0})/3 \qquad (17)$$

The values of σ'_y from CRST and CPT well agree with each other below the depth of 8 m as observed in the figure. Sudden increase and decrease observed in the result of CPT at the depth of 14 m represent the existence of thin sediment of volcano ash fallen about 6,300 years ago (Endo, et al. 1995).

Since the clay is classified as Holocene deposit, the state of overconsolidation can be attributed to aging effect, and the clay can be categorized into normally consolidated aged (NCA) clay (Bjerrum, 1973 and Hanzawa, 1989).

4.2 Consolidation settlement analysis

Dotted lines in Fig. 7 represent the boundaries of sub-layers divided for the analysis in accordance with the change in soil properties. Fig. 8 is an example of e-log σ'_v curve obtained from CRST. Three values of compression indices, C_{c1}, C_{c2} and C_{c3}, are needed to approximate the e-log σ'_v curve, because the compressibility cannot be expressed properly by only one value of C_c after σ'_y. The term "compression index" is used as the inclination of e-log σ'_v curve in this section. Representative CRST result was chosen for the analysis to express the compressibility of each sub-layer. Consolidation constants determined from representative CRST results are listed in Table 1. The value of σ'_b is the stress where the inclination of e-log σ'_v curve moves to C_{c3} from C_{c2}.

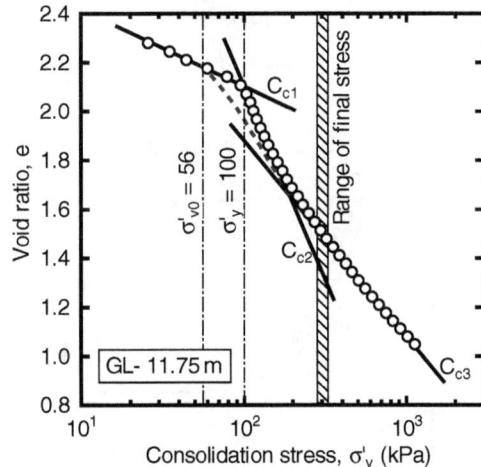

Figure 8. An example of e – log σ'_v curve and its approximation.

Table 1. List of constants for consolidation analysis

Sub-layer	Sample Depth (m)	e_0	σ'_{v0} (kPa)	σ'_y (kPa)	σ'_b (kPa)	C_{c1}	C_{c2}	C_{c3}	$c_{v(NC)}$ (cm²/d)
1	1.35	2.56	6.3	32.3	200.0	0.127	0.960	0.827	80
2	3.65	2.52	17.0	41.5	130.6	0.258	1.205	0.842	90
3	7.55	2.48	35.1	65.1	200.0	0.298	1.272	0.930	100
4	11.75	2.18	56.0	100.0	178.6	0.238	1.747	0.802	100
5	14.75	1.50	71.7	152.5	346.7	0.244	0.813	0.661	70
6	17.75	1.79	89.7	148.8	200.0	0.273	2.414	0.887	60
7	19.75	1.70	101.7	182.8	251.2	0.275	2.246	0.767	80

Stress distribution in the ground induced by the seawall construction was calculated for the location of SP2 and was confirmed that the difference through the depth was insignificant even after rubble mound No. 2. Therefore, one-dimensional consolidation analysis was carried out with finite difference method. The drainage was allowed at both ends of the clay deposit and at the depth where thin volcano ash sediment was found as well as the surface of drain well.

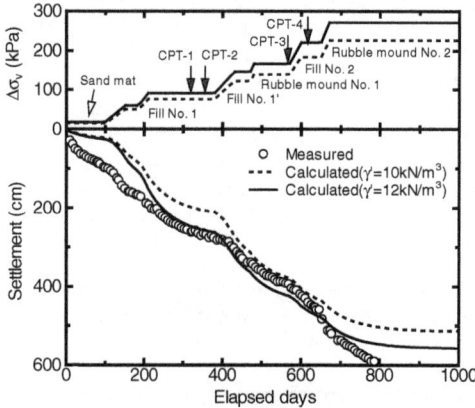

Figure 9. Measured settlement at SP2 and results of the analysis.

The settlement curves from the analysis are presented in Fig. 9 together with the measured values. The predicted settlement shows good agreement with the measured values except the initial stage of the settlement. This discrepancy suggests that e-log σ'_v curve after the installation of sand drain wells becomes something like the dotted line in Fig. 8 due to disturbance.

The actual settlement after the completion of rubble mound No. 2 is much larger than the calculated value. The experience of the first island for the Kansai International Airport revealed that the consolidation settlement of Pleistocene clay layers resting under the Holocene clay starts when the reclamation load reaches about 200 kPa (Endo et al., 1991). The discrepancy in Fig. 9 can be attributed to consolidation settlement of Pleistocene clay layers, which is not considered in the analysis.

4.3 Increase of shear strength

Eq. (17) can be replaced by Eq. (18) when clay moves into normally consolidated state where σ'_y is identical to σ'_v. The value of σ'_v can be obtained easily from consolidation analysis.

$$q_T - \sigma_{v0} - \Delta\sigma_v = 3 \times \sigma'_v \qquad (18)$$

Cone resistance calculated with Eq. (18) is compared with measured one in Fig. 10, where the depth z is normalized by the thickness H. Calculated val-

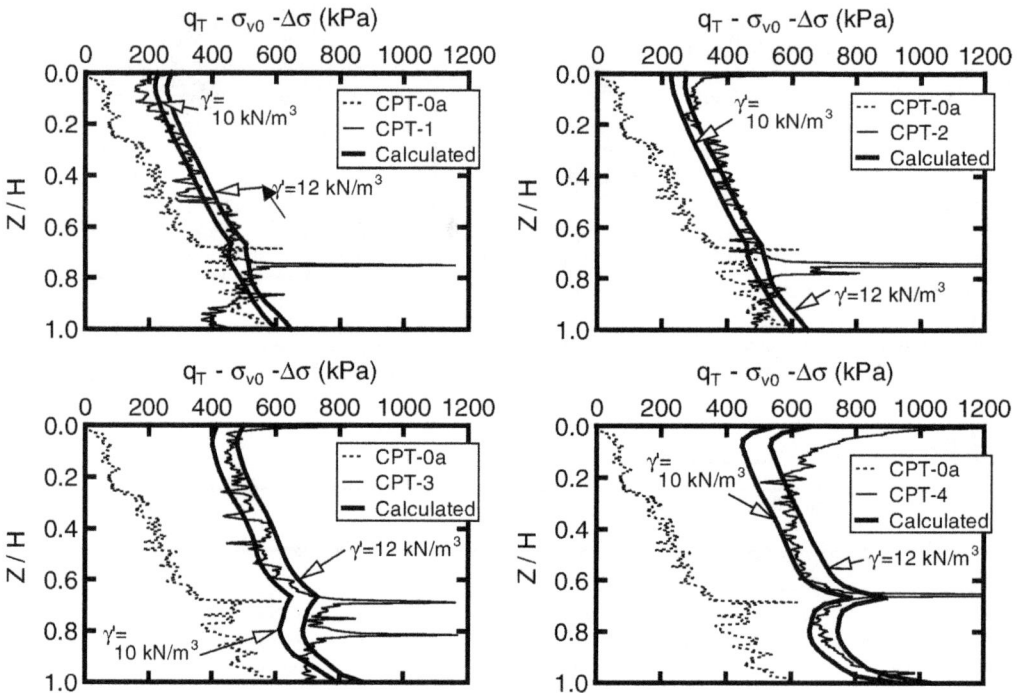

Figure 10. Comparison between measured and analyzed values of cone resistance.

184

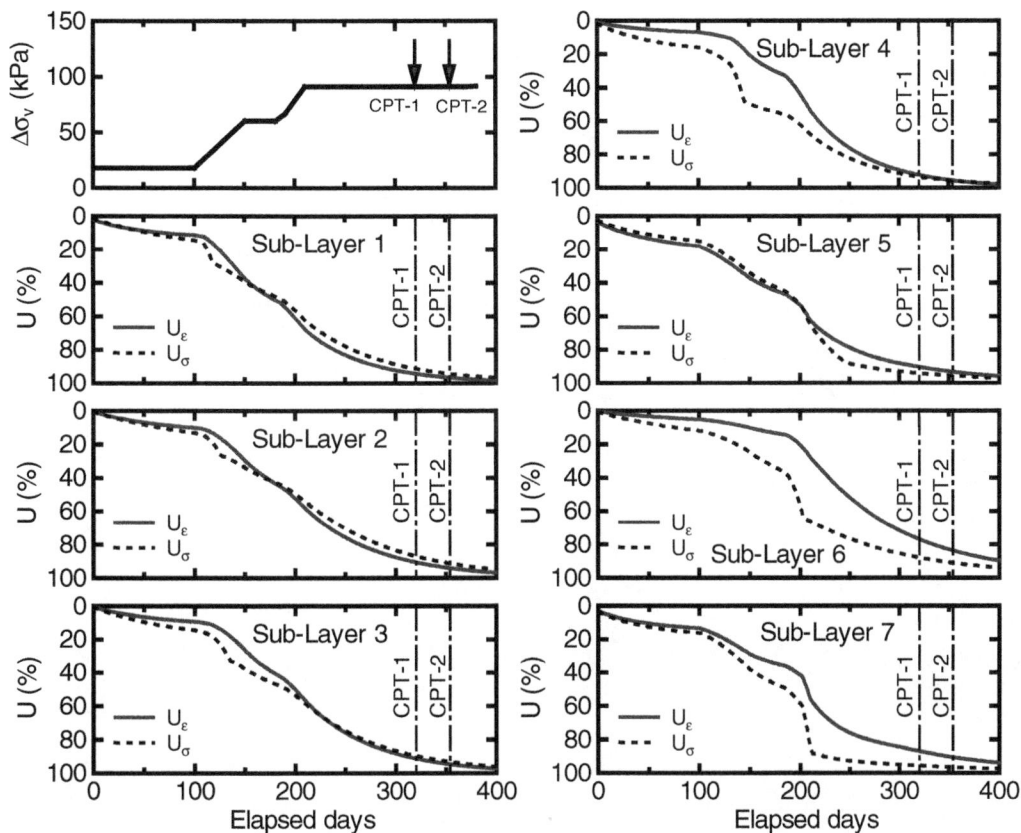

Figure 11. Difference between degree of consolidation in terms of strain and stress calculated for each sub-layer.

ues of cone resistance are corresponding to the center of drain spacing. CPT-0a is obtained before sand mat spreading at location A in Fig. 5. As indicated in the figure, they well agree with each other. Figs. 9 and 10 demonstrate that Eq. (15) works adequately to predict the consolidation with vertical drain.

The measured settlement when CPT-4 was penetrated reaches 4.5 m as presented in Fig. 9. This large settlement suggests that drain wells might be subjected to some extent of deformation and movement from the original position at shallow portion. Therefore, actual position of drain wells was unknown and the cone probe possibly penetrated very near the drain well. The larger value of cone resistance observed in CPT-4 at $Z/H<0.4$ can be attributed to this reason.

Fig. 11 compares the degree of consolidation in terms of strain U_ε and in terms of stress U_σ computed based on Eq. (15) for the loading until Fill No. 1. The values of U_ε and U_σ in the figure are the mean values for each sub-layer. They are almost the same as each other in the upper layers (sub-layer 1 to 4) at the moment when CPT-1 and CPT-2 were performed. On the other hand, U_σ is greater by 5 to 15 % than U_ε in the lower part of the deposit at the

same moments. As demonstrated in Fig. 11, Eq. (15) can analyze consolidation behavior in terms of strain and stress separately.

5 CONCLUSIONS

A non-linear equation for the consolidation with vertical drain was derived and verified through the laboratory test and field observation. Conclusions of this paper are as follows;

1. Assuming $\gamma'=0$, a consolidation equation expressed by Eq. (15) can be given, which has the same form as conventional Eq. (1). Eq. (15) allows using a non-linear stress-strain relation, and consequently, offers the way to evaluate the progress of consolidation in terms of strain and in terms of stress separately.

2. The assumption that γ' is equal to zero distorts Darcy's law. However, Eq. (15) can well explain the consolidation behavior both for reconstituted soil specimen in the laboratory and a naturally deposited Holocene clay found in the construction site of the Kansai International Airport.

3. CRST conducted with a strain rate of 0.02 %/min appropriately gives stress-strain relation for the Holocene clay in this study.

REFERENCES

Bjerrum, L. 1973. Problems of soil mechanics and construction on soft clay and structurally unstable soils, *Proc. 8th ICSMFE*, Vol. 3, pp. 111-159.

Endo, K., Makinouchi, T., Tsubota, K. and Iwao, Y. 1995. Formation process of "Chuseki-sou" in Japan. *Tsuchi-to-Kiso*, Vol. 43, No. 10, pp. 8-12 (in Japanese).

Endo, H., Oikawa, K., Komatsu, A and Kobayashi, M. 1991. Settlement of diluvial clay layers caused by a large scale man-made island. *Proceedings of the International Conference on Geotechnical Engineering for Coastal Development (Geo-Coast '91)*, Yokohama, Vol. 1, pp. 177-182.

Hanzawa, H. 1989. Evaluation of design parameters for soft clays as related to geological stress history. *Soils and Foundations*, Vol. 29, No. 2, pp. 99-111.

Hanzawa, H., Fukaya, T. and Suzuki, K. 1990. Evaluation of engineering properties for an Ariake clay. *Soils and Foundations*, Vol. 30, No. 4, pp. 11-24.

Hanzawa, H. 2000. Use of direct shear and cone penetration tests in soft ground engineering. *Preprints of the International Symposium on Coastal Geotecchnical Engineering in Practice (IS-Yokohama)*, Keynote Address, pp. 25-37.

Mikasa, M. 1963. The consolidation of soft clay —A new consolidation theory and its application—. Kajima-shuppankai Co. Ltd. (in Japanese).

Pradhan, T. B. S., Imai, G., Murata, T., Kamon, M. and Suwa, S. 1993. Experimental study on the equivalent diameter of a prefabricated band-shaped drain. *Proceedings of 11th Southeast Asian Geotechnical Conference*, Singapore, pp. 391-396.

Tanaka, H. 2000. Sample quality of cohesive soils: Lessons from three sites, Ariake, Bothkenner and Drammen. *Soils and Foundations*, Vol. 40, No. 4, pp. 57-74.

Soft Ground Engineering in Coastal Areas, Tsuchida et al. (eds)
© 2003 Swets & Zeitlinger, Lisse, ISBN 90 5809 613 0

Two-dimensional bearing capacity on ultra soft clay

Y. Tanaka & G. Imai
Yokohama National University, Yokohama, Japan

M. Katagiri
Nikkensekkei Nakase Geotechnical Institute, Kawasaki, Japan

ABSTRACT: This paper describes bearing capacity tests of a strip footing on ultra soft clay with various water contents and soil conditions such as consolidated ground or mixed and poured ground. The ratios of bearing resistance q to shear strength τ_v, equivalent to the bearing capacity factor N_c, are investigated. Moreover, movements of the grounds are observed by taking photographs in order to confirm the mode of failure. It is concluded that the deformation behaviors of ultra soft clay grounds are affected on the ground condition of water content and soil condition.

1 INTRODUCTION

Reclaimed land by pump-dredged clay is very weak and its surface has much higher water content than liquid limit. To enable the ground improvement works for further construction on the land, it is necessary to conduct surface soil improvement works. In many of surface improvement works, a sand mat and a sheet net are spread over the surface of reclaimed land to ensure the traficability of working cars. The shallow coast-mixing method is also used. To design these surface improvement works, it is necessary to know the characteristics of shear strength and bearing capacity on such the ultra soft clay ground.

On the shear strength characteristics of ultra soft clay, there have been many researches, for example, Inoue et al. (1990), Tan et al. (1991), Zerik(1997, 1998), Tang et al.(1999) and Fakher et al.(2000).

Tanaka et al. (2000) also investigated the shear strength properties of ultra soft clays using vane shear tests. The results demonstrated that the relationships between shear strength and water content were dependent on the soil conditions that means consolidated and mixed deposits. Therefore, it is considered that the characteristics of bearing capacity depend on the soil conditions.

Generally, the design of surface improvement works is based on the Terzaghi's (1965) theory. The bearing capacity of clay is obtained by following equation;

$$q_u = N_c \cdot s_u \qquad (1)$$

where q_u is the ultimate bearing capacity, s_u is the shear strength of clay and N_c is the bearing capacity factor in terms of cohesion. However, the applicability of Terzaghi's theory for ultra soft clay has not been examined sufficiently yet. Thus, it is important to investigate this applicability.

This paper reports the bearing capacity and the deformation behaviors including failure mechanism of ultra soft clay ground based on the model footing tests of a strip footing. From these experiments, effects of ground condition, such as water contents and soil conditions on the bearing capacity and the failure mechanism of the ground are discussed.

2 TEST MATERIAL AND TEST CONDITION

2.1 Test material

The physical properties of the marine clay used in this study are shown in Table 1.

Table 1. Physical properties of clay.

ρ_s(g/cm^3)	w_L (%)	w_P (%)	I_P
2.62	51	21	30

2.2 Test condition of model grounds

In this study, model grounds of ultra soft clay for two-dimensional bearing capacity tests were prepared for different values of water contents, w, and soil conditions.

Regarding water content, three values are set up between liquid limit, w_L, and twice the liquid limit, $2w_L$.

As far as soil conditions of the model grounds are concerned, two types of grounds are prepared. One is the consolidated ground; the other is the mixed/poured ground. Preparation methods of the grounds are mentioned in the next chapter.

In this study, six cases of footing tests were carried out on various conditions shown in Table 2. In the following figures, the consolidation ground and the mixed/poured ground are called as "Con" and "Mix" for short, respectively.

Table 2. Test conditions of model grounds.

Case	w/w_L	Soil conditions
Case 1-1	1.0	Consolidated ground (Con)
Case 1-2	1.3	Consolidated ground (Con)
Case 1-3	1.8	Consolidated ground (Con)
Case 2-1	1.0	Mixed/poured ground (Mix)
Case 2-2	1.3	Mixed/poured ground (Mix)
Case 2-3	1.8	Mixed/poured ground (Mix)

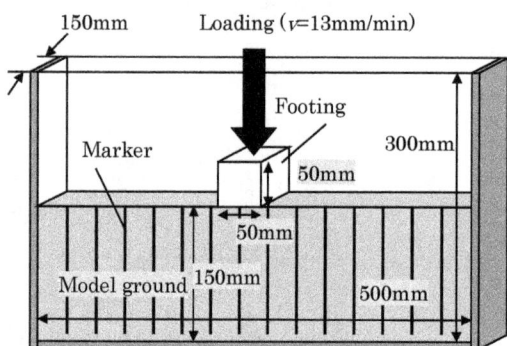

Figure 1. Overview of the model ground for footing test.

3 TEST PROCEDURE

3.1 Model grounds for footing test

Figure 1 shows schematically the model ground for of footing tests on ultra soft clay. The model ground is prepared in the steal container, whose front side is an acrylic plate with 30mm in thickness. The size of the container is 500mm in width, 300mm in depth and 150mm in length. The thickness of the model ground is about 150mm.

3.2 Preparation of consolidated ground

Clay slurry having initial water content, w_0, equal to $2w_L$, is poured into the container at first. Next, the clay slurry is consolidated by a constant pressure, which is applied by an air cylinder through a rectangular plate. Pore water is drained from the bottom and top of the clay slurry during the consolidation process. The end of consolidation is determined by the $2t$ method from the measured settlement of the ground. The applied pressure is shown in Table 3 based on the result of the standard consolidation test on the mixed clay specimen with $w_0 = 2w_L$.

Table 3. Applied pressure for consolidation.

w		$1.0w_L$	$1.3w_L$	$1.8w_L$
p	(kPa)	290	20	5

3.3 Preparation of mixed/poured grounds

Clay slurry well mixed at a relevant water content is prepared beforehand. The mixed clay slurry is then poured into the container. Just after pouring, the footing test is carried out as soon as possible.

In all the cases, linear or point markers are put on the inside plane of the acrylic plate to observe the deformation of the model grounds during the tests before pouring clay slurry.

3.4 Loading of strip footing

The footing tests are carried out as following. A penetration speed of the rectangular foundation is set at 13mm/min to achieve close to be an undrained condition during the penetration. The aluminum foundation is 50mm in width and is fixed to the rod of loading motor. Settlement and applied load are measured by displacement and load transducers during the penetration. Moreover, photographs of the front view of the model grounds are taken by a digital still camera to observe movements of the markers for the deformation of the grounds.

The vane shear tests are also carried out in the model ground after the footing tests. The vane shear strength, τ_v, is measured to obtain the ratio of load intensity, q, to τ_v equivalent to the bearing capacity factor in terms of cohesion, N_c. A center depth of vane blade is set at 2cm below from the surface of the ground. Depth distribution of water contents is also taken from the model grounds.

4 TEST RESULTS

4.1 Shear strength of model ground

Figure 2 shows the curves of vane resistance, τ, and rotation degree, θ, of vane shear tests. Arrows in this figure mean the peak values of τ, respectively. This figure indicates that the peak values are found for the consolidated ground distinctly. For the mixed/poured ground, on the contrary, the peak cannot be seen clearly and θ value at peak τ is lager than that of consolidated ground. It is seemed that brittle failure is observed in the consolidated ground and ductile failure is observed in the mixed/poured ground. From this figure, τ_v is defined as the maximum value of the vane resistance τ.

Figure 3 shows the relationships between vane shear strengths, τ_v, and water contents normalized by liquid limit of the material, w/w_L. This figure also includes the results for sediments consolidated by its own weight at 1G field on laboratory-floor and 30G field in the centrifuge. It is obvious that the shear strength of the consolidated ground is lager than that of the mixed/poured state (▼) with the same w/w_L. The τ_v - w/w_L relationship of the consolidated ground is almost as same as that of the sediments consolidated at 1G and 30G. In this figure, moreover, the result of Tan et al. (1991), which was obtained by the thin plate penetration method, is plotted as a representative of past study. These result is almost as same as that of the mixed/poured ground in this study, since the clay specimen was mixed in the container. These results indicate that the soil condition of the ultra soft clay, that is soil structure, affects the τ_v - w/w_L relationship heavily. Similar results are also obtained by Tanaka et al. (2000).

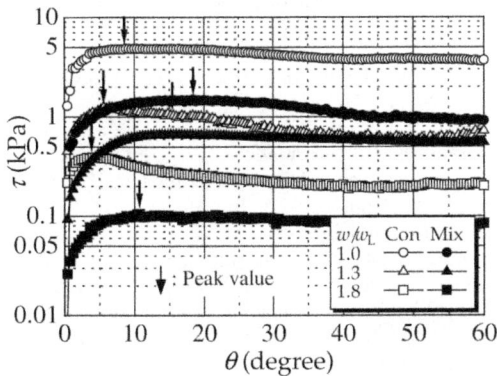

Figure 2. $\tau - \theta$ curves in the vane shear tests.

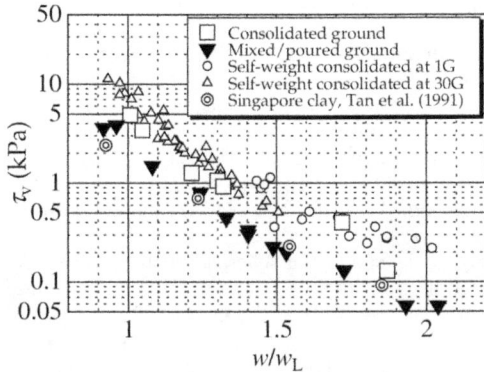

Figure 3. $\tau_v - w/w_L$ relationships in the vane shear tests.

4.2 *q-S/B relationships of footing tests*

Figures 4 and 5 show the load-settlement curves of the footing tests on the consolidated and the mixed/poured grounds, respectively. The vertical axis of the figure is the settlement normalized by width of the footing, and the horizontal axis is the load intensity, q. Values of q are calculated as load divided by cross-sectional area of the strip footing and is corrected for the surcharge effect due to penetration of the footing.

These figures indicate that the load intensity, q, for the consolidated ground is larger than that for mixed/poured ground. In Figure 4, the values of q increases with increase of the S/B and there is a breakpoint (arrows in Figure 4) around $S/B = 0.1$ in each q-S/B curve for the consolidated ground. In the case of mixed/poured ground (Figure 5), however, the break points are not obvious in the q-S/B curves and increases by decreasing the water content. It is confirmed that the differences between these two types of grounds is associated with the relevant failure mechanisms, such as general shear failure or local shear failure.

4.3 *q/τ_v-S/B relationships*

Figures 6 and 7 show the q/τ_v-S/B curves in all cases. The value of q/τ_v is equivalent to the bearing capacity factor, N_c. In general, the theoretical value of N_c for the strip footing is 5.14 for the undrained condition of cohesive soils in Terzaghi's theory. In the cases of consolidated ground, the q/τ_v-S/B curves are broken at around q/τ_v=5 and the q/τ_v-S/B curve is lapped over each other. In the cases of mixed/poured ground, on the other hand, the breakpoint of the

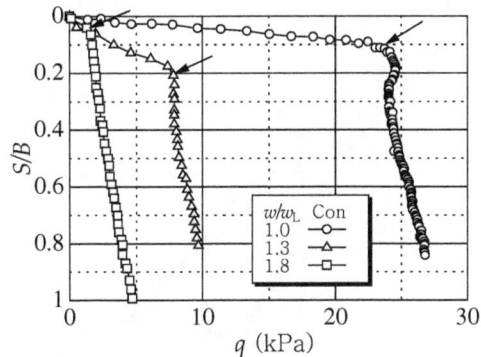

Figure 4. $S/B - q$ curves for footing test on consolidated ground.

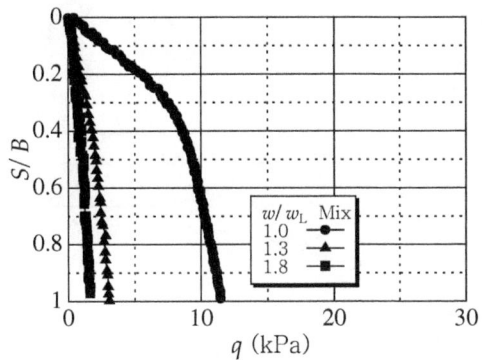

Figure 5. $S/B - q$ curves for footing test on mixed/poured ground.

curve cannot be founded around $q/\tau_v = 5.14$ and q/τ_v increases continuously with increase in S/B.

Figures 8 and 9 show variations of the q/τ_v at S/B = 0.1, 0.2, 0.3, 0.4, 0.5 with w/w_L. Figure 8 shows the q/τ_v values for the consolidated ground are about 5 for various values of w/w_L. On the other hand, Figure 9 shows the q/τ_v values increase with w/w_L and S/B. It is clear that the failure mode for the mixed/poured ground is different from that for consolidated.

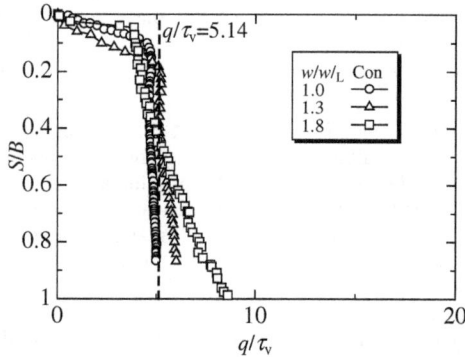

Figure 6. $S/B - q/\tau_v$ curves for consolidated ground.

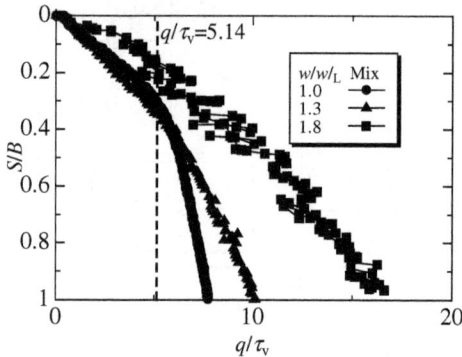

Figure 7. $S/B - q/\tau_v$ curves for mixed/poured ground.

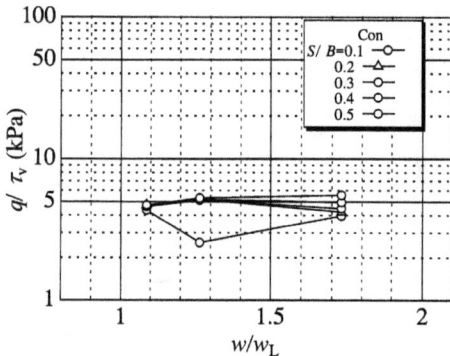

Figure 8. $q/t_v - w/w_L$ relationships for consolidated ground.

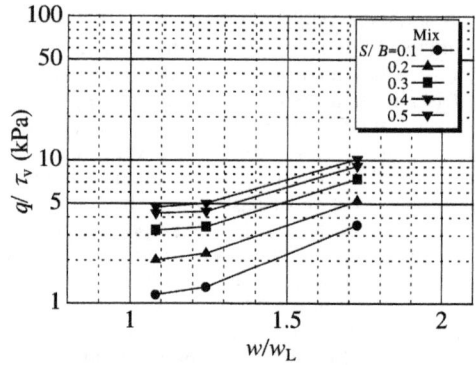

Figure 9. $q/t_v - w/w_L$ relationships for mixed/poured ground.

5 OBSERVATION OF DEFORMATIONS OF MODEL GROUND

Observations of deformation of the model grounds were carried out during the footing tests. The pictures of the front view of the model grounds were taken by a digital still camera.

Figure 10 shows the movements of the markers during the footing tests. In this figure, an area enclosed by broken line means a failure zone evaluated by visual observation. It is clear that the deformations of the grounds are different between the consolidated and mixed/poured grounds. Feature of the deformation in the consolidated ground is heaving of the ground surface around the foundation. Area of the heaving increases with decrease of w/w_L. Moreover, magnitude of the heaving is larger with decrease of w/w_L. It is seemed that the failure of the ground is spread up to the ground surface.

In the case of mixed/poured ground, the deformations of the grounds occur underneath the foundation rather than the surrounding surface ground next to the footing. Area of the movement spread with decrease of w/w$_L$.

Figure 11 shows the schematic pattern of the deformation including the failure of the model grounds of ultra soft clay in both the grounds. In the case of the consolidated ground, the general shear failure occurs during the tests. The local shear failure, on the other hand, is seen in the mixed/poured ground. From consideration of the vane shear tests, the general shear failure occurs for the consolidated ground because of the brittle deformation behavior. For the mixed/poured ground, on the other hand, the local shear failure occurs because of the ductile deformation behavior.

On the water content distributions of the model grounds, changes of water contents are not observed before and after the footing tests distinctly. It is considered that consolidation behaviors are not occurred during penetration of the footing. It is necessary to

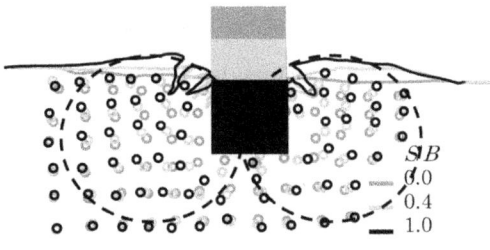

(a) Case 1-1, $w = 1.0 w_L$, Consolidated ground

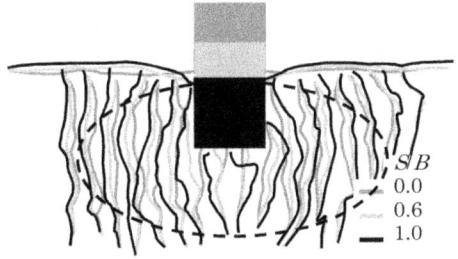

(b) Case 2-1, $w = 1.0 w_L$, Mixed/poured ground

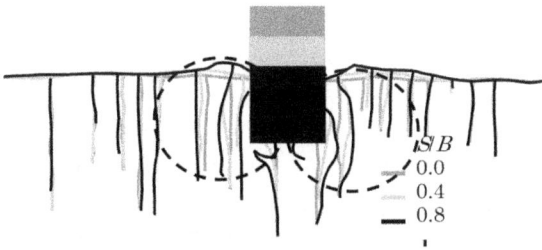

(c) Case 1-2, $w = 1.3 w_L$, Consolidated ground

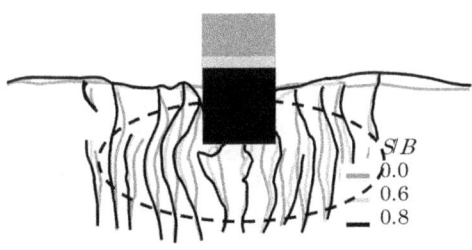

(d) Case 2-2, $w = 1.3 w_L$, Mixed/poured ground

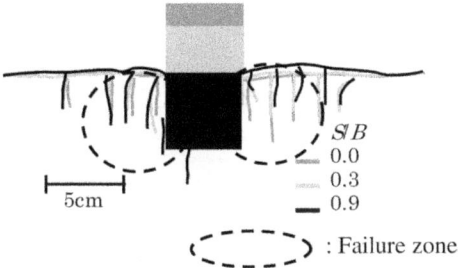

(e) Case 1-3, $w = 1.8 w_L$, Consolidated ground

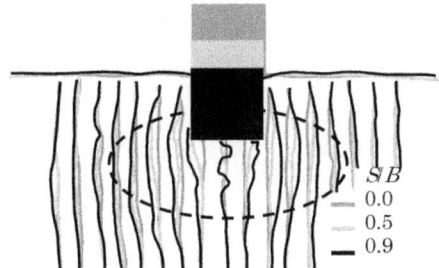

(f) Case 2-3, $w = 1.8 w_L$, Mixed/poured ground

Figure 10. Deformation of the model ground during the footing test.

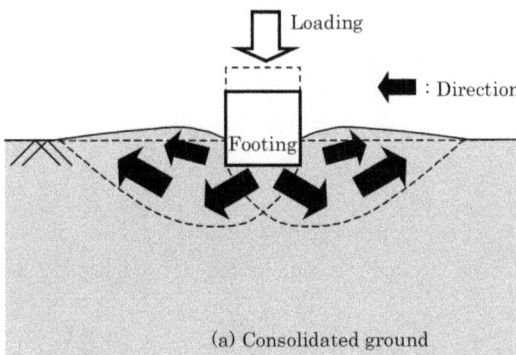

(a) Consolidated ground

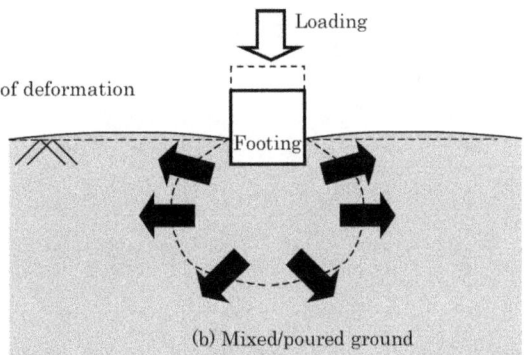

(b) Mixed/poured ground

Figure 11. Overview of the deformation of the model ground.

investigate water contents distributions of the model ground in detail in order to evaluate the drainage condition on ultra soft clay ground for footing test.

6 SUMMARY

A series of the model strip footing test on ultra soft clay is carried out in order to investigate the effect of water content and soil condition on the deformation behaviors such as the ratios of bearing resistance q to shear strength τ_v and the mode of failure. Following conclusions are obtained.

(1) The relationships between τ_v and w/w_L are different between consolidated and mixed/poured grounds. The τ_v - w/w_L relationship for consolidated ground is higher than that of mixed/poured ground at the same water content since these relationships depend on the effect of soil structure heavily.

(2) The q-S/B and q/τ_v-S/B curves of ultra soft clay grounds are different between consolidated and mixed/poured grounds. Values of q/τ_v over $S/B = 0.1$ for consolidated grounds are nearby 5 almost the same as the theoretical values of N_c in Terzaghi's theory. On the other hand, those of mixed/poured grounds increase along with penetration of footing.

(3) It is confirmed that the deformations of the ultra soft clay grounds are different between the consolidated and mixed/poured grounds from the observation.

It is seemed that the consolidated ground has the general shear failure in the footing tests and the mixed/poured ground has the local shear failure.

REFERENCES

Fahker, A., Jones, C.J.F.P and Clarke, B.G.2000. yield stress of super soft clays. Journal of A.S.C.E., 125(6), 499-509.

Inoue, T., Tan, T. S. & Lee S. L. 1990. An investigation of shear strength of slurry clay. *Soils and Foundations*, 30(4): 1-10.

Tan, T. S., Goh, T. C., Inoue, T. & Lee, S. L. 1991. Yield stress measurement by a penetration *method. Canadian Geotechnical Journal*, 28, 517-522.

Tang, Y. X. & Tsuchida, T. 1999. The development of shear strength for sedimentsry soft clay with respect to aging effect. *Soils and Foundations*, 39(6): 13-24.

Tanaka, Y., Imai, G. & Katagiri, M. 2000. Relationships between vane strength and water content of very soft clays. Proc. of International Symposium on Coastal Geotechnical Engineering in Practice, 1: 167-172.

Terzaghi, K. 1965. Theoretical Soil Mechanis, John Wiley and Sons, 1965.

Zreik, D. A., Germaine, J. T. & Ladd, C. C. 1997. Undrained strength of ultra-weak cohesive soils: relationship between water content and effective stress. *Soils and Foundations*, 37(3): 117-128.

Zreik, D. A., Germaine, J. T. & Ladd, C. C. 1998. Effect of aging and stress history on the undrained strength of ultra-weak cohesive soils. *Soils and Foundations*, 38(4): 31-39.

Soft Ground Engineering in Coastal Areas, Tsuchida et al. (eds)
© 2003 Swets & Zeitlinger, Lisse, ISBN 90 5809 613 0

Development of clayey water interception material in a coastal disposal site

K. Yamada, K. Ueno & A. Hada
Penta-Ocean Construction Co., Ltd., Tochigi, Japan

T. Tsuchida & Y. Watabe
Port and Airport Research Institute, Kanagawa, Japan

G. Imai
Yokohama National University, Kanagawa, Japan

ABSTRACT: Many coastal disposal sites have been built by reclaiming soft ground along the coast. For this reason, water interception materials used in impervious seawalls at such disposal sites need to be highly workable at these locations and flexible enough to behave in accordance with ground movements, as well as being impervious and economical. Further, these materials must endure underwater for a long period of time. Directing our attention to the high impermeability of marine clay generated in dredge work, we have endeavored to develop a water interception material based on marine clay.

In this study, we conducted several tests, including a consolidation test and a vane shear test. This was to examine, in terms of soil mechanics, the material properties of a clayey transmutable water interception material produced by adding bentonite and sodium silicate to marine clay. The study has confirmed that the water interception material under discussion maintains its ability to conform flexibly to ground transformation, while satisfying a required coefficient of permeability.

In this study, we also propose a special impervious seawall system using this water interception material. A model was produced based on this seawall system to conduct a permeability test, which has proved the same system to be highly effective.

1 INTRODUCTION

The volume of waste generated in Japan has remained almost flat in recent years. Meanwhile, general waste finally disposed has been decreasing in volume these years thanks to development in recycling and intermediate treatment. In 2000 an estimated 1,100 tons, a decrease from the previous year, of general waste was sent finally to waste disposal sites. Nonetheless, it is indispensable for the country to procure more disposal sites. Current social circumstances are making it increasingly difficult to find new disposal sites, especially on land, due to difficulty gaining the consensus of the local community. In this regard, coastal disposal sites will become more important as an alternative.

Under these circumstances, the Prime Minister's Office and the Health and Welfare Ministry jointly issued an administrative order. Providing technical guidelines and strengthening standards for seepage control work at waste disposal sites, this order requires any impervious system at these sites to meet at least one of the following conditions:

1 The foundation ground of the disposal site must contain a stratum 5 m or more thick with a coefficient of permeability of 1.0×10^{-5} cm/sec or less

(for rock bed, a Lugeon value of 1 or less), or an equally impervious base.
2 Geomembrane must be laid on strata composed of clay or other soils that are 50 cm or more thick with a coefficient of permeability of 1.0×10^{-6} cm/sec or less.
3 Geomembrane sheets must be laid on asphalt/concrete strata that are 5 cm or more thick with a coefficient of permeability of 1.0×10^{-7} cm/sec or less.
4 Double layers of geomembrane must be laid over the surface of bonded textile or similar materials.

As part of sealing work on marine waste disposal sites, several materials have been conventionally used – geomembranes, impervious sheet piles, asphalt mastics, cement-based treated soil, or combination of these. However, these materials alone are unable to deal sufficiently with problems regarding durability, workability, economical efficiency and other requirements.

To cope with the reinforced government order, we have endeavored to develop a water interception material (Clay-Guard Material), a seepage control method (Clay-Guard Method) and a seepage control system. They are intended for use in seepage control work at controllable coastal waste disposal sites. This study describes mechanical characteristics of a transmutable water interception material developed by using marine clay as the main material. It also presents an impervious seawall model using this water interception material and experimentally confirms the seawall's high imperviousness.

2 DEVELOPMENT OF A TRANSMUTABLE WATER INTERCEPTION MATERIAL (CLAY-GUARD MATERIAL)

2.1 Characteristics of a transmutable water interception material (Clay-Guard Material)

In seepage control work at marine disposal sites, several measures are necessary in addition to one aimed at preventing polluted-water leakage out of the disposal site. The site must be workable enough to permit smooth construction of a seepage control system on the sea. The seepage control system needs to be highly transmutable, that is, capable of maintaining a required level of imperviousness in the event of its deformation due to an earthquake. Further, the system must not only be economically efficient but also durable enough to maintain the adequate level of imperviousness for a long period of time. There has so far been no seepage control system that meets all these requirements simultaneously. Focusing attention to the high imperviousness of dredged clay produced in marine works, we have developed what we call Clay-Guard Material, a water interception material based on dredged soil.

The Clay-Guard Material is composed of dredged marine clay. This material contains bentonite added to it to serve as a void conditioner for adjustment of the clay's coefficient of permeability and sodium silicate as a gelling agent to prevent segregation of the material in the pouring process. Although marine clay itself is a highly impervious substance, it is necessary to add water to it as soon as it is dredged in order to facilitate seepage control work. When water is added to it, marine clay may not meet a required coefficient of permeability. Then bentonite is added to the clay to adjust its coefficient. Sodium silicate is also added to the marine clay to reinforce it and prevent material segregation that could occur while the fabricated water interception Material is being poured.

The Clay-Guard material produced by adding bentonite and sodium silicate to marine clay has the following properties:

1 The coefficient of permeability is 1.0×10^{-6} cm/sec or less.
2 The used material does not solidify over time, free from cracks that would otherwise be caused by deformation.
3 Its design strength is modifiable, permitting strength adjustment where necessary.
4 The material maintains high liquidity enough to allow pumping during the pouring process.

5 Material segregation does not occur in underwater pouring.
6 Being natural, the material does not suffer degradation.
7 Dredged soil is recyclable.

2.2 Mechanical characteristics of Clay-Guard Material

To design and construct a seepage control system for a marine disposal site, it is necessary to identify the mechanical characteristics of the water interception Material. Thus, a basic test was conducted on the Clay-Guard Material to examine its mechanical characteristics. This test focused on the permeability property, strength, liquidity and antiwashout underwater properties of the Clay-Guard Material, the values of which were needed to meet the requirements for the water interception material shown in 2.1.

Table 1. Test items

Test items	Testing method	Standard
Flow property	Flow value test	JIS A 313
Strength property	Vane shear test	-
Permeability property	Standard consoulation test	JIS A 1217
	Hydraulic consolidation test	-
Antiwashout underwater property	Antiwashout underwater test	-

2.2.1 Specimen

To make the Clay-Guard Material, pieces of clay were sampled at the Nagoya seaport to be used as the main material. Table 2 shows elemental physical properties of the clay taken from the Nagoya seaport.

Table 2. Elemental physical properties of Nagoya port clay

Elemental physical property		Value	Unit
Wet density		1.571	g/cm^3
Soil particle density		2.652	g/cm^3
Natural water content		77.7	%
Texture	Sand	6.2	%
	Silt	68.6	%
	Clay	25.2	%
Consistency	Liquid limit	65.4	%
	Plastic limit	33.0	%
	Plasticity index	32.4	-
Ignition loss		6.51	%
Organic carbon content		0.66	%

Its particle size distribution is presented in Fig.1. The clay was sieved into particles with a size of 2 mm or less. Then water was added to make them into adjusted slurry with a water content of 110%. A mixture of bentonite (0, 50, 75, 100, 200 and 300 kg/m^3) and sodium silicate (0, 10, 20 and 30 kg/m^3) was added to this slurry, which was then agitated for

5 minutes to be used as a Clay-Guard sample. In the mixture, powder sodium silicate No.2 and three types of bentonite (A, B, and C) were used.

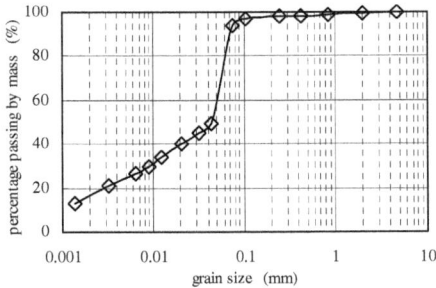

Figure 1. Particle size distribution of Nagoya port clay

2.2.2 Testing method

A flow value test was conducted in accordance with the cylinder method specified in the Testing Methods for Air Mortar and Air Milk under the Japan Highway Public Corporation's standards. An acryl cylinder 80 mm in inside diameter and 80 mm in height was laid on an acryl plate. Clay-Guard samples were placed in the cylinder to be deaerated. After deaeration the cylinder was lifted slowly. Then the samples' lengths were measured in orthogonal directions. The average of the lengths was used as the flow value to calculate the liquidity of the samples.

For a vane shear test, a vessel (200mm × 150mm × 100mm) was filled with samples. The test was conducted immediately, 1 day, 3 days, 7 days, 28 days, and 3 months, after agitation of the samples.

A consolidation test was conducted to examine the consolidation and permeability characteristic of Clay-Guard Material. In the standard consolidation test, the sample was preliminarily consolidated at 4.9 kN/m^2 and then remolded and set inside the consolidation ring to undergo stage loading. The initial consolidation pressure was set at 4.9 kN/m^2. In a standard consolidation test, measurement is only possible at a consolidation pressure of 4.9 kN/m^2 or more. Otherwise it is impossible to identify the coefficient of permeability immediately after agitation of the material. Avoiding this problem, we conducted a hydraulic consolidation test to measure the permeability characteristic of Clay-Guard Material under low stress. Intended solely as an auxiliary test to supplement the standard consolidation test, this hydraulic consolidation test was conducted on only one of the mixtures.

An antiwashout underwater test was conducted in accordance with the Technical Manual for the Lightweight Treated Soil Method. In this test, a sample was mounted in the antiwashout underwater tester and then was poured with a speed-controlled piston into predetermined vessel filled with sea-

water. After pouring, the pH and value and suspended solids of the seawater were measured to identify the antiwashout underwater properties of Clay-Guard Material at pouring. In our test, the pouring speed was adjusted to reflect the actual speed at which the same material would be poured at 50 m^3/hour via an 8-inch tremie pipe on a pump.

2.2.3 Results and discussion

2.2.3.1 Flow value test Fig.2 shows a relation between the amount of bentonite and the flow value. Notice that the flow value decreases as a greater amount of bentonite is added. This suggests that an addition of bentonite reduces both the water content and the void ratio, while the high expensiveness of bentonite decreases the flow value. At the same bentonite level, bentonite A showed a less flow value than B and C. This proved that bentonite A reduces the flow value more effectively than other types of bentonite.

Fig.3 relates the amount of sodium silicate to the flow value. As with Fig.2, this figure shows that the flow value decreases linearly with an increase in the amount of sodium silicate.

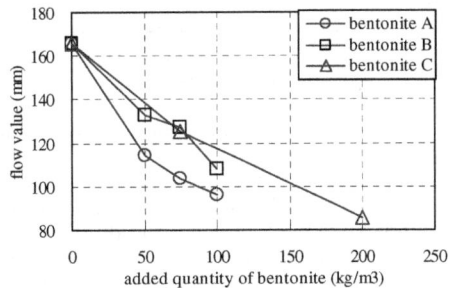

Figure 2. Results of the flow value test (Relation between the flow value and the amount of bentonite added)

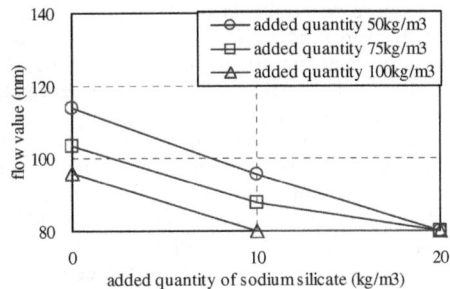

Figure 3. Results of the value test (Relation between the flow value and the amount of sodium silicate added)

It is to be noted, however, that sodium silicate affects the liquidity of the water interception material more strongly than bentonite does, considering that the amount of sodium silicate added was smaller than that of bentonite.

2.2.3.2 Vane shear test Fig.4 shows a relation between strength and curing time at 75 kg/m^3 of bentonite and 10 kg/m^3 of sodium silicate. The figure indicates that the strength of the material surged within a single day after agitation, followed by a slower increase. The value nearly quadrupled in 3 months to 2.5 to 3.5 kN/m^2, suggesting that the sample maintained its gel state without being solidified.

The vane shear test also produced largely the same results as with the flow value test, regarding the difference in bentonite type. At a given length of curing time, the strength of the material was greater in bentonite A than in bentonite B, and also greater in bentonite B than in bentonite C. This confirmed the effect of a difference in bentonite type on the strength of the material.

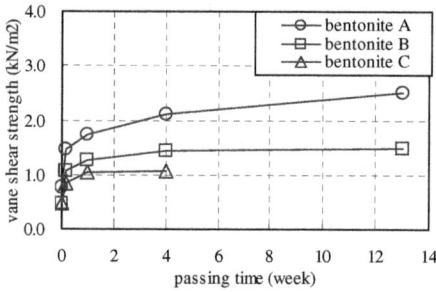

Figure 4. Results of the vane shear test (Relation between vane strength and curing period)

2.2.3.3 Standard consolidation test and permeability consolidation test Fig.5 provides the results of the standard consolidation test to show a relation between the void ratio and the coefficient of permeability of the material that contains 75 kg/m^3 of bentonite but no sodium silicate. Notice that both the void ratio and the coefficient of permeability decrease as more bentonite is added to the adjusted clay. At a given void ratio, bentonite A showed a smaller value than the other types of bentonite. As with the flow value test, this suggests that bentonite A is more expansive than the other types.

Fig.6 shows how the void ratio relates to the coefficient of permeability as a change occurs in the amount of bentonite A. The coefficient of permeability decreased as more bentonite was added at a given void ratio. It was thus concluded that bentonite has a substantial effect on the coefficient of permeability of Clay-Guard Material.

Fig.7 compares the results of the standard consolidation test with those of the hydraulic consolidation test both conducted at 75 kg/m^3 of bentonite with no sodium silicate added. These findings show that, at a theoretical void ratio of 2.5 obtained right after agitation, the coefficient of permeability is below the reference value of k = 1.0 x 10^{-6} cm/sec.

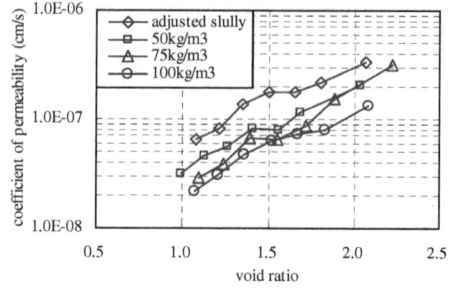

Figure 5. Results of the standard consolidation test I (Relation between the void ratio and the coefficient of permeability)

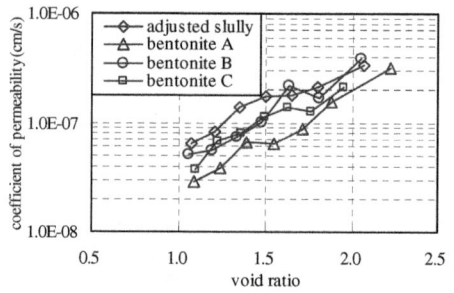

Figure 6. Results of the standard consolidation test II (Relation between the void ratio and the coefficient of permeability)

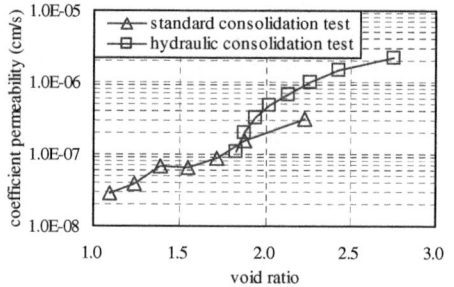

Figure 7. Results comparison between the standard consolidation test and the hydraulic consolidation test

2.2.3.4 Antiwashout underwater test Fig.8 provides the results of the antiwashout underwater test. An addition of sodium silicate confined the suspended solids below the reference value of 100 mg/L, implying that no segregation occurred in the material during the pouring process.

2.2.4 Conclusion

Our tests have proved that the Clay-Guard Material can make a water interception material highly applicable to seepage control work at marine disposal sites. In particular, no other material has so far been found to stay unsolid longer than Clay-Guard Material. This material is expected to retain a required

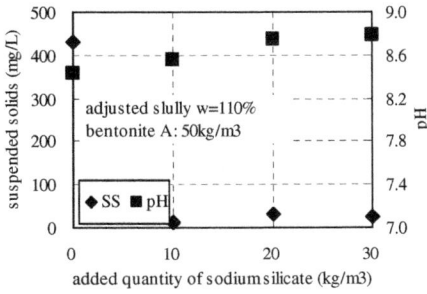

Figure 8. Results of the antiwashout underwater test

coefficient of permeability in the event of an earthquake or any other force working upon the disposal site to deform its seepage control system. In this study, marine clay was used as the main material. In the future we intend to examine the possibility of using other materials. We will also collect more data on the long-term stability of seepage control systems.

3 LARGE-SCALE EXPERIMENT

3.1 Proposed impervious seawall system

In designing an impervious seawall system for marine disposal sites, several factors are taken into consideration – deformation by waves and earthquakes, stability under original-ground conditions, construction period and costs, and workability. To satisfy these requirements, many marine disposal sites have applied geomembranes and impervious sheetpiles to conventionally engineered impervious seawalls, as shown in Figs. 9 and 10. We propose an impervious seawall with a conventional seepage control system using a new water interception material. Fig.11 presents an example. In the proposed system, H-steel sheetpiles (with an inside width of 50 cm or more) treated with expansive waterproof agent on their joints are installed behind an ordinary gravity seawall using caissons. The sheetpiles are then filled with Clay-Guard developed by us as a water interception material. Fig. 12 gives some details of the impervious wall composed of H-steel sheetpiles. Built at a marine disposal site, this impervious wall provides the following advantages:

1 The waterproof agent on sheetpile joints and the water interception material inside sheetpile walls intercept water flows more effectively.
2 The water interception material inside sheetpile walls conforms to any deformation caused by earthquakes or any other force, thus maintaining high imperviousness.
3 Even if the sheetpiles lose their impervious effect due to damage and deterioration in them or their joints, the water interception material itself still ensure imperviousness.

3.2 Transmutability test

In building seepage control systems at marine disposal sites, high priority should be given to their transmutability. In this experiment, we examined effects of the Clay-Guard material on the coefficient of permeability by forcibly displacing the material.

3.2.1 Outline of experiment

Figs. 13 and 14 show an outline of our experiment. As illustrated in Fig. 13, two porous acryl plates were placed inside an earth tank made of steel. The

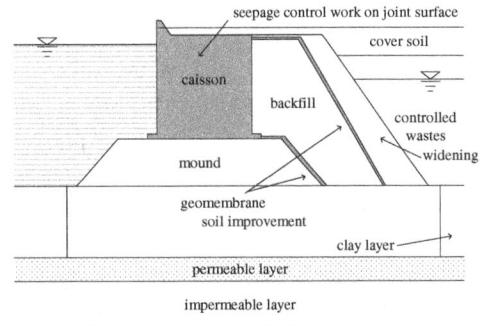

Figure 9. Gravity+double geomembrabe

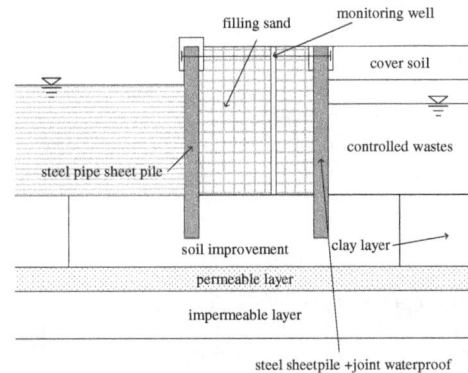

Figure 10. Steel sheetpile+joint waterproof

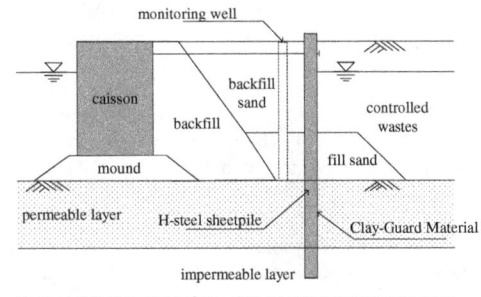

Figure 11. Proposed impervious seawall system

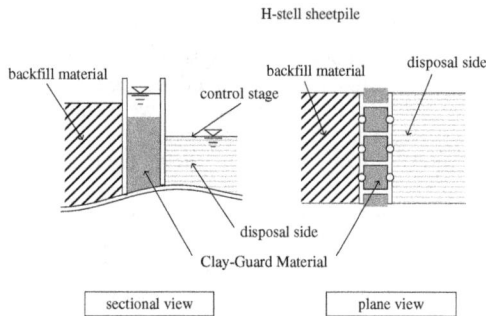

Figure 12. Details of H-steel sheetpile wall

tank was filled with water interception material. The volume of water running horizontally was measured to determine the coefficient of permeability. By installing a continuous thread bolt on top of the tank, it is possible to deform the upper portion of tank, thus generating shearing strain in the water interception material. To prevent potential separation of the material from the steel due to dry shrinkage of the former, a uniform load of 1.0 kN/m² was exerted on top of the material. The following levels of shearing strain were applied to the water interception material: Shearing strain (top displacement/material height) εs: 0%, 1.0%, 2.0% and 4.0%. The clay used was equivalent to those provided in Table 2. Clay-Guard samples were made by adding 75 kN/m² of bentonite A to base clay after adjusting it to a water content of 110%.

3.2.2 Results and discussion

By applying the results of the experiment to Formula (1) obtained from Darcy's Law, we determined the coefficient of permeability.

$$Q = Ak\frac{h}{L}t \qquad (1)$$

where Q = permeability(cm³); A = permeating cross-section(cm²); k = conversion coefficient of permeability(cm/sec); h = water level difference(cm); L = permeating length(cm); t = measuring time (sec). Fig.15 presents the results of the experiment. Even after shearing strain was applied to the water interception material, the coefficient of permeability did not increase, causing no crack in the material. This was because the material contained no solidifying substance, such as cement. But the coefficient of permeability decreased once the shearing strain exceeded 2.0%, probably because deformation enhanced the repletion of the material.

3.2.3 Conclusion

In this experiment, we determined the coefficient of permeability of the Clay-Guard material in the presence of shearing strain. Containing no solidifying substance, such as cement, Clay-Guard was free

from cracks and maintained a required coefficient of permeability. In future experiments, we plan to examine the acceptable volume of deformation.

Figure 13. Summary diagram of the experiment

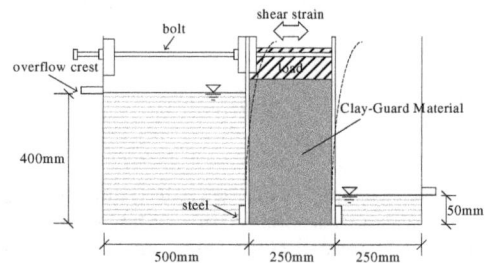

Figure 14. Testing tank

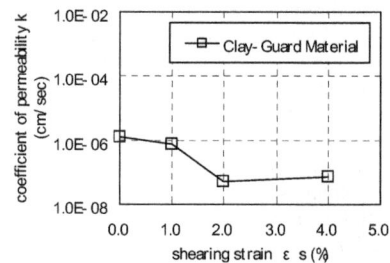

Figure 15. Test results (Relation between shearing strain and the coefficient of permeability)

3.3 Impervious-wall performance experiment

To confirm the effectiveness and impermeability characteristics of the proposed seawall system provided in Figs. 11 and 12, we conducted an experiment using a large-scale seawall model. As shown in Fig.16, H-steel sheetpiles were used in the seawall system.

3.3.1 Outline of experiment

Fig.16 illustrates an outline of experiment. Two types of sheetpiles – sheetpile I and sheetpile II –

198

were installed inside a large earth tank to build two sheetpile walls. Each sheetpile wall had three joints, which were prepared as shown in Table 3. A water-expansive waterproof agent was applied to both end hooks of joints prior to assembly of sheetpiles. After installation of sheetpiles, Cray-Guard was pumped and poured via a small tremie pipe into the sheetpile walls. The Cray-Guard material was prepared as shown in Table 4.

In the experiment, the two compartments separated by sheetpile walls were filled with seawater. By measuring the volume of seawater leaking through sheetpile joints, we calculated the conversion coefficient of permeability of the impervious sheetpile walls. Fig.17 gives some details of the experiment.

3.3.2 Determination of conversion coefficient of permeability

The conversion coefficient of permeability of the sheetpile wall was calculated using Formula (1). Table 5 lists the conditions applied to the calculation. The consolidation displacement volume was corrected in line with consolidation of the Clay-Guard Material.

Table 4. Properties of Clay-Guard Material

Water content of adjusted clay (Nagoya port clay)	110 %
Bentonite A added	75 kg/m³
Sodium silicate added	None

Figure 17. Detail of the experiment

Table 5. Calculation requirements

	A(cm²)	L(cm)	h(cm)
Type I	60.5×h	38	45
Type II	50×h	50	60

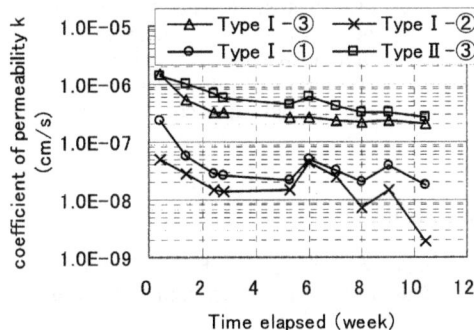

Figure 18. Conversion coefficient of permeability of impervious sheetpile walls

3.3.3 Results and discussion

Fig.18 shows a relation between the conversion coefficient of permeability and measuring time. Sheetpiles II(1) and II(2) do not appear in the graph, because they showed no leakage. Sheetpiles may be considered to be almost completely waterproof if an expansive waterproof agent is working completely all over the sheetpile joints. This has been confirmed in other papers.

For sheetpiles I(3) and II(3), their joints were fed with the new water interception material Clay-Guard, but no waterproof agent was applied to them. The conversion coefficient of permeability recorded about 1.0×10^{-6} cm/s right after pouring and kept decreasing with time, until it almost equaled the coefficient of permeability of the water interception

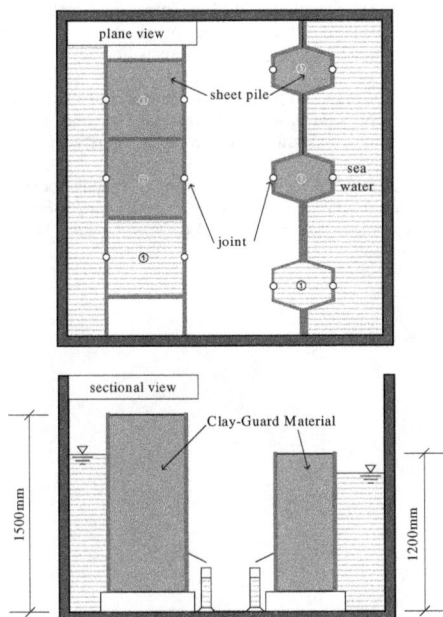

Figure 16. Summary diagram of the experiment

Table 3. Waterproof preparations

	①	②	③
Joint waterproof agent	Used	Used	Not used
Clay-Guard Material	Not used	Used	Not used

material. Thus, we can regard this test as almost equivalent to a permeability test on the Clay-Guard Material using sheetpiles as a mold.

Joint waterproof agent was applied to both sheetpiles I(1) and I(2). Further, sheetpile I(2) was fed with the new water inception material. Both sheetpiles sustained water leakage despite the application of waterproof agent. This implies that the waterproof agent was damaged or peeled off for some reason. Nevertheless, their conversion coefficient of permeability was between 10^{-7} and 10^{-8} cm/s, satisfying the standard value of 1.0×10^{-6} cm/s. Although the coefficient value fluctuated with time for both sheetpiles, it never exceeded the standard value. Especially, for sheetpile I(2), which was fed with the Clay-Guard Material, its conversion coefficient of permeability fluctuated but slowly declined. This suggests that the water interception material refilled the leakage points. Even if its waterproof agent fails to fully function, sheetpile I(2) was considered to have a greater conversion coefficient of permeability than sheetpiles I(3) and II(3), because it contained the Clay-Guard Material.

3.4 Conclusion

Our experiment proved that an impervious sheetpile wall acquires a satisfactory seepage control function when used jointly with H-steel sheetpiles and the Cray-Guard Material. We plan to study the effectiveness of this impervious wall in actual costal environment from various aspects, including its workability.

4 SUMMARY

In this study we clarified the physical properties of the Clay-Guard Material, and examined the imperviousness of the material when it is used in an impervious seawall proposed by us. The results successfully satisfied their requirements. It must be remembered, however, that the water inside a disposal site may behave quite differently from the seawater. In light of this, we need to develop technology for building environmentally secure disposal sites, and at the same time endeavor to expand applications for the Clay-Guard Material.

REFERENCES

Waterfront Vitalization and Environment Research Center. 2000. Manual of Design, Construction and management at a coastal disposal site. Tokyo: Waterfront Vitalization and Environment Research Center

Yamada, K et al. 2002. Proposal of a new impervious sea wall structure in a costal disposal site with the deformation following material; Proceedings of Civil Engineering in the Ocean 2002. Tokyo: Japan Society of Civil Engineers

Ueno, K et al. 2002. Development of the impervious method to follow a deformation in a coastal disposal site (part1); The 37th Japan national conference on Geotechnical Engineerings. Tokyo: The Japanese Geotechnical Society

Hada, A et al. 2002. Development of the impervious method to follow a deformation in a coastal disposal site (part2); The 37th Japan national conference on Geotechnical Engineerings. Tokyo: The Japanese Geotechnical Society

Ueno, K et al. 2002. Development of the impervious method to follow a deformation in a coastal disposal site (part1); The 57th JSCE annual meeting. Tokyo: Japan Society of Civil Engineering

Hada, A et al. 2002. Development of the impervious method to follow a deformation in a coastal disposal site (part2); The 57th JSCE annual meeting. Tokyo: Japan Society of Civil Engineering

Watabe,Y et al 2002.Hydraulic conductivity of seepage control clayey material with consideration of its microstructure;The 47th Geotechnical Symposium,Tokyo:The Japanese Geotechnical Society.

Soft Ground Engineering in Coastal Areas, Tsuchida et al. (eds)
© 2003 Swets & Zeitlinger, Lisse, ISBN 90 5809 613 0

A new construction control method for reclamation work in the Kansai International Airport second phase project

N. Yamane
Osaka Branch, Toa Corporation

T. Fukasawa
Technical Research Institute, Toa Corporation

J. Mizukami
Engineering Department, Kansai International Airport Co., Ltd.

ABSTRACT: Stability and deformation are major geotechnical considerations on soil structures on soft seabed constructed by reclamation. A case study is presented of the application of bathymetric survey to construction control in the Kansai international airport island construction work where by sand drain method is used for soft ground improvement. Effective construction control of such soil structure requires a systematic way of collecting accurate information on progress of consolidation, together with magnitude and distribution of load increment over the improved ground.

1 INTRODUCTION

Generally, the settlement measurement of soil structure constructed on the soft ground is performed using settlement plate. When the settlement plate is not installed, the settlement of the ground can not measure. The settlement of such place where has not installed the settlement plate must be estimated from the settlement of the settlement plate in close vicinity. However, in case of the large-scale reclamation work with the deepwater, there is a problem which can not sufficiently install the settlement plates like the land construction. Therefore, the development of the efficient construction control system in the large-scale reclamation work had been required.

Then, authors developed the construction control system which could grasp thickness of sand fill, settlement of the ground at the site by a newly developed system utilizing bathymetry data. A construction record including unit weight and volume change late of sand fill can be also confirmed by the developed system. Coefficient of consolidation and coefficient of volume compressibility can be calculated based on the thickness of sand fill, settlement of the ground and construction record. In this paper, the case history of applying the developed system to Kansai International Airport 2nd Phase Project is reported.

2 OUTLINE OF CONSTRUCTION WORK

Kansai International Airport, which saw its beginning on September 4, 1994, launched 2nd Phase Project on July of 1999 to supplement its current facilities with an additional runway of 4,000 m. The reclamation area of the 2nd phase airport island reaches 5,450,000 m². Our construction site is located in the north-side in the 2nd phase airport island. The plan of the airport island and our project site is shown in Figure 1. Figure 2 shows A-A cross section in figure 1 as a typical cross section of the seawall. Figure 3 shows the flowchart of construction and soil investigation period. The schematic land development workflow is as follows:

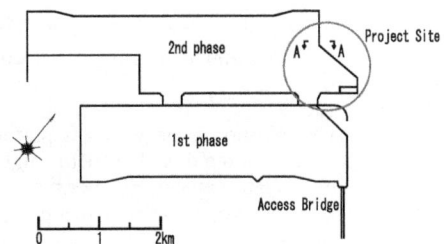

Figure 1. Plan of Kansai International Airport and location of project site

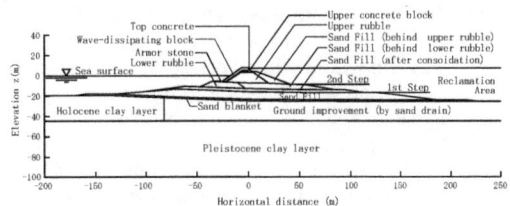

Figure 2. Typical cross section of the seawall (section A-A in Figure 1)

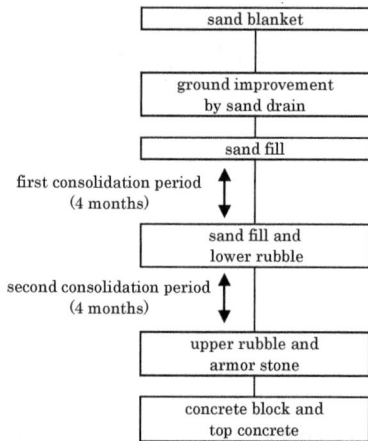

Figure 3. Flowchart of construction and investigation period

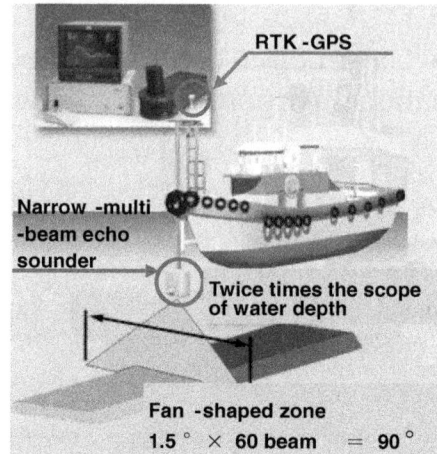

Figure 4. Conceptual scheme of bathymetric survey

1) Spreading of the sand blanket

The 1st part of ground improvement works by the sand drain method is the spreading of a sand blanket. The sand blanket with a layer thickness of 1.5m in 2 steps on the seabed was created by sand spreading barge.

2) Ground improvement by sand drain method

After spreading a thin sand blanket, sand drain piles, 40cm in diameter and about 25m in length, are driven into the alluvial clay layer at 2.5m by 1.6m intervals under the seawall and 2.5m by 2.5m in reclamation area by a sand drain barge.

3) Sand fill

For the 1st step of seawall construction, sand fill with a layer thickness of about 16m is constructed by direct dumping by hopper barge.

4) Rubble dumping

Rubble, which is dumped by sand carrier with grab bucket, serves as the foundation for upper concrete blocks.

5) Armor stone

The weight of an armor stone which is placed outside of upper rubble mound is about 1 ton and it fortifies the rubble against waves. The armor stone is leveled by backhoe above water or manpower of a diver below.

6) Concrete block

Concrete blocks are used to form an earth retaining wall. The concrete blocks, which are produced on land, are transported and placed by crane barge.

To perform reclamation works of an airport island safely, and to prevent uneven settlement, it is necessary to check how the strength of the seabed improved by sand drain method is increasing with the progress of construction works. Therefore, in the 2nd phase construction works, check borings are performed in the end of each consolidation period, and the next stage is to be commenced after it has been

verified that the strength of alluvial clay layer in the current stage has increased.

The 2nd phase island reclamation expected to settlement more than the 1st phase did because the sea area of the reclamation site features a thick layer of soft marine clay under the seabed, in addition to a greater depth.

3 NEW CONSTRUCTION CONTROL SYSTEM BY BATHYMETRIC SURVEY

3.1 Concept of bathymetric survey by narrow multi beam echo sounder

When filling sand is dumped into the seabed, settlement occurs. If this settlement is uneven, a problem will arise when constructing a runway or a building. In order to prevent this uneven settlement, we need to obtain accurate data on the formation of the sand filling which has accumulated on the seabed.

The conceptual scheme of bathymetric survey is shown in Figure 4. The bathymetric survey of the seabed is carried out by a narrow multi beam echo sounder using ultrasonic waves with an acute directivity angle (1.5 degree) and emitting 60 beams at one time to a fan-shaped zone at 90 degree, the depth of seabed can be measured in two dimensions. A positioning survey is performed using RTK-GPS and the effect of survey boat motions such as pitching and rolling is compensated for with correcting instruments.

3.2 Controlling the thickness of the reclamation layer and settlement

The workflow of the bathymetric survey to grasp the layer thickness and settlement is shown in Figure 5.

Actual operational procedures of the bathymetric survey are as follows:

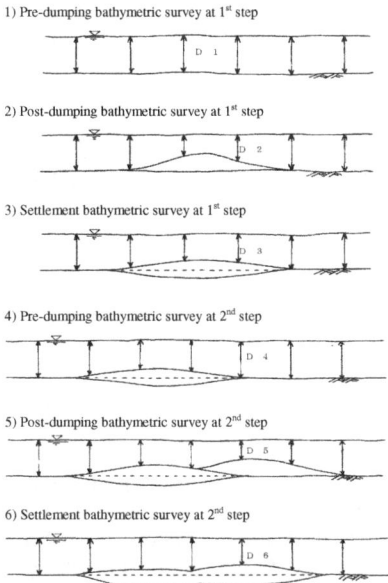

1) Pre-dumping bathymetric survey at 1st step

2) Post-dumping bathymetric survey at 1st step

3) Settlement bathymetric survey at 1st step

4) Pre-dumping bathymetric survey at 2nd step

5) Post-dumping bathymetric survey at 2nd step

6) Settlement bathymetric survey at 2nd step

Figure 5. Workflow of the bathymetric survey to grasp the layer thickness and settlement

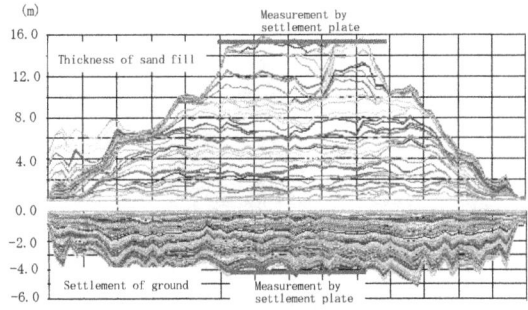

Figure 6. Thickness of the sand fill and settlement by bathymetric survey

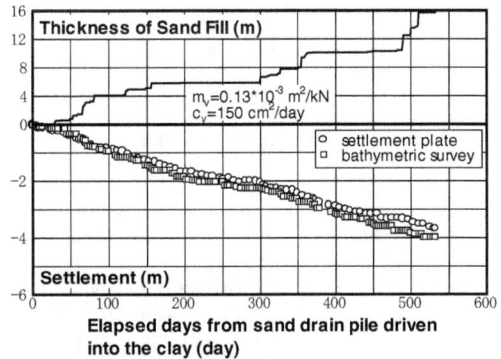

Figure 7. The thickness of sand fill and settlement curves versus times

1) Pre-dumping bathymetric survey

The bathymetric survey of the seabed is carried out as possible as wide area before sand dumping to grasp the current condition.

2) Post- dumping bathymetric survey

After dumping, the bathymetric survey is again carried out to grasp the exact condition of the seabed. The thickness of the reclamation layer is calculated from the difference in these survey 1) and 2).

3) Settlement survey

The Settlement survey is carried out periodically to obtain the settlement of the interval of dumping.

By repeatedly carrying out this process, it is possible to grasp the thickness of the sand fill and settlement.

Figure 6 shows the detailed history of the multiple profiles of the sand fill layers and the corresponding settlement of the ground surface at section A-A obtained from this bathymetric survey. It can be confirmed that the sand fill has been uniformly constructed and neither failure nor excessive settlement occurs beneath the sand fill from this survey data. And the results are good agreement with actual measurement by the settlement plate.

Figure 7 shows the thickness of sand fill and settlement curves versus times obtained from survey data, together with results of settlement plate measurements. The settlement of ground obtained from bathymetric survey show good agreement with actual measurement by settlement plate. Thus, it became possible that the settlement was grasped at any point from the bathymetric survey data. And using this time-settlement curve, coefficient of consolidation and coefficient of volume compressibility can be calculated at any point.

Thickness of sand fill and settlement of the ground are shown in Figure 8 as visually contour figure. These figures enable us to confirm the undulation of thickness of the sand fill and settlement of the ground easily. Thus, by this contour figure local failure and differential settlement can be found.

3.3 Confirmation of construction record (volume change rate and unit weight of sand fill)

In order to calculate the settlement of the ground, it is necessary to convert the thickness of constructed layer into the constructed load. By the developed bathymetric survey system, a construction record such as volume change rate and unit weight of sand fill can be confirmed.

The concept of confirmation of the construction record is shown in Figure 9. Sand fill materials are loaded from a source of sand supply to hopper barge. Weight, volume and water content of the fill materials are measured when it is loaded form a source of sand supply to hopper barge. Accumulated volume, volume change rate and unit weight of the

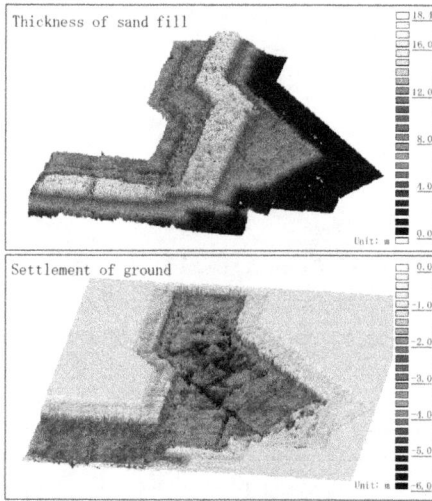

Figure 8. Thickness of sand fill and settlement of ground (three dimension view)

saturated fill materials are calculated based on bathymetric survey data before and after sand dumping.

Figure 10 shows the histogram of the volume change rate and unit weight of sand fill obtained from the bathymetric survey including about 2,000 hopper barge's data. These values change by the construction condition such as water depth of construction site, type of hopper barge, tidal current and so on. The distribution of these data shows the normal distribution and the mean value of the volume change rate is 1.2 and the mean value of saturated unit weight is 12.4 kN/m^3.

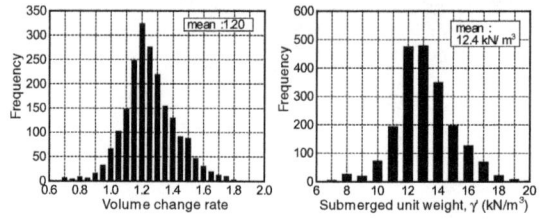

Figure 10. Histogram of the volume change rate and unit weight the concept of confirmation of the construction record

4 CONCLUSIONS

In this paper, the construction control method by newly developed bathymetric survey system was reported. Results of this paper are as follows:

1 The settlement of ground obtained from bathymetric survey show good agreement with actual measurement by settlement plate. Thus, it became possible that the settlement is grasped at any point from the bathymetric survey data.

2 From the bathymetric survey data, the volume change rate and unit weight of sand fill can be confirmed.

REFERENCES

T. Fukasawa, H. Hirabayashi, T. Hosaka and M. MatsudaIn. 2001. Evaluation of thickness of embankment, settlement, of the ground and loading history using echo sounding system.

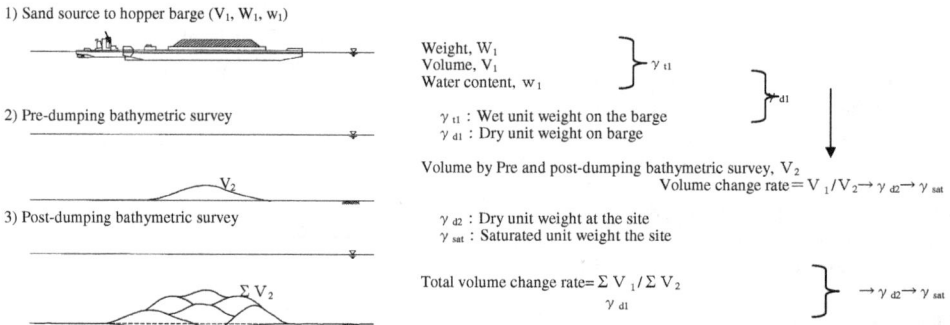

Figure 9. The concept of confirmation of the construction record

Soft Ground Engineering in Coastal Areas, Tsuchida et al. (eds)
© 2003 Swets & Zeitlinger, Lisse, ISBN 90 5809 613 0

Calculation of settlements of foundation soils considering creep

J.H. Yin

Department of Civil & Structural Engineering, The Hong Kong Polytechnic University, Hong Kong, China

ABSTRACT: This paper discusses three approaches to calculation of settlements of foundation soils with creep – two approaches based on Hypothesis B and one based Hypothesis A. Hypothesis A ignores creep compression in the "primary" consolidation period, therefore underestimates the total settlement. Hypothesis B considers creep compressions in both the "primary" and the "secondary" consolidation periods and is logically correct, but faces difficulties in implementation. The author presents a simplified approach to calculation of settlements of soils with creep. A new internet-based calculation tool for calculation of the averaged degree of consolidation and dissipation of excess pore water pressure in 1-D and 2-D with vertical drainage and preloading is introduced. Three example problems are analyzed using the three approaches. Results are compared to each other and/or measured results.

1 INTRODUCTION

Soft soils have been encountered in Hong Kong and in many parts of the world. In Hong Kong, a typical soft soil is soft marine deposits (silty clay or clayey silt) in the seabed (Yin and Graham 1989, 1994). Soft marine soils are often encountered in the construction of reclamation, seawalls near-shore, and foundations on reclaimed land in Hong Kong. The soft marine soils are problematic foundation soils for civil engineering purposes due to low shear strength and high time-dependent compressibility. The skeleton of soft soils exhibits time-dependent stress-strain behavior such as creep, relaxation, strain-rate effects.

Creep is the continuous compression of the soil skeleton under a constant effective stress. Creep of soils is due to mainly (a) viscous squeezing out of adsorbed water in double layers on clay particles and (b) viscous re-arrangement /deformation of clay particles. Adsorbed water is not free water and cannot flow freely under gravity or a hydraulic gradient. Under certain effective stress, however, the adsorbed water will move out slowly. At the same time, the clay particles (plate structure) move closer and re-arrange to a new equilibrium position, resulting further deformation or compression of the soil. This compression occurs slowly with time. According to the mechanism, the creep will occur whenever effective stresses exist in the soils or are acting on the clay particles with adsorbed water, independent of the free porewater in voids or the consolidation process. How to consider the creep in the settlement calculations is the main focus of this paper.

For 1-D straining condition, there are two controversial methods for the calculation of consolidation settlement with creep. The two methods are based on Hypothesis A and Hypothesis B. Hypothesis A assumes that creep occurs after "primary" consolidation. Therefore, the total settlement is equal to primary" consolidation settlement only before t_{EOP} (time at the End-Of-Primary consolidation) and equal to the "primary" consolidation settlement plus "secondary" consolidation (or creep) settlement after t_{EOP}. It is commonly appreciated in the international geotechnical community that Hypothesis A is logically wrong, but simple in calculation, numerically right by good luck (or using larger/ more conservative compression index, for example), and still used by most consulting firms.

Hypothesis B assumes that creep occurs during whole consolidation/ compression process, that is, during and after "primary" consolidation. According to Hypothesis B, the total settlement is equal to "instant" consolidation settlement plus "creep" settlement from the starting time. The latter method is considered to be logically correct, but complicated in calculation. This paper first introduces a fully coupled consolidation analysis approach based on Hypothesis B and using a 1-D elastic visco-plastic model (Yin and Graham 1989, 1992) for soil skeleton. The results are compared with the results from Hypothesis A. Then, the author presents and discusses a simplified method based on Hypothesis B.

2 A RIGOROUS APPROACH BASED ON HYPOTHESIS B USING A 1-D ELASTIC VISCO-PLASTIC MODEL

For 1-dimensional straining (oedometer) condition, the equation based on the condition of continuity is:

$$\frac{k}{\gamma_w}\frac{\partial^2 u}{\partial z^2} = -\frac{\partial \varepsilon_z}{\partial t} \tag{1}$$

where k is the vertical permeability, γ_w is the unit weight of water, u is the excess pore water pressure, z is the vertical co-ordinate, ε_z is the vertical strain and t is time. In Eqn.(1), all assumptions in Terzaghi's 1-D consolidation theory has been used except for the constitutive relationship for the soil skeleton.

Yin and Graham (1989, 1994) proposed a 1-D Elastic Visco-Plastic (1-D EVP) constitutive relationship for the time-dependent stress-strain behavior of soft soils in 1-D straining:

$$\frac{\partial \varepsilon_z}{\partial t} = \frac{\kappa/V}{\sigma_z'}\frac{\partial \sigma_z'}{\partial t} +$$
$$+ \frac{\psi/V}{t_o} exp\left[-(\varepsilon_z - \varepsilon_{zo}^{ep})\frac{V}{\psi}\right]\left(\frac{\sigma_z'}{\sigma_{zo}'}\right)^{\lambda/\psi} \tag{2}$$

where κ/V, λ/V, σ_{zo}', ε_{zo}^{ep}, ψ/V, t_o are six model parameters, all of which can be determined using oedometer creep test data (see Yin and Graham 1889, 1994). Approximate relationships with the conventionally used parameters are:

$$\begin{cases} \kappa \approx \dfrac{C_r}{ln(10)}, \\[2mm] \lambda \approx \dfrac{C_c}{ln(10)}, \ \varepsilon_{zo}^{ep} = \varepsilon_{vc,24} \ (or \ \varepsilon_{vc,EOP}), \\[2mm] \qquad \sigma_{zo}' = \sigma_{c,24}' \ (or \ \sigma_{c,EOP}'), \\[2mm] \psi \approx \dfrac{C_{\alpha e}}{ln(10)}, \ t_o = t_{24} \ (or \ t_{EOP}) \\[2mm] V = 1 + e_o \end{cases} \tag{3}$$

Using the effective stress principle equation of $\sigma_z' = \sigma_z - (u_s + u)$ (u_s is hydraulic static pore water pressure, Eqn.(1) and Eqn.(2) can be combined to solve the consolidation problems. Yin and Graham (1996) proposed using a finite difference method to solve the system of Eqns. (1) and (2).

Fig.1 shows calculated settlement and excess porewater pressure with time using the present method based on Hypothesis B and the old method based on Hypothesis A. It is seen from the figure that at time $t=5\text{x}10^{-6}$min, the settlement is only 13.0 cm based on Hypothesis A, but is 23.0 cm based on Hypothesis B. The difference is 10 cm. This means

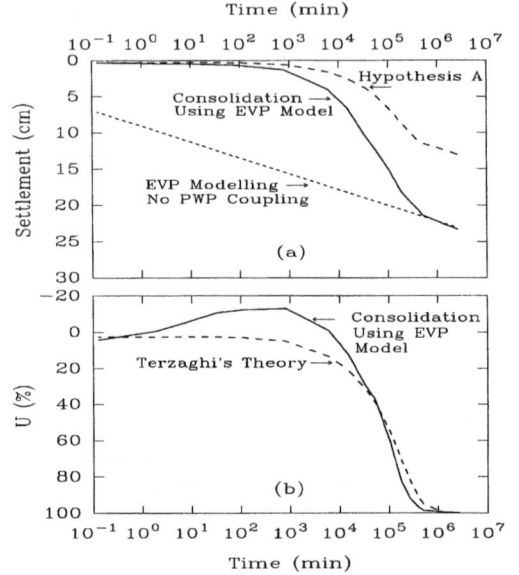

Figure 1. Calculated settlement and excess porewater pressures with time based on Hypothesis A and Hypothesis B (after Yin et al. 1996)

that Hypothesis A underestimates the settlement. The reason is that the creep settlement occurred during the "primary" consolidation has been neglected. From this study, it is found that the settlement larger than that "expected" is likely due to the wrong calculation using Hypothesis A, assuming that other factors are well defined or correct.

3 A SIMPLIFIED APPROACH BASED ON HYPOTHESIS B

Based on the previous study, Hypothesis B is a right approach. However, Hypothesis B normally needs to carry out a fully coupled consolidation analysis using an appropriate constitutive model for time-dependent (creep) of clayey soils (Yin and Graham 1996). This is difficult for practical applications by consultants since programs for such analysis are not readily available. This paper presents a new simplified method based on the principle of Hypothesis B for calculation of settlements of soils with creep. In order to compare with the simplified method, the method based on Hypothesis A is presented here first:

$$S_{totalA} = S_{"primary"} + S_{"secondary"} =$$
$$= \begin{cases} U_v S_f & for \ t < t_{EOP} \\[2mm] S_f + C_{\alpha e} log(\dfrac{t}{t_{EOP}})H & for \ t > t_{EOP} \end{cases} \tag{4}$$

206

where S_{totalA} is the total settlement based on Hypothesis A and U_v is the average degree of consolidation. In Eqn.(1), $C_{\alpha e}$ is the "secondary" coefficient of consolidation defined as $C_{\alpha e} = \Delta \varepsilon_z / \Delta log(t) = C_{\alpha e}/(1+e_o)$. It is noted that $C_{\alpha e} = -\Delta e / \Delta log(t)$. The S_f is the final instant settlement and can be calculated as, for a single layer and for simplicity :

$$S_f = \varepsilon_{inst} H \qquad (5)$$

where $\varepsilon_{inst} = M_v \Delta \sigma_v$ is the instant compression strain (see Fig.4), M_v is the coefficient of volume compressibility, $\Delta \sigma_v$ is the total stress increase, and H is the layer thickness. For soft soil, it is better to use the following equation:

$$\varepsilon_{inst} = \begin{cases} \dfrac{C_c}{1+e_o} log\dfrac{\sigma'_{vo}+\Delta\sigma_v}{\sigma'_{vo}} & (i) \\[2ex] \dfrac{C_r}{1+e_o} log\dfrac{\sigma'_{vc}}{\sigma'_{vo}} + \dfrac{C_c}{1+e_o} log\dfrac{\sigma'_{vc}+\Delta\sigma_v}{\sigma'_{vc}} & (ii) \\[2ex] \dfrac{C_r}{1+e_o} log\dfrac{\sigma'_{vo}+\Delta\sigma_v}{\sigma'_{vo}} & (iii) \end{cases} \qquad (6)$$

where C_r, C_c and e_o are un/reloading compression index, compression index and initial void ratio respectively. The σ'_{vo} and σ'_{vc} are initial vertical effective stress and pre-consolidation stress (or pressure). In Eqn.(5, Case (i) is for normally consolidated soils with $\sigma'_{vo} = \sigma'_{vc}$; Case (ii) is for over-consolidated soils with $\sigma'_{vo} < \sigma'_{vc}$ and total vertical effective stress $(\sigma'_{vo}+\Delta\sigma_v)$ larger than the pre-consolidation pressure; and Case (iii) is for over-consolidated soils with $\sigma'_{vo} < \sigma'_{vc}$, but total vertical effective stress less than the pre-consolidation pressure.

The equation of the simplified method based on Hypothesis B is

$$S_{totalB} = S_{"primary"} + S_{creep} =$$
$$= U_v S_f + C_{\alpha e} log(\frac{t+t_o}{t_o}) H \quad for \ any \ t \geq 0$$
$$= U_v S_f + \frac{\psi}{V} ln(\frac{t+t_o}{t_o}) H \quad for \ any \ t \geq 0 \qquad (7)$$

where S_{totalB} is the total settlement based on Hypothesis B and ψ/V is a creep parameter (Yin and Graham 1989, 1994), defined as $\psi/V = \Delta\varepsilon / \Delta ln(t)$ and is related to $C_{\alpha e}$ by $\psi/V = C_{\alpha e}/ln(10)$. It is noted that the specific volume $V=1+e_o$. The t_o in Eqn.(7) is another creep parameter. In Eqn.(7), the creep strain is

$$\varepsilon_{creep} = C_{\alpha e} log(\frac{t+t_o}{t_o}) \qquad (8)$$

which has definition when $t=0$. The meaning and determination of the parameters are discussed as followings.

The differentiation of Eqn.(8) with time is:

$$\frac{d\varepsilon_{creep}}{dt} = \frac{C_{\alpha e}}{(t+t_o)ln10} \qquad (9)$$

when $t=0$, the initial creep rate is

$$\frac{d\varepsilon_{creep}}{dt}\bigg|_{t=0} = \dot{\varepsilon}_{creep t=0} = \frac{C_{\alpha e}}{t_o ln10} \qquad (10)$$

It is seen from Eqn.(10) that $C_{\alpha e}/(t_o ln10)$ is, in fact, the initial creep rate. It is common that the oedometer test is carried out with 24 hours of loading duration. The compression index C_c (or C_r) can be determined and the instant compression strain ε_{inst} can be calculated using Eqn.(6). The creep settlement is:

$$S_{creep} = \varepsilon_{creep} H \qquad (11)$$

If a creep test is done at Point 1 in normally consolidated range, the data of ε_{creep} vs. log(t) can be best fitted with Eqn.(8). Thus, $C_{\alpha e}$ and t_o can be determined. The best-fitting can be done easily by a trial-error process by knowing that $C_{\alpha e}$ is basically the slope of ε_{creep} vs. log(t) when t is large (say $t>24$ hours for specimen thickness 20mm) and t_o only affects the curvature for $t<24$ hours or even less). The thinner, the test specimen, the better, the determination of t_o. The $C_{\alpha e}$ obtained is the "secondary" consolidation coefficient for the normally consolidated range and t_o is equal to approximately 24hr in our experience.

If a creep test is done in the over-consolidated range, the data of ε_{creep} vs. log(t) can be fitted with Eqn.(8). Thus the two parameters $C_{\alpha e}$ and t_o can be determined. The $C_{\alpha e}$ obtained is the "secondary" consolidation coefficient for the over-consolidated range and is much smaller than that in the normally consolidated range. The t_o shall be determined by the best-curve fitting, but, is probably equal to approximately 24hr in our experience. Thus the initial creep rate $(C_{\alpha e}/t_o/ln10)_{OC}$ in the over-consolidated range is much smaller than $(C_{\alpha e}/t_o/ln10)_{NC}$ in the normally consolidated range. For simplicity, as long as the appropriate creep tests are carried out, the two parameters $C_{\alpha e}$ and t_o can be determined. Yin and Graham (1989, 1994) introduced the concept of "equivalent time" t_e, so that the $C_{\alpha e}$ measured in the normally consolidated range can be used for creep in over-consolidated range and even in the unloading/reloading range.

Eqn.(7) can be used to calculate the total consolidation settlement. This equation includes the creep compression occurring in both the "primary" consolidation and "secondary" consolidations and is an approximation of the fully coupled consolidation analysis presented in the preceding section.

4 COMPARISON OF SETTLEMENTS CACULATED USING HYPOTHESIUS A AND THE SIMPLIFIED METHOD

An example is presented here to explain the proposed simplified method based on Hypothesis B for the calculation of settlements of soils with creep and compared to that using Hypothesis A. The soil profile is shown in Fig.2. Soil parameters are presented in Table 1.

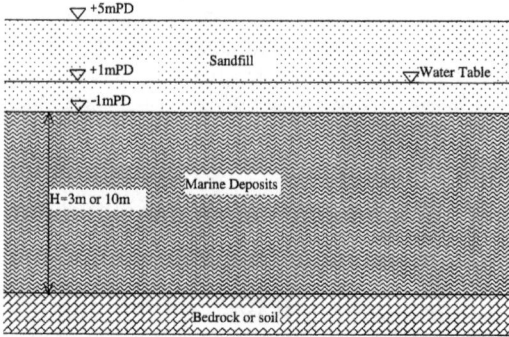

Figure 2. Soil profile used in the example calculation

Table 1. Typical soil parameters

Conventional Parameters	Hong Kong Marine Deposits
k_v (m/sec)	2.5×10^{-10}
$CI=C_c/V$	0.25
$RI=C_r/V$	0.035
$C_{\alpha\varepsilon}=\Delta\varepsilon/log(t_2/t_1)$	0.009
$C_{\alpha e}=\Delta e/log(t_2/t_1)$	0.018
e_o	1
t_o(day)	1
γ_{clay} (kN/m^3)	16
γ_{fill} (kN/m^3)	20
C_v (m^2/year)	0.407 for H=3m
	0.488 for H=10m

In Figure 2, the marine deposits are existing soils. The sandfill is placed on the marine deposits instantly. The effective weight of the sandfill is the uniform loading on the marine deposits. Eqns.(6) and (7) are used to calculate the final settlement S_f of the layer with a thickness H=3m and 10m. The values calculated are presented in Table 2. Eqn.(4) and Eqn. (7) are used to calculate the consolidation settlement of the marine deposits layers with time based on Hypothesis A and Hypothesis B. The results are presented in Fig.3. In the calculation, the U_v is calculated using the solution and chart by Zhu and Yin (1998) or the "Interactive Analysis" tool in the web site: *www.cse.polyu.edu/~civcal/wwwroot/ reclamation/default.htm*, which is developed by Yin and his research team. Fig.4 shows (a) part of a pa-

Table 2. Calculated results for H=3m or 10m (no vertical drains)

Results	H=3m	H=10m
Time for 99% "primary" consolidation	9.9 years (bottom permeable)	97 years (bottom permeable)
σ_{vo} (kPa)	9.3	31.0
$\Delta\sigma_v$ (kPa)	100.4	100.4
$S_{"primary"}$	0.804m	1.569m
S_{creep} for 1 year	0.0692m	0.231m
S_{creep} for 10 years	0.0962m	0.321m
S_{creep} for 100 years	0.123m	0.411m
Hypothesis A: $S_{totalA}=$	0.810m (at t=10 yrs)	1.600m (at t=100 yrs)
Hypothesis B: $S_{totalB}=$	0.900m (at t=10 yrs)	1.970m (at t=100 yrs)
S_{totalA}-$S_{totalB}=$	-0.090m	-0.370m

rameter input page of the "Interactive Analysis", (b) the average degree of consolidation and (c) dissipation of excess pore water pressure for 1-D consolidation analysis of the marine clay layer of 10m thick.

Based on the calculation, it is found that for H=3m and t=10 year, S_{totalA}=0.810m and S_{totalB}= 0.900m for Hypothesis A and Hypothesis B respectively. The difference is S_{totalA}-S_{totalB}=-0.090m. This means that Hypothesis A underestimates the settlement by 0.090m. For H=10m and t=100 year, S_{totalA} =1.600m and S_{totalB}=1.970m for Hypothesis A and

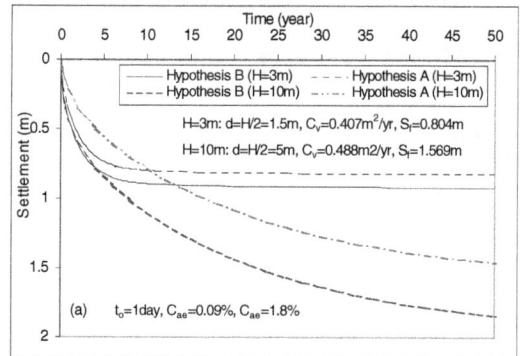

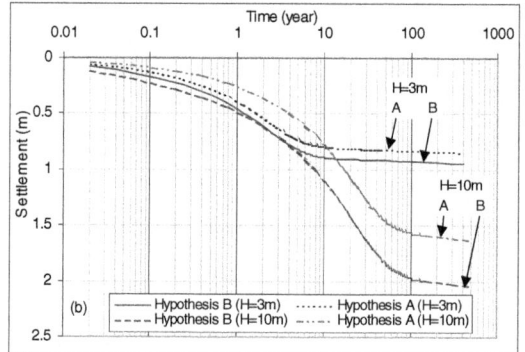

Figure 3. Consolidation settlement with time based on Hypothesis A and Hypothesis B (simplified)

Hypothesis B respectively. The difference is $S_{totalA}-S_{totalB}=$ -0.370m. This means that Hypothesis A underestimates the settlement by 0.370m.

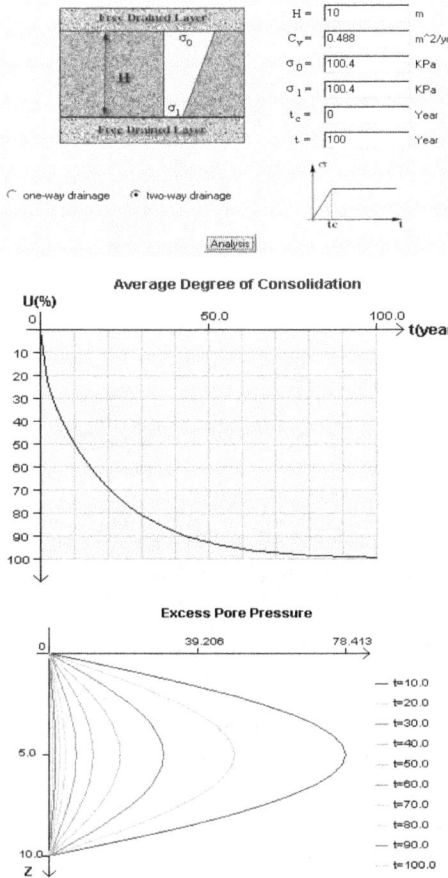

Figure 4. (a) Data input in the "Interactive Analysis" of 1-D consolidation, (b) the average degree of consolidation with time and (c) the dissipation of excess pore water pressure for different time along the clay layer thickness

5 COMPARISON OF SETTLEMENTS OF MEASURED AND CACULATED USING THE SIMPLIFIED METHOD - BERTHIERVILLE TEST EMBANKMENT

Kabbaj et al. (1988) and Lerouril et. al. (1988) reported the results of Berthierville test embankment in Quebec, Canada. Fig.5 shows (a)the soil profile and instrument arrangement, (b) real loading by fill and simplified loading with time, and (c) the measured total pore water pressure at P6 (below the clay layer) and H1 (above the clay layer) and simplified constant water pressures with time.

Those parameters and values used in the simplified calculations of the settlement of the clay layer

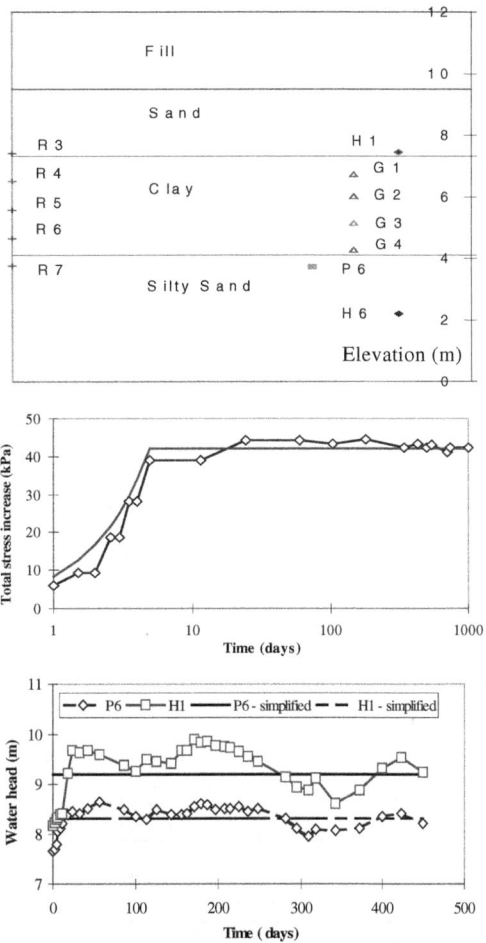

Figure 5. Berthieville test embankment – (a) soil profile and instruments (upper), (b) loading (fill) with time (middle) and (c) the total pore water pressure at P6 and H1

are listed in Table 3. Values of k_1, κ/V, λ/V, and σ'_{vc} are from the publications by Kabbaj et al. (1988) and Lerouril et al. (1988). The t_o shall be 1 day if the compression curve is determined using the compressions after 1 day loading from oedometer tests as discussed in the preceding section. The average values of the total pore water pressure u, the initial vertical total stress σ_{vi} and effective stress σ'_{vi} are calculated at the middle level of the clay layer and using the upper and lower boundary pore water pressure in Fig.5 (bottom). The "primary" consolidated settlement is calculated as $S_{"primary"}$=28.7 cm using Eqn.(5). The creep settlement is calculated using Eqn.(11) as S_{creep}=11.5 cm. The average m_v is as 0.00 1881/kPa.

Lerouril et. al. (1988) reported the vertical permeability k_1 of 1.08x10^{-5} m/hour from pressuremeter test results. The C_{v1} is 5.13 m2/yr. If $k_2=k1/2=$ of 5.4x10^{-6} m/hour. The C_{v2} is 2.57 m2/yr. Eqn.(7) is

209

used to calculated the consolidation settlement. The measured and calculated curves of settlement with time are shown in Fig.6. It is seen that the total final settlement calculated is about the same as the measured. However, the settlement-time curve calculated using C_{v1} is 5.13 m2/yr is much below the measured curve. The settlement-time curve calculated using C_{v2} is 2.157 m2/yr is close to the measured curve. Possible reasons for the difference using C_{v1} is 5.13 m2/yr are: (a) the permeability may be underestimated since the permeability normally decreases the void ratio (or effective stress increase), (b) the external loading and the boundary pore water pressures with time are probably over-simplified.

Table 3. Parameters used in the simplified approach and results

Parameter/variable	Value (or result)
$k_1=$(m/hr)	1.0×8^{-5}
$k_2=$(m/hr)	5.4×8^{-6}
$\kappa/V=$	0.01
$\lambda/V=$	0.184
$t_0=$ (day)	1
$\sigma'_{vc}=$(kPa)	54
$C_r/(1+e_0)=ln(10)\ \kappa/V=$	0.023
$C_c/(1+e_0)=ln(10)\ \lambda/V=$	0.424
u (kPa) (middle clay layer)	29.92
$\sigma_{vi}=$(kPa) (middle clay layer)	69.56
$\sigma_{vi}=$(kPa) (middle clay layer)	39.64
$S_{"primary"}=$(cm)	25.3
$S_{creep}=$(cm) (1000days)	11.5
$m_v=$1/kPa	0.00188
$C_{v1}=$(m^2/yr)	5.13
$C_{v2}=$(m^2/yr)	2.57
$t_c=$ 5 days=0.01366 year	0.01366 year
$S_{totalB}=$ (cm)	36.8

6 REMARKS

Two approaches based on Hypothesis B are introduced. One is a rigorous approach using Yin and Graham's 1-D elastic visco-plastic model and fully coupled consolidation analysis. The second approach is a simplified approach. In connection with the calculation of the average degree of consolidation, which is need in the simplified approach and the traditional approach, an internet-based "Interactive Analysis" tool is introduced.

There examples are presented. The results of the consolidation analysis using Yin and Graham's 1-D EVP model are compared with the test results and results using Hypothesis A. A simple example is analyzed using the simplified approach and Hypothesis A and results are compared. A field case project is also selected and analyzed using the simplified approach and results are compared to the measured results.

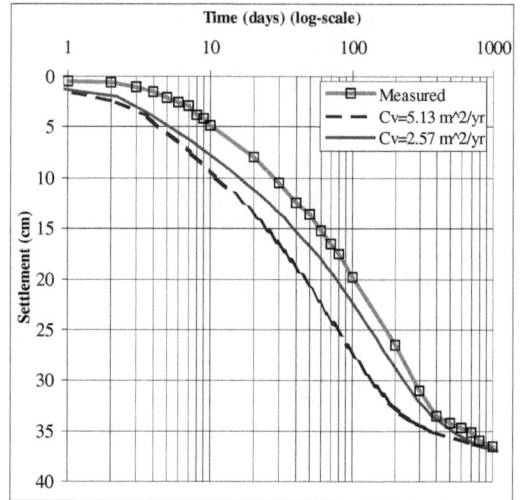

Figure 6. Comparison of measured and calculated settlements with time (for C_v=5.12 m^2/year and C_v=2.57 m^2/year)

The author's views are as follows:
(a) The rigorous approach using Yin and Graham's 1-D EVP model in fully coupled consolidation analysis is more accurate for the prediction of both settlements and excess pore water pressure, especially the phenomenon of "excess pore water pressure when external loading is constant". But this approach is complicated to use. A specialized computer program is needed. This approach is based on Hypothesis B and considers the creep compressions in both "primary" and "secondary" periods.
(b) The simplified approach is an approximate approach and easy to use. This approach is based on Hypothesis B and considers the creep compressions in both "primary" and "secondary" periods.
(c) The traditional approach is based on Hypothesis A and ignores the creep compression in the "primary" consolidation period. Therefore, this approach under-estimates the total consolidation settlement. Care must be taken when using this approach.

ACKNOWLEDGEMENT

Financial supports from a RGC grant and from the grant (H-ZJ73) of The Hong Kong Polytechnic University are acknowledged.

REFERENCES

Leroueil, S., M. Kabbaj, and F. Tavenas, 'Study of the validity of a $\sigma_v - \varepsilon_v - \dot{\varepsilon}_v$ model in in-situ conditions', *Soils & Foundations*, 28(3), 13-25 (1988).

Kabbaj, M., F. Tavenas, and S. Leroueil, 'In situ and laboratory stress-strain relationships', *Geotechnique* 38(1), 83-100 (1988).

Yin, J.-H. and Graham, J., (1989). Viscous elastic plastic modelling of one-dimensional time dependent behaviour of clays. Canadian Geotechnical Journal, Vol.26, 199-209.

Yin, J.-H. and Graham, J. (1994). Equivalent times and elastic visco-plastic modelling of time-dependent stress-strain behaviour of clays. Canadian Geotechnical Journal, Vol. 31. 42-52.

Yin, J.-H., Graham, J., Clark, J.I., and Gao, L. (1994). Modelling unanticipated porewater pressures in soft clays. Canadian Geotechnical Journal, Vol. 31. 773-778.

Yin, J.-H. and Graham, J. (1996). Elastic visco-plastic modelling of one-dimensional consolidation. Geotechnique, 46, No.3, pp.515-527.

Yin, J.-H. and Graham, J. (1999). Elastic visco-plastic modelling of the time-dependent stress-strain behavior of soils. Canadian Geotechnical Journal, Vol.36, No.4, pp.736-745

Zhu, G.F. and Yin, J.-H. (1998). Consolidation of soil under depth-dependent ramp load. Canadian Geotechnical Journal. Vol.35, No.2, pp.344-350.

Zhu, G.F. and Yin, J.-H. (2001). Consolidation of soil with vertical and horizontal drainage under ramp load. Geotechnique, Vol.51, No.2, pp.361-367.

Soft Ground Engineering in Coastal Areas, Tsuchida et al. (eds)
© 2003 Swets & Zeitlinger, Lisse, ISBN 90 5809 613 0

A case history on deformation management of embankment constructed on ultra soft reclaimed land

K. Zen & K. Kasama
Kyushu University, Fukuoka, Japan

K. Egashira
Coastal Development Institute of Technology, Tokyo, Japan

M. Katagiri
Nikken Sekkei Nakase Geotechnical Institute, Kawasaki, Japan

ABSTRACT: An embankment constructed on an ultra-soft reclaimed land had begun to move after its construction due to the uneven force acting on the both sides of embankment. As urgent countermeasures, the counterweight and water level control were adopted. At the same time, the monitoring using the inclinometer was performed and causes of displacement were investigated. In addition, the criteria continuing the safety filling work was prescribed and utilized at the site. In case the horizontal displacement velocity of 20 mm/day was observed successively 3 days, the filling work was prescribed to stop promptly. By such management of displacement, the filling work was successfully completed.

1 INTRODUCTION

In the construction of New Kitakyushu Air Port, the dredged clayey slurry has been utilized for reclamation, disposing it by pump dredgers. As the result of using hydraulic fill, ultra soft ground with high water content of 200% to 300% and with very complicated locality of soil properties was created at the designated air port area enclosed by revetment. Because of limited construction work schedule, rapid construction works on filling and banking had to be executed at several blocks in the area separated by an embankment.

The embankment was constructed at the center of reclaimed area called Area No.1 by the forced replacement method. During the first stage of filling works on both sides of the embankment, gradual displacements of embankment were observed so that the monitoring on movement had been started. At the same time, an allowable management value on displacement for safely continuing construction work was determined. On the basis of estimated mechanism/causes of displacement, urgent countermeasures were applied to restrain the embankment from displacement, continuing the daily construction work.

In this paper, the results of monitoring on the displacement of embankment and the effect of applied countermeasures are presented as a case history on the displacement management of embankment on an ultra soft clay ground.

2 OUTLINE OF THE SITE

The location of central embankment is shown in Figure. 1. The area surrounded by the revetment had been reclaimed with dredged clayey slurry. The central embankment separates the Area No.1 to the west and east blocks at its center.

Figure. 2 shows an example of the cross section of central embankment which was constructed by the forced displacement method starting construction from the both sides of south-most and north-most edges of embankment. The objective position described in this paper is denoted by the cross point between the central embankment and Line No. 60 (X).

Figure. 3 indicates the undrained shear strength of reclaimed layer which was estimated from the Cone Penetration Test (CPT) data by using the equation; $Cu=0.1q_c$ and the Vane Shear Test (VS) results. The location of in-situ sounding test using the CPT and VS is indicated in Figure. 1. Because of the difference of work phases between the west and east blocks, the measured Cu at the east block is 1/4 to 1/2 of that at the west block where the Plastic Board Drain Method (PBD) had been executed. Anyhow, the undrained shear strength is very small at either sides of embankment, indicating the strength of mostly less than $5kN/m^2$.

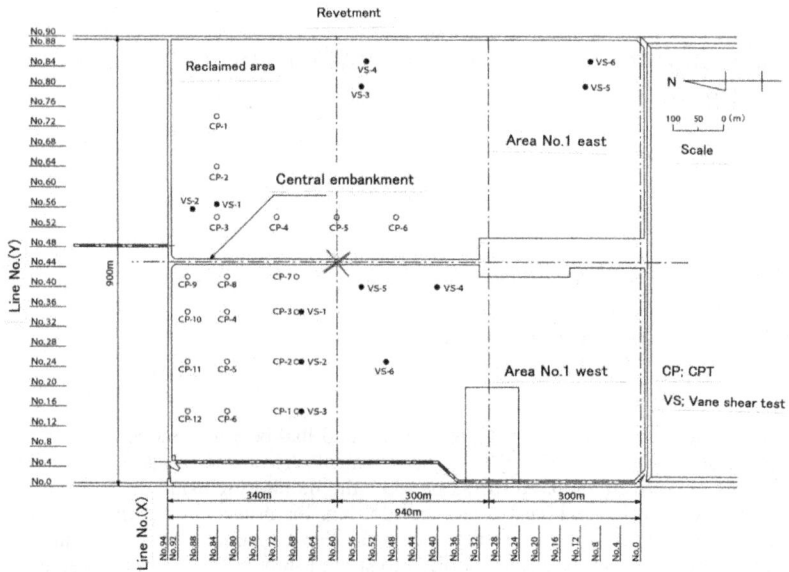

Figure 1. Location

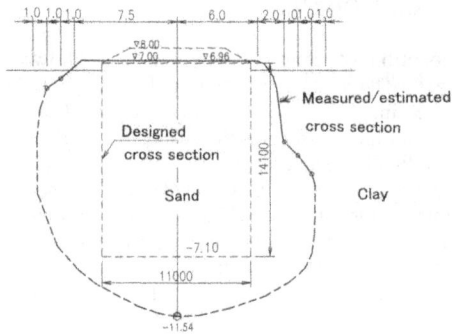

Figure 2. Example of the cross-section of embankment

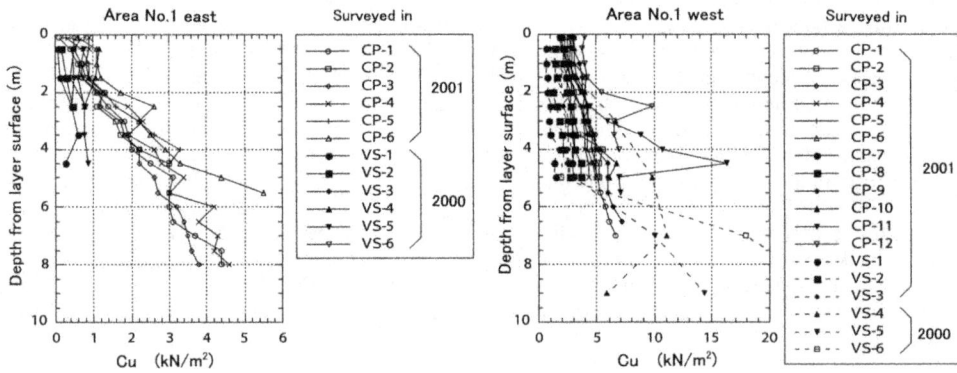

Figure 3. Undrained shear strength

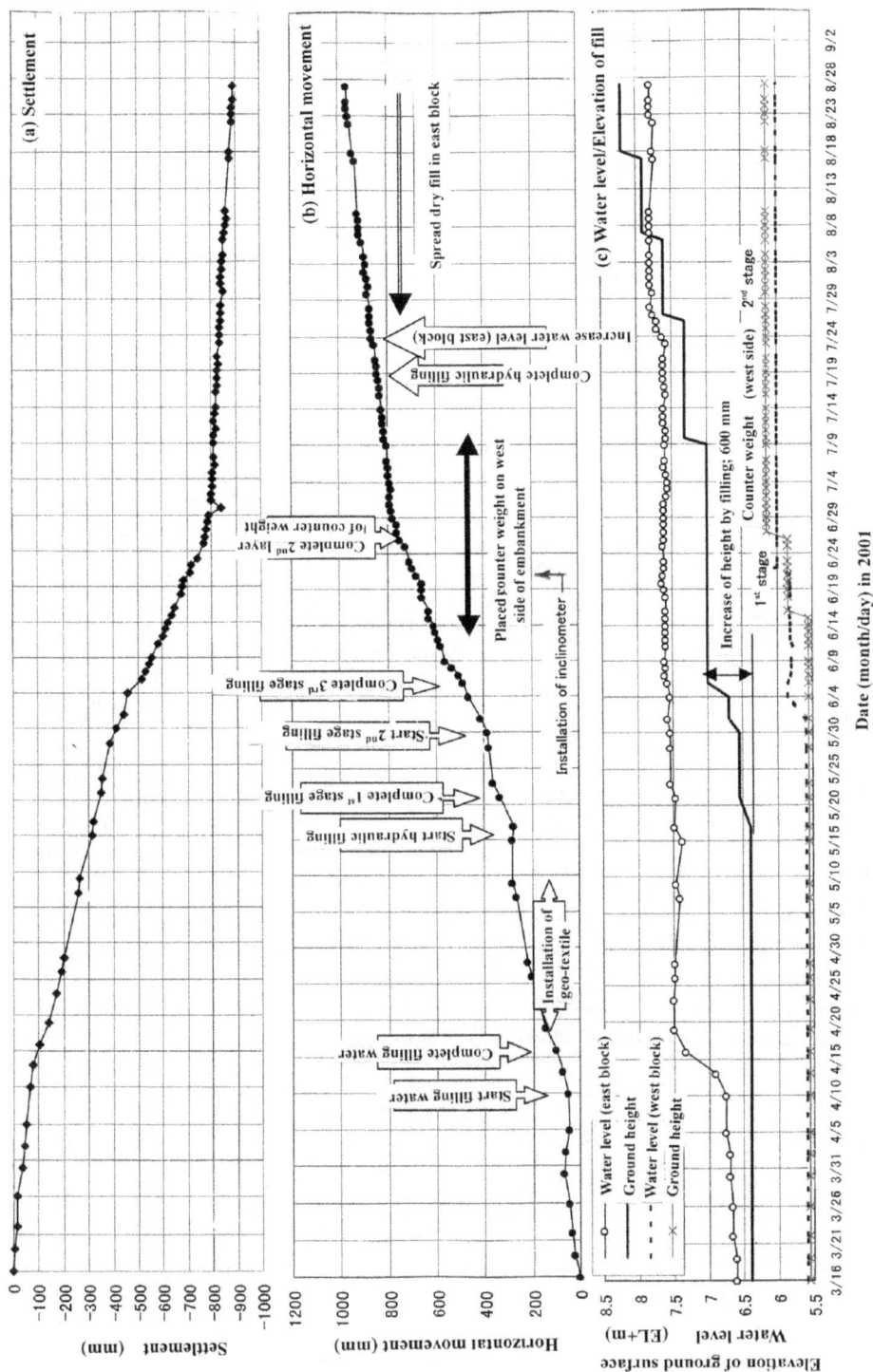

Figure 4. Observed behavior of embankment

215

3 DISPLACEMENT OF EMBANKMENT

3.1 Observed displacement

After completion of central embankment, the construction work for the soil stabilization using the PBD was started on March, 2001 at the west block where the surface treatment had been completed. On the other hand, the surface treatment at the east block had not yet been started but prepared for filling water for the installation of geo-textile. The displacement increased remarkably after starting filling water in the east block on the 10th of April, 2001 as shown in Figure. 4. It was, however, settled by the middle of May in 2001. Then, the hydraulic filling for the sand mat for PBD was started, resulting again increase of the displacement with large velocity. The sand mat was filled with micro-pump in every layer thickness of 150mm up to the second layer and in third layer with 300mm. Then, the displacement was further accelerated after starting the 3rd stage of hydraulic filling. Though this phenomenon seemed not to directly relate to the failure of embankment, it was thought that the failure might happen provided that the accelerated displacement continued further.

3.2 Mechanism of displacements

In order to make clear of the possible causes of displacement, the following phenomena were investigated; i) sliding of embankment, ii) turnover of embankment, iii) circular failure, iv) displacement due to consolidation and v) piping in the embankment. As for the possibility of sliding, it was judged that the possibility was quite small, because the horizontal force was far below the shear strength of clay (alluvial) layer beneath the embankment and the direction of displacement was declined about 45 degree to the ground surface as shown in Figure. 4. The turnover of embankment was confirmed not to occur

because the passive earth pressure acting on the embankment was large enough to resist the driving force. The piping due to the difference of water level, about 2m, between the west and east blocks was checked to find that it was safe enough under the estimated conditions of the site such as soil properties and features of embankment. Figure. 5 indicates the velocity of displacement. The velocity of settlement was measured at the point, 25m far from the embankment to the west (Line No. 60(X), 42(Y)-5) and the velocity of horizontal movement was measured on the central embankment (Line No. 60(X), 45(Y)-0). From Figure.5, it is understood that the velocity of horizontal movement is not always identical to that of settlement but it increases rapidly after starting filling water in east block. This means that the displacement is not due to the settlement by consolidation. The velocity of horizontal movement decreases after the completion of filling water, and it settles nearly zero by the time of starting hydraulic filling on the 15th of May, 2001. Remarkable horizontal movement is observed right after starting hydraulic filling on the 5thof May, 2001. Thus, the displacement was considered to be triggered by the uneven force acting on the embankment from the west and east sides.

Instability of embankment due to the uneven force was analyzed against the circular failure as shown in Figure. 6. The shear strengths for reclaimed clay layers were set $1.4kN/m^2$ and $0.2kN/m^2$ respectively at the west and east blocks and $20kN/m^2$ for clay (alluvial) layer. The internal friction angle for embankment was assumed 20 or 25 degrees. Figure. 6 indicates the result of the circular failure analysis where the internal friction angle of 20 degrees was adopted. In this case, the obtained safety factor against the circular failure was 0.98. For the case of the internal friction angle of 25 degrees, it was 1.20. Since the reclaimed clay repre-

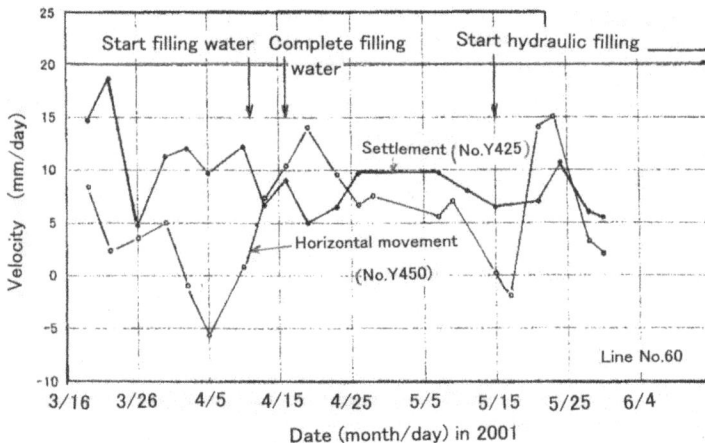

Figure 5. Velocity of displacement

sents fluid-like mud, it may need some discussions whether the circular failure clearly appears or not, the displacement was judged to imply a critical state against the failure attributed to the shear deformation.

3.3 Urgent countermeasures and monitoring

As the displacement velocity increased furthermore after starting 2nd stage of filling at the east block at the beginning of June, 2001 as shown in Figure. 4, countermeasures were urgently applied under the consideration and judgment of the cause of displacement stated above. Namely, the main cause of displacement was attributed to the shear deformation arising from the uneven force between the west and east blocks. Applied countermeasures were; i) making smaller the difference of water levels between the west and east blocks, and ii) placing counter-

weight on the west side of embankment. Firstly, the water level in the west block was increased by stopping the drainage to keep dry ground condition for executing the PBD. However, the filling work at the east block was continued observing the displacement of embankment because of the tight construction work schedule. After the completion of the 3rd stage of hydraulic filling at the east block at the beginning of June, 2001, the displacement of embankment had still continued so that the 300mm thickness of first stage of counterweight was placed on the west side of embankment on the middle of June, 2001. The plane view of counterweight in the first stage is illustrated in Figure. 7, together with the second stage of counterweight mentioned lately.

Right after placing the first stage of counterweight, inclinometers were installed to monitor the deformation in the embankment. Figure. 8 shows the

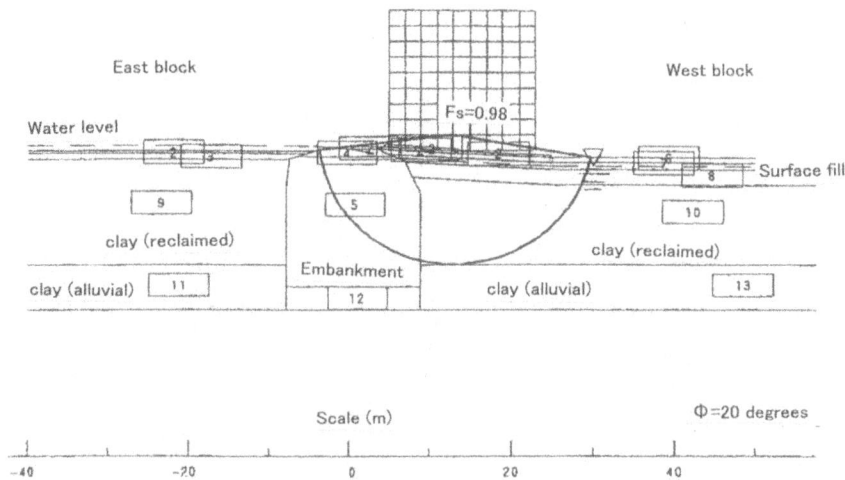

Figure 6. Circular failure analysis (φ=20°)

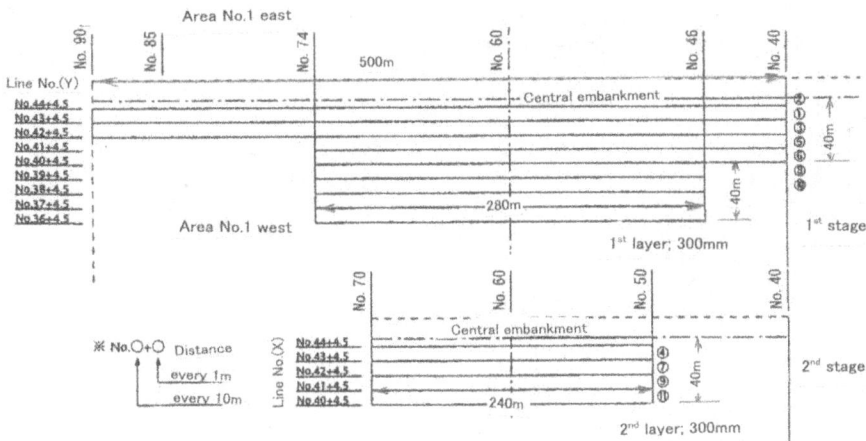

Figure 7. Plane view of counterweight

217

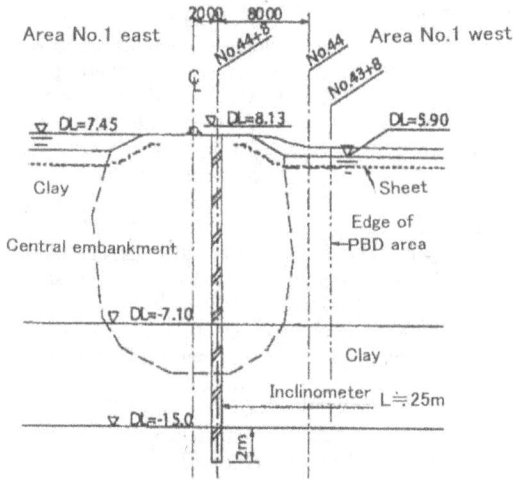

Figure 8. Installation of inclinometer

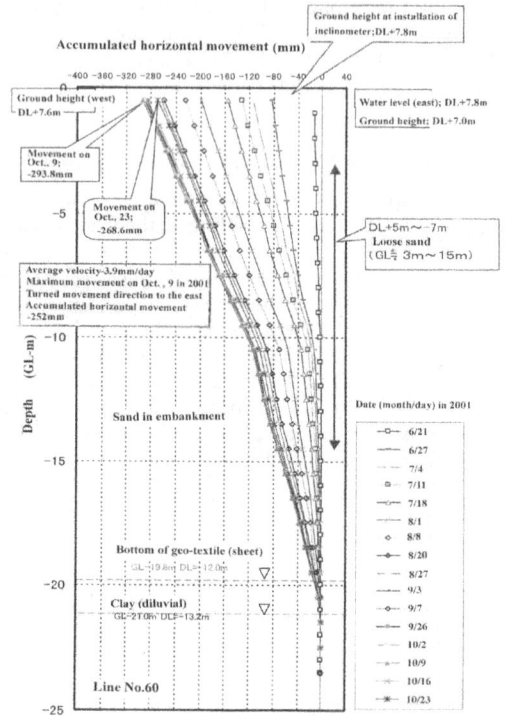

Figure 9. Observed movement

cross-section of embankment where the inclinometer was installed. The measured horizontal movement illustrated in Figure. 9 indicates that the amount of movement becomes larger at shallow depth, showing inflection points at the depth of -10m. Figure. 10 is the N-values by the Standard Penetration Test (SPT) in the embankment investigated at the time when the inclinometer was installed. It is quite interesting to note that the N-values in the embankment were about 5 at the depth of around -10m. This value corresponds to the internal friction angle of 15 to 25 degrees. As the embankment was constructed by the forced replacement method, no compaction was executed, but very loose sandy soil should have been placed to balance the surroundings. Furthermore, the No.60's point corresponds to the connected point of embankment extended respectively from the south and north sides of embankment. This may bring the

lower N-values. The assumed internal friction angle of 20 degrees in Figure. 6 could be considered a reasonable value.

As the displacement had not settled yet after placing the first stage of counterweight as shown in Figure 4, the second stage of counterweight was added at the end of June, 2001 as indicated in Figures 4 and 7. As the result, the horizontal movement drasti-

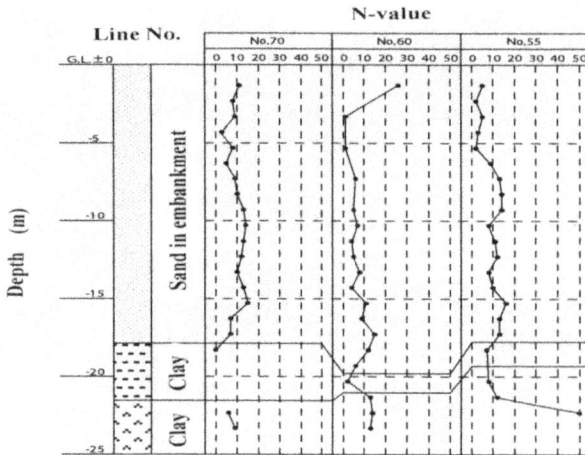

Figure. 10   N-value in the embankment

cally lowered its velocity and settled after the 25th of June, 2001 as shown in Figure. 4, in spite of the increase of thickness of filling at the east block. It may be considered that counterweight works effectively for restraining the embankment from movement.

A little unbalance of forces between the west and east sides of embankment is thought to affect the stability of embankment, since it was constructed on the ultra-soft clay layer by the forced replacement method which held stability under a condition of critical equilibrium on external forces.

4 CRITERIA FOR MANAGEMENT

Because of the tight construction work schedule, the filling work at the east block could not be stopped at the time of the end of May, 2001 when the displacement velocity rapidly increased so that the urgent countermeasures stated above were adopted and the monitoring of displacement was started using the inclinometer. The problem is, however, how to prescribe the criterion on the maximum horizontal displacement per day for continuing the safety construction work. As no data were available on this problem, the existing data shown in Figure. 11 (Kurihara & Ichimito; 1977) were utilized for reference. In Figure. 11, the vertical axis means the lateral deformation of banks per day which induces instability of banks. The horizontal axis is the non-dimensional parameter, where B means the 1/2 of bank width and D_f is the estimated maximum depth of failure plane. According to Figure. 11, instability of banks seems to occur beyond the velocity of 20mm/day, irrespective of the values of B/D_f. Since the deformation of ultra-soft ground was considered far larger than that of soft ground, the minimum value, 20mm/day, in Figure. 11 was adopted to this site. This would be a reasonable when referring to the horizontal movement in Figure. 5 where the velocity is less than 15mm/day. In addition, taking account of the tight construction schedule, the criterion for stopping the filling work was provided as the velocity of horizontal movement of 20mm/day continued successively 3 days.

The results of monitoring indicated that the maximum velocity was about 15mm/day during the monitoring terms up to the 25th of June, 2001. It was, however, decreased to 3mm/day during the successive 3days starting from the 3rd of July, 2001. The change could be attributed to the second stage of counterweight completed placing on the 25th of June, 2001 as shown in Figure. 4. The inclination of horizontal movement seems almost constant, irrespective of the increase of ground height at the east block by filling works. This means that the counterweight worked effectively as expected.

According to Figure. 9, the maximum horizontal movement was 293.8mm which was observed on the

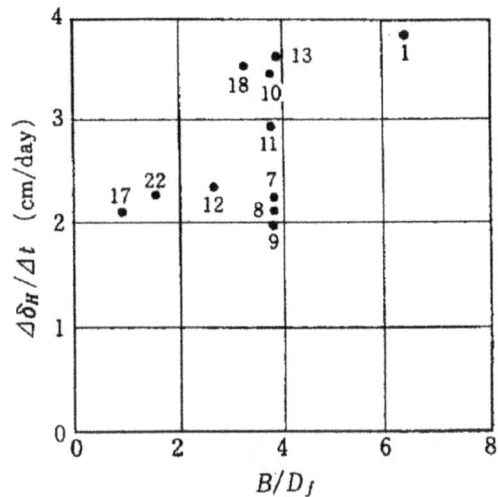

Figure 11. Displacement velocity

9th of October, 2001. After that day, the movement returned to the east side. This reason is thought that the PBD had been started placing at the east block inducing the accelerated consolidation settlement and the preloading had been continued executing at the west block.

5 CONCLUDING REMARKS

An embankment had started moving after its construction by beginning filling water and hydraulic fill into the east block. The measurements of displacement and inclination of embankment were executed as well as the investigation of deformation mechanism and causes. Urgent countermeasures were adopted in order to continue the daily filling work. An allowable displacement velocity was determined as the criteria for continuing works safely. In case the horizontal displacement velocity of 20mm/day was observed successively 3days, the filling work was prescribed to stop promptly. The results of monitoring on the movement of embankment verified that the provided criteria were appropriate as far as this site was concerned. Adopted countermeasures were also confirmed effective to restrain the embankment from movement.

The following conclusions may be drawn from this case study;
(1)The N-value in the embankment constructed by the forced replacement method becomes less than 5 according to circumstances of construction.
(2) The instability against the uneven force acting on the embankment during filling work is to be considered, as the embankment constructed by the forced replacement method on the ultra-soft layer keeps the stability under the critical force equilibrium.

219

(3)The displacement management taking advantage of monitoring was effective to executing safety filling work in the ultra-soft reclaimed land.
(4)The provided criterion on the displacement management of embankment for filling work was appropriate as far as this site was concerned.

REFERENCES

Kurihara, N. & Ichimoto, E.(1977); Practical examples of banking for road construction, Lecture Text, *Kansai Branch*, *JSCE*, pp.71-80 (in Japanese)

Soil improvement methods in coastal area

Soft Ground Engineering in Coastal Areas, Tsuchida et al. (eds)
© 2003 Swets & Zeitlinger, Lisse, ISBN 90 5809 613 0

Application of FGC deep mixing to the braced excavation

K. Azuma
Chigasaki Research Institute, Electric Power Development Co., Ltd., Chigasaki, Japan

K. Ohishi
Nikken Sekkei Nakase Geotechnical Institute, Kawasaki, Japan

T. Ishii
Nikken Sekkei Civil Enginieering Co., Ltd., Tokyo, Japan

Y. Yoshimoto
Civil Engineering Technology Department, Kaihatsu Sekkei consultants Co., Ltd., Tokyo, Japan

ABSTRACT: In earth retaining work undertaken in soft ground, cement-water slurry based deep mixing method ("CDM") is often used to increase the passive earth pressure acting on retaining structure such as sheet pile wall. Due to the high strength of the improved ground by CDM, earth retaining wall is placed in the unimproved ground upon completion of CDM work. The gap between the wall and improved ground is integrated by the other soil improvement such as jet grouting. Jet grouting is much more costly than CDM and the deformation of the earth-retaining walls is accompanied by high pressure injection. One of the special features of the FGC deep mixing method ("FGC-DM") which uses the slurry of three materials, fly ash (F), gypsum (G) and cement (C) is its ability to create uniform improved ground with lower strength than conventional CDM. Thus, the FGC-DM makes it possible to directly drive the earth-retaining sheet piles and to integrate the sheet pile wall with the improved ground. Earth-retaining work using FGC-DM was applied in the construction of a thermal power station. Since this was the first application, a number of technical issues associated with design and construction were to be solved. This report gives details of the design method of earth-retaining structure that employs FGC-DM.

1 INTRODUCTION

In the wet method of deep mixing, CDM, a cement-water slurry is mixed into the soft ground so that firm ground can be prepared by the resulting chemical solidification effect. With this method, if the amount of slurry is insufficient, it is difficult to achieve uniform mixing, and significant variations occur in the strength of the improved ground. The appropriate range of slurry volume to achieve uniform mixture by the current CDM mixing system ranges from 100 to 300 liter per cubic meter of original soil. Therefore, the amount of the slurry is often increased beyond the prescribed level which is determined from the strength requirement.

When CDM is employed to the braced excavation, excessive strength of the improved soil causes the difficulty of driving sheet piles into the improved soil. Therefore, earth retaining wall is placed in the unimproved ground upon completion of CDM work. The gap between the wall and improved ground is integrated by the other soil improvement such as jet grouting. Jet grouting is much more costly than CDM and the deformation of the earth-retaining walls is accompanied by high pressure injection.

The Electric Power Development Co. has long studied the efficient reuse of flyash and gypsum that are the by-products of coal power station and found

that the mixture of flyash, gypsum and cement becomes an alternative for the ordinary cement in the wet method of deep mixing. Figure 1 shows a system of FGC-DM, which employs ordinary CDM mixing machine.

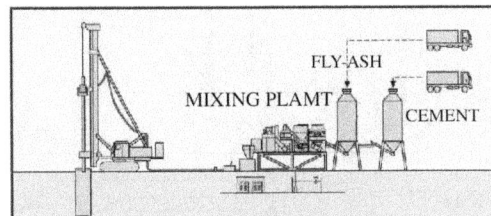

Figure 1. Working system for FGC DM.

The significance of developing the FGC-DM can be summarized by the followings.

• By adding fly ash and gypsum to cement, the FGC-DM makes it possible to increase the total amount of the slurry mixed with the ground in comparison with conventional CDM even when the same amount of cement is used.

For this reason, it enables a uniform improvement at low strength around 0.1 (MN/m^2) – a level at which uniform mixing proved to be difficult to achieve with CDM--and it opens the door to improving ground whose strength ranges widely from low to high. (See Figure. 2).

When uniformly improved ground with a low strength has been prepared as the support of earth-retaining work, it is then possible to drive earth-retaining sheet piles, for instance, directly into the improved ground, thereby obviating the supplementary methods such as jet grouting method which are normally implemented around the sheet piles (See Figure. 3).

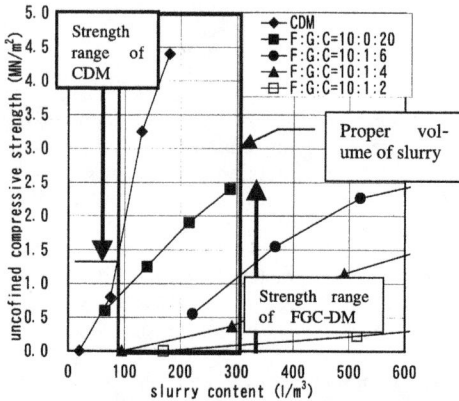

Figure 2. Unconfined compressive strength vs. slurry content.

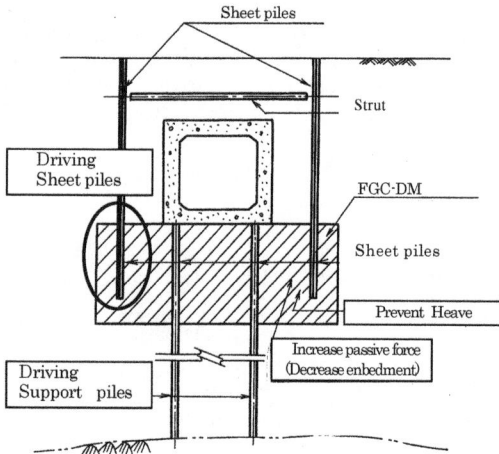

Figure 3. Application of FGC-DM to earth retaining excavation.

2 TECHNICAL ISSUES AND PRELIMINATY INVESTIGATIONS

The major technical issues in the development of FGC-DM and its application to the braced excavation are

- To examine appropriate mix proportion of FGC-DM slurry.
- To confirm the ability of driving retaining structures into FGC-DM improved ground.
- To evaluate passive earth pressure of FGC-DM improved ground in the excavation side.

To examine the appropriate mix proportion of FGC-DM slurry, laboratory tests and field tests were conducted. The effect of F: G: C ratio in the binder, the appropriate binder-water ratio, and the influence of volume of slurry in the actual execution were studied into detail (Asano et al.1996).

To confirm the ability of driving retaining structures into the improved ground, field test was conducted. Sheet piles (type II-IV), Steel pipe piles (D=300, 500 mm), Steel H-beams (H-200,400mm) and Concrete piles (D=300, 400 mm) were driven to the FGC-DM improved ground (6×6m in plan, and 6 m in depth) directly. The unconfined compressive strength of FGC-DM improved ground was about 600 (kN/m^2) sheet piles, steel column piles, H-beams were driven to the improved ground easily by the ordinary vibratory pile driver and oil pressure-hammer without damaging the improved ground. While, the concrete piles could be driven, but ground was damaged and heaving and crack were observed (Azuma et al.1999).

Photograph 1 and 2 show the scene of driving pile to FGC-DM improved ground.

To confirm the passive earth pressure, in situ test of braced excavation down to a depth of 4.5m was executed. According to the field measurement, the passive earth pressure increased from 80 (kN/m^2) to 150 (kN/m^2)as the excavation steps were progressed. The study to evaluate the passive earth pressure of improved ground was continued in the process of designing the actual application that will be described in the following paragraphs.

Photograph 3 shows the scene of in-situ test of braced excavation.

Photograph 1. Scene of driving steel pipe pile.

Photograph 2. Scene of driving concrete pile.

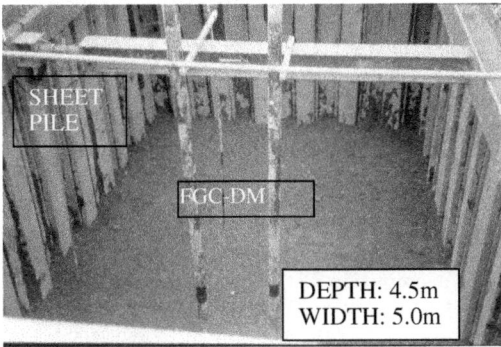
Photograph 3. Scene of in-situ test of braced excavation using FGC-DM.

3 APPLICATION OF FGC-DM FOR EARTH RETAINING

3.1 Construction site

FGC-DM has been applied in the replacement of a power station in Isogo, Yokohama on Tokyo Bay at a location neighboring a sea wall , existing building, public road and gas tanks as shown in Figure 4.

In terms of geology, the top surface of around 3-meter is reclaimed with sand, and the 10 meter underneath is reclaimed ground consisting of dredged clay soil. Underneath the reclaimed soil layer is the alluvial clay layer.

Figure 5 shows a soil profile of the construction site.

Figure 4. Location of braced excavation by FGC-DM.

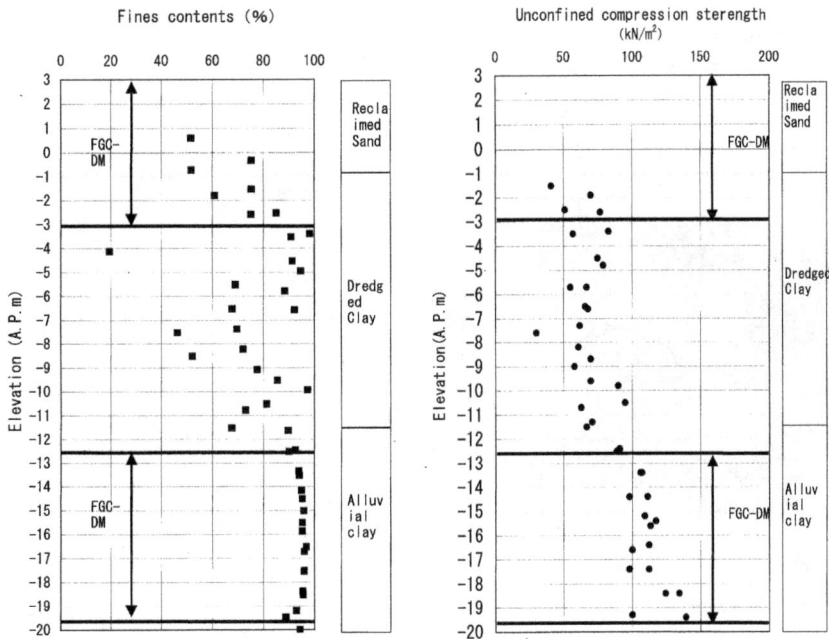
Figure 5. Soil profile of the construction site.

3.2 Outline of earth retaining work

The outline of the braced excavation is shown in Figure 6.

FGC-DM is applied to two different elevations; the surface area on one hand and the base improvement areas on the other. The objective of improving the surface areas is to minimize the displacement of the earth-retaining walls until the upper struts functions whereas the improvement in the excavation bottom are intended to increase the passive earth pressure and prevent bottom heave.

The earth retaining was designed by the ground spring model method. (Refer to Chapter 4.2)

Ninety days after the FGC-DM work was implemented, steel pipe sheet piles and the other piles as shown in photograph 4 were driven directly in the improved ground. Table 1 shows the proportion of binder-water slurry used in the FGC-DM improvement. The excavation was executed in eight stages.

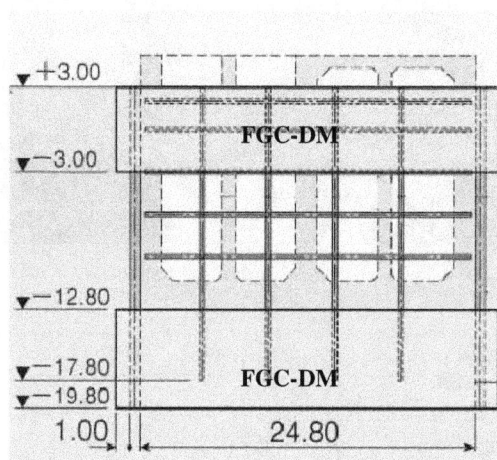

Figure 6. Typical cross section of earth retaining and FGC-DM improvement (at planning).

Photograph 4. Scene of base (at the final excavation).

Table 1. Mix proportion of binder-water slurry

		Amount per 1m³ of soil	Note
Cement	kg	62	F:G:C=
Fly-ash	kg	155	10:0:4
Water	kg	217	W/FGC=100%

4 MEASUREMENTS OF EARTH-RETAINING WORK

4.1 Field measurement

After excavating to a depth of 1 meter beneath the proposed strut positions, struts were installed. The excavation was ultimately undertaken in 8 stages. In order to minimize the displacement of the steel pipe sheet piles after the struts were installed, preloading with a basic load of 9.8 MN per strut was applied. The unconfined compressive strength of the improved ground was about 400 kN/m² for the bottom of improved area (AP -12.6 m to -19.6 m).

Inclinometers, earth pressure cells, water pressure cells are installed on the sheet piles for measurement. Figure 7 shows the results of measurement at the final excavation stage. The maximum displacement in the sheet piles was 12 mm.

An examination of the lateral pressure on the excavation side reveals that the lateral pressure increased with the displacement of the sheet piles on the excavation side (For example, AP -13.5 m.)

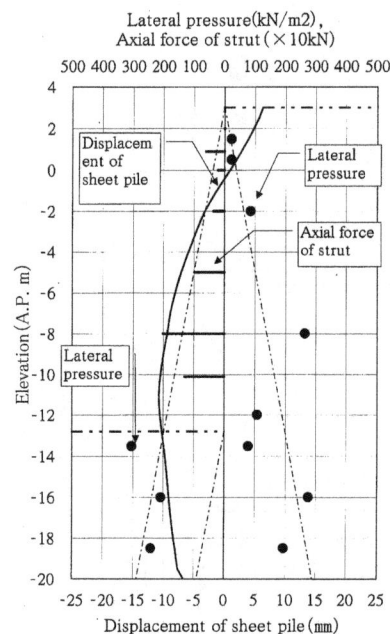

Figure 7. Measurement results (at the final excavation).

In addition, the dotted line on the excavation side in the Figure 7 shows what the values would be if it is assumed that the lateral pressure at rest prior to the commencement of the excavation was still residual. These results indicate that the earth pressure at rest in the excavation side of improvement areas is still residual to some extent even after excavation.

4.2 Design method of earth retaining (braced excavation) in Japan

The ground spring model method is often used for earth retaining (braced excavation) design in Japan. The model for this method is schematically presented in Figure 8. (Nakamura et al.1972)

In this method, the ground on passive side is assumed to be a series of springs that perform elastically when its reaction is smaller than the passive pressure and perfect-plastically when the reaction exceeds the passive pressure.

The elastic reaction in the ground spring model method is generally calculated from the coefficient of subgrade reaction in the horizontal direction, which is specified in the equation below.

$$p = K_h * h$$

$$Kh = \frac{1}{0.3} \cdot \alpha \cdot E_0 \cdot \left(\frac{B_H}{0.3}\right)^{-1} \tag{1}$$

Where p = lateral pressure, Kh = Coefficient of subgrade reaction, h= horizontal displacement of sheet pile, α = Correction factor of elastic modulas E₀= Modulus of deformation, B_H = Loading width.

E_0(modulus of deformation) is inclination of the stress strain curve.

Figure 9 shows the way of calculating E_0 at unconfined compression test.

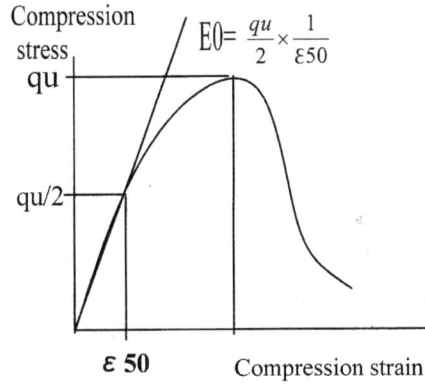

E0 :Modulus of deformation

qu : unconfined compression strength

ε 50: compression strain at qu/2

Figure 9. The way of calculating E₀(Modulus of deformation).

This equation (1) was led by plate loading tests to the loam and the sandy ground. (Yoshinaka et al. 1967)

In this study B_H of 30cm is a base. This equation (1) is for presuming the coefficient of subgrade reaction of every size of B_H. The relation between B_H (=D) and a coefficient subgrade reaction is shown in a Figure 10.

In Japan, B_H is usually taken as 10m in the design of earth retaining. It is because of another studies about the earth retaining.

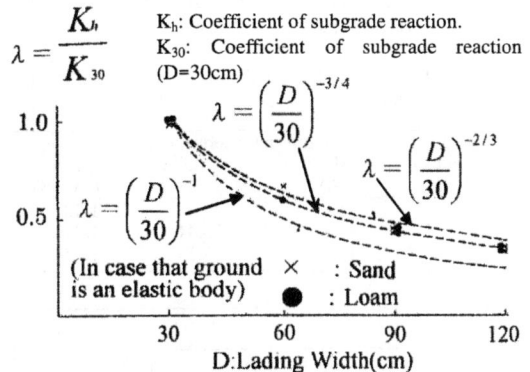

K_h: Coefficient of subgrade reaction.

K_{30}: Coefficient of subgrade reaction (D=30cm)

Figure 10. The relationship between Loading width and coefficient of subgrade reaction.

This coefficient α in the equation(1) is one for unconfined compression test and triaxial test from boring core.

Influences such as the disturbance in core sampling are contained in this coefficient α.

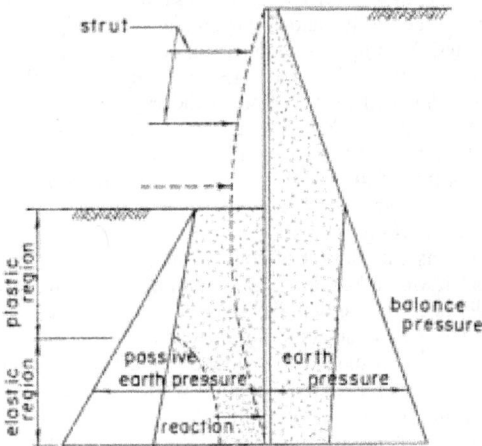

Figure 8. Ground spring method model.

The coefficient α which is gotten from unconfined compression test and triaxial test declines to the value which was gotten from plate loading test. So in this case, coefficient α is taken as 4.

4.3 *Displacement of sheet pile and earth pressure at excavation side*

Equation (1) has been used not only for the standards in the Japan Society of Civil Engineers but for many other design standards as well.

Since the equation(1) was introduced mainly on the basis of research conducted on the loam and sandy soil, the authors examined its applicability to FGC-DM.

The authors compared displacement of sheet pile and the lateral pressure on the excavation side, and the coefficient of subgrade reaction in the horizontal direction was studied.

Figure 11 and Table 2 show the results of the coefficient of subgrade reaction in the horizontal direction from the measurement data. Table 2 also shows the subgrade reaction calculated by Equation (1) by putting α as 4.

In the Figure 11, coefficient subgrade reaction is the coefficient in the regression analysis between displacement of steel pipe pile and subgrade reaction.

For the three points within the FGC-DM improvement range namely, AP -13.5, -16.0 and -18.5m,the results indicate that the coefficient of subgrade reaction in the horizontal direction calculated from the measurement data is approximately one-third of the value calculated by equation (1).

In this calculation, the modulus of deformation of the unconfined compression test result at corresponding depths is used.

Table 2. Comparison of measured and calculated coefficient of subgrade reaction

Elevation of measurement (A.P +m)	Measurement (MN/m³) (A)	equation(1) (MN/m³) (B)	Ratio ((A)/(B))
-13.5	15.3	44.09	1/2.8
-16.0	11.5	28.44	1/2.5
-18.5	11.6	33.73	1/2.9

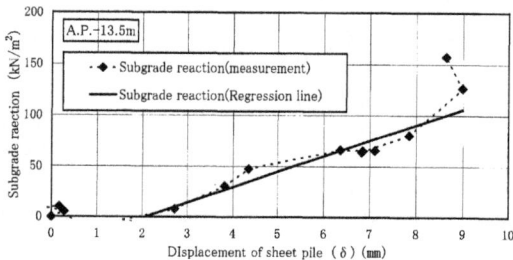

Figure 11. The relation between displacement of sheet pile and measured subgrade reaction.

5 EVALUATION OF THE COEFFICIENT OF SUBGRADE REACTION

5.1 *Study based on field measurement*

A comparison of the coefficient of subgrade reaction in the horizontal direction calculated from the equation (1) with the measurement values calculated on the basis of the on-site measurement results revealed that the measurement values were considerably lower than--about 30 per cent of--the theoretical values calculated using the equation (1). Possible reasons for this include the correction factor α and the characteristics of the materials in the improved ground.

Applicability of correction factor α in the equation(1) is introduced mainly on the basis of research conducted on the loam soil. About α, plate loading tests, unconfined compression tests and borehole load tests were carried out, and various tests data were obtained.

So, upon examination of the applicability of α on the basis of the various test data obtained for FGC-DM improved ground, the authors realized that the correction factor used in the equation (1) can still be applicable in the FGC-DM.

By these examination,α is taken as 4

Generally speaking, cement-stabilized soil is a brittle material. The behavior of brittle materials up until their failure resembles that of elastic bodies, the coefficient of subgrade reaction in the horizontal direction can be expressed by the following equation (2) if the ground is assumed to be an entire elastic object.

$$Kh = \frac{1}{0.3} \cdot \alpha \cdot E_0 \cdot \left(\frac{B_H}{0.3}\right)^{-1} \qquad (2)$$

Figure 12 shows the relation between subgrade reaction measured by earth pressure cells installed on the sheet piles and displacement of sheet pile calculated by inclinometers installed along the sheet piles at AP. – 12 m and – 13.5m. The slope angle of solid line and broken line in the Figure 11 are the coefficient of subgrade reaction in the horizontal direction obtained from equation (1) and (2).

Figure 13 shows the results of unconfined compression tests using samples in the vicinity of the corresponding depths.

In this calculation of coefficient of subgrade reaction, α was taken as 4, E_0 (modulus of elastic) was calculated by the way of Figure 9.

Among the stress strain curves for AP -12.0m in Figure 13, some show evidence of a brittle failure behavior and a clear yield point while the other show evidence of a ductile failure behavior. This is believed to be the result of variations in the stirring and mixing of the slurry where AP -12.0 m approaches the limit of the target improvement range.

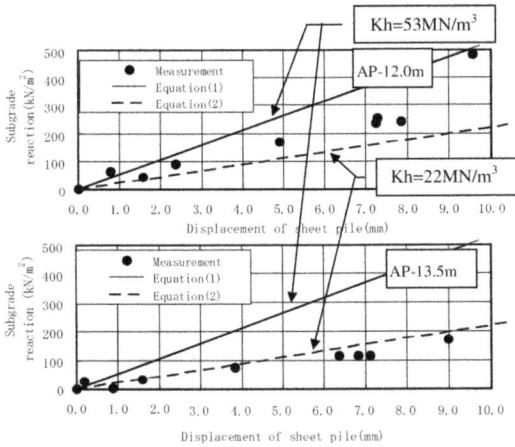

Figure 12. Coefficient of the subgrade reaction obtained from earth retaining measurement.

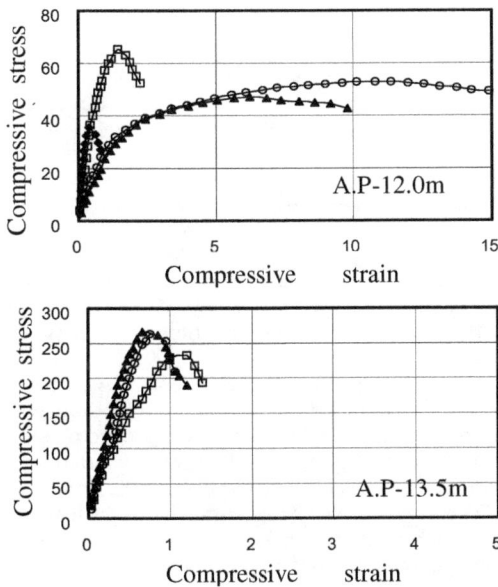

Figure 13. Stress strain curve of unconfined compression tests.

It can be said therefore that overall the material has characteristics midway between the two. In the same way, it is clear that the coefficient of subgrade reaction in the horizontal direction measured from Figure 13 is positioned midway between the plastic object (equation (1)) and elastic object (equation (2)).

It is clear that the stress strain curves for AP -13.5 m indicate brittle failure behavior and a clear yield point, and that the coefficient of subgrade reaction in the horizontal direction is close to that of an elastic body (equation (2)).

5.2 Study based on centrifuge model tests

Authors carried out centrifuge model tests in advance, in order to examine a coefficient of subgrade reaction of improved ground. The centrifuge test results were compared with the on-site measurement result.

5.2.1 Test condions

Using a model (scale: 1/40) such as the one shown in Figure 14 for the centrifuge model tests, the authors conducted horizontal load tests using a 40G centrifuge model. Using a flexural rigidity of El as an index, models of steel pipe sheet piles with a diameter of 700 mm were made using 9.7 mm aluminum plates. The sheet piles were given pin supports at their bottom ends and connected to bi-directional load cell, and the perpendicular and horizontal reactions of the bottom ends of the sheet piles under a horizontal load were measured. In order to ascertain the deformation behavior of the sheet piles accompanying the horizontal load and the earth pressure acting thereupon, strain gauges were positioned on the sheet piles. The characteristics of the test cases and FGC-DM improved ground are shown in Table 3.

In the discussions given below, the prototype values are all used for the results of the centrifuge model tests.

Table 3. Strength and modulus of deformation

	Unconfined compression strength (kN/m^2)	Modulus of deformation (kN/m^2)
Case-1	196	17400
Case-2	392	65200

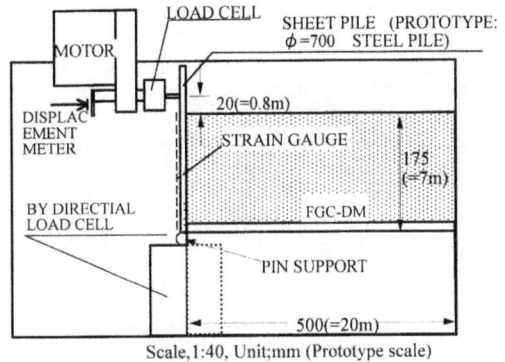

Scale,1:40, Unit;mm (Prototype scale)

Figure 14. Model setup for centrifuge test.

5.2.2 Results of tests

Figure 15 shows the correlation between the earth pressure increment $\Delta\sigma_H$ and the horizontal displacement d_H of the sheet piles accompanying the horizontal load. Here, d_H indicates the horizontal displacement of the sheet piles at the positions where $\Delta\sigma_H$ was applied, and it is calculated from the

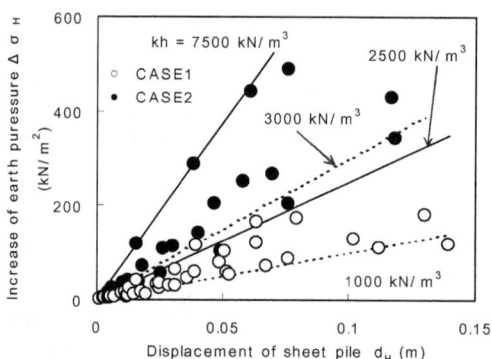

Figure 15. Coefficient of the subgrade reaction obtained from centrifuge model tests.

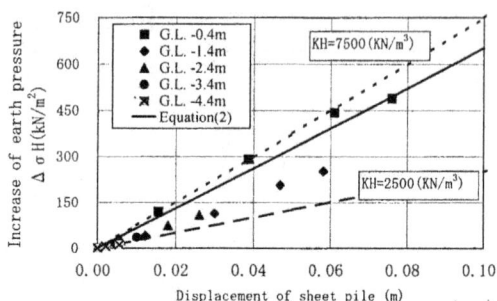

Figure 16. Coefficient of subgrade reaction based on centrifuge model tests (case 1).

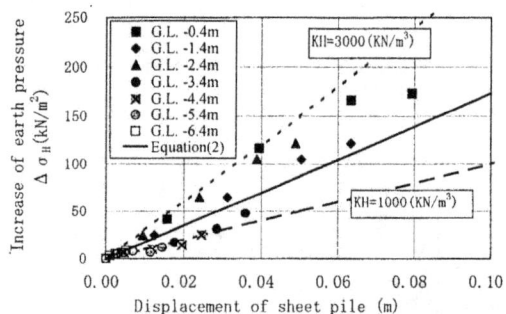

Figure 17. Coefficient of subgrade reaction based on centrifuge model tests (case 2).

deflection of the sheet piles which is in turn calculated from the bending moment. A comparison between cases 1 and 2 shows that the earth pressure increment was naturally greater for the same horizontal displacement case 2 where the strength and rigidity of the improved ground were higher. Furthermore, when the coefficient of subgrade reaction in the horizontal direction k_H is calculated as a coefficient correlated to $\Delta\sigma_H$ and d_H, k_H equivalent to 1,000 to 2,500 kN/m^3 is yielded in case 1 and k_H equivalent to 3,000 to 7,500 kN/m^3 is in case 2.

The coefficient of subgrade reaction in the horizontal direction calculated from the centrifuge model test was compared with the coefficient of subgrade reaction on-site measurement.

The correction factor(α)used in the equation (1) and (2) from unconfined or triaxial compression tests among the correction factors are correction factors that apply when boring cores are used. In contrast to this, the FGC improved ground used for the centrifuge model tests were samples prepared by mixing in the laboratory. As such, the authors reasoned that since there was no loosening at the sampling, the evaluations should be made without taking these correction factors into consideration.

The correlation between the amount of horizontal displacement of the sheet pile and the earth pressure measured on the centrifuge model tests is shown in Table 4 and Figure 16 and 17.

Table 4. Comparison between results of calculating coefficient of subgrade reaction and actually measured values in the centrifuge model test

	Equation(1) (A) (kN/m^3)	Equation(2) (B) (kN/m^3)	Measurement (C) (kN/m^3)
Case-1	4200	1670	1000-3000
Case-2	15700	6270	2500-7500
	Reduction ratio of measurement		
	C/A	C/B	
Case-1	1/4.2-11.7	1/1.7-1/0.7	
Case-2	1/5.2-1/2.1	1/2.1-1/0.8	

In these cases, modulus of deformation (E0) was calculated by the way of Figure 9.

B_H is taken as 10m on the ordinary Japanese design method.

Figure 16 and 17 show that, in both cases 1 and 2, the coefficient of subgrade reaction in the horizontal direction for the FGC-DM improved ground obtained from the centrifuge test results agrees approximately with the value calculated using equation (2).

These results also agree with what is presented in "5.1.2 Examination on subgrade reaction of FGC-DM with on-site measurement results.

6 APPLICABILITY OF GROUND SPRING METHOD

In view of the fore-mentioned coefficient of subgrade reaction in the horizontal direction, the authors studied the applicability of the ground spring model method described in 5.1 equation (2).

Table 5 lists the condition for analysis.

In addition, lateral pressure at rest was also taken into consideration in this analysis. (Azuma et al. 2002).

Among the results of the analysis, Figure 18 shows the displacement of the steel pipe sheet pile calculating by equation (1) and (2).

By re-examining the coefficient of subgrade reaction in the horizontal direction by equation (2), the calculated displacement of the steel pipe sheet pile was close to the actually measured value.

Table 5. Condition of analysis

	Elevation (AP+.m)			Shear strength (kN/m^2)	Wet density (t/m^3)
Improved	+3.0	-	+1.0	250	1.53
	+1.0	-	-1.0	120	
Unim-proved	-1.0	-	-2.0	27.5	1.58
	-2.0	-	-3.0	9.4	
	-3.0	-	-5.0	8.1	
	-5.0	-	-7.0	22.7	1.69
	-7.0	-	-9.0	49.6	
	-9.0	-	-11.0	183.6	
Improved	-11.0	-	-12.8	260.0	1.54
	-12.8	-	-18.0	260.0	1.48
	-18.0	-	-19.8	90.0	

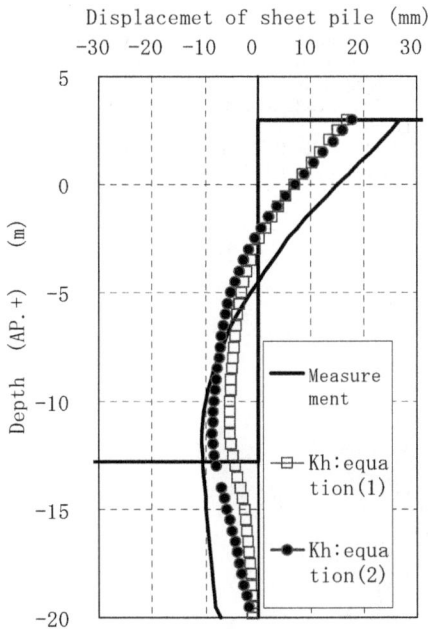

Figure 18. Comparison of measured and calculated displacement of sheet pile.

7 DESIGN FLOW FOR EARTH RETAINING WITH FGC-DM

Design of deep braced excavation with FGC-DM follows the conceptual flow chart shown in Figure.19.

As the brade excavation is a complicated interaction of the retaining structure and the surrounding soil even the natural homogeneous ground, the design becomes an iterative process.

Given the soil profile and the requirement for excavation, the size of earth retaining structure (enbedment of sheet pile, strength of FGC-DM etc) is assumed.

If necessary in this stage, the standardized laboratory test result and database on the previous similar experience will help these assumption.

Design is executed by the ground spring model method. In this stage, coefficient of subgrade reaction should be calculated by equation (2).

If the deformation and stress of sheet pile is acceptable, it is necessary to confirm the stress of FGC-DM.

Especially FGC-DM is typical brittle material,so stress must be existed in the elastic range.

As the merit in the case of applying an FGC-DM to the earth retaining, it is mentioned that sheet piles can be driven on the direct to improvement ground.

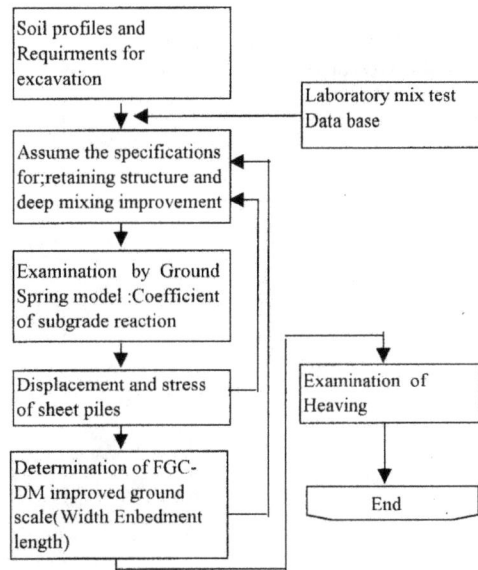

Figure 19. Design flow chart of earth retaining (braced excavation) with FGC-DM.

Therefore, it is desirable to be compatible not only the obtaining the shear strength and rigidity for proper coefficient of subgrade reaction in horizontal direction but also drivability of sheet piles to the FGC-DM improved ground.

In parallel with the examination of the retaining structure, the heaving of the excavation bottom should be examined . There exist a variety of calculation methods to examine the bottom heave.

In this case, heaving of the bottom was examined in centrifuge model test and numerical analysis. (Ohishi et al.1999) (Ishii et al.2001).

8 CONCLUSIONS

1. FGC-DM was successfully applied to improve the base of braced excavation to minimize the displacement of the earth-retaining sheet piles and prevent heave of the bottom.

2. Field measurements undertaken for evaluating the coefficient of subgrade reaction which is a critical parameter for ground spring model. The measured coefficient values are lower than those calculated by traditional equation in Japan.

3. A new calculation equation for the coefficient of subgrade reaction to the FGC-DM ground has been established on the basis of the on-site measurement results and centrifuge model tests for the coefficient of subgrade reaction in the horizontal direction.

ACKNOWLEDGEMENT

Authors acknowledge the advice provided by Dr. K. Takahashi, Executive Director, Port and Airport Research Institute

REFERENCES

Asano, J., Ban, K., Azuma, K. and Takahashi, K.(1996). "Deep Mixing Method of soil stabilization using coal Ash." Proceeding of Is-Tokyo '96, The 2nd International Conference on Ground Improvement Geosytems: 393-398: Rotterdam: Balkema.

Azuma, K., Noguxchi, S., Kurisaki, K. and Takahashi, K. (1999) "Earth retaining excavation using sheet pile and self supported earth retaining wall by Deep Mixing Method Using Coal Ash." Proceeding of 13th International Symposium on Use and Management of Coal Combustion Products: 3: 90-1 - 90-22

Azuma,k., Takeuchi.G., Takahashi.K.,Yoshimoto.Y., Ri.R., Ishii,T(2002): "The relation between the lateral pressure on site and inner excavation about the earth retaining structure using deep mixing soil stabilization method using coal ash" Proc.of 37th Japan National Conference on Geotechnical Enginnering,pp1603-1604. (in Japanese)

Ishii,T., Ohishi, K., Katagiri, M., Saitoh,K. and Azuma, K. (2001) "Stability analysis of braced excavation with base improvement." Proceeding of 10th International Conference on Computer Methods and Advanced in Geomechanics..

Nakamura.H., Nakazawa,A.(1972) "Stress calculation of earth retaining structures during construction" Soil and Foundations,Vol.12,No.4,pp95-103(in Japanese)

Ohishi, K., Katagiri, M., Saitoh, K. and Azuma, K. (1999) "Deformation behavior and heaving analysis of deep excavation" Proceeding of Is-Tokyo '99, Geotechnical Aspects of underground construction in soft ground: 693-698: Rotterdam: Balkema.

Yoshinaka.R.(1967) "Coefficient of subgrade reaction "Civil Engineering Journal,Vol.10,No.1,pp33-37(in Japanese)

Soft Ground Engineering in Coastal Areas, Tsuchida et al. (eds)
© 2003 Swets & Zeitlinger, Lisse, ISBN 90 5809 613 0

Deformation of ultra-soft soil

V. Choa
Nanyang Technological University, Singapore

M.W. Bo
SPECS Consultants Pte. Ltd., Singapore

ABSTRACT: Reclamation on ultra-soft soil is extremely difficult due to its low strength and high compressibility. Slow and steady application of load is required in order to maintain the stability of the foundation. Since soil is too soft, conventional sampling and site investigation are not feasible. A special twist sampler to collect samples and a Gamma-Gamma probe for in-situ density measurements were utilized. The conventional oedometer cannot be used to determine deformation parameters, a hydraulic type of consolidation equipment is used instead. Step loading or constant rate of strain tests can be carried out with this equipment. Large settlement usually occurs which does not follow Terzaghi's small strain theory. The large settlement, which takes place during the initial stage of deformation, occurs with little gain in effective stress. The prediction of deformation in terms of magnitude and rate of settlement are possible with a new set of equations. This paper presents a case study of deformation of an ultra-soft soil after reclamation.

1 INTRODUCTION

Due to a shortage of land in coastal cities, reclamation works are being carried out to expand the land. Areas with favourable foundation for reclamation are becoming scarce. As such reclamation works have to be carried out on non-favourable foundation such as waste pond, slurry pond and recently formed estuary deposits which are still undergoing self-weight consolidation. Since such deposits are very soft with high compressibility and negligible shear strength, reclamation on them are extremely difficult. Special techniques are required for reclamation of this type of deposit. In addition to the difficulties in reclamation on such foundation, it also contributes large settlement upon application of the additional load. Large settlement usually occurs with little effective stress gain. This type of large strain deformation does not comply with Terzaghi's small strain theory. This paper describes the deformation of ultra-soft soil and proposes a theory to explain this behaviour.

2 ULTRA-SOFT SOIL

For the purpose of this paper, an ultra-soft soil is defined as a soil with a liquidity index of greater than unity, i.e. a soil with a moisture content greater than its liquid limit. The strength of such soil is extremely low and may not have measurable effective stress in the in-situ condition. The soil may still be undergoing self-weight consolidation. A tall column of this type of soil if left undisturbed will commence to develop a structure during the self-weight consolidation process. However it will take a long period of time for this type of soil to possess a measurable effective overburden pressure. The bulk density of such soil generally varies between 11 to 13 kN/m^3.

3 DEFORMATION OF ULTRA SOFT SOIL

There are various types of deformation of soil such as plastic flow, elastic deformation, shear deformation, undrained creep, primary compression, secondary compression and liquefaction. Most theories of deformation cover deformation of either cohesive or cohesionless soil but there are some theories, which cover cohesive and frictional soils. Some types of deformation are stress dependent and some are time dependent. Generally the majority of large strain deformations are associated with cohesive compressible soils. In 1923, Terzaghi proposed a time independent linear-elastic model of compression behaviour for a low permeable thin layer of soil. Early researchers have however concentrated mainly on naturally sedimented clay deposits, which are either normally consolidated or over-consolidated.

The deformation behaviour of the ultra soft soil due to additional load is different from that of the normally or over-consolidated soil. The prediction of magnitude and time rate of deformation using Terzaghi's consolidation theory may not be appropriate (Bo et al. 1997a, 1997b and 1999). Terzaghi's consolidation theory leads to an under-estimation of the magnitude and an over-estimation of degree of consolidation. Since saturated soil has two phases viz water and solid, the model used consists of a viscous and elasto-plastic component which follows fluid and solid mechanics respectively. Since ultra-soft soils are in a state of under consolidation the slurry-like soil behaviour can be explained by a combined viscous and plastic deformation. There could be two scenarios of deformation. In the first scenario, the ultra-soft soil will start with viscous deformation and then switch to plastic deformation. The transition is likely to be smooth. This model can be explained by a spring and dashpot model shown in Figure 1a in which viscous deformation is represented by a dashpot and plastic deformation by a spring. The viscous deformation is likely to be time dependent and irreversible. Both viscous and plastic deformations may be linear within its own stage. In the second scenario, the deformation of slurry like viscous material starts with visco-plastic behaviour and at a certain point it smoothly switches to plastic deformation. This second scenario is illustrated in Figure 1b. Idealized e log σ'_v curve compared with that of natural soil is shown in Figure 2.

The compression index in the soil stage is likely to be equivalent to the intrinsic compression index of natural clays. However the compression index of the soil in the ultra-soft stage will be greater than the intrinsic compression index. The ultra-soft soil has three compression indices viz C^*_{C1}, C^*_{C2} and C^*_{C3} where C^*_{C3} is the same as the intrinsic compression index. It is proposed that for predicting consolidation of ultra-soft soil, a large strain coefficient of consolidation (C_F) be used instead of the conventional coefficient of consolidation (C_V) (Bo 2002 and Bo et al. 2003).

4 SITE INVESTIGATION FOR ULTRA-SOFT SOIL

Since ultra-soft soil is too soft for conventional methods of site investigation, sampling was carried out with a twist sampler as shown in Figure 3. Field vane shear tests were carried with bigger dimension blades due to the low shear strength. Bulk density was measured with a Gamma-Gamma Probe. Site investigations had to be carried out from a floating pontoon rather than a jack-up barge. Figure 4 & 5 shows density contour and density profile determined by Gamma- Gamma probe at the slurry pond in the Changi East reclamation project.

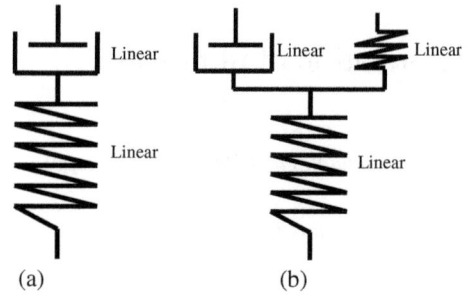

Figure 1. Visco-plastic model for ultra-soft soil.

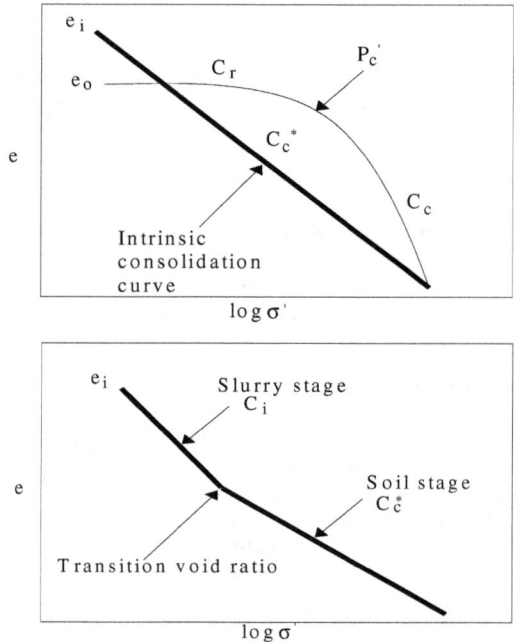

Figure 2. Comparison of e-log σ' curves for natural soil and ultra-soft slurry-like soil.

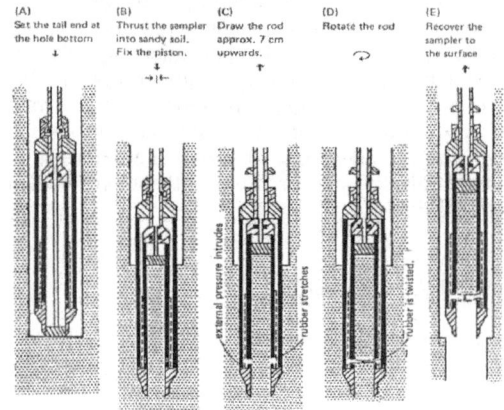

Figure 3. Twist sampler.

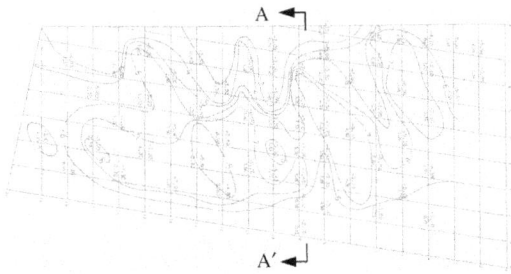

Figure 4. Contour map showing depth to the density 1.5 g/cm³.

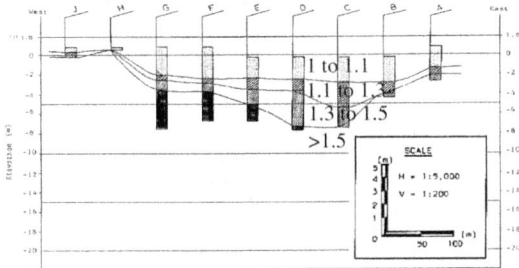

Figure 5. Density profile along Section A-A'.

5 CHARACTERIZATION OF ULTRA-SOFT SOIL

Basically ultra-soft soil does not have structure and is like slurry. Therefore, although the samples collected with the twist sampler are disturbed, the determination of the soils in-situ parameters is not affected. Physical properties of samples can be determined as the same way as that determined for natural soil. It may not be necessary to carry out strength test since ultra-soft soil does not have measurable strength. The essential parameters to be measured are the compressibility and consolidation parameters. Conventional oedometer tests may not be appropriate because the soil is too soft. A hydraulic consolidation cell such as a Rowe cell is a more suitable equipment to characterize the compression and consolidation of the ultra-soft soil. The compression and consolidation parameters obtained from a step-loading hydraulic consolidation cell test is shown in Figure 6.

However it took several days to complete a set of test because the permeability of the ultra-soft soil is extremely low. Alternative consolidation tests are the constant rate of loading (CRL) and constant rate of strain (CRS) tests using the hydraulic consolidation cell. However these tests require appropriate rates of loading and strain in order to

obtain the correct $e \log \delta'_v$ curve. Details can be found in Bo (2002). Figures 7 & 8 shows $e \log \delta'_v$ curves obtained from CRL and CRS tests respectively. Based on the comparison with EOP test for the particular soil tested, 1 kPa/800sec was found to be a suitable rate of loading for the CRL test and 0.01% strain rate was a suitable rate of strain for the CRS test. Figures 9 & 10 shows the permeability and large strain coefficient of consolidation parameters obtained from the CRL and CRS tests.

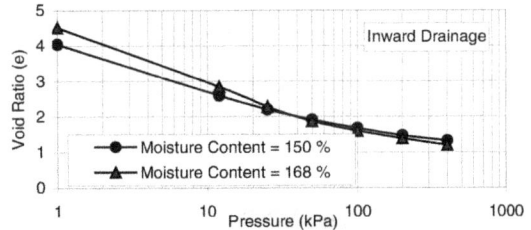

Figure 6. Void ratio vs. pressure test results from hydraulic consolidation cell for two different moisture contents.

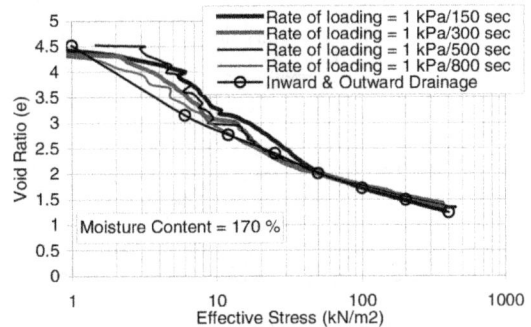

Figure 7. Comparison of e-log σ' curves from various loading rate CRL tests with step loading hydraulic cell tests.

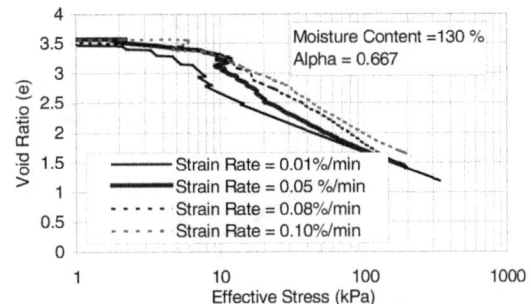

Figure 8. Comparison of e-log σ'ᵥ curves from CRS tests with various strain rate and conventional 24 hour test curves.

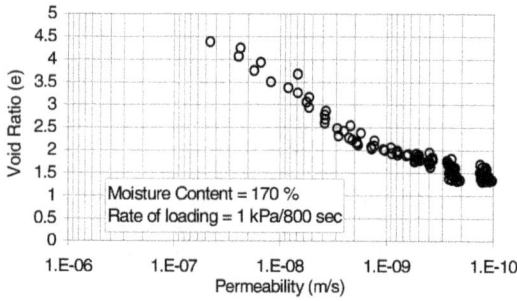

Figure 9a. Void ratio versus permeability.

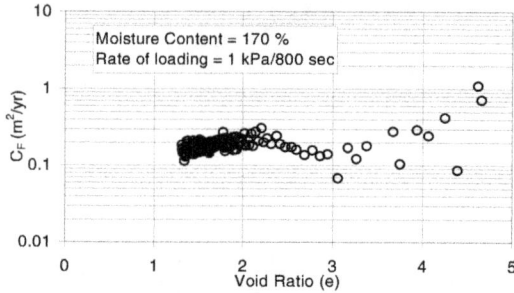

Figure 9b. C_F versus void ratio.

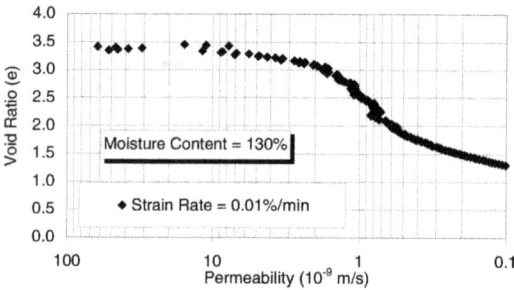

Figure 10a. Permeability measured from CRS test.

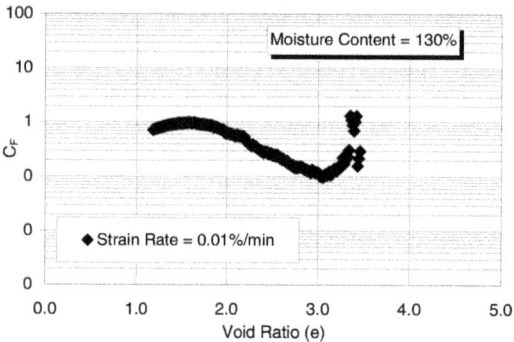

Figure 10b. Void ratio vs. coefficient of consolidation (C_F) from various CRS test.

6 PREDICTION OF ULTRA-SOFT SOIL DEFORMATION

After obtaining the compression and large strain consolidation parameters from the appropriate consolidation test, the magnitude of settlement can be predicted by Equation (1) proposed by Bo (2002).

$$S = \frac{H_o}{1+e_i}\left(C_{c1}^* + C_{c2}^* + \left(C_{c3}^* \times \log\frac{\sigma_f^{'}}{100}\right)\right) \qquad (1)$$

where H_o is initial thickness of soil; e_i is initial void ratio; $\sigma_f^{'}$ is final stress; and C_{c1}^*, C_{c2}^* and C_{c3}^* are various compression indices.

The time rate of consolidation can also be predicted by Gibson's large strain theory using large strain coefficient of consolidation. The time factor curves shown in Figure 11 can be used for predicting of time rate of settlement for both single and double drainage conditions. Using the C_F obtained from compression test, time factor can be calculated from:

$$T = \frac{C_F t}{D_e^2} \qquad (2)$$

where T is the time factor; t is time ; and D_e is the drainage path.

A program using the explicit finite difference method has also been developed. Details can be found in (Bo, 2002).

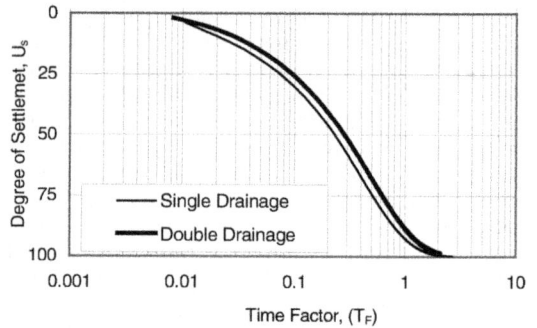

Figure 11. Time factor curves covering both ultra-soft to soil range for single and double drainage.

7 CASE STUDY

7.1 Pilot embankment and other test areas

In order to investigate the deformation and pore pressure dissipation of an ultra-soft soils, a pilot embankment was carried out at the Changi East reclamation project where reclamation and soil improvement was carried out on an ultra-soft soil. Details of the reclamation on a slurry-like soil has been described in Bo et al. (1998). Soil instrument clusters were installed after filling to +4 mCD. Instruments installed in a typical cluster included surface settlement plates, deep settlement gauges, pneumatic piezometers, electric vibrating-wire piezometers, open-type piezometers and water stand-pipes. The deep settlement gauges and piezometers were installed at various elevations in the soil. The arrangement of each cluster in plan and elevation are shown in Figures 12.

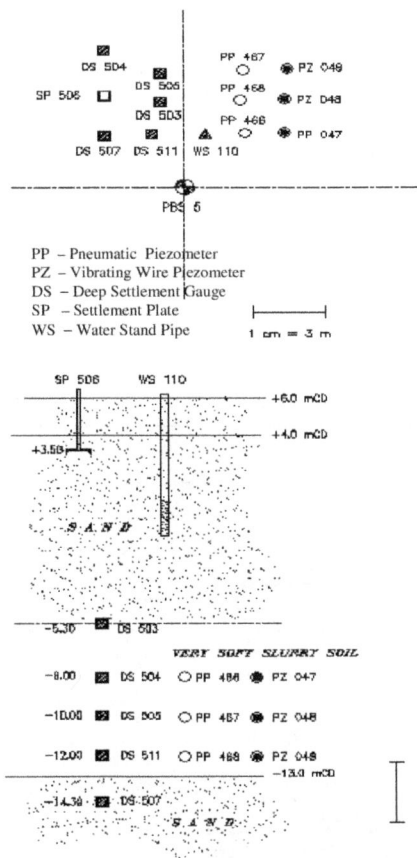

Figure 12. Instrument layout and installed elevations at siltpond area.

7.2 *Deformation behaviour of the siltpond slurry*

Instruments in the pilot area registered a total settlement of 2.5 m within a one year period. This settlement is equivalent to a strain of 30%. This settlement did not take into consideration the settlement that occurred due to sand filling up to elevation +4 mCD prior to the installation of the instruments.

It was noticed that although the foundation soil was settling, no pore water pressure dissipation was recorded up to eleven months in the top two piezometers and up to two weeks in the bottom most piezometer which was installed close to the drainage boundary. Only after eleven months did the pore water pressures in the top two piezometers start to dissipate and the soil began to gain strength. Actually there was a small reduction of excess pore pressure during the initial 11-month period. This was not due to pore pressure dissipation, but rather the submergence of the sand fill. The observation of no effective stress and strength gain despite the occurrence of large settlement had been confirmed by various in-situ and laboratory tests (Bo et al., 1997 a & b). Figure 13 shows pore pressure and settlement measurement at the pilot embankment.

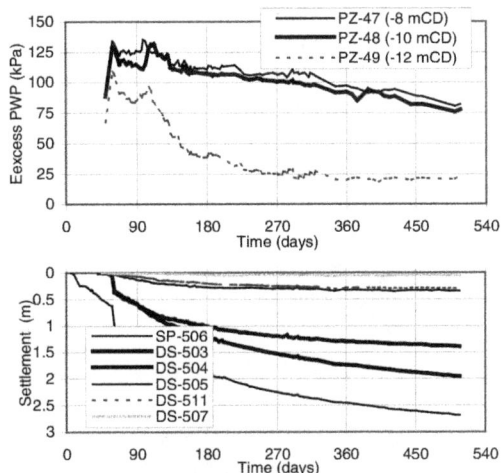

Figure 13. Pore pressure and settlement measurement at pilot embankment.

7.3 *Verification of the proposed large strain deformation model*

The proposed large strain deformation was verified by two case studies. One is at siltpond pilot area explained in Section 7.2. Another is area with largest settlement in the same siltpond. At the siltpond pilot area, the top 5.3 m of material was in a slurry state and the bottom 2.5 m had already become a Terzaghi soil and was in a normally consolidated state. The settlement of both layers were studied and analyzed. Figure 14 shows the profile of soil together with the geotechnical parameters of the siltpond pilot area.

7.3.1 *Determination of compression indices*
The compression indices at the different log cycles of pressure were obtained from consolidation tests. The final void ratios for the relevant additional load were calculated using Equations 1. The compression indices, void ratios at the liquid limit and void ratio at 10 kPa for the ultra-soft upper layer are summarized in Table 1. The compression indices of the normally consolidated lower soil layer are also included in Table 1. Table 2 shows the final expected

load at various construction stages and the predicted settlement at the end of each stage together with the total cumulative settlement. For the ultra-soft soil, the initial effective stress is taken as 1 kPa and for the normally consolidated soil the initial stress is calculated using $\sigma' = \gamma h$. Submergence of the fill due to settlement was taken into consideration in the estimation of the final expected pressure. The void ratio and thickness of the compressible layer were updated at each stage of loading.

-5.2m CD

Ultra-soft Soil
$\overline{W}$ =130%
e_i = 3.484
Gs = 2.68
LL = 78%
PL = 25%
Clay Content = 70 - 80%

-10.50m CD

Normally consolidated Soil
$\overline{W}$ = 80%
e_i = 2.144
Gs = 2.68
LL = 78%
PL = 25%
Clay Content = 70 - 80%

-13.00m CD

Figure 14. Profile of soil at siltpond pilot area.

Table 1. Summary of compressibility parameters for ultra-soft upper and lower soil layer at siltpond area

Parameters	Ultra-soft upper layer	Lower soil layer
Void ratio at liquid limit (e_L)	2.09	-
Void ratio at 10 kPa (e_{10})	2.20	-
C^*_{c1}	1.28	-
C^*_{c2}	0.91	0.91
C^*_{c3}	0.49	0.49

Table 2. Summary of predicted settlement under various stages at siltpond pilot area

	Stages		
	I	II	III
Additional load (kPa)	65.20	105.45	186.45
Load duration (month)	2.00	28.70	16.53
Ultimate settlement for upper layer (m)	2.174	2.416	2.585
Settlement at time (t) for upper layer (m)	0.472	2.408	2.584
Ultimate settlement for lower layer (m)	0.63	0.72	0.806
Settlement at time (t) for lower layer (m)	0.13	0.717	0.805
Predicted total settlement at time (t)	0.602	3.125	3.389
Measured total settlement at time (t)	0.25	3.10	3.40

The calculated final settlement was in close agreement compared with the measured settlement. The time rate of settlement was predicted using the proposed finite difference model. Construction stages and duration of load in various stages were modeled as shown in Figure 15. The pore pressure measured at large settlement area is shown in Figure 16.

Various C_F values were applied to the different loading stages. The applied C_F values for each stage are shown in Table 3. Several passes of vertical drains at two metre square spacing were installed. This was because after the vertical drains have performed for a certain duration, the large strains and buckling of the drains would have made them virtually ineffective and the installation of another round of drains was required.

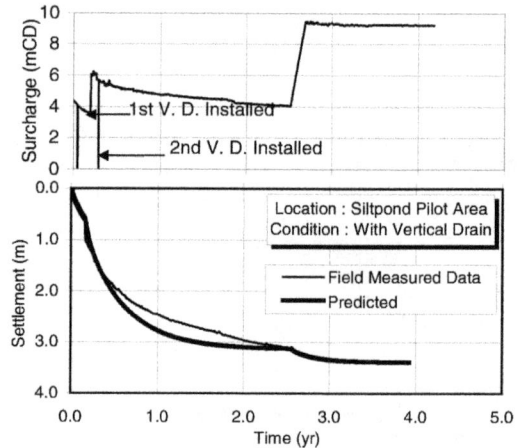

Figure 15. Construction sequence for pilot area and verification using data from siltpond pilot area.

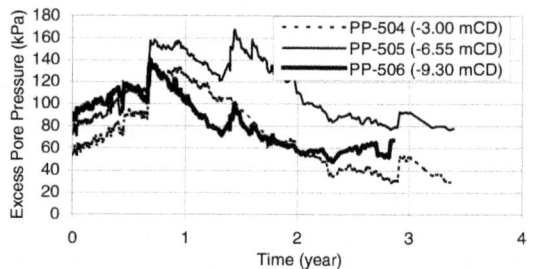

Figure 16. Excess pore pressure versus time at large settlement area.

Table 3. Summary of applied C_F values in prediction of time rate of settlement in siltpond pilot area

Stages	C_F (m²/yr.)	
	Upper Layer	Lower Layer
1	0.0287	0.0539
2	0.0903	0.1264
3	0.1733	0.1827

238

The time rate of settlement were calculated for the two separate layers and combined to give the total settlement. It can be seen in Figure 15 that the time rate of settlement predicted using the proposed model closely match the measured data.

7.4 Verification using data from area with the largest deformation in the main works

Since verification of the reliability of the proposed equations and model using one set of test data may not be sufficiently conclusive, prediction was made for another area in the main works where the largest settlement occurred. The physical parameters and compression indices of the study area is shown in Table 4. Figure 17 shows the soil profile and geotechnical parameters of the area. In this case, the compression indices for the various pressure ranges were obtained from the proposed equations using the measured liquid limits (Bo 2002). The computed settlements were found to be in close agreement with the measured settlement (Table 5). The construction sequence is shown in Figure 18.

Table 4. Summary of compressibility parameters for ultra-soft upper and lower soil layer at siltpond large settlement area

Parameters	Ultra-soft soil	Lower soil layer
Void ratio at liquid limit	2.09	-
Void ratio at 10 kPa (e10)	2.20	-
$C*c1$	1.28	-
$C*c2$	0.912	0.92
$C*c3$	0.50	0.50

Table 5. Summary of settlement under various stages at siltpond large settlement area

	Stages				
	I	II	III	IV	V
Additional load (kPa)	52.8	72.8	93.2	128.6	159.6
Load duration (month)	9.0	2.0	6.0	19.0	5.5
Ultimate settlement for upper layer (m)	3.579	3.974	4.277	4.533	4.678
Settlement at time (t) for upper layer (m)	2.030	2.700	3.697	4.530	4.651
Ultimate settlement for lower layer (m)	1.491	1.748	2.068	2.278	2.422
Settlement at time (t) for lower layer (m)	1.491	1.748	2.072	2.278	2.424
Predicted total settlement at time (t)	3.521	4.448	5.767	6.810	7.075
Measured total settlement at time (t)	3.40	4.30	5.40	6.70	6.95

The final expected loads and settlements at various stages are shown in Table 5. The measured and predicted time rate of settlement are in close agreement as shown in Figure 18. The slight variation of time rate of settlement in each stage may be due to a slight variation of the applied C_F values at each stage with the field C_F values. The applied C_F values in each stages are shown in Table 6. It can be seen that C_F values of the lower layer clay for this area is very different from those in the pilot area. In reality, C_F values varies with applied total and effective stress at each step. It should be noted that C_F values of the lower clay layer in this area for various stages are very different from the C_F values in the pilot test area due to the differences in stress levels. The slight variation in settlement magnitude at each stage could also be due to the slight variation of estimated additional load at each stage.

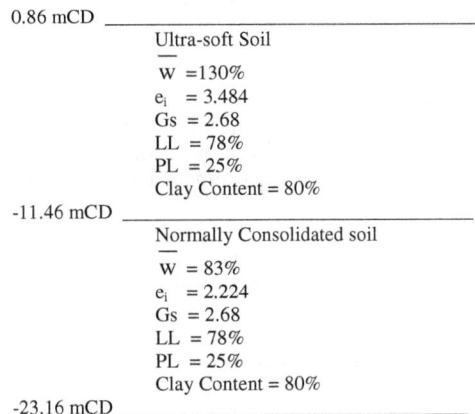

0.86 mCD _____

Ultra-soft Soil
w =130%
e_i = 3.484
Gs = 2.68
LL = 78%
PL = 25%
Clay Content = 80%

-11.46 mCD _____

Normally Consolidated soil
w = 83%
e_i = 2.224
Gs = 2.68
LL = 78%
PL = 25%
Clay Content = 80%

-23.16 mCD _____

Figure 17. Soil profile and geotechnical parameters of the largest settlement area.

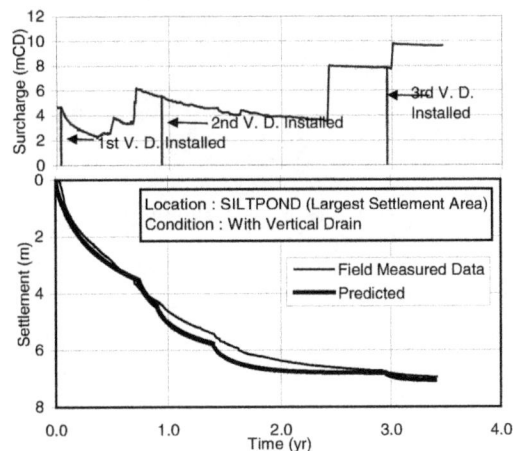

Figure 18. Construction sequence for siltpond and verification using data from the largest settlement area in the main work.

Table 6. Summary of applied C_F values used in prediction of time rate of settlement area

Stages	C_F (m²/yr)	
	Upper layer	Lower layer
1	0.0465	0.2758
2	0.1018	0.3605
3	0.0957	0.5492
4	0.1488	0.7155
5	0.1943	0.7805

8 CONCLUSIONS

- Ultra-soft soil with moisture content is greater than the liquid limit has extremely low strength and is often still undergoing self-weight consolidation.
- Deformation of these soils are very different from normally consolidated soils and large settlement with little or no pore pressure dissipation may be experienced in the early stages of deformation.
- Site investigation of these soils require special equipment such as the Gamma-gamma probe and twist sampler.
- Compression tests can be carried out in a hydraulic consolidation cell. The compressibility and consolidation parameters of the ultra-soft soil can be determined by step incremental loading, constant rate of loading (CRL) or constant rate of strain (CRS) tests.
- Prediction of the magnitude and time rate of consolidation can then be made using the equations and time factor curves presented in this paper.
- The proposed equations and model have been validated with case studies and found to be reliable.

ACKNOWLEDGEMENT

Mr. Joel Z. Indedanio and Mr. Hla Shwe of Hyundai Engineering and Construction Co. Ltd. are greatly appreciated for their assistance in preparing this paper.

REFERENCES

Bo Myint Win, Arulrajah, A. and Choa, V. 1997a. Large deformation of slurry-like soil due to additional load. *International Conference on Foundation Failure,* May 1997, Singapore, pp. 289-296.

Bo Myint Win, Arulrajah, A. and Choa, V. 1997b. Large deformation of slurry-like soil. *Deformation and Progressive Failure in Geomechanics*, Asoaka, A., Adachi, T. and Oka, (eds) Rotterdam: Balkema, pp. 437-442.

Bo Myint Win, Arulrajah, A., Choa, V. and Na Y. M. 1998. Land reclamation on slurry-like soil foundation. *Problematic Soils*, Yonagisawa, Morota and Mitachi (eds), pp. 763-766. Rotterdam: Balkema.

Bo Myint Win, Arulrajah, A., Choa, V. and Na, Y. M. 1999. One-dimensional compression of slurry with radial drainage. *Soils and Foundations, Japanese Geotechnical Society,* November 1999, Vol. 39, pp. 9-17.

Bo Myint Win 2002. Deformation on ultra-soft soil. *Ph.D. Thesis, Nanyang Technological University*, Singapore.

Bo M. W., Choa, V. and Wong, K. S. 2002. Compression test on slurry with small scale consolidometer, *Canadian Geotechnical Journal*. April 2002, Vol. 39, pp. 388-398.

Bo M. W., Wong, K. S., Choa, V. and Teh C. I. 2003. Compression tests of ultra-soft soil using an hydraulic consolidation cell. *The Geotechnical Testing Journal*, (On print).

Soft Ground Engineering in Coastal Areas, Tsuchida et al. (eds)
© *2003 Swets & Zeitlinger, Lisse, ISBN 90 5809 613 0*

Long-term strength change of cement treated soil at Daikoku Pier

M. Ikegami & T. Ichiba
Kanto Regional Development Bureau, Ministry of Land, Infrastructure and Transport, Yokohama, Japan

K. Ohishi & M. Terashi
Nikken Sekkei Nakase Geotechnical Institute, Kawasaki, Japan

ABSTRACT: This paper presents results of a research on long-term characteristics of cement treated soil. The purpose of this investigation has two aspects concerning change of properties with time in a long time span. One is to confirm long-term strength of the cement treated soil 20 years after construction. The other is to investigate the deterioration at the boundary surface of the treated soil. The results are that the present strength inside the treated ground is 2.1 times greater than 20 years ago, while at the periphery the strength reduction is observed up to a depth of 30 to 50mm from the surface of treated soil.

1 INTRODUCTION

Deep mixing method is a kind of soil stabilization technique using cement and/or lime as a binder, constructing cement treated soil block, column or w all in soft ground. It was developed in 1970's as cement deep mixing (CDM) method using cement-water slurry in Japan. Dry jet mixing (DJM) method using dry cement powder was also developed in Japan after several years following CDM method. These soil stabilization methods have been applied in practice for construction in soft ground for 25 years.

Many researches have been carried out to investigate mechanical characteristic of cement treated soil. As to long-term properties of treated soil, especially long-term strength, there are many reports showing strength increase with time. In most of these reports, however, laboratory test and field verification were conducted on the treated soil samples aged less than a couple of years. Taking account of the life of structures, it is not sufficient testing period for assurance of the long-term performance. So far, several research groups have reported the long-term test results on in-situ treated soils exceeding 10 years period (Terashi et al. 1992; Hayashi et al. 2001; Inagaki et al. 2001).

On the other hand, some reports points out the possibility of deterioration (strength reduction) of the treated soil in the long term (Terashi et al. 1983, Saitoh, 1988, Kitazume et al. 2003). In the studies, laboratory mixed soil specimens exposed to outer environments, such as seawater or untreated clay, were tested to investigate deterioration progress from surface with time. However, long-term change of mechanical, physical or chemical characteristic of treated soil is not entirely made clear yet.

In this research, in-situ treated soil samples were retrieved from the treated soil mass 20 years after construction and a series of test was conducted to investigate mainly strength change with time in 20 years from construction to present. There are two aspects for the purposes of this study. One is to investigate the strength increase inside the treated soil mass where outer environment is considered to have not affected. The other is to investigate the deterioration at the periphery of the treated soil exposed to untreated original clay ground.

2 OUTLINE OF THE TEST SITE

The test site is at Daikoku Pier, Port of Yokohama in Tokyo bay (see Figure 1). It is a seawall construction surrounding new reclaimed land, which construction was started in 1971 and reclamation was completed in 1990. There are 25 berths in this pier. The foundation ground for 9 berths, called T1 ~ T9 berth, were improved by CDM method as seawall foundation. This is the site where the wet method of deep mixing (CDM method) was first applied in large scale by the Port and Harbor Bureau, Japanese Ministry of Transport (currently, Ministry of Land, Infrastructure and Transport). Therefore, intensive study on the cement treated soil was conducted to evaluate the quality and the properties during the construction, which is the reason the authors selected the site for investigating long-term characteristics of cement treated soil. The outline of these berths is shown in Table 1.

The test site is located at the end of T2 berth. The original ground at T2 berth was thick clay layer from

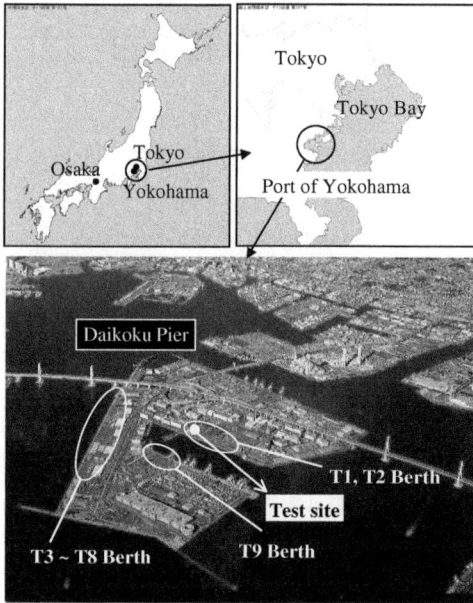

Figure 1. Location of Daikoku Pier and the test site.

Table 1. Outline of T1 ~ T9 berths

Berth	Length (m)	Depth (m)	Construction period of ground improvement
T1, T2	480 (@240)	12	1976 ~ 1981
T3 ~T8	1110 (@185)	10	1779 ~ 1883
T9	240	12	1985 ~ 1988

Table 2. Properties of the alluvial clay layer

Elevation (m)	Water content (%)	Wet density (t/m^3)*[1]	Unconfined compressive strength (kPa)*[2]
~ -23	80 ~100	1.50	$q_u = 5.7H$
-23 ~ -38	50 ~ 60	1.68	$q_u = 8.6 + 4.9H$
-38 ~ -50	75 ~ 85	1.53	$q_u = 64.7 + 2.7H$

*[1] Average value, *[2] $H = 0$ at El. -12 m

elevation -12 to -70 m overlying bedrock of mudstone. The layer from sea bottom to around El. -50 m is soft alluvial clay and diluvial clay is beneath it. Properties of the alluvial clay are shown in Table 2. The clay deposit may be divided into three layers according to physical properties.

The ground improvement was applied to increase the shear strength of soft alluvial clay for stability of seawall. The binder used was the ordinary Portland cement and the binder content of 160 kg/m^3 was recorded in the design document. The deep mixing machine used had eight shafts as shown in Figure 2. Typical cross section of the improved ground in T2 berth is shown in Figure 3. The clay ground was improved to depth of El. -49 m as massive block with 57 m in length and 35.9 m in thickness. The im-

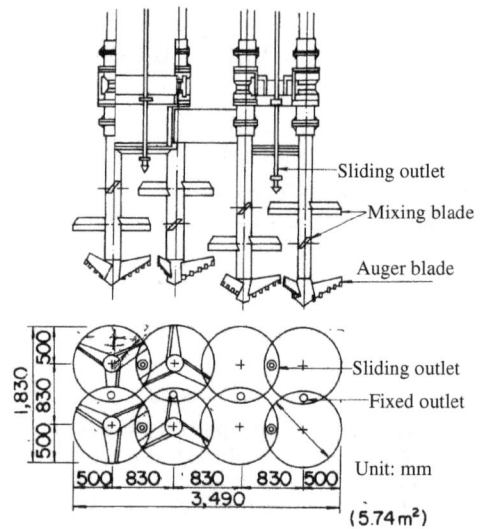

Figure 2. Mixing equipment used in this construction.

proved ground is a floating-type structure, which bottom remains in the soft ground.

Three to six months after the ground improvement execution in 1981, a number of core samples were retrieved at this site and physical properties, strength and calcium (Ca) content were measured.

3 TEST PROCEDURE

In this research in 2001, after 20 years of construction, four borings are conducted to obtain undisturbed core samples for the investigation of long-term characteristics of the treated soil. The layout of the borings in cross section is shown in Figure 3. Two borings (Bor. No.1 & No.2) are inclined at angle of 20 degrees from vertical, by which continuous

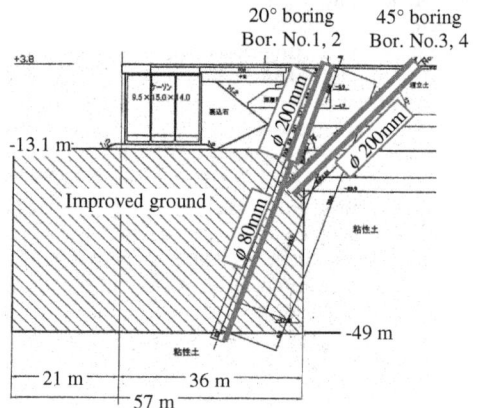

Figure 3. Cross section of the treated ground and boring.

core samples with 80 mm in diameter are retrieved from top to bottom of the treated soil. Other two borings (Bor. No.3 & No.4) inclined 45 degrees from vertical are conducted to retrieve the core samples with 200 mm in diameter at the side boundary between treated soil mass and original ground as shown in Photo 1.The depth of retrieving these core samples at the side boundary is about El. -18 m.

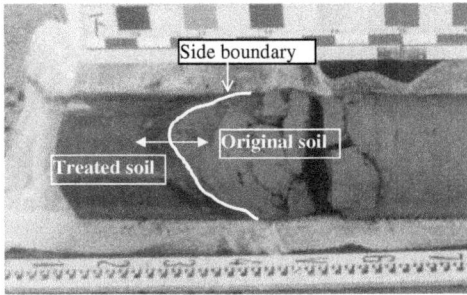

Photo 1. The core sample of Bor.No.3 at side boundary of the treated ground.

Figured 4 shows the location of borings at the present ground level in plan. Open triangles represent points of the vertical borings conducted in 1981, referred to as Check bor. No1 ~ No.5. The borings in 2001 represented by solid circle are inclined from vertical as shown already in Figure 3 and are located near the Check bor. No.4 and 5 conducted in 1981. To investigate long-term characteristics of the treated soil, test results in 2001 are compared with those in 1981.

The core samples taken by Bor. No1 and No.2 are tested to determine the water content, wet density and unconfined compressive strength of treated soil

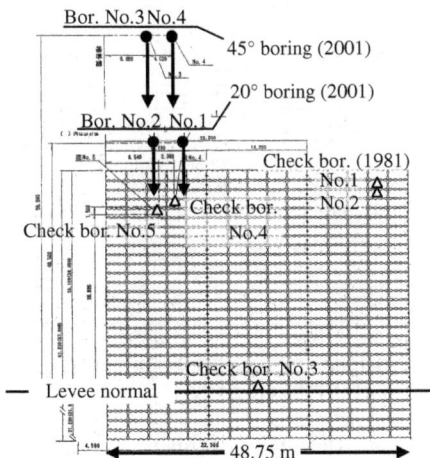

Figure 4. Plan of boring point in 1981 and 2001.

inside the treated soil mass to which outer environmental condition has negligible influence.

The core samples by Bor. No.3 and No.4 are tested to determine whether deterioration has occurred or not at the periphery exposed to the original clay ground. Total Ca content in soil was measured on the treated and the original soil by atomic adsorption spectrometry. Strength was measured by the needle penetration test to investigate the strength distribution in detail. In this test as shown in Photo 2, a sewing needle with a diameter of 0.84 mm was penetrated into the treated soil sample at a constant speed of 3 mm/min. The needle penetration tests are conducted on several specimens of Bor. No. 1 and 2 before unconfined compression tests, and the penetration resistance at penetration depth of 5 mm, Q_N, is correlated to the unconfined compressive strength q_u. The relationship between Q_N and q_u is determined as q_u (kPa) = 76.4Q_N (N).

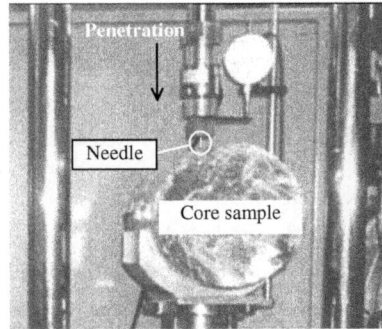

Photo 2. Needle penetration test on core sample.

4 LONG-TERM PROPERTIES INSIDE THE TREATED GROUND

Figure 5 and 6 compare the water content and wet density distributions with depth between 1981 (open circles) and 2001 (open and solid triangles). These physical properties have not changed in the last 20 years. The distributions with depth can be divided into three layers of upper, intermediate and lower layer, which reflect the original soil profile as shown in Table 2.

The strength distributions with depth are compared between 1981 and 2001 as shown in Figure 7. The strength in 1981 is based on Check bor. No.4 and No.5, which were conducted near the present Bor. No.1 and 2 as shown in Figure 4. In 1981, the average field strength throughout the depth was 6.3 MPa in unconfined compressive strength. After 20 years in 2001, the average field strength is 13.2 MPa. About 2.1 times strength increase is observed in these 20 years although no change is found in the physical properties.

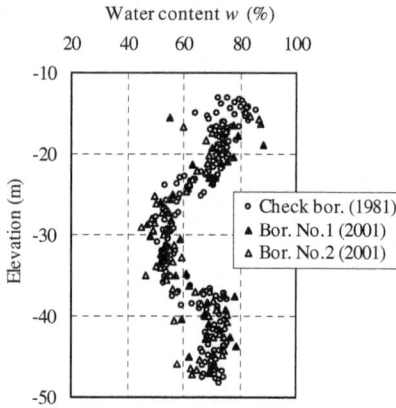

Figure 5. Water content distribution with elevation.

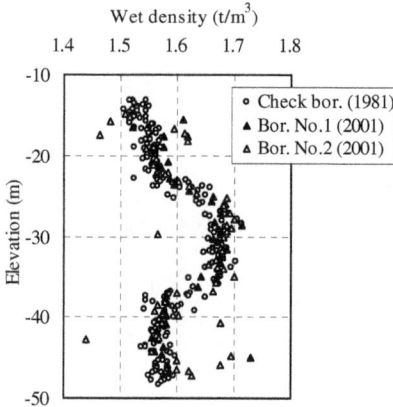

Figure 6. Wet density distribution with elevation.

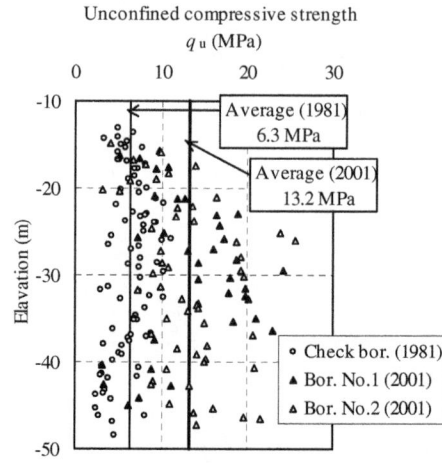

Figure 7. Strength distribution with elevation.

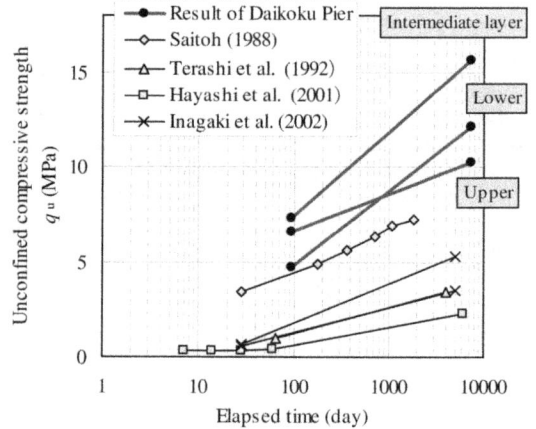

Figure 8. Long-term strength increase with time.

Table 3. Summary of researches on long-term strength

Research	Original soil	Binder	Treatment pattern / site
This research	Clay	OPC	Block / in situ
Saitoh	Clay	OPC	In mold / laboratory
Terashi et al.	Clay	Lime	Column / in situ
Hayashi et al.	Clay	BSC	Column / in situ
Inagaki et al.	Peat and clay	SC	Tangent column / in situ

OPC: Ordinary Portland cement
BSC: Blast Furnace Slag cement type B
SC: Special cement-type hardening agent

The strength at the test site is plotted versus time as shown in Figure 8. On the basis of physical property distributions shown in Figure 5 and 6, the treated soil can be divided into three layers, upper layer from El. −13.1 m to -24 m, intermediate layer from -24 m to −37 m, lower layer from -37 m to −49 m, so that the strength increase is shown for each layer. The strength in 1981, 93 days after construction, is based on Check bor. No.4 and 5, which were conducted near the present Bor. No.1 and 2. In Figure 8, similar research results on long-term strength are plotted together. The summary of the previous researches referred is shown in Table 3. Although various binders and treatment patterns were used, the strength of treated soil indicates steady increase up to more than 10 years. As regarding this test site, the rate of strength increase from 93 days to 20 years is 1.6 times for upper layer, 2.1 times for intermediate layer, 2.6 times for lower layer. The overall average of these for the present study is around 2.1.

5 DETERIORATION AT THE PERIPHERY OF THE TREATED GROUND

Strength and Ca content distribution at the periphery of the treated soil samples are shown in Figure 9. The strength in q_u is estimated by needle penetration tests, using q_u (kPa) = 76.4Q_N (N) relationship. The lines shown as "average of inside" indicate respectively the average value of strength in 2001 and Ca content in 1981 inside the treated soil at upper layer. The "average" strength inside the treated soil is the average in the range from El. -13.1 m to -24 m on the basis of Bor. No.1 & 2. The "average" Ca content in 1981 is also the average in the same range of depth. Strength decrease is confirmed to a depth of 30 to 50 mm from the boundary surface. At a depth of more than 30 to 50 mm, the strength coincides with that inside the treated soil in 2001. Ca content distribution is in good agreement with strength distribution. Ca content inside the treated soil is equal to that measured 20 years ago and starts to decrease at a depth of 30 to 50 mm from the boundary. For the treated ground at Daikoku Pier, the deterioration depth can be estimated as 30 to 50 mm judging either by strength or Ca content distribution.

Figure 10 shows Ca content distribution across the boundary between treated and original soil at Bor. No.3. Ca content in the treated soil decreases toward the side periphery of treated soil. In contrast Ca content in the original ground increases toward the side periphery. It is evident that Ca leaching phenomenon from treated soil to original soil occurred at the surface of the treated soil. Ca leaching phenomenon may be one of factors to cause the deterioration of treated soil, judging from reduction both in strength and Ca content as shown Figure 9.

To estimate the rate of deterioration with time, the field data at Daikoku Pier is plotted versus time with laboratory test results by Terashi et al. (1983) in Figure 11. Terashi et al. (1983) conducted a series of laboratory test on treated soil samples. The laboratory mixed specimens were prepared in molds, 2 hours after mixing removed from the molds, then exposed to different environments. Solid squares are the data of specimens cured in artificial seawater and open squares are those cured in untreated clay. The progress of deterioration depth in logarithm scale is almost in the linear proportion to logarithm time.

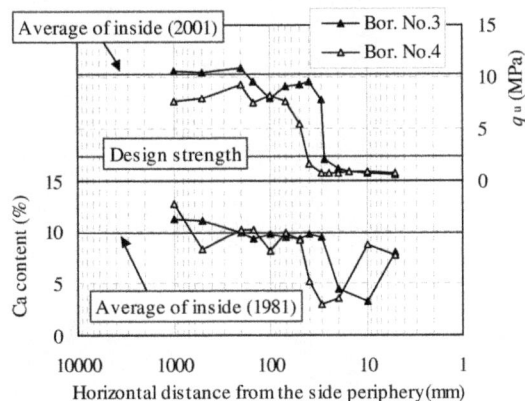

Figure 9. Comparison of strength and Ca content distribution.

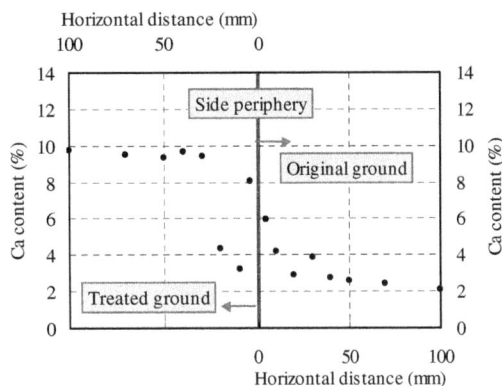

Figure 10. Ca content distribution across the boundary at boring No.3.

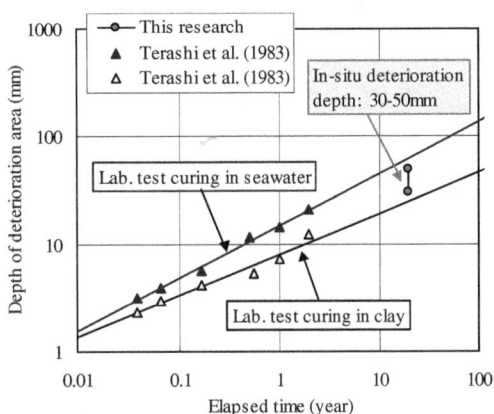

Figure 11. Depth of deterioration with time.

6 CONCLUSIONS

Concerning long-term property change of treated soil, two aspects should be taken into account. One is the strength increase with time in the long term and the other is the possibility of deterioration such as the strength reduction at the boundary surface of the treated soil. At the test site in Daikoku Pier, to investigate these aspects a series of test was conducted on the treated soil samples retrieved from the improved ground constructed 20 years ago. The conclusions are drawn as follows,

1. No change of physical properties of treated soil, water content and wet density, was detected at least by the standardized geotechnical test procedure.
2. The strength inside the treated soil mass increases with time. The average strength shows about 2.1 times increase in the past 20years, from 93 days after construction to present.

3. At the boundary surface of treated soil, deterioration is confirmed as strength reduction. The progress of deterioration in the past 20 years is around 30 to 50 mm in depth from the boundary.
4. There is a possibility that Ca leaching from treated soil is one of the factors to cause the deterioration.

REFERENCES

Kitazume, M., Nakamura, T., Terashi, M. and Ohishi, K. 2003. Laboratory tests on long-term strength of cement treated soil, *Proc. of Grouting and Ground Treatment 2003*, (to be submitted).

Hayashi, H., Nishikawa, J., Egawa, T., Terashi, M. and Ohishi, K. 2001. Long-term strength of cement treated column by deep mixing method, *Proc. of 56th Annual Meeting of JSCE* (in Japanese).

Inagaki, T., Fukushima, Y., Nozu, M., Yanagawa, Y. and Kasahara, Y. 2002. Quality of deep mixing column for organic clay under highway embankment after 10 years, *Proc. 37th Japan National Conference on Geotechnical Engineering* (in Japanese).

Saitoh, S. 1988. Experimental study of engineering properties of cement improved ground by the deep mixing method, *Ph.D. Thesis*, Nihon niversity (in Japanese).

Terashi, M., Tanaka, H., Mitsumoto, T., Homma, S. and Ohashi, T. 1983. Fundamental properties of lime and cement treated soils (3rd Report), *Report of Port and Harbour Research Institute*, 22(1)1, 69-96 (in Japanese).

Terashi, M. and Kitazume, M. 1992. An investigation of the long-term strength of a lime treated marine clay, *Technical Note of the Port and Harbour Research Institute*, 732, 14 (in Japanese).

Soft Ground Engineering in Coastal Areas, Tsuchida et al. (eds)
© 2003 Swets & Zeitlinger, Lisse, ISBN 90 5809 613 0

Effects of negative pressure and drain spacing in the horizontal drain method

S.S. Kim
Department of Civil and Environmental Engineering, Hanyang University, Ansan, Korea

H.Y. Shin
Department of Civil Engineering, Chungang University, Seoul, Korea

ABSTRACT: In this paper, the laboratory test results with middle-sized soil box test modeling the in-situ installing of horizontal drains are discussed for the estimation of the optimum negative pressure. The test was carried out in the different vacuum pressure conditions together with the measurement for the settlement and volume change of drained water by the installed drains during the consolidation process. After the test, the water content was measured to both directions of lateral distance from the drain and depth of the soil, to find out the distribution of ground improvement and strength enhancement. From the analysis on the distribution of water content, the gradual application of vacuum pressure from lower to higher level by pre-determined stages starting from low vacuum pressure is found to be effective and desirable. In the comparison of the degrees of consolidation with elapsed time, the calculated value by the prediction method based on the Barron's conventional theory showed a good agreement with the measured value. With this, it is positively considered that the applicability of the prediction method base on Barron's theory to the practical design of horizontal drains can be justified in calculating drain spacing and consolidation period.

1 INTRODUCTION

The horizontal drain method is introduced as a new method to stabilize the surface of marine dredged clay layer. As presented in Figure 1, it is performed by laying drain into the very soft ground composed through inputting the dredged clay soil on the pond. After then, it pursues large consolidation improvement effects in a short time by extracting a large volume of void water in the dredged soil by acting vacuum pressure using a vacuum pump on the drain.

To apply 'Horizontal Drain Method' on the surface stabilization of very soft dredged soil and trafficability, it is recognized that vacuum pressure and drain spacing are one of the most important factors to be considered. In this study, we tried to identify the influence of the vacuum pressure and drain spacing on horizontal drain method through laboratory soil box test.

2 THEORETICAL REVIEW

In vertical drain method, as drain sections are shaped as panels, an analysis is complicated according to size and shape, if direct application is tried.

To simplify the analysis, equivalent diameter of drain, d_w is used, and the efficiency of which has been proved in various literature.

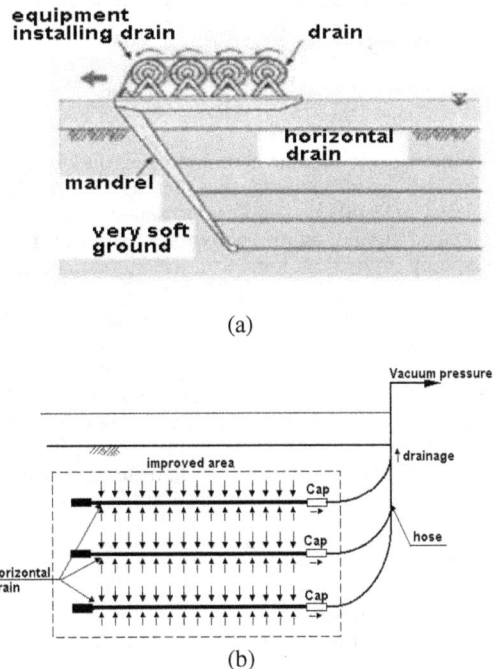

(a)

(b)

Figure 1. Schematic diagram of Horizontal Drainage using vauccum pressure: (a) installing method, (b) mechanism of horizontal method.

In horizontal drain method, as there has not been a theoretical equation, theory of vertical drain method was applied in the calculation of drain spacing, installing method and equivalent diameter.

2.1 Equivalent Diameter of Drain

In the drain used in this study, as a core and a filter are heat-bonded with each other as shown in Figure 2, there is no pore water flow through the joint of core and filter. Therefore, the equation (1) provided by Rixner et al. (1986) was used to calculated equivalent diameter.

$$d_w = \frac{(B+h)}{\pi} \tag{1}$$

where, B = width of drain
h = thickness of drain

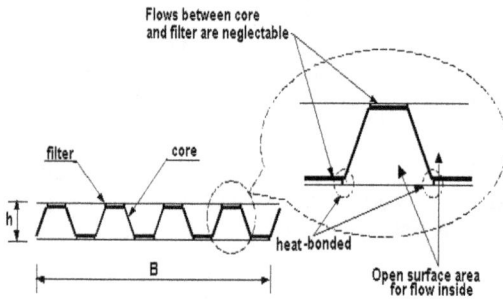

Figure 2. Explanation of open surface area.

2.2 Drain Spacing

In this study, horizontal drains were distributed in square shape following the method proposed by Kjellman(1948). To simplify the calculation of drain installed ground, an interpretation was performed by deciding influential circle of drain, d_e, which is equivalent diameter of cylindrical soil mass, when horizontal drainage occurs by single equivalent circular drain. Figure 3 shows spacing and efficient diameter of drain. In this Figure, pore water occurring inside of the square is introduced to drain located in the center. If converting this drain area to effective influential circle with same width, there will be following relations between this diameter and average space between drain (Kjellman, 1948).

$$\frac{\pi d_e^2}{4} = d_m^2 \rightarrow d_e = 1.13 \cdot d_m \tag{2}$$

where, d_m: average drain spacing
d_e : effective diameter of drain board
d_w : equivalent diameter of drain board

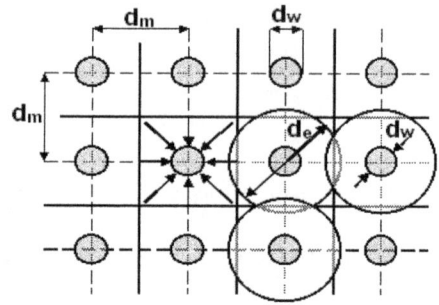

Figure 3. Arrangement and effective influence diameter of drain.

2.3 Duration Time for Vacuum

In horizontal drain method, activation region of vacuum pressure can be governed by the initial spacing of drain. In this study, Barron's equation (1948) is applied to calculate the degree of consolidation and an approximated solution.

(1) Duration time for vacuum

$$t_U = \frac{d_e^2}{C_v} T_h \tag{3}$$

(2) Degree of consolidation

$$U = 1 - \exp\left(\frac{-8T_h}{F(n)}\right) \tag{4}$$

where, T_h = Time factor
C_v = Coefficient of consolidation
U = Degree of consolidation
$$F(n) = \frac{n^2}{n^2-1} \ln(n) - \frac{3n^2-1}{4n^2}$$
$$n = \frac{d_e}{d_w}$$

To design the horizontal drain method, usually an approximated solution of Barron(1948) regarding hollow circular diameter consolidation theory is used. However, as shown in Figure 4, as settlement is progress, effective influential circle becomes oval. Therefore boundary condition of constant diameter that is applied in vertical drain method cannot be applied here. Thus, in this study, the following effective influential circle has been calculated by correcting average spacing of drain (Shinsa, 1988).

$$d_m = \sqrt{\frac{d_0^2 + d_f^2}{2}} \qquad (5)$$

$$d_e = 1.13 \cdot d_m \qquad (6)$$

where, d_0 = initial spacing of drain
d_f = final spacing of drain
d_m = average spacing of drain
d_e = influence diameter of drain

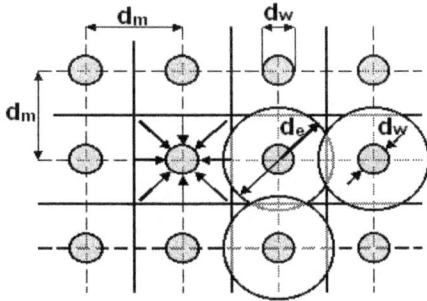

Figure 4. Average Drain Spacing and Effective Influence Diameter of Drain.

2.4 Vacuum Pressure

Vacuum pressure delivered to very soft ground through horizontal drain buried in the ground using vacuum pump is shown smaller than that in vacuum cell. Shinsa(1988) explained the causes about the reduction of vacuum pressure as follows ;

1) Difference between the location of vacuum pump and water level (head loss of location)
2) Resistance due to friction of drain, and/or bending (head loss of channel)
3) Absorption of air from ground plumbing or outside

Therefore, the amount of vacuum pressure acted on very soft clay in the field shall be set considering this kind of loss. However, laboratory experiment performed in this study has very small range comparing to that of actual reclaimed land. Therefore the difference due to the location of vacuum pump and water level is fairly small. Additionally, as the top of the cell was capped to prevent absorbing outside air, vacuum reduction effect was not to be considered.

3 LABORATORY TEST

3.1 Test Scope

In this study, two kinds of laboratory soil box tests were performed to investigate characteristics of consolidation of horizontal drain spacing. Among them,

middle-sized soil box test is for modeling consolidation characteristics of soil mass where single drain is installed (so called unit cell test). During the test, settlement characteristics and consolidation velocity in the center of the soil box are investigated and change in water content after completion of the experiment is measured. On the other hand, the large-sized soil box test was performed to identify the complicated coupled consolidation of ground where drain has installed with multi-layers. To analyze optimum spacing of horizontal drain, we implemented drain spacing similar to actual ground by using miniature drain.

3.2 Soil Characteristics

Sample used in this study was marine clay collected from Jinhae and Yeocheon in the south seashore of Korea. Samples were collected in disturb. Alien substances are removed by sieve analysis using No. 10 sieve in wet condition. Additionally, seawaters which has been transported from the sample collection area is added to the sample, to increase degree of saturation and produce high water content sample. Characteristics of the samples used in this study are shown on Table.1 and Table. 2.

Table 1. Soil properties used in this study (from Jinhae)

Classification	Properties	Classification	Properties
Soil classification	CH	Liquid Limit (LL)	56.3%
Percent passing No.200 sieve	93%	Plastic Limit (PL)	20.6%
Specific Gravity	2.67	Plastic Index (PI)	36.7

Table 2. Soil properties used in this study (from Yeocheon)

Classification	Properties	Classification	Properties
Soil classification	CH	Liquid Limit (LL)	56.3%
Percent passing No.200 sieve	93%	Plastic Limit (PL)	20.6%
Specific Gravity	2.67	Plastic Index (PI)	36.7

3.3 Characteristics of Drain

In Figure 5, shows sections of drain used in this study. Drains used in large-sized soil box test have the section of 100mm (with) × 5mm (thickness) and is named Type O-5. It contains 800-1,000cm^3/sec of discharge capacity under hydraulic gradient i=0.01 conditions. Type O-5 is composed of a castle type core and heat melting-adhesive type filter. In this study, to maintain equilibrium with drain spacing, miniatured drain such as the dimension of 25mm (with) × 10mm (thickness) was used.

On the other hand, in middle-sized soil box test, original section of drain is used (which is named

(a)

(b)

Figure 5. Horizontal drains used in this study; (a) Type O-5 (width 100mm × height 5mm), (b) Type O-10 (width 100mm × height 10mm).

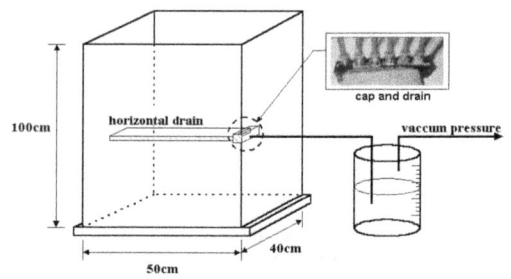

Figure 6. Schematic diagram of middle-sized soil box.

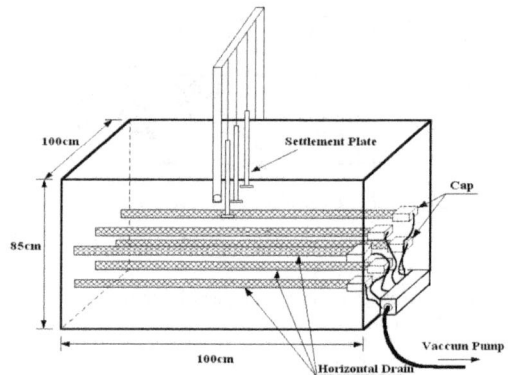

Figure 7. Schematic diagram of large-sized soil box.

Type O-10). Type O-10 has the section of 100mm in width and 10mm in thickness. Discharge capacity is 1,200 – 1,600 cm^3/sec for hydraulic gradient i=0.01.

3.4 Test Equipment

(1) Middle-sized soil box
Figure 6 present a schematic diagram of middle-sized soil box manufactured applying frequently used horizontal consolidation device. Soil box was made of acryl so that surface settlement can be identified by bare eye. The dimension of the box was 500mm × 400mm × 1,000mm. Ends of drain were free so that they could be relocated downward as the samples consolidate. Additionally, in the end of drain, a rubber tube was connected to measure pore water exhausted through drain by vacuum pressure. The upper part of soil box was capped with vinyl sheet to minimize influence of evaporation.

(2) Large-sized soil box
Figure 7 shows large-sized soil box consolidation test device. It is made of steel frame of square section. The size is 1,000mm × 1,000mm × 850mm. Connection device to each drain is same as in middle-sized soil box. As this soil box is made of steel frame and it is impossible to identify surface settlement in bare eyes, a settlement plate was installed on top of the box and a wire was connected to settlement gauge in order to identify the settlement amount.

3.5 Test Procedure

(1) Middle-sized soil box
Experiments were performed to find out the effect of change in vacuum pressure. Settlement and drainage were measured with the elapsed time. Test conditions were initial sample height was 90cm and initial water content was 250±25% as shown in Table 3. Numeric value difference of each condition was from sample mixing and placing procedures and it is very difficult to set these values accurately. As settlement begins after slurry sample is added, vacuum pressure was adopted approximately 24 hours, when self-weight consolidation began.

Table 3. Test conditions for middle-sized soil box

	Initial water content (w_0)	Initial surface height (H_0)	Vacuum pressure
CASE 1	267%	87.8cm	20kPa
CASE 2	278%	86.8cn	40kPa
CASE 3	230%	89.3cm	80kPa

(2) Large-sized soil box
Figure 8 shows drain spacing condition in large-sized soil box. Miniature drains were distributed in 2

layers and 3 columns. Each end of drain were fixed with clippers and connected to drainage tube for extraction of pore water.

In Table 4, conditions of large-sized soil box test are presented. Initial water content of slurry samples is about 300%. Experiments were performed for total 5 kinds of vacuum pressure to identify the best vacuum pressure condition inducing the best improvement effects. ; no-pressure, constant 80kPa of pressure, and step loading of vacuum pressure controlled by gradual increment of pressure such as 20kPa, 40kPa and 80kPa.

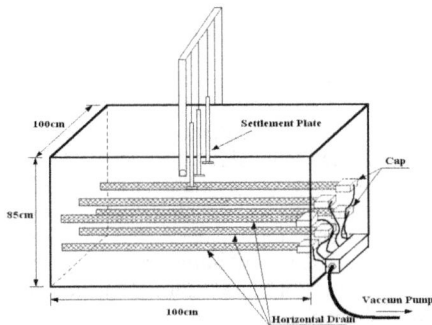

Figure 8. Location of horizontal drains and settlement gauge.

Table 4. Test conditions for large-sized soil box

	Initial water content (w_0)	Initial height (H_0)	Spacing ratio (n)	Equivalent diameter (d_w)
Step loading, 20, 40, 80kPa of vacuum pressure	301%	80cm	25	15mm

4 TEST RESULTS AND ANALYSIS

4.1 Results from the Middle-sized Soil Box Test

(1) Settlement and drainage
Table 5 shows the middle-sized soil box test condition. It shows cumulated drainage extracted from horizontal drain located in the center of the soil box, initial height of surface layer right after the application of vacuum pressure and the final height of surface layer after 25 day (Consolidation is considered over 90%).

Table 5. Results from middle-sized soil box test

	Initial surface Height (H_0)	Final surface height (H_f)	Cumulated drainage	Settlement
CASE 1	87.8cm	56.6cm	34,369cm³	30.9cm
CASE 2	86.8cm	53.6cm	39,520cm³	33.2cm
CASE 3	89.3cm	56.0cm	49,360cm³	33.3cm

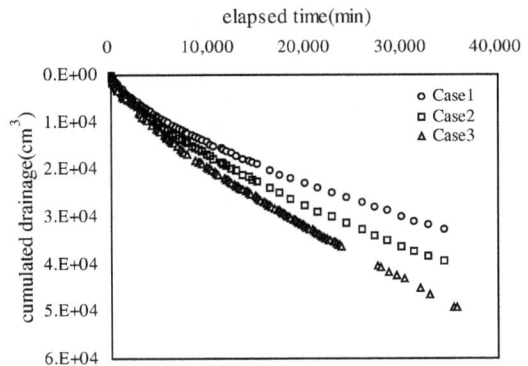

Figure 9. Cumulated drainage with elapsed time.

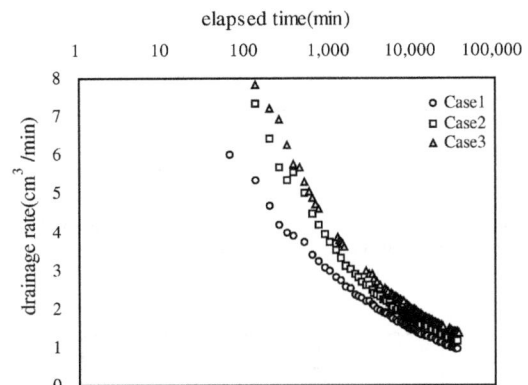

Figure 10. Change in drained water rate with time.

Figure 9 shows cumulated drainage according to time in each condition respectively to investigate the influence of vacuum pressure. It was identified that vacuum pressure and cumulated drainage were increased proportionally.

Figure 10 shows drain velocity with time. As shown in the Figure, initial drain velocity of pore water extracted from drain was less than 10cm³/min under all conditions. Considering the discharge capacity of Type O-10, well-resistance effects can be ignored, because pore water amount at in-laboratory experiments is very small and collected pore water can be drained very easily. However, the length of the drain used in this experiment was 50cm, which was shorter than field conditions (100-200m), it is needed to calculate minimum probable discharge capacity according to installation length of drain through detail study using in-situ construction and theoretical equation.

(2) Effect of vacuum pressure
Figure 11 compares settlement (drain/soil box sectional width) back-calculated from surface settlement and drain according to time. While surface settlement showed similar tendency, back calculated

251

settlement was 74% of real settlement. The cause of this difference was that settlement by self-weight consolidation and vacuum pressure occur simultaneously. Additionally, the bigger vacuum pressure was, the smaller the difference between real settlement and back-calculated settlement was estimated. From this result, we can judge that self-weight consolidation settlement is accelerated, when vacuum pressure is big.

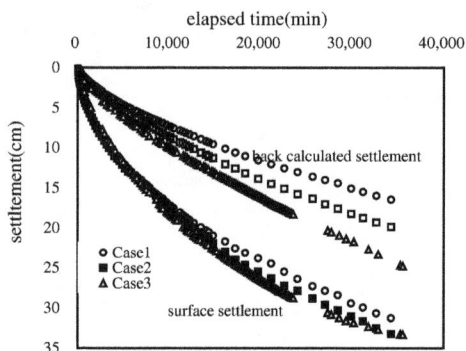

Figure 11. Comparison of surface ground settlement and back calculated settlement from drained water with time.

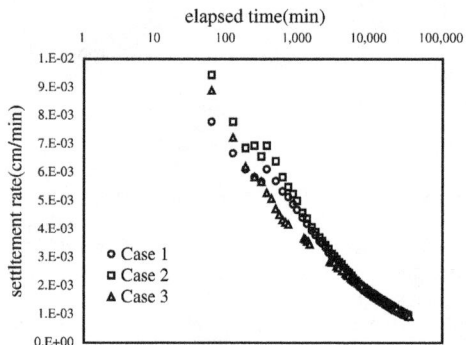

Figure 12. Surface settlement rate with elapsed time.

(3) Results from the step-loading test
To identify the effects of step loading condition, it was compared with the experiment result of 80kPa vacuum pressure loading test. In this comparison, surface settlement and drainage with time, drain velocity and vertical water content distribution after completion of the test were identified.

Figure 13 shows surface settlement with time. We can find out that there is some difference in settlement for two loading conditions. However, after approximately 30,000 minutes, while 80kPa of vacuum pressure loaded model showed near end of settlement, step loading conditioned one showed continuous settlement. The difference of initial settlement is probably because initial vacuum pressure is small in step loading. It implies that as time goes by, settlement in two models will be closer.

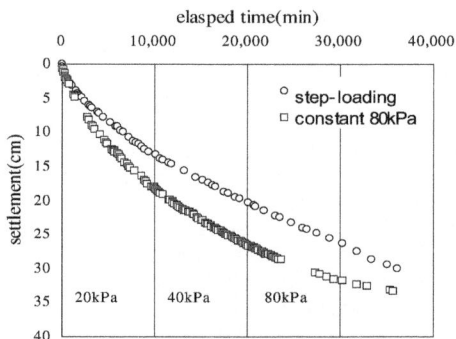

Figure 13. Surface settlement with elapsed time.

Figure 14 shows settlement velocity according to time. Excluding the initial settlement velocity by initial cumulated drain, the settlement velocities of two conditions become identical, as the vacuum pressure in the step loading is approaching to 80kPa. Figure 15 and Figure 16 shows the relationship between cumulated drain and drain velocity. These show that excluding the influence of initial vacuum pressure difference, the result of two conditions become identical as time elapse.

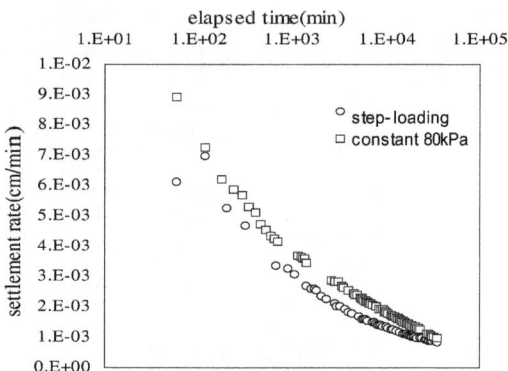

Figure 14. Surface settlement rate with elapsed time.

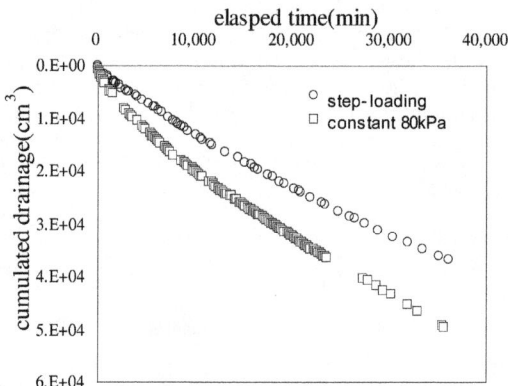

Figure 15. Cumulated drainage with elapsed time.

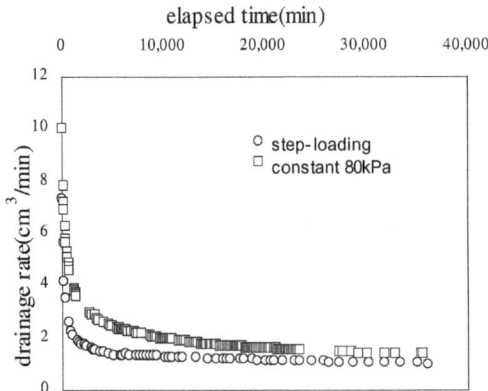

elapsed time(min)

Figure 16. Drainage rate with elapsed time.

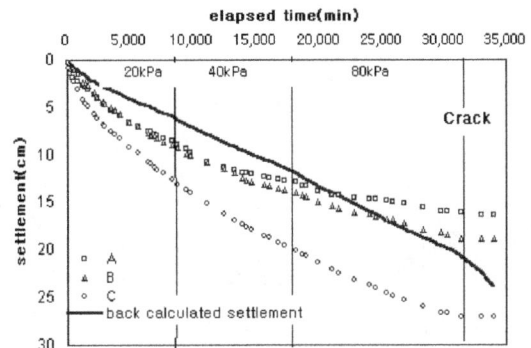

Figure 18. Cumulated drainage and drainage rate with elapsed time.

(4) Water content distribution

Figure 17 shows final average water content distribution by horizontal distance from drain. While they show similar water content near drain, the gradient against the step loading condition becomes smaller as the distance from drain is bigger.

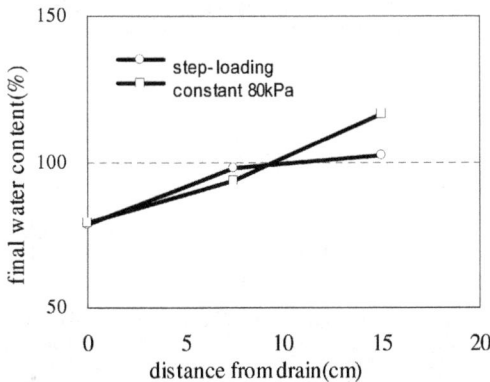

Figure 17. Distribution of final water content with distance from drain.

A restriction in applying the horizontal drain method is that vacuum pressure efficiency is getting smaller as the distance from drain is greater. It is probably because of the decreased channel for pore water due to filter cake or clogging. However, according to this study, step loaded condition can be very effective although it takes a relatively long time to improve ground soil.

4.2 Results from the Large-sized Soil Box Test

Figure 18 shows the large sized soil test with 2 layers and 3 columns of drain. It shows drain and drain velocity according to time. In the experiment for this study, there were cracks in the surface of the drain 23 days after test initiation. The experiment was terminated as it was judged that surface water could

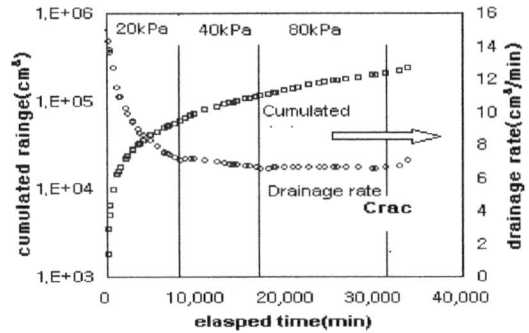

Figure 19. Change in settlement with elapsed time.

penetrate to the sample. When seeing Figure 18, drain velocity is increasing after 30,000 minutes. We can guess that surface water can flow in at this stage.

Figure 19 shows settlement by location and by layer according to time. In the Figure A, B and C shows layer settlement measurer installed in 60cm, 40cm and 20cm from surface. As shown in the Figure, the settlement was the biggest in A (27.1cm), B (18.8 cm) and C (16.3cm) in descending order. It means that the efficiency of horizontal drain method reduced as installation depth gets bigger. It is probably because that clay layer on the top of drain acts as confined pressure and reduces discharge capacity of drain.

4.3 Estimation of Drain Spacing

In this section, consolidation curve calculated from theoretical equation of Barron(1948) with the actually measured consolidation curve to estimate effective installing spacing in horizontal drain installation was compared and analyzed. Actually measured consolidation curve is estimated result using Asaoka's method. Table 6 shows constant value and the final settlement calculated from the constant.

Initial drain spacing (d_0=33cm) before improvement was getting decreased with the vacuum consolidation and reached d_f= 23.4cm, the spacing after experiment. It means that we cannot ignore the change in drain spacing according to that in layer thickness when considering consolidation velocity of the dredged soil ground.

Table 6. Final settlement and coefficient(β) obtained by Asaoka's method

Values obtained by Asaoka's method		β_1	β_0	Final settlement, S_f (mm)
Large-sized Soil box	A	0.9261	1.2837	173
	B	0.9392	1.2846	211
	C	0.9487	1.6512	322

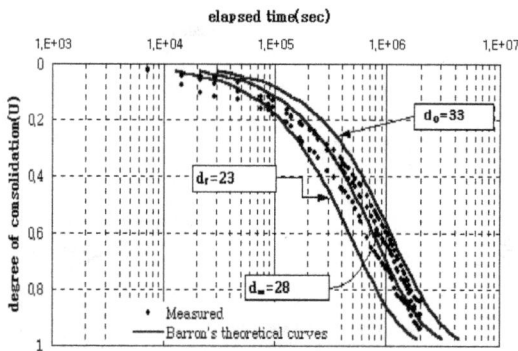

Figure 20. Comparison of theoretical and measured consolidation curve.

In this study, Barron's consolidation theory was applied as an approximated solution and at the same time Shin's proposal that consolidation velocity is proportional with square root of the layer thickness was used to calculate average value of drain spacing.

Figure 20 compares consolidation degree of Barron's theory and actual measurement result. As Figures show, actually measured consolidation curve is well consistent with average spacing (d_m) acquired from formula (5) provided by Shinsa(1988). Therefore it is possible to apply Barron's consolidation theory to calculate consolidation velocity of horizontal drain method applied ground. Additionally, it means that we need to consider the change in layer thickness and the change in drain spacing.

5 CONCLUDING REMARKS

In this study in-lab model soil box tests were performed to investigate the effects of vacuum pressure and drain spacing. The conclusion of this study is as follows;

1. In the result of middle-sized soil box test, well-resistance effect can be ignored, as pore water was too small to influence discharge capacity of drain.
2. Although the efficiency of horizontal drain method was bigger with larger vacuum pressure, it is decreasing with the depth of spacing. It is because soil in the upper layer of the drain acts to drain as restraint loading and reduce discharge capacity of the drain.
3. Step loading needs longer development time than the method to put large vacuum pressure sporadically. However to get even development effects in horizontal drain method, step loading is very effective by loading low vacuum pressure initially and incrementing it gradually.
4. When calculating average installing space of drain using Shin(1988)'s formula, it was nearly identical to the consolidation result of Barron's equation and actual settlement. It is judged that there will not be many problems to apply Barron's equation to the design of horizontal drain method, if the change in drain spacing according to settlement progress is properly considered.

REFERENCES

Kim, J.K. (2001) Effect of vacuum pressure and drain spacing in horizontal drain method, Paper of Master, Chungang Univ., Seoul, Korea.
Kim, S.S. (1994) Initial application of paper drain on development of soft ground in Korea, Trace of Korean Geotechnical engineering, Korean Institute of Geotechnical Engineering Press.
Kang, M.S. (1998) Development of vertical drain method considering of influence factors, Paper of Ph. D., Chungang Univ., Seoul, Korea.
Asaoka, A. (1978) Observation procedure of settlement prediction, Soils and Foundation, Vol. 18, No. 4.
Barron, R.A. (1948) Consolidation of fine-grained soils by drain wells, Trans. ASCE, Vol. 113, No. 2346, pp. 718-742.
Rixner, J. J., Kremer, S. R. and Smith, A. D. (1986) Prefabricated vertical drains, Vol. II : Summary of Research Effort", FHWA, Research Report No. FHWA/RD-86/169, Washington.
Shinsha, H., Matunaga, S., Watari, Y. and Satou,H.(1988) Consolidation of a clayey layer installed horizontally drains by vacuum(part 3), The 23th Japan National Conference on Soil Mechanics and Foundation Engineering, Miyazaki, pp. 2129-2132.
Shinsha, H. et. al (1991) Dewatering of Dam deposited soil by lateral drain, Annual Report of Penta-Ocean Construction Institute of Technology, Vol. 20, pp. 63-68.

Soft Ground Engineering in Coastal Areas, Tsuchida et al. (eds)
© 2003 Swets & Zeitlinger, Lisse, ISBN 90 5809 613 0

Design and performance of fibredrain in soil improvement projects

S.L. Lee
National University of Singapore

G.P. Karunaratne
Formerly, National University of Singapore

ABSTRACT: The effective coefficient of consolidation C_h is a function of the hydraulic conductivity and compressibility of the clay as well as the field performance of prefabricated vertical drains (PVD). The effective discharge capacity of PVD is influenced by the deformation of the core and the filter, clogging of the filter and kinking of the PVD during the consolidation process. The design of soil improvement projects using PVD is a function of the effective C_h, the final settlement, the drain spacing as well as the lateral pressure on the drain. Prefabricated vertical drains should possess sufficient tensile strength to withstand installation stresses associated with thick clay deposits, as well as the densification of granular soil layers at the surface by heavy tamping. Fibredrain is a PVD that is biodegradable and ecologically harmonious. Several selected projects in East and Southeast Asia are discussed to illustrate the field performance of Fibredrain.

1 INTRODUCTION

Infrastructure construction on soft clay leads to long-term settlement problems. These are best dealt with by pre-consolidating the clay until the anticipated settlement under the design load is enforced prior to superstructure construction. To achieve the desired degree of consolidation within the project duration prefabricated vertical drains (PVD) are normally used. The associated surcharge intensity should be designed to take the post-construction settlement due to secondary compression into account.

The excess pore pressure in the soft clay should dissipate by pore water permeating through the filter cover into the PVD core, which conveys the water axially to the drainage boundary.

Both axial and filter permeability of the PVD are important in this regard. The filter cover should satisfy two requirements simultaneously, cross-plane water transmission and soil retention. The cross plane filter permeability should be large enough to take advantage of the pervious inclusions that may exist in natural soil deposits. The reduction of axial discharge flow capacity caused by the deformation of the core and the filter, the clogging of the filter and the kinking of the PVD in the consolidation process must be taken into account in the design of the soil improvement project.

During installation in thick deposits of soft clay PVDs are subject to tension as the mandrel is withdrawn. Adequate tensile strength in the PVD filter and core to withstand installation stresses is therefore essential.

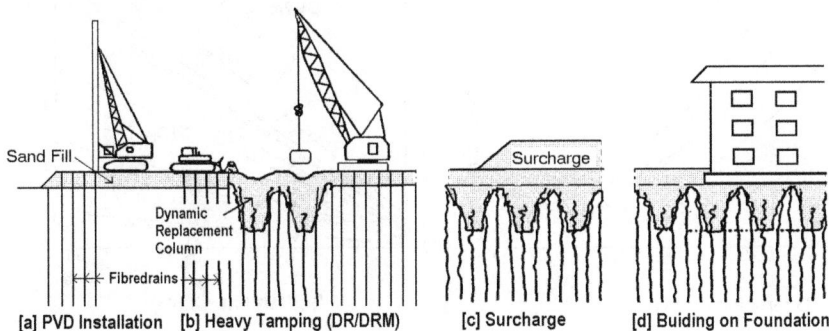

Figure 1. Sequence of ground improvement prior to building construction on soft clay.

In many soil improvement projects, to improve surface bearing capacity, recently placed granular fills have to be densified while the PVDs are employed for soft subsoil consolidation under the fill and surcharge, as shown schematically in Figure 1. Fill soils, even with a high ground water table, can be effectively treated with the application of high energy tamping. The PVD passing through the fill soils should be able to withstand the above high energy tamping without significant reduction to its discharge flow capacity. In addition, adequate tensile strength and flexibility are important for a PVD to withstand the sliding and heaving of soil layers in such treatment projects.

The interaction of PVD with the soft soil layer to be consolidated determines its field performance. The field discharge capacity of PVD could be quite different from that determined in the laboratory depending upon the testing procedure. Reduction factors discussed by Koerner (1997) demonstrate this aspect of divergence between laboratory and field performance. This paper deals with a simple rational design method for estimating the required field discharge capacity of a PVD as well as the corresponding effective C_h and drain spacing for estimating the time of consolidation.

2 DISCHARGE FLOW CAPACITY

2.1 *Required Discharge Capacity, Q_r*

The average degree of consolidation Ū of a clay deposit under surcharge accelerated by PVD installation can be determined by (Hansbo, 1981)

$$\overline{U} = 1 - e^{\left[\frac{-8C_h t}{D^2 \mu}\right]} \tag{1}$$

where c_h = effective coefficient of consolidation with horizontal flow, D = influence diameter of a vertical drain, t = time, $\mu \cong \ln(D/d) - \frac{3}{4}$, and d = equivalent drain diameter = (a+b)/2, where a and b are the width and the thickness of the PVD cross section respectively (Welsh, 1983).

The average degree of consolidation Ū may be taken as the ratio of current settlement s to the final settlement s_f, i.e.,

$$\overline{U} = s/s_f \tag{2}$$

Assuming that all parameters in Equation 1 are relatively constant, differentiation of Equation 2 with respect to time in view of Equation 1 gives the rate of settlement as

$$\frac{\partial s}{\partial t} = s_f \frac{8c_h}{D^2 \mu} e^{\left[\frac{8c_h t}{D^2 \mu}\right]} \tag{3}$$

The maximum rate of settlement occurs at the early stage of consolidation; taking t = 0, Equation 3 yields

$$\left(\frac{\partial s}{\partial t}\right) = \dot{s}_0 = \frac{8s_f C_h}{D^2 \mu} \tag{4}$$

where $\dot{s}_0$ is the initial ground settlement rate. If A = $\pi D^2/4$ denotes the influence area of a PVD, then the required discharge flow rate Q_r is estimated by

$$Q_f = A\dot{s}_0 = 2\pi s_f C_h/\mu \tag{5}$$

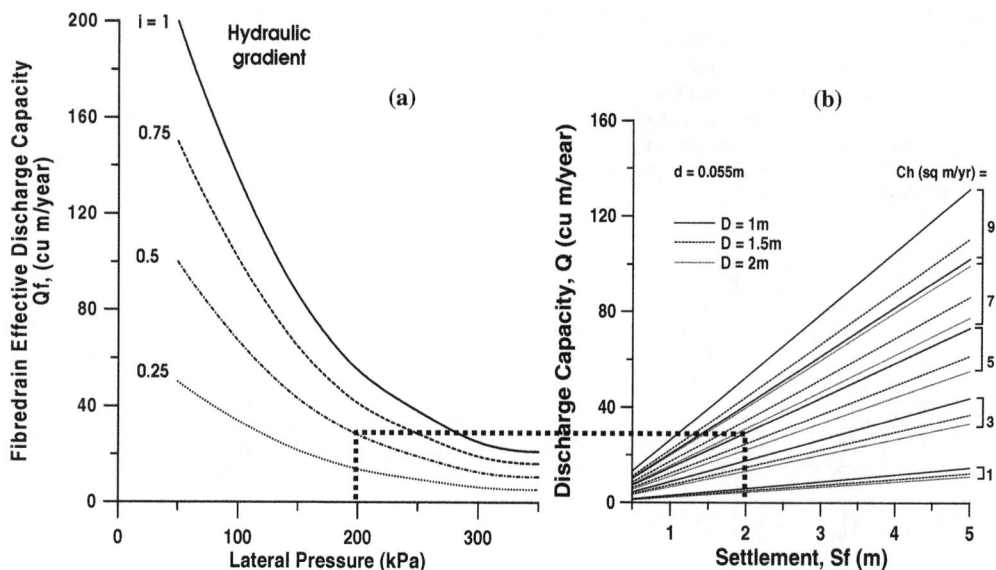

Figure 2. Design charts for Fibredrain

Figure 2(b) shows the variation of Q_r against s_f for various C_h and D/d values based on Equation 5 assuming d = 0.055m. It can be seen that the larger the final settlement S_f, which depends on the thickness and compressibility of the clay deposit, and C_h, which is characterized by the compressi-bility and the hydraulic conductivity of the clay as well as the performance of the PVD, the larger is the required Q_r. It should be noted that reducing D leads to the increase of Qr.

2.2 *Effective Discharge Capacity of PVD*

The effective discharge capacity of PVD can be determined by field tests, or it can be deduced from laboratory tests. Depending on how close is the simulation of the laboratory test to field conditions, the effective discharge capacity Q in the field should be deduced from the laboratory values Q' by

$$Q = Q'/R \qquad (6)$$

where R denotes a reduction factor suggested by (Koerner, 1997) in the form

$$R = F_i F_d F_c F_b F_k \qquad (7)$$

Koerner suggested that Q' be determined from short-term laboratory tests in accordance with ASTM D4716. In Equation 7 F_i represents the reduction factor due to intrusion of the filter in to the core space (F_i = 1.5 – 2.5), F_d is due to creep deformation of the core (F_d = 1.0 – 2.5), F_c refers to chemical clogging of the filter and core space (F_c = 1.0 – 1.2), F_b is due to biological clogging of the filter and core space (F_b = 1.0 – 1.2) and F_k is due to kinking of the PVD under axial deformation caused by the compression of the clay layer (F_k = 1.0 – 4.0).

3 PROPERTIES OF FIBREDRAIN

Most PVDs are manufactured with a polymeric core and a geotextile filter sleeve. Fibredrain, a PVD developed at the National University of Singapore, will be illustrated as a case study in the following. It consists of four axial coir strands enveloped within a filter comprising two layers of jute burlap (Lee et al, 1995). Jute in raw form, after preliminary treatment, is spun into yarns and woven into burlap, which is used for the two filter layers. Coconut coir used in the core is spun into strands after preliminary processing. Figures 3 and 4 show a bale of Fibredrains delivered on site and a schematic view of Fibredrain respectively. It has shown satisfactory performance in many projects in Southeast Asia and Japan involving treatment of soft marine and fluvial clays, as well as peaty clays, and in some cases in conjunction with surface densification using high energy tamping.

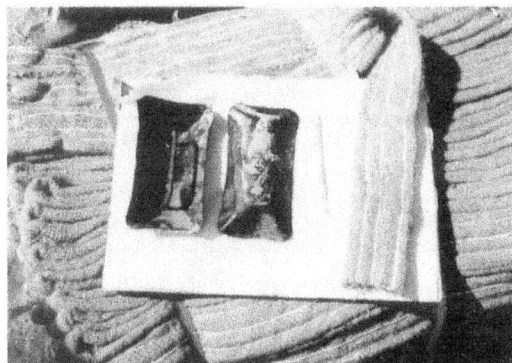

Figure 3. Bale of Fibredrains and anchor shoes.

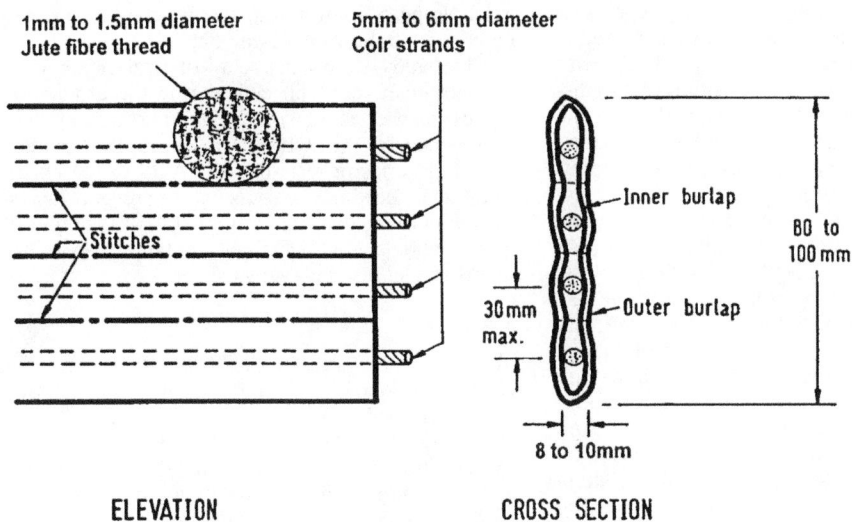

ELEVATION   CROSS SECTION

Figure 4. Schematic view of Fibredrain.

257

From the environmental point of view, the energy consumption of jute and coir products in the production stage, transformation stage and utilization/disposal stage is much lower than that needed for synthetic products (De, 1994). Raw materials for the production of Fibredrain are natural fibres extracted from jute plants and ripened coconuts both of which are renewable resources. Jute plants and coconut trees grow in soil with nutrients such as nitrogen, phosphorous and trace elements, as well as water, sunlight and carbon dioxide. No air pollutants and/or water pollutants are generated during this process. The production of jute fibres and coconut coir are labour intensive requiring land and manpower as the main inputs. The solid waste generated during the production as well as the utilization /disposal stages is basically organic and biodegradable and in some instances reusable as building materials. When used in stabilizing soft subsoil, Fibredrain decomposes biologically after the consolidation process. The decomposed products are organic wastes with traces of methane and carbon dioxide, which are non-polluting. It is a green product, environmentally friendly and ecologically harmonious. Some of the laboratory tests conducted to examine the properties of Fibredrain are reported in the following, details of which can be found in Karunaratne et al (1999).

3.1 Axial Discharge Capacity

To evaluate the discharge capacity of Fibredrain a 200mm long specimen was installed at the center of a 100mm diameter cylinder of Singapore marine clay ($w_n = 65\%$, $w_L = 70\%$, $w_P = 40\%$). A 10mm thick coarse sand layer was provided at both ends of the clay cylinder, which was enclosed within a large diameter triaxial membrane. The discharge capacity was measured by connecting the drainage ends to a constant head tank. Lateral pressure was increased in stages and volume change in the clay cylinder was measured followed by the determination of the discharge capacity under each pressure (Lee et al, 1995). Figure 2a shows the mean discharge capacity of Fibredrain under several hydraulic gradients plotted against argument of lateral pressure.

Referring to Equation 7, kinking is not a problem for Fibredrain (Miura et al, 1995a) and hence F_k for Fibredrain is close to 1.0. For short duration projects F_c is low or close to 1.0 (Koerner, 1997). Mlynarek (1998) and Mlynarek and Rollin (1995) reported that biological clogging is not a problem if the apparent opening size (AOS) is large. AOS of Fibredrain is in the range of 200-600 μm when measured individually, hence F_b for Fibredrain is 1.0. The reduction factors due to the deformation of core and intrusion of the filter into the core space, $F_i \times F_d$, due to lateral pressure, have been taken in to account in the testing procedure.

3.2 Filter Permeability and Clogging Potential

The large AOS serves to tap pervious layers and lenses in clay deposits. The double filter layer, which has a smaller effective AOS, intercepts the clay slurry generated immediately after installation of Fibredrain in the clay as the mandrel is withdrawn. The cross-plane permeability of the filter was measured under a constant head across a stack of four identical burlaps fastened across an open end of a circular cylinder (Lee et al, 1994). The cross-plane filter permeability determined in this way for clean water was better than 10^{-5} m/s. This is equivalent to the hydraulic conductivity of fine sand, which helps in tapping natural drainage layers such as sand and silt seams, lenses and other pervious paths. Clay slurry with water content from 65% to 600% was passed through four burlap layers mentioned above under a pressure head of 0.5 m of water. About 2 litres of marine clay slurry at water content of 260% to 600% flowed out within 15 minutes. For Singapore marine clay ($w_L = 70\%$) at water content of 65%, the passage of water was slower but clear from the beginning.

This scenario is similar to the field installation stage of PVD as the mandrel is withdrawn when disturbed slurry clay under high water pressure tends to rush through the filter into the core. The four burlap filter layers were removed carefully and cleaned in an ultrasonic agitator to extract the clay embedded in each burlap layer separately. Gravimetric analysis showed that soil particles retained on the outermost burlap layer and the immediately next inner layer, but no particles were detected in the third and fourth burlap layers (Figure 5a). The water passing through the four burlap layers was examined similarly to be free of any soil particles (Lee et al, 1994). The retention of soil particles on the two burlap layers under axial permeability test with marine clay at 65% water content is shown in Figure 5b.

This study shows that clay of near liquid limit will not enter the drain core during the installation process as well as the consolidation process but will be retained by the two burlap layers. It is evident from this series of tests that the AOS of the burlap filter layers need not be too small to prevent clay intrusion in to the core, but larger AOS of the order of 200 – 600 μm is beneficial in tapping pervious seams and lenses in clay deposits.

3.3 Kinking

Miura et al. (1995a) compared the effective coefficient of consolidation, C_h, of Fibredrain and a plastic drain identified as PD in Ariake clay ($w_L = 86 - 97$%, $I_p = 48 - 54\%$, $W_n = 105 -134\%$, $G_s = 2.59 - 2.62$) by installing both separately in 500 mm diameter and 500 mm high consolidation cells. On back analysis, at consolidation pressures of 98 kPa and 294 kPa, the effective C_h was found to be 1.35 m^2/yr and 2.19 m^2/yr with PD, and 9.2 m^2/yr and 10.7m^2/yr with Fibredrain respectively. Figure 6 shows the deformed shapes of the two drains after carefully removing the clay around the drains at the end of consolidation (Miura et al, 1995a). PD has deformed significantly resulting in kinking and decreased in axial drainage capacity.

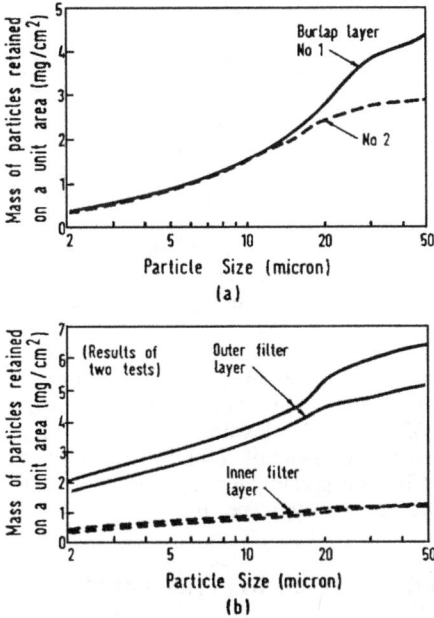

Figure 5. Distribution of particles retained on filter layers: (a) filter and (b) axial permeability tests.

The deformation of Fibredrain, on the other hand, was largely confined to increase in thickness and longitudinal compression of the drain without kinking. Axial compression in the Fibredrain is largely manifested as an increase in cross sectional area due mainly to the unwinding characteristics of jute fibers in the filter layer as well as the core. The resulting increase in cross-sectional area and the decrease in filter opening size enhance the unclogged water flow into the drain. The above experimental results show that the effective C_h of the same clay under the same consolidation pressures is function of the performance of different PVD.

Aboshi (2001) reported the consolidation of Hiroshima clay ($W_L = 49\%$, $W_P = 21\%$, $C_C=0.36$, $P_C=300$ kPa) at 15m depth and clay from Kojima Bay, Japan ($W_L = 64\%$, $W_P = 22\%$, $C_C=0.53$, $P_C=78$ kPa) in 300 mm diameter laboratory consolidation cells with Fibredrain and PD. Pore pressure measurement in Fibredrain test showed zero excess pore pressure at the end of primary consolidation whereas the PD test showed 5m of water head remaining at a respective axial strain of 24% and 19%. The measurement of axial permeability at the end of the test

was reported to be 7×10^{-9} m/sec for Fibredrain and zero for PD, illustrating the effect of kinking. The examination of the drain materials after the tests showed that Fibredrain had increased its thickness without kinking, whereas the PD had developed three to four kinks which completely curtailed the axial flow capacity.

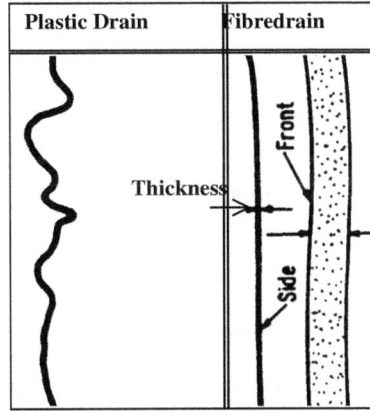

Figure 6. Deformed shape of PD and Fibredrain (Miura et al, 1995b).

3.4 Tensile Strength

Figure 7 shows the tensile strength of Fibredrain in air-dry condition, under conditions of soaking in water for 4 days, exposing for 50 days outdoor and after consolidating in Ariake clay for 126 days (Miura et al, 1995b). The tensile strength ranged from 8.96 kN (916 kg), 8.45 kN (860 kg), and 7.56 kN (771 kg) to 2.04 kN (208 kg) for the respective cases. The strength deterioration with time in the Ariake clay shows the biodegradability of the fibers. It should be pointed out that the deterioration in strength does not imply a reduction in drainage capacity as observed in Pantai Mutiara reclamation project in Jakarta, Indonesia where Fibredrain performed satisfactorily for more than two years (Lee et al, 1988). Because of high tensile strength, (6.8 kN with 8.7% strain at rupture), the robustness and the flexibility of the core and the jute filter layers, Fibredrain can withstand installation stresses, high energy tamping while retaining its drainage functions unimpaired. As the mandrel is driven into the ground during the installation process, soils intrude into the mandrel through the mandrel-anchor shoe interface and develop significant tensile forces in the PVD, as the mandrel is withdrawn. In recent field measurement of installation stresses, a tensile force of 1.4 kN on the filter of a PVD was recorded at 30m depth of installation (Karunaratne, 2002). It is imperative that PVD should possess sufficient tensile strength to withstand these tensile forces.

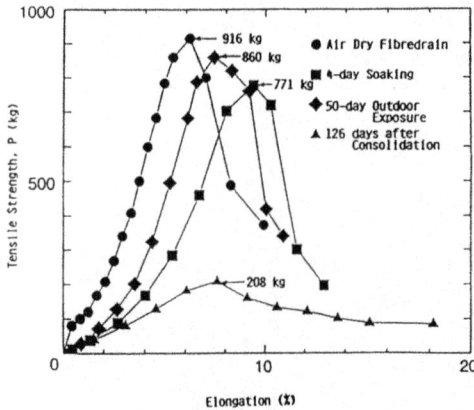

Figure 7. Tensile test and biodegradability of Fibredrain (Miura et al, 1995b).

4 DESIGN OF SOIL IMPROVEMENT

In the design of soil improvement work using Fibredrain, S_f can be estimated from the results obtained from soil investigation and design loads. The value of Q_r can be obtained from Figure 2b as a function of S_f for various values of influence diameter D and the effective C_h and compared with the discharge flow capacity given in Figure 2a, which depends on the lateral pressure corresponding to the depth of installation and the design load and surcharge. The hydraulic gradient set up along the PVD may be estimated by the intensity of design load and surcharge in terms of metres of water divided by the length of flow path along the drain.

The design of spacing to achieve a certain degree of primary consolidation, say 80% under surcharge, which is equivalent to 100% primary consolidation under design load, within a prescribed time is determined, as usual, by means of Equation 1 using the influence diameter D and effective C_h determined from Figure 2b. For example, under a lateral pressure of 200 kPa, i = 0.5 and a settlement S_f equal to 2m, the effective C_h is 5 m^2/yr with an influence diameter of 1.5m. The time of consoli-dation by Equation 1 corresponding to $\bar{U}$=80%, S_f = 2m, C_h= 5 m^2/yr, D=1.5m is t = 84 days. It should be emphasized that the time of consolidation cannot be reduced by decreasing D as this leads to larger Q_r, which will be greater than the effective dis-charge capacity of the drain under the lateral pressure and hydraulic gradient. In most cases, the effective C_h is smaller than the C_h of the soil determined in the laboratory (Rowe, 1968; Head, 1985). If the latter is smaller, it should be used in Equation 1 to determine the time of consolidation.

Back analysis of field settlement-time data obtained in Singapore marine clay using a plastic PVD (Choa et al, 1979) showed effective C_h values in the range of 2 to 10 m^2/yr. Figures 2a,b show that for i = 0.5 and lateral pressure of 200 kPa, the values of effective Ch corresponding to settlement of 3, 2 and 1 m are 3, 5 and 7 m^2/yr respectively.

5 FIELD MONITORING IN HIROSHIMA, JAPAN

Yoshida et al. (1995) reported the installation of Fibredrain in Hiroshima under the jurisdiction of Hiroshima Prefecture Government. The site was under-

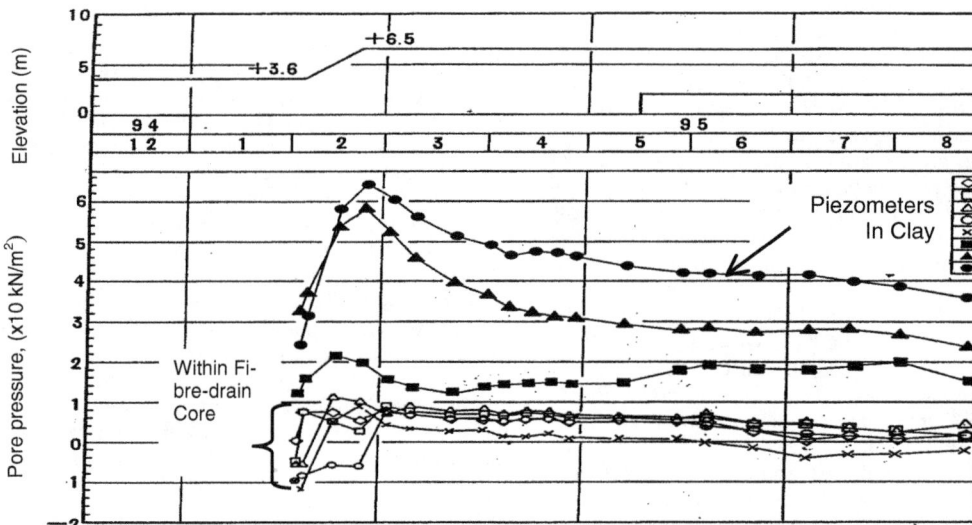

Figure 8. Piezometric variations within Fibredrain and in the surrounding clay, Hiroshima, Japan (Yoshida et al, 1995)

260

lain by about 15 -18 m of Hiroshima Clay (Wn = 90%, C_c = 1.30, e_0 = 2.40, ρ_t = 15 kN/m^3, q_u = 120 kN/m^2). Fibredrain was installed at a spacing of 1.1m and sand drains at 2.5m spacing for comparative assessment. A considerable excess pore pressure remained in the clay at the end of 8 months in the sand drain area (Aboshi, 1999).

Figure 8 shows the monitoring results of piezometers installed in the clay in the Fibredrain area at El -9.2m, -12.5m and -19m, where the clay occurs between El -8.5m and -25m. Piezometers were also installed at El -10.6m, -13.6m, -16.6m, -19.6m and -22.6m within the Fibredrain core to investigate the well resistance. Within the same eight-month period, the piezometers in the Fibredrain core showed only an excess pore pressure not exceeding 10-kPa indicating that well resistance is insignificant and the axial flow capacity is more than adequate.

6 TREATMENT OF EX-MINING LAND NEAR KUALA LUMPUR, MALAYSIA

Fibredrain was used in conjunction with high energy tamping for the treatment of ex-mining land for a housing project (Lee et al, 1989) for which a safe bearing pressure of 60 kN/m² with a factor of safety of 2.5 was required. The site was composed of loose mine tailing soils, with N values less than 2 in clays and between 2 -10 in sand. To treat the deep seated clay varying from 9 m to 21 m, Fibredrains were installed first at 1.5 m square spacing, followed by high energy tamping with a 15-tonne pounder falling freely from heights between 10 m - 25 m with 6 to 12 blows per pass for three passes over a 6m x 6m grid, together with a fourth ironing pass imparting an energy of 225 to 250 tm/m² for dynamic compaction in the sandy areas. The energy intensity was increased to 270 to 335 tm/m² for dynamic replacement (DR) in the clayey deposits A surcharge fill of 4.5 m was placed and the resulting settlement was observed to taper off in 60 to 70 days as shown in

Figure 9. Figure 1 illustrates the concept of this treatment. Drains and surcharge combined with high energy tamping enforced anticipated total settlement of the order of 1 m. This case record shows that using Fibredrains, surcharge and high energy tamping, a safe bearing capacity adequate for 5-storey residential houses can be easily achieved. For highway embankment construction, this treatment method can be easily adopted.

7 PEATY SOIL TREATMENT IN SINGAPORE

This site was a waterlogged land with peaty clay and fluvial deposits. In a series of field trials conducted in 1983 to evaluate the feasibility of stabilizing some 7.8 m of peaty clay underlain by 5.6 m of fluvial clay (Lee et al, 1984), Fibredrains were installed at 2.2 m square spacing from a 1 m thick sand blanket. Subsequently, dynamic replacement and mixing method (DRM) was employed in treating the upper layer of the peaty clay. DRM consisted of dropping a 15 tonne pounder through a height of 15 - 20 m in six passes. A surcharge equivalent to 3.7 m of well-rolled clayey sand fill was placed subsequently for verification. The settlement in the area treated with Fibredrains0 and DRM levelled off within 6 months compared with areas using drains and surcharge method and surcharge only where the settlement continued at the same rate, as illustrated by the load settlement curves in Figure 10.

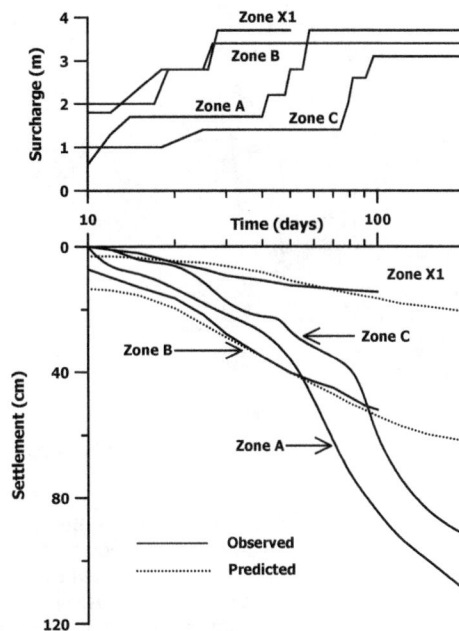

Figure 10. Surface settlement of different treated areas under surcharge at Bishan Depot, Singapore.

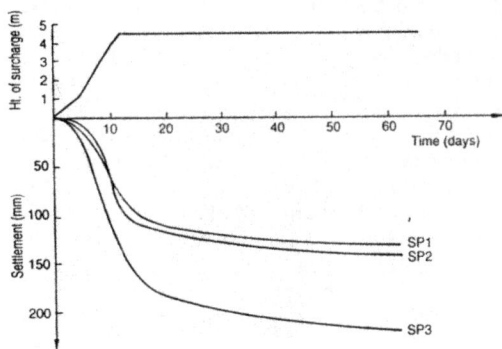

Figure 9. Load-settlement-time behaviour of ex-mining site near Kuala- Lumpur, Malaysia.

8 RECLAMATION ON SEABED CLAY AT PANTAI MUTIARA, JAKARTA, INDONESIA

Pantai Mutiara, a 90 ha residential-cum-recreational development, is being constructed on land which was reclaimed at the waterfront in the Pluit area of Jakarta. The project site, located immediately north of metropolitan Jakarta, extends about 1500m from the old shoreline spanning a width of 650m to 750m as shown in Fig 11.

Part of the existing site was reclaimed for approximately 400m from the original shoreline during 1979 to 1981. Area beyond this reclaimed land was reclaimed in three phases to a proposed formation level of El. +2.6m. Phase IA of the reclamation works consists of filling an area designated as 'Block R' (Fig 11) from the end of 1986 to 1988, where the seabed varied from El. 0m to El. -1.2m. Phase IB consisting of Block S and Block Z, where the seabed varied from El. -1.5 to El. -2.5m, was completed in 1994 using dredged clay. Phase II, comprising Blocks T, U, V, W, X and Y, started in August 1994 and was completed in 1997. In each phase, the sequence of reclamation work involves construction of perimeter dykes, placement of fill, soil improvement using vertical drains and surcharge, dredging of internal canals and marina, and construction of canal/marina revetment. Temporary dykes were also constructed to allow phase development of the project.

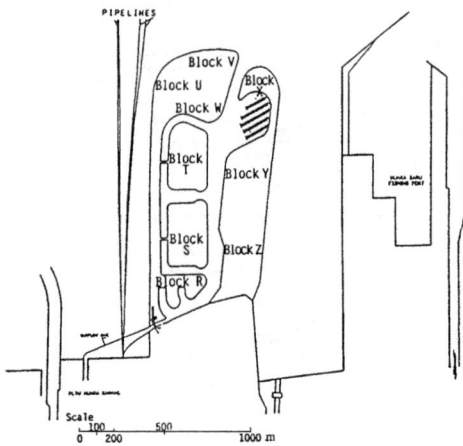

Figure 11. Site plan of Pantai Mutiara reclamation in Jakarta, Indonesia.

A typical soil profile across the north-south direction of the proposed development is shown in Fig 12. The seabed at the site varies from El. 0m to El. -5m and the proposed formation level is El. +2.6m. The site is underlain by a 16m to 18m thick layer of very soft to medium stiff silty clay followed by interbedded layers of silts and sands of dense to very

dense consistency and very stiff silty clays. The upper 12m of the deposit are highly plastic silts and organic clay with w_n = 70% - 150%, w_L = 60 to 130, I_P = 30 - 70, void ratio e = 1.5 - 3.5 and C_c = 0.7 - 1.2. Between El. -12m and El. -18m, the soft to medium stiff silty clay is relatively less compressible with w_n = 50% - 90%, void ratio, e ≅ 1.7, C_c = 0.5. The SPT N value is 0 in the very soft silty clay, 2 to 15 blows/30 cm in the soft to medium stiff silty clay layer and larger than 50 blows/30 cm in the dense to very dense sand/silt strata. The cone resistance rarely exceeds 1 MPa until 12m depth. Thereafter, it increases to about 2 to 3 MPa between 12m and 18m, and exceeds 3 MPa below 18m depth. The vane shear strengths increased from 2 kPa near the seabed to about 14 kPa at 10m depth. Between 12m and 18m depths, the shear strength values vary from 20 to 40 kPa and below 18m depth, the soil was too stiff for the vane apparatus.

The preconsolidation pressures from consolidation tests confirmed that the clayey soils are mostly normally consolidated with C_u/p' ratio of about 0.22 to 0.25.

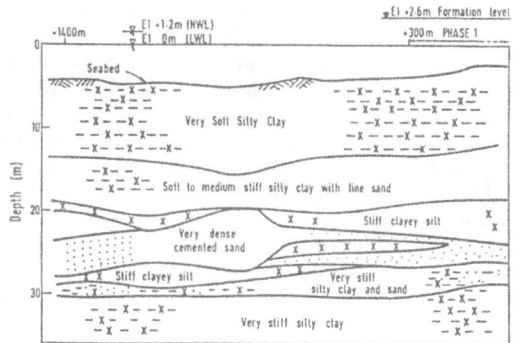

Figure 12. Soil profile at Pantai Mutiara reclamation in Jakarta, Indonesia.

Settlement of the filled areas will occur due to the consolidation of the underlying soft sediments. To provide construction stability during fill placement and to minimize post-construction settlement to an acceptable magnitude, a system of vertical drains with preloading was adopted. Initially, imported vertical drains were installed but due to devaluation in the local currency, they were replaced by locally manufactured Fibredrains at the request of the developer because of lower material costs. More than 350,000m of Fibredrains (80 mm x 10 mm) were installed in a 2 m square grid over an area of about 4 ha in Blok R to a depth of about 16 m to 18 m. Due to the non-availability of suitable materials, fill placement was interrupted and the reclamation works stretched to more than 15 months. Following Lee et al (1988) the back calculated Ch for this soft clay was obtained as 3.8 m²/yr using Equation 1.

For an estimated m_v of about 1.97 m²/MN, which is based on an ultimate settlement of 75 cm under an applied load of 1.4 m of fill of unit weight 17 kN/cu m, the average k_h is obtained as 2.3×10^{-9} m/sec by $k_h = C_h \, m_v \, \gamma$, where γ is the unit weight of water. Based on the observed maximum rate of settlement, the flow through a drain due to the compression of clay volume within the tributary area of a drain yields a rate of 4.7 m³/yr for the single drainage condition. After 12 months to 15 months of drain installation the fill level was raised by 1 m. The ground settlement accelerated at all depths as expected and the effective presence of the drain was observed. In-situ vane tests indicated $C_u = 20$ kN/m² within the mid depth of clay corresponding to average vertical effective stress p'=80 kN/m². Based on $C_u/p' = 0.25$, nearly 90 to 95% consolidation of clay under the applied load is complete.

In Phase II, with the reclamation of about 17 to 20 ha of land with hydraulically dredged seabed clay from the vicinity of the site, to increase the undrained shear strength of the underlying sea bed clay to about 22 kN/m² at El. -6 m at the periphery of the reclamation so as to facilitate seawall construction without instability, an estimated degree of consolidation of 90% was needed. With an average C_h of about 9 m²/yr, a drain spacing of 1.4 m was adopted to achieve this degree of consolidation in 3 months with an appropriate surcharge. Subsequently increase in undrained shear strength and ground settlement was observed (Lee et al, 1988).

In this reclamation, excavation for drainage channels in an area where Fibredrain was installed about eight years earlier, there was no trace of its existence, which verified the biodegradability and hence ecological friendliness of Fibredrain.

Fig 13 shows a typical section of the western breakwater at the seabed level of El. -4.5m. The dyke was designed to retain fill materials inside the reclamation areas and at the same time protect the reclaimed land from wave forces. Prior to construction of the dykes, bamboo clusters consisting of 7 bamboo poles, length of 8m and diameter of 80 to 100mm, were inserted vertically into the soft seabed at 1.5m spacing with the aid of a floater guide until the top of the clusters are at the seabed. These bamboo clusters have two Fibredrains wrapped around (Fig 14) the centre pole of the cluster. The bamboo is tied together by jute ropes and bamboo pins. A bamboo raft comprising 7 layers of bamboo mat was placed on top of the clusters with the aid of a crane mounted on a pontoon. Each bamboo pole has a nominal 1 ton tensile strength.

Sand was then spread over the rafts to a height of 0.30m. For temporary dykes, sandbags were placed on top of this reinforced raft-sand layer. For permanent dykes, the raft-sand layer was covered with a geotextile (Polyfelt TS700) before continuing the build-up with graded quarry stone. Two rock mounds comprising 200-kg quarry stone were constructed gradually with a slope of 1:1.5. The next stage consisted of placing the geotextile on the centre portion between the two rock mounds after which sand was placed up to El. -1.0 m. After a period of time for the underlying soft clay to gain strength from dissipation of excess pore pressure through Fibredrains wrapped in the bamboo clusters, quarry stones of 15 to 50 kg were placed in a configuration as shown in Fig 13. To provide protection against wave forces, a 2m thick layer of precast tetrapods weighing 1.6 tons each were placed on the external sides of the dyke up to a crest height of El. +4.0m. There was a time lapse between each stage to allow the underlying soft clay to gain strength.

During the construction of the dyke, the bamboo raft acts as a tension element, which increases the dyke stability. The main influence of the bamboo clusters is to increase the length of a potential rupture surface during the fill operation and the Fibredrain is to accelerate the gain in strength of the underlying soft clay. Stability analysis was carried out for each stage and the minimum factor of safety

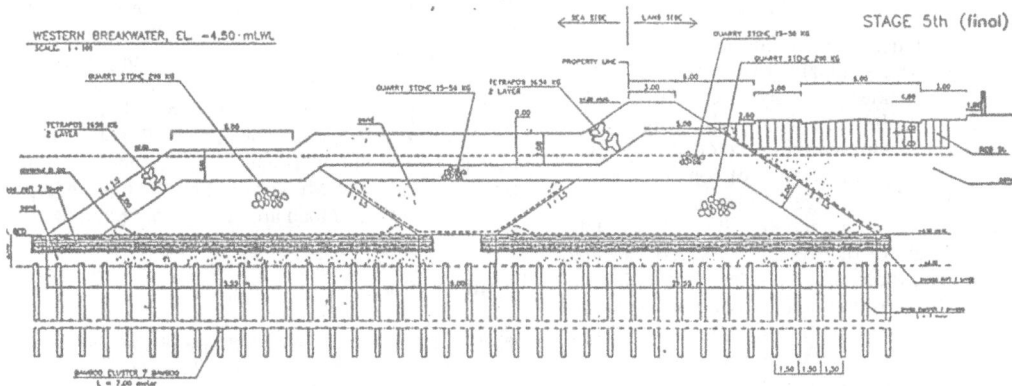

Figure 13. A typical section of the western breakwater.

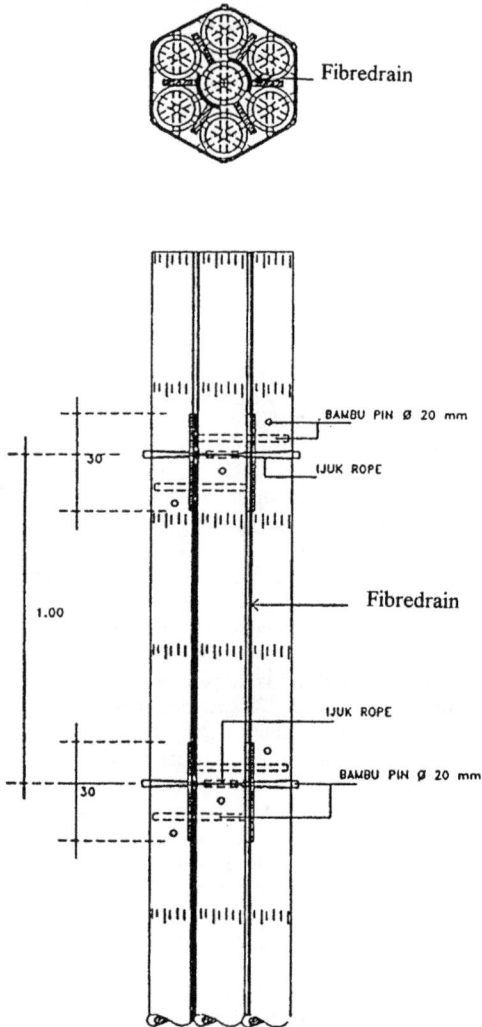

Figure 14. Bamboo cluster with Fibredrain.

9 CONCLUSIONS

The effective coefficient of consolidation C_h is a function of the hydraulic conductivity and compressibility of the soil as well as the field performance of the PVD, which is effected by the deformation of the core and the filter, the clogging and kinking of the drain. In field monitoring, the slope of the settlement-time curve in the early stage of consolidation multiplied by the influence area per drain gives the effective discharge capacity of the PVD-soil system under the surcharge.

For Fibredrain, the effective discharge capacity in the field is given in Figure 2a as a function of lateral pressure. The required discharge capacity is shown in Figure 2b as a function of the final settlement S_f for various values of the influence diameter D and the effective C_h. With given lateral pressure and final settlement S_f the value of D and C_h can be determined in Figure 2. With these values of D and C_h, the time of consolidation can be calculated by Equation 1. The time of consolidation cannot be reduced by decreasing D as this leads to increase in Q_r. It should be emphasized that Fibredrain continues to function even at an axial strain of 24% (Aboshi et al, 2001). In soil improvement projects where the PVD was observed to be under-performing (Bo et al, 1997) the effective discharge capacity of the PVD should be examined with particular reference to the effect of kinking under axial strain caused by settlement and clogging. The damage caused by installation stresses may be important for PVD with lower tensile strength. For Fibredrain the effect of these factors is not significant as discusses earlier.

Fibredrain is environmentally friendly in the production and transformation processes, biodegradable in the utilization phase and hence ecologically harmonious.

ACKNOWLEDGEMENTS

The cooperation of Associate Professor M. A. Aziz, Drs S. D. Ramaswamy and N. C. Das Gupta in the development of Fibredrain is gratefully acknowledged. The advice rendered by Professor H. Aboshi, and the support given by Dr T. Inoue, both of Fukken Co Ltd., Hiroshima, Japan as well as the contribution of Mr Ludi Bone, P.T. Indonesia Nihon Seima, Jakarta and Mr T. Murakami, Amano Corporation, Onomichi, Japan are deeply appreciated.

REFERENCES

Aboshi, H. (1999). "On some problems of consolidation and soil stabilisation in soft clays." *Proc., the Seng-Lip Lee Symp. on Innovative Solutions in Structural and Geotechnical Engineering*, Bangkok, Thailand, 241-250.

was 1.10 for a deep-seated failure. The factor of safety will increase gradually with time due to consolidation of the underlying clay layers. The stability of the dyke with the reclamation fill behind it was also analysed and the minimum factor of safety was found to be about 1.20 at the end of construction.

During the initial stages of constructions the bamboo clusters will support the dyke. At the later stages, the weight of the dyke will exceed the bearing capacities of the bamboo clusters. The latter will subside and the surrounding clay will commence the consolidation process with consequent increase in shear strength. From previous measurements of completed dyke, the settlement is estimated to be in the order of 1.5m to 2.0m.

Aboshi, H., Sutoh Y., Inoue, T. and Shimizu, X. (2001). "Kinking deformation of PVD under consolidation settlement of surrounding clay." *Soils and Foundations*, 41(5), Oct 2001

Bo, M. W., Arulrajah, A. and Choa, V. (1997), Performance verification of soil improvement work with vertical drains", *Proc., 30th Year Anniversary of the Southeast Asian Geotechnical Society, Bangkok,* 1-191 – 1-203

Choa, V., Vijiaratnam, A., Karunaratne, G. P., Ramaswamy S. D. and Lee S. L. (1979). " Consolidation of Changi marine clay of Singapore using flexible drains", *Proc., 7th ECSMSFE, Brighton*, 3, 29-36

De R. N. (1994). "Jute and Environment", *Business Standard Ltd*, New Delhi, India

Hansbo, S. (1981). "Consolidation of fine grained soils by prefabricated drains." *Proc., 10th Int. Conf. on Soil Mech. And Found. Engrg.,* Publications Committee of ICSMFE, ed., A. A. Balkema, Rotterdam, The Netherlands, 677 - 682.

Head, K. H. (1985), *Manual of Soil Laboratory Testing*, Pentech Press, London

Karunaratne, G.P., Chew, S. H., Leong, K.W., Wong, W.K., Lim, L.H., Yeo, K.S. and Hee, A.M. (2002). "Installation Stresses in Prefabricated Vertical Drains", Submitted for Publication in *J. Geot. and Geoenv. Engrg,* ASCE

Karunaratne, G.P., Lee, S. L., Aziz, M.A., Yong, K.Y. and Soehoed, A. R. (1999). "Fibredrain for soil improvement". *Proc., the Seng-Lip Lee Symp. on Innovative Solutions in Structural and Geotechnical Engineering,,* Bangkok, Thailand, 261-275

Koerner, R.M. (1997). *Designing with geosynthetics*, Fourth Edition, NJ: Prentice Hall.

Lee, S.L., Karunaratne, G.P., Aziz, M.A., and Inoue, T. (1995):"An environmentally friendly prefabricated drain for soil improvement". *Proc.B.B. Broms symp. on geotechnical engineering*, Singapore, 13-15 December. 1-9

Lee, S.L., Karunaratne, G.P., Das Gupta, N. C., Ramaswamy, S. D. and Aziz, M. A. (1994). "Natural Geosynthetic Drain for Soil Improvement". *Geotextiles and Geomembranes*, London, U.K., 13(6-7), 457-474

Lee, S.L., Karunaratne, G.P. and Yong, K.Y. (1988). "Performance of Fibredrain in Pantai Mutiara". *Proc. seminar on ground improvement application to Indonesian soft soils*, Cawang, Indonesia

Lee, S.L., Yong, K.Y., Tham, K.W., Singh, J. and Chen, W.P. (1989). Treatment of ex-mining land by Fibredrains, surcharging and high-energy impact. *Proc. Symp on application of geosynthetic and geofibre in South Asia,* Petaling Jaya, Malaysia, 5-18 to 5-22.

Lee, S.L., Lo, K.W., Karunaratne, G.P. and Ooi, J. (1984). "Improvement of peaty clay by dynamic replacement and mixing." *Proc. JSSMFE-NUS-AIT Seminar on soil improvement and construction techniques in soft ground*, Singapore, 208-214.

Mlynarek, J. (1998). Panel discussion on filtration and drainage, *5th Int. conf. on geotextiles, geomembranes and related products*, Singapore, 4, 1383-1385

Mlynarek, J. and Rollin, A. L. (1995). "Bacterial clogging of geotextiles – overcoming engineering concerns." *Proc. Geosynthetics '95*, 1, 177 – 188.

Miura, T., Tou, M., Murota, H. and Bono, M. (1995a). Large Scale Consolidation test on drainage characteristics of Fibredrain, (In Japanese). *Proc. of annual regional meeting of JSCE,* Kyushu, Japan

Miura, T., Tou, M., Murota, H. and Bono, M. (1995b). The basic experiment on permeability characteristics of Fibredrain, (In Japanese). *Proc. of annual regional meeting of JSCE,* Kyushu, Japan

Rowe, P W (1968), "Influence of geological features of clay deposits on the design and performance of sand drains", *Proc Institution of Civil Engineers*, London, Paper 70583

Welsh, J. K. (1983). "Soil Improvement – A Ten Year Update", *Proc. Symp. Placement and Improvement of Soils*, Geotechnical Special Publication No 12, ASCE

Yoshida, Y., Hamada, K., Sakimori, H and Goto, H. (1995). "Effectiveness of soil stabilization of Fibredrain method." (In Japanese), *Proc. of annual regional meeting of JSCE*, Kyushu, Japan

Soft Ground Engineering in Coastal Areas, Tsuchida et al. (eds)
© *2003 Swets & Zeitlinger, Lisse, ISBN 90 5809 613 0*

Cone resistance of sand compaction pile in soft clay with different area replacement ratio

J. Takemura
Asian Institute of Technology, Thailand

T. Mizuno
NTT communications, Tokyo, Japan

ABSTRACT: Cone penetration tests were conducted in model SCP grounds with different area replacement ratio, 0, 30, 50, 70 and 100%, in a geotechnical centrifuge using a miniature cone penetrometer with diameter of 3mm. Tip resistance and shaft friction were measured in the test. Pressuremeter tests were also conducted in the SCP model ground in the centrifuge to examine the horizontal constrain effect in terms of coefficients of subgrade reaction of SCP composite ground with different area replacement ratio under cylindrical expansion condition. From the tests the effect of area replacement ratio on the cone tip resistance and shaft friction was confirmed, that is, the higher the replacement is, the larger the resistances. It was also found from the pressuremeter tests that the horizontal coefficient of subgrade reaction of SCP composite ground is affected by the area replacement ratio.

1 INTRODUCTION

Sand Compaction Pile Method (SCP) is one of the most common ground improvement techniques in Japan for both soft clay and loose sand. For the application to the soft clay it is primarily used to prevent the failure and reduce the settlement of structure founded on the ground. This method can be categorized according to the area replacement ratio (A_s), namely, high replacement ratio SCP: As > 0.7; medium replacement ratio SCP: 0.4<As<0.7; low replacement ratio SCP: As < 0.4 (Japan Port and Harbor Association 1989). Kanda & Terashi (1990) reviewed 85 case records about the near shore soil improvement works using SCP. From the comprehensive survey, they reported that the high replacement ratio SCP was commonly applied in port harbor construction works, especially for the foundation of gravity type caisson structures, like quay wall and breakwater. This is mainly because the high replacement ratio SCP can provide greater stability, less deformation and a shorter construction time than the low replacement ratio SCP. Although the high replacement ratio SCP has these advantages and huge application records in Japan, it is not yet popular in developing counties, e.g., South East Asian Region due to its high construction cost. Therefore, SCP with lower replacement ratio seems to have greater prospect.

Although precise assessments are required for both short term stability and long term settlement in the application of low replacement ratio SCP (Rah-

man et al., 2000), in the current design practice the short time stability is the critical condition which determines the improvement condition of SCP, like improvement area and replacement ratio. In the stability analysis on the structure founded on SCP improved soft clay, the resistance in the sand piles' portion is usually dominant in the total resistant force or moment even for relatively low replacement ratio (Kanda and Terashi 1990). Therefore the evaluation of friction angle of the sand pile, ϕ', is the most important in the stability analysis of SCP. In-site confirmation of sand pile is normally conducted by the standard penetration test (SPT) after installing SCPs and the friction angle of the sand pile is evaluated using empirical relations between N-value and ϕ' value which have been derived from a vast data on SPT in sand. However, the conditions of SPT for uniform sand and a sand pile surrounded by composite ground, that is, clay improved by SCP, are different as shown in Figure 1. The horizontal pressure induced by the penetration is larger for the former than the latter, which may give different N value even for the same density of the sand. Furthermore it can be expected that this horizontal constraint effect depends on the replacement ratio. Figure 2 shows the N-values observed in SCP with different replacement ratios of 25% and 70% at a same construction site in Hiroshima harbor. Although the N-values for A_s=70% show some scatter, the dependency of N-values on replacement ratio can be clearly seen, that is, the higher A_s, the higher N-values. This trend of N-value on replacement ratio results in

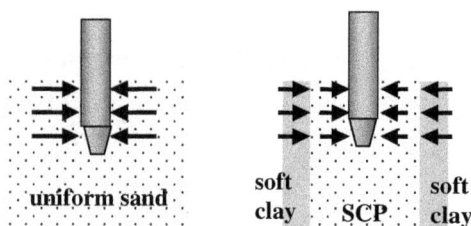

Figure 1. Horizontal constraint in sand and SCP during penetration test.

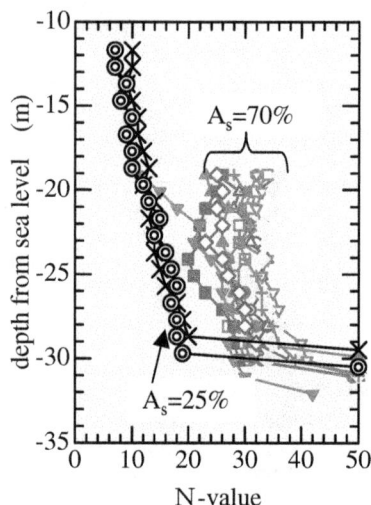

Figure 2. N-values measured at center of SCPs with As =25% and 70% at Dejima in Hiroshima Harbor. (Third Bureau of Harbor Construction, MOT 2000).

the lower friction angle for the lower replacement ratio adopted in the current design guideline (Japan Port and Harbor Association 1989).

If the lower horizontal constraint effect for the low replacement ratio is the main reason for the trend of N-value, the current design guideline could underestimate φ' value for the SCP with lower replacement ratio. However, the horizontal constrain effect is also the case during SCP installation, which may create denser sand piles for the high replacement ratio SCPs than the low ones, which is the background of the current design guideline. Because it is difficult to directly measure φ' value of the sand in SCP, a proper answer to this question has not been given yet. One of the possibilities to give an insight on this argument is static cone penetration test, because it can be easily conducted in practice and theoretical interpretation, like cavity expansion theory, is easier than SPT (e.g., Houlsby & Hitchman 1988, Yu & Houlsby 1991, Chen & Juang 1996, Salgado et al. 1997). Centrifuge model tests are also very useful in the study of SCP improved soft clay (Nakase & Takemura 1989).

In this research, in order to study the effect of replacement ratio on the cone resistance, cone penetration tests were conducted in model SCP grounds with different area replacement ratio, 0, 30, 50, 70 and 100%, in a geotechnical centrifuge using a miniature cone penetrometer with diameter of 3mm. Tip resistance and shaft friction were measured in the test. Pressuremeter tests were also conducted in the SCP model ground in the centrifuge to examine the horizontal constrain effect in terms of coefficient of subgrade reaction of SCP ground with different replacement ratio under cylindrical expansion condition.

2 CENTRIFUGE MODEL TESTS

2.1 Soils used in the tests

Soil used for making the sand compaction piles and sand drains in the model was Toyoura sand. Clay used for the model ground was Ariake clay collected from Kumamoto bay. The collected clay was remolded and sieved through 2mm mesh and stored in plastic containers at Tokyo Institute of Technology (TIT). Physical and mechanical properties of those materials are shown in Table 1.

Table 1. Material properties

(a) Ariake clay

Specific gravity: G_s	2.67
Liquid limit: w_L	71.1 %
Plastic linit: w_P	38.0%
Plasticity index: I_p	33.1
Compressibility index: C_c	0.52
Void ratio e at 1kPa in NCL	2.807
Permeability k at 100kPa	10^{-9}m/sec
c_u/p	0.42

(b) Toyoura sand

Specific gravity: G_s	2.64
Maximum void ratio, e_{max}	0.974
Minimum void ratio, e_{min}	0.609
Relative density of SCP	80%

2.2 Model preparation

Mark III centrifuge of TIT (Takemura et. al. 1997) was used in the tests. Three different diameters of SCP (D_s) were used: 15mm, 30mm and 60mm. Two model containers with different inner dimensions were used in the tests. The container for the model ground with SCP of D_s=15mm and 30mm was 500mm in length, 150mm in width and 350mm in height and the container for the model with the largest diameter SCP had the same length and height as the other one but different width of 292mm. Figure 3 shows the schematic illustration of the test setup. Coarse silica sand was placed at the bottom of the strong box and compacted to the thickness of 20mm to form a bottom drainage layer and a filter paper was laid on it. De-aired clay slurry with water content at about 1.5times the liquid limit was then poured into strong box to the depth of about 260mm.

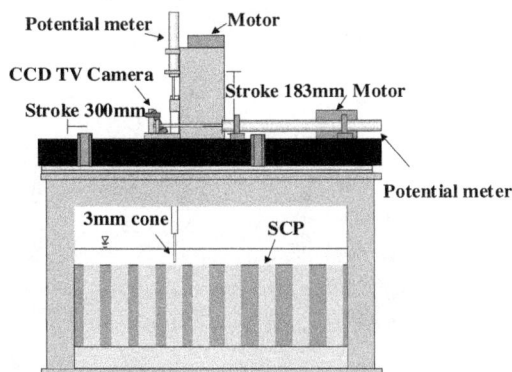

Figure 3. Test setup for cone penetration test.

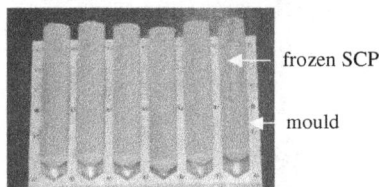

Figure 4. Frozen sand pile made by metal mould: Ds=30mm.

Figure 5. Cross sectional view of model SCP ground after test: Ds=30m.

The specimen was consolidated under 5kPa preconsolidation pressure using a bellofram cylinder for about 10 days. After completion of the preconsolidation, 7mm diameter vertical sand drains (SD) were made at the location where model SCPs would be installed. The relative density of SDs was about 30%. The function of SDs was to accelerate the first centrifugal consolidation, in which normally consolidated clay with increasing strength and stiffness with depth was prepared. The model clay with SDs was then mounted on the centrifuge and the first centrifugal consolidation was conducted under 100g for the model SCP ground with D_c=15mm and 30mm, and 50g for that with D_c=60mm. With the aid of SDs, 90% consolidation in the first centrifuge test was completed in 17 hours for the longest case. After the first centrifugal consolidation, thickness of the model ground was 150mm which corresponding to 15m and 7.5m in the prototype scale under the centrifuge acceleration of 100g and 50g employed in the tests respectively.

Metal mould was used to prepare the model SCPs. Saturated sand piles were made in the mould at relative density about 80% and frozen at -40^{0}C. Figure 4 shows the frozen SCPs in the half of metal mould for D_s=30mm. In the installation of the model SCP, vertical holes were first made in the ground by extracting clay columns together with SD using thin walled steel tube with the same diameter of the SCP from predetermined positions through a thick acrylic template and then the frozen piles were installed in the holes. Figure 5 shows a photo of the sand piles in the ground taken after test. In the actual construction of SCPs, the SCPs are driven by displacement, which gives quite severe disturbance to the surrounding clay. Ng et al. (1998) developed an inflight SCP installer for their centrifuge model tests to investigate the effect of actual installation process on the mechanical behavior of the SCP ground. Although the effect of the disturbance is one of the most crucial aspects on this problem, in this study above mentioned technique was adopted to make precise arrangement of SCPs and to avoid the disturbance of the clay so that the initial conditions of the clay and SCP could be specified clearly. Details in installing the SCP and SD have been presented by Rahman et al. (1998).

In the model SCP ground with D_s=60mm, a model pressuremeter with 5 pressure cells (60mm in diameter and 25mm in height) was inserted in a hole of the arrangement of SCPs as shown in Figure 6.

Each cell is connected to a separate hydraulic cylinder with build-in pressure transducers with thin-flex tube. The displacement of hydraulic piston and pressure of the cylinder correspond to the expansion and pressure of the cell. Details of the model pressuremeter were given by Okamura et al. (1998). After completing the model SCP ground, 2D actuator with the miniature cone penetrometer, which had a 3mm diameter rod of 50mm length from the tip (see. Figure 7), was placed on the model container as shown in Figure 3. The second centrifugal consolidation was then conducted with the same centrifugal acceleration in the first centrifugal consolidation.

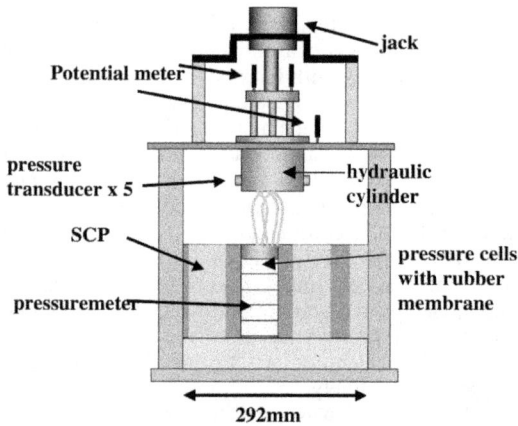

Figure 6. Test setup for pressuremeter test.

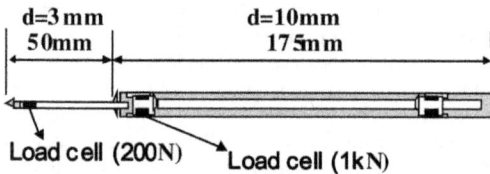

Figure 7. Model miniature cone penetrometer.

2.3 Test procedures and conditions

Having confirmed the completion of the second centrifugal consolidation, miniature cone was penetrated to the depth of 100mm at the center of SCPs in a row indicated in Figures 8. After penetrations in the first row, the centrifuge was once stopped, the position of the actuator was adjusted to the next row for the further penetration, and centrifugal consolidation was again conducted before conducting the penetration in the next row. Beside the SCP ground, the cone penetration tests were conducted in clay ground after completion of the first centrifugal consolidation and in uniform saturated sand with D_r=80%. Two different penetration rates were employed in the tests, i.e., 0.2mm/sec and 2.2mm/sec. During the penetration test, the axial loads at the tip and at the end of 3mm diameter rod in the cone penetrometer were measured by the load cells 1 and 2 shown in Figure 7. The tip resistance q_c is obtained from the load at the tip (load cell 1) and average shaft friction at the penetration depth z, $\tau_{sm}(z)$, is evaluated from the following equations:

$$q_c = \frac{Q_1}{\pi d_c^2 / 4} \tag{1}$$

$$\tau_{sm}(z) = \frac{Q_2 - Q_1}{\pi d_c z} \tag{2}$$

where Q_1 and Q_2 are loads measured by load cell 1 and 2 respectively, and d_c is diameter of the cone.

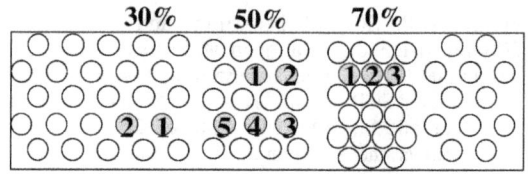

Figure 8(a). Model 1 (D_S/d_C=5, D_S=15mm, 100G).

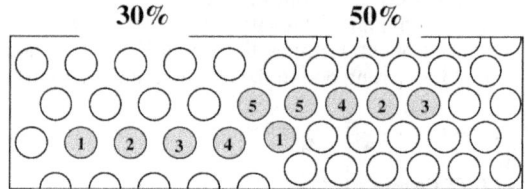

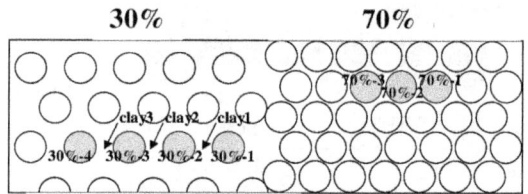

Figure 8 (b). Model 2-1&2 (D_S/d_C=10, D_S=30mm, 100G).

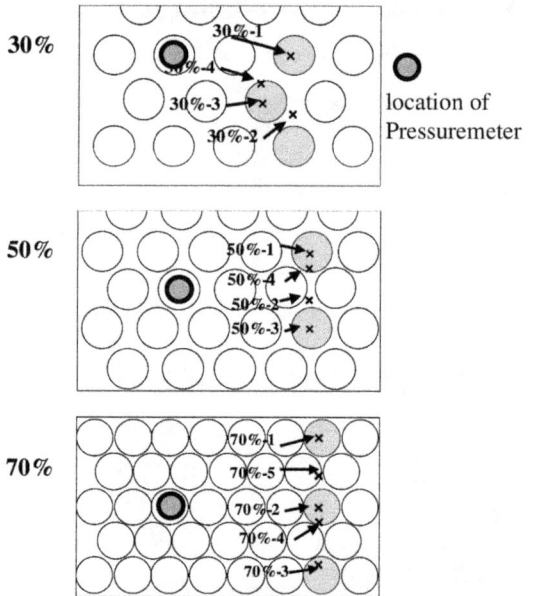

Figure 8 (c) Model 3-1,2 &3 (D_S/d_C=20, D_S=60mm, 50G).

In the SCP grounds with D_s=60mm, pressuremeter test was conducted at the location shown in Figure 8(c) before conducting the cone penetration test. During the pressuremeter tests, displacement of the pistons and pressure in the cylinders were measure, which gave the strains of cavity and cavity expan-

270

sion stresses at different depths. The rates of cavity strain employed are 0.2%/min for the clay, the uniform sand and SCPs with A_s=30%, and 0.38%/min for SCPs with A_s=50% and 70%. In the following chapter, test results and discussion are all given using prototype scale.

3 TEST RESULTS AND DISCUSSION

3.1 Cone penetration test

Variations of the tip resistance and the average shaft friction with penetration depth observed in the model SCP ground with D_s=3m and D_s/d_c=10 are shown in Figures 9. Penetrations were conducted more than twice with the same As and penetration rate for all the test cases. The tip resistances and shaft frictions with same conditions show some difference in all the cases as shown in the figures. Positioning of the cone at the center of SCP was manually done seeing the monitor of image taken by an on-board CCD camera. However the actual position of the cone penetration was possibly apart from the center to some extent because of two reasons. One is the unclearness of the top view of sand piles and the other is the difficulty to install the sand piles in a precise straight line (see Figure 10). As the resistance is expected to decrease with increasing the apartness, the maximum tip resistance and shaft friction are considered to be the most reliable ones as the representative of those at the center of SCP.

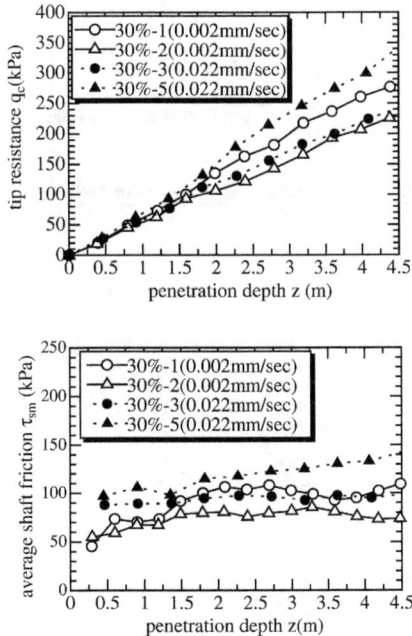

Figure 9. Variation of tip resistance and average shaft friction with penetration depth: D_s=3m, A_s=30%.

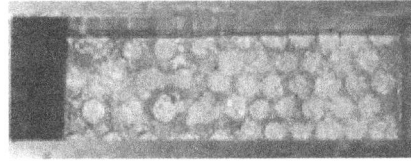

Figure 10. Top view of model SCP ground: Ds=30mm, left A_s=30%, right A_s=50%.

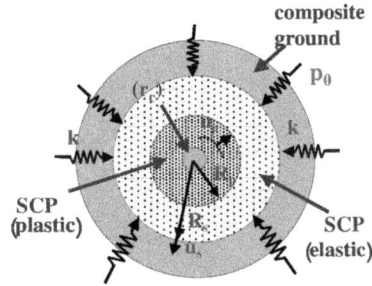

Figure 11. Cavity expansion model for cone penetration in SCP.

From the figures it can be also seen that the slower the penetration rate is, the lower the tip resistance and shaft friction. It is contradict to the trend in clay where the cone resistance increases with decreasing penetration rate due to the effect of consolidation by partial drainage near the cone. The cone penetration into SCP can be modeled by cavity expansion in finite area as shown in Figure 11. The sand-clay boundary expands radially resulting the subgrade reaction from the surrounding composite ground. The difference in the expansion pressure with different replacement ratio may be considered by the subgrade reaction. In the sand portion the fully drained condition is satisfied in the normal range of the penetration rate and partial drained condition near the sand-clay boundary depends on the penetration rate. As the radial displacement at the sand-clay boundary is not large for the normal sand–cone diameter ratio as shown in Figure 12, expansion volume could be compensated by the volume decrease due to consolidation in the clay more for the slower penetration rate than the higher one, resulting in the lower subgrade reaction for the former than the latter.

Figures 13 show variations of the tip resistance and average shaft friction with penetration depth for various replacement ratios in the model SCP grounds with D_s=3m and D_s/d_c=10. From these figures, the effect of penetration rate on the cone resistances of SCP can be also seen but not the case for clay in the penetration rate difference adopted in the tests. Regardless of the penetration rate, the resistances tend to increase with replacement ratio and this trend is clearer for the shaft friction than the tip resistance. Ratios of the tip resistance and average

271

shaft friction to those in the uniform sand are plotted against the replacement ratio for different penetration depths. The tip resistances and shaft frictions used in the figures are the observed maximum values in the tests with the same conditions for SCP ground and the average values for the uniform sand. The effect of replacement ratio on the cone resistances is clearly confirmed in these figures. However, as the sand pile-cone diameter ratio increases,

the resistances also increase reducing the difference from those in the uniform sand. From Figure 12 it can be known that the cylindrical cavity, which might be related to the shaft friction, gives larger radial displacement than the spherical cavity, which can be related to the tip resistance. This suggests that the tip resistance of SCP could be close to that of the uniform sand with smaller D_s/d_c than the shaft friction. However, this cannot be seen in the figures. One of the possible reasons about it may be the difference between actual deformation mechanism and the assumed cavity expansion because of the relatively shallow measured depth. One of the possible reasons about it may be the difference between actual deformation mechanism and the assumed cavity expansion because of the relatively shallow measured depth. In order to discuss more about this point, more precise tests with less error are required.

3.2 Pressuremeter tests

Figures 15 shows the relationships between cavity strain, $\Delta r/r$ (r is the initial cell radius and Δr the radial displacement of the cell perimeter), and cell pressure increment measured at different cell depths. The value in parentheses shows the vertical effective stress at the middle depth of each cell. During the pressuremeter tests, unloading and reloading were conducted as shown in the figures. Due to some in-

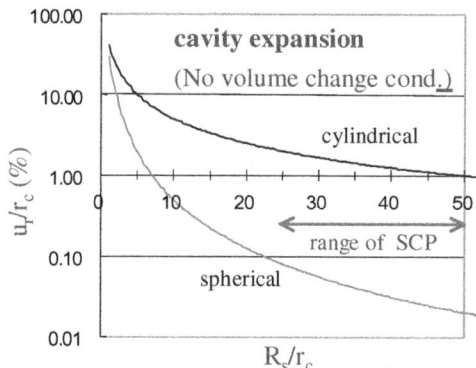

Figure 12. Relationship between radial displacement at SCP perimeter (u_c) and SCP radius (R_s) normalized by cone radius (r_c) for no volume change conditions.

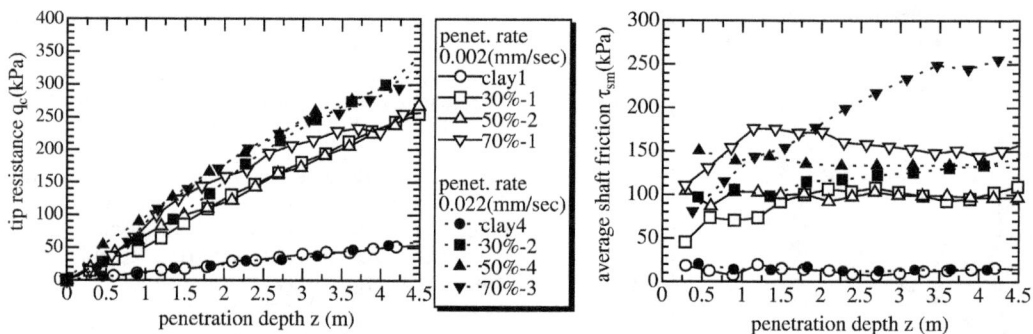

Figure 13. Variation of tip resistance and average shaft friction with penetration depth for various A_s: D_s=3m, D_s/d_s=10.

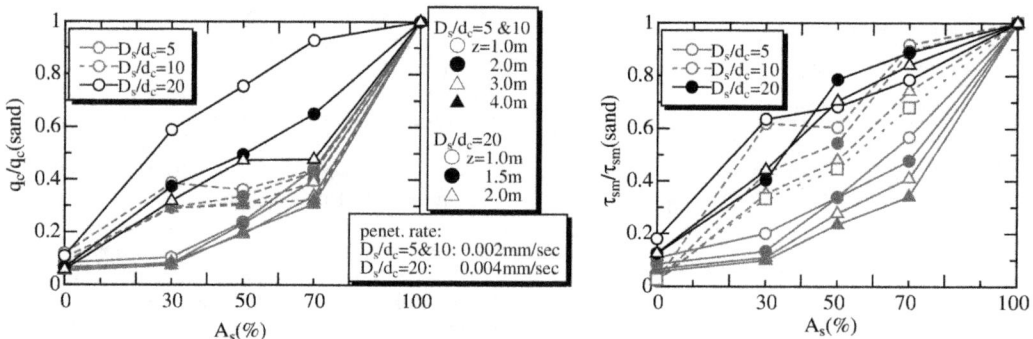

Figure 14. Plots of ratios of tip resistance and average shaft friction to those in uniform sand against replacement ratio.

272

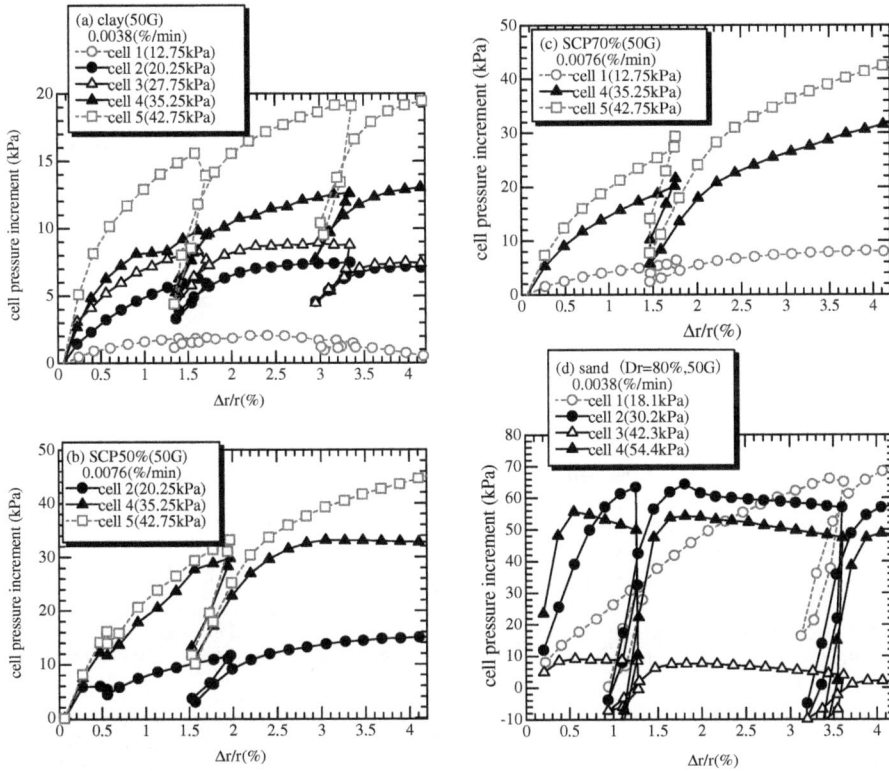

Figure 15. Relationship between cavity strain and cell pressure increment in pressuremeter tests.

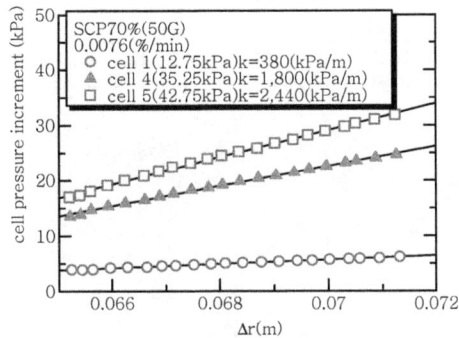

Figure 16. Relationship between radial displacement and cell pressure increment.

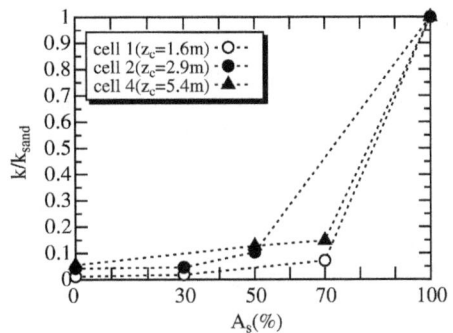

Figure 17. Relationship between ratio of coefficient subgrade reaction to that of uniform sand and replacement ratio.

evitable gap between rubber membrane of the cell and the surrounding clay, reliable relationship could not be obtained especially in the initial loading portion. However, in the reloading process, linear relationships were observed in some cells as shown in Figure 16. From these straight lines coefficients of subgrade reaction, k, were obtained.

Figure 17 shows relationship between the ratio of coefficient of subgrade reaction (k) to that of the uniform sand (k_{sand}) and replacement ratio. From this figure it can be seen that the coefficient of subgrade reaction increases with increasing replacement ratio. However, the trend is not so clear as that seen in the tip resistance and shaft friction in the cone penetration tests. Furthermore there is marked difference between k values of the SCP grounds and the uniform sand. These difference between the effects of replacement ratio on the cone resistances and coefficient of subgrade reaction could be attributed to two reasons. One is that the inevitable gap between the cell and the surrounding clay impairs the effect on the latter. The other is that the loading rates in the pressuremeter test were even smaller than the cone penetration test, increasing the consolidation and reducing the reaction from the surrounding clay as discussed in the previous section. More accurate test and some numerical or analytical simulations are required, in order to have a right answer to these arguments.

4 CONCLUSIONS

In this research, cone penetration tests and pressuremeter tests were conducted in model SCP grounds with different area replacement ratio in a geotechnical centrifuge using a miniature cone penetrometer and a model pressuremeter. The effect of replacement on the tip resistance and shaft friction ratio was clearly seen, the higher the replacement is, the larger the resistances. From the pressuremeter tests, it was also found that the horizontal constrain effect in terms of the coefficient of subgrade reaction of SCP composite ground was affected by the replacement ratio. These observations suggest that the current practice using empirical relation of φ' value to N-value in uniform sand has a possibility of providing underestimation of φ' value of sand pile surround by SCP composite ground especially for the low replacement ratio. However, beside the replacement ratio, the cone penetration resistances and the subgrade reaction are affected by many factors, such as, sand pile-cone diameter ratio, penetration rate or loading rate, stiffness of the clay portion. In order to examine the effects of these factors, further study is strongly recommended using both experimental and analytical approaches.

REFERENCES

Chen, J. W. & Juang, C. H. 1996. Determination of drained friction angle of sands from CPT. J. Geotech. Engrg., ASCE, 122(5): 374-381.
Houlsby,G.T. & Hitchman,R. (1988) Calibration chamber tests of a cone penetrometer in sand. Geotechnique, 38(1): 39~44.
Japan Port and Harbor Association 1989. Technical guideline and commentary for harbor facilities. (in Japanese)
Kanda, K. & Terashi, M. 1990. Practical formula for the composite ground improved by sand compaction pile method. Technical Note of PHRI, No.669. (in Japanese)
Nakase, A. & Takemura, J. 1989. Stability of clays improved by sand compaction piles. Technical Report of Dept. Civil Engrg, Tokyo Institute of Technology, 40: 1-18.
Ng, Y. W., Lee, F. H., and Yong, K. Y. 1998. Development of an in-flight Sand Compaction Piles (SCPs) installer. Proc. Intl. Conf. On Centrifuge 98, 1, Tokyo: 837-842.
Okamura, M, Sahara, F., Takemura, J. & Kusakabe, O. 1998. Load-settlement behavior of shallow footings on compressible sand deposits. Proc. Intl. Conf. On Centrifuge 98, 1, Tokyo: 435-440.
Rahman. Md. Z., Takemura, J., Kouda, M., Yasumoto, K. 1998. Stability and deformation of soft clay improved by SCP with low replacement ratios. Proc. 13th SEAGSC, 1, Taipei: 393-398.
Rahman. Md. Z., Takemura, J., Kouda, M., Yasumoto, K. 2000. Experimental study on deformation of clay improved by low replacement ratios SCP under backfilled caisson loading. Soils & Foundations, 40(5): 19-35.
Salgado, R., Mitchell, J.K. & Jamiolkowski, M. 1997. Cavity expansion and penetration resistance in sand. J. Geotech. And Geoenvironmental Engrg., ASCE, 123(4): 344-354.
Takemura, J., Ishii, H., Okamura, M., Takahashi, A. & Kusakabe, O. 1997. Centrifuge model test of embankment on soft clay with sheet piles wall reinforcement against settlement of adjacent ground. Technical Report, Dept. of Civil Engrg., Tokyo Institute of Technology, 55: 61-78 (in Japanese).
Third Bureau of Harbor Construction, MOT, 2000. Internal report.Yu, H. S. & Houlsby, G. T. 1991. Finite cavity expansion in dilatant soil: loading analysis. Geotechnique, 41(2): 173-183.

Soft Ground Engineering in Coastal Areas, Tsuchida et al. (eds)
© 2003 Swets & Zeitlinger, Lisse, ISBN 90 5809 613 0

Improvement of soft ground applying vacuum consolidation method by expecting upper clay layer as sealing-up material

H. Yoneya, T. Shiina & H. Shinsha
Penta-Ocean Construction Co., Ltd., Institute of Technology, Tochigi, Japan

ABSTRACT: This method is a kind of vacuum consolidation method. While it is necessary to cover the foundation surface with a seal sheet in the conventional vacuum consolidation method, this proposed method eliminates the necessity of using one by taking advantage of the sealing performance of a clay layer. This paper reports the results of our experiments on indoor and field models. These results have shown that our method is useful.

1 INTRODUCTION

Renewed focus has been directed in recently years toward the application of the vacuum consolidation method to soft-ground improvement. This is because three advantages have been recognized in this method: (1) no banking is required, (2) the method does not cause shear failures from such events as circular slip, because a load generated by vacuum pressure works inwards in all directions, and (3) when banking is to be used, it can be formed rapidly because its excess pore water pressure is offset against negative pressure.

Meanwhile, sediment generated in dredge work has been disposed of at landfills. While it is becoming increasingly difficult to procure new sediment-disposal sites, there is growing need for reducing the volume of sediment by consolidating it so that the limited capacity of existing disposal sites can be utilized more effectively. Best fit for these needs, the vacuum consolidation method without using banking is expected to be an important consolidation-based soil improvement approach.

The vacuum consolidation method using capped vertical drains is a process in which a clay ground is improved by vacuum consolidation while using its top layer about 1 m thick for negative-pressure sealing. Under conventional vacuum consolidation methods, it is necessary to cover the soil ground surface with a sealing sheet. They have frequently failed in the presence of a thick sand outer layer or the sites under water. The new method, on the other hand, dispenses with a sealing sheet by taking advantage of the sealing performance of clay, thus making itself highly applicable to conventional vacuum consolidation works.

This paper introduces the new vacuum consolidation method using capped vertical drains and describes its features and effectiveness based on the results of laboratory tests and field works. It also clarifies the behavior of improved soil and effects of improvement, while summarizing problems to be solved in the future.

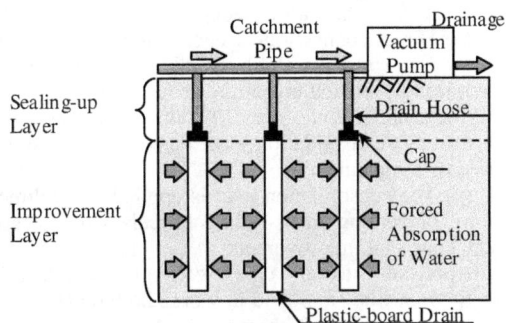

Figure 1. Outline of the method

2 OUTLINE OF THE METHOD

The vacuum consolidation method using capped vertical drains has the following features:

(1) A clay layer 1 m to 2 m in thickness is used as a negative-pressure sealing-up layer, eliminating the necessity to employ a sealing sheet. Thus this method is especially useful at sites under water or on the thick permeable outer layer, where it is almost impossible to lay a sealing sheet or embed its edges in the ground.

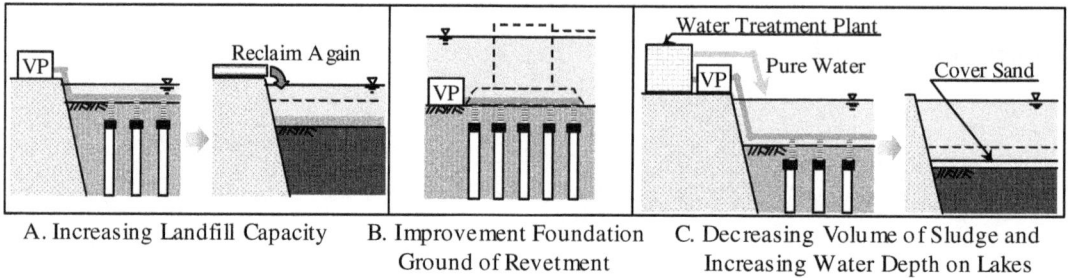

A. Increasing Landfill Capacity   B. Improvement Foundation   C. Decreasing Volume of Sludge and
                            Ground of Revetment   Increasing Water Depth on Lakes

Figure 2. Examples of application

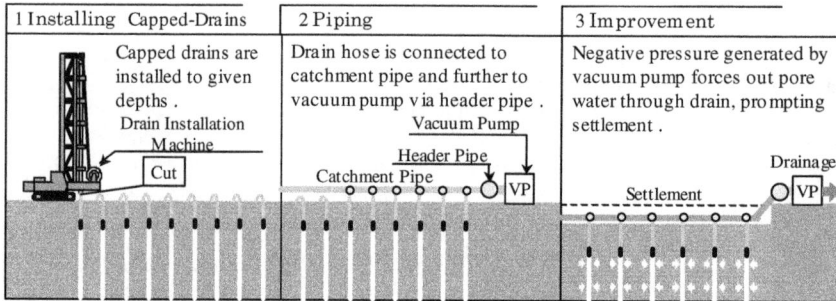

Figure 3. Work flow (on land)

(2) If a permeable layer is found below the outer layer in a preliminary bowling survey, it is possible to apply a waterproof seal to the relevant part of drains prior to the start of work.

(3) Direct connection of each drain to the catchment pipe provides high sealing performance, permitting the vacuum pump to fully exert a negative pressure of up to 50 to 70 kN/m². In the event of insufficient consolidation load, banking may also be used to complement it.

(4) The consolidation load is applied in all directions, preventing shear failure (circular slip) to ensure consistent improvement.

Applicable to all types of soft ground, this method may be employed in works both on land and under water. Figure2 shows an example of application of this method. The method is effective not only for applications under the conventional vacuum consolidation method but also for new applications and sites that have defied the conventional method. Figure3 illustrates the work flow under the new method.

3 LABORATORY MODEL TEST

3.1 *Outline*

Under the vacuum consolidation method using capped vertical drains, clay in the top layer of the improvement area is used as a negative-pressure sealing-up layer. Thus, it is essential to determine the thickness of the sealing-up layer accurately. To examine the ability of clay to maintain sealing performance, we conducted a laboratory test using a model.

Figure 4 presents the testing equipment used in our test. Clay was fed to a cylindrical earth tank. And the capped drain was installed to a given depth. A negative pressure of 50 kN/m² was applied to three sealing-up layers – 0.5, 1.0 and 1.5 m in thickness.

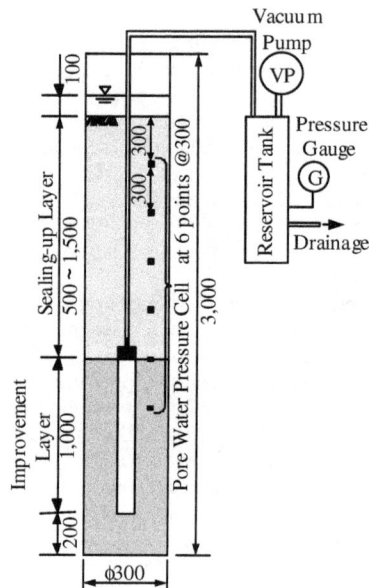

Figure 4. Outline of the testing equipment

276

3.2 Results

In the case of the sealing-up layer 1.5 m thick, six pore water pressure cells were placed in the soil. Figure5 provides the isochrone of pore water pressure of the sealing-up layer. The graph indicates that the pore water pressure of the sealing-up layer gradually came to form a triangular distribution.

Figure6 shows a theoretical isochrone for the triangular stress distribution. The curves in Figures5 and 6 behaved precisely in the same manner.

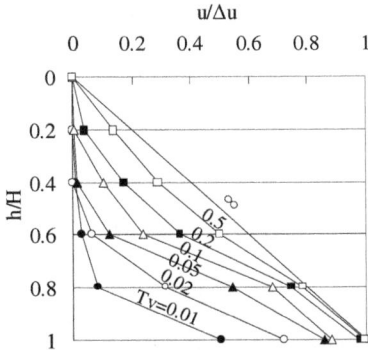

Figure 5. Isochrone (Actual measurement)

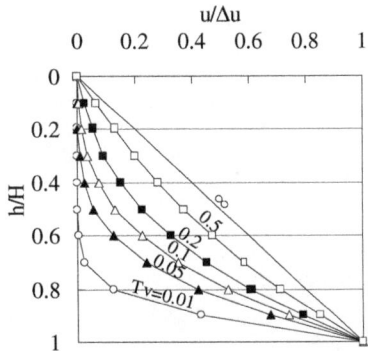

Figure 6. Isochrone (Theoretical value)

Figure7 gives a water content distribution after the test. The water content remained unchanged in the top layer (w=80%), but declined in lower layers. The water content of the improvement area was 55% to 60%, almost the same at all points on the area. Within the dotted line in the figure is a calculated distribution. This distribution was calculated from the e-logp curve in some prior tests on the assumption that a uniform negative pressure of 50 kN/m^2 would be exerted over the improvement area and forms a triangular distribution in the sealing-up layer. (Figure8)

The calculated distribution quite precisely reflects the measurements.

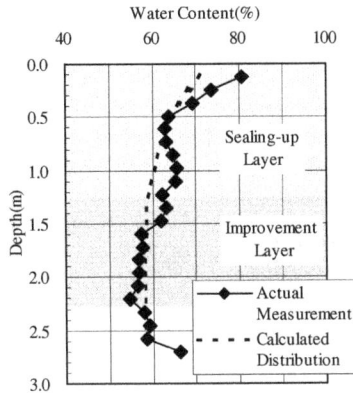

Figure 7. Water content distribution

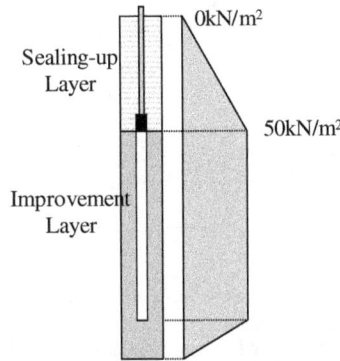

Figure 8. Assumed stress distribution

Figure9 compares the estimated settlement curve with the measurements. The final settlement level was determined using the Cc method by plotting consolidation pressure values represented in a distribution shown in Figure8. The total settlement curve of the ground was estimated by combining a theoretical settlement curve of the sealing-up layer (with a triangular distribution of stress) and that of the improvement area (Barron's solution).

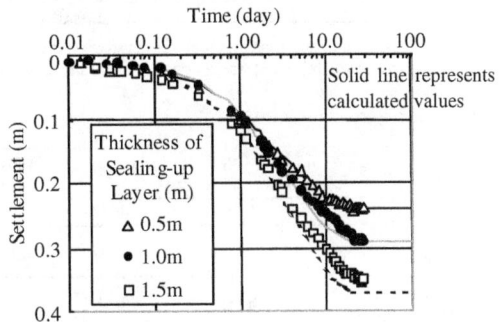

Figure 9. Settlement behavior

Figure10 presents settlement measurements and total displacement volume (for a sealing-up layer thickness of 0.5 m). The total displacement volume was calculated by diving the flow rate by the plane area of the earth tank. Initially, the displacement and settlement values moved side by side, but slowly the former increased relative to the latter, with displacement continuing at a constant speed. This happened probably because a seepage flow emerged from the earth surface. In this experiment we used a large vacuum pump, which prevented negative pressure acting upon the improvement area from declining.

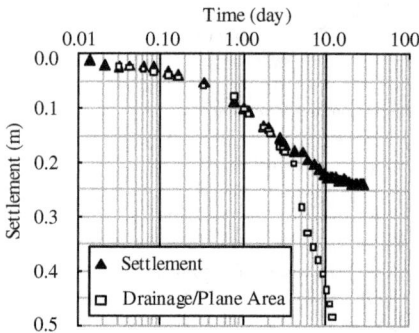

Figure 10. Settlement and total displacement (Sealing-up layer thickness of 0.5)

The results for the three cases show that a seepage flow appears later and displacement volumes become smaller, as the sealing layer becomes thicker. Table1 provides a relation between the time taken for seepage flow to appear and the time factor and the coefficient of permeability. In each case, time factor Tv of the sealing layer was about 0.1 for the number of days taken for a seepage flow to appear.

3.3 Summary

In the test using a model to identify the ability of clay to maintain sealing performance, we obtained the following findings:

(1) The negative pressure on the sealing-up layer started to slowly increase from near the drain, finally forming a triangular distribution. The estimated settlement value favorably matched the measured value when a triangular distribution was produced for the negative pressure on the sealing-up layer.

(2) In the initial phase of consolidation, the displacement volume equaled the settlement level. Then, at $Tv \doteqdot 0.1$, a seepage flow was generated, causing displacement at a given speed. However, the pump capacity was large enough relative to the displacement volume to prevent a decrease in working negative pressure.

4 CONSTRUCTION EXPERIMENT (1)

4.1 Outline

This section describes the results of application of our method in order to prevent uneven settlement when building a pnewmatic caisson. In the construction, vacuum consolidation and surcharge by banking were implemented.

Figure11 provides an overview of the ground under experiment. Improvement by consolidation was attempted on a highly organic soil layer 2.0 m to 5.8 m in thickness that lay across the improvement area. Table2 shows the physical properties of the ground. With a water content as high as 400%, the highly organic soil layer was high in fiber.

Table 1. Relation of appearance of a seepage flow to time factor and coefficient of permeability

Case	Thickness of Sealing-up Layer	Time taken for Seepage flow to Appear	Time Factor Tv	Coefficient of Permeability (cm/sec)	
				k_1 *1	k_2 *2
1	0.5m	3days	0.12	$3.1*10^{-6}$	
2	1.0m	12days	0.12	$1.7*10^{-6}$	$7.0*10^{-7}$
3	1.5m	23days	0.10	$2.0*10^{-6}$	

*1: Calculated using the water head difference as negative pressure and the distance of permeation as sealing-up layer thickness.
*2: A falling head permeability test was conducted using sample soil after its water content was adjusted to 80%.

Table 2. Ground conditions

Soil Layer	Depth (layer thickness)	N Value	Water Content(%)	Cu (kN/m²)
Waste soil	~1.9m(1.9m)	3~4	-	-
Highly Organic Soil	~6.9m(5.0m)	2~4	400	19.1
Sand within Volcanic Ash Soil	~8.8m(1.9m)	8~11	46	174.0
Volcanic Cohesive Soil	~14.0m(5.2m)	9~25	80	108.0

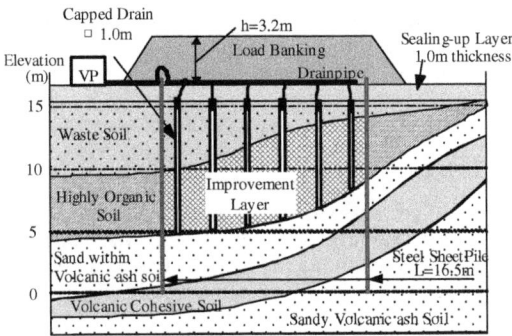

Figure 11. Outline

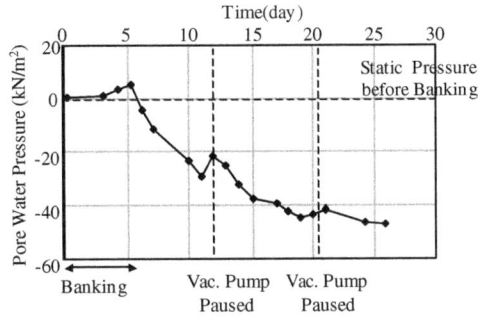

Figure 13. Change in pore water pressure with time

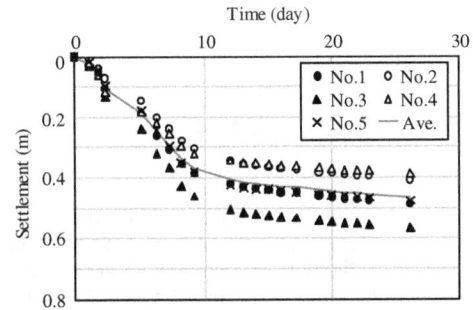

Figure 14. Settlement behavior

The improvement area, about 600 m², was enclosed with steel sheetpiles. Capped drains 7 m to 13 m long were used, depending on their positions on the ground, depth of which had been identified in a preliminary survey. The drains were installed to make 1-m squares. On installation of drains and pipes, a vacuum pump was activated. After two days of trial operation, banking was laid up to a thickness of 2.7 m at a daily banking rate of 0.9 m.

During banking, measurements were taken on the settlement of the improvement surface, displacement from the drain tank, applied load, and pore water pressure in the improvement ground (Figure12).

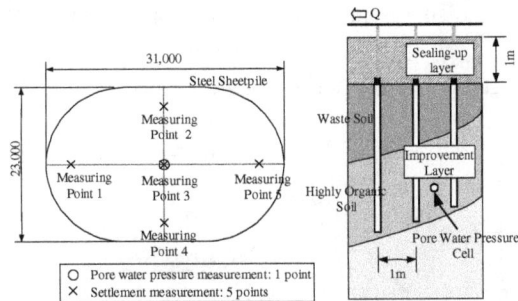

Figure 12. Measuring points

4.2 Results

Figure13 shows the behavior of the pore water pressure of clay between drains. The initial pore water pressure was set at 0 kN/m², that is, the static pressure prior to banking. The pore water pressure increased at first, but began to decline and finally reduced to -50 kN/m².

Figure14 illustrates settlement behavior. The average settlement was about 50 cm 30 days after the start of banking. The degree of consolidation during construction was estimated using the hyperbolic fitting method. In about 30 days following the start of banking, the value reached 90% at each measuring point. At this point of time, the maximum difference in residual settlement was judged to remain at about

2 cm, and thus we terminated the improvement process by stopping negative pressurization.

Table3 shows changes in water content and strength after improvement. A combined pressure of 116 kN/m² was applied in our experiment with banking plus negative pressure. The rate of increase in strength m (=$\Delta Cu/\Delta p$) as a result of improvement by consolidation was about 0.3. Meanwhile, the water content declined from 400% to 340%, confirming an improved effect. Still, some samples showed almost no change in water content. In the case of highly organic soil soil, it is generally difficult to determine water content precisely with a single experiment because results tend to vary widely due to uneven sampling.

Table 3. Improved effect

Property	Before Improvement	After Improvement
w(%)	396.2	339.2
Cu(kN/m²)	19.1	57.1

4.3 Summary

Our soil improvement experiment was conducted using the vacuum consolidation method together with banking, with the following results:

(1) The pore water pressure decreased from its initial value, that is, static pressure before banking, to the applied negative pressure of -50 kN/m^2, suggesting that a given pressure (banking plus negative pressure) was acting upon the ground.

(2) The degree of consolidation was controlled during construction using the hyperbolic fitting method. The improvement process was terminated once the degree of consolidation reached 90%. The improvement area settled by an average of 50 cm.

(3) The strength (cohesiveness) of the soil increased to 38 kN/m^2, with an increase in strength m = 0.3, thus showing an improvement.

After soil improvement by consolidation, the caisson was rebuilt. Almost no uneven settlement was found in it, confirming that this method would enable effective soil improvement.

5 CONSTRUCTION EXPERIMENT (2)

5.1 *Outline*

This section introduces a construction experiment where our proposed method was employed in order to accelerate settlement of an unconsolidated clay ground. Figure15 shows an outline and the columnar section of the target improvement ground, and Table 4 the results of our soil test on the clay layer.

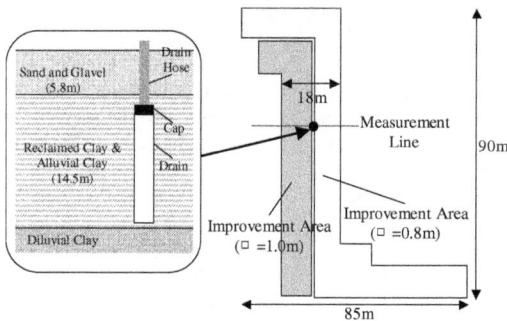

Figure 15. Outline

Table 4. Ground conditions

Layer	Depth	Cc	Cv(cm^2/day)	w(%)
1	7.0-7.8m	0.75	60	80.0
2	10.0-10.8m	1.15	30	62.1
3	13.0-13.8m	1.08	50	72.9
4	16.0-16.8m	1.17	30	78.4
5	19.0-19.8m	1.06	30	90.7

The ground stayed unconsolidated, still retaining excess pore water pressure in clay. There was much concern that a difference in level might be caused between the structure to be built in the future which will have the pile foundation and its surrounding un-

consolidated area. The clay layer was about 15 m thick, with a 6-m-thick surface composed of sand and gravel. The improvement area was about 2,600 m^2. Drains had been installed in 0.8-m squares at some parts and 1.0-m squares at others.

5.2 *Results*

Figure16 presents working negative pressure near the vacuum pump. Figure17 shows pore water pressure on drain tip and in the improvement area. A negative pressure of 50 to 70 kN/m^2 continuously worked until the termination of the improvement work. The pore water pressure on drain tip decreased by about 80 kN/m^2 immediately after the start of improvement. The pore water pressure dropped more than the working negative pressure probably because excess some pore water pressure remained in the ground. The pore water pressure of clay in the improvement area gradually declined to about 80 kN/m^2.

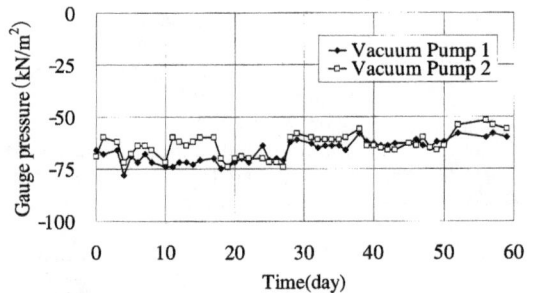

Figure 16. Change in working negative pressure with time

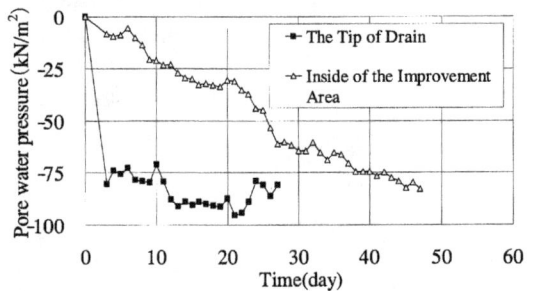

Figure 17. Change in pore water pressure with time

Figure18 illustrates how the improvement area changed its cross-sectional form. On the whole, the area settled to form a U-shape. Even at a point 15.0 m away from an edge of the improvement area, the ground settled by about 10 cm. Figure19 shows changes in horizontal displacement near the edge of the improvement area. The clay layer was horizontally displaced by up to about 30 cm on the improvement side. This was presumably attributable to inward isotropic stress generated during improvement by negative-pressure consolidation.

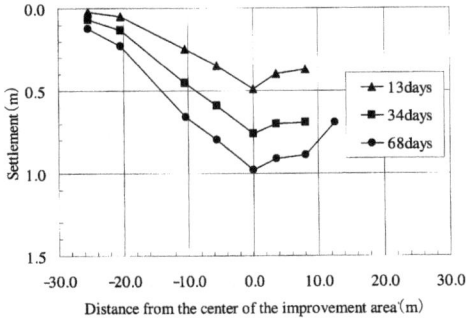

Figure 18. Cross-sectional change

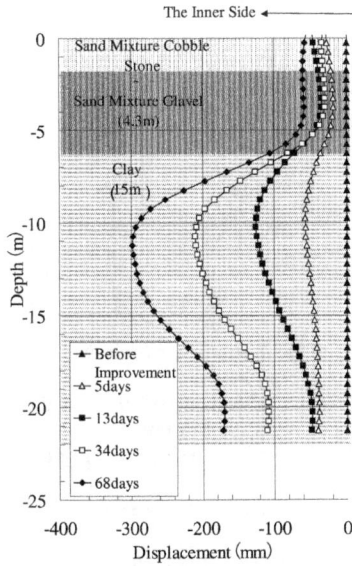

Figure 19. Horizontal displacement

Figure20 provides a settlement curve plotted along the longitudinal measurement line. At the end of application of negative pressure, the ground had settled by 0.91 m in the center of the improvement area. To compare measurements with the time-settlement curve obtained by the Cc method and

Barron's solution, combined with the results of a consolidation test, actual settlement gradually decreased to about 70% of the calculated value for the same period of time in the center of the improvement area. In our test, the area of improvement was small, compared with the depth of improvement. In such cases, actual settlement may be less than one-dimensional settlement due to inward isotropic improvement.

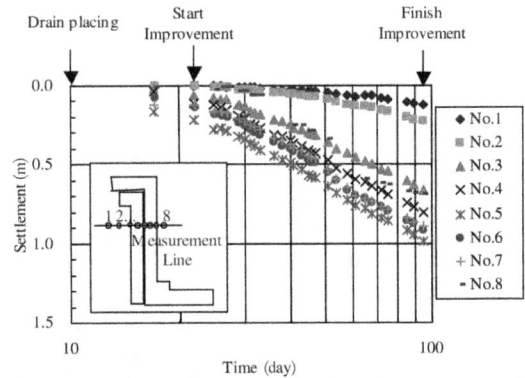

Figure 20. Settlement curve on the longitudinal measurement line

Analysis by the finite element method (FEM) was conducted to see how the improvement area settles and is transformed in accordance with a change in the width-depth ratio of the area. The mesh used in analysis is shown in Figures21 and 22. In the analysis, the soil constant was computed based on the results of a soil test on the layer in which samples were taken from five different depths, while assuming that the sand layer on the surface was elastic (Table5).

The negative pressure generated by vacuum consolidation was calculated at an assumed pore water pressure of 80 kN/m² below static pressure. Analysis was conducted on four cases of an improvement depth of 15 m–(1) one-dimensional model, (2) improvement width of 10 m, (3) improvement width of 18 m and (4) improvement width of 30 m.

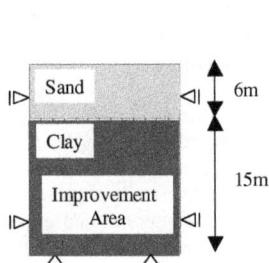

Figure 21. Mesh analysis(One-dimensional)

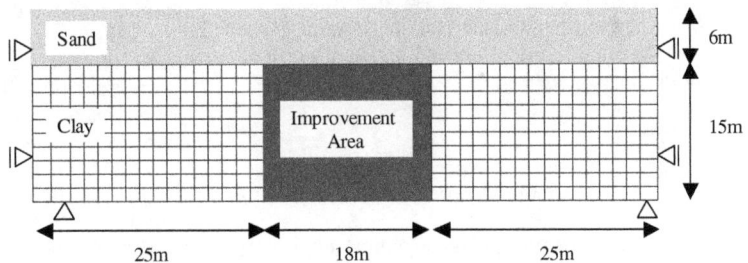

Figure 22. Mesh analysis (Two-dimensional)

Table 5. Material properties

Layer	Depth	λ	κ	M	k (cm/sec)
1	6-9m	0.53	0.053	1.2	1.0×10^{-6}
2	9-12m	0.50	0.050	1.2	1.0×10^{-7}
3	12-15m	0.47	0.047	1.2	1.0×10^{-7}
4	15-18m	0.51	0.051	1.2	1.0×10^{-7}
5	18-21m	0.46	0.046	1.2	1.0×10^{-7}

Table 6. Final settlement

Thickness of Improvement Area	Width of Improvement Area	Final Settlement in The Center of Improvement Area S_f
15m	One Dimension	1.29m
15m	10m	0.76m (59%)
15m	18m	0.95m (74%)
15m	30m	1.29m (100%)

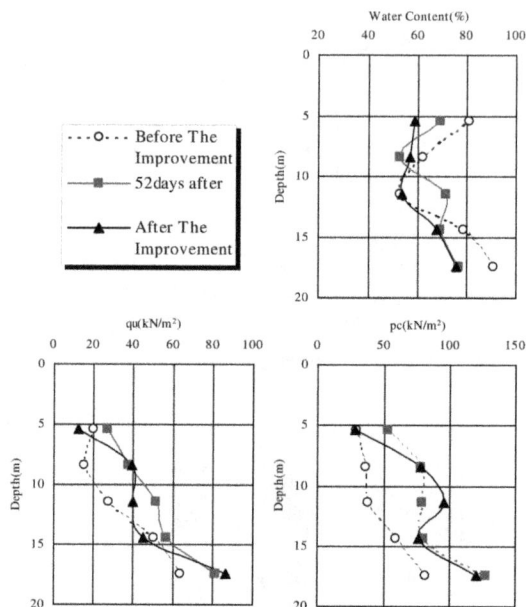

Figure 23. Improved effect

Table6 shows the final settlement in the center of the improvement area for each case. Compared with one-dimensional consolidation, which recorded a final settlement of 1.29 m, settlement reduced in partial improvement as the improvement width decreased. At an improvement width of 30 m, the final settlement equaled the corresponding value for one-dimensional consolidation, while at an improvement width of 18 m, the final settlement was declined to 74%. This suggests that, when the improvement width is small relative to the improvement depth, settlement may decrease in the improvement area.

Figure23 shows the result of drilling examination in an improvement area. The examination was carried out before and after the improvement and 52 days after the start of the improvement.

There is a clay layer of which the water content is increasing in comparison with that at the preliminary survey, it might be caused by a difference of a sampling position and the disturbance by sampling.

Also, a clay layer deeper than 15m has the coefficient of consolidation cv as small as 20~25cm²/day. So, compared with the upper clay layer, progress of consolidation is slow.

The unconfined compressive strength has increased by about 20 kN/m² in general, although there is a part with little increase. The rate of increase in strength m (=ΔCu/Δp) is about 0.2~0.3, this is considered to be a standard increase tendency of marine clay in Japan.

5.3 Summary

Our ground-improvement experiment using the vacuum consolidation method produced the following results:

(1) Applied negative pressure almost constantly stayed at 60 kN/m². The pore water pressure on the drain tip decreased immediately after operation was started. The pore water pressure finally decreased in the improvement area to almost the same level as on the drain tip.

(2) In settlement, the following events were recognized:
1. A decrease in settlement relative to theoretical one-directional settlement.
2. Horizontal displacement into the improvement area.
3. Occurrence of settlement outside the improvement area.

FEM analysis also proved that, when improvement width is small relative to improvement depth, settlement tends to decrease in the center of the improvement area.

(3) A decrease in settlement is caused probably by distortion due to isotropic vacuum consolidation.

6 CONCLUSION

In this report, we have presented the results of a laboratory test and a field experiment that employed the vacuum consolidation method using capped drains. The findings proved that this method is effective in reducing soil volumes.

The vacuum consolidation method produces a stress distribution different from one seen in the surcharge method because it generates a consolidation load to work on drains in all directions. It is reported that the upper layers of the improvement area are under isotropic stress, while the lower layers undergo K_0-consolidation (one-dimensional consolidation), restricted on lateral pressure [3]. Still there are several problems to be solved in designing this method:

(1) A convincing stress distribution must be established to identify an interface between K_0-consolidation and the consolidation by isotropic stress.

(2) The method needs to be compatible with seepage consolidation phenomena.

(3) It is required to clarify the range of decreasing settlement due to effects of the surrounding non-improvement area

(4) It is also necessary to identify surrounding areas that settle in line with consolidation progressing from the non-improvement area to the improvement area.

In the future, we will endeavor to develop technology for large-scale construction.

REFERENCES

Watari, Y., Shinsha, H. and Hayashi, K., 1984. Experiment on the Vacuum Consolidation Method Using Plastic Drains. *Ground and Construction*. vol.2, No.1: 33-40.

Takano, Y., Shinsha, H., Watari, Y. and Sato, H., 1988 In Situ Consolidation Test of a Clay Layer Installed Vertically Drains by. *Proc. of the 23rd Japan National Conference Soil Mechanics and Foundations Engineering*. 2133-2134.

Suzuki, S., Umezaki, T., Kawamura, T and Iizuka, T., 2002 Characteristics of Vacuum Consolidation and Undrained Shear of Saturated Clay under Anisotropic Stress State. *Proc. of the 37th Japan National Conference on Geotechnical Engineering*. 239-240.

Soft Ground Engineering in Coastal Areas, Tsuchida et al. (eds)
© 2003 Swets & Zeitlinger, Lisse, ISBN 90 5809 613 0

Research on the coefficient of earth pressure at rest (K_0) for a CPG improved ground

K. Zen
Department of Civil Engineering, University of Kyushu, Fukuoka, Japan

M. Ikegami & T. Masaoka
Kanto Regional Development Bureau, Ministry of Land, Infrastructure and Transport, Yokohama, Japan

T. Fujii & M. Taki
Fukken Company Limited, Consulting Engineers, Hiroshima, Japan

ABSTRACT: The authors have been continuously doing research on the ways of application of compaction grouting (CPG) technique to the sandy soil ground-improvement against liquefaction under seismic forces. From the authors' previous experimental works, it is found that the CPG increased N-value, bulk density, coefficient of earth pressure at rest (K_0), and cyclic shear strength $(\sigma_d/\sigma_c')_{20}$, proving itself sufficiently counteracting against liquefaction. The design method that has been proposed in the previous study for the purpose of preventing liquefaction in CPG technique was based on the same principle that was used in the sand compaction pile method (SCP). This proposed design procedure takes into consideration the increased liquefaction strength accompanied by the increased bulk density, but it is without consideration of the increased liquefaction strength that results from the increase in K_0. In order to confirm the effect of K_0 on liquefaction, the results from the various geotechnical investigations, carried out before and after improvement works, were re-evaluated. This paper examines some ways of introducing K_0 to the design on the basis of the actual measurements of K_0 at the improved ground.

1 INTRODUCTION

Compaction grouting (CPG) is known to increase the coefficient of earth pressure at rest K_0 and the soil bulk density, which in turn increase the liquefaction strength; however, a design method that takes into account such increased coefficient K_0 has yet to be established. Experimental CPG with different improvement rates was carried out in the Tokyo International Airport Project to determine and compare pre- and post-CPG coefficients of earth pressure at rest. As it was anticipated that the coefficient at the center differs from that at the edge of a ground-improved area due to different boundary conditions, the effect of CPG at the edge of the improved area was examined.

2 EXPERIMENTAL WORK

2.1 *Strata and soil profiles of ground*

The experimental CPG was carried out in the soil under an old taxiway of the Tokyo International Airport. The stratigraphy of the improve area area and the soil properties of the main stratum to be improved were as follows.

1) Stratigraphy: The top layer was a compacted subgrade about 1.5 m thick, and lying under the subgrade was a fill layer 0.8 to 1.3 m thick. Lying further below was a 7.0 to 8.2 meter-thick sandy layer of dredged sludge, which was the main layer to be improved and called A_{s0}. Lying below it was a clayey layer A_{C1}. The groundwater level was in the fill layer. Accordingly, grout columns 9 m long were driven downward from the depth of 2.8 m below the ground surface.

2) Soil properties of A_{s0}: Table 1 shows the soil properties of the pre-CPG layer A_{s0}. Figure 1 shows the grain-size distribution curves of the layer A_{s0}, revealing its fines content F_c (finer than 75 µm) was 10-50%. This means that the sandy soil had the grain sizes nonuniformly distributed.

Table 1. Physical properties of A_{s0}.

ρ_s(g/cm^3)	D_{50}(mm)	I_p(%)	U_c	e_0	e_{max}	e_{min}
2.71	0.2	7.5	11.8	0.90	1.53	0.80

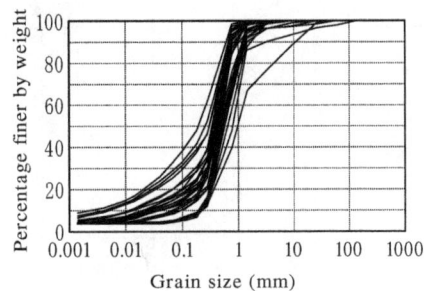

Figure 1. Grain size distribution curves of A_{s0}.

2.2 Experimental cases

The experimental CPG was carried out adopting four cases shown in Figure 2. Cases 1, 2, and 3 were designed to examine the relationship between the rate of improvement (displacement rate) a_s and the effect resulting from the improvement (improvement effect). Cases 2 and 4 were meant to conduct the comparison of the ground behaviors between two different grouting methods (bottom up method and top down/bottom up method).

2.3 Soil tests

The following in-situ tests and soil-property tests were conducted before and after the experimental CPG. See Figure 2 as for the test locations.

The parameters tested were the design parameters used in Japan for the prediction of liquefaction and important criteria to estimate the improvement effect resulting from CPG.

1) Standard penetration test (SPT) to measure depth-wise N-values at one-meter interval

2) Soil property test to determine bulk density ρ_t, grain-size distribution, fine-grain content F_c, plasticity index I_p, etc.

3) Density log in borehole to measure depth-wise density values continuously using a radioisotope

4) Self-boring pressure meter test (SBP) in soil to measure coefficients of earth pressure at rest K_0 in situ to evaluate the confining effect of grout columns' lateral pressure on the lateral movement of soil

5) Cyclic undrained triaxial test (liquefaction test) of soil to determine the cyclic shear strength $(\sigma_d/\sigma_c')_{20}$ by using undisturbed soil specimens. The cyclic shear stress ratio is the value of $\sigma_d/2\sigma_c'$ when liquefaction occurs at the 20th loading under the condition that the double amplitude of axial strain DA is 5% (σ_d is the amplitude of cyclic deviator stress; σ_c', the isotropic confining pressure).

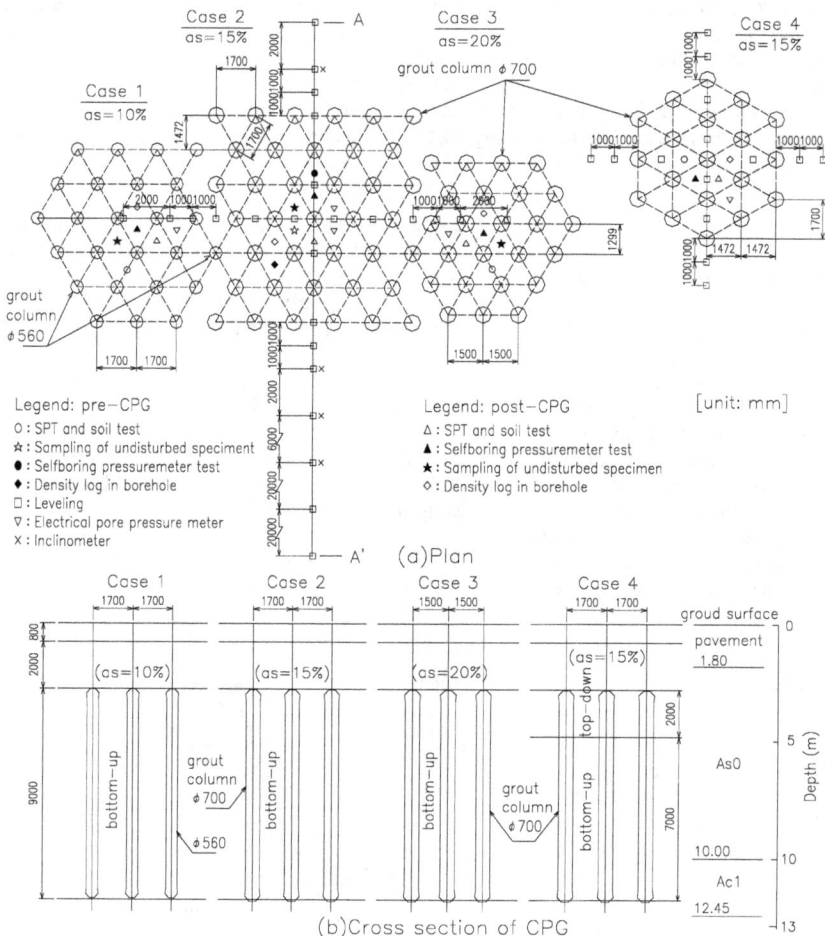

Figure 2. Details of experimental work.

286

3 RESULTS OF EXPERIMENTAL WORK

The following soil parameters are directly related to the design criteria for against liquefaction in Japan.

3.1 N-values

Figure 3 shows the depth-wise distribution of pre- and post-CPG N-values of each case. The fines content F_c is also plotted in the figure. It is apparent from the figure that the N-values were increased as a result of the CPG. It appears that as the degree of compaction and the ground improvement rate (displacement rate) a_s increase, the increment of the

N-value increases. Although the effect of the fine grain content F_c on the compaction degree or the N-value is uncertain from these data, it was ascertained that the sandy soil was compacted by the CPG.

3.2 Soil parameters

Figure 4 shows the depth-wise distribution of pre- and post-CPG ρ_t, K_0, and $(\sigma_d/\sigma_c')_{20}$ of each case.

1) Bulk density ρ_t of soil: The bulk density was increased by the CPG in every case. Because of the fact that the measured values of bulk density were dispersed depth-wise and the layer A_{S0} was inhomogeneous, it was difficult to evaluate the increment of the bulk density quantitatively.

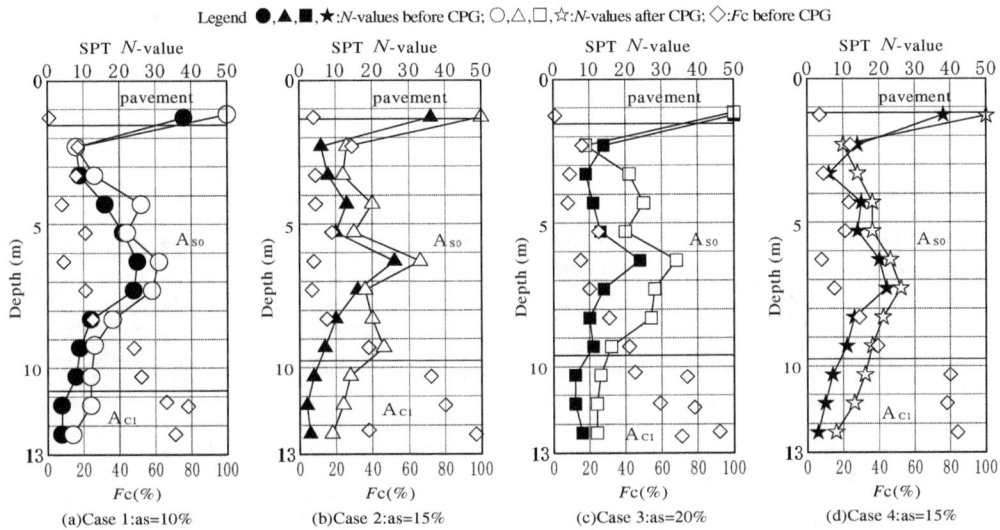

Figure 3. N-values before and after CPG.

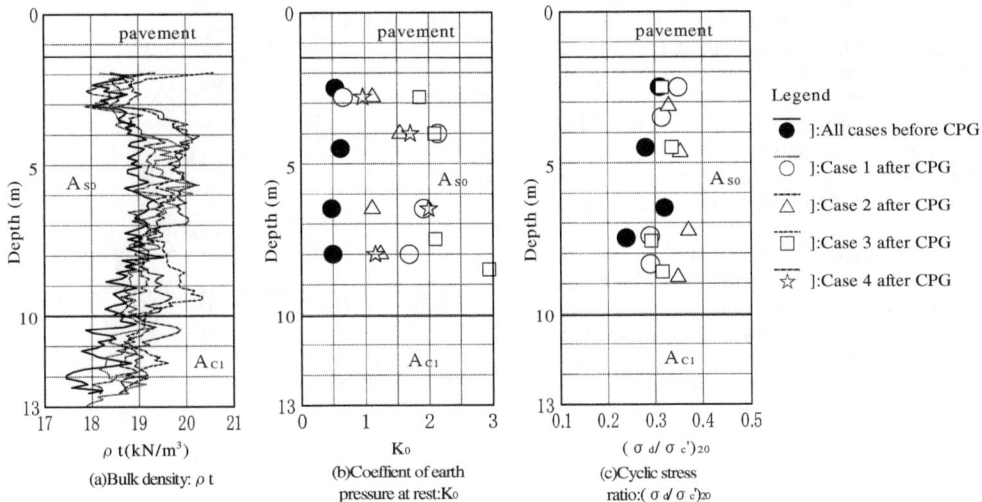

Figure 4. Soil properties before and after CPG.

2) Coefficient of earth pressure at rest K_0: The pre-CPG coefficient ranged from 0.45 to 0.60; the post-CPG coefficient, 0.65 to 2.95. As the pre-CPG mean coefficient of the layer A_{S0} had a common value of 0.5, these measured values seem to be reasonably accurate. Accordingly, the increase of K_0 was due to the increase of lateral pressure caused by the CPG. The dispersion of K_0, when taken into account the increment of K_0 as a result of the CPG, is estimated to be 0.5 to 1.5.

3) Cyclic shear strength $(\sigma_d/\sigma_c')_{20}$: This was apparently increased by the CPG, but the increment of the ratio was not so salient as those of the N-value and K_0. The reason behind it may be the effects of the isotropic consolidation ($K_0 = 1.0$) in the cyclic undrained triaxial test. It is generally known that as K_0 increases, the cyclic shear strength increases. Therefore, the cyclic shear strength shown in Figure 4 (c) may be underestimated.

4 EVALUATION OF INCREMENT OF N-VALUE WHEN K_0 INCREASES

Figure 5 shows the corrected pre- and post-CPG N-values of each case. The correction was made by the method proposed by Tokimatsu, Yoshimi et al. The corrected N-value increased from about 20 before the CPG to 30–40 after the CPG. As shown in the figure, a tendency is seen that the higher the improvement rate (displacement rate), the larger the increment of the N-value will be. Figure 6 and Figure 7 show the pre- and post-CPG cyclic shear strength $(\sigma_d/\sigma_c')_{20}$ and coefficient of earth pressure at rest K_0, respectively. The CPG increased R_{l20} by 0.05–0.1 to 1.2–1.3 times the pre-CPG value, whereas the CPG increased K_0 by 0.5–1.5 to 2–4 times the pre-CPG value. In Figure 7, the K_0 values determined one and a half years after the CPG and those determined three years after the CPG are indicated, too. The latter values remained nearly the same level as the former ones.

Of the increment of the corrected N-value of Figure 5, the portion attributable to the increment of bulk density was estimated by using the relationship between the corrected N-value and cyclic shear strength by Tokimatsu, Yoshimi et al., and the increment of the corrected N-value of Figure 5 minus the portion attributable to the increment of bulk density was assumed to be attributable to the increment of K_0. Figure 8 shows the bulk density-related increment and the K_0-related increment. As shown in the figure, the bulk density-related increment is 0.5–4.0, or 20–30% of the gross increment, whereas the

K_0-related increment is 3–15, or 70–80% of the gross increment. Figure 9 shows the relation between ΔK_0 (increment of K_0) and ΔN_{k0} (increment of N-value due to ΔK_0). A relatively good correlation was observed between them and their correlation is represented by the equation below.

$$\Delta K_0 = 0.143 \Delta N_{K0} \tag{1}$$

Figure 10 shows the relationship between the improvement rate a_s and K_0. As shown in the figure, the pre-CPG K_0 is about 0.5 and K_0 tends to increase as the improvement rate increases. The tendency may be affected by the initial void ratio and the fines content; however, it was ascertained that an improvement rate of 10% or more brings about a K_0 of 1.0 or more under the soil conditions of the present study.

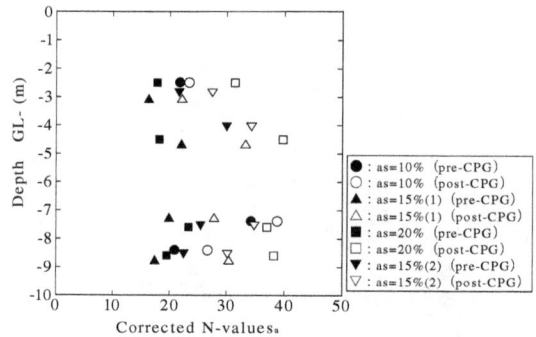

Figure 5. Pre-and post-CPG corrected N-values.

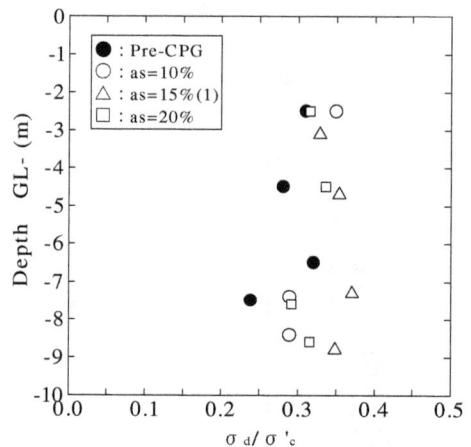

Figure 6. Pre- and post-CPG σ_d/σ'_c.

288

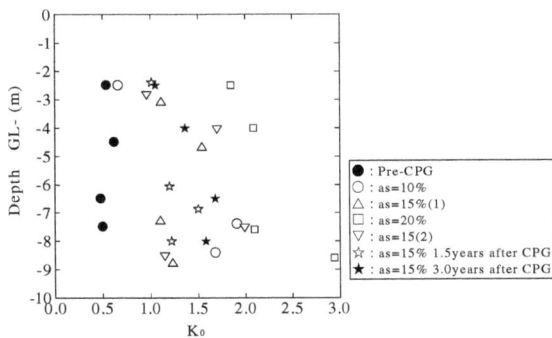

Figure 7. Pre-and post –CPG K_0.

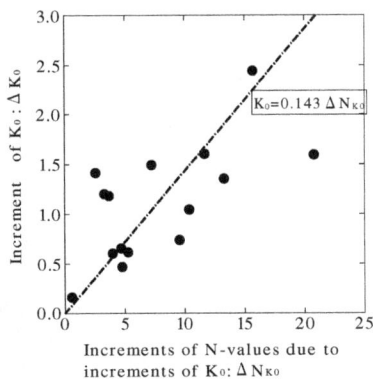

Figure 9. Relationship between increments of K_0 and N-value.

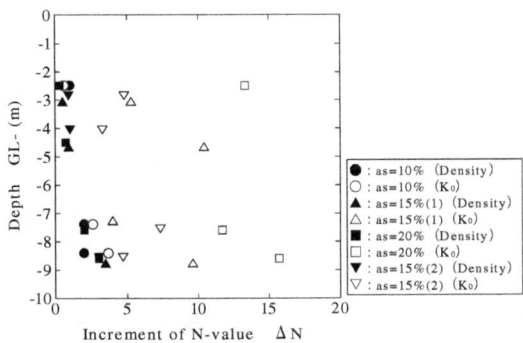

Figure 8. Increments of N-values.

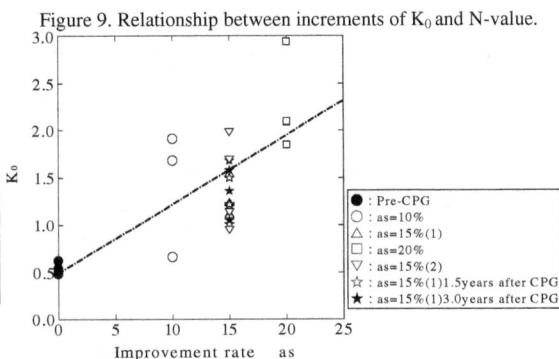

Figure 10. Relationship between as and K_0.

Table 2. Soil properties of stratum to be improved and specifications of CPG.

Stratum	N-value	Fines content Fc	Plasticity index Ip	Improvement rate a_s	Diameter of grout column	Interval
Bs	4 - 9	40 - 52%	29.1 - 37.8	11%	68.3mm	1.7m
As1	2 - 3	39 - 66%	0 - 11.3	15%	69.2mm	1.7m
As2	5 - 21	21 - 54%	0 - 12.2	8%	50.5mm	1.7m

5 IMPROVING EFFECT AT EDGE OF IMPROVEMENT AREA

The standard penetration test, self-boring pressure meter test, cyclic undrained triaxial test (liquefaction test), and physical-property test of soil were conducted at the four spots shown in Figure 11 to ascertain the improvement effect at the edge of the improved area. The edge tests were conducted at a distance of 85 cm from the center of the outermost grout column.

Table 2 shows the soil properties of the layers to be improved and the specifications of CPG. The layers are of silty sand as shown in the table. Figure 12 shows a sectional view of strata of the area to be improved and pre- and post-CPG N-values. According to a geologic survey by boring, the A_{s1} layer was discontinuous and the B_s and A_{s2} layers were almost horizontal and continuous. The N-value was over 50 at several spots, it is supposed to be due to the fact that the grout spreading in the form of veins was struck.

Figures 13-15 show the relationship between the distance from the edge of the improvement area and soil parameters. The N-values of Figure 13 are equivalent ones (N_{65}) free of the effects of the pressure of structures on the ground and the fines content F_c. As shown in Figure 13, the equivalent N-value N_{65} at the edge was 5-10 lower than that at the first or outermost grout column. The equivalent N-value was fairly even in the improved area within the outermost grout columns. As shown in Figure 14, the

289

cyclic shear strength $(\sigma_d/\sigma'_c)_{20}$ at the edge and that in the improved area within the outermost grout columns were almost the same. The pre-CPG K_o values as well as the post-CPG ones are shown in Figure 15. As shown in the figure, a tendency was observed that the K_o value increased as the testing point moved from the edge toward the center of the improved area. The K_o value was below 1.0 on the average in the zone from the edge to the second grout columns, whereas the same was over 1.0 on the average within the third grout columns. Thus, the changes of K_o value occurred around the third grout column, or a distance of four meters from the edge; however, the K_o value was over 0.8 even at the edge.

As for the N_{65} and R_{l20}, salient difference of improvement effect between the edge and the center of the improved area was not observed. It is supposed to be due to the fact that the fines content F_c of the soil was as much high as 50%.

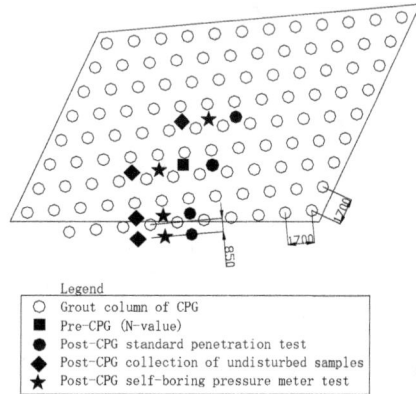

Legend
○ Grout column of CPG
■ Pre-CPG (N-value)
● Post-CPG standard penetration test
◆ Post-CPG collection of undisturbed samples
★ Post-CPG self-boring pressure meter test

Figure 11. Locations of test.

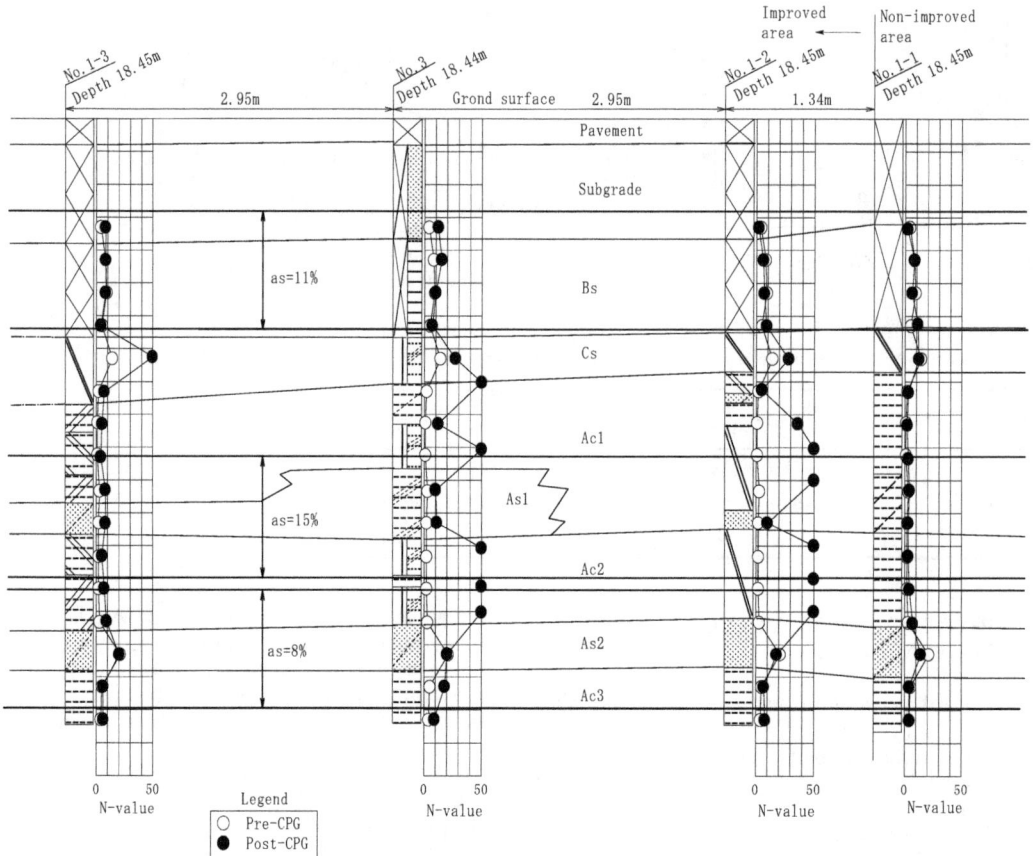

Figure 12. Section of stratum and pre- and post-CPG N-values.

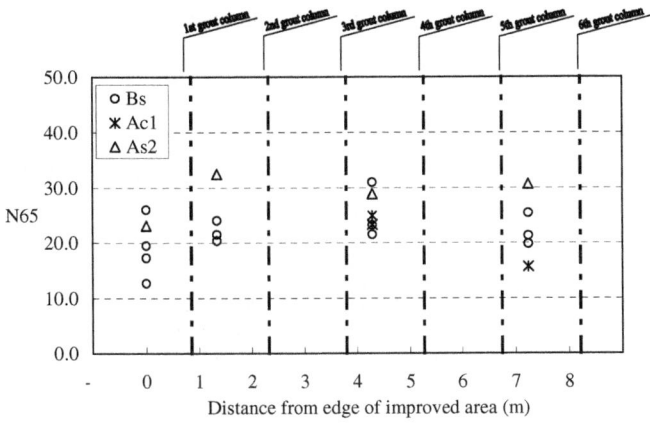

Figure 13. Relationship between N_{65} and distance from edge of improved area.

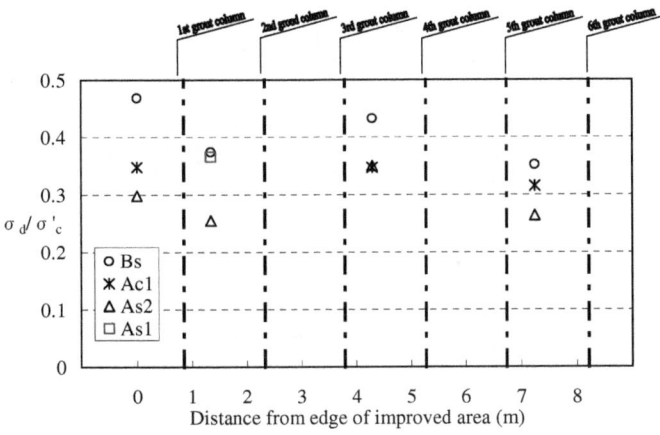

Figure 14. Relationship between σ_d/σ'_c and distance from edge of improved area.

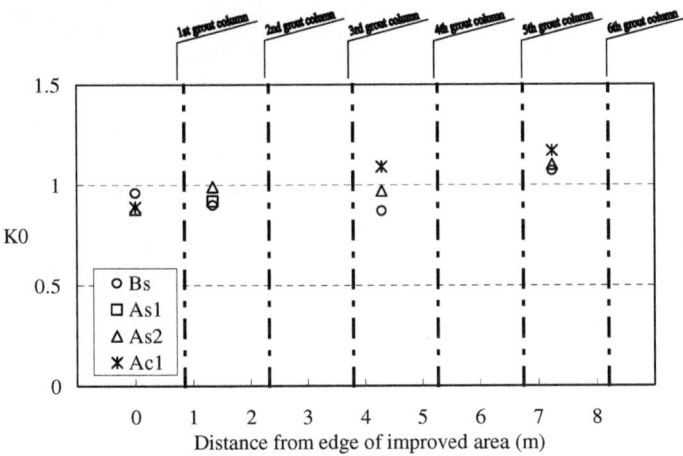

Figure 15. Relationship between K_0 and distance from edge of improved area.

6 CONCLUSIONS

The findings derived from the present study are as follows.

- The K_o values of three years after the CPG were as high as those immediately after the CPG.
- The increment of K_o accounted for 70-80% the increment of the N-value.
- If the improvement rate is 10% or more, a K_o value of 1.0 or more can be secured.

The improvement effect of the CPG differed from the edge to the center of the improved area. This tendency was most salient in the case of K_o, K_o being around 0.8 at the edge and over 1.1 at the center of the improved area.

REFERENCES

S. Yamaguchi, D. Kozawa, M. Arata, H. Matsumoto, M. Taki and Y. Kanno 2000. Design and construction method of compaction grouting as a ground-improving technique against liquefaction. *Coastal Geotechnical Engneering in Practics*, pp.557-560.

K. Tokimatsu and Y. Yoshimi 1983. Empirical correlation of soil liquefaction based on SPT N-value and fines content. *Soils and Foundations,* vol. 23, No. 4, pp.56-74.

Reuse of dredged soils & behavior of coastal structures under earthquake

Soft Ground Engineering in Coastal Areas, Tsuchida et al. (eds)
© *2003 Swets & Zeitlinger, Lisse, ISBN 90 5809 613 0*

Applicability of CPT for construction control in coastal area

T. Fukasawa
Technical Research Institute, TOA Corporation, Yokohama, Japan

ABSTRACT: It is essential for Cone Penetration Test (CPT) application to consider the cone factor, which correlates the shear strength of cohesive soil with the cone resistance, since CPT only provides index parameter such as cone resistance. Aiming for the extensive use of CPT, this paper firstly describes the cone factors based on theoretical analysis and empirical correlation through in-situ tests. The paper also discusses the cone factors obtained from various types of laboratory tests and correlation between cone factors, plasticity index, over consolidation ratio and rigidity index. Then, an example of the effective use of the CPT as a means of construction control of a seawall on consolidating soft ground improved by sand drain method is presented. It can be concluded through this study that the cone factors obtained from in-situ tests, which are not relate to plasticity index, over consolidation ratio and rigidity index, are fall within the 7~18 range and cone factor for the ground under consolidation is the same as that for naturally deposited soil.

1 INTRODUCTION

Soft cohesive soil is deposited in thick layers in the coastal areas of Japan. The ground investigation is conducted prior to the construction of structures on such soft ground to examine various civil engineering problems. The quality of such investigations affect significantly on the amount of construction cost.

Up to now, field vane test (FVT) has been extensively used in Western Europe and North America, and unconfined compression test (UCT) has been often used in Japan for undisturbed samples, as a test method for determining undrained shear strength (S_u) of the cohesive soil. The strengths obtained from these tests are influenced by local factors such as the soil properties, the level of technical expertise and so on. The actual application of the data has been left to the discretion of experienced engineers in each region or country. As the reduction of construction cost is needed, it is required to establish a test method, which provides data with necessary and sufficient accuracy at a low cost throughout the whole process from the investigation to the construction control.

In CPT, cone resistance, sleeve friction and pore water pressure are continuously measured at a regular time interval, while a cone probe at the end of rod is pushed into the ground at a fixed speed. It has been pointed out that advantages of the CPT are (i) the simplicity and cost effectiveness of the testing method, (ii) the measured information is readily available on site and (iii) index of strength obtained

from this test is unaffected by soil types. The following is a comparison of the cost between CPT and UCT in cohesive soil. As far as the cost of the test itself is concerned, the cost of CPT is about 1/5 of that of UCT, which needs undisturbed samples using thin-wall sampling after boring. The test period of CPT is about 1/5~1/10 of that of UCT. Furthermore, CPT provides continuous data in direction of depth whereas the data of UCT is usually taken at an interval of 1m. However, in order to interpret the CPT data to deduce design soil parameters, a correlation between CPT data with other either in-situ or laboratory test results has to be established. Many researchers have so far attempted to deduce S_u of cohesive soils from measured cone resistance by utilizing a correlation between S_u and cone resistance for a given site (e.g. Tanaka and Tanaka (1996), Konrad and Law (1987)). Estimation of S_u from CPT using cone resistance is commonly made from the following equation.

$$S_u = (q_t - \sigma_{vo})/N_{kt} \qquad (1)$$

where q_t is corrected cone resistance for pore water pressure effects, σ_{vo} is in-situ total vertical stress and N_{kt} is termed cone factor.

It is noted that the cone factor is important when studying the cone penetration mechanism and interpreting the results of CPT applied to cohesive soil ground.

This paper describes the cone factors obtained from previous studies based on theoretical analysis and empirical correlation through tests. Then, the results of site investigation with CPT on Japanese marine clays are shown. The paper also discusses the cone factors obtained from various types of laboratory tests and correlation between cone factors, plasticity index (I_p), over consolidation ratio (OCR) and rigidity index (I_R). Furthermore, an example of the effective use of the CPT as a means of construction control of a seawall on consolidating soft ground improved by sand drain method is presented. The paper finally shows that relationship between the cone resistance and S_u is not affected by the state of consolidation of the ground.

2 THEORETICAL ANALYSIS OF CONE FACTOR

Numerous interpretations of cone factor are based on either theoretical analysis or empirical correlation

through tests. The theoretical analysis method may be classified into the following five categories:
1 Bearing capacity theory
2 Cavity expansion theory
3 Strain path theory · Steady state approach
4 Incremental finite-element method
5 Analytical and numerical approaches using linear and non-linear stress-strain relationships

The theoretical analysis results in the following equation of the cone resistance (q_c) and the S_u, with the cone factor (N_c). σ_i is the in-situ total stress; the overburden pressure (σ_{vo}), the horizontal stress (σ_{ho}) and the mean stress ($\sigma_{mean} = (\sigma_{vo} + 2\sigma_{ho})/3$ are used.

$$q_c = N_c \cdot S_u + \sigma_i \qquad (2)$$

Figure 1 shows the correlation between the theoretical solution of cone factor and I_R reported by many researchers. In the Figure, the cone resistance to obtain the cone factor is q_t (q_c) $-\sigma_{vo}$ and the theoretical solution using σ_{mean} and σ_{ho} was excluded. In

Table 1. Cone factors obtained from previous studies

Country	Site	Cone factor	S_u	q	Investigation depth	Thickness of clay layer	Plasticity index	Reference
Canada	Champlain	9~18	FVT	q_c	10~27m	10~20m	6~45	Tavenas et al. (1982)
	Champlain	14~16	FVT	q_c	10m	10m	5~37	Roy et al. (1982)
	Champlain	7.5~11.5	FVT	q_t	20~25m	20m	5~40	Konrad and Law (1987)
	Champlain	10~17.5	FVT	q_t	5~8m	5~8m	10~40	La Rochelle et al. (1988)
	Vancouver	8~9	CAUC	q_t	23m	–	9~16	Rad and Lunne (1988)
USA	Chicago et al.	8~25	UCT	q_c	21m	21m	15~42	Eid and Stark (1998)
	California	8.5~16.5	UU	q_c	–	–	8~54	
	Massachusetts, Louisiana	9~12	FVT	q_c	–	–	22~70	Jamiolkowski et al. (1982)
Scandinavia	Norway, Sweden, Denmark	8~22	FVT	q_c	12~35m	11~33m	5~55	Lunne and Eid (1976)
Norway	Drammen, Onsoy	12~19	FVT	q_c	20m	16~19m	10~36	Lacasse and Lunne (1982)
	Onsoy et al.	12~21	$S_{u(LAB)}$	q_t	–	–	3~50	Aas et al. (1986)
	Drammen et al.	8.5~21	CAUC	q_t	7~25m	–	4~46	Rad and Lunne (1988)
Denmark	Niverod, Niva	4~12	FVT	q_c	7m	7m	10~15	Denver (1988)
	Yoldia et al.	8.5~12	CAUC	q_t	1.25~2.4m	–	7~137	Luke (1995)
UK	Easington et al.	12.5~20.5	UU	q_c	16~22m	16~22m	10~25	Nash and Duffin (1982)
	Cowden et al.	13~30.5	Tri	q_t	18~20m	16~20m	12~54	Powell and Quarterman (1988)
	Cowden, Brent	12~27	CIUC	q_t	15~18m	–	16~49	Rad and Lunne (1988)
Ireland	Athlone	6.5~35	FVT	q_c	11m	10m	10~63	O'riordan et al. (1982)
	Belfast	10.5~18.5	FVT	q_t	11m	7m	14~56	
	Belfast, Cavan	4~14	FVT	q_t	6~9m	5~7m	10~75	Faulkner et al. (1998)
North Sea	Sleipner, Gullfaks	9~19.5	CAUC	q_t	33~50m	18~28m	18~32	Lunne et al. (1985)
	Troll, Brage	8~24	CAUC	q_t	30~42m	–	19~41	Rad and Lunne (1988)
Korea	Pyeongtaek, Youngam	8.5~32	FVT,CIUC	q_t	9~16m	9~16m	22~31	Lee et al. (1998)
Japan	Kurihama et al.	9~14	FVT	q_t	14~40m	11~40m	20~150	Tanaka and Tanaka (1996)
		8~16	UCT	q_t				
Singapore		14.5~21	CFVT	q_t	32m	30m	35~75	Dobie (1988)
Hong Kong	Chek Lap Kok et al.	15.5~21	FVT	q_t	28m	28m	28~39	Newman et al. (1995)
Thailand	Bangkok	11~17	CFVT	q_t	17~21m	14~19m	41~78	Shibuya et al. (1998)
		7.8	DST	q_t	21m	19m	41~78	
Itary	Porto Tolle	8~16	FVT	q_c	30m	10m	30	Jamiolkowski et al. (1982)
Brazil	Rio de Janeiro	11~27	CAUC	q_t	10m	–	60~87	Rad and Lunne (1988)
	Rio de Janeiro	8~18	FVT	q_t	11m	8m	40~200	Almeida (1998)

Cone factor:$N_k, N_{kt} = (q - \sigma_{vo})/S_u$; FVT:field vane shear test; CFVT:field vane shear test with correction;
UU:unconsolidated undrained triaxial compression test; CIUC:consolidated isotropically undrained triaxial compression test;
CAUC:consolidated anisotropically undrained triaxial compression test; UCT:unconfined compression test;
$S_{u(LAB)}$:average undrained shear strength, =$(S_{uc}+S_{uD}+S_{uE})/3 \fallingdotseq S_{u(D)}$; Tri:triaxial test (D98mm)

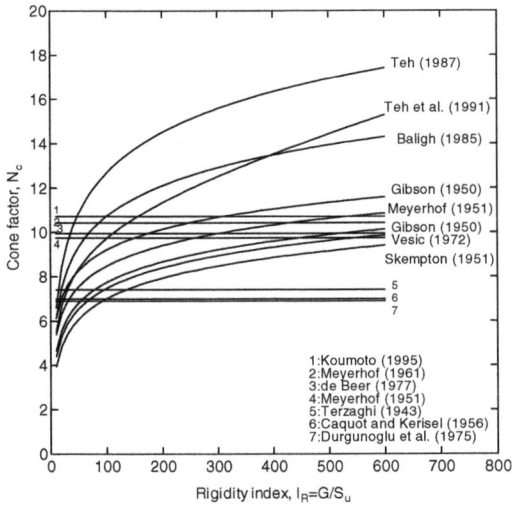

Figure 1. Relation between cone factor and I_R

this paper, the cone factor obtained from the theoretical solutions explained earlier is defined as N_c, and those from tests are defined as N_k $(=(q_c-\sigma_{vo})/S_u)$ and N_{kt} $(=(q_t-\sigma_{vo})/S_u)$.

Many analysis models have been developed to theoretically explicate the cone penetration mechanism. However, as the ground transformation around the cone is complex during penetration, and the in situ is peculiar, for example, in the non-homogeneity of sedimentary stratification. Thus, a conventional solution is yet to be made.

3 CONE FACTOR OBTAINED FROM EMPIRICAL CORRELATION

The theoretical analysis of the cone penetration process is difficult. Therefore, attempts have been made to empirically obtain the correlation between the cone resistance and S_u. In Western Europe and North America where the CPT is advanced, numerous reports on cone factors have been submitted in various countries since 1970's. Table 1 shows the cone factors gained through tests using the electric cone penetrometer in Western Europe, North America, East and Southeast Asia. Fukasawa and Kusakabe (2001) show that cone factors listed here reflect local characteristics of each region and country and none of them are dependent on I_p. Cone factor is considered to change with different S_u values. Mesli (2001) suggests that it is useful to attempt a direct correlation between cone resistance and S_u mobilized in full-scale in stability situations. Cone factor is also affected by the level of technical expertise. It may be noted that the cone factor needs to be examined with reference to the most up-to-date data available at the time.

In Figs. 2 and 3, N_{kt} values are plotted versus I_p and OCR of the cohesive soil at 10 locations in Japan shown in Table 2. In order to obtain N_{kt}, three types of shear strength are used; namely $S_{u(DST)}$ from direct shear test (DST), $S_{u(FVT)}$ from FVT and $q_u/2$ from UCT. They are described as $N_{kt(DST)}$, $N_{kt(FVT)}$ and $N_{kt(UCT)}$. Figure 4 shows the frequency distribution of N_{kt}. Figures (a), (b) and (c) show the frequency distribution of $N_{kt(DST)}$, $N_{kt(FVT)}$ and $N_{kt(UCT)}$, respectively, and (d) shows their sum total. The number of samples (n), the mean values (mean) and the coefficient of variation (V) are also included in the figures. The dispersion of N_{kt} is the smallest in DST, and that of N_{kt} in FVT and UCT is large. The values of N_{kt} of the cohesive soil of 10 different locations in Japan are 7~15 (Mean=10.8) in DST, 7~16 (Mean=11.6) in FVT and 8~18 (Mean=13.0) in UCT; as illustrated in Figs. 2 and 3, they are constant regardless of I_p or OCR. This means that, unlike the $q_u/2$ and $S_{u(FVT)}$, the cone resistance is not affected by the soil property. However, it is necessary to check the local cone factors, as they do exist, in case of detailed investigations. Figure 5 indicates the relationship between the cone factor from field investigations and I_R. The figure includes N_c ob-

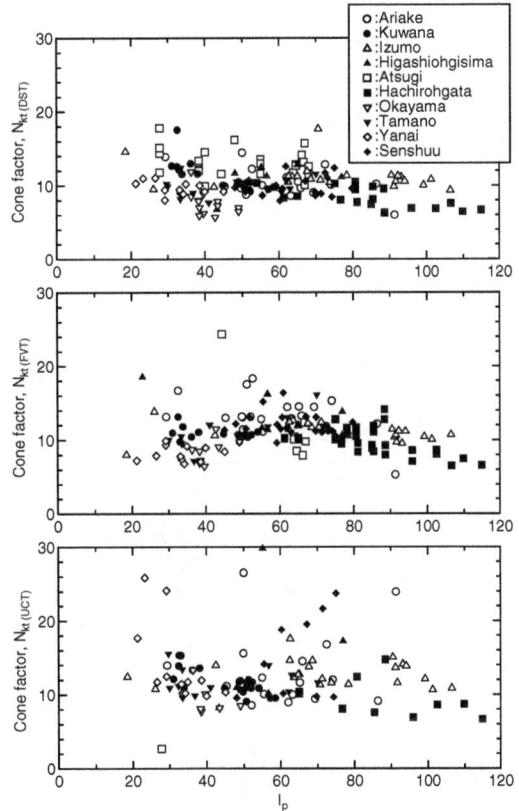

Figure 2. Relation between cone factor and I_p

tained from the theoretical methods previously explained and the measurement values of the cone factors shown in Table 1 are also added. Within the I_R's range of 15~250, no clear correlation is observed between the I_R and the cone factor obtained from the in situ test results.

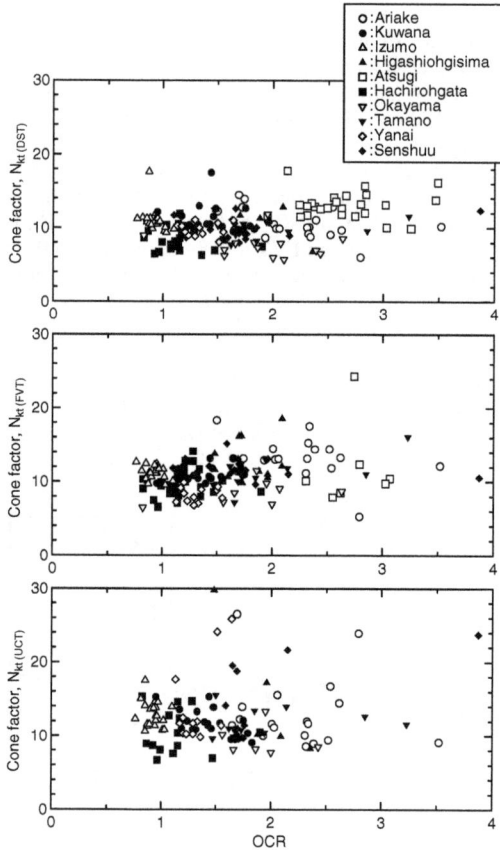

Figure 3. Relation between cone factor and OCR

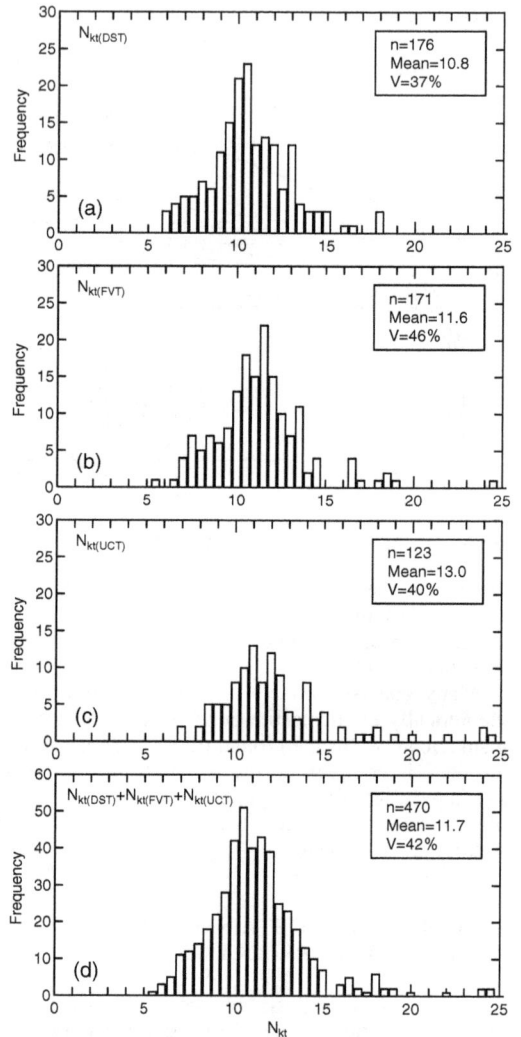

Figure 4. Frequency distribution of cone factor

Table 2. Investigation site (10 locations in Japan)

Site	Ip	OCR	Thickness of clay layer
Ariake	45~91	1.7~3.5	18m
Kuwana	31~59	1.2~1.8	20m (below 10m thick of sand layer)
Izumo	18~26 63~107	0.8~1.1	27m (below 8m thick of sand layer)
Higashiougishima	30~62	1.5~2.4	15m (below 5m thick of sand layer)
Atsugi	28~68	2.0~3.5	18m
Hachirohgata	29~152	0.9~1.9	40m
Okayama	36~49	1.5~2.6	7m (below 4m thick of sand layer)
Tamano	30~57	1.5~3.2	13m
Yanai	21~45	1.1~1.6	13m (below 6m thick of sand layer)
Senshuu	45~80	1.4~4.0	21m

4 APPLICATION OF CPT FOR CONSTRUCTION CONTROL

Stability and deformation are major geotechnical considerations on soil structures on soft seabed constructed by reclamation, typically with a soil improvement method such as vertical drain method. Effective construction control of such soil structures requires a systematic way of collecting of accurate information about the gain in soil strength and the progress of consolidation. A common testing method for evaluating the gain in strength of the improved ground in Japan is UCT using undisturbed samples. In-situ tests are not typically used. The gain in soil strength is evaluated by comparing the unconfined compression strengths after and before the ground

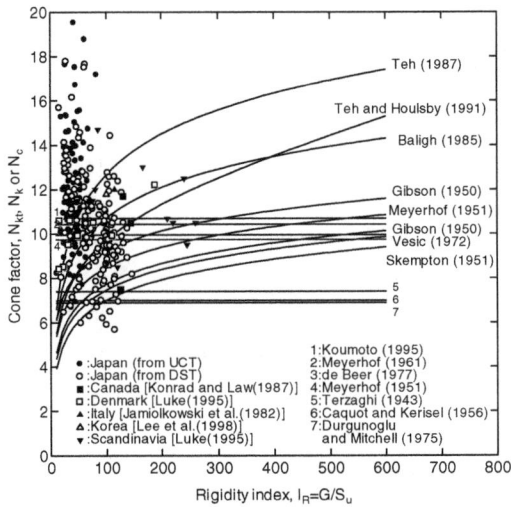

Figure 5. Relation between cone factor and I_R from in-situ test

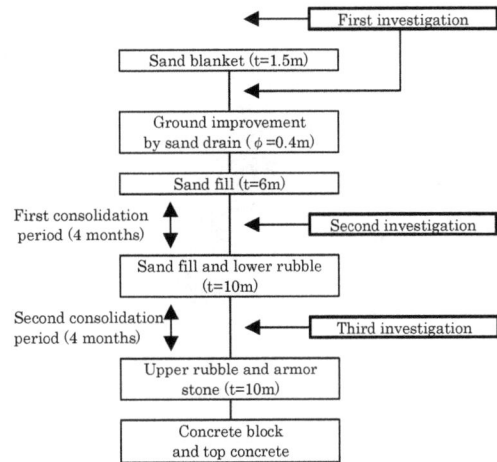

Figure 7. Flowchart of construction and investigation period

improvement. The average degree of consolidation is calculated by comparing of the measured settlement, to the estimated final settlement. Since the soil strength and the degree of consolidation are separately evaluated as the above, actual construction control of soil structures on improved ground is rather complicated. It is therefore desirable and more practical if a simple in-situ testing method can be adopted, which could simultaneously give information about the gain in strength as well as the degree of consolidation for the purpose of construction control. In this study, as a speedy and economical means for construction control, CPT was extensively used to monitor the gain in strength of improved ground and the degree of consolidation during the period of construction.

4.1 An outline of the project

The seawall construction is a part of Kansai International Airport 2nd phase project. Figure 6 shows a typical cross section of the seawall. A soft Holocene

clay layer extends to the depth of about –45 m as shown in Fig. 6. The seawall embankment is composed of rubble mounds and sand fills. Construction sequence of the seawall is illustrated in Fig. 7, in which the time required for consolidation between each stage is indicated. Soil investigation was carried out at three different stages, which are also shown in Fig. 7. Sand piles of about 21 m long were driven into the Holocene clay layer at intervals of either 1.6 m×2.5 m right under the seawall or 2.5 m×2.5 m in the reclamation side in a square pattern. The sand pile reaches at thin sand layer below the Holocene clay layer. The increase in vertical effective stress due to the first stage of the sand fill ($\Delta\sigma'_{(1st)}$) was about 75 kN/m^2 at the center of the seawall. The second stage of the sand fills further increased the vertical effective stress ($\Delta\sigma'_{(2nd)}$) for about 100 kN/m^2.

4.2 Soil properties at the site

The first investigation was conducted prior to the construction. Table 3 tabulates various types of test conducted in this study and their testing methods.

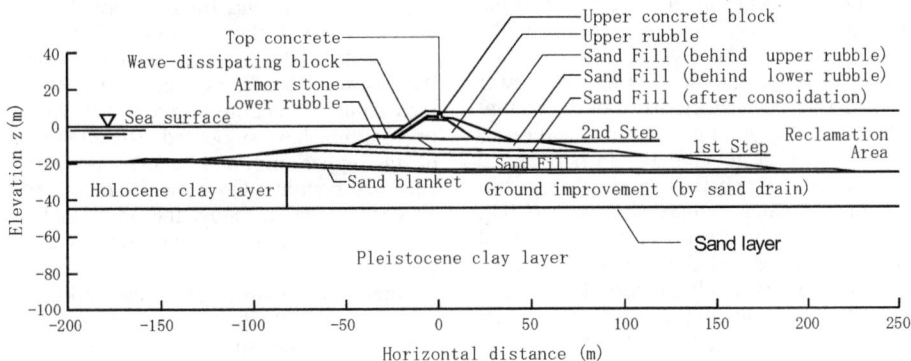

Figure 6. Typical cross section of the seawall

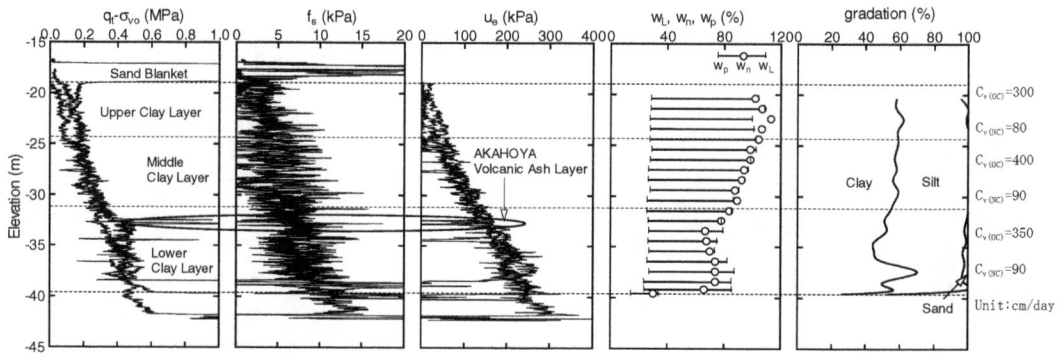

Figure 8. Results of the first investigation

Table 3. Testing methods conducted in this study

Items			Test method	Derived values
First investigation	In situ	CPT	JGS 1435	q_t-σ_{vo}, f_s, u_d
		FVT	JGS 1411	$S_{uf(FVT)}$
	Laboratory	UCT	JGS 0511	q_u
		DST-1	JGS 0560 recompression stress: σ'_{vo}	$S_{uf(DST)}$
		CIU	JGS 0523 recompression stress: 2/3 σ'_{v}	$S_{u(CIU)}$
		DST-2	JGS 0560	$S_{u(DST)}$, $S_{un(DST)}/\sigma'_{v}$
		Consolidation test	JGS 0412	c_v, m_v, C_c
		Physical test	each method	ρ_s, ρ_t, w_n, w_L, w_p
Consolidation period	In situ	CPT	JGS 1435	q_t-σ_{vo}, $\Delta\sigma$, f_s, u_d
		Dissipation test with CPT	measuring pore water pressure	U
	Laboratory	DST-1	JGS 0560 recompression stress: σ'_{vo}	$S_{uf(DST)}$
		UCT	JGS 0511	q_u

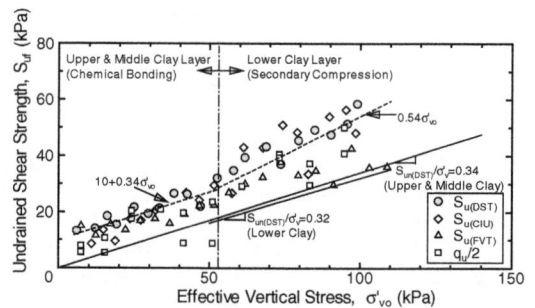

Figure 9. Relation between S_{uf} and σ'_{vo}

Figure 8 shows the profiles of the CPT data with depth of Holocene clay before the construction. The average thickness of the Holocene clay at the construction site is 21.2 m. The cone resistance in terms of q_t-σ_{vo}, the sleeve friction (f_s) and the excess pore water pressure (u_e) response, together with consistency limits and the grain size information are shown. Coefficient of consolidation (c_v) is also described in the Fig. 8. As indicated in Fig. 8, the Holocene clay layer may be divided into three layers: the upper clay, the middle clay and the lower clay, by considering the characteristics of the natural water content (w_n), the liquid limit (w_L), u_e and c_v.

In-situ undrained shear strengths (S_{uf}) obtained from various shear tests, DST, isotropically consolidated undrained triaxial compression test (CIU), FVT and UCT, are plotted versus effective vertical stress in Fig. 9, together with the undrained strength increment ratio in normally consolidated state (S_{un}/σ'_{vc}) as obtained from DST-2. The in-situ direct shear strength ($S_{u(DST)}$) in Fig. 9 may be expressed as a function of the effective vertical stress as indicated

in the figure. The intersection of these two equations approximately corresponds to the effective vertical stress at the depth of –31 m, which corresponds to the boundary between the middle and lower layers.

It has been recognized that S_{uf} varies with types of the shear test adopted, because of strength anisotropy and strain rate dependency of S_u. Values of S_{uf} plotted in Fig. 9 varies depending on the type of shear test, but it is noted that most of the data lie above the S_{un}/σ'_{vc} line. It is therefore considered that all the three clay layers at the construction site are over-consolidated state. The clay layers at the construction site have no historical evidences of experiencing overburden pressures. It is thus considered that the clays are classified as normally consolidated aged clay (NCA clay). Hanzawa and Adachi (1983) discussed that NCA clays have been subjected to either chemical bonding or secondary compression, or their combined action. Figure 9 is a typical example of what they have suggested. The additional strength of the upper and middle clay is mainly due to chemical bonding. Once the effective vertical stress exceeds a certain value, the additional strength of lower clay develops stemmed from secondary compression.

Shear strength for design use mobilized in full-scale in stability situations of soft clay ($S_{u(mob)}$) are often determined from shear strength obtained from laboratory tests, which are modified by a correction factor. Various recommendations have been made and some of which are given below.

$$S_{u(mob)} = 0.85 S_{u(DST)} \quad \text{(Hanzawa (1992))} \qquad (3\text{-}1)$$

$$= 0.75 S_{u(CIU)} \quad \text{(Tsuchida et al. (1989))} \quad (3\text{-}2)$$

$$= q_u/2 \qquad (3\text{-}3)$$

$$= S_{u(FVT)} \qquad (3\text{-}4)$$

Using the relationships in Eq. 3, the cone resistance versus $S_{u(mob)}$ are plotted in Fig. 10, from which a good correlation is found between the cone resistance and $S_{u(mob)}$, with the cone factor of 12 as is given in Eq. 4.

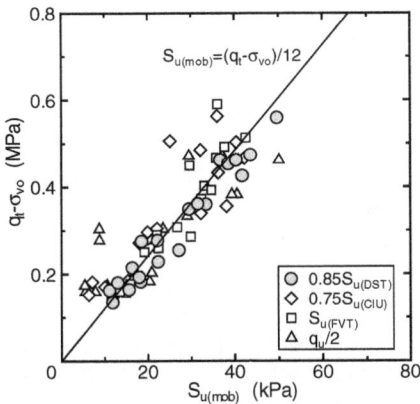

Figure 10. Relation between q_t-σ_{vo} and $S_{u(mob)}$

$$S_{u(mob)} = (q_t - \sigma_{vo})/12 \qquad (4)$$

The value of 12 for the cone factor is in good agreement with the data on several Japanese clays, as previously mentioned.

In order to predict strength increment of the improved ground, S_{un}/σ'_{vc} is in need. The mean values of S_{un}/σ'_{vc} for design use, which can be calculated from S_{un}/σ'_{vc} shown in Fig. 9 and Eq. 3-1 are 0.29 for the upper and middle clay, and 0.27 for the lower clay, respectively.

4.3 Strength increase and degree of consolidation in the ground improved by sand drain

To examine the change in S_u of the consolidating clay ground improved by sand drains, CPT and laboratory tests on undisturbed samples were conducted during the second and third soil investigations. During the consolidation, investigations were performed at three months and four months after the start of the consolidation period. Results of the investigation are shown in Fig. 11 with respect to cone resistance, $S_{u(mob)}$ and the degree of consolidation with depth. The data obtained from the first investigation before construction is also shown. The cone resistance in terms of q_t-$(\sigma_{vo}+\Delta\sigma)$ is presented versus depth in Fig. 11 (a). It can be confirmed that cone resistance steadily increases with the progress of consolidation. It is also noticed in the figure that at the second soil investigation, the values of cone resistance below the depth of –40 m show virtually no increase since the first soil investigation before construction. The reason for this is that in-situ vertical effective stress increased by the first stage of the sand fill $(\Delta\sigma_{(1st)} \doteqdot 75 \text{ kN/m}^2)$ is nearly equal to the consolida-

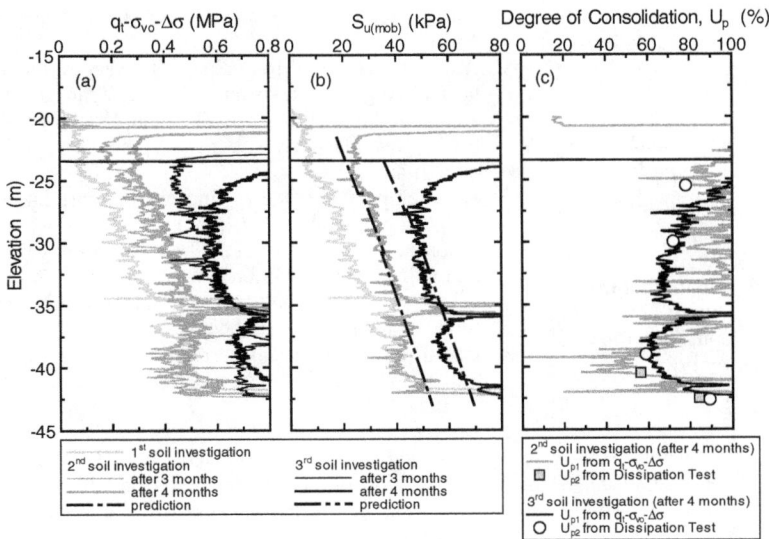

Figure 11. Results of CPT during consolidation period

tion yield stress of the clay around that depth. This observation is consistent with the suggestion of Hanzawa and Adachi (1983), that is to say that S_u does not exhibit a significant increase until effective vertical stress reaches at consolidation yield stress. Figure 11 (b) shows $S_{u(mob)}$ calculated from cone resistance using Eq. 4. It is evident that $S_{u(mob)}$ increases with the progress of consolidation, the same trend as cone resistance with time. In Fig. 11 (b), the predictions of $S_{u(mob)}$ with depth are also drawn by dotted lines based on the discussions in the previous section. It is seen that these predictions agree well with the results of CPT both for the second and third soil investigation, confirming that $S_{u(mob)}$ can be accurately estimated from CPT during the process of consolidation.

In practice, the average degree of consolidation (U_ε) is used for the purpose of construction control for structures on soft clay, where the average degree of consolidation is calculated from the settlement of the ground measured by the settlement plate, representing the average value of the degree of consolidation throughout the clay layer in terms of strain. In CPT, the degree of consolidation is obtained by measuring the dissipation of excess pore water pressure with time while the cone remains stationary (U_{p1}) and by comparison with cone resistance after sand fill and increased effective vertical stress (U_{p2}). The dissipation tests were conducted at four depths during the second investigation and at two depths during the third investigation, respectively, shown in Fig. 11 (c). U_{p1} and U_{p2} are the degree of consolidation, which are based on the stress. It can be seen in Fig. 11 (c) that the values of U_{p1} are close to the values of U_{p2}. The average values of U_p and U_ε were 84 % and 87 % at the end of the first consolidation period (second soil investigation), and were 73 % and 77 % at the end of the second consolidation period (third soil investigation). The observed relationship between U_p and U_ε is the same as the one demonstrated by Mikasa (1963).

4.4 Correlation of cone resistance with undrained shear strength of consolidating ground

The data of $q_u/2$ during consolidation are plotted against cone resistance in terms of $q_t-(\sigma_{vo}+\Delta\sigma)$ in Fig. 12, together with the results obtained from the first investigation. It is seen in the figure that the degree of scattering increases with increasing undrained shear strength due to consolidation. But the data fall within the lines given by Eq. 5.

$$q_u/2 = (q_t - (\sigma_{vo} + \Delta\sigma))/(7 \sim 18) \qquad (5)$$

It is thus considered that the relationship between cone resistance and undrained shear strength is not affected by the state of consolidation of the ground. Although the scatter of our data indicates slightly different from data of Tanaka and Sakagami (1989),

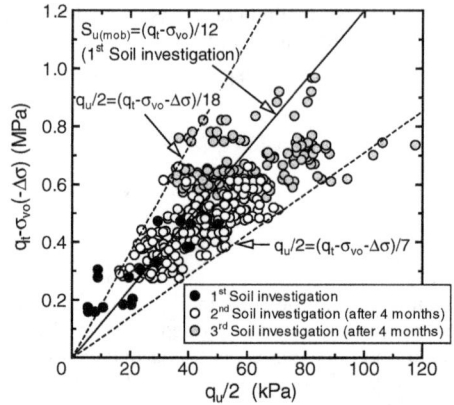

Figure 12. Relation between $q_t-\sigma_{vo}-(\Delta\sigma)$ and $q_u/2$ during consolidation period

the result of investigation is practically identical with the one reported by them. In CPT, S_u of the ground under consolidation is made accurately obtained from a relationship between cone resistance and S_u from the result of first investigation before construction.

5 CONCLUSIONS

The following conclusions can be made from this study:

1 N_{kt} of the cohesive soil from 10 locations in Japan with the I_p values within the range of 20 and 115 and the OCR values between 0.8 and 3.5 is constant, regardless of I_p and OCR values. N_{kt} from DST is between 7 and 15 (mean value = 10.8), N_{kt} from FVT between 7 and 16 (mean value = 11.6) and N_{kt} from UCT between 8~18 (mean value = 13.0).

2 Although affected by values of shear strength normalizing cone resistance, N_{kt} in Japan generally falls within the scope of 7 and 18. N_{kt} or N_k of the low plasticity soil in Northern Europe is quite high.

3 An attempt was made to find out the correlation of the cone factor with I_R from the theoretical solution and the in-situ test. The result shows that no evident correlative tendency is observed in the values between I_R and the cone factor within the scope of I_R (15~250) in the cohesive soil.

4 Based on the empirical correlation, gain in S_u of the ground during construction is accurately evaluated by CPT.

5 Degree of consolidation of the ground during construction can be estimated by CPT. The degree of consolidation obtained from dissipation test agrees well with the one calculated from cone resistance.

6 Relationship between the cone resistance and S_u is not affected by the state of consolidation of the ground. N_{kT} for the ground under consolidation is the same as that for naturally deposited soil.

7 The result of this study clearly shows that CPT is a very useful investigation method and has high applicability for construction control.

REFERENCES

Aas,G., Lacasse,S., Lunne,T. and Hoeg,K. 1986. Use of in situ tests for foundation design on clay. Proceedings of the ASCE Specialty Conference In Situ'86, Use of In Situ tests in Geotechnical Engineering, 1-30.

Almeida,M.S.S. 1998. Site characterization of a lacustrine very soft Rio de Janeiro organic clay. Proceedings of the First International Conference on Site Characterization-ISC'98, 2, 961-966.

Baligh,M.M. 1985. Strain path method. Journal of Soil Mechanics and Foundation Division, ASCE, 111(9), 1180-1136.

Caquot,A. and Kerisel,J. 1956. Traite de mecanique des sols. Imprimerie Gauthier-Villars, Paris.

De Beer,E. 1977. Static cone penetration testing in clay and loam. Sondeer Symposium, Utrecht.

Denver,H. 1995. CPT in Denmark-National Report. Proceedings of International Symposium on Cone Penetration Testing, CPT'95, 1, 55-61.

Dobie,M.J.D. 1988. A study of cone penetration testing in the Singapore marine clay. Proceedings of the first International Symposium on Penetration Testing, 2, 737-744.

Durgunoglu,H.T. and Mitchell,J.K. 1975. Static penetration resistance of soil: Analysis. Proceedings of ASCE specialty Conference on In Situ Measurement of Soil Properties, ASCE, 1, 151-171.

Eid,H.T. and Stark,T.D. 1998. Undrained shear strength from cone penetration test. Proceedings of the First International Conference on Site Characterization-ISC'98, 2, 1021-1025.

Faulkne,A.R., Lehane,B.M. and Farrel,E.R.I. 1998. Cone penetration testing in Irish soils. Proceedings of the First International Conference on Site Characterization-ISC'98, 2, 1033-1038.

Fukasawa,T. and Kusakabe,O. 2001. A history of development of cone penetration tests with reappraisal of interpretation methods and its applicability to clay soils. Technical Report, Department of Civil Engineering Tokyo Institute of Technology, 64, 23-60.

Gibson,R.E. 1950. Discussion of Wilson,G. The bearing capacity of screw piles and screwcrete cylinders. Journal of the Institution of Civil Engineers, 34, 382.

Hanzawa,H. 1992. A new approach to determine soil parameters free from regional variations in soil behaviour and technical quality. Soils and Foundations, 32(1), 71-84.

Hanzawa,H. and Adachi,K. 1983. Overconsolidation of alluvial clays. Soils and Foundations, 23(4), 106-118.

Jamiolkowski,M., Lancellotta,R., Tordella,L. and Battaglio,M. 1982. Undrained strength from CPT. Proceedings of the Second European Symposium on Penetration Testing, 2, 599-606.

Konrad,J.M. and Law,K.T. 1987. Undrained shear strength from piezocone tests. Canadian Geotechnical Journal, 24, 392-405.

Koumoto,T. 1995. On the foundation methods for soft Ariake clay. Journal of JSIDRE, 63(6), 587-592.

Lacasse,S. and Lunne,T. 1982. Penetration tests in two Norwegian clays. Proceedings of the Second European Symposium on Penetration Testing, 2, 661-669.

La Rochelle,P., Zebdi,M., Leroueil,S., Tavenas,F. and Virely,D. 1988. Piezocone tests in sensitive clays of eastern Canada. Proceedings of the First International Symposium on Penetration Testing, 2, 831-841.

Lee,S.J. and Kim,M.M. 1998. Estimations of geotechnical properties from piezocone penetration tests in Korea. Proceedings of the First International Conference on Site Characterization-ISC'98, 2, 1099-1104.

Luke,K. 1995. The use of cu from Danish triaxial tests to calculate the cone factor. Proceedings of International Symposium on Cone Penetration Testing, CPT'95, 2, 209-214.

Lunne,T., Christoffersen,H.P. and Tjelta,T.I. 1985. Engineering use of piezocone data in North Sea clays. Proceedings of the 11th International Conference on Soil Mechanics and Foundation Engineering, 2, 907-912.

Lunne,T. and Eide,O. 1976. Correlations between cone resistance and vane shear strength in some Scandinavian soft to medium stiff clays. Canadian Geotechnical Journal, 13, 430-441.

Mesri,G. 2001. Undrained shear strength of soft clays from push cone penetration test. Geotechnique, 51(2), 167-168.

Meyerhof,E. 1961. The ultimate bearing capacity of wedge-shaped foundations. Proceedings of the 5th International Conference on Soil Mechanics and Foundation Engineering, 2, 105-109; and 3, 193-195.

Meyerhof,G.G. 1951. The ultimate bearing capacity of foundations. Geotechnique, 1, 301-332.

Mikasa,M. 1963. The consolidation of soft clay, - a new consolidation theory and its application -. Kajima Shuppankai, Tokyo. (in Japanese)

Nash,D.F.T. and Duffin,M.J. 1982. Site investigation of glacial soils using cone penetration tests. Proceedings of the Second European Symposium on Penetration Testing, 2, 733-738.

Newman,R., Wood,R., Berner,P., Covil,C. and Ng,N. 1995. CPT testing at Hong Kong's new airport at Chek Lap Kok. Proceedings of International Symposium on Cone Penetration Testing, CPT'95, 3, 227-239.

O'riordan,N.J., Davies,J.A. and Dauncey,P.C. 1982. The interpretation of static cone penetration tests in soft clays of low plasticity. Proceedings of the Second European Symposium on Penetration Testing, 2, 755-760.

Powell,J.J.M. and Quarterman,R.S.T. 1988. The interpretation of cone penetration tests in clays, with particular reference to rate effects. Proceedings of the First International Symposium on Penetration Testing, 2, 903-909.

Rad,N.S. and Lunne,T. 1988. Direct correlations between piezocone test results and undrained shear strength of clay. Proceedings of the First International Symposium on Penetration Testing, 2, 911-917.

Roy,M., Tremblay,M., Tavenas,F. and La Rochelle,P. 1982. Development of a quasi-static piezocone apparatus. Canadian Geotechnical journal, 19, 180-188.

Shibuya,S., Hanh,L.T., Wilailak,K., Lohani,T.N., Tanaka,H. and Hamouche,K. 1998. Characterizing stiffness and strength of soft Bangkok clay from in-situ and laboratory tests. Proceedings of the First International Conference on Site Characterization-ISC'98, 2, 1361-1366.

Skempton,A.W. 1951. The bearing capacity of clays. Proceedings, Building Research Congress, London, I, 180-189.

Tanaka,Y. and Sakagami,T. 1989. Piezocone testing in underconsolidated clay. Canadian Geotechnical Journal, 26, 563-567.

Tanaka,H. and Tanaka,M. 1996. A site investigation method using cone penetration and dilatometer tests. Technical Note of the Port and Harbour Research Institute Ministry of Transport, Japan, 837, 1-52 (in Japanese).

Tavenas,F., Leroueil,S. and Roy,M. 1982. The piezocone test in clays: Use and limitations. Proceedings of the Second European Symposium on Penetration Testing, 2, 889-894.

Teh,C.I. 1987. An analytical study of the cone penetration test. PhD thesis, Oxford University, Oxford, U.K.

Teh,C.I. and Houlsby,G.T. 1991. An analytical study of the cone penetration test in clay. Geotechnique, 41(3), 17-34.

Terzaghi,K. 1943. Theoretical soil mechanics. John Wiley & Sons, Inc.

Tsuchida,T., Mizukami,J., Oikawa,K. and Mori,Y. 1989. New method for determining undrained strength of clayey ground. Report of the Port and Harbour Research Institute Ministry of Transport, Japan, 28(3), 81-145 (in Japanese).

Vesic,A.S. 1972. Expansion of cavities in infinite soil mass. Journal of the Soil Mechanics and Foundation Division, ASCE, 98(3), 265-290.

Soft Ground Engineering in Coastal Areas, Tsuchida et al. (eds)
© 2003 Swets & Zeitlinger, Lisse, ISBN 90 5809 613 0

Prediction method of liquefaction-induced settlement of remedied embankment with deep mixing method

M. Okamura, M. Ishihara & K. Tamura
Public Works Research Institute, Tsukuba, Japan

ABSTRACT: In order to mitigate liquefaction-induced embankment settlement, remedial measures are often implemented for existing embankments. This paper describes calculation method for settlement of embankment remedied with deep mixing method due to foundation liquefaction. Firstly, an empirical relation between crest settlement and displacement of remedied zone was established. Then, a new calculation method for displacement of remedied zone was developed. In the method, a macroscopic failure surface and a plastic displacement potential in the general load space are considered to evaluate the subgrade reaction force from foundation ground. The method is capable of calculating not only horizontal, vertical or rotational displacement alone, but also their combined effect. The method was validated through comparison with centrifuge test results of embankments resting on loose saturated sands subjected to strong base shakings. The calculated displacement components, that is vertical, horizontal and rotational displacement at the end of shaking, compared well with those measured.

1 INTRODUCTION

Although river dikes in Japan have often been damaged during past large earthquakes, any earthquake effects were not taken into account in the design practice of river dikes, because it is usually easy to restore in a short term. The Hyogoken-nambu earthquake of 1995 hit 6 m high dikes of Yodo river causing them to subside by as much as three meters. The highly urbanized hinterlands were in real danger of an overflow due to dike settlement. After this earthquake, the Ministry of Construction started a remediation program against liquefaction-induced failures for vulnerable dikes.

It was reported that large settlement of dikes during past earthquakes was more or less associated with liquefaction of foundation soils. Countermeasures in the remediation program for river dikes aim at mitigating crest settlements by providing containment for the deformation of the liquefiable foundation soils by forming remedied stiff zones in liquefiable soil layer under embankment toes (Adalier et al., 1998; Matsuo and Shimazu, 1998). Ground improvement technique by the deep mixing method is often used for this purpose.

In the current practice, countermeasures are designed so that factors of safety for the bearing capacity, sliding and rotation failure modes and for failure of the remedied zone itself are higher than certain values. This design procedure may assure that dis-

placement of remedial countermeasures as well as deformation of countermeasures itself are limited, but does not directly assess crest settlement. Since a minimum requirement for river dikes is that the crest of dikes remains higher than a river water level, crest settlement should be directly considered in design procedures.

In this study a practical method is developed for predicting crest settlement of embankment which is remedied with the deep mixing method. The method is validated through comparisons of predicted results with centrifuge test observations. In this paper, the improved zone by the deep mixing method is assumed as a rigid block and deformation of the improved zone itself, which may be termed as the internal stability, is out of the scope.

2 CREST SETTLEMENT

From observations of a series of centrifuge model tests on embankment resting on loose saturated sand with and without countermeasure, Okamura and Matsuo (2002a) demonstrated that the following three major factors contributed the crest settlement; i) shear deformation of embankment due to horizontal deformation of underlying liquefied sand, ii) lateral deformation of loose sand layer and iii) contractive volume change of loose sand under embankment. These factors are schematically illustrated in Fig.1.

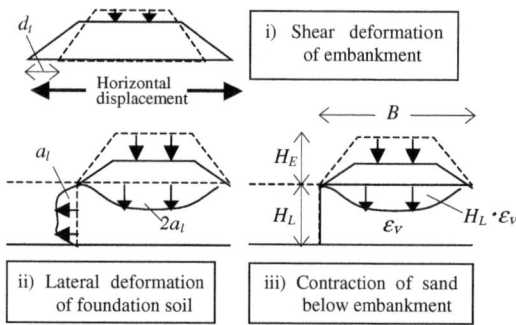

Figure 1. Three major factors contributing crest settlement

The factors ii) and iii) are associated with settlement of the embankment base, while factor i) is associated with a decrease in the embankment height. They concluded that the crest settlement could be reasonably estimated by the following equation,

$$S_c = c_1 \left(\frac{2a_l}{B} + \varepsilon_v H_T \right) + c_2 \frac{2d_l H_E}{B} \qquad (1)$$

where, S_c = crest settlement, a_l = area of lateral deformation of the improved zone, ε_v = volumetric strain of liquefied soil beneath embankment, H_L = depth of liquefied layer, d_l = lateral deformation at embankment toe, B = width of embankment base, H_E = height of embankment and c_1 and c_2 = constants. The constants, c_1 and c_2, may depend on unevenness or curvature of the settlement profile along the embankment base and heterogeneity of strain in the embankment. Centrifuge test results indicated, however, that they can be regarded essentially as constants (c_1 = 1.86 and c_2 = 1.25) irrespective of several influential factors including size and shape of embankment, depth of liquefied layer, thickness of unsaturated soil layer at ground surface, and base acceleration time history.

With regard to the volumetric change of liquefied sand, an empirical chart proposed by Ishihara and Yoshimine (1992) is ready to be used in practice. In the following sections of this paper, we concentrate on the displacement of the improved zone, i.e., the

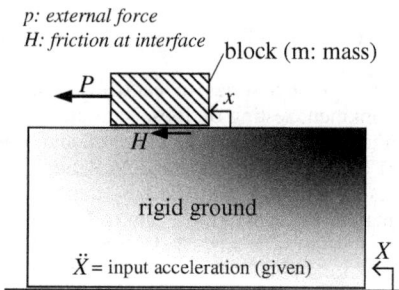

Figure 2. Single degree-of-freedom problem

area of lateral displacement and horizontal displacement at toes.

3 DISPLACEMENT PREDICTION OF IMPROVED ZONE

3.1 Single degree-of-freedom problem

Consider a rigid block resting on ideal rigid level ground as shown in Fig.2, with the admissible movement of the block to be sliding only. For a given time history of external horizontal force, P, acting on the block and of horizontal acceleration of the rigid ground, the motion of the block is described by an equation of motion as follows:

$$m\ddot{x} = P + H - m\ddot{X} \qquad (2)$$

where X = horizontal displacement of the ground, x = relative displacement of the block with regard to the ground, H = frictional force at the interface between the block and the ground and m = mass of the block. Sliding displacement of the block, x, can be calculated by integrating equation (2) twice, in much the same way as the sliding block method originally proposed by Newmark (1965). Design earthquake motion or the input acceleration $\ddot{X}$ will be given in design procedures. The forces P and H have to be determined properly.

The sliding block analogy as well as the model shown in Fig.2 is only available to a single degree-of-freedom problem and can calculate sliding displacement alone. But, generally, the movements of a block on a ground in reality cannot result from purely sliding, subsidence or rotation but their combined effect. Not only the sliding displacement but also rotation is of primary importance, since horizontal displacement at the top of blocks is one of the determining factors of the embankment settlement.

In the following sections, a method which is capable of calculating fully coupled displacement, that is vertical, horizontal and rotational displacement, of a block under earthquake loadings is described, in which a macroscopic constitutive law for the entire foundation soil-block system is considered. The method is applied for predicting fully coupled displacement of improved zone under embankment toes.

3.2 Fully coupled displacement prediction

Figure 3 depicts a schematic illustration of a block as a countermeasure beneath the embankment toe resting on a bearing stratum of a rigid-perfectly plastic media. Assuming that the direction of the input acceleration to the bearing stratum as well as the earth pressure acting on the side of the block from the liquefied soil are horizontal alone, the motion of the block relative to the bearing stratum can be expressed by the equations of motion as follows:

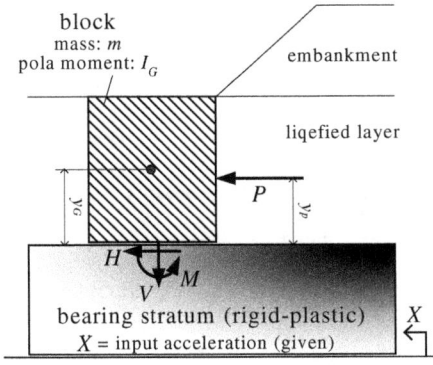

Figure 3. Forces acting on improved stiff zone

$$-V + mg = m\ddot{y} \quad \text{(vertical)} \quad (3)$$

$$P - H - m\ddot{X} = m\ddot{x} \quad \text{(horizontal)} \quad (4)$$

$$P \cdot y_p - M = m(\ddot{x} \cdot y_G) + \left(I_G + m y_G^2\right)\ddot{\theta} \quad \text{(rotational)}(5)$$

in which θ = rotation of the block, I_G = polar moment of inertia of the block about the center of gravity, y_G = the height of the center of gravity from the base, P = earth thrust from liquefied layer, y_p = height of acting point of P, and y and x are vertical and horizontal displacement at the center of gravity relative to the bearing stratum. The subgrade reaction force acting on the base of the block can be divided into three components of vertical and horizontal forces, V and H, and moment with regard to the center of the base, M. Provided that the subgrade reaction forces (V, H and M) and horizontal force on the side of the block from the liquefied soil layer, P, are properly determined, displacement of the block (x, y, and θ) under the action of a given base acceleration ($\ddot{X}$) can be readily obtained by integrating equations (3) – (5).

3.3 Earth pressure acting on side of block

Typical centrifuge model observations on earth pressures acting on the both side of the block, that is on the free field side and the embankment side, from the liquefied sand layer are given in Fig.4 (Okamura et al., 2003). In the figure, the base acceleration and excess pore pressure ratio time histories are also presented. The test was conducted at 50g environment and results were presented in prototype scale. It can be observed that the broken lines in the figure corresponding to the residual earth pressure was almost constant after the liquefaction condition (100% excess pore pressure ratio) was reached. Deviation from the residual pressure during liquefaction was essentially proportional to the acceleration of the block and the pressures on the embankment side and the free field side are in the opposite phase. Thus, the earth thrust from liquefied layer on the side of

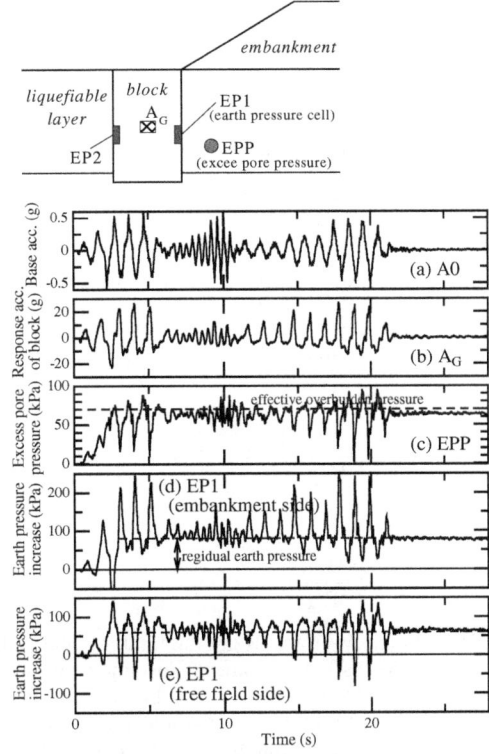

Figure 4. Observed responses of block as liquefaction countermeasure beneath embankment toe

the block, P, in equations (4) and (5) has the relation,

$$P = m_a\left(\ddot{X} + \ddot{x}\right) + P_r \quad (6)$$

where, P_r = the residual earth thrust and m_a = is added mass. Equation (4) becomes,

$$\ddot{x} = \ddot{X} - \frac{H}{m_a - m} + P_r \quad (7)$$

The profiles of the residual earth pressure, P_r, and the added mass, m_a, will be reported in detail and given in a simple chart for easy reference (Okamura et al., 2003).

3.4 Subgrade reaction

For a rigid block on a level ground, the failure envelope of the ground in a V-H-M/B general load space has been analyzed and many investigators have proposed mathematical expressions of the envelope (Murff, 1994; Georgiadis and Butterfield, 1988; Nova and Montrasio, 1991; Bransby and Randolph, 1998; Ukritchon et al., 1998; Martin and Houlsby, 2000). To preserve dimensional homogeneity the moment, M, is divided by the block width, B. One of the simplest expressions of the failure envelope for a

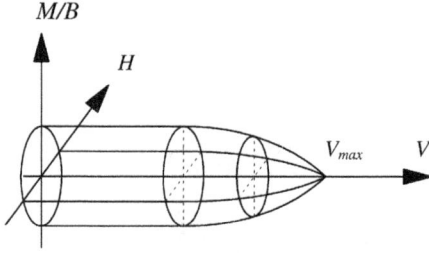

Figure 5. Failure envelope in general V-H-M/B load space

uniform clay deposit is of the form shown in Fig. 5 and equations (8) and (9).

for $V \geq V_{max}/2$

$$F = \left(\frac{H}{H_{max}}\right)^2 + \left(\frac{M}{M_{max}}\right)^2 - 4\left(1 - \left(\frac{V}{V_{max}}\right)^2\right)^2 = 0 \quad (8)$$

for $V < V_{max}/2$

$$F = \left(\frac{H}{H_{max}}\right)^2 + \left(\frac{M}{M_{max}}\right)^2 - 1 = 0 \quad (9)$$

The symbols V, H and M are the total vertical and horizontal forces and moment sustained by the ground, respectively, thus the subgrade reaction forces. H_{max} is the ultimate horizontal load when $M = 0$ and $V = 0$, M_{max} is the ultimate moment when $V = 0$ and $H = 0$ and V_{max} is the maximum vertical load capacity. The reference point of the moment is the center of the block base. H_{max}, V_{max} and M_{max} are often taken as c_uB, c_uN_cB and $c_uN_cB/2$, respectively, where c_u is undrained strength of the clay and N_c is the bearing capacity factor. Fig. 5 shows the cross sections of the failure envelope in the M/B-V plane ($H = 0$) and in the H-V plane ($M = 0$) to be constant for $V<V_{max}/2$ and parabolas for $V>V_{max}/2$. The cross section of the failure envelope in M/B-H plane at a constant V is an ellipse with an axis ratio of $N_c/2$.

In order to describe the displacement characteristics of the block under combined loading, the flow rule has been examined, in which a macroscopic constitutive law for the entire soil-block system was considered. In a manner consistent with the theory of plasticity, in which the ratio of incremental plastic strain is related to the state of stress, "plastic displacement potential", Q, is defined. The incremental displacement vector in the work-conjugate displacement space, $\dot{s}$, is assumed to be orthogonal to the plastic displacement potential. Thus we have,

$$\dot{s} = \{\delta y, \delta x, B\delta\theta\} = \lambda \frac{\partial Q}{\partial R} \quad (10)$$

in which $R = \{V, H, M/B\}$ is the subgrade reaction vector and λ is a constant. In this study associated plasticity is assumed so that the failure envelope also describes the plastic displacement potential defining

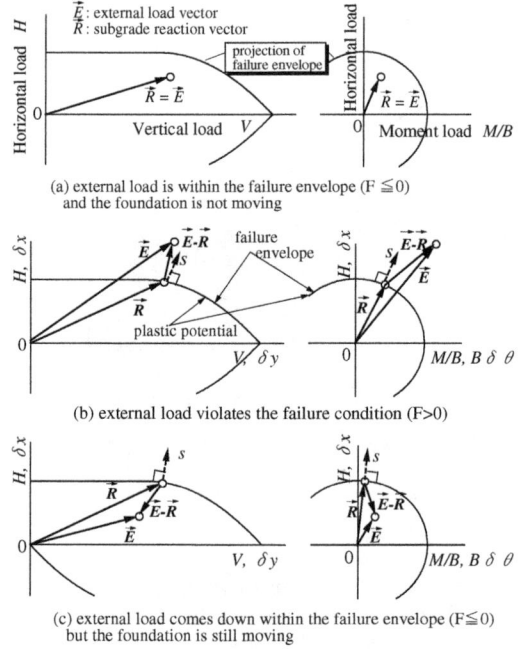

(a) external load is within the failure envelope (F $\leqq$ 0) and the foundation is not moving

(b) external load violates the failure condition (F>0)

(c) external load comes down within the failure envelope (F$\leqq$0) but the foundation is still moving

Figure 6. External load, subgrade reaction force in general load space and superimposed incremental displacement vectors

the relative magnitudes of the incremental plastic displacement during failure.

Computational procedure of displacement of the block resting on the bearing stratum having the failure envelope, F, and plastic displacement potential, Q, in the form of Equations (8) and (9), and subjected to an external load at time t, $E_{(t)} = \{V_{e(t)}, H_{e(t)}, M_{e(t)}/B\}$ as illustrated in Fig. 6, is as follows. The external load, E, includes all the loads acting on the block except for the subgrade reaction force on the base, $R_{(t)} = \{V_{(t)}, H_{(t)}, M_{(t)}/B\}$.

Step 1: Divide the time period into n equal segments, t_1, t_2, $\cdots$, t_i, $\cdots$, t_n, start from t = 0 when the block stops under the equilibrium condition of $E = R$, as shown in Fig. 6(a). The block does not move relative to the ground ($\dot{s} = 0$ and $s = 0$) and the equilibrium holds ($E = R$) while external load vector E stays within the failure envelope.

Step 2: Time is increased by a small increment, $t_i = t_{i-1} + \Delta t$, where $\Delta t = t_n/n$. Examine the external load $E_{(ti)}$ to determine whether it exceeds the failure envelope or not. If $E_{(ti)}$ stays within the failure envelope, the block does not move at this time step. Step 2 is repeated with increasing time step by step until $E_{(ti)}$ violates the failure condition.

Step 3: Once $E_{(ti)}$ exceeds the failure envelope the subgrade reaction force is no longer the same as the external load and the difference in the forces ($E_{(ti)}$-$R_{(ti)}$) accelerates the block. In Fig.6, the directions of the displacement increment are plotted on the same diagram as the states of loads.

The reaction force vector R has to be determined so that R is on the failure envelope and the direction of the vector $\dot{s} = \lambda \cdot \partial Q / \partial R$ coincides with that obtained from equation (11). Knowing the relative velocity $\dot{s}_{(t_{i-1})}$ and displacement $s_{(t_{i-1})}$ at the preceding time step, relative velocity and displacement of the block at this i-th time step can be derived by integration of the relative acceleration.

$$\ddot{s}_{(t_i)} = \left\{ \ddot{y}_{(t_i)}, \ddot{x}_{(t_i)}, B\ddot{\theta}_{(t_i)} \right\};$$

$$\dot{s}_{(t_i)} = \int_{t_{i-1}}^{t_i} \frac{\ddot{s}_{(t_i)} + \ddot{s}_{(t_{i-1})}}{2} \, dt + \dot{s}_{(t_{i-1})} \qquad (11)$$

$$s_{(t_i)} = \int_{t_{i-1}}^{t_i} \frac{\dot{s}_{(t_i)} + \dot{s}_{(t_{i-1})}}{2} \, dt + s_{(t_{i-1})} \qquad (12)$$

This calculation is repeated until the block velocity drops to zero, even after the external load E returns within the failure envelope (Fig. 6 (c)). Step 2 is returned to when the block stops. More detailed information about the calculation procedure was given by Okamura and Matsuo (2002b).

The necessary input data for this calculation includes a bearing load for vertical central loading (V_{max}), width of the base B, mass m and polar moment of inertia of the block, I_G, and acceleration time histories of the bearing stratum.

4 VERIFICATION OF THE METHOD

The proposed method described above was utilized to simulate the centrifuge model embankment with countermeasure subjected to a strong base shaking (Okamura and Matsuo, 2002a). In this section, the centrifuge model is briefly reviewed first and the method is validated through the comparison of computational results with model observations.

4.1 Centrifuge test

Figure 7 illustrates the centrifuge model. The model was constructed in a rigid container with internal dimensions of 1500 mm long, 300 mm wide and 500 mm deep. A clay stratum was formed on the bottom of the container by consolidating de-aired kaolin clay slurry to the pressure of 200 kPa. This made the whole clay stratum to be lightly overconsolidated condition through the course of the centrifuge test.

Acrylic blocks which have the same unit weight as that of surrounding loose sand layer were placed on the clay layer. The acrylic blocks did not accurately simulate the treated soil by the deep mixing method but rather they were intended to be stiff and strong as compared with surrounding soils. Then, a uniform sand layer of 160 mm deep was constructed by compacting Edosaki sand to a relative density of

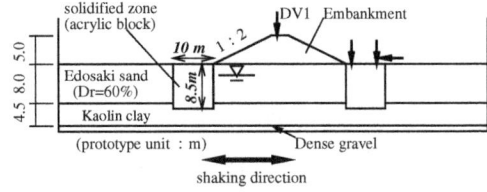

Figure 7. Centrifuge model configuration (Okamura and Matsuo, 2002a)

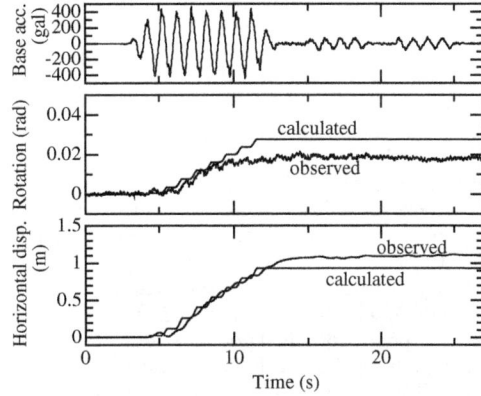

Figure 8. Comparisons of calculated displacements with centrifuge test observations

60%. The sand layer was saturated under a vacuum of -92 kPa. The water level was set 35 mm below the soil surface and a 100 mm high model embankment was constructed on the foundation soil by compacting wet Edosaki sand to a dense condition.

The undrained strength of the clay was estimated to be 47 kPa based on the measured water contents after centrifuge test and the stress history of the clay.

The model was brought up to 50 g centrifugal environment and one-dimensional horizontal shaking was imparted along the model long axis. This model simulated a 5 m high embankment founded on a 8 m deep loose sand deposit with 10 m wide solidified zones at both sides of embankment. In the following discussion, all comparisons are made in prototype units unless otherwise mentioned.

The residual earth thrust, P_r, and the added mass, m_a, in equation (6) for this model were 92.0 kN and 135 t, respectively.

4.2 Comparison of computational results with test observation

The measured displacement time histories of the block in the model are compared with those calculated from the proposed method in Fig.8, together with imparted base acceleration. Vertical load due to the self-weight of the block was smaller than the half of the bearing capacity force ($V_{max}/2$) and the vector $\dot{s}$ was always normal to the δy axis in Fig.6, calculated settlement of the block was zero. This is consistent with the test result that the observed settlement of the block was very small, of the order of 0.01 m. It appears that the calculated residual horizontal displacement and residual rotation agreed quite well with those observed.

It can be seen in the figure that the displacement in the calculation increased stepwise, while the observed displacement comprised both cyclic and permanent residual displacement, indicating that the block moved back and forth cyclically with gradually accumulating residual displacement. This difference may be attributed to the fact that the rigid-perfectly plastic response of the foundation soil is assumed in the calculation. The cyclic displacement observed in the model, however, is relatively small and can be negligible for the practical design purpose.

It can be concluded that the proposed method can be an effective tool to assess displacements of solidified zone by deep mixing method beneath embankment toes.

Finally, Crest settlement was calculated from equation (1), with the block displacement calculated from the prediction method, and with volumetric strain of liquefied sand derived from the empirical chart (= 3.0%) proposed by Ishihara and Yoshimine (1992), was 1.76m. This agreed well with the observed crest settlement of 1.65m.

5 CONCLUSION

River dikes in Japan have often been damaged during earthquakes due to the foundation liquefaction. Remedial measure has been constructed In order to assess settlement of embankments remedied with the deep mixing method, a new calculation method for displacement of improved zone was developed. In the method, a macroscopic constitutive low for the entire solidified zone-foundation soil system is considered to evaluate the subgrade reaction force from the bearing stratum. The method is capable of calculating not only horizontal, vertical or rotational displacement alone, but also their combined effect. The method is validated through comparison with centri-

fuge test results of a embankment resting on a loose saturated sand subjected to strong base shaking. The calculated displacement components compared quite well with those measured. The proposed method can be an effective tool to assess displacements of solidified zone by the deep mixing method beneath embankment toes.

REFERENCES

Adalier, K., Elgamal, A.-W. and Martin, G. R. 1998: Foundation liquefaction countermeasures for earth embankment, *J. of Geotechnical and Geoenvironmental Engineering, ASCE*, 124(6), 500-517

Bransby, M. F. and Randolph, M. F. 1998: Combined loading of skirted foundations, *Geotechnique* 48(5), 637-655

Georgiadis, M. and Butterfield, R. 1988: Displacements of footings on sand under eccentric and inclined loads, *Canadian Geotechnical Journal* 25(1), 199-212

Ishihara, K. and Yoshimine, M. 1992: Evaluation of settlements in sand deposits following liquefaction during earthquakes, *Soils and Foundations* 32(1), 173-188

Martin, C. M. and Houlsby, G. T. 2000: Combined loading of spudcan foundations on clay: laboratory tests, *Geotechnique* 50(4), 325-338

Matsuo, O., and Shimazu, T. 1998: Design and construction manual for countermeasures against liquefaction-induced river dike failure, *Technical Memorandum of Public Works Research Institute* (in Japanese).

Murff, J. D. 1994: Limit analysis of multi-footings foundation systems, *Proc. 8th Int. Conf. Comput. Methods. Adv. Geomech. Morgantown* 1: 223-244

Newmark, N. M. 1965: Effects of earthquakes on dams and embankments. *Geotechnique* 15(2): 139-160

Nova, R. and Montrasio, L. 1991: Settlements of shallow foundations on sand, *Geotechnique* 41(2), 243-256

Okamura, M. and Matsuo, O. 2002a: Effects of remedial measures for mitigating embankment settlement due to foundation liquefaction. *J. of Physical Modelling in Geotechnics.* 2(2): 1-12

Okamura, M. and Matsuo, O. 2002b: A displacement prediction method for retaining walls under seismic loading. *Soils and Foundations.* 42(1): 131-138

Okamura, M., Ishihara, M. and Tamura, K. 2003: Seismic earth pressure on rigid wall in liquefied sand, *Soils and Foundations* (in preparation)

Ukritchon, B., Whittle, A. J. and Sloan, S. W. 1998: Undrained limit analyses for combined loading of strip footings on clay, *J. of Geotechnical and Geoenvironmental Engineering*, ASCE, 124(3), 265-276.

Soft Ground Engineering in Coastal Areas, Tsuchida et al. (eds)
© 2003 Swets & Zeitlinger, Lisse, ISBN 90 5809 613 0

Sea reclamation with dredged soft soil for Central Japan International Airport

T. Satoh
Central Japan International Airport Co. Ltd., Japan

M. Kitazume
Port and Airport Research Institute, Japan

ABSTRACT: Recently, Pneumatic Flow Mixing Method, is developed for sea reclamation and land development, in which dredged clay is mixed with relatively small amount of cement in a transporting pipe. The soil mixture forms several separated mud plugs in the pipe, and is thoroughly mixed by means of turbulent flow during the transporting. The mixture transported and deposited at reclamation site gains relatively large strength so that no additional soil improvement is usually required. This method is expected to provide an economical and rapid construction for sea reclamation, and is applied to construct a man-made island for Central Japan International Airport at Nagoya area, where about 10 million cubic meters of dredged soil is stabilized by the method. In this article, the mechanical properties of the treated soil, the execution technique of the method and its application to the Airport island construction project are described briefly.

1 INTRODUCTION

Many man-made islands have been constructed in Japan to obtain enough plain land area for airport, electric power plant and so on. These islands require huge amount of relatively high quality soil for sea reclamation. Recently it becomes more difficult to obtain such a soil with reasonable expense because of strict environmental protection, which requires use of poor quality soil. On the other hand, a huge amount of soft soil is dredged at many ports every year to maintain enough sea route and sea berth. These soft soils are dumped at disposal sites in coast area. It becomes more difficult to construct the disposal area for dredged soil and subsoil, because of environmental restriction and economical reason.

These circumstances promote to use dredged soft soils as a reclamation material. Recently, a new soil improvement technique has been developed in Japan, named as Pneumatic Flow Mixing Method, in which dredged soft soil is mixed with small amount of stabilizing agent, usually cement, in a pipe during transporting by compressed air and is deposited for sea reclamation. The mixture of dredged soil and stabilizing agent forms many separated mud-plugs in the pipe, and is thoroughly mixed during transporting by a help of turbulent flow generating within the plug. The soil mixture deposited and cured at site can gain relatively high strength so that no additional soil improvement is required for constructing structures on it. The method requires only a stabilizing agent supplier facility to existing pneumatic fa-

cilities. As no mixing blade is used to mix soil and agent and the soil improvement can be performed continuously, this method is expected for construction of man-made island within relatively short period and more economically.

Authors have started a research project together with Ministry of Land, Infrastructure and Transport (former Ministry of Transport) to study the applicability of Pneumatic Flow Mixing Method for sea reclamation through investigations of treated soil properties, construction techniques of sea reclamation, executing ability, quality control and assurance and so on. In the research project, laboratory mixing tests, centrifuge model tests and field tests were performed (Kitazume, 1997, 1998, Kitazume, et al., 1999, 2000, Makibuchi, et al., 1999, Ministry of Transport, 1999). These research efforts show the high applicability of the method and promote several constructions (Porbaha, et al., 1999a, 1999b, Horii, et al., 1999, Iwatsuki, et al., 1998). Recently this method is applied to construct a man-made island for Central Japan International Airport at Nagoya area in Japan. The construction project has started in 2000 and will be completed in 2005, in which the total of 10 million m^3 of dredged soil is stabilized by this method to construct a part of the airport island.

In this article is described briefly the development of Pneumatic Flow Mixing Method and its application to Central Japan International Airport Construction Project as well as the mechanical properties of the treated soil, the execution technique of the method.

2 PNEUMATIC FLOW MIXING METHOD

2.1 Mechanism of the method

Transporting soft soil in a pipe without any amount of air requires high pressure due to the friction generating on an inner surface of pipe. When relatively large amount of compressed air is injected into the pipe together, the soft soil is separated into small blocks by the injected air in the pipe, as schematically shown in Fig. 1. The separated soil block, called as a 'plug', is forwarded to an outlet by a help of compressed air. The forming plug composed of soil and air block has a function to reduce the friction on a pipe inner surface and in turn can reduce considerably the air pressure required to transport. The formation of the plug is dependent upon the mixing ratio of soil and air, and the pile diameter.

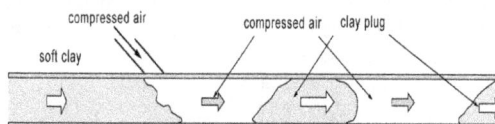

Figure 1. Schematically view of plug flow.

2.2 Characteristics of clay plug

The characteristics of clay plug measured in field tests are summarized in Table 1 (Ministry of Transport, 1999). It can be found that the clay plugs with an average volume of 0.41 m^3 are transported at average speed of about 11 m/s and an average interval of about 7 seconds.

Table 1. Characteristics of clay plug

	Test results
Test condition	
soil volume	210 m^3/hr
water content	132.5 %
liquid limit	70.5 %
cement added	38 kg/m^3
Test results	
plug speed	10.9 (1.6– 25.0) m/sec
plug volume	0.41 (0.23–0.52) m^3
plug length	4.3 (0.23– 5.4) m
plug interval	7.1 (1.3– 30.3) sec

The field tests also revealed that these characteristics are almost independent upon the transported soil volume and amount of cement. As the clay plugs are transported very high speed in the pipe, the turbulent flow is generated within the plugs due to the friction on the inner surface of pipe. The turbulent flow provides to mix the clay and the stabilizing agent thoroughly. Previous research efforts found that thorough mixing can be obtained in the condition of Reynolds number, $Re = uD/v$ of 500 to 3,000, where u is the plug speed, D is the pipe diameter and v is the viscosity of plug. It is also known that at least 50 m to 100 m of transporting distance is necessary to ensure sufficiently mixing.

2.3 Facility

Figure 2 shows a group of barges for one kind of the Pneumatic Flow Mixing Method available in Japan, which includes a pneumatic barge, a stabilizing agent supplier barge and a placement barge. In Figure, dredged clay on the soil transport barge is loaded into the hopper on the pneumatic barge at first, and is transported by a help of compressed air to the reclamation site. Stabilizing agent, usually cement is then injected to the soft soil on the stabilizing agent supplier barge and they are thoroughly mixed during the transporting in the pipeline. Cement in either slurry or dry form is added to the soil, while the slurry form is common in Japan.

There are two major types of the method according to where the stabilizing agent is injected; *compressor addition type* and *line addition type*. In the former type, the stabilizing agent is injected to the soft soil before the compressed air is injected into the pipe. In the later type, on the other hand, the stabilizing agent is put into the pipe after the air injection. In the both techniques, the soil mixture is allowed to be placed to reclamation site through a cyclone on the placement barge, which functions to release the air pressure transporting soil plugs. A tremy pipe is usually used to place the soil mixture under seawater not to entrap seawater within the soil, which can cause considerably decrease of the treated soil strength.

There are several variations in the stabilizing agent injection techniques and transporting techniques by various construction firms to improve the mixedness of clay and agent (Porbaha, et al., 1999, Horii, et al., 1999, Yamane, et al., 1998). One is additional equipment installed along the pipeline to detect the plug location and its volume accurately to inject cement slurry directly on clay plugs, and the others are the techniques where the diameter or shape of pipe is changed locally.

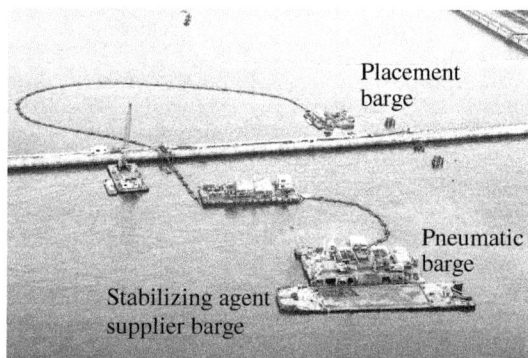

Figure 2. Group of barges.

3 CHARACTERISTICS OF TREATED SOIL

3.1 *Effect of amount of agent on strength increase*

Figure 3 shows the relationship between the amount of cement and the unconfined compressive strength, qu (Ministry of Transport, 1999). The two types of field strength are plotted in the figure; treated soil placed on land in Fig. 3(a) and under seawater in Fig. 3(b). The amount of cement is defined as a dry weight of cement per one cubic meter of soil to be treated. The treated soil named as 'laboratory' means one manufactured and cured in a laboratory according to the Standard by Japanese Geotechnical Society (Japanese Geotechnical Society, 2000). It is found that the 'laboratory' specimen shows almost linear increase in the strength with increasing the amount of cement. In the figure, several kinds of field treated soil strength are also plotted together with the laboratory treated soil. The specimen named as 'mold' is the soil and cement mixture obtained at the outlet of the pipe and cured in a laboratory according to the Standard. The specimens named as 'core' and 'L core' are the treated soils sampled *in-situ* which is cured in the field condition. The specimen of 'core' is 5 cm in diameter and 10 cm in height, while the 'L core' is 50 cm in diameter and 100 cm in height.

It is found in Fig. 3(a), in the case of placement on land, that the unconfined compressive strengths of all the treated soils increase almost linearly with the increase of amount of stabilizing agent irrespective of the specimen type. But it is found that the 'laboratory' samples show the highest strength among the test specimens. The field treated soils also show large increase in the strength. But it is found that the 'L core' sample shows the lowest strength, which might be due to relatively large scatter in the strength within the specimen. It can be emphasized that at least about 40 kg/m^3 of cement is required to obtain a certain strength gain in the field, while the 'laboratory' shows about 100 kN/m^2 in the same amount of cement. In the case of placement under seawater as shown in Fig. 3(b), the strength of the field treated soil 'core' is relatively small com-

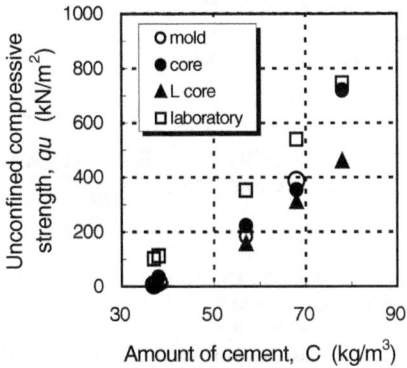

(a) placement on land

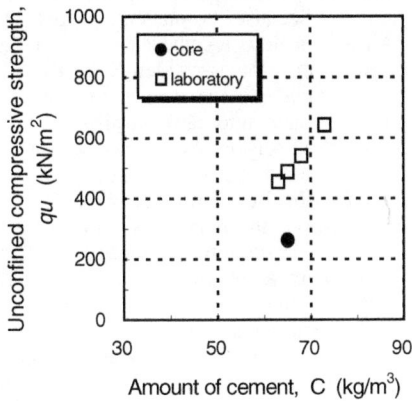

(b) placement under seawater

Figure 3. Unconfined compressive strength and amount of cement.

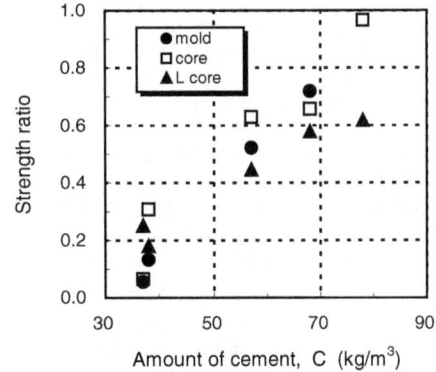

(a) placement on land

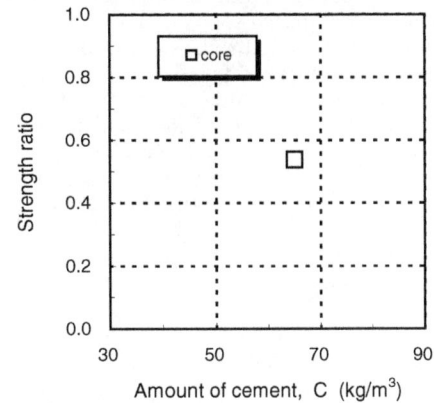

(b) placement under sea water

Figure 4. Strength ratio and amount of cement.

pared with the 'laboratory' specimen. This can be due to seawater entrapped within the treated soil during placement.

In order to compare tdhe strength of field with the laboratory treated soil in detail, Figure 4 shows the strength ratio of the field treated soil to the 'laboratory' soil (Ministry of Transport, 1999). It is found that the strength ratios are not constant but increase with increasing the amount of cement irrespective of specimen. This means that the small amount of cement is not mixed well with the soil in the field. The strength ratio of 'core' specimen is about 0.5 to 0.8 in the case of placement on land when the amount of cement is about 60 to 70 kg/m^3, which is a target amount of cement in this method. Figure 4(b) also shows the strength ratio of the field 'core' sample placed under seawater. The strength ratio of the treated soil under seawater is about 20 % lower than that placed on land. This phenomenon can be due to increase of water content of treated soil by entrapping seawater during placement. This also emphasizes that the best care should be paid during the placement of treated soil not to entrap seawater. The strength of 'L core' is lower than that of 'core' and the strength ratio is about 0.7, which is probably the influence of the relatively large scatter in strength within the specimen. This strength ratio is almost same as a previous research on the treated soil manufactured by the Deep Mixing Method (Futaki, et al., 1996). This relation indicates that the treated soil with scatter in strength can be evaluated as a whole by the average strength on small sized samples.

3.2 *Effect of amount of agent on strength deviation*

It is well known that the strength deviation is much dependent upon the amount of cement, the manufacture techniques and the size of specimen. Figure 5 summarizes the relationship between the coefficients of deviation of unconfined compressive strength, *qu* against the amount of cement. In the figure, several field manufactured treated soils are plotted together with the laboratory treated soil. It is found that the coefficient of deviation of the 'laboratory' treated soil is relatively small with the order of 15 %, and almost constant irrespective of the amount of cement. This means that quite uniform mixing can be obtained according to the Japanese standard (Japanese Geotechnical Society, 2000). The field manufactured treated soils, on the other hand, indicate relatively large coefficient of deviation comparing to the 'laboratory' specimen. It is found that the coefficient of variation of the field specimen with small size, shown as 'core' in the figure, is about 35 % irrespective of the amount of stabilizing agent, which is almost same order to that of the Deep Mixing Method (Hosomi, et al., 1996). But the specimen with large size, 'L core', shows almost same coefficient as the small sized specimen as far as the

amount of cement exceeds about 50 kg/m^3 but increases in the coefficient to about 60 % when the amount of cement decreases to 38 kg/m^3. This phenomenon indicates that the relatively small amount of cement cannot be distributed thoroughly within the treated soil by the Pneumatic Flow Mixing Method.

The phenomena shown in Figs. 3 to 5 emphasize that at least about 50 kg/m^3 of cement is necessary to obtain some amount of field strength gain with relatively high uniformity.

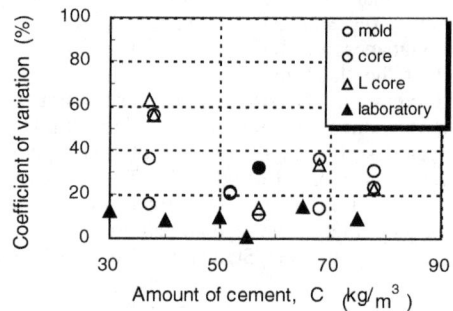

Figure 5. Strength deviation of field manufactured treated soil against amount of cement.

4 SEA RECLAMATION FOR CENTRAL JAPAN INTERNATIONAL AIRPORT

4.1 *Outline of construction project*

Central Japan area has about 20 million population and a quarter manufactured product shipment of the whole nation. Recent rapid increase of passenger and cargo in this area requires increase of the current international airport capacity. However there is not enough land space available to expand the current airport facilities, because the current Nagoya International Airport is now located in high populated suburban area with a lot of residential houses. This circumference provides a new airport construction at coastal area. The new International airport, Central Japan International Airport, is planned to construct on a man-made island at 2 to 3 kilometers offshore at Tokoname City in Aichi prefecture (Fig. 6).

The plane area of the man-made island is about 4.7 million m^2 while the total amount of soil for reclaiming the island is about 56 million m^3. It is rather difficult to obtain the whole amount of reclamation soil from mountainous area with reasonable expense because of strict environmental protection. In order to reduce the amount of mountainous soil, and to promote recycling of dredged soil, dredged soil excavated at Nagoya port is used as a reclamation material after treatment by the Pneumatic Flow Mixing Method.

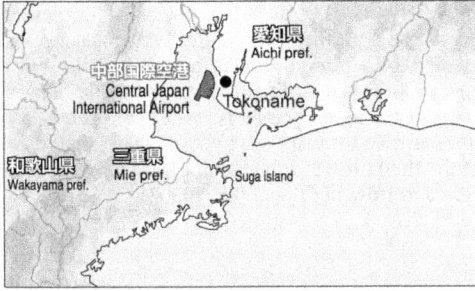

Figure 6. Location of Central Japan International Airport.

4.2 Ground condition at construction site

The sea depth of the construction site is about 6 m in average, 3 m at the shallowest to 10 m at the deepest. There is an eroded flat seabed and a buried valley between the new island and the opposite land. Almost all area of the new airport is constructed on a relatively good foundation, called as Tokoname layer. At the old valleys, alluvial sand and clay layers are laid alternatively. The alluvial sand layer has medium strength with SPT value of 14 to 15 in average. The alluvial clay layer is a soft layer and has about 70 kN/m² in unconfined compressive strength, qu. The diluvial clay layer and the sand layer, on the other hand, have high SPT value of 13 to 43. These soil conditions are preferable to construct the new airport comparing Kansai International Airport or Tokyo International Airport. A total of 12 km long sea revetment is constructed with gravel mound along the airport island. A part of revetment foundation soil is improved by the sand compaction pile method to assure the sufficient bearing capacity of the sea revetment and also to reduce the consolidation settlement.

4.3 Sea reclamation and design of treated soil

The field strength of the treated soil is easily controlled by changing the amount of cement added and/or the water content of soft soil. Figure 7 shows the design flow of mixing condition of treated soil.

(1) At first of the design flow, the design strength of the treated soil, quf is determined to obtain the required CBR value and so that no consolidation settlement takes place in the ground. In the former criteria, the minimum CBR value of 2 is regulated by the Ministry of Transport, while the design CBR value is set to 3 in this project. According to preliminary laboratory tests on the relationship between CBR value and unconfined compressive strength, the required unconfined compressive strength, qu is obtained as 80.1 kN/m². In the latter criteria, the preconsolidation pressure of the treated soil, which is found in the preliminary tests to be about 1.25 times higher than the unconfined compressive strength, should be larger than the overburden pressure at the

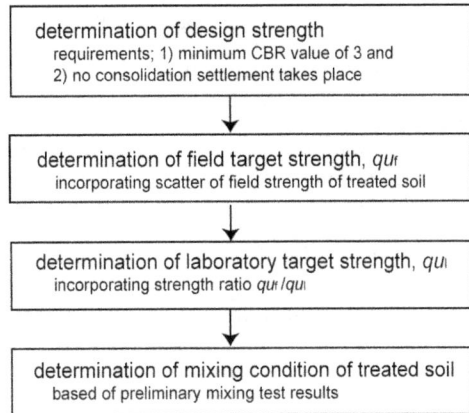

Figure 7. Design flow of mixing condition on treated soil.

bottom of the island. The required unconfined compressive strength for this criterion is obtained as 118 kN/m². As the result of these procedures, the design strength of the treated soil, quf should be equal or higher than two required strengths and is determined as 120 kN/m².

(2) It is usual that the field treated soil has a lot of scatter in the strength. Based on the preliminary test results, as shown in Fig. 5, the target average field strength, quf is then determined as 157 kN/m² so that more than 75 % of the field treated soil strength exceeds the quf = 120 kN/m².

(3) As shown in Fig. 4, the field strength, quf is relatively smaller than the laboratory strength, qul due to the difference in mixing and curing condition. For the influence of the strength ratio between the field and laboratory strengths, the average strength of laboratory treated soil is determined by $qul = quf / 0.5$; as $qul = 314$ kN/m².

(4) The soft soils are excavated at 10 sites in Nagoya Port and their soil properties are much different each other. A series of preliminary laboratory tests was carried out on these soils to obtain the optimum mixing conditions for the target laboratory strength. Figure 8 shows a typical example of test results, in

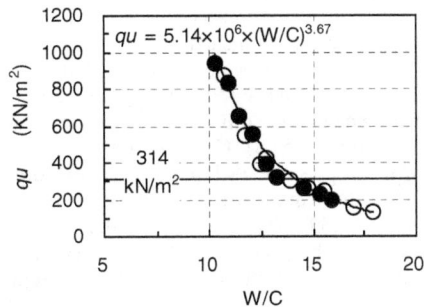

Figure 8. Typical example of relationship between unconfined compressive strength and the ratio of water to cement.

which the vertical and horizontal axes are the unconfined compressive strength, q_u and the water and cement ratio, W/C, respectively. The amount of water in the W/C ratio is defined as whole water including both soil and cement slurry. The figure shows that the q_u is almost inversely proportional to the W/C ratio but the unique relation can be obtained for each dredged soil.

4.4 Construction of the island

Amount of soil for reclaiming the island is about 56 million m^3, out of which 46 million m^3 of soil is mountainous soil and 10 million m^3 of soil is dredged soil (Fig. 9). The dredged soil is treated by the Pneumatic Flow Mixing Method to make high quality soil for reclamation. Three sets of pneumatic execution system operate at the construction site to complete constructing the man-made island within about 15 months, which include a pneumatic barge, a cement supply barge and a placement barge (Figs. 10 and 11). Each system has a maximum capacity of 1,000 m^3/h of manufacturing treated soil. These machines are able to construct about 25,000 m^3 of the treated soil per one day.

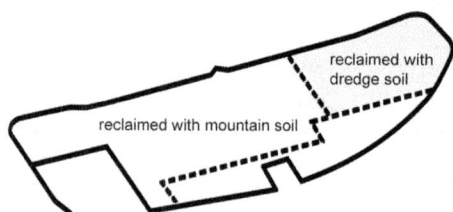

Figure 9. Airport island reclamation plan.

Dredged soil transported from dredged site is poured into the hopper on the pneumatic barge at first. Since there is not any dumped area for functioning a buffer of the soil, the close cooperation between the excavation side and reclamation side. Then the soil is mixed with some amount of seawater to adjust its initial water content of about 110 %. The soil is transported forward by the sand pump and its density is measured by means of the γ ray density meter. The water content of the soil is calculated by the measured density to obtain the amount of cement slurry to be added by the preliminary test results (Fig. 8). Compressed air of 390 to 490 kN/m^2 in pressure is injected into the pipe to form a plug flow and to transport the soil to the cement supply barge. The diameter of the pipeline at the cement supply injection is locally large of 152 cm compared to the other of 76 cm to improve mixedness due to temporary dis-forming the plug flow (Porbaha, et al., 1999). The cement slurry is manufactured at the batching plant whose water and cement ratio is 100 %. After injecting the cement slurry to the soil, the soil mixture is then transported as a plug by the compressed air toward the outlet along 1500 m long pipeline. At the outlet on the placement barge, a cyclone is installed to release the air pressure and transporting energy. The treated soil mixture then flows down to the outlet by its own weight and flow out the outlet (Fig. 12).

Figure 10. Pneumatic execution system operating the site.

Figure 11. Placement barge.

In the construction procedure, the treated soil is placed on the seabed to form the first layer of about 4 m in thickness. During the construction of this layer under sea level, a tremy pipe is used so that negligible seawater can be entrapped within the soil mixture that can reduce the treated soil strength considerably. After several weeks' curing to ensure the strength gain of the first layer, another soil mixture is placed on it to the design level of about +2.5 m D.L., where the best care is also paid not to entrap seawater and/or air babble within the mixture.

4.5 Quality control and assurance

During the execution, the density and the amount of dredged soil are measured every 20 seconds in order to control the water and cement ratio (W/C) and the amount of cement mixed, which feed back for quality control of the mixture. For the treated ground, field cone tests and unconfined compressive strength tests on the core sample are performed every 25,000 m² and 40,000 m², respectively. Figure 12 shows a typical measured data of unconfined compressive strength on core samples. It is found that there are a lot of scatters in the measured data, even if the execution is controlled with high quality. The figure shows that the treated soil strength is about 100 to 400 kN/m², which is relatively higher than the expected value of 157 kN/m². This phenomenon indicates that the reduction factor of 0.5 (strength ratio between field and laboratory strengths) is somewhat conservative in the actual construction with the quality control. Further discussion will be done to reduce some amount of cement to obtain the design strength more precisely from the viewpoints of technical, risk management and economics.

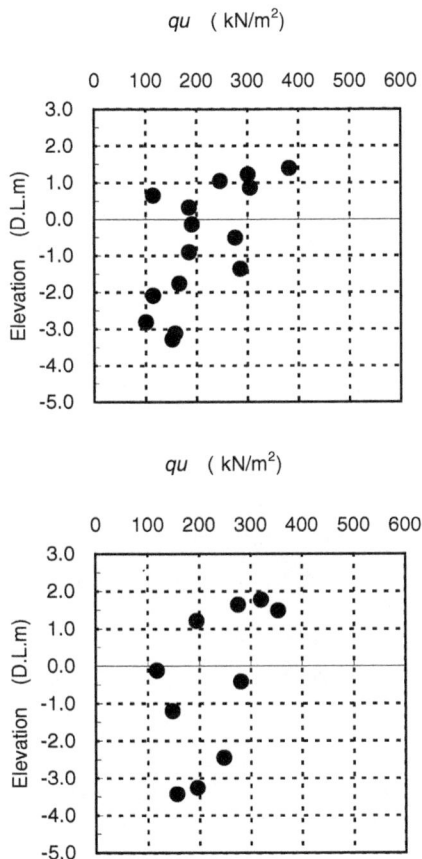

Figure 12. Unconfined compressive strength in field.

For the environmental protection, seawater from the reclamation site, which might be wasted, is pumped out through the wastewater treatment plant after controlling hydrogen exponent (pH) and suspended solids (SS). A contaminant test is also carried out on the treated soil every 400,000 m³ to ensure that no harmful contaminant is expanded.

5 CONCLUDING REMARKS

The research reported herein forms a part of an ongoing research effort to investigate the applicability of the Pneumatic Flow Mixing Method for sea reclamation and back filling waterfront retaining structures. A lot of laboratory and field tests confirm that the Pneumatic Flow Mixing Method has relatively high applicability for construction of man-made island, sea reclamation and back filling. Authors expect that the man-made island construction of Central Japan International Airport will be completed successfully and efficiently. A lot of experiences and know-how in execution technique and quality control and assurance accumulated in the construction will be published in near future to promote further development of the method.

Authors sincerely show their thanks to Ministry of Land, Infrastructure and Transport (former Ministry of Transport), Japan Dredging Reclamation Engineering Association and Toa, Kumagai, Nishimatsu, Mitsui, Tokura Joint Venture for their great assistance for development of the method and execution of laboratory and field tests.

REFERENCES

Horii, R., Shinsha, H. & Fujio, Y. Plant for the pneumatic flow mixing method. *Journal of Kensetsu no Kikaika*, 1999, 30-35, (in Japanese).

Hosomi, H, Nishioka, S & Takei, S. Method of deep mixing at Tianjin Port, People's Republic of China, *Proc. of the 2nd International Conference on Ground Improvement Geosystems*, 1996, 491-494.

Iwatsuki, T., Kamiyama, Y., Hashimoto, F., Yanai, E. & Masuyama, T. Effective cement-mixing method for mud transport using a compressed-air mixture pipeline, *Annual Journal of Hydraulic Engineering*, Japan Society of Civil Engineerers, 1998, (42) 655-660, (in Japanese).

Japanese Geotechnical Society, Practice for making and curing stabilized soil specimens without compaction, JGS T 0821-2000, Japanese Geotechnical Society, (in Japanese).

Kitazume, M. Centrifuge model tests on stability of embankment improved by cement, *Proc. of 32nd Annual Conference of Japanese Geotechnical Society*, 1997, 2429-2430, (in Japanese).

Kitazume, M. Centrifuge model tests on stability of improved embankment, *Proc. of 33rd Annual Conference of Japanese Geotechnical Society*, 1998, 2257-2258, (in Japanese).

Kitazume, M., Matsubara, Y., Matsuura, T., Hayashi, K., Shinohara, K., Oomori, K., Kaneshiro, T., Hoshi, H. & Kojima, T. Field test on soil admixture stabilization with suppressed pH agent, *Proc. of 34th Annual Conference of Japanese Geotechnical Society*, 1999, 805-806 (in Japanese).

Kitazume, M., Yamazaki, H. and Tsuchida, T. Recent soil admixture stabilization techniques for port and harbor constructions in Japan- deep mixing method, premix method, light-weight method-, *Proc. of the International Seminar on Geotechnics in Kochi*, 2000, 23-40.

Kitazume, M., Yoshino, N., Shinsha, H., Horii, R. and Fujio, Y. Field test on pneumatic flow mixing method for sea reclamation, *Proc. of the International Symposium on Coastal Geotechnical Engi-neering in Practice*, 2000, (1) 647-652.

Makibuchi. M., Yoshino, N., Kitazume, M. & Okano, K. Strength characteristics of the clay ground improved with a small amount of cement, *Proc. of 34th Annual Conference of Japanese Geotechnical Society*, 1999, 807-808. (in Japanese).

Ministry of Transport, Pneumatic Flow Mixing Method, Yasuki Publishers, 1999, 157, (in Japanese).

Porbaha, A., Hanzawa, H. & Shima, M. Technology of air-transported stabilized dredged fill, Part 1: Pilot study, *Ground Improvement Journal of ISSMGE*, 1999, 49-58.

Porbaha, A., Hanzawa, H. & Shima, M. Technology of air-transported stabilized dredged fill, Part 2: Quality assessment, *Ground Improvement Journal of ISSMGE*, 1999, 59-66.

Yamane, N., Taguchi, H., Fukaya, T., Dam, K.L., Kishida, T. & Iwatsuki, T. Strength Characteristics of cement-treated soil using compressed air-mixture pipeline. *Proc. of 34th Annual Conference of Japanese Geotechnical Society*, 1998, 2253-2254. (in Japanese).

Soft Ground Engineering in Coastal Areas, Tsuchida et al. (eds)
© 2003 Swets & Zeitlinger, Lisse, ISBN 90 5809 613 0

Behavior of caisson with extended footing quay wall during ordinary and earthquake conditions

T. Sugano
Port and Airport Research Institute, Japan

Y. Shiozaki
Applied Technology Research Center, NKK Corporation, Japan

T. Tanaka & K. Ebihara
Japan Science and Technology Corporation, Japan

ABSTRACT: A gravity type quay wall is made of a caisson or other gravity retaining structure placed on the subsoil as a foundation. The major stability systems are based on the mass of the wall and the friction at the bottom of the wall and the bearing capacity of foundation against the reaction force between the wall bottom and foundation. To maintain the stability of gravity type quay wall on soft soil condition, a new type of caisson was developed, which has extended footing. To investigate the behavior of a caisson with extended footing quay wall during static and dynamic conditions, a series of centrifuge model test and underwater shake table test was conducted. The behavior of caisson with extended footing quay wall during the tests were made by comparing the conventional reinforced concrete caisson type quay wall tests and it showed small differences.

1 INTRODUCTION

A caisson with extended footing using steel - concrete composite structures (HB Caisson) has been developed to decrease the intensity of reaction force distribution between caisson bottom and its foundation, as shown in Figure 1 and Figure 2.

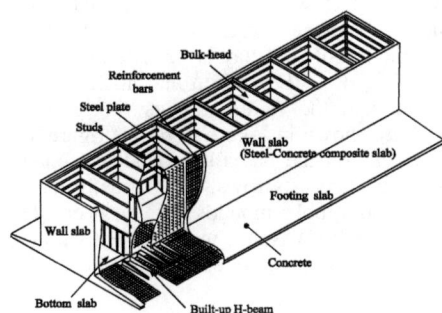

Photo 1. Steel skeleton of HB Caisson

Figure 1. Overview of a HB Caisson

Photo 2. Complete the concrete works

The HB Caisson is composed of outside-wall slabs, bottom slabs and footing slabs, all made of steel-concrete composite. By using steel members such as Built-up H-beams, the footing can be extended.

The snap shots of a steel skeleton work and the HB Caisson just before shipping out to Shin-Okitsu quay in Shimizu port are shown in Photo. 1 and Photo. 2 respectively. (Sugano et al., 2002)

Schematic comparison of a conventional reinforced concrete caisson (RC Caisson) and the HB Caisson is shown in Figure 2. The HB Caisson is characterized by the narrow body due to extended footings both seaside and landside. (Shiozaki et al., 2000)

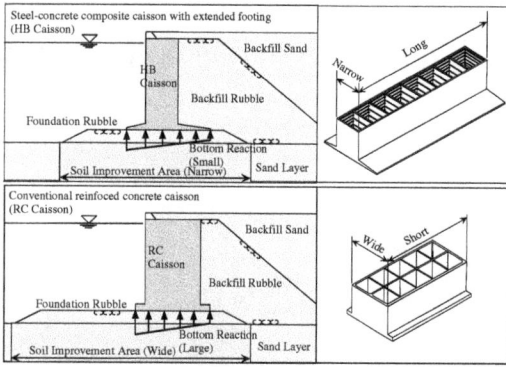

Figure 2. Comparison of HB and RC Caissons

The design standard on port and harbour structures in Japan (Goda et al., 2002) assumes that the backfill rubble above the landside footing is a part of gravity structure resisting against active earth pressure as shown in Figure 3. In this figure, the shaded part is assumed as wall body of a gravity type quay wall. Therefore, by extending the landside footing, the caisson width can be reduced.

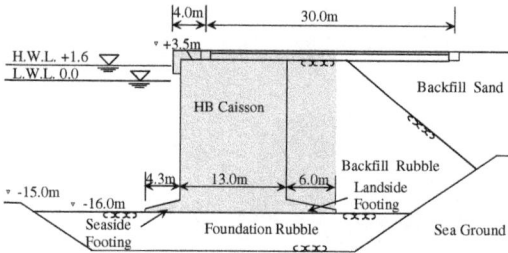

Figure 3. Cross section of HB Caisson type quaywall (Shin-Okitsu Quay)

Due to extension of the seaside footing, the intensity of bottom slab reaction force distribution is decreased. For that reason, soil improvement area is also reduced. The weight of HB Caisson is lighter than an equivalent RC Caisson because of long footing effect, whereas the construction cost of HB Cais-

son is more expensive than RC Caisson because of the HB Caisson use many steel members. However, in severe construction conditions such as soft subsoil or deep design depth, the total cost including soil improvement and caisson setting could be reduced.

A series of RC Caisson and HB Caisson model tests was conducted with following similar conditions, to investigate the influence of extending landside footing and the backfill rubble, especially the behavior of backfill rubble above the landside footing, a) Design seismic coefficient was $k_h=0.15$. b) Water depth was -7.5 m for static test, and −10 m for dynamic test as of prototype scale.

2 STATIC CONDITION

To investigate the static stability system of both RC and HB caissons on the soft subsoil foundation, centrifuge tests were conducted as shown in Figure 4. The dimensions in this figure are for a model scaled to 1:50 of a prototype structure. In the subsoil region a thin soft clay layer ($c=6$ kN/m^2) was placed. The cohesion of soil for the soft clay layer was selected in order to induce circular sliding failure at applied centrifuge acceleration arrives up to 50 G.

The soft clay layer was made from artificial bentonite powder, which water content of 120 %. To construct the model, No.6 Soma-sand for backfill and Grade 7 crushed gravel for foundation rubble and backfill rubble were used.

The time histories of centrifuge acceleration and horizontal displacement of the caisson top are shown in Figure 5. At first the caisson top moved toward seaside direction, however around 25 G state, suddenly moved landside direction due to the large shear deformation of soft clay layer. The failure initiated at 25.3 G for HB Caisson and at 27.9 G for RC Caisson, and the failure modes (i.e. residual conditions) are shown in Figure 4. In this figure, the observed cracks on the backfill ground and on the seabed are indicated as arrows, and the deformation of caissons and rubble mounds can be seen. The failure mode of HB Caisson is in agreement with RC Caisson's failure mode. These results indicate that

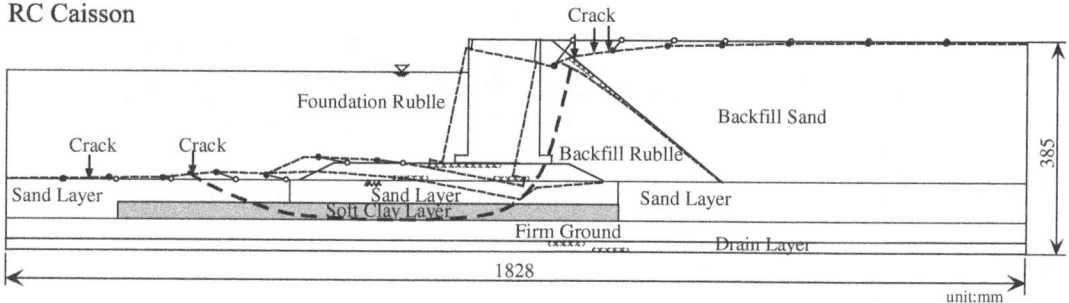

Figure 4. Cross section of centrifuge model and residual displacement

320

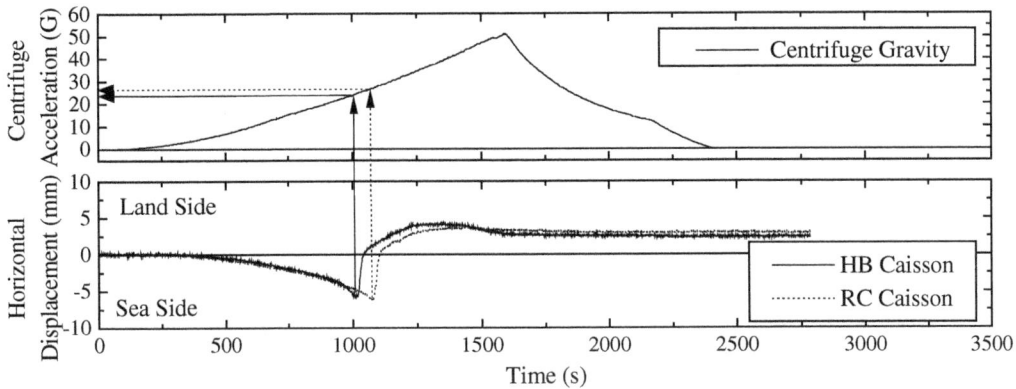

Figure 5. Time histories of applied centrifuge acceleration and horizontal displacement of the caisson

the performances of the RC and HB Caissons under static condition are almost the same.

3 DYNAMIC CONDITION

Underwater shake table tsts provide an opportunity to investigate the behavior of HB and RC caissons during an earthquake. The experimental techniques adapted here were the same as those developed by Sugano (et al., 1995, 1996) to analyze the mechanism of collapse of the gravity type quay walls damaged by the 1995 Hyogoken-Nambu earthquake.

The models were built within a hollow rectangular model container placed on the shake table. The model container had dimensions of 1500 mm deep by 2500 mm wide by 3500 mm long. The 1:20 model caisson set was installed in the model container on the shake table, which was set in the middle bottom of a water pool 2 m maximum deep and 13 m by 13 m wide to simulate the effect of sea water. The front end (i.e. sea side in model quay wall) of the container above the model seabed level was open to the water pool.

The similitude in 1 G field for soil-structure-fluid system (Iai 1989, Iai and Sugano 1999) was adopted for the underwater shake table tests. Due to the fact that the shear modulus of sand in small strain level is proportional to the square root of the confining pressure, the scaling relation includes a scaling factor for strain as shown in Table 1.

The cross sections of the test model are shown in Figure 6. In the tests, three units of model caissons were installed along the quay wall face line for each HB and RC caissons to prevent boundary effect such as frictional resistance between the model and the model container wall. The caisson in the middle was used for monitoring accelerations, pore water pressures and displacements. The model caisson was made of thick aluminum plates and filled with dry sand to adjust up to the dry density of the model

Table 1. Scaling factors for shake table tests

Quantities	Scaling factor	Scaling factors for 1/20 model
Length	λ	20.0
Time	$\lambda^{0.75}$	9.46
Acceleration	1	1
Displacement	$\lambda^{1.5}$	89.4
Stress	λ	20
Strain	$\lambda^{0.5}$	4.47

caisson 21 kN/m³. The foundation subsoil was idealized in the model tests by firm sand layer, which relative density of 90 %. The rubble mound and the backfill rubble were modeled using Grade 4 and Grade 6 crushed gravel respectively.

Several tests were carried out using the "Hachinohe" motion, which is used as a standard design input motion for port structures in Japan, and the peak values of the shake table acceleration (i.e. Input motion) were reached to 170 Gal, 255 Gal and 340 Gal each step. To observe the large deformation mode of quay walls at failure state, the sinusoidal input test was conducted with 5 Hz frequency and 300 Gal acceleration for 10 cycles.

For each of the Hachinohe 340 Gal cases for RC and HB caisson quay wall models, the dynamic performances did not reveal large differences as shown in Figure 7. Also, the failure mode of the RC and HB caisson after shaking can be seen in Figure 6. The dynamic acceleration responses of the back fill rubble and the HB caisson are shown in Figure8. Figure 8 demonstrates that the measured backfill rubble motion above the landside footing is approximately similar to the HB caisson motion during shaking.

4 CONCLUDING REMARKS

A series of model tests was reviewed in order to discuss the performance of HB Caisson quay walls. Major conclusions obtained are as follows.

321

HB Caisson

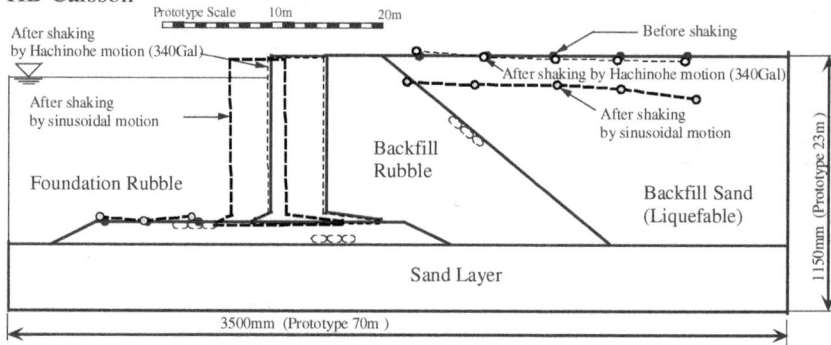

RC Caisson

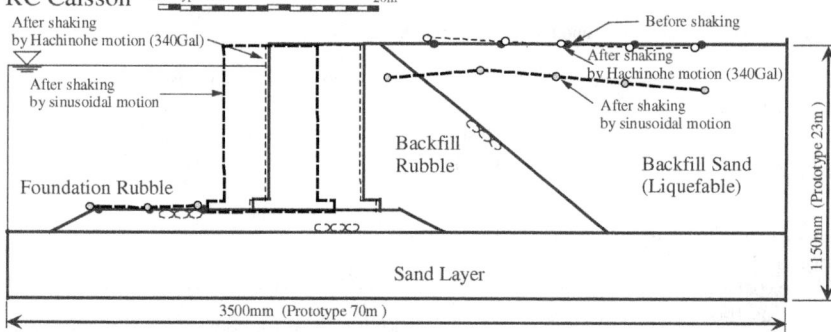

Figure 6. Cross section of the HB and RC caisson model

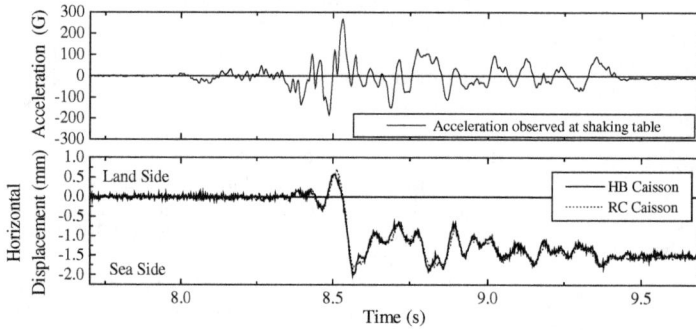

Figure 7. Time histories of input motion acceleration and horizontal displacement of the caisson top

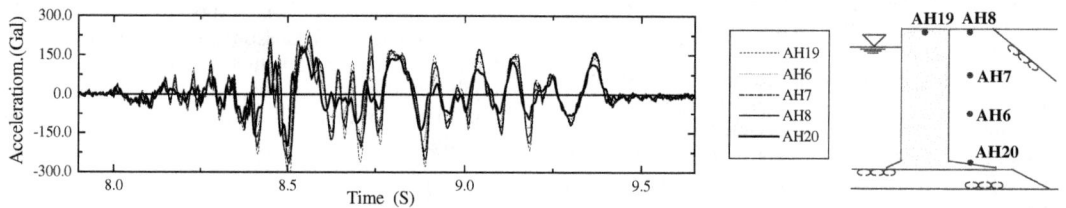

Figure 8. Time histories of the response of model (HB Caisson)

(1) The failure mode of HB Caisson and conventional RC Caisson on soft subsoil in static condition were the same.

(2) The HB Caisson body and backfill rubble above the landside footing moved together, and the backfill rubble above the landside footing was working as the effective weight.

(3) The seismic performance of HB Caisson and RC Caisson during underwater shake table tests indicate similar motions and residual displacements.

(4) The test results indicating that the current design code is applicable for the HB Caisson design.

As the conclusion was derived from only the model tests, more study must be necessary for appropriate design method of the HB Caisson quay wall.

REFERENCES

Goda, Y., Tabata, T. and Yamamoto, S. eds., 2002, Technical Standards and Commentaries for Port and Harbour Facilities in Japan, The Overseas Coastal Area Development Institute of Japan.

Iai, S., 1988, Similitude for shaking table tests on soil-structure-fluid model in 1G gravitational field, *Report of the Port and Harbour Research Institute*, PHRI, Vol.27, No.3, 3-24.

Iai, S. and Sugano, T., 2000, Shake Table Testing on Seismic Perfomance of Gravity Quay Walls, *Proc. of the 12th World Conference on Earthquake Engineering*, WCEE, Paper No.2680.

Shiozaki, Y., Sugano, T., Yamamoto, S., Tanaka, T. and Sekiguchi, K., 2000, Experimental Study of the Behavior of Hybrid (Steel-Concrete Composite) Caisson-type Quay Walls during Earthquakes using an underwater shaking table, *Proc. of the 12th World Conference on Earthquake Engineering*, WCEE, Paper No.1518.

Sugano, T., Fujii, A. and Shiozaki, Y., 2002. Application of the Hybrid-Caisson with Extended Footing to High Seismic Resistance Quays, *Tsuchi-to-Kiso*, JGS, Vol.50, No.4, Ser.No.531, 4-6. (in Japanese)

Sugano, T., Morita, T., Mito, M., Sasaki, T. and Inagaki, T., 1996, Case Studies of Caisson Type Quay Wall Damage by 1995 Hyogoken-Nanbu Earthquake, *Proc of the 11th World Conferrence on Earthquake Engineering*, WCEE, Paper No.765.

Sugano, T., Mito, M. and Oikawa, K., 1995, Mechanism of damage to port facilities during 1995 Hyogo-ken Nambu Earthquake, *Technical Note for the Port and Harbour Research Institute*, PHRI, No.813, 207-252. (in Japanese).

Soft Ground Engineering in Coastal Areas, Tsuchida et al. (eds)
© 2003 Swets & Zeitlinger, Lisse, ISBN 90 5809 613 0

Evaluating undrained cyclic strength of gravelly soil by shear modulus

Y. Tanaka
Central Research Institute of Electric Power Industry, Chiba, Japan

ABSTRACT: The possibility of evaluating undrained cyclic strength of natural gravelly deposits using shear moduli was investigated. Through this study, the following conclusions are drawn. 1) It was found that cyclic shear moduli at relatively large strain amplitude were closely related with undrained cyclic strengths, irrespective of the effective confining pressure, grain size distributions and whether or not the samples are intact, whereas there seems no unique relationship between initial shear moduli and undrained cyclic strengths. 2) It was found that the $G/G_0 \sim \gamma$ relation of a intact sample can be approximated by that of the reconsitited sample depending on the initial shear modulus of the intact sample. Based on this fact and the relationship mentioned in 1), a simplified procedure to evaluate undrained cyclic strength of gravelly soil by initial shear modulus of the ground and shear modulus of the disturbed sample was proposed.

1 INTRODUCTION

Recently, accurate evaluation of seismic stability of gravelly ground is required further as compared with before because important structures are being constructed on gravelly ground. In assessing the seismic stability of gravelly layers, it is necessary to evaluate their undrained cyclic strengths. However, high-quality intact gravelly samples often need to be taken using in-situ freezing sampling because the undrained cyclic strength of gravelly soils is susceptible to sample disturbance, though in-situ freezing sampling is much more expensive than conventional tube sampling. Thus a simplified method to evaluate the undrained cyclic strength of gravelly soils by using the LPT (Large Penetration Test)(Kaito et al., 1971) blow count has already been proposed (Tanaka et al., 1989; Tanaka et al., 1991; Tanaka et al., 1992). However, penetration tests generally seem to have some limitations in terms of grain size of the gravelly soil layer because the penetration probe can not be driven into the gravelly ground when the probe hits a very large particle even if the ground itself is very loose.

Figure 1 shows the results of the LPT conducted for the gravelly debris avalanche deposit which liquefied during Hokkaido-Nansei-Oki earthquake of July 12, 1993 (Tanaka et al., 1994). The Large Penetration Test (LPT), which was firstly introduced by Kaito et al.(1971) for testing gravelly soils, was carried out. The penetration resistance of LPT, N_L-value, is defined as the blow counts for a 30-cm

drive by freely dropping a 100-kg hammer from a height of 150 cm. The maximum external diameter and the internal diameter of the penetration probe of LPT are 73 mm and 54 mm, respectively. In Fig.1, N_L-value value is extraordinarily large at an approximate depth of G.L. - 4 m. The extraordinarily large value is attributable to the existence of a very large gravel particle.

To overcome this difficulty, it would be thus useful to find another index and to establish its correlation with the undrained cyclic strength of gravelly

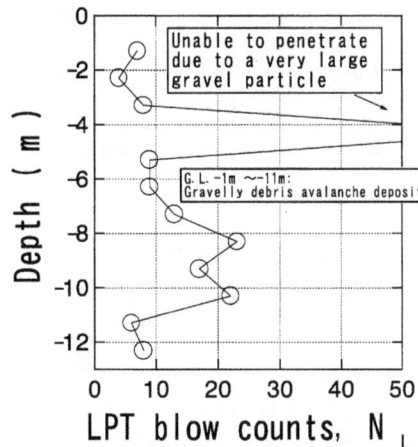

Figure 1. A case study of Large Penetration Test for a gravelly ground which liquefied during Hokkaido-Nansei-Oki Earthquake of July 12, 1993. (Tanaka et al., 1994)

soils. In the present study, the validity of the shear modulus as an alternative index to evaluate the in-situ undrained cyclic strength of gravelly soils is investigated.

There are several precedents in literature on evaluation methods for undrained cyclic strengths of gravelly soils using shear wave velocity.

Tokimatsu et al.(1988) suggested that there is a unique relationship between undrained cyclic strengths and normalized shear modulus for several kinds of sands and gravelly soils, irrespective of the effective confining pressure and whether or not the samples are intact.

$$G_N = \frac{G_f}{F(e_{min}) \cdot (\sigma'_m / P_1)^{\frac{2}{3}}} \qquad (1)$$

where,

$$F(e_{min}) = \frac{(2.17 - e_{min})^2}{1 + e_{min}} \qquad (2)$$

e_{min}: Minimum void ratio, σ'_m : Mean effective stress, P_1 : =98kPa, G_f: Initial shear modulus calculated by the following equation.

$$G_f = \rho_t V_s^2 \qquad (3)$$

V_s: In-situ shear wave velocity, ρ_t: Wet density of soil

Since minimum void ratio, which is used in Eq.(2), of gravelly soils seems to vary depending on the method of compaction, it seems to be a subject for future study to establish standard method for determining minimum void ratio of gravelly soils. Furthermore, reportedly, the Hardin - Richart's formula (Hardin and Richart, 1963), which is the origin of Eq.(2), can not be applied to gravelly soils, when the void ratio of which is less than 0.5 (Tanaka et al., 2000b). This also seems to be another subject for future study.

Hatanaka et al.(1997b) investigated the relationship between undrained cyclic strength of undisturbed samples and in-situ shear wave velocity, implying that for Holocene gravels, a fairy good correlation between undrained cyclic strength and shear wave velocity can be found, whereas there seems no unique correlation between undrained cyclic strength and shear wave velocity for the data of Pleistocene gravels.

Andrus and Stokoe (2000) investigated the relationships between shear stress ratio generated in the ground during earthquakes and shear wave velocity modified in terms of effective overburden pressure based on field performance data. They also proposed curves for assessment of liquefaction of uncemented Holocene-age gravels as boundary between liquefied cases and nonliquefied cases. In general, undrained cyclic strength of dense or Pleistcene gravelly soils can not be determined based on field performance data because there is almost no case of liquefaction.

According to the review of precedents in literature described above, it was found that there was no appropriate method to evaluate undrained cyclic strength of dense or Pleistcene gravelly soils. Thus, in this study, the evaluation method of undrained cyclic strength of gravelly soils including dense or Pleistcene ones is investigated.

In this paper, the undrained cyclic strength and cyclic shear modulus of intact gravelly soils and their reconstituted samples were measured by large triaxial tests basically under their in-situ effective confining pressure. 'Intact sample' or 'undisturbed sample' means a sample obtained by the in-situ freezing method. 'Reconstituted sample' means a sample which is reconstituted at the same density as the undisturbed sample which was used for the undrained cyclic strength test. 'Disturbed sample' means a sample , which has the same grain size distribution as the gravelly soil layer, obtained in disturbed state. The undrained cyclic strength is defined as the stress ratio required to reach the double amplitude of axial strain, 2% or 2.5%, in 20 loading cycles. R_s(DA=2% or 2.5%, N_c=20) or R_s denote the undrained cyclic strength.

2 ELATIONSHIP BETWEEN SHEAR MODULUS AND UNDRAINED CYCLIC STRENGTH

Reportedly, there is a unique relationship between undrained cyclic strength of gravelly soils and G_v / σ_{mv}' (G_v : Secant shear modulus at maximum volumetric strain in triaxial compression test, σ_{mv}': Mean effective stress at maximum volumetric strain in triaxial compression test), irrespective of the effective confining pressure, grain size distributions and whether or not the samples are intact (Tanaka et al. 1992). Since it is difficult to evaluate in-situ values of G_v and σ_{mv}', relationships between undrained cyclic strength and various kinds of shear modulus are investigated.

2.1 Relationships between undrained cyclic strength and initial shear modulus

Since the initial shear modulus can be obtained more easily than G_v by site investigation, it is worth investigating the relationship between undrained cyclic strength and initial shear modulus instead of G_v. Figs. 2(a) and 2(b) show the relationship between undrained cyclic strength and initial shear modulus, where G_f denotes initial shear modulus calculated from the shear wave velocity and G_0 denotes the initial shear modulus by the cyclic triaxial test and σ_c' denotes effective confining pressure.

Though there seems to be some correlation in Figs. 2(a) and (b), it is not sufficient for accurate evaluation of undrained cyclic strength (Tanaka et al. 1992).

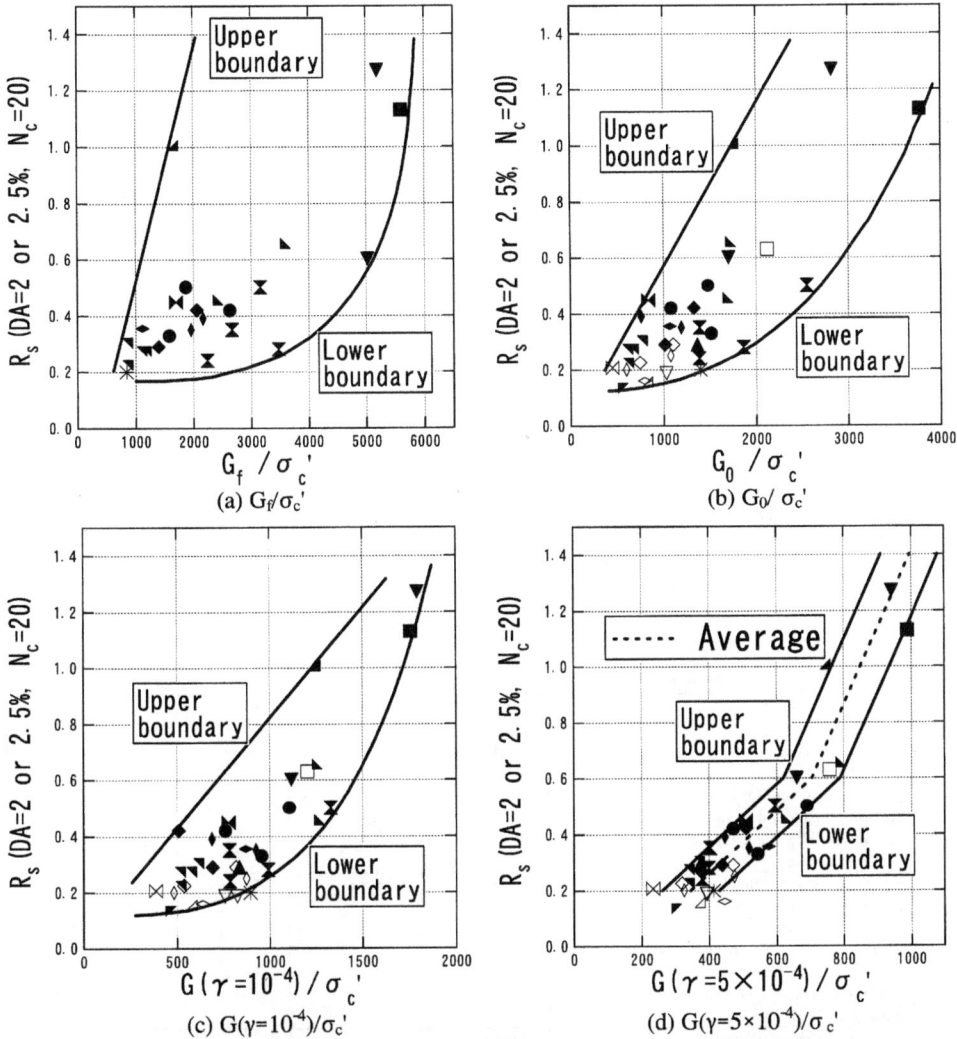

Figure 2. Undrained cyclic strength VS. various kinds of shear moduli (See Table 1 for symbols)

Data in Fig.2(a) seem to vary more widely than those in Fig.2(b). This indicates that G_0 is more appropriate index for evaluation of undrained cyclic strength than G_f, though G_0 and G_f, which should theoretically be equal. There are several reports saying that G_f equals G_0 of undisturbed samples (Goto et al., 1987; Goto et al., 1992; Hatanaka et al., 1988). However, on the contrary, some reports say G_f possibly differs from G_0, which is the cause of the difference between Fig.2(a) and Fig.2(b)(Tanaka et al., 2000 ; Tanaka, 2001). Thus, causes of difference between G_f and G_0 are discussed in the following.

The possible causes of the difference between G_f and G_0 are itemized as follows:

1 G_0 of samples was affected due to sample disturbance during sampling even by the in-situ freezing method.

2 Since axial strain of the sample was mostly measured externally, G_0 was underestimated due to bedding error.

3 Since shear wave velocity by in-situ velocity logging was affected by anisotropy and heterogeneity of the gravelly layer, G_f was overestimated.

Firstly, the cause 1) shown above is discussed. Figure 3 shows the relationships between undrained cyclic strength of undisturbed samples and that of reconstituted samples. Figure 3 indicates that undrained cyclic strengths of undisturbed samples are larger than those of reconstituted samples without exception. This means that undisturbed samples are not affected by sample disturbance considerably. However, it is impossible to demonstrate that undisturbed samples are barely affected by sample disturbance. Even if undisturbed samples are affected by

327

Table 1. Symbols in Fig. 2

Symbols	Sites	Mean grain size (mm)	Maximum grain size (mm)	Fines content (%)	Effective confining pressure (kPa)
●	A-site gravel (Tanaka et al.,1988; Tanaka et al., 1990)	0.3~2.5	20~40	< 2.0	69~118
▲	Reclaimed gravel in Rokko island （Tanaka et al., 2000a)	0.5~8.0	40~100	10~30	98~245
■, □	K-site gravel (Tanaka et al., 1988)	15~35	100~150	1~3.2	98
◆, ◇	T-site gravel （Tanaka et al., 1991)	3.0~20	50~200	0.9~8.1	176~225
▼, ▽	KJ-site gravel (Tanaka et al., 1991)	3.0~7.0	40~60	1~15	157
✖	H-site gravel (Tanaka et al., 2000b)	0.4~15	20~200	10~30	69~245
✳	Gravelly debris avalanche deposit (Tanaka et al., 1994), (Kawai et al., 1994)	4.0~20	10~30	5~10	49
◤, △	Mandano gravel (Shamoto et al., 1986 ; Goto et al, 1987)	2.0	15~35	<1.0	118
◆, ◇	Tokyo gravel (Hatanaka et al, 1988)	5.6~19.5	80~90	0.81~11.8	294
◆, ◊	Tone-river gravel (Goto et al, 1992)	2.0~30	50~90	< 1.0	127~186
◣	KF-site gravel (Uchida et al, 1997)	12~37	106~125	0.2~0.6	108~226
◤	Reclaimed Masa ground at Port Island (Improved ground) (A society for inveatigating the ground damaged during the 1995 Great Hanshin Earthquake, 1998)	0.6~4	20~120	2~10	76~186
▶	Reclaimed Masa ground at Port Island (Unimproved ground) (A society for investigating the ground damaged during the 1995 Great Hanshin Earthquake, 1998; Hatanaka et al, 1997)	1.7~3.7	37~102	3.9~8.6	108
▶◀, ⋈	Tadotsu site gravel (Watabe et al, 1991)	3~17	70~150	< 3	186

Solid symbols and * symbol mean undisturbed samples recovered by the in-situ freezing method, while open symbols mean reconstituted samples.

disturbance to a certain degree, it is meaningful to establish the relationship between undrained cyclic strength and shear modulus because the effect of disturbance on undrained cyclic strength and shear modulus has a tendency to compensate each other.

Secondly, the cause 2) mentioned above is discussed. It is pointed out that the externally measured axial strain in triaxial tests is considered to underestimate the stiffness due to the effect of bedding error (B.E.) between the ends of the specimen and the cap and pedestal of the triaxial cell (Tatsuoka and Shibuya, 1992). Figure 4 shows the results of cyclic triaxial tests of sandy soils and gravelly soils conducted for investigating the effect of bedding error (Tanaka et al., 2000). Figure 4 indicates that the un-

derestimation of the stiffness due to the bedding error is about 15% or smaller over a wide range of grain size. Since the effect of the B.E. is small in triaxial tests of gravelly soils, the effect of B.E. on underestimation of shear modulus obtained by the triaxial test is neglected in this paper.

Finally, the cause 3) mentioned above is discussed. There are some reports describing that initial shear modulus evaluated by the results of in-situ velocity logging coincides with that identified by inversion analysis using records of earthquake observation (For example, Sato et al.(1996)), whereas the author reported a case study describing that the two kinds of initial shear moduli do not coincide each other (Tanaka, 2001). Figure 5 shows the distribu-

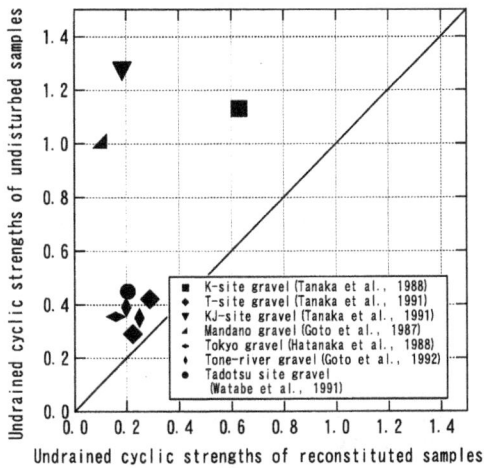

Figure 3. Undrained cyclic strengths of undisturbed samples VS. undrained cyclic strengths of reconstituted samples

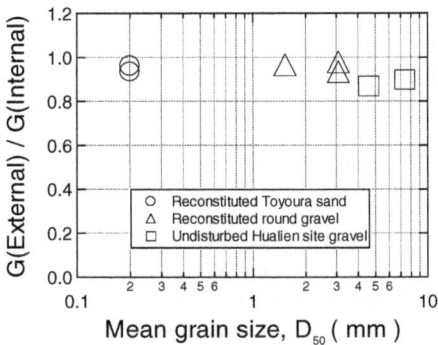

Figure 4. Effect of grain size on underestimation of initial shear modulus due to bedding error (Tanaka et al, 2000)

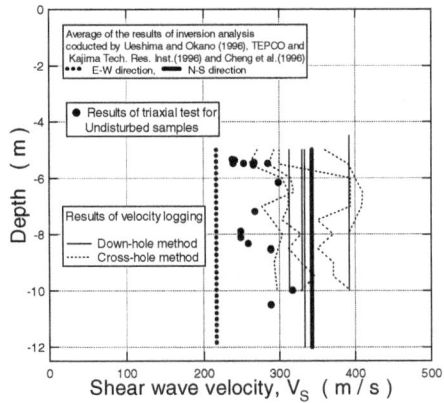

Figure 5. Distribution of shear wave velocity evaluated by various methods (Tanaka, 2001)

tion of shear wave velocity measured by in-situ velocity logging and identified by inversion analysis using records of earthquake observation. The shear wave velocity measured by in-situ velocity logging varies widely from 270 to 410 m/s. Shear wave velocity identified by inversion analysis, which is also plotted in Fig.5, shows obvious azimuth dependency. Figure 5 shows that shear wave velocity identified from N-S component of earthquake records is about 340 m/s, whereas that from E-W component is about 220 m/s.

The author insisted that the difference in shear wave velocity shown in Fig.5 should be attributable to anisotropy and heterogeneity of the gravelly soil layer (Tanaka, 2001). Thus, for the gravelly soil layer shown in Fig.5, shear wave velocity should be determined considering both anisotropy and heterogeneity. Unlike shear wave velocity by velocity logging, shear wave velocity by inversion analysis is hardly affected by heterogeneity because the wavelength determined by the predominant frequency of

earthquakes is significantly larger than the gravel particles. Moreover, the average of shear wave velocity identified from N-S component and E-W component in Fig.5 is about 280 m/s, which approximately equals shear wave velocity converted from initial shear moduli by cyclic triaxial tests as shown in Fig.5. Thus, if shear wave velocity by in-situ velocity logging coincides with that by inversion analysis, shear wave velocity by in-situ velocity logging can be used for evaluation of undrained cyclic strength. However, for the anisotropic gravelly soil layer as shown in Fig. 5, the average of shear wave velocity identified from orthogonal components should be attached importance.

The discussion described above is summarized as follows:

1 It is meaningful to establish the relationships between undrained cyclic strengths and shear modulus of intact samples because intact samples do not seem to be affected considerable sample disturbance.

2 Initial shear modulus used for evaluating undrained cyclic strengths can be evaluated by velocity logging or inversion analysis, if both evaluated results are the same. If initial shear modulus velocity logging differs from that by inversion analysis, initial shear modulus by inversion analysis should be attached importance.

2.2 *Relationship between undrained cyclic strength and shear modulus at large strain amplitude*

Since undrained cyclic strength is a mechanical property at large shear strain, the shear modulus at relatively large strain such as G_v mentioned above may be more closely related to undrained cyclic strength than initial shear modulus. Thus, relationships between undrained cyclic strength and shear modulus at large strain amplitude are investigated herein. Figure 2(c) and 2(d) show the relationships

329

between undrained cyclic strength and $G(\gamma=10^{-4})/\sigma_c'$ and $G(\gamma=5\times10^{-4})/\sigma_c'$, respectively. $G(\gamma=10^{-4})$ and $G(\gamma=5\times10^{-4})$ are defined as shear moduli at shear strain amplitude, $\gamma=10^{-4}$ and $\gamma=5\times10^{-4}$, respectively. Both $G(\gamma=10^{-4})$ and $G(\gamma=5\times10^{-4})$ were calculated from the average initial shear moduli of the samples used for the undrained cyclic strength test and degradation curves of shear modulus obtained from cyclic deformation tests. Expectedly, the scatter of the data in Fig. 2(c) is smaller than that in Figs. 2(a) and 2(b). Furthermore, the scatter of the data in Fig. 2(d) is smaller than that in Fig. 2(c). The relationship between undrained cyclic strength and $G(\gamma=5\times10^{-4})$ $/\sigma_c'$ is an almost unique relationship irrespective of sampling site, confining pressure, and whether or not the sample is intact.

2.3 Expression of effect of confining pressure on undrained cyclic strength

It is well known that the undrained cyclic strength of gravelly soils and sands decreases with an increase in the effective confining pressure beyond the in-situ effective overburden pressure (Tatsuoka et al., 1981; Kokusho et al., 1983; Yoshimi et al., 1984; Kon-no et al., 1987; Tanaka et al., 1992). This trend should be evaluated for evaluation of undrained cyclic strength of dense gravelly soil because this trend is remarkable for dense gravelly soils and sands.

Tanaka et al. (1992) classified the effect of effective confining pressure on undrained cyclic strength of undisturbed samples into two groups according to the relation between the effective confining pressure and the effective overburden pressure.

Group A: Two or more different stress values, which equal or exceed the in-situ effective overburden pressures, are selected as effective confining pressures for undisturbed samples which are taken from a specific depth (Tatsuoka et al. 1981; Kokusho et al. 1983; Yoshimi et al. 1984).

Group B: In-situ effective overburden pressure of each specimen, which were obtained at various depths in homogeneous soil layers are selected as effective confining pressures (Kon-no et al. 1987; Furuta and Konno 1988).

The experimental results from Group A are useful for estimating the change in the in-situ undrained cyclic strength due to the increase in the effective overburden pressures caused by construction, such as filling embankments, while the experimental results from Group B are important when estimating the effect of overburden pressure on the in-situ undrained cyclic strength of deep soil layers.

Tanaka et al. (1992) showed that the effects of effective confining pressure of both Group A and Group B can be unified and expressed by the following equation.

$$R_s = 0.15 + (R_{s1} - 0.15) \cdot \left(\frac{\sigma_c'}{P_1} \right)^{-0.65} \qquad (4)$$

where, R_{s1}: stress ratio R_s at $\sigma_c'=P_1$ in Eq. (4).

On the other hand, the effect of effective confining pressure on shear modulus can be evaluated only by conducting cyclic triaxial test in the laboratory. Generally, for sandy gravels and sands, the effect is usually expressed by the following equation.

$$\frac{G_0}{P_1} = A_1 \cdot \left(\frac{\sigma_c'}{P_1} \right)^{n_1} \qquad (5)$$

where, n_1 is a constant and is about 0.5.

Generally, the relationships between G/G_0 and γ/γ_r of sands and gravelly soils are independent of effective confining pressure, whereas the relationships between G/G_0 and γ are affected by the effective confining pressure (Tanaka et al., 2000b). γ_r is called reference strain defined as γ when $G/G_0=0.5$. The effect of effective confining pressure on γ_r is usually expressed by the following equation.

$$\gamma_r = A_2 \cdot \left(\frac{\sigma_c'}{P_1} \right)^{n_2} \qquad (6)$$

where, n_2 is a constant which can be approximated by $1-n_1$ (Tanaka et al., 2000b).

Figure 6 shows relationships between G/G_0 and γ/γ_r for various kinds of gravelly soils, indicating that the values of $G(\gamma=5\times10^{-4})/G_0$ is in the region of $0.20 \sim 0.55$. The solid line in Fig.6, which shows an average relationship, is calculated by the following equation.

$$\frac{G(\gamma)}{G_0} = 0.5 \cdot \left(\frac{\gamma}{\gamma_r} \right)^{-0.43} \qquad (7)$$

Substituting Eqs.(5) and (6) into Eq.(7), the following equation can be derived.

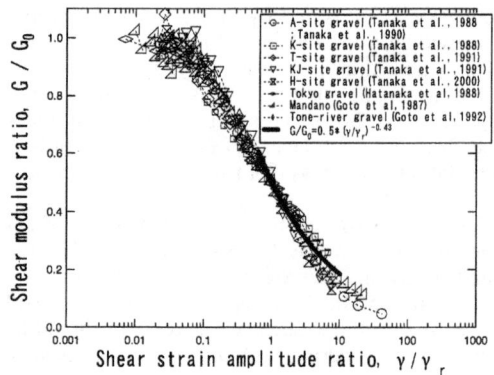

Figure 6. Shear modulus ratio VS. shear strain amplitude of various kinds of gravels

$$\frac{G(\gamma)}{\sigma_c'} = 0.5 \cdot (\gamma)^{-0.43} \cdot A_1 \cdot A_2^{0.43} \cdot \left(\frac{\sigma_c'}{P_1}\right)^{n_1 + 0.43n_2 - 1} \qquad (8)$$

Equation (8) can be rewritten as follows:

$$\frac{G(\gamma)}{\sigma_c'} = \left\{\frac{G(\gamma)}{\sigma_c'}\right\}_{\sigma_c' = P_1} \cdot \left(\frac{\sigma_c'}{P_1}\right)^{n_1 + 0.43n_2 - 1} \qquad (9)$$

Substituting $n_1 = n_2 = 0.5$ as representative values, Eq.(9) of $\gamma = 5 \times 10^{-4}$ can be further derived as follows:

$$\frac{G(5 \times 10^{-4})}{\sigma_c'} = \left\{\frac{G(5 \times 10^{-4})}{\sigma_c'}\right\}_{\sigma_c' = P_1} \cdot \left(\frac{\sigma_c'}{P_1}\right)^{-0.285} \qquad (10)$$

The solid lines in Fig.7 show the effect of effective confining pressures on undrained cyclic strength. The solid lines were calculated using Eq.(10) and an average relationship between R_s and $G(\gamma = 5 \times 10^{-4})/\sigma_c'$ shown in Fig.2(d). Calculated results by Eq.(4) are also plotted as broken lines in Fig.7. Though the assumptions mentioned above are rather rough, the broken lines agree well with the solid lines in Fig.7.

Thus, it was proved that the effect of confining pressure on undrained cyclic strength can be considered properly by using the relationships between R_s and $G(\gamma = 5 \times 10^{-4})/\sigma_c'$ shown in Fig.2(d).

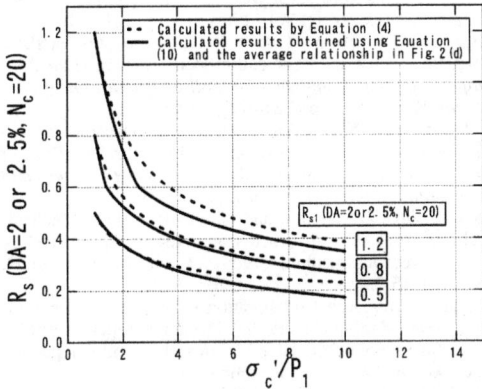

Figure 7. Relationship between undrained cyclic strength and effective confining pressure

3 A SIMPLIFIED METHOD FOR EVALUATION OF UNDRAINED CYCLIC STRENGTH OF GRAVELLY SOIL BY INITIAL SHEAR MODULUS OF THE GROUND AND THE $G/G_0 \sim \gamma$ RELATIONSHIP OF THE DISTURBED SAMPLE

3.1 Evaluation of in-situ shear modulus $G(\gamma = 5 \times 10^{-4})$

To make use of the relationship between undrained cyclic strength and $G(\gamma = 5 \times 10^{-4})$ shown in Fig.2(d), it is necessary to estimate the in-situ value of $G(\gamma = 5 \times 10^{-4})$. Since in-situ initial shear modulus can

be evaluated by in-situ velocity logging or inversion analysis as described previously, it is necessary to estimate $G(\gamma = 5 \times 10^{-4})/G_0$ for estimating $G(\gamma = 5 \times 10^{-4})$. The relationships between G/G_0 and γ of intact samples are reportedly similar to that of reconstituted samples (Hatanaka and Uchida, 1995; Tanaka et al., 1995). Thus, $(G(\gamma = 5 \times 10^{-4})/G_0)_{intact} / G(\gamma = 5 \times 10^{-4})/G_0)_{reconst}$ is plotted in Fig.8 against G_0/σ_c' of intact samples.

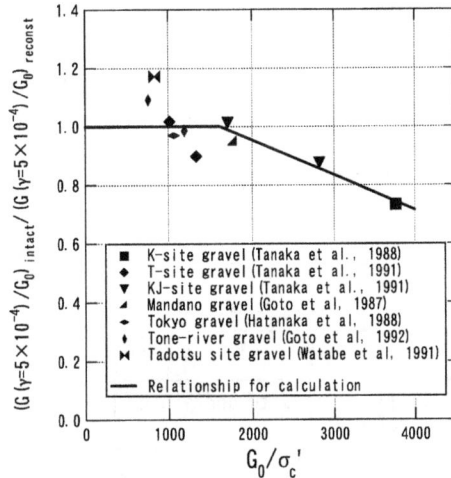

Figure 8. Relationship between $G(\gamma = 5 \times 10^{-4})/G_0)_{intact}/ G(\gamma = 5 \times 10^{-4})/G_0)_{reconst}$ and G_0/σ_c'

$(G(\gamma = 5 \times 10^{-4})/G_0)_{intact}$ and $G(\gamma = 5 \times 10^{-4})/G_0)_{reconst}$ are defined as $G(\gamma = 5 \times 10^{-4})/G_0$ of intact samples and reconstituted samples, respectively. The value of $(G(\gamma = 5 \times 10^{-4})/G_0)_{intact} / G(\gamma = 5 \times 10^{-4})/G_0)_{reconst}$ decreases as the value of G_0/σ_c' increases. It is reasonable to consider that the value of G_0/σ_c' increases, as the gravelly soil layer becomes older. Thus, the trend in Fig.8 mentioned above is attributable to the difference in soil skeleton due to the aging effect, such as cementation, consolidation and strain history.

The solid line in Fig.8 is used for evaluation methods for undrained cyclic strength of gravelly soils described in the next part of this paper.

3.2 A procedure to evaluate undrained cyclic strength by the simplified method

Figure 9 shows a procedure for evaluating the undrained cyclic strength, which would enable the undrained cyclic strength of gravelly soils to be estimated using in-situ shear modulus and results of cyclic deformation tests conducted for disturbed samples reconstituted at in-situ density. Moreover, in Fig.10, measured results are compared with calculated results obtained through the procedure. The calculated results are in agreement with the measured results.

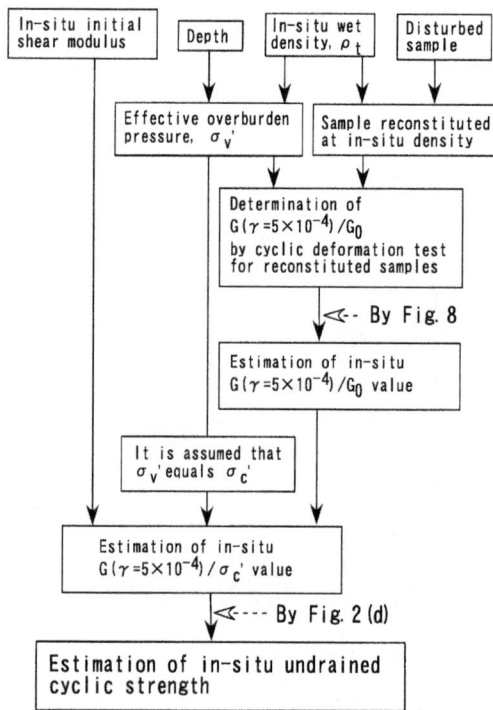

Figure 9. Procedure to evaluate undrained cyclic strength of gravelly soils by in-situ initial shear modulus and results of cyclic deformation tests for reconstituted samples

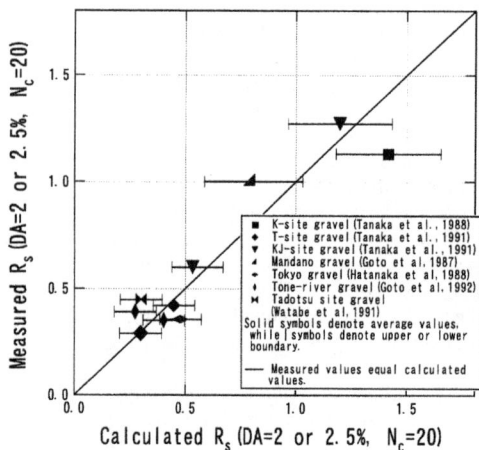

Figure 10. Estimated undrained cyclic strengths by in-situ initial shear modulus and results of cyclic deformation tests VS. measured undrained cyclic strengths

4 CONCLUSIONS

Relationships between undrained cyclic strengths and shear moduli were investigated on intact samples of gravelly soils sampled by in-situ freezing method and on their reconstituted samples. The possibility of evaluating undrained cyclic strength of natural gravelly deposits using shear moduli was investigated. Through this study, the following conclusions are drawn.

1   It was found that cyclic shear moduli at relatively large strain amplitude were closely related with undrained cyclic strengths, irrespective of the effective confining pressure, grain size distributions and whether or not the samples are intact, whereas there seems no unique relationship between initial shear moduli and undrained cyclic strengths.

2   It was found that the $G/G_0\sim\gamma$ relation of a intact sample can be approximated by that of the reconstituted sample depending on the initial shear modulus of the intact sample. Based on this fact and the relationship mentioned in 1), a procedure to evaluate undrained cyclic strength of gravelly soil by initial shear modulus of the ground and shear modulus of the disturbed sample was proposed.

REFERENCES

Andrus,R.D. and Stokoe II,K.H. (2000): Liquefaction resistance of soils from shear-wave velocity, *Journal of Geotechnical Engineering*, Vol.126, No.11, pp. 1015 - 1025.

Furuta, I. and Konno, M. (1988): Relationship between shear strength of soils and basic soil properties, *Proc. of the 23rd Japan National Conference on Soil Mechanics and Foundation Engineering*, pp.681–682, (in Japanese).

Goto,S., Shamoto, Y. and Tamaoki,K. (1987): Dynamic properties of undisturbed gravel sample by In-situ frozen, *Proc. of the 8th Asian Regional Conference of Soil Mechanic and Foundation Engineering*, Vol.1, pp.233-236

Goto,S., Suzuki,Y. Nishio,S. and Oh-oka,H. (1992): Mechanical properties of undisturbed Tone-river gravel obtained by In-situ freezing method, *Soils and Foundations*, Vol.32, No.3, pp.15-25

Geotechnical Research Collaboration Committee on the Hanshin-Awaji Earthquake (1998) : Dynamic properties of Masado fill in Kobe Port Island improved by soil compaction method, Final report of the committee.

Hardin,B.O. and Richart,F.E. (1963) : Elastic wave velocities in granular soils, *J. of SMF, ASCE*, 94, SM2, pp.353 – 369.

Hatanaka,M., Suzuki,Y., Kawasaki,T. and Endo,M.(1988): Cyclic undrained shear properties of high quality undisturbed Tokyo gravel, *Soils and Foundations*, Vol.28, No.4, pp.54-68

Hatanaka, M. and Uchida, A. (1995) : Effect of test methods on the cyclic deformation characteristics of high quality undisturbed gravel samples, Static and Dynamic Properties of Gravely Soils, *Proc. of sessions sponsored by the Soil Properties Committee of the Geotechnical Engineering Division of the A.S.C.E.*, pp.136 - 151.

Hatanaka,M., Uchida,A. and Ohara,J. (1997a) : Liquefaction characteristics of a gravelly fill liquefied during the Hyogo-ken Nanbu Earthquake, *Soils and Foundations*, Vol.37, No.3, pp.107 - 115.

Hatanaka,M., Uchida,A. and Suzuki,Y. (1997b) : Correlation between undrained cyclic shear strength and shear wave velocity for gravelly soils, *Soils and Foundations*, Vol.37, No.4, pp.85 - 92.

Kaito,Y., Sakaguchi,S., Nishigaki,Y., Miki,K. and Yukami,H. (1971) : Large penetrationtest, *Tsuchi-to-kiso, JSSMFE*, Vol.19, No.7, pp.15-21(in Japanese)

Kawai,T., Tanaka,Y., Kokusho,T. and Seo,K. (1994) : Dynamic strength of gravelly debris avalanche deposit which liquefied during Hokkaido-Nansei-Oki earthquake of July 12, 1993, *Proc. of 49th Annual convention of Japan Society of Civil Engineers*, Division III, pp.628 – 629(in Japanese).

Kokusho,T., Yoshida,Y., Nishi,K. and Esashi,Y. (1983) : Evaluation of seismic stability of dense sand layer, Part1 Dynamic strength characteristics of dense sand, Report No.383025, Central Research Institute of Electric Power Industry (in Japanese).

Kon-no,M., Furuta,I., Sawada,S. and Sakuma,N. (1987): Relationship between cyclic undrained shear strength of soils and basic soil properties, *Oyo Technical Report*, No.9, pp.1-19 (in Japanese).

Sato, K., Kokusho, T., Matsumoto, M. and Yamada, E. (1996) : Nonlinear seismic response and soil property during strong motion, Special issue on geotechnical aspects of the January 17 1995 Hyogoken-nambu earthquake, *Soils and Foundations*, pp. 41 - 52.

Shamoto,Y., Nishio,S., Baba,K., Goto,S., Tamaoki,K. and Akagawa,S. (1986) : Cyclic stress strain behaviour and liquefaction strength of diluvial gravels utilizing freezing sampling, *Proc. of Symposium on deformation and strength characteristics of gravelly soils and their testing method*, pp. 89 - 94(in Japanese).

Tanaka,Y., Kudo,K., Yoshida,Y., Kataoka,T. and Kokusho,T. (1988) : A study on the mechanical properties of sandy gravel, – Mechanical properties of undisturbed sample and its simplified evaluation –, Report No.U88021, Central Research Institute of Electric Power Industry(in Japanese).

Tanaka, Y., Kudo, K., Yoshida, Y., Nishi, K., Aida, M. and Suzuki, H. (1990): "On the applicability of various sampling methods to the gravelly ground," Report No.U90046, Central Research Institute of Electric Power Industry, (in Japanese).

Tanaka,Y., Kokusho,K., Kudo,K. and Yoshida,Y. (1991) : Dynamic strength of gravelly soils and its relation to the penetration resistance, *Proc. of 2nd International Conference on Recent Advances in Geotechnical Earthquake Engineering and Soil Dynamics*, Vol.1, pp.399–406.

Tanaka,Y., Kudo,K., Yoshida,Y. and Kokusho,T. (1992) : Undrained cyclic strength of gravelly soil and its evaluation by penetration resistance and shear strength, *Soils and Foundations*, Vol.32, No.4, pp. 128–142.

Tanaka,Y., Kokusho,T., Okamoto,T., Kusunoki,K., Kawai,T., Suzuki,K., Kataoka,T., Tohda,S., Abe,S. and Honsho,S. (1994) : Investigation in cases of liquefaction of gravelly debris avalanche deposit occurred during Hokkaido-Nansei-Oki earthquake of July 12, 1993 (Part 1), – Geotechnical investigation and liquefaction assessment –, Report No.U94007, Central Research Institute of Electric Power Industry, (in Japanese).

Tanaka,Y., Kokusho,T. and Kudo,K. (1995):Cyclic strength evaluation of gravel by shear modulus, *Proc. of the 10th Asian Regional Conference on Soil Mechanics and Foundation Engineering*, Vol.1, pp.91 - 94.

Tanaka,Y., Kanatani,M., Hataya,R., Sato,K., Kawai,T. and Kudo,K. (2000a) : Evaluation of liquefaction potential of gravelly soil layer based on field performance data, *Journal of Japan Society of Civil Engineers*, No.666/III-53, pp.55 - 72(in Japanese).

Tanaka,Y., Kudo,K., Nishi,K., Okamoto,T., Kataoka,T. and Ueshima, T. (2000b) : Small strain characteristics of soils in Hualien, Taiwan, *Soils and Foundations* , Vol.40 , No.3 , pp.111 - 125.

Tanaka, Y. (2001): Modeling anisotropic behavior of gravelly layer in Hualien, Taiwan, *Soils and Foundations*, Vol.41, No.3, pp.73 - 86.

Tatsuoka, F., Iwasaki, T., Tokida, K. and Kon-no, M. (1981): Cyclic undrained triaxial strength of sampled sand affected by confining pressure, *Soils and Foundations*, Vol.21, No.2, pp. 115 - 120.

Tatsuoka, F. and Shibuya, S. (1992): Deformation characteristics of soils and rocks from field and laboratory tests, *Report of the Institute of Industrial Science*, The University of Tokyo, Vol.37, No.1, (Serial No.235).

Tokimatsu,K., Yoshimi,Y. and Maeda,S. (1988): Prediction of liquefaction strength of sandy soils using S-wave velocity, *Proc. of the 23rd Japan National Conference on Soil Mechanics and Foundation Engineering*, pp.711 - 712(in Japanese).

Uchida,A., Hatanaka,M. and Iizuka,S.(1997):Undrained cyclic shear strength and post undrained cyclic shear volumetric behavior of high-quality undisturbed gravel, *Trans. of 14 th SMiRT*, K03/4, pp.135-142.

Watabe,M., Hayashi,M., Ishihara,K., Akino,K., Iizuka,S., Ukita,T., Yamazaki,T., Konno,T., Suzuki,S., Kitazawa,K., Suzuki,Y., Matsuda,T., Mori,K., Nagai,K. (1991):Large scale field tests on quaternary sand and gravel deposits for seismic siting technology, *Proc. of 2 nd International Conference on Recent Advances in Geotechnical Earthquake Engineering and Soil Dynamics*, Vol.I, pp.271 - 289.

Yoshimi, Y., Tokimatsu, K., Kaneko, O. and Makihara, Y. (1984): Undrained cyclic shear strength of a dense Niigata sand, *Soils and Foundations*, Vol.24, No.4, pp. 131 - 145.

Soft Ground Engineering in Coastal Areas, Tsuchida et al. (eds)
© *2003 Swets & Zeitlinger, Lisse, ISBN 90 5809 613 0*

Utilization of cement treated soft dredgings with special working ship

Y.X. Tang & Y. Miyazaki
Kanmon Kowan Kensetsu, Co., Ltd., Shimonoseki, Japan

T. Tsuchida
Port and Airport Research Institute, Yokosuka, Japan

ABSTRACT: Recently, how to effectively reduce and reuse soft dredgings has become an important matter in coastal engineering. Other than disposal of them into reclamation area, the dredgings after proper cement treatment usually shows good characteristics, useful for various port structures. Special working ships, pre-installed with cement treatment system, have been introduced into practice.

This paper presents compressive strength characteristics of the cement treated soils, then reports two applications in practice, discusses underwater casting method of this material and finally proposes a ground improvement method by use of the introduced ships.

1 INTRODUCTION

How to effectively reduce and reuse soft dredgings has become an important matter in coastal engineering. Other than disposal of them into reclamation area, the dredgings after proper cement treatment show good characteristics, suitable for various purposes of marine structures. Practical reuses of soft dredging are presented in this paper.

At first, an empirical, simple but useful correlation among compressive strength, cement content and water content is proposed. Variance property of compressive strength is interpreted on the basis of cement mixing degree and fluctuation in water content. This information results in an estimate for the variance coefficient of compressive strength near 0.3, which is in agreement with the values actually observed in projects.

Next, working ships special to cement treatment and placement of the soft dredging is exhibited. Such ships have been introduced in practice, aiming to provide a recycled geomaterial with high quality and to reduce construction cost. The working ships are equipped with cement treatment system, able to handle the soft dredging at a rate of 300 m³ per hour. The improved dredging has already found applications in many marine construction projects. In fact, they are frequently used as reclamation, backfill and liner materials for various purposes. More than 1.7 million m³ of dredging has been constructed at 11 sites by this method, two examples in Nagoya port are presented in detail.

Finally, a ground improvement method by use of the introduced technique is proposed instead of sand

replacement method. The proposed method will be conducted through dredging, improving and replacing processes. This method is supposed to eliminate the disposal of dredging, and no need of good sand.

2 COMPRESSIVE STRENGTH OF CEMENT TREATED DGREDGING

2.1 *Estimate of compressive strength*

Strength of cement treated soils is confirmed usually by means of unconfined compression test. Many factors, such as cement content, water content, curing period, temperature, casting method etc. will affect the compressive strength. The farmer two are considered as the major factors. That is to say, the compressive strength increases with increasing cement content and decreases with increasing water content (Tang et al, 2000, 2001, Miyazaki et al, 2001).

The author investigated about 30 cases of cement blending tests on various soft soils, and found that unconfined compressive strength q_u can be empirically expressed as Equation 1:

$$q_u = \frac{K(C - C_0)}{(G_S w/100 + 1)^2} \quad (1)$$

where, w and C are water content and cement content in mixed soil. G_S is specific gravity of soil grain. K is a strength coefficient and C_0 is regarded as the minimum of cement content required. In the case that cement content C actually mixed is less than the minimum value C_0, probably no improving effect could be expected. Figure 1 gives a fitting example

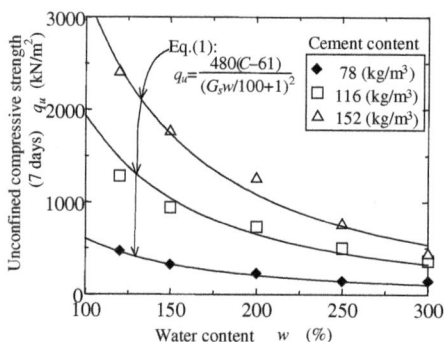

Figure 1. Cement blending test on Yamaguchi clay

of Equation 1 to blending test result for Yamaguchi clay.

Figure 2 presents 4 cases analyzed, in which unconfined compressive strengths q_u^* empirically estimated by Equation 1 and q_u actually measured are compared. It can be seen that they are well correlated, and the correlation coefficients r are generally larger than 0.97. In Equation 1, K and C_0 are unknown constants, which are determined from the obtained result of blending test. Moreover, prior to actual blending test, the parameters K and C_0 can be roughly given for a typical soil as bellows.

$K = 150 \sim 400$ kN·m/kg (7 day)

$K = 250 \sim 600$ kN·m/kg (28 day)

$C_0 = 30$ kg/m^3

2.2 Characterization of strength variance

As stated above, cement content and water content are regarded to be the substantial factors, though there may be various other reasons. If variances in cement content and water content are determined, the authors believe that the variance in compressive strength can be evaluated.

Variance in cement content depends on degree of mixing of cement with soil, and variance in water content reflects water content fluctuation of dredging. The evaluating method for the variance of compressive strength is derived from the relation of Equation 1 (Tang et al, 2001, 2002):

$$V_{qu} = \frac{\sigma_{qu}}{\overline{q}_u} = \left[\left(\frac{\overline{C}}{\overline{C} - C_0} V_C \right)^2 + \left[\frac{2 G_s \overline{w}/100}{G_s \overline{w}/100 + 1} V_w \right]^2 \right]^{\frac{1}{2}} \quad (2)$$

where, V_C, V_w are variance coefficients of cement content and water content respectively, while $\overline{C}$ and $\overline{w}$ are their averaged values.

At a project site in Nagoya port, the authors observed the water content of original dredgings at a

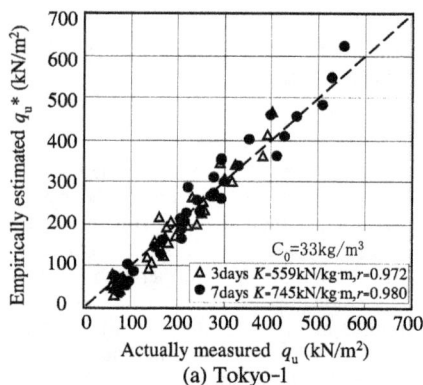

(a) Tokyo-1

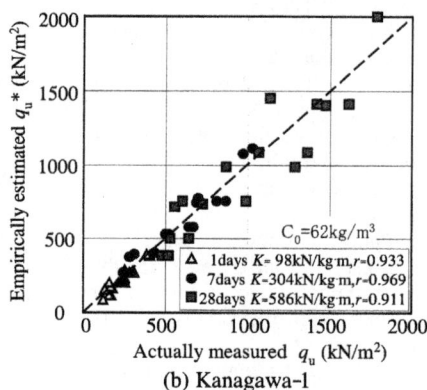

(b) Kanagawa-1

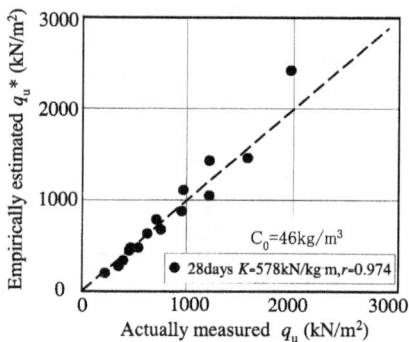

(c) Miyagi

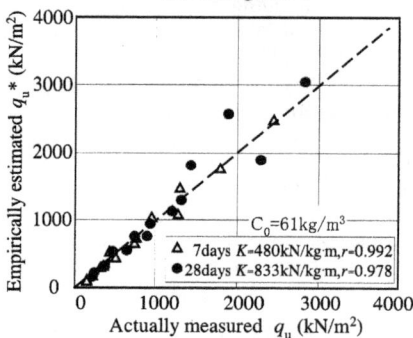

(d) Yamaguchi-1

Figure 2. Comparison of estimated and measured unconfined compressive strengths

frequency of every two days, and daily sampled the cement treated soils. The cement treated samples were cured underwater for 7 days and 28 days, and subjected to unconfined compression test. After the tests, the compressed sample showing the middle strength was selected as the representative, and its cement content was assayed.

Figures 3 and 4 show the histograms obtained for unconfined compressive strength and for cement content. On the basis of Figure 4, we can obtain $\bar{C}$ =76 kg/m^3 and V_C=0.15. The observation of water content results $\bar{w}$=102 % and V_w=0.15. Let us assume C_0=30 kg/m^3, G_S=2.68, Equation 2 yields the variance coefficient V_{qu} for unconfined compressive strength, equal to 0.33. Comparing with the result in Figure 3, it can be said that Equation 2 reasonably interprets the inherent process from the variances in cement and water contents to the variance of unconfined compressive strength.

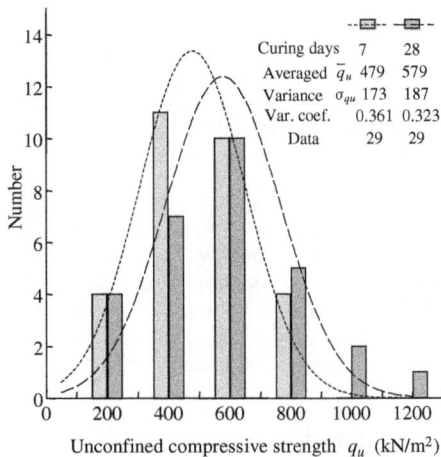

Figure 3. Distribution of compressive strength

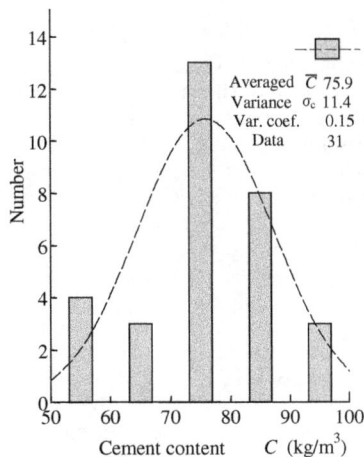

Figure 4. Distribution of cement content

In addition to site A in Nagoya port, the variance properties are also examined for other projects at site B, C and D. At those sites, water content of the dredgings and unconfined compressive strengths for the treated samples cured within molds and for the undisturbed core samples taken from *in-situ* grounds were ordinarily measured. However, the information of cement content actually mixed in the treated soils has not been measured. Considering the mixing and casting processes are identical with site A, the variance of cement content in all the projects is assumed to be the same, the variance coefficient from site A is straightly adopted for the other sites. Meanwhile, the designed cement contents are used as the averaged values of actual cement contents.

Table 1 summarizes the variance properties of water content, cement content and unconfined compressive strength. It is seen that the variance coefficient of cement content is about 0.15, while that of water content ranges from 0.08 to 0.28 and most probably to be 0.16. This information results in an estimate for the variance coefficient of compressive strength near 0.3, which is in agreement with the values actually observed in projects. Therefore, the variance coefficient, V_{qu}^*, evaluated by use of the relationship of Equation 2, essentially explains with the strength variance by unconfined compression test.

Table 1. Variance properties of water content, cement content and compressive strength

Project	Site A	Site B	Site C-1	Site C-2	Site D
$\bar{w}$ (%)	102	82	180	133	71
V_w	0.15	0.16	0.17	0.28	0.08
$\bar{C}$(kg/m^3)	76	110[*1]	50[*1]	50[*1]	50[*1]
V_C	0.15	0.15[*2]	0.15[*2]	0.15[*2]	0.15[*2]
C_0(kg/m^3)	30	25	20	20	25
G_S	2.68	2.68	2.68	2.68	2.71
V_{qu}^*	0.33	0.29	0.38	0.50	0.32
$V_{qu(7)}$	0.36	--	0.29	0.46	0.30
$V_{qu(28)}$	0.32	0.24	0.27	0.41	0.32
$V_{qu(core)}$	--	--	0.42	0.55	0.30

[*1] Designed cement content
[*2] The same value with site A

3 SPECIAL WORKING SHIPS

Photo 1 shows the largest working ship special to the cement treatment for soft dredging. Important specification for the special ship is listed in Table 2. This ship is equipped with 5 fundamentals: (a) prehandling system for the crude dredging; (b) cement slurry plant; (c) soil cement mixing apparatus; (d) oil piston pump unit and (e) spreader.

The crude dredging is dug and dumped into a hopper, here large bulks of soil are crumbled, while

Photo 1. Overview of cement treatment working ship

Table 2. Specification of special working ship

Capacity	Handling ability	300 m³/h
	Conveying distance	500 m
	Water depth	-10 ~ +15 m
Processing	Back hoe	4.5 m³
	Hopper	30 m³
	Cement milk plant:	
	Cement consumption	50 ton/h
	Cement water ratio	0.5 - 5.0
	Soil cement mixer	0-34 rpm
Transportation	Oil piston pump	φ 500, L 3500 (pair)
	Max. push pressure	7 MPa
	Pipe diameter	350 mm
	Pipe material	Steel / Rein. rubber
	Spreader length	56 m
Positioning	Spud	Square pillar × 2
	Placing point	Differential GPS
	Sounding	Sonar / lead wire
Ship size	Length	58.0 m
	Width	22.4 m
	Depth	5.1 m
	Draught	2.1 m

dissimilar substances, such as waste tires, wires, or concrete bulks are removed.

Cement slurry plant mixes cement with water and stocks the cement slurry in a large agitator. This plant will consume cement by 50 ton per hour at its maximum rate, and the water cement ratio can be arbitrarily changed between 0.5 and 5.

The cement slurry is supplied to the soil cement mixing apparatus, to which the pre-handled soil is sent at the same time. This mixing apparatus can produce treated soil by 300 m³ per hour at its nominal rate. The treated soil is passed to an oil piston pump unit, which was specially developed to convey soft soils with an ability of 600 m³ per hour. With the help of this high power piston pump, the treated soil is moved to the casting points through a transport pipe. The casting points are reached by swing-ing the spreader, which bears the transport pipe. According to work condition, placing points can also be reached through the transport pipe which can be extended as long as 500 m, floating on the water surface or laying on the ground. The positions of the casting points are determined by use of differential GPS, and underwater sounding is carried out with a sonar device or a lead wire.

In projects that the amount of dredging to be treated is greater than about 20,000 m³, the special working ship is likely the optimal alternative in view of the construction cost. Besides, since all the fundamentals are systematically installed on the ship, we need not to care much about the pre- and post-procedures, significantly reducing the labor works.

4 PRACTICAL REUSES OF CEMENT TREATED DREDGING WITH SPECIAL WORKING SHIPS

4.1 Practices of cement treated dredging

The special working ships have acted at 11 sites for more than 28 projects. As shown in Table 3, total volume of cement treated dredging conducted by the special working ships reached 1.7 million m³. Usually, the artificially improved soils are used for reclamation, covering soft ground, widening dike and embankment without compaction, etc. Advantages of using the special working ship can be explained that it provides high quality cement treated soil efficiently, because all the equipments are pre-installed on the same ship. The ships were manufactured to be able to mix cement with dredging, convey and cast it at natural low water content, making it possible to build embankment directly by use of the cement treated soft dredging. Here is to be presented two applications in Nagoya port.

338

Table 3. Practical utilization of cement treated dredging with special working ship in coastal area

Site	Project name	Number	Volume (m³)	dredging Soil	dredging w (%)	Cement (kg/m³)	$q_{u(28)}$ (kN/m²)	Usage	Notes
1	Reclamation in Isozaki port	1	40,766	clay	196.6	70	368	trafficability	
2	Construction of Minami-honmoku wharf	3	154,051	clayey	97.2	90	1009	leak protect	slope 1:3
3	Reclamation in Ishinomaki port	3	97,645	clayey	254.0	60	122	reclamation	
4	Restore of Kobe port	2	16,915	clay	103.0	70	451	backfill	
5	Trial ground improvement, Kitakyushu	1	6,888	silt	120.4	80	466	trafficability	
6	Construction of Shinkaimen site	8	376,288	clayey	172.0	50	330	waste site	$k<10^{-6}$cm/s
7	Reclamation in Nagoya port	3	153,619	clayey	101.2	80	869	widening	slope 1:3
8	Reform of Ohi jetty	1	34,677	clay	102.2	70	409	backfill	
9	Runway movement, Iwakuni	2	673,000	silt	110.3	50	301	foot protect	slope 1:5
10	Environment reform, Tachibana	1	49,246	clay	51.6	50	205	waste site	slope 1:3
11	Land development, Kanda port	3	120,330	clay	200.0	80	456	widening	slope 1:3
	Total	28	1,723,424						

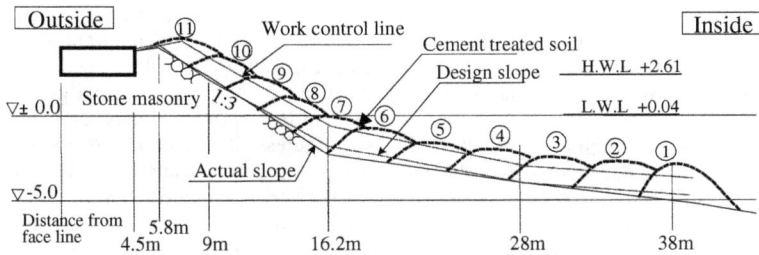

Figure 5. Placement of treated soil along the slope

4.2 Widening dike

Nagoya port island is constructed mainly to provide a disposal space for the dredging during water channel maintenance. The dike of the third phase has been completed, and it is planned to discharge the dredging by 3 million m³ every year, into this space. To avoid the muddy water seeping out of the dike, it is planned to place a cement treated layer inside the dike by use of the dredging.

Figure 5 shows a section of the dike. The gradient at the shoulder is 1:3, while at the toe is between 1:5 to 1:8. More than half of the slope is underwater in the typical section. Along the slope, a seeping protection was demanded to place thicker than 1.0 m. Cement amount to be mixed with the dredging was prescribed at 80 kg/m³, and the unconfined compressive strength at 7 days' curing was expected to be greater than 400 kN/m². Actually, samples taken during construction showed an averaged strength of $q_{u(7)}$=490 kN/m² when cured for 7 days.

Photo 1 is the whole view of construction when placing the treated dredging under water. The working ship is moored outside the dike, and cement treated dredging is conveyed over the dike through the long spreader. The placement of cement treated soft soil began from the toe of the slope, climbing towards the shoulder step by step. The weakness of the treated dredging limited the height possible to

place on the slope at one step. Figure 6a illustrates the stability analysis in the present case. A simple relation was evidenced approximately valid in the actual field.

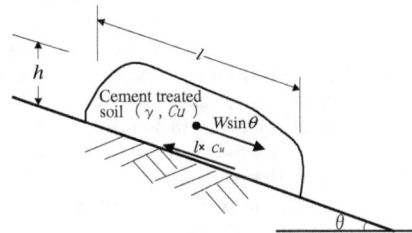

Figure 6a. Stability of soft soil independently placed on a slope

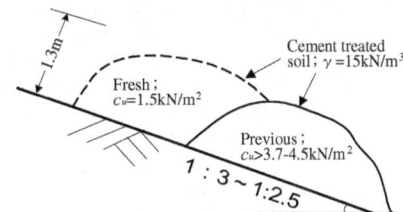

Figure 6b. Stability of soft soil as is placed on a slope step by step

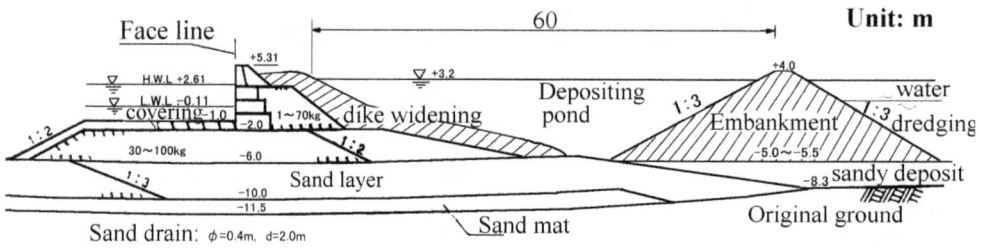

Figure 7. Configuration of embankment and widened dike

$$h = \frac{c_u}{\gamma_t \sin \theta} \qquad (3)$$

here: h is thickness possible to place (m).

c_u is undrained shear strength (kN/m²).

γ_t is unit weight (kN/m³).

θ is angle of slope.

Another stability problem is whether or not the previously placed cement treated soil possesses enough shear strength to sustain the load by the successively placed one. This situation is illustrated in Figure 6b. Stability analyses by circular arc method showed that shear strength needed to be greater than 3.7 to 4.5 kN/m² for the treated soil previously placed. The interval between the two placing steps was so controlled that the previous soil showed shear strength greater than 5 to 6 kN/m², measured by use of a hand vane tool.

4.3 Embankment inside reclamation

As large amount of dredging was discharged into the damping space, surplus muddy water occurs simultaneously in Nagoya port island. The muddy water needs to be dealt with before being released out of the dike. Thereby, it was planned to build a temporary embankment so as to form a depositing pond for the suspended particles to settle. The rectangular depositing pond is about 60 m in width and 195 m in length. Figure 7 shows the configuration of the widening structure and the present embankment. A sand layer, lying beneath the dike structure, occasionally appears upon original soft ground.

A site investigation was conducted using vane shear test, as shown in Figure 8. The soft sediment of recently deposited dredging had an undrained shear strength of c_u=0.2 to 1 kN/m², while the original ground shows an undrained shear strength of c_{u0}=2.3 kN/m² at the depth 8.5 m, gradually increasing with depth. The artificial sand layer laid between the sedimentary and original deposits.

On the conditions above it is asked to build the embankment as high as 10 m, without previous improvement of the ground. The geomaterial available for the embankment is the dredging after cement treatment as the construction of widening dike. A stability analysis was conducted using the circular arc method. The analysis gave stable results providing the lower part of the embankment developed undrained shear strengths greater than 50 kN/m² (i.e. q_u>100 kN/m²). Such strengths are fully expectable when the treated soils are placed and cured for few days. Bearing capability of the original ground to the embankment was also examined. It was found that the added load due to the embankment slightly surpasses the bearing capability if the undrained shear strength is assumed c_u=2 kN/m² for original ground while the existence of the inhomogeneous sand layer is neglected. Nevertheless, it is considered possible to establish such a structure because of little unit weight difference between the cement treated dredging (γ=15 kN/m³) and the sedimentary dredging (γ=13 kN/m³), as well as the solidification of the cement treated soil.

The embankment was built from a depth of 5.0 to 5.5 m, by replacing sedimentary dredging with the cement treated dredging. Unlike the case of widening dike as shown in Figure 5, even the long spreader equipped on the working ship could not reach the position of the embankment. Here additional transport pipe was mounted, connecting the working ship outside the dike to the placing pontoon inside the dike. By this way, the conveying distance was extended to 220 m. The transport pipe floated on the water, flexibly mobile in correspondence with the movement of the placing pontoon.

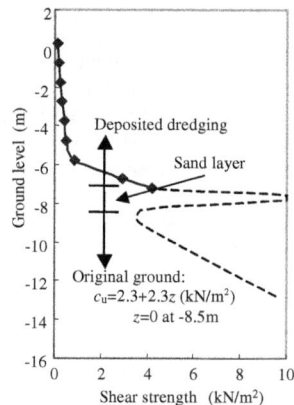

Figure 8. Profile of shear strength

340

Figure 9 shows the section of the completed embankment, roughly coinciding with designed section. Actually, the maximum displacement at embankment shoulder was observed about 50 cm in vertical direction and 80 cm in horizontal direction.

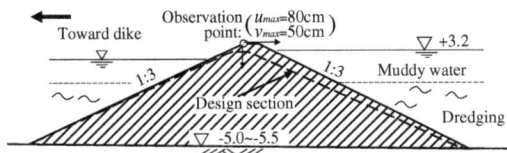

Figure 9. Completed embankment by cement treated dredging

5 UNDERWATER CASTING TECHNIQUES

Excavated underwater, the soft dredging is usually reused around coastal area. This means that the dredging shall probably be cast underwater. In fact, more than two thirds of total volume shown in Table 3 was constructed submerged.

Generally, casting cement mixed soil underwater is more difficult than that above the water surface. If the material separates seriously underwater, not only the designed strength cannot be obtained, but also the surround seawater maybe tainted by muddy or alkaline water generated. Whether or not the material separation occurs depends on soil property and casting method. The author carried out several laboratory and field tests on this matter (Miyazaki et al, 2001, Tang et al, 2002).

Figure 10 shows the strength comparison for a clayey soil treated with cement and cast under and above water in laboratory. It can be seen that reduction in compressive strength caused from underwater casting is insignificant. Meanwhile a field test was conducted on a cohesionless soil, which was cement treated and then cast underwater. In this case the casting pipe was suspended 30 to 80 cm from the bottom. It was observed that the soil separated intensely due to its weak cohesion, when dropped into water. Figure 11 shows the compressive strengths cast underwater and that prepared by normal method. Evidently, soil separation virtually decreased compressive strength, and we must avoid such a phenomenon.

On the basis of the experimental knowledge, the casting method in practice is controlled in such a manner that no cement treated soil is allowed to fall underwater. Figure 12 explains the casting situation. After adequate mixing with cement, normal clayey dredging shows certain cohesion. The cement treated dredging is filled fully into the transportation pipe, pushed by a powerful oil piston pump, and spewed out of the casting pipe underwater. Since the casting pipe is erected merely apart from the ground, the soil

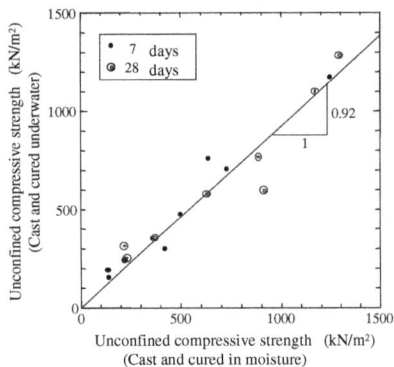

Figure 10. Effect of casting condition on unconfined compressive strength

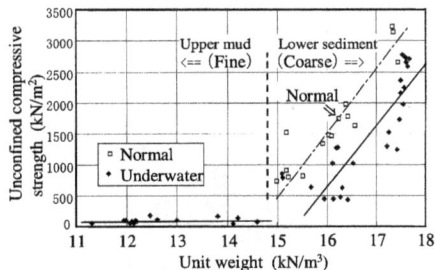

Figure 11. Reduction of unconfined compressive strength due to material separation underwater

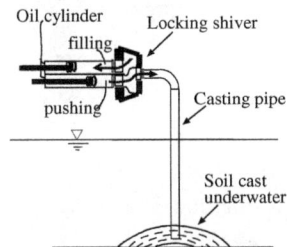

Figure 12. Method adopted to avoid material separation

is kept out of touch with water until underwater casting is finished.

Figure 13 presents the results of water examination conducted during constructions. Some observed results in laboratory and field tests are also plotted in the figure, where the distance between casting and water sampling points are in log scale. Except project 1 for its particular condition, examinations in project 2 to 4 yielded very low turbidities ranging within 1 to 10 ppm, and steady pH values fluctuating between 7.8 to 8.3. These are essentially identical with natural seawater environment. Casting test D gave the similar result. This field was carried out in a shipbuilding dock, so the scale had been large enough to reproduce the real construction.

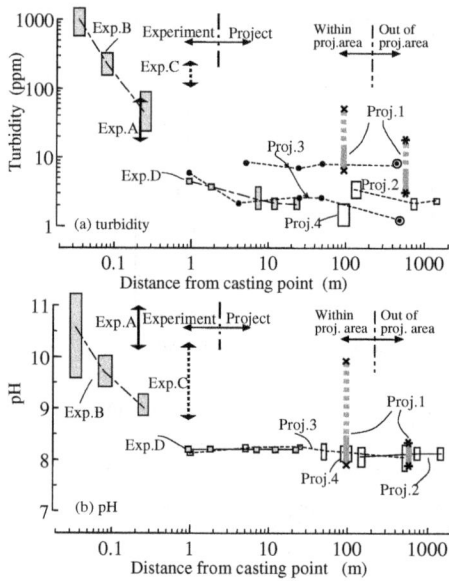

Figure 13. Environmental impacts on surrounding seawater due to underwater casting of cement treated soil

6 PROPOSAL OF A GROUND IMPROVEMENT METHOD

To improve a soft ground, a traditional method of replacing soft soil with sandy material is popularly used. This method is used both on land and in marine practices. However, it becomes more and more difficult to dispose the soft soil excavated, recently, and the same is to supply good sandy material for replacement. So there comes an idea to reuse the soft soil at the place it occurred.

Figure 14 shows the working flow proposed for soft ground improvement in marine projects. The working processes are: -1) excavation of target soft ground; -2) examination of designed section; -3) cement treatment of soft soil and replacement underwater; -4) finish. It is noticed from the figure that there is a surplus of dredging between processes -1) and -3). In conventional marine works, the surplus dredging can be contained with several idle barges. But when the surplus bulk turns to be too large (i.e. excess of 10,000 m^3), it might be better to stock it temporarily at a proper place.

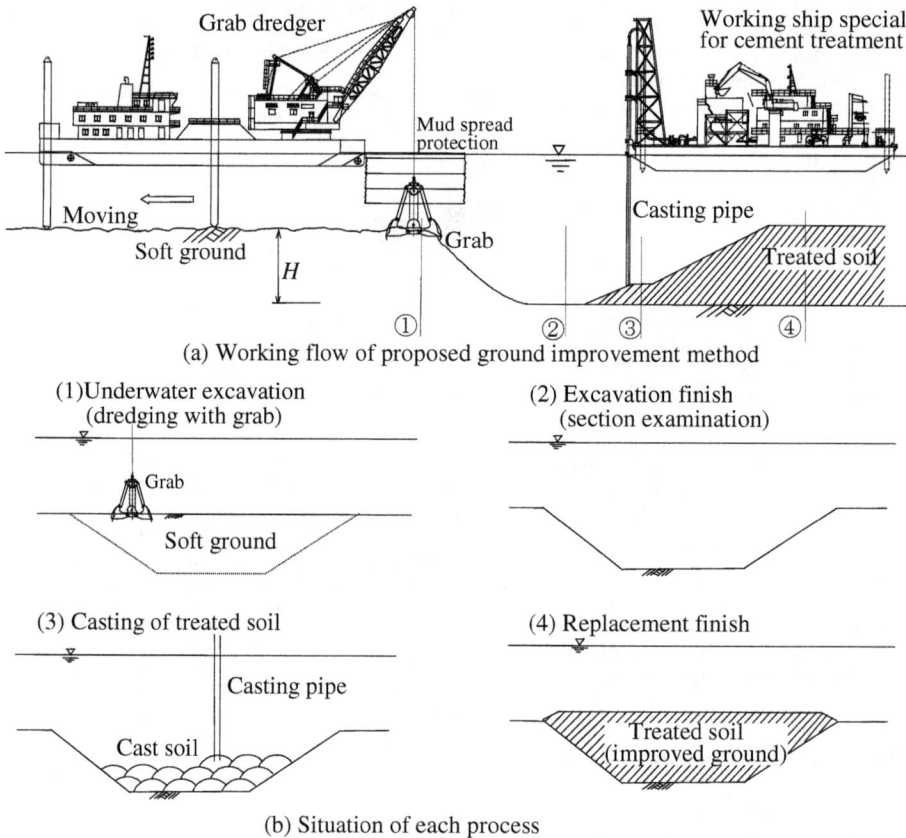

(a) Working flow of proposed ground improvement method

(1) Underwater excavation (dredging with grab)

(2) Excavation finish (section examination)

(3) Casting of treated soil

(4) Replacement finish

(b) Situation of each process

Figure 14. Proposal of ground improvement by dredging, cement treating and replacing method

Although there are various subjects to be studied toward practical application, the proposed ground improvement method should be a feasible way to reduce the waste disposal of dredging. In fact, a trial cost estimation showed that this method is able to compete with the method of replacement with sandy material, supposed the cost for dredging disposal is taken into consideration. It is also of interest to apply such method to construction of impervious seawall for waste reclamation, because the cement treated dredging generally shows an order of permeability smaller than 1×10^{-6} cm/sec.

7 CONCLUSIONS

Soft dredging will continue to occur with water channel maintenances or coastal constructions. There are strong needs to make the best use of this geo-material, because less sites are available for new reclamations in Japan. Practices involving reuse of soft dredging presented in this paper can summarized as bellows:

1 By cement treatment, the soft dredging can be changed into a useful geomaterial. The strength property depends mainly on cement content and water content. Accordingly, the quality related to strength can be evaluated from the viewpoints mixing degree and water content fluctuation. It is shown that the variance coefficient of compressive strength for treated dredging is normally near 0.3.
2 Special working ships have been introduced in practical engineering. Since all the necessary equipments are systematically installed, the ships offer high quality treated soils, possible to serve in various structures in coastal area. Two application examples in Nagoya port are presented.
3 Underwater casting techniques are discussed. It is emphasized that underwater separation of the cement treated soils must be avoided. This phenomenon depends on soil property and casting method. It is shown that soil separation can be controlled unless the soil is roughly dropped underwater.
4 With the help of the introduced working ship, improvement of soft ground can achieve by dredging the soft soil, treating it with cement and returning the treated soil to the ground.

REFERENCES

Miyazaki, Y., Tang, Y.X., Ochiai, H., Yasufuku, N. & Omine, K. 2001, A correlation among unconfined compressive strength, cement content and water content for cement treated dredgings, *Technology reports of Kyushu Univ.*, Vol.74, No.1, 1-8. (in Japanese)

Miyazaki, Y., Tang, Y.X., Ochiai, H., Yasufuku, N. & Omine, K. 2001, Strength characteristic of cement treated soils and environmental impacts on surrounding water due to underwater casting, *Technology reports of Kyushu Univ.*, Vol.74, No.2, 99-106. (in Japanese)

Tang, Y.X., Miyazaki, Y. & Tsuchida, T. 2000. Advanced reuses of dredging by cement treatment in practical engineering. *Coastal Geotech. Eng. in Practice*. Nakase & Tsuchida (Ed), 725-731. *Proc. intern. symp. Sept. 2000, Yokohama*, Rotterdam: Balkema.

Tang, Y.X., Miyazaki, Y. & Tsuchida, T. 2001. Practices of reused dredgings by cement treatment. *Soils and Foundations, JGS*, Vol.41, No.5, 129-143.

Tang, Y.X., Miyazaki, Y., Ochiai, H., Yasufuku, N. & Omine, K. 2002, Environmental impacts on seawaer due to casting cement treated soil underwater, *JSCE Jour., Geotech. Eng.*, No.708/III-59, 211-220. (in Japanese)

Tang, Y.X., Miyazaki, Y., Ochiai, H., Yasufuku, N. & Omine, K. 2002, Quality assessment related to strength property of cement treated dredging, *Technology reports of Kyushu Univ.*, Vol.75, No.1, 1-8. (in Japanese)

Technical standard for port facilities, 1999, *The Japan Port & Harbour Association*. (in Japanese).

Soft Ground Engineering in Coastal Areas, Tsuchida et al. (eds)
© 2003 Swets & Zeitlinger, Lisse, ISBN 90 5809 613 0

Construction of seawall for waste landfill in Tachibana Port

T. Tsuchida & Y. Watabe
Port and Airport Research Institute, Yokosuka, Japan

H. Yuasa
Tokushima Prefectural Government, Tokushima, Japan

Y. Oda
Corporation for Advanced Transport and Technology, Tokyo, Japan

ABSTRACT: A new seawall structure for waste landfill has been constructed in Tachibana Port, Tokushima Prefecture, which was the first one in port after Ministry of Health and Welfare revised the technical standard of structures for waste disposal facility in 1998. The design and construction of the seawall were summarized, as follows: 1) According to BEM analysis, it can be said that SCP method will not spoil the ground properties as impermeable soil layer, when the replacement ratio is less than 50%. 2) As a barrier to prevent the seepage, two polyvinyl chloride (PVC) sheets and cement treated clay layer were constructed. According to the settlement-deformation analysis, PVC sheets and cement treated soil will be stable in the long term. 3) Because most of the work had to be carried out underwater, new techniques, including on-ship heat bonding of PVC sheets and direct casting of cement treated clay on slope, had been developed and successfully carried out.

1 INTRODUCTION

The volume of domestic wastes and industrial wastes, which have to be disposed in the controlled disposal site, is increasing continuously. For most of the local communities, the construction of final waste disposal site is an urgent need. However, it is getting extremely difficult to find a new disposal site inland because the agreement of the residents near the site would hardly be obtained due to a lot of problems of waste disposal site such as, transportation noise, flying dust, bad smell and especially potential risk of ground water pollution. In these circumstances, the waste disposal sites located in seaport are getting more important.

In 1998, Ministry of Health and Welfare revised the technical standard of waste disposal facility in order to secure the safety to the environment nearby. In the revised standard, the barrier to prevent the water containing hazardous materials from inside the controlled disposal site into the open sea must be doubled based on the failsafe concept. A new seawall structure for controlled waste landfill has been constructed in Tachibana Port, Tokushima Prefecture, which is the first one constructed in seaport after the revision of the standard. As constructed under seawater, the structures of waste disposal site in port area have engineering problems different from those located inland. This report introduces the design and construction method of the seawall structure in Tachibana Port Project, especially on the design of barrier consisting of 2 impermeable PVC (polyvinyl

chloride) sheets and cement treated dredged soil, and techniques of on-ship heat-bonding of PVC sheets, pavements of PVC sheets under the tidal change and the direct under water casting of cement treated clay.

2 WASTE DISPOSAL SITE IN PORT AREA AND ITS TECHNICAL PROBLEMS

An important role of seaport in Japan is to provide a waste disposal site in urban area. Fig.1 shows the change of total volume of disposed waste in Japanese port. As shown in Figure, in 1990-1995, about 18,000,000m³ of wastes were disposed in port area every year, and the volume of waste is increasing continuously.

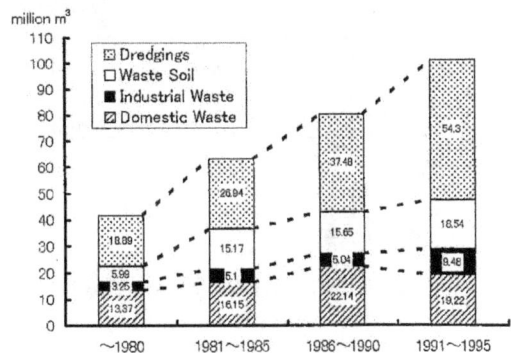

Figure 1. Volume of waste disposed in port area

Fig.2 shows the area of waste disposal site in the ports of 3 major bays, and the ports of other areas. The area of waste disposal sites located in port range from 0.1 million m^3 to 3 million m^3, and they are much larger than those located inland, most of which are less than 0.1 million m^3. As shown in Fig.2, in the metropolitan area located in 3 major bays, the waste disposal sites in port have larger areas and capacities, and are taking 60 % share of the total volume of waste disposal in these cities. In 1999, 57 billion yen were paid for the construction of seawall for waste disposal site in Japanese ports.

As shown in Fig.1, disposed waste consists of the dredging, waste soil transported from inland construction site, industrial wastes and domestic wastes. As the dredging and the waste soil are not hazardous, they are disposed in the *stable disposal site*, where no special measures are carried out in the environmental point of view. The industrial wastes and the domestic wastes are disposed in the *controlled disposal site,* where the wastes must be surrounded by impermeable soil layers or equivalent barriers. As mentioned in introduction, the waste disposal sites located in port is getting more important and at the same time, the higher level of the safety to environment are urgently requested.

In the revised technical standard by Ministry of Health and Welfare (1998), the impermeable soil layer is defined as "clay layer of more than 5m thickness whose permeability is less than 10^{-5} cm/sec" and the equivalent barriers are defined as follow:

1 clay layer of more than 50cm thickness whose permeability is less than 10^{-6}cm/sec, and to lay a impermeable sheet
2 asphalt concrete layer of more than 5cm thickness whose permeability is less than 10^{-7}cm/s and to lay a impermeable sheet
3 to lay two impermeable sheets
4 installation of sheet pile up to the impermeable soil layer
5 installation of wall whose thickness is larger than 50 cm and the permeability is less than 10^{-6} cm/sec

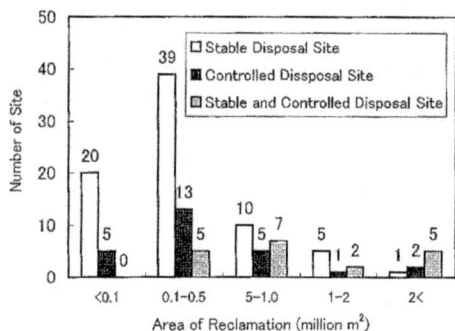

Figure 2. Scale of waste disposal site in port area

Most of waste disposal sites in port have been constructed on soft clay layer, because thick alluvial clay layer is laid at the seabed in most Japanese ports and they can be considered as impermeable soil layer of permeability less than 10^{-5}cm/sec. Accordingly a major engineering problem is how to design and construct the barrier of seawall structure.

Fig.3(a) is a typical cross-section of the seawall of controlled disposal site constructed before 1998, where an impermeable PVC sheet was laid on the back fillings. Fig.3(b) is another seawall where the steel sheet pile was installed as a barrier.

According to the revised standard, impermeable sheets must be double layered based on the failsafe concept, and in case of installation of steel sheet pile, the more considerations must be taken for the impermeability of the joints.

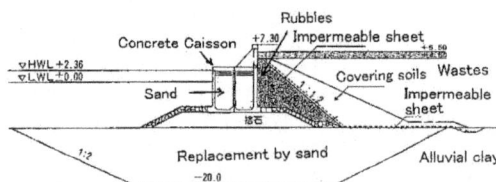

Figure 3(a). Typical cross section of seawall where PVC sheet was used as a barrier

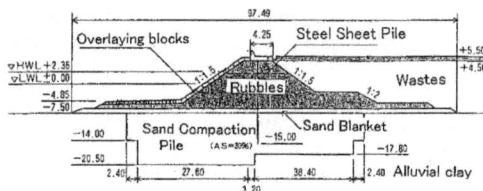

Figure 3(b). Typical cross section of seawall where the steel sheet pile was installed as a barrier

3 CONSTRUCTION OF WASTE DISPOSAL SITE IN TACHIBANA PORT, TOKUSHIMA PREFECTURE

Tachibana port is located in the central part of Tokushima prefecture (Fig.4), and is playing an important role as an industrial center for the south of the prefecture. As a large scale coal power plant project is located in the port, Kokatsu-Goto district development project has been proceeded in order to construct industrial complex related to the seaport, including waste disposal facility. The planned total capacity of the waste disposal site is 630,000 m^3, consisting of domestic wastes of 290,000m^3, industrial wastes of 178,000m^3, waste soil of 90,000m^3 and dredging 72,000m^3.

The geotechnical condition is shown in Fig.5. A 15-16m thick alluvial clay layer is laid on a 3m thick

Figure 4. Location of Tachibana Port, Tokushima

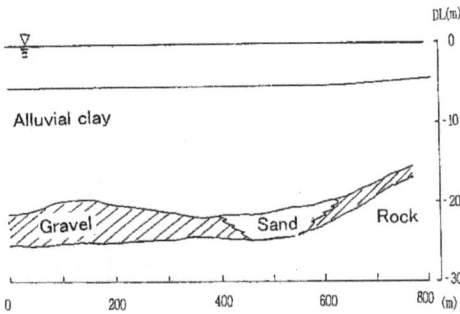

Figure 5. Soil condition of construction site

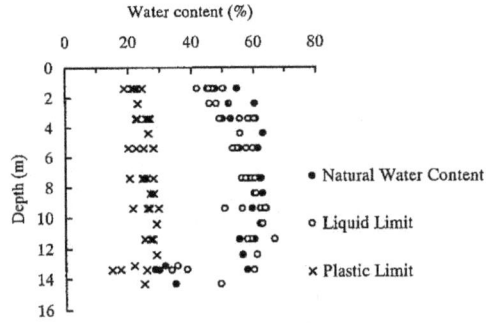

Figure 6(a). Water content and consistency limits

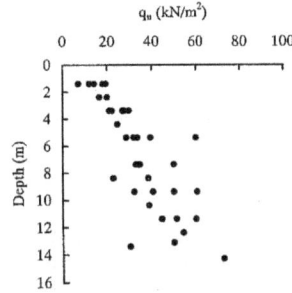

Figure 6(b). Unconfined compression strength

layer of sand and gravel. The water content, consistency limits and unconfined compression strength of clay are shown in Fig.6(a) and Fig.6(b), which indicating that the clay is apparently normally consolidated condition. As the coefficient of permeability obtained in laboratory consolidation test ranges between 10^{-6} and 10^{-7} cm/sec, the clay layer can be an impermeable soil layer.

The plane view of waste disposal site are shown in Fig.7(a). The site is constructed successively to the existing coal power plant. The seawall of 600m lengths was newly constructed, and only the barrier consisting of PVC sheets and cement treated soil was constructed at the boundary of the coal power plant. The cross-section of seawall and the barrier were shown in Fig.7(b), which were designed as follows;

1 the seawall made of rubbles were constructed on the soft clay ground which was improved by Sand Compaction Method (SCP) of 30% replacement ratio.

2 The inside of seawall was covered by two polyvinyl chloride (PVC) sheets of 3 mm thickness. To protect the lower sheet from the damage due to the edge of rubbles, the surface voids of rubble were packed by small stones and below the PVC sheet a 10mm thick unwoven geotextile was laid. On the top of the second PVC sheet, a 10mm

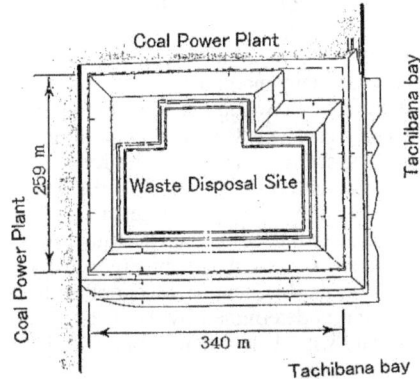

Figure 7(a). Plane view of waste disposal site

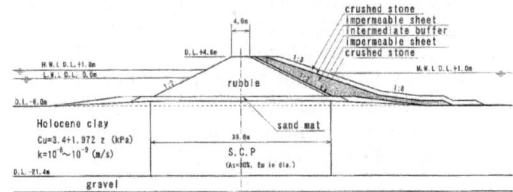

Figure 7(b). Cross-section of seawall and barrier

thick unwoven geomenbrane and the 1m thick crushed stone were laid to cover the sheet against the dumping of wastes.

347

3 Between the PVC sheets, cement treated dredged soil was utilized for an intermediate buffer zone. The intermediate zone is aimed to relieve the stress of PVC sheet when the consolidation settlement takes place by reclamation of wastes. And in the case that the sheet was damaged by some reasons, it is expected for the low permeability of treated soil to prevent the leakage. Also the intermediate zone is helpful in sinking the sheets by its weight.

Recently, as impermeable barrier for waste disposal site, the high density polyethylene (HDPE) sheets are commonly used, however, in the case of underwater work, HDPE sheets of large width, whose unit weight is smaller than 1.0, is extremely difficult to be bonded and be laid at the seabed. This is the reason to use PVC sheet.

4 SOME CONSIDERATIONS OF DESIGN

4.1 *Permeability change due to existence of sand compaction piles*

As most of waste disposal sites in port are constructed on normally consolidated clay, soil improvement work is inevitable. Although SCP method is common and economical improvement method, the increase of permeability of clay ground due to the installation of sand piles must be taken into consideration.

To evaluate this effect, the seepage analysis with the boundary element method (BEM) was carried out with the model shown in Fig.8, where k_2 and k_1 are permeability coefficient of sand column and the in-situ clay, respectively, and A_s is the replacement ratio defined as area rario. The result of the analysis is shown in Fig.9, where the permeability of improved ground is evaluated by the equivalent permeability coefficient k^*. As shown in Fig.9, the ratio k^*/k_1, which means the increase of permeability by the sand column is determined by A_s, if k_2/k_1 is larger than 10, and with A_s less than 70%, k^*/k_1 is less than 10.

Considering that the permeability of marine clay is ranged from 10^{-6} cm/sec to 10^{-7}cm/sec, it can be said that the improvement by SCP method will not spoil the ground properties as impermeable soil layer whose permeability should be less than 10^{-5}cm/s. However, as there can be some errors on the location or the width of sand columns in the construction practice, the replacement ratio of foundation for waste disposal site should be less than 50%.

4.2 *Stress-strain characteristics and permeability of cement treated soil*

In the waste disposal site constructed inland, a clay liner consisting of compacted clay is usually used as preserving soil for impermeable sheet. However, in

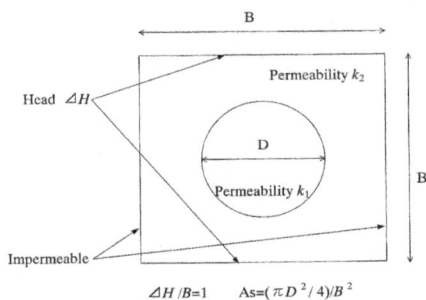

$\Delta H /B=1$ $As=(\pi D^2/4)/B^2$

Figure 8. BEM Analysis for ground improved by SCP

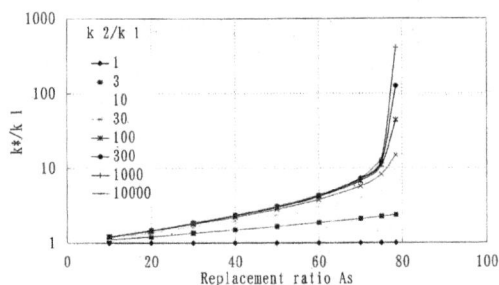

Figure 9. Replacement ratio and the increase of equivalent permeability of ground improved by SCP

the case of underwater construction, compaction is not available because the soil is fully saturated. Hence, instead of compacted clay liner, other materials such as cement treated soil have to be used.

Although the cement treatment is usually aimed to increase the shear strength of soft soil, the aim of treatment in this case is to relieve the stress of PVC sheet and to prevent the leakage when the sheet was damaged by some accidents. Accordingly, the cement treated soils are expected to have more ductile stress-strain characteristics with the least shear strength necessary to keep the designed shape and to have lower permeability.

Fig.10 is the stress-strain relationship of cement treated Tachibana bay clay (Watabe et. al, 2000), that are obtained in consolidated undrained triaxial test (CIU test) with 49kPa isotropic consolidation pressure. As shown in Fig.10, when the treated soil is sheared with a certain consolidation and confining stress, the treated soil fails with a peak at 1.5-3.0 % axial strain, and after the peak the shear strength does not decrease remarkably, even some show strain hardening behavior.

Fig.11 is e-log p and e-log k relationship of cement treated Tachibana bay clay. Since the cement treated soil hardened in a high water content and a large void ratio, its permeability was larger than that of untreated soil at the same consolidation pressure. However, the permeability of treated soil was smaller than that of untreated soil at the same void ratio. The reason is considered that the ettringite and silicic acid calcium hydrate developing with cement hydration reaction clog the void in the soil and prevent the water flow (Watabe et.al, 2000).

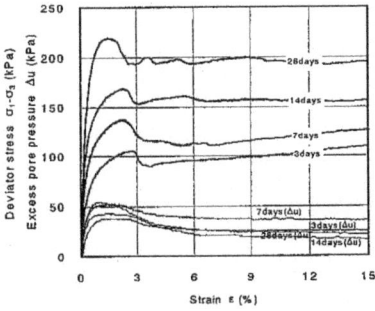

Figure 10. Stress-strain relationship obtained by CIU test

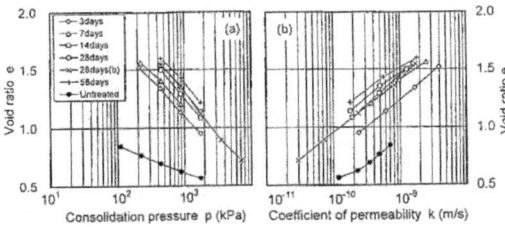

Figure 11. *e*-log *p* and *e*-log *k* relationship of cement treated Tachibana bay clay

4.3 *Seepage analysis of seawall*

To study the effect of cement treated soil layer, the seepage analysis was carried out. Fig12 shows the conditions of analysis. For the simplicity, the permeability of alluvial clay is assumed to be 10^{-7} cm/s and for following 3 cases, the calculation was carried out:

Case 1: inside of the sea wall is covered by sand layer of $k = 10^{-2}$ cm/sec

Case 2: inside of the seawall is covered by cement treated soil of $k = 10^{-7}$ cm/sec

Case 3: inside of the sea wall is covered by impermeable by PVC sheet

Fig.13 is the result of the analysis, showing the volume of seepage water from inside of seawall into outside. Case 2 in Fig.13 shows that even if the PVC sheets does not work at all, the effect of cement treated soil to prevent the seepage is large compared to Case 1, where the seawall is covered by sand. The barrier consisting of 2 PVC sheets and the intermediate cement treated soil is considered to be quite safe to the risk of leakage.

4.4 *Finite element analysis of settlement and deformation of ground*

Most of waste disposal sites in port, as in the case of Tachibana bay, have been constructed on soft thick

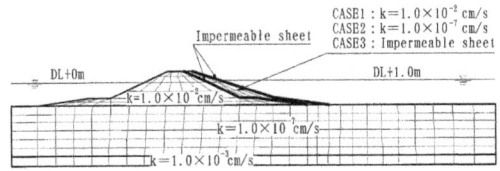

Figure 12. Model of seepage analysis

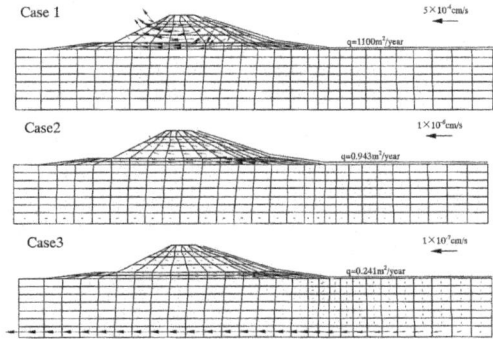

Figure 13. Results of seepage analysis

clay layer to make it as the impermeable soil layer. Accordingly, as the waste landfill proceeds, the consolidation settlement and the lateral shear deformation will take place, and they must be taken into consideration in the design of seawall.

Fig.14 shows the model and the calculation conditions of finite element analysis of seawall. For the alluvial clay, Sekiguch-Otha's visco-plastic model was used; sand pile and gravel layers and cement treated soil were assumed to be elastic materials. The waste landfill is to continue for about 10years after the construction of seawall, and calculation was carried out from the start of seawall construction to 10,000days (27 years) after that.

Fig.15 (a) shows the deformation of seawall, and Fig.15 (b) is the settlements at the surface of seabed. About 1.8m consolidation settlements took place inside the waste disposal site, while the settlement of seawall was reduced to about 50 cm due to the improvement by sand compaction piles.

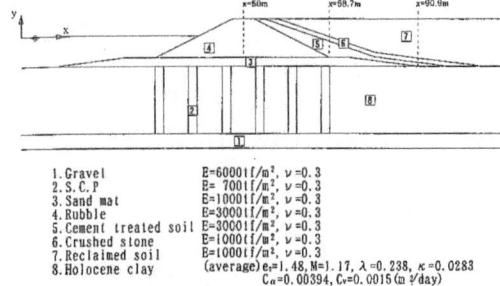

Figure 14. Model and conditions of FE analysis

Based on the analysis, predicted deformations of PVC sheets with time are plotted in Fig.16. As the maximum extension of sheets is 2%, the sheets have no risk to be broken due to the ground deformation. Fig.17 shows the largest principal strains of cement treated soil layer with time. As shown in Fig.17, the maximum strain is about 0.7% at 10,000 days, and it is considered that the cement treated soil between PVC sheets will be stable during and after the waste landfill.

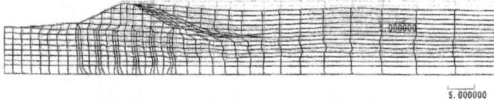

Figure 15(a). Deformation of seawall

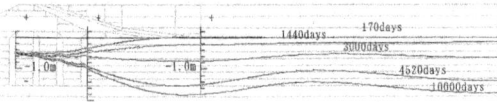

Figure 15(b). Predicted Settlement

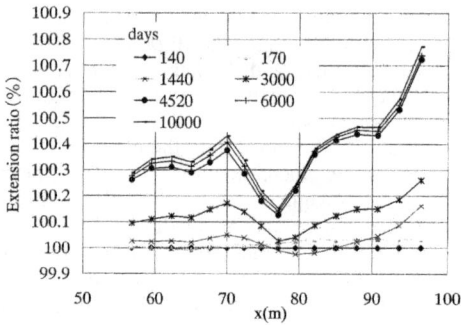

Figure 16. Extension rate of PVC sheets

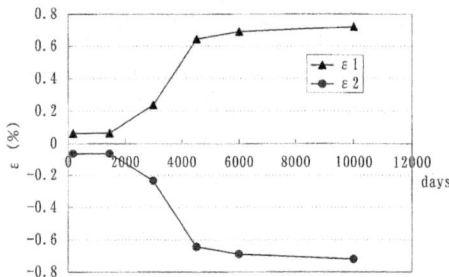

Figure 17. Principal strains of cement treated soil

5 CONSTRUCTION WORKS

The construction work started in January 1999 and finished in March 2000. Fig.18 shows the flow of construction work.

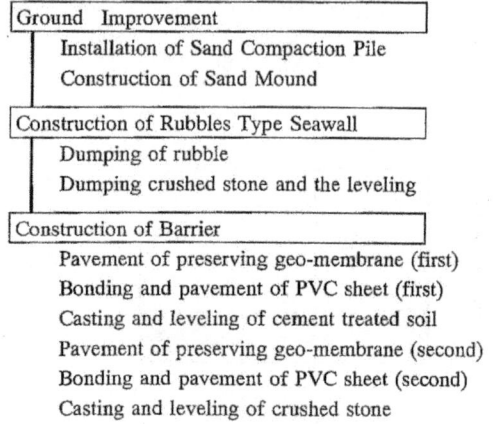

Figure 18. Flow of construction work

5.1 Sand compaction pile

The foundation of seawall was improved by SCP method of 30% replacement ratio. Fig.19 shows the settlements of seabed, measured with the settlement plates during the construction of seawall. As shown in Fig.18, the measured settlements were 60-70cm at 100days and were slightly larger than 50cm calculated by FEM. The reason will be that the sand piles were assumed to be linearly elastic in FEM analysis, while the actual sand piles were not elastic but more deformable.

5.2 Pavement of PVC sheets under water

Total area of PVC sheets was 100,000m^2 and all of them were bonded in-situ to be a piece of sheet. The procedure of works on PVC sheet is as follows:
1 PVC sheets, 6m width, 60m length and 3mm thickness, were rolled up in the factory and were shipped to the site.
2 The 6 rolls of PVC sheets were moved on the construction vessel, and spread out on board every 7m length. By bonding the 6 sheets of 6m

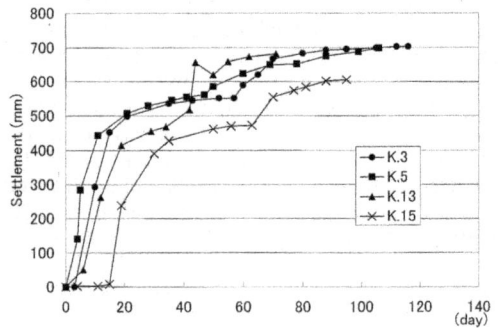

Figure 19. Measured settlements of ground improved by SCP method

width and 5 overlapping sheets of 1m widths, a piece of PVC sheet of 40m widths and 7m lengths were made up (Photo 1). The on-ship bonding was carried out with the heat-bonding device taking 25mm bonding width (Photo 2).

3 The sheet was pushed out to the sea and was spread with floats. Pulling up both sides of 40m width sheets on the other vessel, the heat bonding was carried out to make all of them one piece of by PVC sheet.

4 The sheet was fixed at the top of the seawall and the seabed.

Table 1. Index properties of dredged clay

Wet density:	1.67
Particle density:	2.70
Natural water content(%):	54
Liquid limit (%):	49
Plastic limit (%):	23
Plasticity index:	26
Sand content (%):	32
Silt content (%):	39
Clay content (%):	29

Photo 1. PVC sheets on the construction vessel

Photo 3. Special working ship for cement treatment

termined by the water content of clay, and the volume of cement is influential on the long-term strength. Based on the stability analysis of slope, the initial water content of dredged soil was determined to be $1.4w_L$ and the cement is 50kg/m^3.

The total volume of cement treated soil was 63,000m^3 and the construction work was carried out by Special vessel for cement treatment (construction capacity is 250m^3/hour, Photo 3). The surface flatness of cement treated soil layer was investigated by red and the leveling and modifications to keep the irregularity within ±30cm were carried out by works of long back hoe.

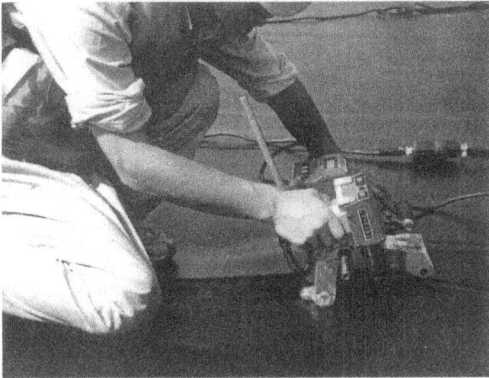

Photo 2. Heat bonding of PVC sheet on ship

5.3 Design and performance of cement treated soil

The index properties of dredged soil in Tachibana bay is shown in Table.1. As shown in Fig.7 (b), the cement treated soil layer must have a initial shear strength to be stable on the slope of 1:3 to 1:8 gradients, on the other hand, considering the requirement as the barrier, it is expected that the cement treated soil is not brittle but deformable. Therefore the volume of cement should be a minimum essential.

Fig.20 shows the shear strength of cement treated soil with time, where Portland cement was used. The initial strength of cement treated clay is mainly de-

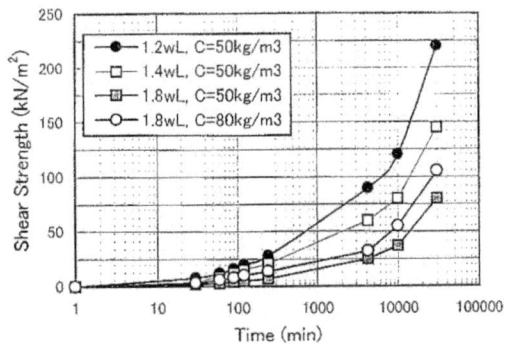

Figure 20. Shear strength of cement treated soil with time

351

6 SUMMARY AND REMARKS

A new seawall structure for waste landfill has been constructed in Tachibana Port, Tokushima Prefecture, which was the first one constructed in seaport after Ministry of Health and Welfare revised the technical standard of structures for waste disposal facility in 1998. In this project, the design and construction method of a new type of seawall structure for waste landfill were developed and can be summarized as follows:

1 As the seawall was constructed on the soft clay layer of 15-18 meter thickness, the sand compaction pile method of 30% replacement method was used for ground improvement. According to BEM analysis, it can be said that SCP method will not spoil the ground properties as impermeable soil layer, when the replacement ratio is less than 50%.

2 As double barrier system to prevent seepage from inside of the seawall, two polyvinyl chloride (PVC) sheet of 3mm thickness and cement treated clay layer of 1 – 3 meter thickness were constructed. By the seepage analysis, the effect of cement treated soil between PVC sheets was discussed. According to the settlement-deformation by FE analysis, it is considered that PVC sheets and cement treated soil will be stable in the long term.

3 Because most of the work had to be carried out underwater, new techniques, including on-ship heat bonding of PVC sheets and direct casting of cement treated clay on slope, had been developed and successfully carried out.

On the waste disposal site in coastal area, the authors think more research and development are necessary. For example, although the tidal change is only 80 cm in Tachibana bay, the works of PVC sheets were much influenced and were troubled by it. It may be difficult to do without some improvements where the tidal change is much larger. Finally, the authors sincerely appreciate the members of Soil Mechanics Laboratory in PHRI for their corporations.

REFERENCES

Watabe, Y., Furuno,T., Tsuchida,T. and Yuasa,H. (2000):The utilization of cement treated soil in the waste reclamation landfill and it is characteristics of shear, Proceedings of 35th Japan National Conference on Geotechnical Engineering, JGS, pp.1227-1228, (in Japanese).

Furuno,T., Watabe, Y., Tsuchida,T. and Yuasa,H. (2000):The utilization of cement treated soil in the waste reclamation landfill and it is characteristics of consolidation, Proceedings of 35th Japan National Conference on Geotechnical Engineering, JGS, pp.1251-1252, (in Japanese).

Watabe,Y., Tsuchida.,T., Furuno.T. and Yuasa,H. (2000): Mechanical characteristics of a cement treated dredged soil utilized for waste reclamation landfill, International Symposium on Coastal Geotechnical Engineering in Practice(IS-Yokohama).

Soft Ground Engineering in Coastal Areas, Tsuchida et al. (eds)
© 2003 Swets & Zeitlinger, Lisse, ISBN 90 5809 613 0

Use of lightweight treated soil method in seaport and airport construction projects

T. Tsuchida & M.S. Kang
Port and Airport Research Institute, Yokosuka, Japan

ABSTRACT: A lightweight treated soil (LWTS) method has recently been developed to reuse dredged soils as an artificial lightweight geomaterials, and the density ranges from 1.0 to 1.2 g/cm^3, for coastal construction projects. There are two types of lightweight soils, foam treated soil (FTS) and bead-treated soil (BTS). The slurry of dredged soil or waste soil is mixed with cement and air foam, or cement and EPS (expanded polystirol) beads whose diameters are 1 - 3 mm, respectively. Since 1996, LWTS has been applied in several seaport and coastal airport projects. The slice method was developed to calculate the earth pressure at earthquake when LWTS is used as backfill of quaywall. The four cases in Kobe Port, Kumamoto Port, Ishikari Bay New Port and Tokyo International Airport were presented, where lightweight soil was used to reduce the earth pressure or consolidation settlement.

1 INTRODUCTION

In Japanese seaports, soft clayey soils of about 6 million m^3 are dredged annually and are dumped in waste-dumping sites enclosed by seawalls. As the lack of waste-dumping sites has become a serious problem for many ports, the demand for the reuse of dredged soils in port construction work has increased. Recently, Tsuchida et al. (1996) developed the lightweight treated soil (LWTS) method, in which the chemically treated dredged clay is reused as a filling material in port and harbor works (Tsuchida, 1999).

There are two types of lightweight treated soils made from dredged soil. One is a foam treated soil (FTS), and the other is a beads treated soil (BTS). The slurry of dredged soil is mixed with stabilizing agent and air foam or expanded polystirol (EPS) beads with the diameters of 1~3 mm, respectively. The wet density of the lightweight soil is usually 1.0 g/cm^3 over sea level and 1.1~1.2 g/cm^3 below sea level. The conventional design shear strength ranges from 200 kN/m^2 to 400 kN/m^2.

The LWTS method was firstly applied for the rebuilding of a quay wall damaged by Hanshin-Awaji Great Earthquake in 1995. Since then, the method has been used successfully in several seaports and airports projects in Japan. One of the most popular usages of LWTS method is backfilling of a gravity type quay wall for the purpose of the reduction of seismic earth pressure. In this case, the lightweight treated soil was directly placed underwater using a pusher pump and a tremie pipe. Because of the density change and the larger risk of washout in deep water, the LWTS method is recommended in seawater shallower than -3m. However, considering the tidal change, in some ports it is necessary to place the lightweight soil at a deeper sea level.

2 ENGINEERING PROPERTIES OF LIGHTWEIGHT TREATED SOIL

Engineering properties of foam treated soil and beads treated soil have been investigated by Tsuchida et al. (1996, 1999). Here, the outlines of geotechnical properties of those lightweight soils are described briefly.

2.1 *Density and fluidity*

Density of light-weight treated soil can be controlled between 0.6 and 1.5 g/cm^3 by induced amount of air foam or polystirol beads, and water content. The density of foam treated soil increases slightly during placing and hardening. Therefore, the design of density must be done by accounting for the increase of density from 0.05 to 0.1 g/cm^3.

Because clayey soil to be mixed with FTS or BTS has relatively high water content, the treated soil has high liquidity at the mixing stage and then looses liquidity quickly. The liquidity of treated soil is much influenced by the initial water content of soil and the amount of stabilizing agent. Typical liquidity properties of FTS and BTS are shown in Figure 1(a), in which flow values measured by the Standard JHS A 313 are plotted against the initial water content normalized by the liquid limit (water content at liquid limit is 76.1%). The figure includes the flow values of treated soils placed at the atmosphere and in the water.

As shown in the figure, the flow value of FTS placed at the atmosphere increases almost linearly with the water content. The flow value of FTS placed in the water, however, is small and maintains almost constant irrespective of the initial water content. This is due to the effect of buoyancy. Similar properties are seen in BTS in Figure 1(b), where the flow value at the atmosphere increases very rapidly with the increase of initial water content.

Figure 2 shows the relationship between the flow value of FTS and elapsed time, in which the soil is manufactured with the initial water content of 190% and amount of air foam of 187 l/m^3 (Coastal Development Institute, 1999). The figure indicates that the fluidity of treated soil rapidly decreases with increasing elapsed time to a third of the initial value. Accordingly, FTS should be transported and placed at construction site after mixing.

2.2 Strength, deformation and compression properties

The strength of lightweight treated soil is given by the solidification due to stabilizing agents, such as cement. Stress-strain curves in unconfined compression tests of FTS are shown in Figure 3. The figures show that the compressive stress increases rapidly with increasing the axial strain and reaches the peak at the strain of $1 - 2$ %. The secant modulus E_{50} can be estimated from the unconfined compressive strength q_u with the following equation;

$$E_{50} = 100 \text{ to } 200 \ q_u$$

Figure 4(a) and (b) show a relationship between amount of cement and q_u of FTS and BTS, respectively, showing that q_u value increases almost linearly with the increase in amount of cement. The increasing ratio of q_u is larger for soils with relatively low initial water content. The amount of cement necessary for FTS and BTS depends on the soil type

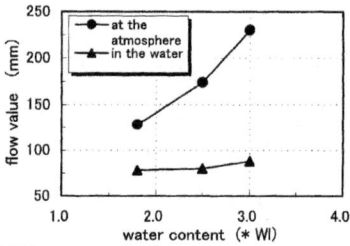

(a) Foam treated soil

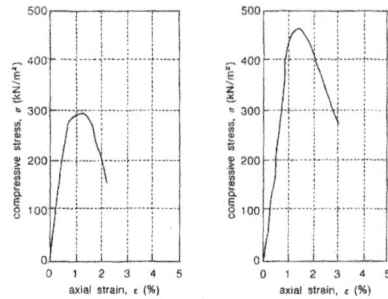

Figure 3. Stress-strain curve of foam treated soil

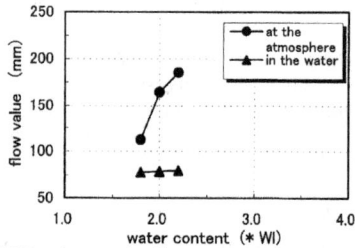

(b) Beads treated soil

Figure 1. Flow value and water content of soil

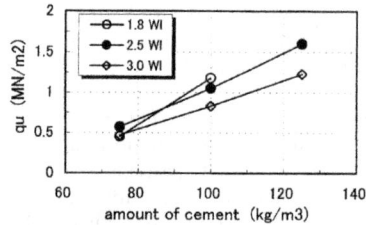

Figure 4 (a). Amount of cement and q_u (28day, Foam treated soil)

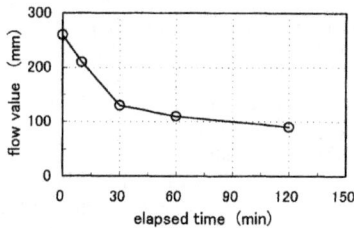

Figure 2. Fluidity of FTS with elapsed time

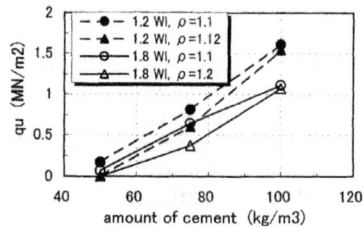

Figure 4 (b). Amount of cement and q_u (28day, Beads treated soil)

354

and the initial water contents, ranging from 50kg/m^3 to 140 kg/m^3.

Figure 5 shows the axial stress-strain curve obtained in one dimensional consolidation test. The curves are characterized by a sharp bend at the consolidation yield stress p_y. The ratio of p_y and q_u is about 1.4-1.6.

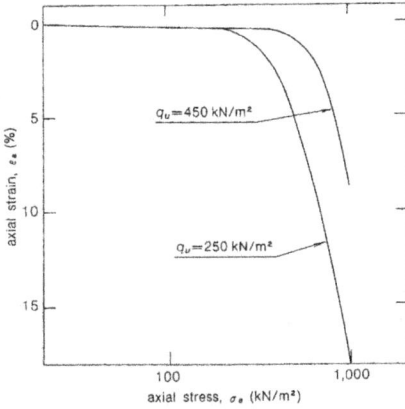

Figure 5. Compression curve of FTS in 1-D consolidation test

3 CALCULATION OF EARTH PRESSURE BY SLICE METHOD AND THE DESIGN OF QUAYWALL

An important problem to use lightweight treated soil for backfilling of quaywall structure is how to calculate the earth pressure at earthquake. Mononobe-Okabe's equation, which is based on the equilibrium of forces with the seismic coefficient method, has been used in the conventional design standards. However the equation is applicable to semi-infinite ground of horizontal deposits, and is not available when the solidified block, such as FTS or BTS are used for backfilling of wall. Tsuchida et al (2000) developed the slice method, which is common in slope stability problem, to calculate the earth pressure at earthquake for complicated composition of backfilling.

The slice method for calculating the earth pressure resultant force is obtained as follows. The slip surface behind the wall is assumed as Figure 6. The earth pressure resultant force is calculated based on a equilibrium of the weight, buoyancy, slip surface shear force, and seismic force on each slice.

The definitions of symbols in the figure are as follows:

P: earth pressure resultant force
α_i: angle of slip surface
W_i: total mass of the slice
W_i': effective mass of the slice
T_i: shear force on the slip surface

V_i: vertical force acting on the right side of the slice
k_h: lateral seismic coefficient
δ: angle of friction on the wall
l_i: length of the slip surface of the slice
N_i: normal force on the slip surface
E_i: lateral force acting on the right side of the slice

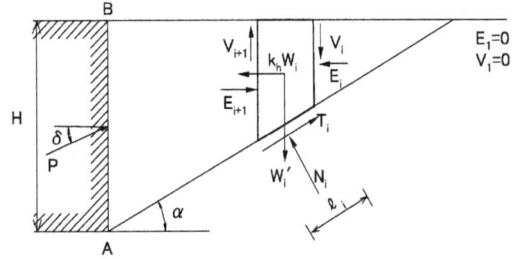

Figure 6. Calculation of the earth pressure resultant force by the slice method

The equilibrium of these forces is shown in Figure 7. The vertical and lateral equilibrium is established by the following equations.

$$W_i' + \Delta V_i = T_i \sin \alpha_i + N_i \cos \alpha_i \quad (1)$$

$$W_i k_h + \Delta E_i = T_i \cos \alpha_i - N_i \sin \alpha_i \quad (2)$$

where

$$\Delta V_i = V_i - V_{i+1}, \, \Delta E_i = E_i - E_{i+1}$$

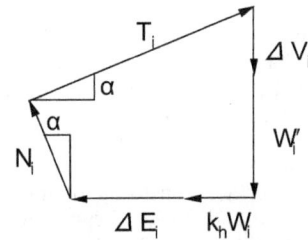

Figure 7. Equilibrium of forces working in each slice

Using safety factor F_s, cohesion and frictional angle on slip surface of each slice c_i, ϕ_i, the following equation on shearing force T_i is obtained;

$$T_i = (c_i l_i + N_i \tan \phi_i)/F_s \quad (3)$$

Erasing N_i with equations (1) and (2) gives:

$$T_i = (W_i' + \Delta V_i)\sin \alpha_i + (W_i k_h + \Delta E_i)\cos \alpha_i \quad (4)$$

Also by erasing N_i with equations (1) and (3), we have:

$$T_i = \frac{c_i l_i + (W_i' + \Delta V_i) \cdot \sec \alpha_i \cdot \tan \phi_i}{F_s \cdot (1 + \tan \alpha_i \cdot \tan \phi_i / F_s)} \quad (5)$$

As the earth pressure takes place at $F_s = 1$, the following equation on lateral force E_i is obtained as follows;

$$\Delta E_i =$$

$$\frac{c_i l_i \cdot \sec\alpha_i + (W_i' + \Delta V_i)(\tan\phi_i - \tan\alpha_i)}{1 + \tan\phi_i \tan\alpha_i} - W_i k_h \qquad (6)$$

The equilibrium of forces of all the slices in horizontal direction is given as follows;

$$\sum \Delta E_i = E_1 - E_{n+1} = -P\cos\delta \qquad (7)$$

By assuming that the direction of the resultant forces working on both sides of each slices is equal to the direction of the resultant force of earth pressure, the following equation can be used;

$$V_i = E_i \tan\delta \qquad (8)$$

By substituting equations (6) and (8) into the equation (7), the resultant force of earth pressure can be calculated for a linear slip surface passing through the bottom of wall, point A in Figure 6 by the following equation;

$$P\cos\delta =$$

$$\sum \frac{W_i k_h + \{-c_i l_i \cdot \sec\alpha_i + W_i'(\tan\phi_i - \tan\alpha_i)\}/X}{1 + (\tan\alpha_i - \tan\phi_i)\tan\delta / X} \qquad (9)$$

where, $X = 1 + \tan\phi_i \cdot \tan\alpha_i$

Calculations of earth pressure of P with varying α, the maximum value of P is the resultant force of the active earth pressure and α at that time is the angle of rupture.

When lightweight treated soil is backfilled, the solidified bloc as shown in Figure 8 is usually placed on the ordinary soil. The following three modes of rupture are hypothesized to calculate the active earth pressure in this condition.

Mode I: Linear slip surface cutting the inside of lightweight treated soil.

Mode II: Bilinear slip surface consisting of linear surface in ordinary soil and vertical hypothetical crack in lightweight treated soil. The shear resistance on the vertical surface of the interior of the lightweight treated soil is not considered.

Mode III: Tri-linear slip surface consisting of linear surfaces in ordinary soils and the boundary between the lightweight treated soil and ordinary soil.

For each angle α, the resultant forces of earth pressure are calculated with all three failure modes, and the maximum value is taken as the representative on the angle. The maximum value of P is the resultant force of the active earth pressure and α at that time is the angle of rupture.

Assuming that the distribution of earth pressure does not change by the height of wall, earth pressure p can be calculated by dividing the difference between the resultant forces of earth pressure P of ad-

joining layers by the depth difference (see Figure 9), as follows;

$$p = \frac{P_{d_{i+1}} - P_{d_i}}{d_{i+1} - d_i}$$

where, p is the earth pressure between AB in Figure 9, and P_{d_i} is the total earth pressure when the depth of the bottom of wall is d_{i+1}.

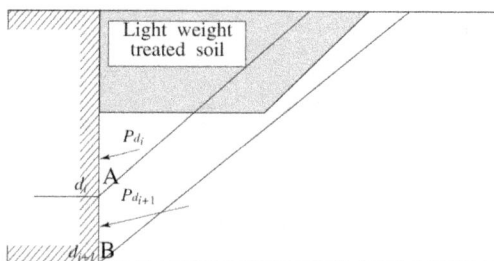

Figure 8. Earth pressure in the Slice Method

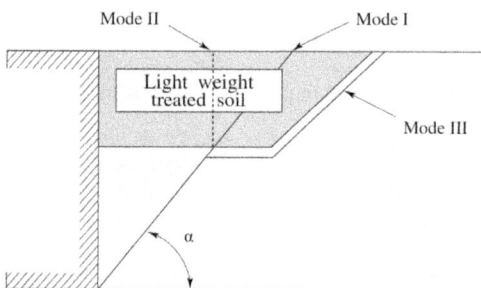

Figure 9. Three failure modes when a slip surface passes through light weight treated soil

Figure 10 shows the effect of earth pressure reduction due to use of LWTS, which is calculated by the slice method. In the figure, r^* is a ratio of the resultant force of active earth pressure when LWTS is used, to that of the conventional type filled by the rubbles and the sands. k_h is the design horizontal seismic coefficient, S is the volume of LWTS for 1m length of the quay wall and H is the height of quaywall. As shown in Figure 10, the earth pressure is reduced more with the increase of the replacement

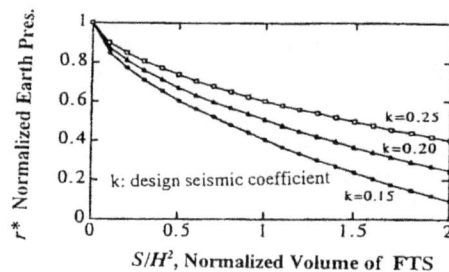

Figure 10. Effect of earth pressure reduction due to use of FTS

ratio of LWTS, S/H^2. In the case of k_h=0.2, which is common in Japanese major ports, the replacement ratio S/H^2 will be about 40% in order to get the 30 % reduction of earth pressure.

4 CASE IN KOBE PORT

4.1 *Overview*

The caisson type quaywall of 183m length at the second port Island in Kobe Port was damaged by Great Hanshin-Awaji Earthquake 1995. The lateral movements of caissons were as much as 0.8m to 3.5m in the seaside, and the settlements were 1.1m to 2.5 m, while the inclination of the caissons were less than 3 degree. In the reconstruction work of the quaywall, the reuse of the moved caissons were strongly requested to reduce the cost and the time, Further, the design seismic coefficient k_h of damaged facilities had to be increased from 0.18 to 0.20 after the reconstruction.

To satisfy these conditions, the backfilling of the caissons was replaced by foam treated soil (FTS), to reduce the design earth pressure by 30%. The design density of FTS was 1.00t/m³ in the areas above sea level and 1.20 t/m³ in the area under sea level. The design shear strengths FTS was 100kN/m² (Wako, T. et al., 1998).

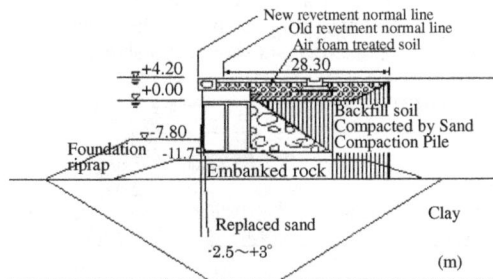

Figure 11. Design cross section of Kobe Port - Port Island 2

The cross section for reconstruction was determined as shown in Figure 11, where FTS was filled at the level from –1.0 m to +3.5 m and the total volume used were 22,000m³. As the average sea level on the site were +1.7m, more than half of the FTS was placed in the sea water, and the others were in the atmosphere condition. The sand layer beneath FTS was improved by the Sand Compaction Pile Method (SCP) to prevent the liquefaction in the earthquake time.

4.2 *Construction work*

The clay with natural water content of 122% and the liquid limit of 97%, which was dredged near by for the other reconstruction work, was mixed with seawater, cement (140kg per 1m³ of FTS) and air foam

(0.279m³ per 1m³ of FTS above sea level, and 0.196m³ per 1m³ FTS below sea level). Figure 12 is a construction system, which consists of the following machinery units (Wako et al., 1998);

a. unit for water-mixing and producing a clay slurry of uniform water content
b. unit for transportation of clay slurry
c. unit for mixing the slurry with cement and air foam
d. unit for casting at the site

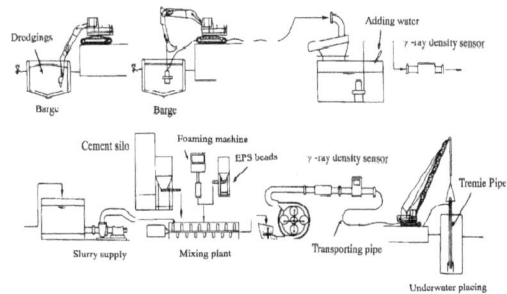

Figure 12. Construction system of LWTS method

The important point of the construction work was to maintain the density of FTS constant. It was found that about 30% of the air foam was disappeared during mixing, transportation and the placing, because the strength of the air foam was weakened by some organic that was contained in the dredged clay. For the compensation of lost air foam, it was necessary to increase the volume of air foam at the mixing unit. If the volume of air foam was too large and the density of FTS became less than 1.03 g/cm³ of seawater density, the placed FTS floated in the sea making the serious segregation of material and the sea water pollution. Accordingly, the density of FTS was continuously measured by the on-line Gamma ray density devices which were installed between the units. The density of FTS could be controlled within the ±0.05g/cm³ of design value. Figure 13 shows the placing of FTS at site.

Figure 13. Placing of foam treated soil

4.3 Investigation of density and strength after construction

Because FTS was cast in the seawater, whether the density is kept with time or not is important (Tsuchida et al., 1999). The author investigated the FTS used in the old wharf structure and observed the slight increase of the density under sea level (Tsuchida et al., 1999).

Figure 14 is the wet densities of FTS which were easured with core samples taken by boring after 1, 4, 7,10 and 22 months after construction. As shown in Figure 13, the measured densities of FTS above sea level were slightly larger than the design value of 1.0 ton/m^3 (1.0g/cm^3) and below the sea level, the densities were smaller than the design. The change of the density with time was not observed. The increase of the density above sea level seems to be due to the shrinkage with the initial hardening of the material. Figure 15 is the unconfined compression strengths of the cores with depth. Although the large scatters were observed, the average of the unconfined strength increased from 400kN/m^2 in 1 month to 600kN/m^2 Pa in 22 month. It could be concluded

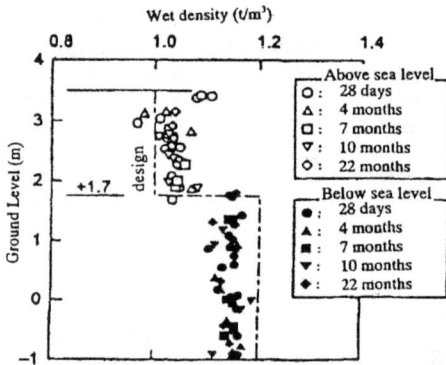

Figure 14. Wet density of FTS measured by Boring core after the construction

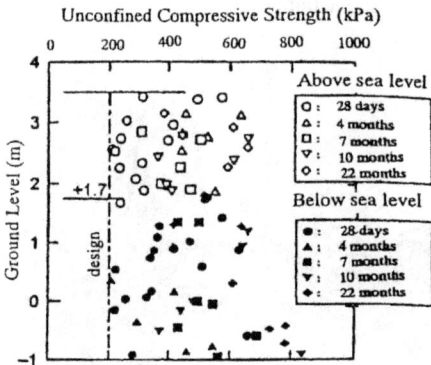

Figure 15. Unconfined Compression Strength of Foamed Treated Soil

that the mechanical properties of FTS used for backfilling of quaywall were in the stable conditions.

5 FIELD TEST IN KUMAMOTO PORT

5.1 Overview

Kumamoto Port is located on the Ariake Sea on Kyushu Island. For the construction of a new quay wall to the depth of 10m below the sea level in Kumamoto Port, it was known that the usage of lightweight soil for the backfill would reduce the construction cost by 20~25%. However, due to the tidal changes of as much as 4.5m in this port, it was necessary that the lightweight soil should be placed at a depth of 10m underwater at high tide. The full scale placing test of both FTS and BTS was carried out at a testing site with 10m water depth (Satoh, et al., 2001).

5.2 Underwater placing of lightweight treated soil

Ariake clay is well known for typical soft sensitive clay in Japan. The natural water content, liquid limit and plastic limit were 84.0%, 63.8% and 26.9%, respectively. The target value for the unconfined compressive strength is 200kN/m^2, and the target value for the wet density of lightweight treated soils is 1.1t/m^3 immediately after mixing, and 1.2t/m^3 at 28days after placing, respectively. For the above target values, the mix proportion for the lightweight treated soils is decided and shown in Table 1.

Table 1. Mix proportion of LWTS in Kumamoto Port

	FTS	BTS
Dry soil	367kg(136liter)	368kg(137liter)
Water	611 kg	625kg
Lightweight material	8.9kg(208liter)	6.9kg(215liter)
Cement	80-100kg	80-100kg

Underwater placing tests of lightweight treated soils were performed inside the cells of concrete caisson which were under construction in Kumamoto Port to avoid the possible separation of beads by spreading around the construction site. The inside of the concrete caisson was divided into 4 sections and sub-divided into 9 testing zones, as shown in Figure 16. The field placing tests at the 9 testing zones were designated as Case 1 to Case 9 in order. The total quantity of lightweight treated soils in the test was 860m^3, and the testing duration was 5 days.

During the placing of lightweight treated soils under seawater, a sample of seawater in the caisson was taken from the middle part of the placed surface and the top of the caisson. The pH value in the caisson increased from the initial value of 8.2 to 8.5~9.8 due to the placing of lightweight treated soil, and the density of suspended substances (SS) was increased to 160mg/l at maximum. These data represent the

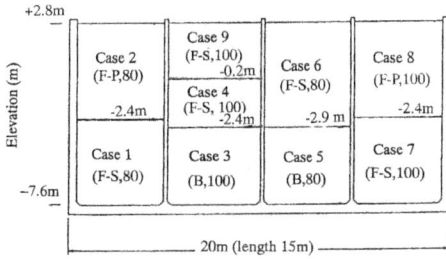

+2.8m

Elevation (m)

Case 2 (F-P,80)	Case 9 (F-S,100) -0.2m	Case 6 (F-S,80)	Case 8 (F-P,100)
	Case 4 (F-S, 100) -2.4m		
-2.4m		-2.9 m	-2.4m
Case 1 (F-S,80)	Case 3 (B,100)	Case 5 (B,80)	Case 7 (F-S,100)

-7.6m

20m (length 15m)

*F-S:Surface active type air foam, F-P:Protein type air foam, B:EPS beads

** 80, 100: unit cement content 80kg/m³,100kg/m³

Figure 16. 4 sections and sub-divided into 9 testing zone in caisson

water inside the caisson and at 1~3 m distance from the placed materials.

It should be noted that, when the lightweight treated soil is placed under water, some protection to the sea environment is necessary, although the increase of pH and the density of SS were observed in only a limited area near the placed material. Water samples were also taken from 10 sampling points, 20~50m away from the testing site and no significant changes from the initial values on water quality

were observed because the tests were carried out inside the concrete caisson.

5.3 Wet density after placing

The wet density of BTS immediately after mixing meets the target value of 1.1 t/m³ while the wet densities of FTS were slightly larger than the target value. This means that a part of air foam was lost during mixing. The variation range of the wet density immediately after mixing for each case is shown in Figure 17 as a shadow area. In the same figure, the measuring range of the wet density of the lightweight treated soils after transportation is also shown except for Case 9. The wet density after transportation was measured using samples of lightweight treated soils taken at the tremie pipe after transporting.

The wet densities after transportation were larger than the wet density immediately after mixing. This is due to the pressure in the transportation pipe and the separation and loss of lightweight agents during placing. The maximum pressure in the transportation pipe was 180~300 kN/m², which may cause the compression of foam according to Boyle's law and the irrecoverable compression of EPS beads. During the underwater placing, it was observed that some of

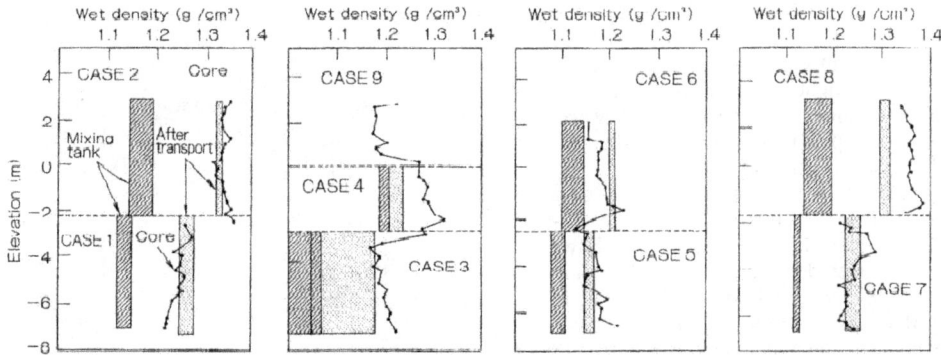

Figure 17. Wet density after mixing, transportation and 28-day curing

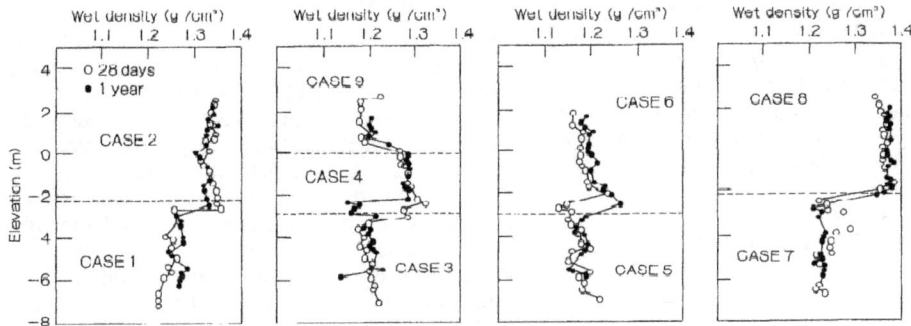

Figure 18. Wet density of 28-day samples and 1-year samples

359

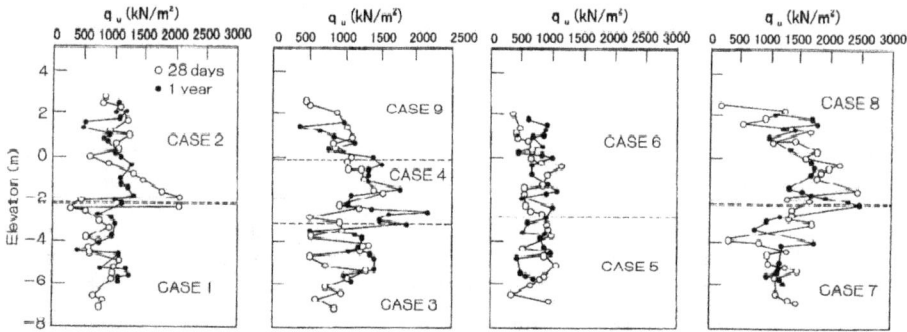

Figure 19. Unconfined compressive strength with depth

the foam in FTS and the EPS beads in BTS were separated from the treated soil and drifted up to the water surface. The EPS beads were not allowed to spread around the construction site, but were carefully collected and re-added into the mixing plant; the volume of this was about 2~3% of the total beads mixed.

For investigating the change in wet density of the treated soils after placing, the core samples of the lightweight treated soils after curing of 728 days (28-day samples) were collected by boring. In Figure 17, the comparison with the wet density after mixing and transportation is shown. It can be seen that the wet density of the lightweight treated soil of 28-day samples is not different from the wet density after transporting. A slight decrease in wet density was seen in 28-day sample of Case 6, while the wet density of 28 -day samples was slight larger than the wet density after transportation in Cases 4 and 8.

To investigate the long-term quality change, core samples of the lightweight treated soils were collected 1 year after construction. Figures 18 shows the comparison of wet density of the lightweight treated soils after 28 days and 1 year with the depth. It can be seen that there is no significant difference in wet density between the underwater samples of 28 days and 1 year after the construction.

5.4 Strength after placing

Compressive strength is also an important index for quality control of lightweight treated soils. The unconfined compressive strength q_u was obtained using 28-day core samples and 1-year core samples. The q_u values with depth are shown in Figure 19. The unconfined compressive strengths were much larger than the target value of 200 kN/m^2.

6 CASE IN ISIKARI BAY NEW PORT

6.1 Overviews

Ishikari Bay New Port is located near Sapporo, the capital city of Hokkaido Island. The steel-sheet pile

quaywall, which was constructed in 1982, was damaged by the progress of corrosion, and the improvement was made for increasing the earthquake-resistant capacity. The foam treated soil was used to reduce the earth pressure. Figure 20 shows the cross section for the improvement. A part of the backfill of the steel-sheet-pile quaywall was replaced by FTS whose wet density was 1.3 t/m^3. The replaced area was 5.25 m in depth, 16.00 m in width, and the improvement by the sand compaction pile method and the gravel drain pile method were also carried out in the sand layers surrounding FTS as a countermeasure against liquefaction (Hirasawa, et al., 2000).

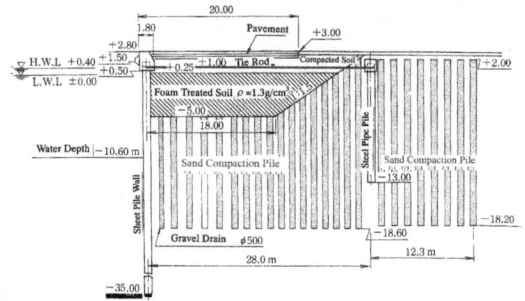

Figure 20. Cross section of quaywall in Ishikari Bay New Port

6.2 Use of bentonite and low temperature experiments

As the excavated soil did not contain much fines (sand 90%, silt and clay 10%), a series of mixing experiments were conducted to investigate the feasibility of underwater placement. As the result of a washout resistant test with the placing velocity of 20cm/s, it was known that without addition of fine, FTS placed underwater were quickly segregate and the density became more than 1.3 t/m^3. To improve the consistency of FTS, bentonite of was mixed with the excavated sand. The mixing condition determined with the laboratory test is shown in Table 2.

As the construction work is to be carried out under the temperature from –5°C to 5°C in January, the

Table 2. Mixing condition of FTS

Dry Soil	393 kg/m^3	Seawater	393 kg/m^3
Cement	200 kg/m^3	Air foam	368 liter/m^3
Bentonite	98 kg/m^3		

effect of low temperature on FTS was a problem. The curing test was conducted in a thermo-siatic chamber of –5°C, 0°C and 5°C, and the flow value, wet density and unconfined compressive strength were measured in the chamber. Flow values were measured immediately after mixing and five minutes later to examine the influence of low temperatures on the fluidity of FTS.

When the outside temperature was –5°C, it was known that foaming would be difficult because the foaming nozzle would freeze and the on-site temperature control would be necessary to keep the temperature of foaming equipment about 5°C.

Figure 21 shows the relationship between the mixing temperature of FTS and the flow value. As shown in the figure, the effect of temperature on the flow value is small and seems to be almost negligible in the case of higher than 0°C.

Figure 22 shows the relationship between the density change $\Delta\rho_t$ ($=\rho_{t28}-\rho_{t7}$) and the kneading temperature, where ρ_{t28} is the density of 28-day cure sample and ρ_{t7} is the density of 7-day cure sample. As shown in the figure, $\Delta\rho_t$ increased when the temperature was lower than 0°C. It was also known that the temperature at the mixing and curing did not have a major influence on the strength of FTS.

6.3 Construction

The construction work of FTS was conducted during the period between January 27 and February 16, 1999. The placement of FTS was carried out by dividing the 31m length of the site into three sections with the steel plates, and the depth of 5.25 m into four layers. The total volume of placement was 2,550m^3.

Based on the low-temperature experiment results the entire plant, excluding the cement and bentonite silos was wrapped with a cold protection cover to keep the internal temperature at 5°C or higher. When seawater was used, its temperature was raised to 5°C by a heater. The outside air temperature was below zero through most of the placement period, and the lowest temperature was 9°C, observed on February 3.

Figure 22 shows the wet density and the unconfined compressive strength of samples collected by boring from the FTS application site 28 days after placement. All the samples collected from each elevation satisfied the target value of ρ_t (1.3g/m^3 or less). As the water depth increased, ρ_t tended to increase due to the influence of water pressure.

As a summary of the case in Ishikari Bay Port, when the excavated sand was used as material for FTS, the addition of adequate amount of bentonite

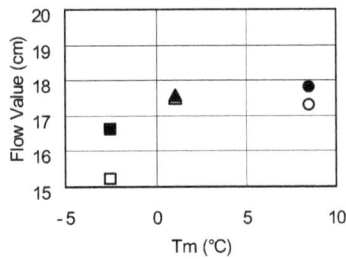

Figure 21. Flow value and mixing temperature

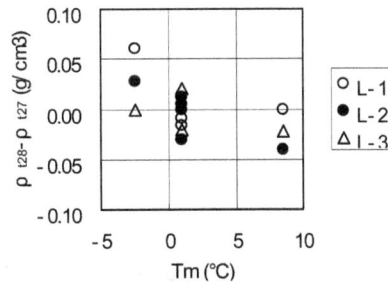

Figure 22. Density change and mixing temperature

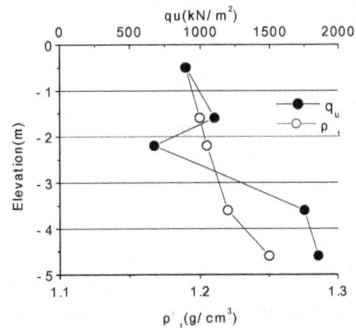

Figure 23. Wet density and unconfined compressive strength after placing

Figure 24. Underwater placing in winter

was necessary to prevent the segregation in seawater. And, even at low temperatures, it was possi-

361

ble to apply FTS by conducting appropriate temperature control).

7 CASE IN TOKYO INTERNATIONAL AIRPORT

7.1 Use of FTS in Tokyo International Airport

Since 1984, the Offshore Expansion Project of Tokyo International Airport (Haneda) has been conducted by Ministry of Transport. At present the project entered Phase III, in which a new runway, taxiways, apron and a new passenger terminal building have been constructed on soft clays whose thickness are 35~40m in Holocene layer and 10~15m in Pleistocene layer. In this project, the total amount of 90,000m³ of FTS and BTS was used at the six sites for the purpose of reduction of consolidation settlement or earth pressure. The Tokyo International Airport and the location of the sites are shown in Figure 25.

Figure 26 shows the cross-section of Case-1, where FTS and BTS of 1,500 m³ were used for the back-filling of the revetment (Tsuchida, et al., 2000). FTS and BTS were made by mixing the waste clay slurry, which was caused by the construction of shield tunnel in the airport, with cement, air foam and EPS beads. The density and unconfined compressive strength of lightweight soil in the design were 1.2 g/cm³ and 200 kN/m², respectively.

To investigate the ground properties of the lightweight soils, the core sampling and the laboratory tests were carried out as well as in-situ plate loading test, CBR test, and Falling Weight Deflectmeter (FWD) test, 2-6 months after the construction. According to test results, the properties of FTS and BTS satisfied the range for the quality control in spite of the large variance of the properties of waste soil.

7.2 Use of FTS for reduction of consolidation settlement in construction of new taxiway

The new parallel taxiway was constructed on the west side of the existing A-Runway, having a length of 2,000 m and a width of 30 m. For the construction of the taxiway, 2~3m height embankment was carried out on the sites where the railroad tunnel, the road tunnels, and the utility conduits were already constructed underground. Because the new taxiway had not been included in the master plan of the project, the load of the embankment had not been taken account for the existing underground structures. Figure 28 shows a typical cross section of ground improvement for the taxiway constructed on the railway shield tunnels. As the Holocene layer is loose enough to be liquefied in the case of designed earthquake, the layer was improved by the deep cement mixing method. If the lightweight treated soil is not

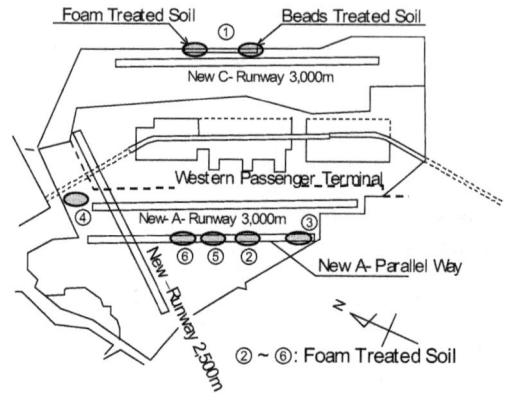

Figure 25. Tokyo International Airport and the sites where LWTS was used

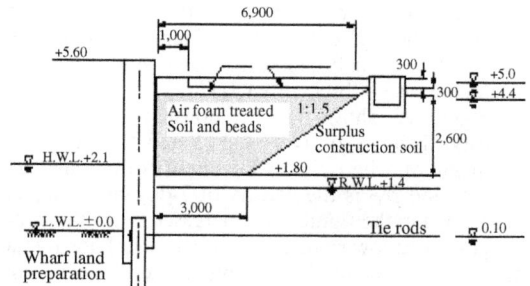

Figure 26. Cross-section of revetment

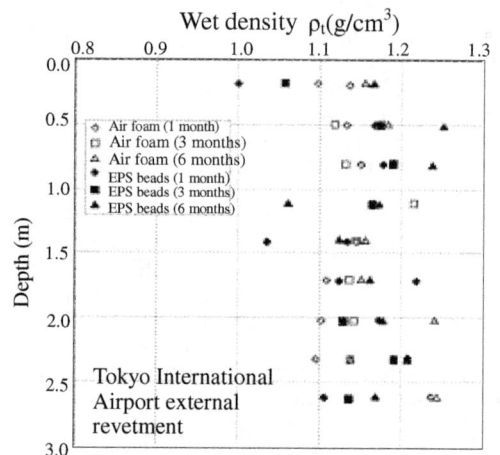

Figure 27. Wet density with FTS and BTS

used, the predicted consolidation settlement of clays beneath the tunnels was 64 cm and it was much larger than that allowable for the structural safety of tunnel. By using the FTS as shown in the figure, the increment of overburden stress was reduced by 70%,

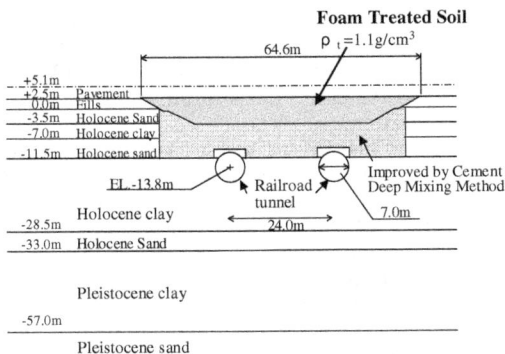

Figure 28. Cross-section of parallel taxiway constructed on railroad tunnel

and predicted consolidation settlement was reduced to 19cm (Tsuchida, 2001).

Figure 29 shows the sight of construction work of FTS, which was made by mixing air foam and water with the excavated soil in the airport. The excavation and the placing of FTS were carried out systematically, keeping the change of overburden stress on the tunnel within a limited range. The sight after the construction of FTS is shown in Figure 30.

Figure 31 shows the measured vertical movement of the rail plane monitored during the construction. As shown in the figure, the measured movement was kept less than 20mm and agreed well with the prediction.

7.3 Grand expansion project of Tokyo International Airport and the merits of LWTS method

Recently, the new "Grand expansion" project has been emerged to counter the international airport competition in Asian region.

Figure 32 shows one of the likely plans, where a new runway (the fourth runway) is constructed, increasing the capacity of taking off and landing of the airport by more than 40%. However, there are some engineering problems as follows:

Figure 29. Placing of FTS

Figure 30. Completion of FTS work

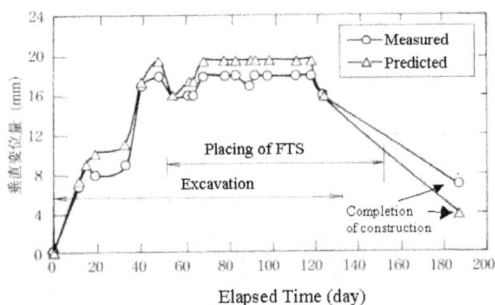

Figure 31. Measure and predicted movement of rail plane during excavation and placing of FTS

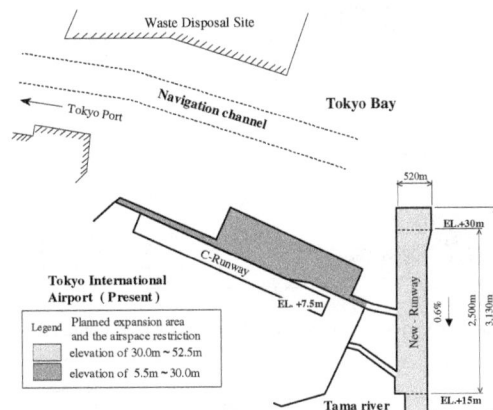

Figure 32. Planned Grand excavation of Tokyo International Airport

- As the site of new runway is close to the navigation channel to Tokyo Port, the elevation of new runway must be from 15m to 30m over the sea level in order to take a sufficient clearance to container ships.
- Although a large scale of the improvement of soft clays are necessary, the time slot for use of the improvement vessels is limited due to the airspace restriction around the present C-Runway.

363

Figure 33 shows a typical cross section of seawall in which the improvement is carried out by Sand Compaction Pile and the reclaimed by mountain sand. To support the height of 15m, the area necessary for the improvement are 37m in depth and 140m in width. A cross section when FTS is used for the backfill is shown in Figure 34. As shown in the figure, by virtue of lightweight of fills, it is possible to reduce the width of improvement drastically and this is important because of the limitation of time slot for improvement work. Further, 3m^3 of FTS can be made of 1m^3 of marine clay, which will be produced by dredging for the construction of new navigation channel.

As the study of the Grand expansion Project has just started in 2001, there are so many studies to be carried out, including intensive geotechnical investigations. It can be said that the project will provide challenging topics for soft ground engineering.

8 CONCLUDING REMARKS

In the construction work on soft ground, the use of lightweight geotechnical material is an effective method to save the cost and construction time. However, in the coastal area, because the buoyant force due to the sea level change and the inertia force in earthquake time must be taken into consideration, the use of lightweight material has been difficult. In this report, a lightweight treated soil (LWTS) method and the case studies in the coastal projects are introduced.

A lightweight treated soil method has recently been developed to reuse dredged soils as an artificial lightweight geomaterials, and the density ranges from 1.0 to 1.2 t/m^3, for coastal construction projects. There are two types of lightweight soils, foam treated soil (FTS) and bead-treated soil (BTS). The slurry of dredged soil or waste soil is mixed with cement and air foam, or cement and EPS (expanded polystirol) beads whose diameters are 1 - 3 mm, respectively. Since 1996, the lightweight treated soil method has been applied in several seaport and coastal airport projects in Japan.

An important problem to use LWTS for backfilling is how to calculate the earth pressure at earthquake, because the conventional Mononobe-Okabe's equation is not available when the block type material such as the formed treated soil are used for backfillings of the wall. To solve the problem, the slice method was developed to calculate the earth pressure at earthquake.

The four cases in Kobe Port, Kumamoto Port, Ishinomaki Port and Tokyo International Airport were presented, where lightweight soil were used to

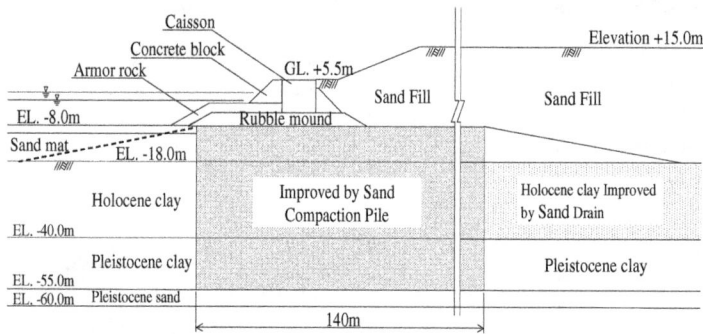

Figure 33. Cross-section of seawall reclaimed by mountain by mountain sand

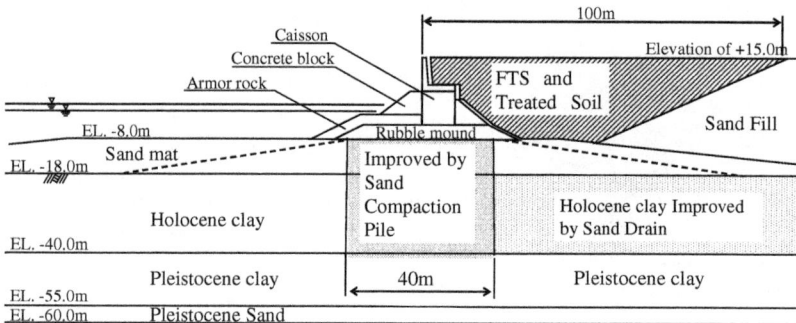

Figure 34. Cross section of seawall backfilled by FTS

reduce the earth pressure or consolidation settlement.

In the case of Kobe quays, FTS was filled between the level −1.0 m and +3.5 m with the total volume was 22,000 m^3 to reduce the active earth pressure in earthquake by 30%. The increase of the density below sea level was almost negligible and the average of the unconfined strength increased from $400kN/m^2$ in 1 month to $600kN/m^2$ in 22 month.

In the cases of Kumamoto Port, a full scale field placing test of lightweight soils was carried out at sites 10m below the sea level in Kumamoto Port with the purpose of investigating the material properties of lightweight soils placed under deep water. Especially the change of density through the process of mixing, transportation, placing and hardening was examined in detail. It was found that the lightweight soil method can be applied under sea water of −10m, although a part of the total volume of mixed foam or mixed EPS beads is swept away during the construction process. The wet-density of lightweight soil measured 1 year after the construction was almost the same as that of a 28-day specimen, while the 1 year strength was 40 % larger than the 28-day strength.

In the case of Ishinomaki Port, the steel-sheet pile quaywall, constructed in 1982, was improved for increasing the earthquake-resistant capacity. The backfill was replaced by FTS whose density was 1.3 g/cm^3. As the excavated sand was used as material for FTS, the addition of adequate amount of bentonite was necessary to prevent the segregation in seawater. The construction work was carried out under the temperature from −5°C to 5°C in January, and it was found that it was possible to apply FTS by conducting appropriate temperature control.

In Tokyo International Airport, the total amount of 90,000m^3 of FTS and BTS was used at the six sites for the purpose of reduction of consolidation settlement or earth pressure. The new parallel taxiway was constructed on the sites where the railroad tunnel were already constructed underground. As the predicted consolidation settlement of clays beneath the tunnels was much larger than that allowable for the structural safety of tunnel, a part of the soil was replaced by FTS and the increment of overburden stress was reduced by 70%. The excavation and the placing of FTS were carried out systematically, keeping the change of overburden stress on the tunnel within a limited range. The measured movement was kept less than 20mm and agreed well with the prediction. In newly studied *Grand Expansion* project, the LWTS method is considered one of the effective methods to solve the engineering problems, such as the extremely high embankment on soft clay, and the airspace restriction.

The author would like to express his deep gratitude to the staff of Soil Mechanics and Geo-environment laboratory, Port and Airport Research Institute in preparing the manuscript

REFERENCES

Hirasawa, M., Saeki, S., Kodama, S., Yakuwa, T., and Tsuchida, T.(2000): Development of light-weight soil using excavated sand and its application for harbor structures in cold regions, *Coastal Geotechnical Engineering in Practice*, Proceedings of International Symposium on Coastal Geotechnical Engineering in Practice, Balkema, Vol.1, pp.599-604.

Satoh, T., Tsuchida, T., Mitsukuri, K. and Hong, Z. (2001): Field placing test of lightweight treated soil under seawater in Kumamoto Port, Soils and Foundations, Vol.41, No.5, pp.145-154.

Tsuchida, T., Takeuchi, D., Okamura T. and T. Kishida (1996). Development of lightweight fill from dredgings. *Environmental Geotechnics*, Proceedings of the Second International Congress on Environmental Geotechnics, Balkema, pp.415-420.

Tsuchida, T.(1999): Development and use of foamed treated soil in port and airport project, *Report of Port and Harbour Research Institute*, Vol.38, No.2, pp.131-167, (in Japanese).

Tsuchida, T. Fujisaki, H., Makibuchi, M., Shinsya, H., Nagasaka, Y.and Hikosaka, M. (2000): Use of light-weight treated soils made of waste soil in airport extention project, Journal of JSCE, No.644, VI-46, pp.13-23 (in Japanese).

Tsuchida, T., Kikuchi, Y., Yamamura, K., Funada, K and Wako, T. (2001): Slice Method for Earth Pressure Analysis and its Application to Light-Weight Fill, Journal of the Japanese Geotechnical Society, Vol.41, No.3 (in Japanese).

Tsuchida, T.(2001):Settlement of Pleistocene clay layer in coastal area, the reason, prediction and measure, Keynote Lecture, Proceedings of the 3rd International Conference on Soft Soil Engineering, HongKong, pp.67-80.

Wako, T., Tsuchida, T., Matsunaga, Y., Hamamoto, K., Kishida, T and Fukasawa, T.(1998): Use of Artificial Light Weight Materials for Port Facility, Journal of JSCE, No.602, VI-40,pp.35-52.

Evaluation for the stability of coastal structures

Soft Ground Engineering in Coastal Areas, Tsuchida et al. (eds)
© *2003 Swets & Zeitlinger, Lisse, ISBN 90 5809 613 0*

Osaka Port Yumeshima tunnel project

K. Ikeda, K. Kusachi, M. Miyata, K. Kawasaki, T. Esaki, A. Hirota & M. Ushiro
Kinki Regional Development Bureau, Ministry of Land, Infrastructure and Transport, Japan

ABSTRACT: In this paper, we first introduce an overview of the plan and structure of the Osaka Port Yumeshima Tunnel that is currently under construction. Then, we discuss the method for estimating the consolidation settlement of diluvial clay layers at the construction site, as well as the results of our study on the new flexible tube joint structure we developed for immersed tunnels to solve the technological problems, that was decided on after ground settlement estimation.

1 INTRODUCTION

The City of Osaka is currently developing and constructing international physical distribution centers and other state-of-the-art facilities in its waterfront area including Yumeshima Island, an artificial offshore island in Osaka Bay, in order to expedite the "Techno Port Osaka" project that the city drew up to commemorate the centennial anniversary of its incorporation as a city. The top plan view of the Osaka Port design for the future is shown in Fig.1.

In the Osaka Port design, an undersea tunnel (Osaka Port Yumeshima Tunnel) was planned and designed in 1999 to support harbor transportation activities for these distribution centers and as well as personnel flow, both of which are estimated to increase in volume and number in the future. As shown in Fig.1, this tunnel will serve as a link connecting Sakishima Island and Yumeshima Island. Construction of this tunnel started at the site in 2000 under the execution management of the Japanese Government, with the target completion date set at FY2007.

As lands have been reclaimed in deepwater areas in Osaka Bay in recent years, the seabed has received increased land load. Prolonged consolidation of the diluvial clay layer is observed in these areas. The Yumeshima Tunnel is being constructed between Sakishima Island, a recently-constructed artificial island being exposed to primary consolidation, and Yumeshima Island, that is under reclamation at present. Therefore, this tunnel will be exposed to large residual ground settlement due to the prolonged consolidation. In the design of this tunnel, one of the most challenging tasks was how to design

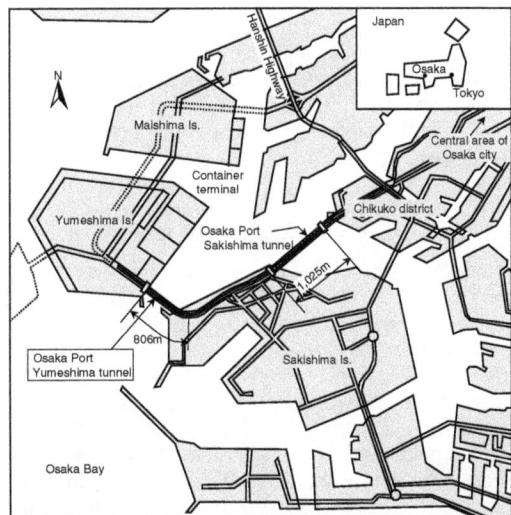

Figure 1. The top plan view of the Osaka Port design and the location of Osaka Port Yumesima Tunnel

and construct the tunnel in this bad ground condition.

In this paper, we first introduce an overview of the plan and structure of the Osaka Port Yumeshima Tunnel that is currently under construction. Then, we discuss the method for estimating the consolidation settlement of diluvial clay layers at the construction site, as well as the results of our study on the new flexible tube joint structure we developed for immersed tunnels to solve the technological problems, that was decided on after ground settlement estimation.

2 GENERAL DESCRIPTION OF YUMESHIMA TUNNEL DESIGN

2.1 Introduction

The longitudinal sectional view, and cross sectional view of the Yumeshima Tunnel are given in Figs. 2 and 3. This tunnel consists of an immersed tunnel section, access areas on the ground, and vertical shafts (ventilation station separation system). Its overall length is approximately 2.1 km (immersed tunnel section: approx. 0.8 km, access roadway sections on the ground: approx. 1.3 km). In the longitudinal view, the immersed tunnel section crosses underneath the 15-m deep main sea channel planned for large ships.

For the tunnel construction, tube immersion method was chosen because this method has various advantages over bridging method or shielding tunnel method. The primary reason for this is that the immersion method enables us to give a shorter length of the tunnel, which is including access areas, to cross the sea channel between two islands.

The tunnel tube section consists of 8 tube elements (length of each tube element: approx. 100 m). For constructing the approach sections on the ground (access roads), a large-scale open cut method is used to excavate the ground to the maximum depth in excess of 25 m. Among these main tunnel components, we discuss only the immersed tunnel section in this paper.

2.2 Structure of Immersed Tunnel Section

Tunnel tube structures can be classified into "reinforced concrete (RC) structure," "RC and steel shell structure," and "steel-concrete composite structure." Until around 1990, tunnel tubes using an RC structure or RC and steel shell structure were used for most of the immersed tunnels constructed in Japan (see Fig. 4). In RC and steel shell structure, the steel shell is not counted on as a structural member. However, many recent tunnels use tunnel tubes of steel-concrete composite structure because the composite structure enable us to reduce construction costs/periods and to secure construction workability (see Fig. 5). As well known, various technological developments have been made for this type of structure.

Steel-concrete composite structures are further classified into open sandwich structure and full sandwich structure. In the open sandwich structure, the outer shell of the tube is used as a water resistant steel plate and also as a steel member alternative to reinforcing bars. The inner portion of the tube slab is constructed of an RC structure. The full sandwich structure is a composite structure in which both the outer and inner shells are constructed of steel members in place of reinforcing bars (see Fig. 5)

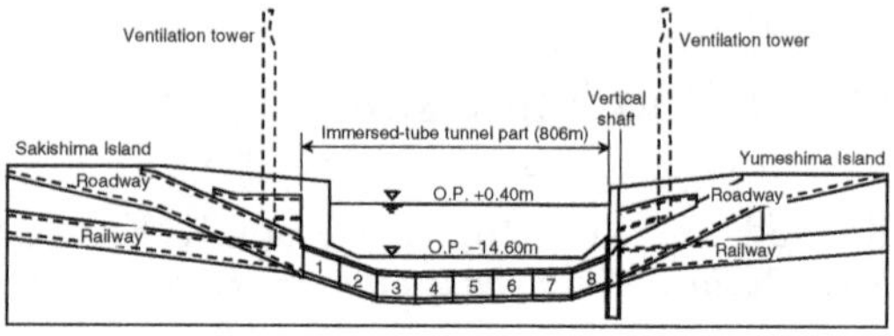

Figure 2. Longitudinal sectional view (Osaka Port Yumeshima Tunnel)

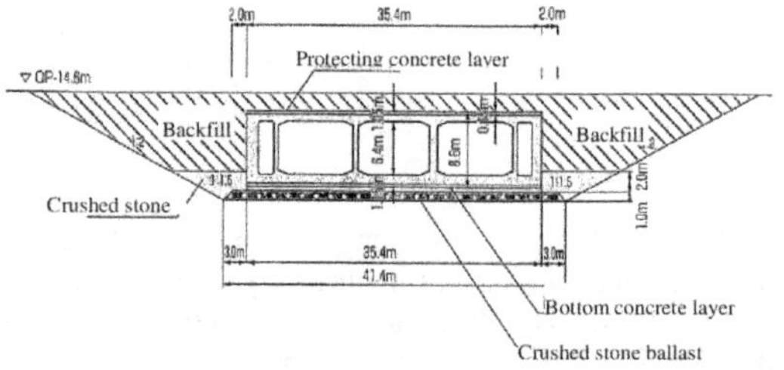

Figure 3. Cross sectional view of immersed tunnel section (Osaka Port Yumeshima Tunnel)

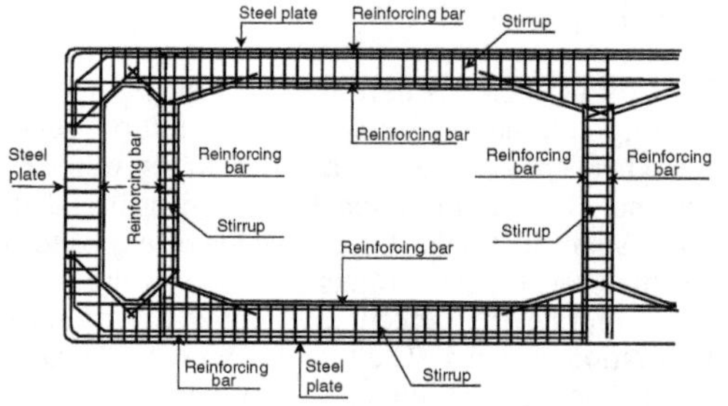

Figure 4. RC structure of immersed tunnel tube element

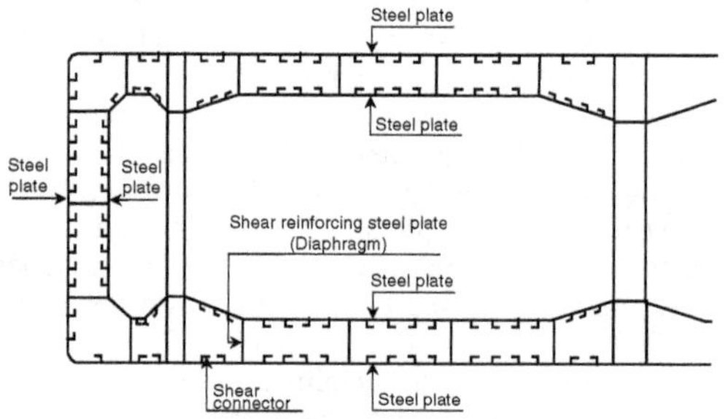

Figure 5. Steel-concrete composite structure of immersed tunnel tube element (full sandwich structure)

The Osaka Port Sakishima Tunnel is the first immersed tunnel in Japan consisting of tunnel tube elements of steel-concrete composite structure. In this tunnel, the floor slab and sidewall of the tube elements are of open sandwich design and the roof slab and center wall are of RC. Tunnel tubes of this structure design were also used the New Kinuura Tunnel. The Kobe Port Minatojima Tunnel is the

World's first immersed tunnel to use tunnel tubes of full sandwich structure with the exception of one tube element. The roof slab and sidewall are of full sandwich design, while the center wall and the partition wall are of RC. Tunnel tubes of this structure were also used for the Okinawa Naha Port Submerged Tunnel. For this tunnel, all members including the floor slabs are of full sandwich design.

The tunnel tube structure of the Yumeshima Tunnel was determined from the standpoint of construction period and cost. As a result, full sandwich structure was selected for the roof slab, sidewall, partition wall, and center wall, while open sandwich structure was selected for the floor slab. The standard cross section of the immersed tunnel tube elements for the Yumeshima Tunnel is shown in Fig. 6. Similarly to the Sakishima Tunnel, the Yumeshima Tunnel adopts a railway center system in which a railway runs along the tunnel centerline and roads run on both sides of the railway. Consisting of two lanes, each road is used for one-way traffic. Emergency evacuation walkways are provided besides the roadway. Each immersed tunnel tube element measures 35.4 m in width and 8.6 m in height. The effective dimension (height) of the tube inner space is 6.4 m. This dimension was determined from the dimension required for the roads, for which the construction gage (tall container), roadbed, and ground settlement margin were taken into account.

3 ESTIMATION OF GROUND SETTLEMENT

An immersed tunnel has the feature of reduced apparent specific gravity of the tunnel tube section due to buoyancy, thus less tube supporting force is required from the ground. From this standpoint, tunnel construction by immersion is advantageous when constructing tunnels on soft grounds. However, in designing an immersed tunnel that will be constructed on a soft ground, residual ground settlement (differential settlement) that will occur in the future, should be accurately estimated in addition to the tunnel bearing capacity of the ground.

Since the Yumeshima Tunnel is being constructed on freshly reclaimed ground, as has been already discussed, there is concern about the conspicuously uneven ground settlement that is estimated to occur around the seawalls in the future. In this paper, we discuss an outline of the method for estimating consolidation settlement of the ground that we applied in the tunnel design, as well as the results. However, note that the estimation results we are going to discuss here are the ones we obtained at the initial stage of the tunnel design, that have been partially modified after the load conditions, construction sequences, etc. were reviewed.

he seabed of Osaka Bay consists of a soft alluvial clay layer on the sea bottom and a thick diluvial clay

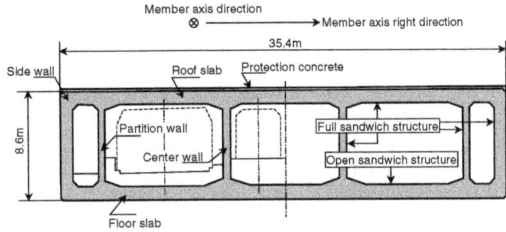

Figure 6. Standard cross sectional view of immersed tunnel element

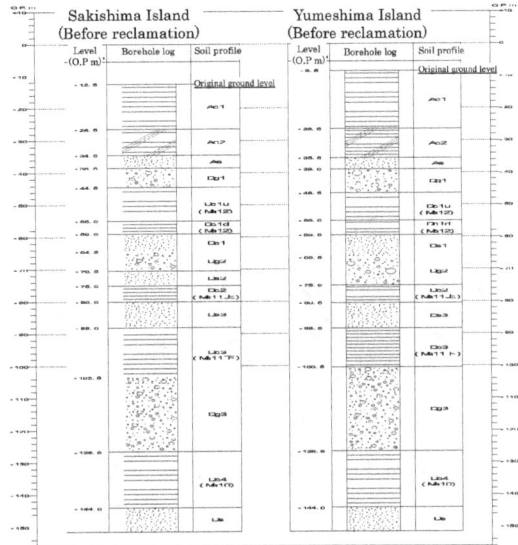

Figure 7. Soil profile (Sakishima Is. and Yumeshima Is.)

layer below the alluvial clay layer. Sandwiched between these layers is a sandy soil deposit. The diluvial clay layer is composed of marine deposits defined by Ma numbers, and is said to have pseudo over-consolidation characteristics due to age effect. The composition of the soil layer measured by drilling through the mud at some spots near the tunnel construction site is shown in Fig. 7. The diluvial clay layer will be exposed to a high pressure due to reclamation load. When considered from the data acquired during the construction of the Kansai International Airport, the boundary of ground consolidation is estimated to reach a depth-below-water-level of approximately OP - 150 m. Based on the above investigation result, we chose Ma12 deposit, Ma11 deposit, and Ma10 deposit for the ground settlement estimation. We excluded the alluvial clay layer located below the access road section from the scope of settlement estimation, because the alluvial clay layer in this area had been improved (by deep mixing treatment method) so that residual settlement will not occur for 100 years (design service life) after the opening of this tunnel to traffic.

As the method of ground settlement estimation, we used a quasi-single dimensional consolidation analysis method. This method incorporates Terzaghi's primary consolidation theory into the secondary consolidation theory (settlement proportional to log (t)). The latter theory can express consolidation settlement of creeping characteristics. For the stress calculation (load condition) in the soil, we used Boussinesq's that assumes reclaimed land as a semi-limitless horizontal deposit. By using this theory, we are able to consider three-dimensional distribution of stresses, when estimating complicatedly fluctuating ground stresses due to excavation, tunnel tube setting, back filling of the trench, and other construction works.

We set such analysis parameters as consolidation yield stress Pc; compressibility index Cc (in normal consolidation zone); volume compressibility coefficient m_v (in over-consolidation zone); and secondary consolidation coefficient ε_a as follows:

For the consolidation yield stress Pc, which divides consolidations into over-consolidation and normal consolidation according to stress, we removed the effect of rate dependency from the constant strain rate consolidation test results to determine the values that would be obtained from a standard consolidation test. For the consolidation index Cc, we set a value for the steepest slope used for a standard consolidation test. In setting the volume compressibility coefficient m_v for the $Ma12$ deposit, we used a variable m_v model where m_v changes at a point near Pc. For the $Ma11$ and $Ma10$ deposits, we set constant values. Fig. 8 shows the relationship between the volume compressibility coefficient and Pc we determined from actual measurements, and Fig. 9 shows the relationship between the secondary consolidation coefficient and Pc. To set the secondary consolidation coefficient ε_a, we referred to the results of a long-period consolidation test that had been conducted for a non-disturbed test block sample in the Osaka Port, together with the actual ground settlement measured in the same site (at the ventilation tower constructed on the Nanko exit side of the Sakishima Tunnel).

For the starting point of ground settlement estimation, sea bottom condition without consolidation should be used. Based on this concept, we set the ground settlement initiation time at the time point immediately after land reclamation. Since ground settlement after completion of tunnel tube installation is crucial, we estimated the overall residual settlement during about 100 years (estimated to be Mar. 2108) from completion of the tunnel (which is scheduled to be in Dec. 2004). Fig. 10 shows the distribution of ground settlement along the longitudinal cross section (longitudinal axis) of the tunnel that we estimated on the basis of the quasi-one dimensional settlement analysis method. This figure indicates that ground settlement of approximately 1.0 m will be caused at the access road section during 104 years

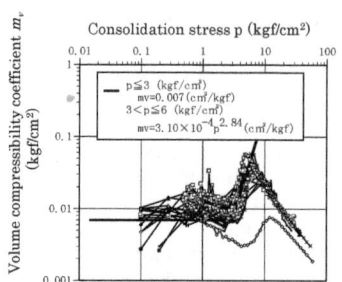

(a) Dc1(Ma12: Upper)

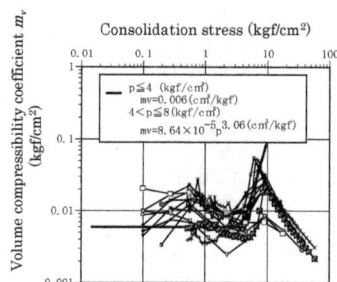

(b) Dc1(Ma12: Lower)

Figure 8. Relationship between volume compressibility coefficient m_v and consolidation stress P

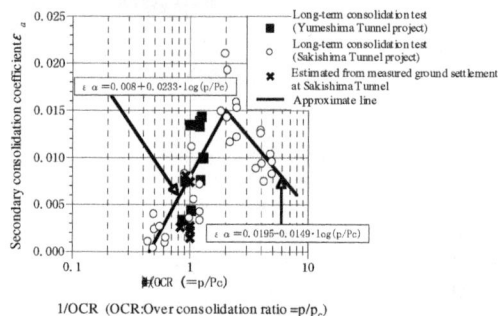

1/OCR (OCR:Over consolidation ratio =p/p$_c$)

Figure 9. Relationship between the secondary consolidation coefficient ε_a and consolidation stress P

after completion of tunnel tube installation. In the tunnel tube section, however, the ground will hardly settle since there is no conspicuous load that will be applied to this section. Substantially large differential settlement is predicted near the seawall. As the result of the ground settlement estimation, it became clear that the effect of differential settlement on the tunnel tube and tube joints must be carefully studied during the tunnel design.

4 DEVELOPMENT OF FLEXIBLE JOINT

Immersed tunnel tube elements are usually connected by flexible joints to minimize shearing force resulting from earthquakes, differential ground settlement, etc. Flexible joints consisting of a rubber gasket and connection cable (conventional joints) are usually used for immersed tunnels. Reliable dewatering performance and flexibility are required to the flexible joints. Tunnel tube elements are deformed differently from each other depending on their location. Therefore, we should determine the performances actually required of the individual joints.

We calculated the joint displacement at the immersed tunnel tube section. In the calculation, we took into account dead load, temperature, residual ground displacement (in both vertical and horizontal directions), seismic load, etc. as the load conditions, and determined the joint displacement as the sum of the ordinary displacement and the displacement obtained from seismic response analysis. We used an overall framework analysis model of the tunnel's longitudinal cross section for the ordinary displacement calculation, while we used an overall analysis model of a lumped mass system for the seismic displacement calculation.

Both vertical and horizontal displacements are required as the residual ground displacement to be inputted to the overall framework model. The reason that horizontal displacement is required is because horizontal ground displacement gives large influence on openings between adjacent tube elements. Excessively large openings will deteriorate the dewatering performance of tube joints. As the differential vertical settlement for the calculation of openings between tube elements, we selected the safer side of the two values: one obtained from the quasi one-dimensional settlement analysis, which has been already discussed in this paper, and the other obtained by a two-dimensional FEM-based settlement analysis, which we performed additionally for this study. For the horizontal displacement, we used the result obtained from the two-dimensional FEM-based settlement analysis for areas near the sea walls. Though we do not discuss the two-dimensional FEM analysis in detail here, we confirmed that the ground displacement estimated by the FEM analysis method has the same tendency as that obtained by the quasisingle one-dimensional analysis.

The above analysis results showed that the displacement of tube joints in the tunnel axial direction will reach approximately three times the maximum allowable displacement of the conventional joints (Gina #148 type joints), which is approximately 80 mm. This indicates that the conventional joints cannot assure sufficient dewatering performance for the immersed section of the Yumeshima Tunnel. Thus, it became necessary to develop a flexible tube joint of a new structural design capable of absorbing such a larger displacement. The outline of the newly developed joint, called crown seal joint[3)4)], and the results of its performance confirmation tests are discussed below.

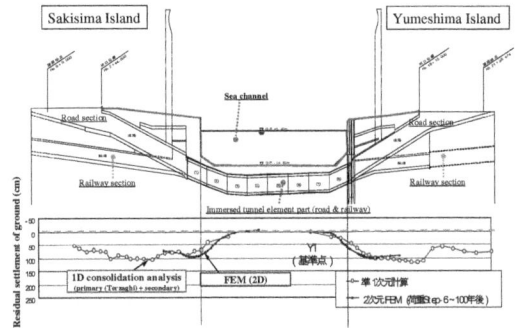

Figure 10. Distribution of ground settlement along the longitudinal cross section (longitudinal axis) of the tunnel

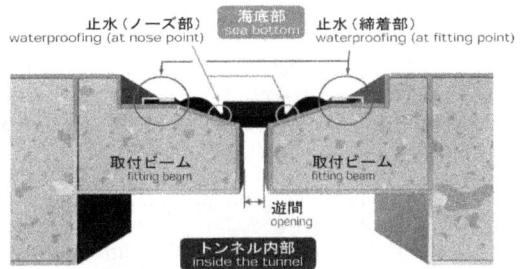

Figure 11. Schematic figure of crown seal joint

Table 1. Principal components and their function of crown seal joint

Component	Principal function
Crown seal rubber strip (primary dewatering rubber strip)	Keeps water out of its adhesion point. Also employs water pressure to stop water at nose point. The elasticity of rubber itself allows deformation of joint (in both axial and shear directions).
Secondary dewatering rubber strip	Its W-profile end assists primary dewatering function of joint.
Stopper cable	Shares axial tension, if axial tensile deformation exceeds allowable limit.
Fitting beam	Engages with crown seal rubber strip and secondary dewatering rubber strip. Transmits compressive force to immersed tube body, if axial compressive deformation exceeds allowable limit.
Shock absorbing rubber block	Prevents local stress generation in fitting beam, even when axial compressive deformation causes adjacent beams to contact each other.
Dust proof plate	Protects joint from mud, sand, and other foreign substances.
Shear key	Shares shearing force if shearing deformation exceeds allowable limit.

4.1 Structure of crown seal joint

The structure of the newly developed flexible joint is schematically illustrated in Fig. 11. The name of this joint, "crown seal joint," is derived from the crown-shaped center portion of the primary dewatering rubber piece that ensures dewatering performance of the joint. Principal components and performance of the crown seal joint are given in Table 1. The conventional joints use the compression characteristics of the rubber gasket to ensure the required dewatering performance, whereas the new joint is structured to depress its dewatering rubber strip tightly against the circumference of the tube element to ensure a sufficient dewatering performance.

The following are the major advantages of the crown seal joint.

1 It can be deformed freely in both the tunnel axial direction and radial direction (hereinafter referred to as shear direction) by preliminarily providing an opening (a gap in the tunnel axial direction) between the adjoining fitting beams. Thus, it can absorb displacements between tunnel tube elements and substantially reduce shearing force in the immersed tubes.

2 Doubled dewatering effects can be expected. To describe it in more detail, water pressure acting on the crown portion depresses the nose points (protrusions provided on both sides of the face to be adhered to the fitting beams) against the fitting beams. This creates a failsafe dewatering performance (called self sealing performance), in addition to its intrinsic dewatering capability (primary dewatering).

3 The basic structure of this joint cannot prevent the adjacent tunnel tube elements from displacing (opening) in excess of its allowable limit. To make this joint function reliable by removing this shortcoming, an additional member (stopper cable) is attached to control the tensile directional displacement of the joint within an allowable limit. This member does not prevent free deformation of the joint, as long as the joint displacement is within the allowable limit.

A detailed illustration of the crown seal rubber strip is shown in Fig. 12, and its principal components and their function are listed in Table 2. The crown seal rubber strip is a rubber seal consisting of a crowned portion in the center and sidewalls on both sides, and provides primary dewatering, the main role of the joint. The crowned portion has sufficient hardness to eliminate the chance of undesired deformation (resulting in falling in opening between tube elements, etc.). It also has a suitable flexural rigidity to ensure self-standing (self-supporting) characteristics. The gripping mechanism (a metal bracket that uses the principle of leverage) of the sidewalls has the same configuration as that of Ω-profile rubber seals, which have been widely used because of their high dewatering performance. Inheriting the superior a adhesion characteristics of the Ω-profile rubber seals, the sidewalls exhibit desired functional reliability. The dewatering rubber strip of crown seal joint is impregnated with (single directional weave) and ordinary fiber (plain weave) to reinforce its tensile strength.

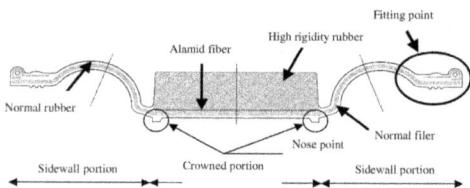

Figure 12. Schematic figure of crown seal rubber

Table 2. Principal components and function of crown seal rubber

Component	Principal function
High hardness rubber strip (hardness: 70 deg.)	Covers most part of crowned portion. Increased hardness improves self-standing (self-supporting) feature of crowned portion.
Normal hardness rubber strip (hardness: 50 deg.)	Covers part of sidewalls, nose points, and crowned portion. Hardness, 50 deg., was determined to give dewatering feature to nose points and flexibility to sidewalls.
Alamid fiber	Impregnated in the tension side of crowned portion in single-directional weave pattern to limit deformation.
Ordinary fiber (Nylon)	Impregnated in sidewalls and crowned portion in plain weave pattern to offer multi-directional flexibility.

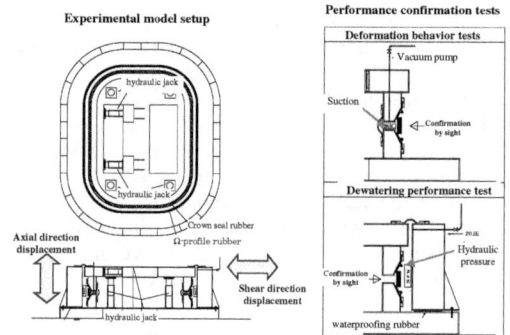

Figure 13. Three-dimensional model test setup

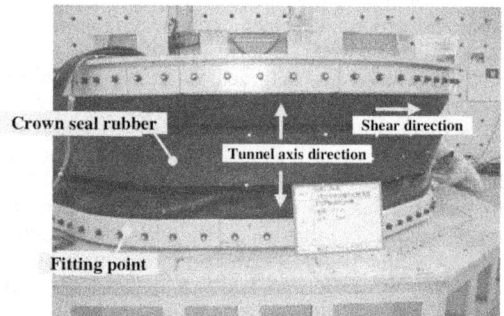

Figure 14. Photograph of three-dimensional model test setup

4.2 Performance Confirmation Tests

a) Outline of the Tests

The structural design concept of crown seal joint was proposed as a new technology that can flexibly handle large displacement while maintaining dewatering performance. Because this tube joint is based on an innovative structural design concept that was under development at that time, it had not yet been put into practical use. No confirmation had been made whether or not the performance of this joint would be sufficient for the Yumeshima Tunnel. Under the above circumstances, we carried out performance confirmation tests of full-scale crown seal rubber strips, to evaluate appropriateness for use in the Yumeshima Tunnel. For the tests, which were intended to evaluate the performance (deformation and dewatering performances) of this joint as a means for ensuring water tightness of the

374

Yumeshima Tunnel, we carried out two-dimensional and three- dimensional model tests. Here, we discuss only the three-dimensional model test. The three-dimensional model test setup is outlined in Fig. 13 and Fig. 14. We prepared a ring-shape three-dimensional model simulating the overall structure of the crown seal joint, and applied loads to simultaneously deform the ring-shape crown seal rubber molding in the axial and shear directions to test its deformation behavior and dewatering performance.

b) Test Results
To check the deformation behavior of the test specimen, we operated a hydraulic jack installed inside the test setup to deform the specimen 0 to 300 mm in the axial direction and 0 to 150 mm in the shear direction, and visually observed its deformation pattern. At the same time, we measured the deformation using photogrammetric method. As a result, the specimen responded flexibly to the displacement of the joint without exhibiting any abnormal deformation behavior such as longitudinal buckling (so-called "spanworm movement") or snaking (see Fig. 14). Fig. 15 shows the deformation of the specimen under a deformation load applied in the shear direction. This figure shows that the specimen maintains the neutrality of crowned portion in both axial (opening is equal to axial displacement in this case) and shear directions, which is essential for self-sealing.

To assess the specimen's dewatering performance, we enclosed the specimen with an outer counter (steel wall) and filled the space between the specimen and the counter with water. Following this procedure, we applied water pressures of up to 100, 200, and 300 kPa to the specimen by using a pump, differential pressure regulator, and other necessary devices. While applying deformation loads in the same manner as that for the deformation behavior observation test, we observed the specimen visually for leakage of water. For reference, 300 kPa is the maximum estimated pressure to the tunnel. For the self-sealing performance, we supplied water to the sidewall area through a communicating tube and checked the nose points for leakage of water. As the result, no water leaked from the sealed portion, even after the specimen was displaced by 0 to 300 mm in axial direction and 0 to 150 mm in shear direction. These tests confirmed that the self-sealing function of the specimen is effective as long as the joint displacement stays within 0 to 250 mm in axial direction and 0 to 100 mm in shear direction (see Fig. 16).

As discussed above, the tests have confirmed that the newly developed crown seal joint can successfully meet the performance required of the immersed tube joints for the Yumeshima Tunnel, and a decision was made to apply this technology to this tunnel, the first time it has been used in Japan.

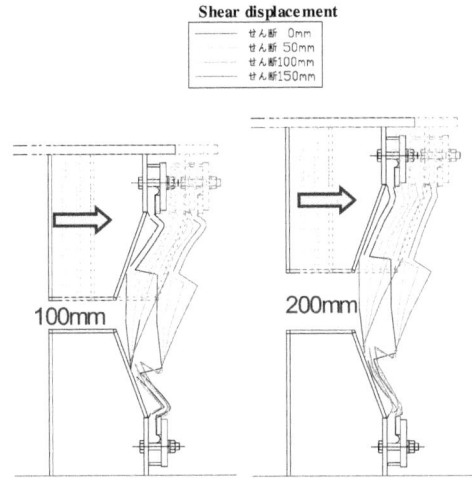

(a) Initial opening = 100mm   (b) Initial opening = 200mm

Figure 15. Deformation of specimen under a deformation applied in the shear direction

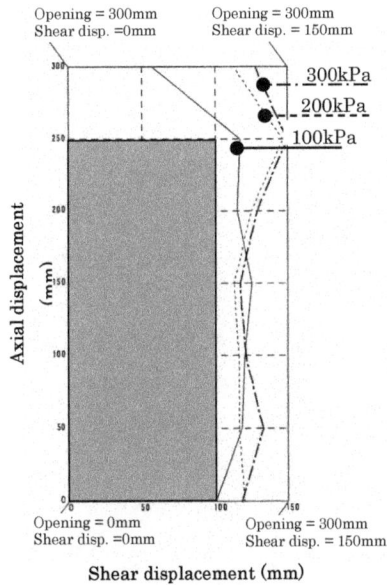

Figure 16. Effective range of self-sealing function

Before concluding our discussions, note that the crown seal joint was developed under a joint project involving the Kinki Regional Development Bureau (Ministry of Land, Infrastructure and Transport), Port and Airport Research Institute (independent administrative agency), Sumitomo Rubber Industries, Ltd., Penta-Ocean Construction Co., Ltd., Oriental Consultants Co., Ltd., and Prof. O. Kiyomiya (professor at Waseda University).

5 CONCLUDING REMARKS

Like the Sakishima Tunnel, the Yumeshima Tunnel is also expected to play an important role as part of the traffic network organically connecting various functions of Osaka Port. As discussed in this paper, technological problems originating from settlement of soft ground have to be resolved to systematically organize the transportation network in the Osaka Bay area. Kinki Regional Development Bureau of the Ministry of Land, Infrastructure and Transport will positively contribute to the systematic organization of the transportation network in this area by taking necessary measures including new technology development.

Soft Ground Engineering in Coastal Areas, Tsuchida et al. (eds)
© 2003 Swets & Zeitlinger, Lisse, ISBN 90 5809 613 0

Recent research into the fundamental behaviour of piles in clay

R.J. Jardine
Imperial College, London, England

ABSTRACT: Recent research by a team from Imperial College London has focused on understanding the fundamental behaviour of piles driven in sands and clays. The work has led to new design procedures that are used widely in offshore practice and to surprising new discoveries concerning the effects of time on piles driven in sand. The presentation made to the Nakase Symposium will outline some early results from new research into the effective stress behaviour of piles bored in clay and into the effects of long term ageing for steel piles driven in clay. The results demonstrate the shortcomings of existing predictive approaches.

Soft Ground Engineering in Coastal Areas, Tsuchida et al. (eds)
© 2003 Swets & Zeitlinger, Lisse, ISBN 90 5809 613 0

Proposal of new design method for soft landing breakwater with piles

Y. Kikuchi
Port and Airport Research Institute, Independent Administration Institute, Yokosuka, Japan

ABSTRACT: A soft landing breakwater was originally proposed for the site where ground condition is not good and wave condition is not sever. The first design method is proposed in 1991. It was much simplified for limited conditions. That is the reason why the structure is used only in Kumamoto Port. Nowadays even in Kumamoto Port, the wave condition of the construction site is more sever and the limitation of existing method is appeared. In this paper, the results of the field loading test done in 1989 are examined to discuss the characteristics of lateral resistance of a soft landing breakwater with piles. Finally, this paper and proposes new design method.

1 INTRODUCTION

Research of a soft landing breakwater started because of a construction project of the Kumamoto port. The area of Kumamoto port is located in the middle of Ariake sea. The area is covered with thick soft clay layer and the wave condition is rather mild. If using a conventional gravity type breakwater in such an area, huge ground improvement is needed to support the weight of the structure. If the weight of the breakwater is light, no ground improvement is needed and construction cost can be inexpensive. This is the reason why new type breakwater is developed.

The feature of the new type breakwater called soft landing breakwater is that the self weight is

light. The lateral resistance of this breakwater is according to cohesion between base plate and clay surface. This is the original idea. Figure 1 shows the two types of soft landing breakwater. Figure 1a) is a original type called flat type. If the lateral resistance is not enough to resist against wave force, piles are used to improve resistant capacity. It is called piled type (figure 1b)). The mechanism of lateral resistance of this type is rather complex, because not only base plate but piles resist to horizontal load and existing of piles may change cohesion between base plate and ground surface.

The design method for a sot landing breakwater with piles was proposed in 1991. Supported by this code, the breakwater of Kumamoto port was constructed quickly and safely. But nowadays the de-

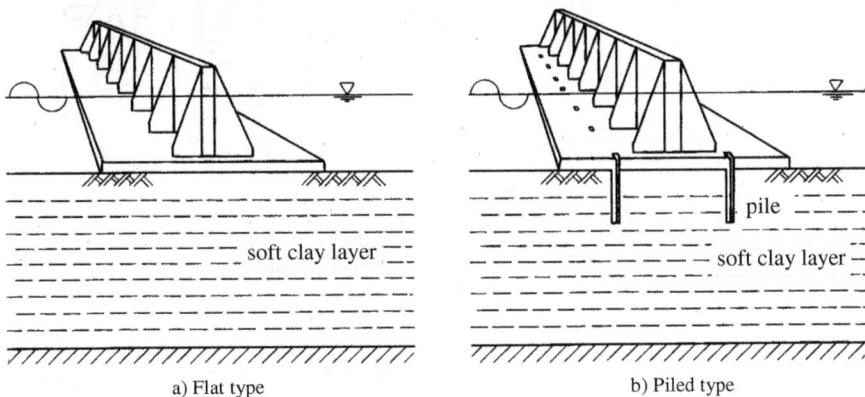

a) Flat type   b) Piled type

Figure 1. Two types of soft landing breakwater.

sign method is known to be too conservative, because the breakwater experienced in severe storm several times and no damage was found. Size of the breakwater will be extensively larger when the depth of the construction site is deeper.

In this paper, the mechanism of the lateral resistance of a soft landing breakwater with piles is discussed according to several laboratory tests and field test results.

2 EXISTING DESIGN METHOD FOR SOFT LANDING BREAKWATER

The existing design method proposed in 1991 is rely on the field test and laboratory test done in from 1982 to 1990.

First field loading test was conducted in 1985(Kuchida et al., 1986). In this case, flat type model and piled type model were horizontally and statically loaded. Figure 2 shows the plan view of piled type model. As seen in figure 2, embedded length of piles used in this model is 4m. A piled type initially designed has short and rigid piles. The

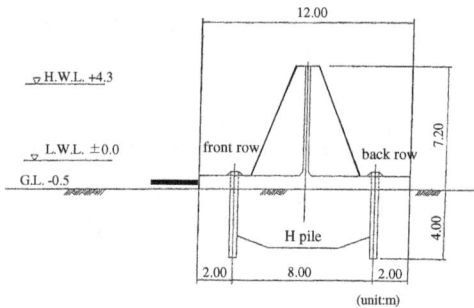

Figure 2. Model soft landing breakwater with piles used in the field test done in 1985.

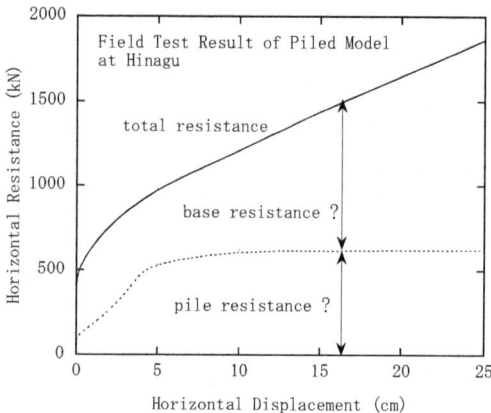

Figure 3. Relation between horizontal displacement and horizontal resistance in the first field loading test.

model used in the test was designed full scale with twelve piles, because the model had been planned to be tested long term durability tests in the construction site. According to the limitation of measuring facilities, bending moment distributions were measured in only four piles. In the test, lateral resistance of piles were estimated by these measurements. Figure 3 shows the relation between horizontal displacement and horizontal resistance. The solid line shows total resistance of the model and the dotted line shows the estimated horizontal resistance of piles. We might conclude from figure 3 that in the beginning of the loading, piles and base can resist to horizontal load, but later resistance of piles are saturated and base resistance increase according to horizontal displacement increasing.

Broms (1965) showed one simple calculation method for ultimate lateral resistance of a pile. Figure 4 shows the lateral ultimate resistance of a short pile with fixed pile head. According to this method, ultimate lateral resistance of a short and head fixed pile will be

$$P_{Broms} = 9c_u B(l - 1.5B) \tag{1}$$

where P_{Broms}: ultimate lateral resistance of a short pile with fixed pile head, c_u: shear strength of ground(homogeneous), B: width of a pile, L: embedded length.

The measurement results of bending moment distribution of piles are shown in Figure 5. From this figure, the moment distribution of a pile in back row at the maximum horizontal load quite fits to the moment distribution according to Broms' method. However, moment distribution of a pile in the front row shows less load is applied to the piles.

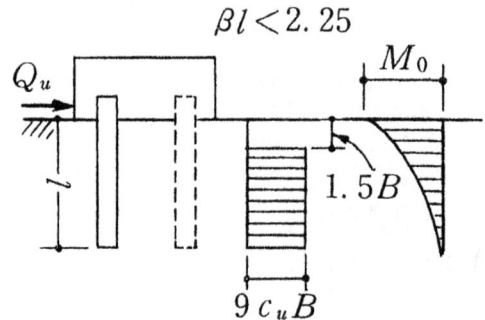

Figure 4. Ultimate lateral resistance of rigid pile with pile head fixed proposed by Broms.

These test results appeared that 1) a short, rigid pile can show ultimate lateral resistance, 2) ultimate lateral resistances of piles in back row is the same order as it according to Broms' method, but ultimate lateral resistance of piles in front row is negligible. This conclusion gives the following design equation for estimating the lateral resistance of piles.

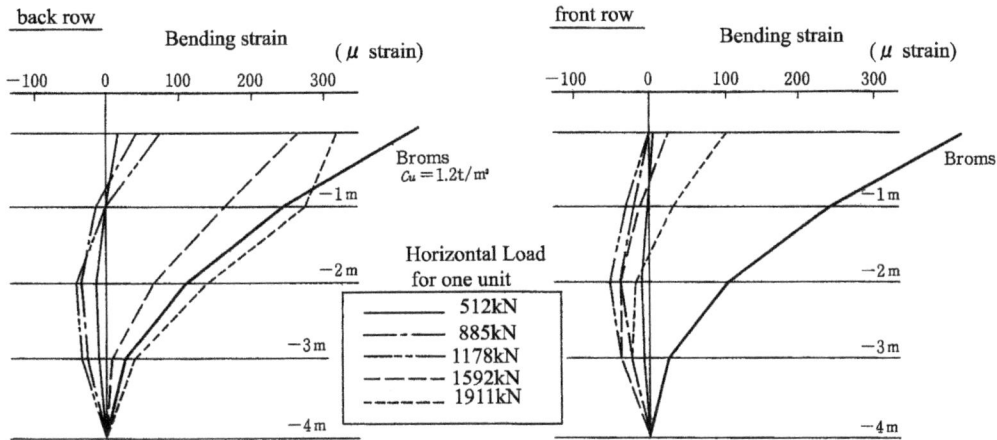

Figure 5. Bending moment distributions of the test result are compared to Broms' equation. Left side show the back row's result and right side the front row.

$$P = \frac{n}{2} \cdot 9c_u B(l - 1.5B) \qquad (2)$$

where P: total lateral resistance of a piled type model, n: number of piles of a structure.

Several series of laboratory tests(Kikuchi et al., 1990) gave important results concerning to a piled type model. Figure 5 shows a laboratory test result which was conducted on the model clayey ground consolidated in 10kN/m². In this series, thickness of piles were changed for varying embedded length ratio E_r which is defined as a ratio of embedded length l and a depth of a first 0 point of the bending moment l_{m1}. Piles are called short pile if E_r is less than 1. In the figure, a ultimate resistance P_y is defined that the first yield load of load displacement curve. Figure 6 shows P_y is as the same amount as the equation (2). But all the piles in laboratory tests

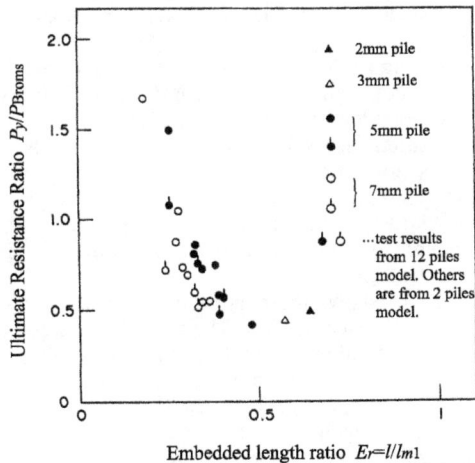

Figure 6. Ultimate resistance ratio P_y/P_{Broms} changes according to embedded length ratio E_r.

show the almost equal resistance and the mode of deflection was much far from the assumption of Broms' method.

3 FIELD LOADING TEST

We had unknown for the resistance mechanism of piled type model with long piles and we had no information about the mechanism against repeating load. Another field test program was conducted in 1989. This tests result was not included in the existing design procedure, because the procedure is restricted only for with short piles.

3.1 Test procedure

The tests were conducted in Kumamoto port in 1989. In the series five tests were done. First one is for check the subgrade reaction of the ground by a single pile test. Three static loading tests were conducted for checking the effect of embedded length of piles with piled type model. In these models, the length of a model is one sixth of a full scale structure and has two piles. Flexural rigidity of the pile is 23 MNm². and width of the pile is 25cm. Embedded length of piles were selected in 5m, 15m, 25m where as the depth of the first 0 point of the bending moment l_{m1} is almost 10m. Stage repeated loading test was done in the model with 25m length piles. Figure 7 shows the plan of the model. In this case, the length of the model was 1.5m, width was 10m. In this test, base plates were set, then piles were installed. Fourteen days of curing time was employed after set the model. The distance of two piles were selected in 5m.

In the tests, bending moment distribution, axial force distribution of piles, displacement, inclination of base plate and applied load were measured.

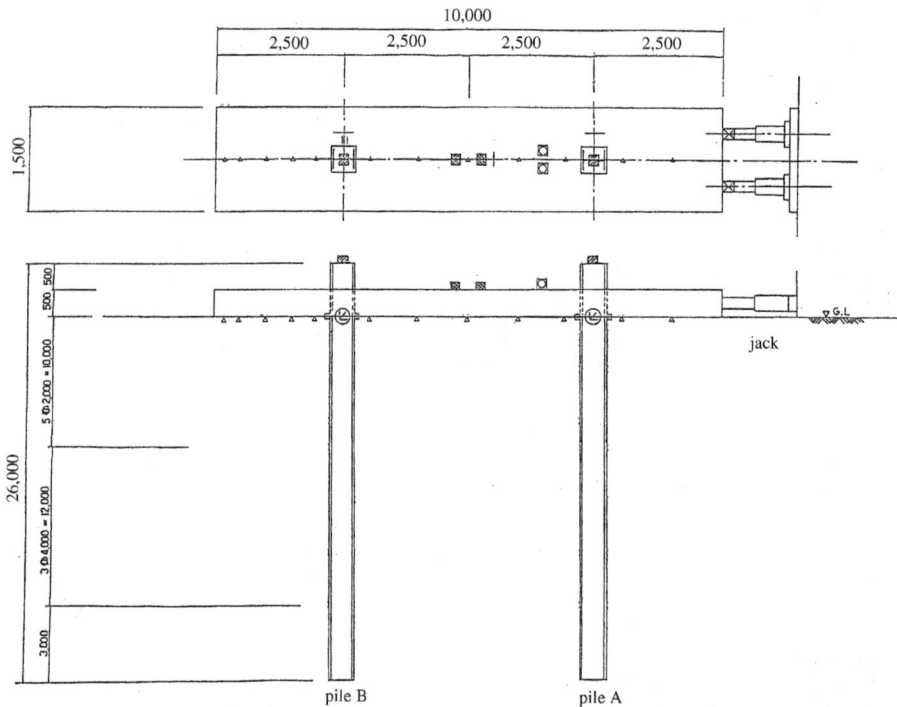

Figure 7. Model used in field test done in 1989. This figure shows the piled model with the longest piles. Right side jacks push the model statically to the left.

3.2 Test results

3.2.1 Static loading test

After the single pile test, the ground can be modeled in S type ground of PHRI method. The subgrade reaction model can be written as follows;

$$p = k_s \cdot x \cdot y^{0.5} \tag{3}$$

where p: subgrade reaction, k_s: coefficient of subgrade in S type model of PHRI method, x: depth, y: deflection. The test result gave $k_s = 150 kN/m^{3.5}$. The test results given by cases from 2 to 4 show that subgrade reaction for piles used in the soft landing structure is also modeled by S type of PHRI method. Figure 8 show the comparing results of the pile behavior and the model.

Figure 9 shows the displacement of the structure against horizontal load. Long piles embedded in 25m was used in case 2, medium length piles embedded in 15m in case 3, and short length piles embedded in 5m in case 4. This figure shows that case 4 shows the smaller resistance than that of other cases. Cases 2 and 3 show almost the same resistance. Arrows directed upward show where the first yield point was appeared in top of one of the piles. Double arrows directed downward show where the second yield point was appeared in middle depth of one of the piles. It is interesting that the first yield point appeared in the same level of resistance.

Figure 10 shows the relation between inclination of the base and horizontal load. Inclination was small when embedded length of pile is long, but large when embedded length is small.

Figure 11 shows the distribution of the base contact pressure to the ground. In this figure, the change of contact pressure from the start of the loading is shown. Only in case 4, the contact pressure of the base is about 30 kN/m^2 which is almost the ultimate bearing capacity calculated by Davis and Booker's theory.

According to the shear strength of ground, maximum pull out capacity of a pile is 44, 350, 850 kN for embedded length of 5m, 15m, 25m respectively. The measurement result of axial forces of piles show that piles embedded 5m were fully mobilized the pull out resistance (figure 12).

Theses results show that embedded pile length is not enough long leads rotational failure of the soft landing structure.

Figure 13 shows the relation between horizontal load and resistance of piles. The shear force applied the pile heads are estimated from the distribution of the bending moment. There are some variance in the test results, but a ratio of resistance of piles to horizontal load is small in the beginning of the loading and it goes to unity at the end of loading. This result means that lateral resistance of the structure is ac-

382

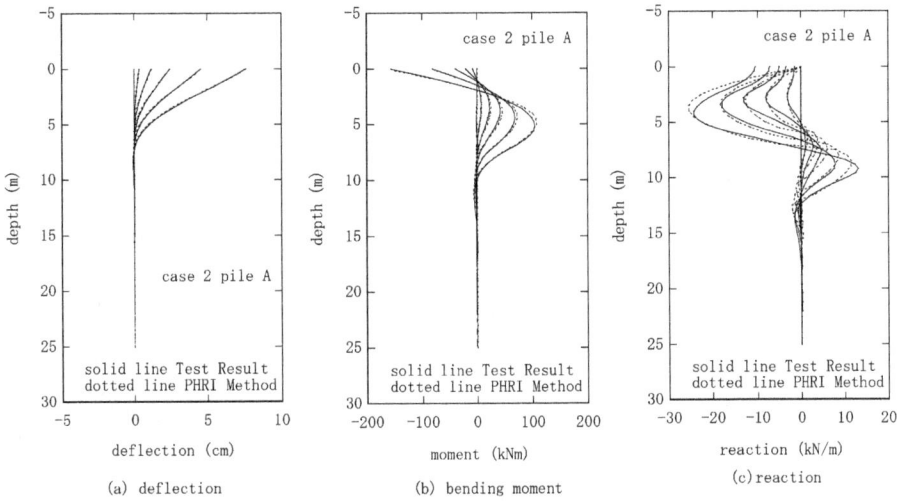

Figure 8. Comparison between test results and PHRI model.

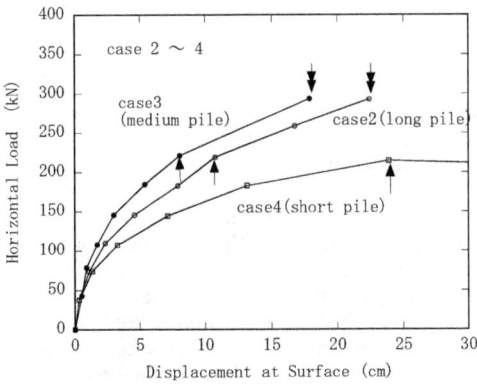

Figure 9. Relation between displacement and horizontal load of piled models.

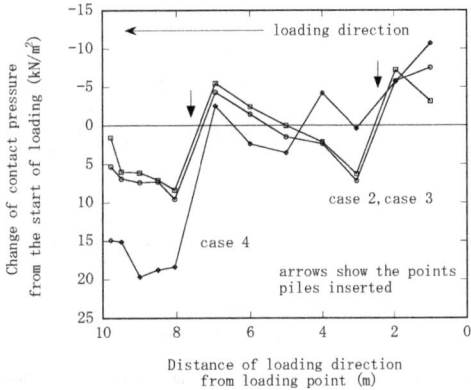

Figure 11. Change of contact pressure at final loading stage from the start of the loading.

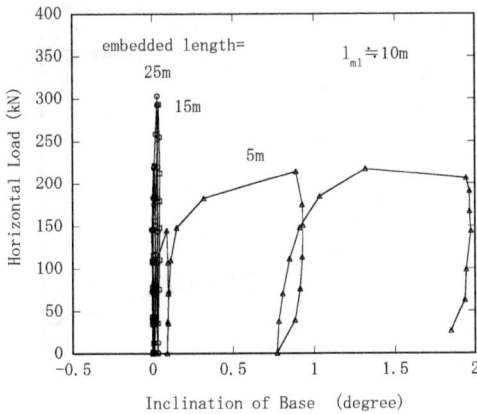

Figure 10. Relation between inclination of the base and horizontal load of piled models.

cording to lateral resistance of piles when the structure loaded to fail. As shown in figure 3, former field test results show opposite conclusion. But the results of the series of laboratory tests agree with the conclusion from figure 13. Former field test results may have some problem.

3.2.2 Repeated loading test

Repeated loading test was conducted in the same size of the model as shown in figure 5. Load was applied reciprocally with return period of 5 sec. At first, maximum load of 60 kN was applied 50 cycles. Then that of 80 kN 30 cycles. Finally that of 120kN was applied about 100 cycles. Figure 14 shows the relation between the cycles and the maximum displacement of the crown of the base plate of each cycle. The figure shows displacement increase according to the number of cycle increases. Especially

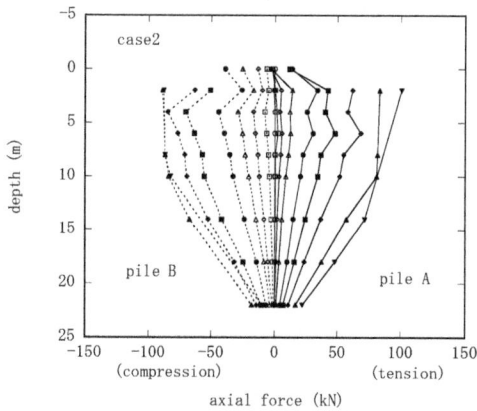

a) embedded length is 25m

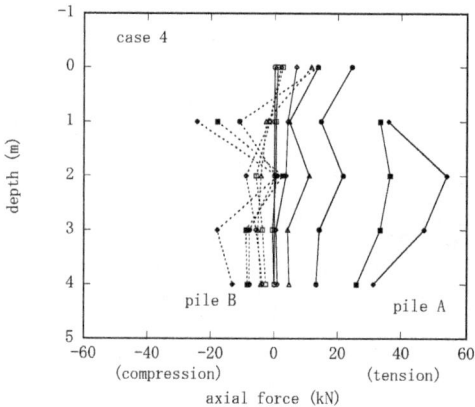

b) embedded length is 5m

Figure 12. Axial force distribution of piles in piled type model.

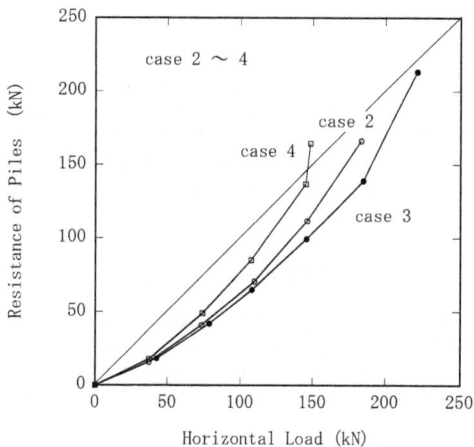

Figure 13. Relation between horizontal load and lateral resistance of piles.

increment of displacement is large when the maximum repeated load is 120 kN. There are two reason why the displacement increase according to increment of cycles. One is that the shallow part of the ground around the pile is weaken according to repeated load and it leads the reduction of the coefficient of subgrade reaction. The other is reduction of the resistance of the base plate. It makes the increment of shear force to the piles. Figure 15 show the effect of the former reason. Solid lines show the distributions of deflection, bending moment, subgrade reaction of the first cycle and dotted lines for last cycle. As seen in figure 15 c), shallow part of subgrade reaction decrease with increment of load cycles. It leads increment of maximum bending moment and deflection as shown in figure 15 a),b). As already mentioned that subgrade reaction of this ground for the first time of loading is modeled in S type of PHRI method, we try to present the reduction of subgrade reaction by using the change of the coefficient of subgrade reaction. Here we define the coefficient of the subgrade reaction ratio R_k as follows;

$$R_k = k_{tn} / k_{t1} \tag{4}$$

where k_{tn}: a coefficient of subgrade reaction when nth cycle's maximum load applied, k_{t1}: a coefficient of subgrade reaction when first maximum load applied.

Figure 16 shows the example of the change of R_k according to cycle. This figure shows that R_k decrease with increment of loading cycles and final reduction of R_k depends on the depth. Similar plots are done for each loading level. Following conclusions are found; 1)reduction of R_k stops when number of cycles are more than around 25; 2) final reduction of R_k differs according to load intensity; 3) final reduction of R_k differs according to depth. I simply make a model of the reduction of R_k in the following way.

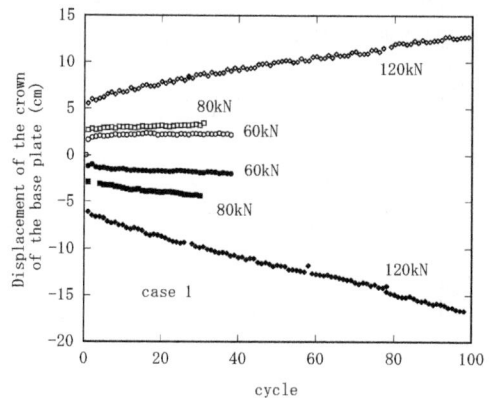

Figure 14. Displacement of the crown of the base plate increase with increment of loading cycles.

384

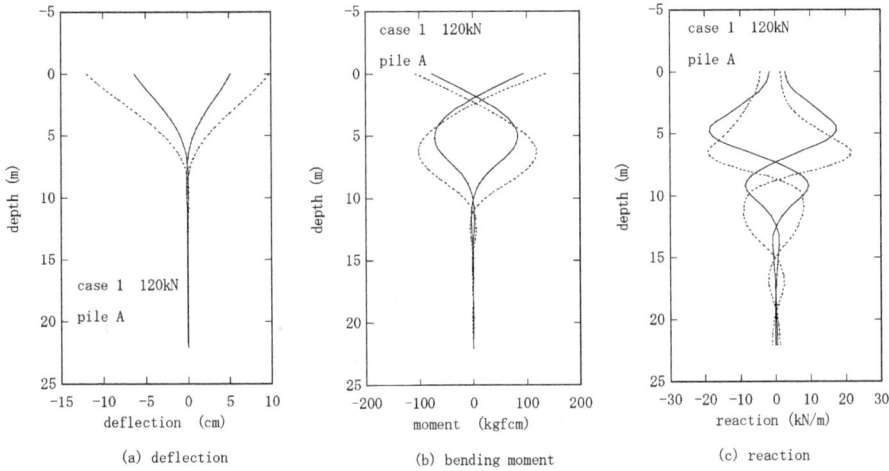

(a) deflection   (b) bending moment   (c) reaction

Figure 15. Change of pile reaction according to repeated load.

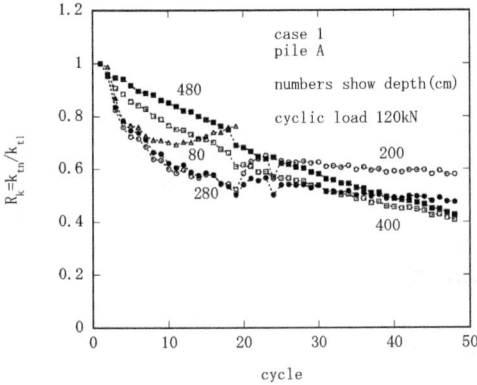

Figure 16. Coefficient of subgrade reaction ratio changes according to number of cycles of repeated load.

$$R_k = 1 - \alpha\beta\gamma \qquad (5)$$

where α: a parameter affected by load intensity, β: a parameter affected by number of loading cycles, γ: a parameter affected by depth.

Parameters α, β, γ are defined as follows from the test results;

$$\alpha = 3.75 y_0 / B$$

$$\beta = \begin{cases} \dfrac{n}{25} & (n \le 25) \\ 1 & (n \ge 25) \end{cases} \qquad (6)$$

$$\gamma = \begin{cases} 1 & (0.25 l_{m1} \ge x) \\ \dfrac{0.5 l_{m1} - x}{0.25 l_{m1}} & (0.5 l_{m1} \ge x \ge 0.25 l_{m1}) \\ 0 & (x \ge 0.5 l_{m1}) \end{cases}$$

where y_0: deflection of the pile at ground surface, B: pile width, n: number of loading cycle.

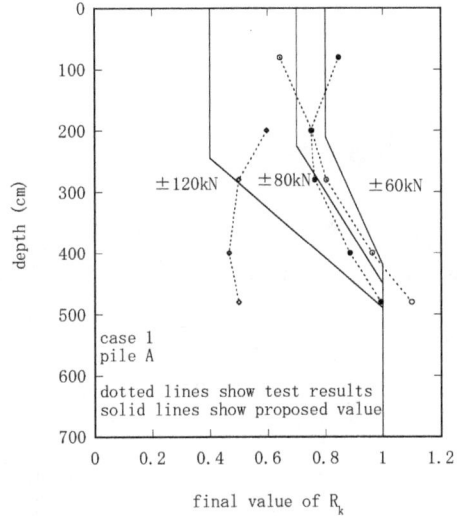

Figure 17. Distribution of the change of coefficient of subgrade reaction ratio.

Figure 17 shows the comparison of the test result and the reduction model.

4 NEW DESIGN METHOD

Followings are the important point differ from the existing design method:

1 The model of subgrade reaction is changed to PHRI method. Both rows of piles work in full ability for the lateral resistance.
2 Pullout resistance and push in resistance of piles are taken into consideration.
3 Fixity of pile head is taken into consideration for the design of cross section of piles. This paper

385

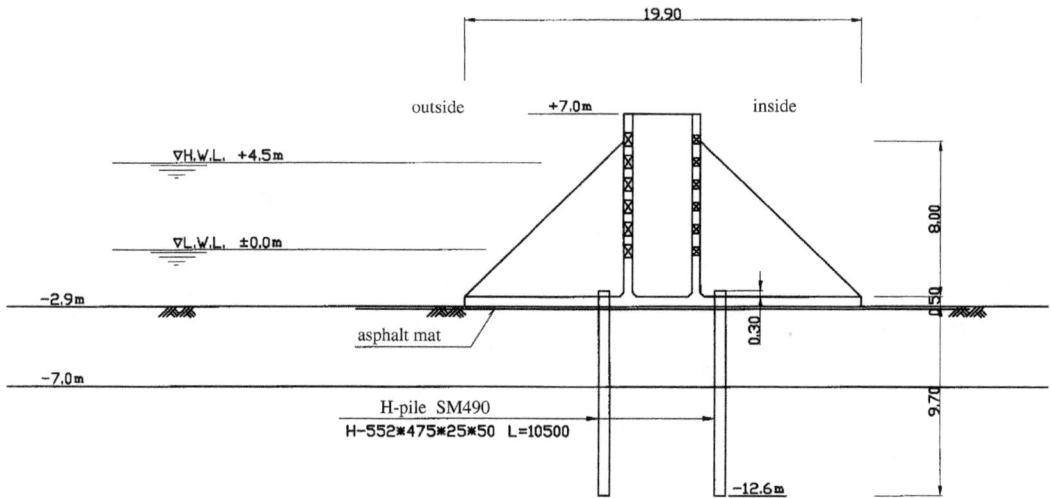

Figure 18. Breakwater designed by existing method.

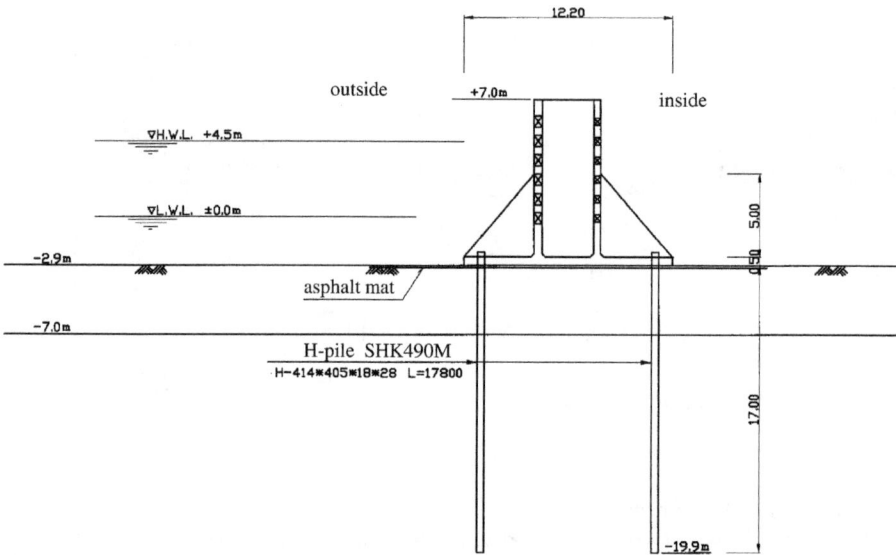

Figure 19. Breakwater designed by proposed method.

never mentions this point. But all the test result show that pile heads are never fully fixed, but fixity of it is around 0.7 for average of all the piles.

4 The reduction model of the coefficient of subgrade reaction for repeated load is proposed.

To compare the new design method and existing design method, a trial design is made. In this case, the breakwater will be constructed at the depth of C.D.L. -2.9m, C.D.L +4.5m of high water level, C.D.L. +7.0m of the crown of the breakwater. Wave condition for design will be as follows; considering return period is 50 years; 4.0m is the design wave height; 5.5sec for the wave cycle; wave direction is

rectangle to the structure. Return period of 2 years are considered to take into consideration of the effect of repeated load. Wave condition for that will be 2.0m of wave height and 5.0sec of wave return period. Ground condition will be alluvial clay layer with shear strength of $3.3 + 1.41z$ (kN/m^2, z=0m where C.D.L.0.0m). The coefficient of earthquake will be 0.17. Safety factors of the structure are 1.5 for sliding, 1.2 for return, 1.5 for bearing capacity. Allowable strength of steel will be 190 MN/m^2.

Figures 18 and 19 show the result of the design. Figure 19 is designed followed by new method where as figure 18 by existing method. If the breakwater is constructed in the depth of C.D.L-2.3m, the width of it is 14.9m and embedded length of piles are 9.6m by existing method. It means that only the difference of 60cm depth gives completely different design result by using existing method. It proves the limitation of existing design method.

Although the structure designed by existing method has 19.9m of width, 12.6m of embedded length of piles, it by new method has 12.2m of width, 19.9m of embedded length of piles. As construction cost will be much affected by the weight of the upper structure in this case because of machinery restriction, the structure shown in figure 19 proves to be much cheaper than that in figure 18.

5 CONCLUSIONS

In this paper, the lateral resistance of a soft landing breakwater with piles is examined and the new design method is proposed and compared with existing design method. Followings are the main conclusion of this paper;

1 PHRI method is suit for the subgrade reaction model for the lateral resistance of piles used in a soft landing breakwater.
2 Pullout and push in resistances of piles affect much to the lateral resistance of the structure.
3 The reduction method of coefficient of subgrade reaction for repeated load is proposed.
4 The structure designed by proposed method, which is more rational than existing one, proves economical if the structure is constructed in deeper area.

REFERENCES

Broms, B.B. (1965) "Design of laterally loaded piles." Proc. of ASCE, 91(SM3), 79-99.
Kikuchi, Y., Takahashi, K., Nakamura, R., (1990) "Horizontal loading tests on models of soft landing comb-shaped structures." Technical report of Port & Harbour Research Institute, 679. (in Japanese)
Kuchida, N., Ogura, K., Hosokawa, Y., Nakano, T. (1986) "Field tests on the development of soft landing breakwater - horizontal loading tests and its results-." 41st Annual meeting of JSCE, III, 137-138. (in Japanese).

Soft Ground Engineering in Coastal Areas, Tsuchida et al. (eds)
© 2003 Swets & Zeitlinger, Lisse, ISBN 90 5809 613 0

Stability analysis of a ground-sheet pile system on the soft ground

K. Mizuno
Wakachiku Construction Company Ltd., Tokyo, Japan

T. Tsuchida
Independent Administrative Institution, Port and Airport Research Institute, Yokosuka, Japan

ABSTRACT: In this paper, the stability of a ground-sheet pile system was quantitatively evaluated using the elasto-plastic finite element method. First, on the case in which plastic deformation was generated in the outside of sheet pile, the sensitivity analyses about the effect of the FEM analysis conditions on safety factor were carried out. The analytical results of FEM were compared with the conventional slip circle methods of slices. As the result, the safety factor of FEM was greatly changed with the finite element mesh size and the friction assumption between the sheet pile and the ground. Second, on the case in which the slip surface passes the inside of sheet pile since the bending rigidity of sheet pile is relatively smaller than the rigidity of the ground, the applicability of FEM was examined. It was proven that the bending rigidity of sheet pile greatly affected the stability of this system and the elastic modulus of the ground had to be appropriately given in order to evaluate the safety factor quantitatively.

1 INSTRUCTION

It is necessary to carry out the slope stability when a sheet-pile quaywall are planned on the soft ground. It is general to evaluate the stability of this system by the conventional slip circle method of slices. In this method, slip surface is assumed passing the outside of sheet pile, and the stability analysis is carried out only on the slip surface which passes under the lower end of sheet pile. This assumption is based on the empirical knowledge that it does not need to consider the stability for the slip surface which passes the inside of sheet pile if the rigidity of sheet pile is decided based on a balance of the earth pressure.

However, it is uncertain on the validity of this assumption when the new material is used. For example, they are the cases in which the lightweight treated soil was used as a back-filling material and the cases in which they are designed in the sheet pile of which the rigidity is comparatively low. And, it is present state that it can not adopt the approach of extending the depth of sheet pile in order to resist the sliding.

From such background, it is important to quantitatively examine the stability for the sliding which passes the inside of sheet pile and the resistance force of the sheet pile must be considered in the calculation. In this study, the stability of a ground-sheet pile system was quantitatively evaluated using elasto-plastic finite element method.

First, on the case in which plastic deformation of the ground was generated in the outside of sheet pile, the sensitivity analyses about the effect of the FEM analysis conditions on the safety factor were carried out. The analytical results of FEM were compared with the conventional slip circle methods of slices. Here, this parametric analysis is called "Series-A".

Second, on the case in which the slip surface passes the inside of sheet pile since the bending rigidity of sheet pile is relatively smaller than the rigidity of the ground, the applicability of FEM was examined. This parametric analysis is called "Series-B".

2 STABILITY ANALYSIS PROCEDURE USING FINITE ELEMENT METHOD

In the slope stability analysis using FEM, it is assumed that the ground is the elasto-plastic material and the failure criterion follows the Mohr-Coulomb's. Shear strength of the ground (τ_f) is shown as below.

$$\tau_f = c + \sigma \tan \phi \qquad (1)$$

Safety factor is calculated by the prediction procedure called "shear strength reduction method (Kobayashi, 1984)". The coefficient that reduces shear strength by equation (1) is defined as the reduction rate (*F*). Stress analysis is carried out for the imaginary shear strength in which equation (1) is divided by F. In this procedure, the divergence of calculation is considered as the failure of the system. Therefore, it is possible to evaluate the largest reduction rate (F) as the safety factor of the system (F_s) in cases where the calculation converges. The theory of imaginary viscosity method (Zeinkicwicz and Cormeau, 1974) is adopted as stable calculation method. The geotechnical general purpose program GeoFem (CDIT, 1997), which built in this theory, was used in this study.

3 PARAMETRIC STUDIES (SERIES-A)

3.1 Analytical model

The sectional view of a sheet pile quaywall (tie rod type) used in this analysis is shown in Figure 1. The front water depth is $z=-12.6$m and the back height of sheet pile is $z=+3.0$m. The range of vertical modeling area is $z=-40$m and the range of horizontal modeling area is ± 70m from the normal line of quaywall. Numerical simulations were carried out under the ground condition of three types in this study. Type-1 is the sand-clay model, and Type-2 is the clay model, and Type-3 is the sand model. It is Type-1 model as shown in Figure 1. Table 1 shows the material properties of each model. The elastic modulus was calculated in following equations.

$$E_{s(sand)} = 700N \quad (kN/m^2) \qquad (2)$$

$$E_{s(clay)} = 210c_u \quad (kN/m^2) \qquad (3)$$

and N is a N-value of standard penetration test.

The boundary condition of bottom was set for horizontal and vertical displacement to be fixed and the boundary condition of side was set only for horizontal displacement to be fixed. And, the anchor point of tie rod was set for horizontal displacement to be fixed. The sheet pile was modeled as a linear beam element with 3 nodes, and the interaction (friction) between the ground and the sheet pile was modeled as a linear joint element with 6 nodes. It has already been confirmed that the plastic deformation, which passes through the outside of sheet pile, was generated when the bending rigidity of sheet pile was assumed as shown in Table 1.

3.2 Analytical cases

The purpose of these analyses is to parametrically examine the effect of the subdivision size of finite element mesh, the consideration of the joint element between the ground and the sheet pile, and the shear stiffness of the joint element on the analytical results, safety factor. Table 2 shows the analytical cases in Series-A. The FE mesh with the coarse size (Case-A and Case-B) is shown in Figure 2. Figure 3 shows the mesh with the fine size (Case-C1~Case-

E2). The properties of both meshes are shown in Table 3.

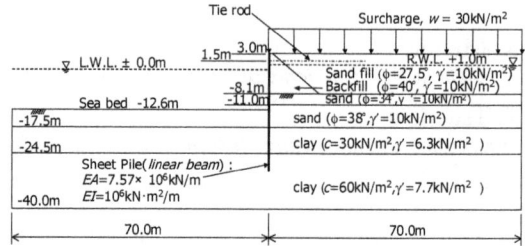

Figure 1. The sectional view of sheet pile quaywall (Type-1: Sand-Clay)

Table 1. Material properties of each model

Model Type		1:Sand-Clay	2:Clay	3:Sand
(Soil-1)	γ (kN/m³)	10.0	6.0	10.0
-11.0m	c_u (kN/m²)	-	30	-
∫	ϕ (degrees)	38	-	30
-17.5m	E_s (kN/m²)	14700	6300	14700
	ν	0.28	0.40	0.33
(Soil-2)	γ (kN/m³)	6.3	6.3	10.0
-17.5m	c_u (kN/m²)	60	60	-
∫	ϕ (degrees)	-	-	30
-24.5m	E_s (kN/m²)	12600	12600	14700
	ν	0.40	0.40	0.33
(Soil-3)	γ (kN/m³)	7.7	7.7	10.0
-24.5m	c_u (kN/m²)	150	150	-
∫	ϕ (degrees)	-	-	30
-40.0m	E_s (kN/m²)	31500	31500	14700
	ν	0.40	0.40	0.33
Sheet pile		$EA = 7.57*10^6$ kN/m		
(linear beam element)		$EI = 10^6$ kN·m²/m		

Table 2. Analytical cases (Series-A)

Case No.	Model Type	FE Mesh Roughness	Properties of Joint (linear joint element)	
			K_n (kN/m²)	K_s (kN/m²)
A	Sand-Clay	Rough	–	–
B	Sand-Clay	Rough	10^7	10^{-2}
C1	Sand-Clay	Fine	10^7	10^{-2}
C2	Sand-Clay	Fine	10^7	10^4
D1	Clay	Fine	10^7	10^{-2}
D2	Clay	Fine	10^7	10^4
E1	Sand	Fine	10^7	10^{-2}
E2	Sand	Fine	10^7	10^4

K_n: Normal Stiffness
K_s: Shear Stiffness

Table 3. Properties of FE mesh

Mesh type	Num. of node	Num. of element
Coarse	1346	441
Fine	5157	1704

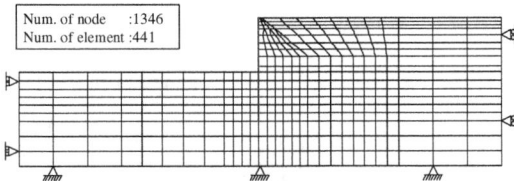

Figure 2. Finite Element mesh (Coarse subdivision)

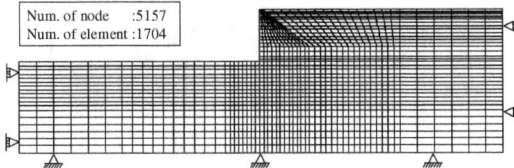

Figure 3. Finite Element mesh (Fine subdivision)

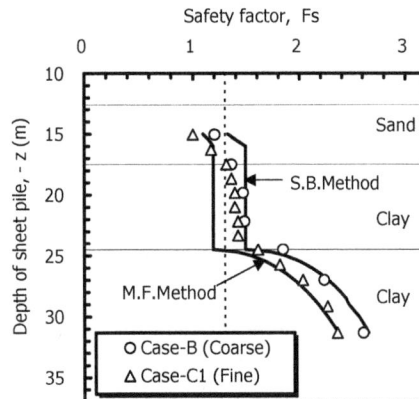

Figure 4. Effect of the subdivision mesh size (Sand-Clay)

Case-A is the model that does not consider the joint element, and it means that the friction between the ground and the sheet pile sufficiently works.

By comparing Case-B with Case-C1, it is possible to examine the effect of the roughness of mesh division on the safety factor. And by comparing Case-C1 with Case-C2, it is possible to examine the effect of the shearing stiffness of the joint element on safety factor. The comparison of Case-D1 and Case-D2 or the comparison of Case-E1 and Case-E2 in which the ground condition changes is also same.

$K_s=10^{-2}$ kN/m^2 in Table 2 means that it is calculated with the friction not essentially working between the ground and the sheet pile. On the other hand, $K_s=10^4$ kN/m^2 is the value which is equivalent to the shearing modulus of the ground (G_s) calculated from the elastic modulus of the ground (E_s) shown in Table 2, and it means that the friction sufficiently works between the ground and the sheet pile.

The analysis was carried out with changing the length of sheet pile. The longest length of sheet pile is $L=20$m, which is set to the lower clay layer (Soil-3). The shortest length is $L=2.5$m. The following Figure 4~Figure 7 show the relationship between the safety factor and the length of sheet pile. In addition, the safety factor calculated by the conventional slip circle methods of slices is shown in each figure. The upper limit of the slip circle method was calculated by the simplified Bishop's method (S.B.Method) and the lower limit is the value by the modified Fellenius' method (M.F.Method).

3.3 Results of FEM

3.3.1 Effect of the subdivision size of FE mesh

Figure 4 shows the comparison of Case-C1 and Case-B, and it shows the effect of the subdivision size of FE mesh on the analytical result.

The safety factor of Case-C1 was greatly calculated in comparison with the safety factor by the conventional M.F.Method when the depth of sheet pile was set in the upper clay layer (Soil-2). However, when the depth of sheet pile reached the lower clay layer (Soil-3), the safety factor of Case-C1 was almost equivalent to M.F.Method. It is generally said that the accuracy of stability analysis by FEM is improved when the mesh subdivision becomes fine. In present analysis, the effect on the safety factor is merely examined for two kinds of meshes. As far as it is concerned within this analysis, it was shown that the safety factor by FEM could be calculated the equivalent value to M.F.Method if the mesh is modeled in the fineness as Case-C1.

3.3.2 Effect of the shearing stiffness of joint element

The effect of the shearing stiffness of the joint element (K_s) on the safety factor was examined in the sand-clay deposit. Here, extreme two cases were assumed. One is the case in which the friction between the ground and the sheet pile works fully (Case-C2). K_S was made to be 10^4 kN/m^2 in order to express this condition analytically. Another is the case in which the friction does not work and K_S was made to be 10^{-2} kN/m^2 (Case-C1). The mesh for both cases is a fine mesh shown in Figure 3.

Figure 5 shows the comparison of Case-C1 and Case-C2. It has already been explained that the safety factor of Case-C1 is approximately equal to the conventional M.F.Method. As in Figures 5, it is shown that the safety factor of Case-C2 is approximately equal to S.B.Method. The ratio of the safety factor of Case-C1 to Case-C2 ($F_{s(C1)}/F_{s(C2)}$) was 0.84~0.96. The difference of the safety factor in both cases increased with extending the length of sheet pile.

The result of Case-A was also shown in Figure 5. Case-C2 and Case-A are the substantially equal model that assumed the same boundary conditions. The difference in the subdivision of mesh seems to be a cause on calculating large safety factor of Case-A comparing with Case-C2.

3.3.3 Effect of the ground condition

The effect of the friction between the ground and the sheet pile on the analytical result was examined for cohesive ground and sandy ground.

Figure 6 shows the Type-2 model in which the ground was changed to cohesive soil and Figure 7 shows the Type-3 model of sandy soil. Generally, there is no difference in the safety factor between the slip circle methods for the cohesive soil ground. However, as in Figure 6, the difference in the safety factor between two methods is slight. It seems to be a cause to modeling the backfilling as ϕ-material. For the sandy soil ground, the difference in the safety factor between two slip circle methods was increased, and the ratio of the safety factor of M.F.Method to S.B.Method was from 0.75 to 0.88.

As in Figure 6 and Figure 7, it is proven that FEM shows the tendency that is approximately equal to slip circle method on the relationship between the depth of sheet pile and the safety factor. The same observation applies to the results of the sand-clay ground.

For the cohesive soil ground model shown in Figure 6, Case-D2 in which the friction worked fully calculated the safety factor that is approximately equal to S.B.Method. And, Case-D1 in which the friction did not work calculated the safety factor that is lower than M.F.Method, when the depth of sheet pile was set to the lower clay layer (Soil-3). However, when the depth of sheet pile was set to the middle clay layer (Soil-2), the safety factor of Case-D1 was placed within two slip circle methods.

For the sandy soil ground model shown in Figure 7, Case-E2 in which the friction worked fully calculated the safety factor that is approximately equal to S.B.Method. The ratio of the safety factor of Case-D2 to Case-D1 ($F_{s(D1)}/F_{s(D2)}$) was from 0.88 to 0.95. Similarly, $F_{s(E1)}/F_{s(E2)}$ was from 0.79 to 0.84.

3.4 Failure mode of the ground

Figure 8 and Figure 9 show the displacement vector in the ground failure for the model that made the depth of sheet pile to be $z=-27$m. As in both figures, Case-C1 that assumed the friction between the ground and sheet pile to be zero, shows the wedge failure-mode in comparison with Case-C2. Both cases show the failure mode that passes the lower end of sheet pile, since the bending rigidity of sheet pile is relatively stiffer than the rigidity of the ground. It is considered that the difference in the safety factor between two cases is produced by such slight difference in the failure mode.

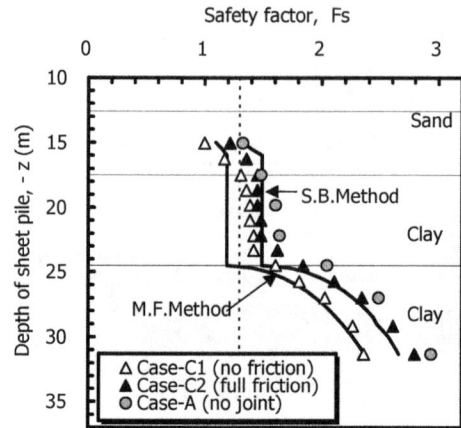

Figure 5. Effect of the shearing stiffness of the joint element (Sand-Clay)

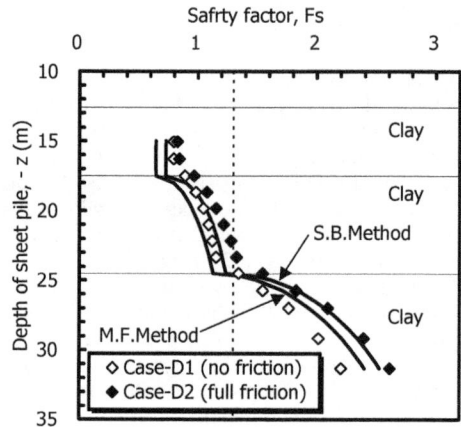

Figure 6. Effect of the shearing stiffness of the joint element (Clay)

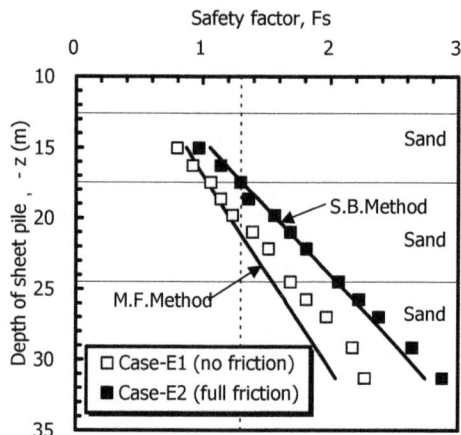

Figure 7. Effect of the shearing stiffness of the joint element (Sand)

392

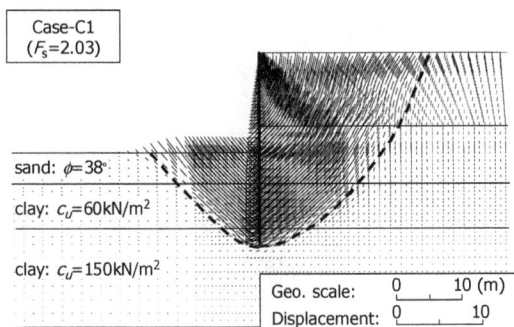

Figure 8. Displacement vector (Case-C1)

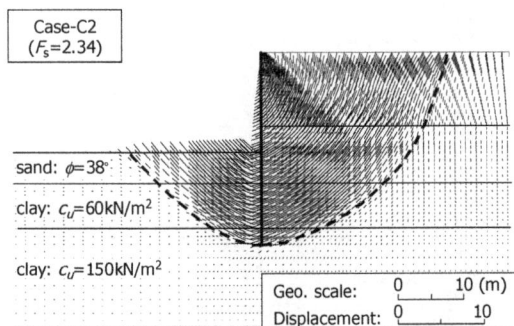

Figure 9. Displacement vector (Case-C2)

4 PARAMETRIC STUDIES (SERIES-B)

4.1 Analytical model and cases

For the ground model of three types shown in Table 1, the sensitivity analyses were carried out in order to examine the relationship between the bending rigidity of sheet pile and the safety factor of this system. The depth of sheet pile was also common on three ground models, and it was made to be $z=-27$m. And, the interaction of the ground and the sheet pile was assumed the friction sufficiently working.

The bending rigidity range of sheet pile was changed from $EI=1$ MN·m^2/m to $EI=10^4$ MN·m^2/m. This value was decided considering the lower limit per 1m of steel sheet pile (SP I-type) and the upper limit per 1m of one steel-pipe sheet pile ($\phi2,000*t25$).

4.2 Results of FEM

Figure 10, Figure 11 and Figure 12 respectively show the relationship between the bending rigidity of sheet pile and the safety factor of the ground-sheet pile system for three models. Since the sheet pile was modeled by using the linear beam element, the safety factor calculated here means the failure of the ground.

As in Figure 10, the safety factor takes constant value 2.34 from $EI=10,000$ MN·m^2/m to $EI=100$ MN·m^2/m and decreases with the bending rigidity of sheet pile at $EI<100$MN·m^2/m. The similar tendency is shown in Figure 12 in which plotted the case of the sandy ground. As in Figure 11 for the cohesive soil, the safety factor suddenly decreases from 2.09 to 1.42 at $EI=15$ MN·m^2/m.

Figure 13 and Figure 14 show the displacement vector for the cohesive soil model.

As in Figure 13 which shows the case of $EI=100$MN·m^2/m, the sheet pile sufficiently demonstrates the sliding resistance against the deformation of the ground, since the rigidity of sheet pile is relatively stiffer than the rigidity of the ground. As the result, it is considered that the failure mode, which passes through the lower end of sheet pile, becomes a critical condition. As in Figure 14 which shows the case of $EI=5$MN·m^2/m, the failure mode which passes the inside of sheet pile becomes a critical condition because of the flexibility of the sheet pile. The safety factor seemed to decrease as the result.

The rapid decrease of the safety factor in Figure 11 is caused by the discontinuity of the strength of the cohesive soil and the sliding of the ground at the shallow layer.

4.3 Safety factor considering the yield of sheet pile

Since the stability analysis by FEM mentioned above does not take into account the yield of sheet pile, it was permitted that the sheet pile was infinitely transformed. Therefore, the safety factor calculated by this procedure was decided only by the failure of the ground, and it is unrelated to the yield of sheet pile.

In the stability problem of the ground-sheet pile system, it seems to be rational to evaluate the safety factor as the failure of the ground or the yield of sheet pile. In this study, this evaluation procedure was given as follows.

1 The sheet pile is modeled as a linear beam element, and the prediction procedure of safety factor is not different from subsection 4.2
2 The shear strength reduction rate of the ground when the sheet pile reached the yield stress is defined as the safety factor for the yield of sheet pile.
3 This is calculated from the relationship between the shear strength reduction rate and the bending stress of sheet pile that required one after another by procedure 1.
4 In comparison with the safety factor that is obtained by usual calculation procedure and the safety factor by the above method, the small value is evaluated as a safety factor of the ground-sheet pile system.

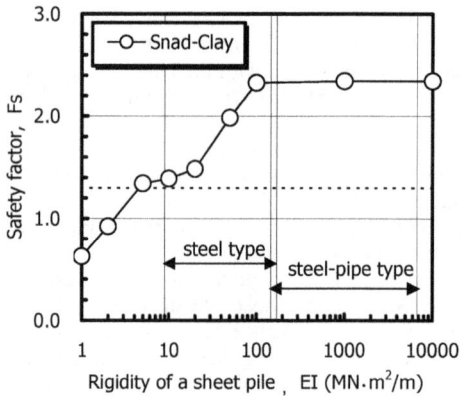

Figure 10. Effect of the bending rigidity of sheet pile (Sand-Clay)

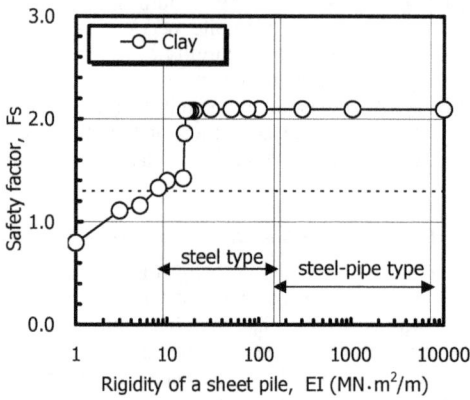

Figure 11. Effect of the bending rigidity of sheet pile (Clay)

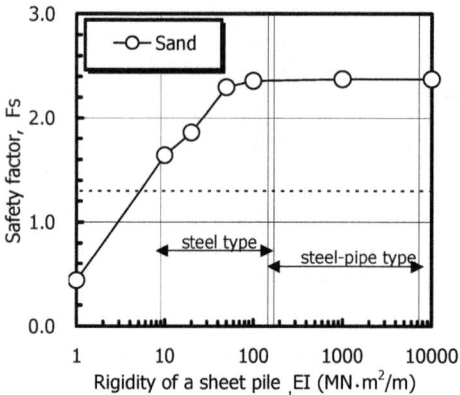

Figure 12. Effect of the bending rigidity of sheet pile (Sand)

Figure 15 shows the shear strength reduction rate and the maximum bending stress of sheet pile in FEM calculation procedure with the bending rigidity was EI=1000MN·m^2/m. From Figure 15, the safety

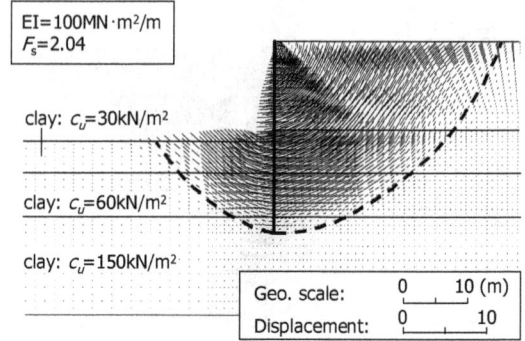

Figure 13. Displacement vector (Clay, EI=100MN·m^2/m)

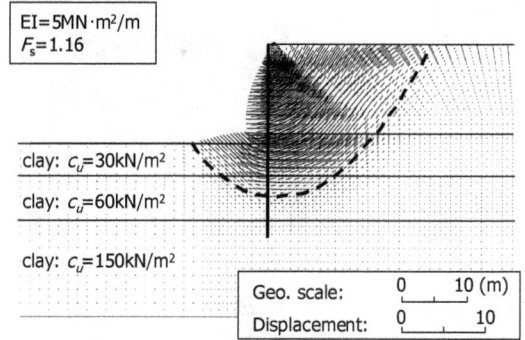

Figure 14. Displacement vector (Clay, EI=5MN·m^2/m)

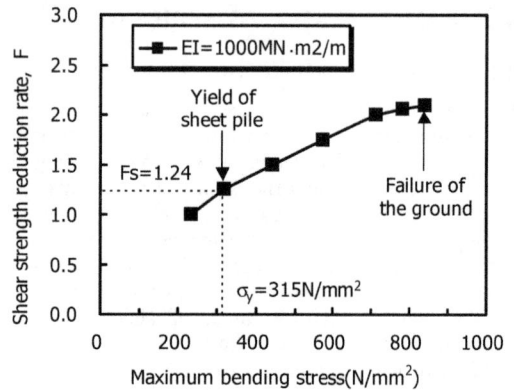

Figure 15. Relationship between the shear strength reduction rate and the bending stress of sheet pile (EI=1,000MN·m^2/m)

factor when the sheet pile reaches the yield stress is calculated as F_s=1.24.

Figure 16 shows the safety factor of cohesive soil ground based on the above method. As in Figure 16, it was judged that the sheet pile reached the yield, before the failure of the ground is occurred except for the case of EI=10^4 MN·m^2/m.

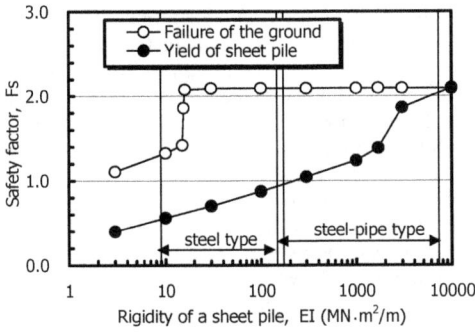

Figure 16. Safety factor considering the yield of sheet pile

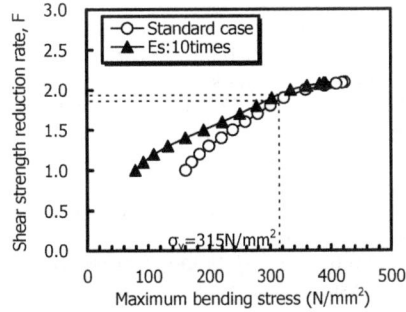

Figure 17. Effect of the elastic modulus of the ground on the safety factor (EI=10000MN·m²/m)

4.4 *Effect of the elastic modulus of the ground*

It seems to produce large plastic deformation of the ground with approaching the failure condition, because elasto-plastic model of Mohr-Coulomb's failure criterion is assumed for soils in this analysis. Accordingly, the excessive force may affect the sheet pile and the evaluation of safety factor may be improperly carried out.

Such problem in the FEM was examined analytically in detail (Tsuchida and Mizuno, 2002). As the result, it was confirmed that the large plastic deformation was not generated in the ground, when the bending stress of sheet pile reached the yield stress. However, it is considered that the safety factor is dependent on the elastic modulus of the ground under such condition. Here, the results of examining the effect of the elastic modulus of the ground on the evaluation of safety factor are described.

Figure 17, Figure 18 and Figure 19 show the relationship between the shear strength reduction rate and the maximum bending stress of sheet pile. These cases in which the bending rigidity are to be 10,000MN·m²/m, 3,000MN·m²/m, and 1,000 MN·m²/m are respectively shown.

Analysis case shown in Figure 16 was made to be a standard case (legend: closed triangle). The elastic modulus of the ground for this case has already been shown at Table 1. The elastic modulus of another case (legend: open circle) was assumed to be 10 times in comparison to the standard case.

These figures indicate that the safety factor on the failure of the ground is not almost affected by the elastic modulus of the ground, even if the elastic modulus is assumed 10 times. However, the maximum bending stress that affects sheet pile decreased and the safety factors which is decided by the yield of sheet pile were greatly calculated.

Figure 20 shows the effect of elastic modulus of the ground on the safety factor for the yield of sheet pile. This parametric analysis showed that the elastic modulus of the ground changes the safety factor and the change of it was from 1.1 times to 1.2 times for the 10 times the elastic modulus of the ground.

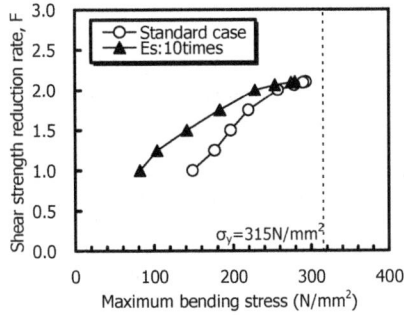

Figure 18. Effect of the elastic modulus of the ground on the safety factor (EI=3000MN·m²/m)

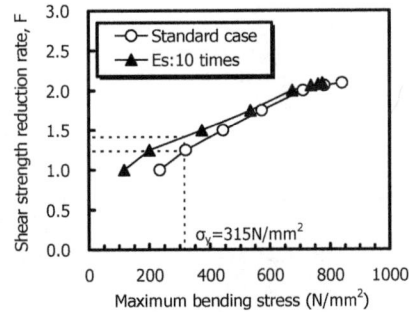

Figure 19. Effect of the elastic modulus of the ground on the safety factor (EI=1000MN·m²/m)

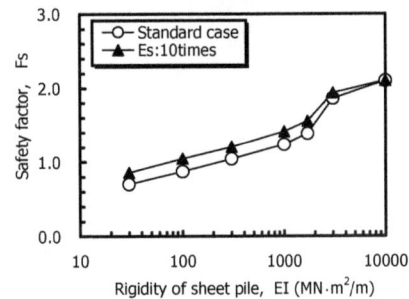

Figure 20. Change in safety factor due to the elastic modulus

5 CONCLUSIONS

In this study, the stability of a ground-sheet pile system was quantitatively evaluated using the elastic-plastic finite element method. The conclusions derived from this study are as follows.

1 The difference in the safety factor between the fine finite element mesh and the coarse mesh was examined. The safety factor calculated by the fine mesh was smaller than the coarse mesh and the value was almost equal to the safety factor calculated by conventional slip circle method of slices (modified Fellenius' method).

2 The friction property between the ground and the sheet pile was evaluated using the linear joint element with the shear stiffness and the normal stiffness. The analytical result, which assumed that there is no friction between the ground and the sheet pile, agreed almost with the safety factor calculated by modified Fellenius' method. And, the analytical result in which the friction sufficiently worked agreed almost with the safety factor by simplified Bishop's method. The similar tendency was confirmed in the analytical results of various ground conditions.

3 It was shown that the safety factor of the ground-sheet pile system decreased with decreasing of the bending rigidity of sheet pile. This reason seems to be able to explain in the relative relationship between the bending rigidity of the sheet pile and the rigidity of the ground. In other words, the failure mode of the ground in the shallower depth becomes critical condition, if the bending rigidity of sheet pile is relatively smaller than the rigidity of the ground.

4 The method to evaluate the safety factor of the ground-sheet pile system rationally was shown. In this method, it was important to appropriately evaluate the elastic modulus of the ground.

REFERENCES

Coastal Development Institute Technology, 1997. GeoFem user's manual (in Japanese).

Kobayashi, M., 1984. Stability analysis of geotechnical structures by finite elements, Report of the port and harbour research institute, Vol.23, No.1,: 83-101(in Japanese).

Tsuchida, T. and Mizuno, K. 2002. Bearing capacity analysis and stability analysis by "GeoFem", Technical note of the port and airport research institute, No.1023 (in Japanese).

Zeinkicwicz, O. C. and Cormeau, I. C., 1974. Visco-Plasticity - Plasticity and Creep in Elastic Solids - A Unified Numerical Solution Approach, Int. Journal for Numerical Method in Engineering 8: 821-845.

Soft Ground Engineering in Coastal Areas, Tsuchida et al. (eds)
© 2003 Swets & Zeitlinger, Lisse, ISBN 90 5809 613 0

Settlement of an embankment placed on the old seabed in a polder

H. Ohta
Tokyo Institute of Technology, Tokyo, Japan

A. Iizuka
Kobe University, Kobe, Japan

ABSTRACT: The deformation and the stress state within a specified domain of soft clay foundation loaded by an embankment are analysed by using a soil/water coupled finite element program incorporated with an elasto-viscoplastic constitutive model. The embankment is placed on the old sea-bed in a polder. In addition to laboratory tests usually carried out for the design purpose, some in-situ tests are carried for the research purpose. All the test data are intentionally interpreted aiming at seeking a reasonable procedure for parameter determination. Thus specified parameters are found to be in good accordance with an accidental failure of a slope in the polder and with the settlement of the embankment. This leads to the conclusion that the use of in-situ tests in specifying the material parameters is promising in the application of finite element methods in engineering practice.

1 INTRODUCTION

In the author's understanding, any kind of the mechanical behaviour of a foundation, loaded either by human activities such as construction works or through the agency of natural environment, should be analytically treated as an initial/boundary value problem in which a set of governing equations are to be solved under a set of initial and boundary conditions. The reaction of a foundation to the external agencies is governed by a set of governing equations where constitutive equations for the soil skeleton and the pore water flow have to be incorporated. An incremental elasto-viscoplastic model proposed by Sekiguchi and Ohta (1977) is used in this investigation to describe the mechanical response of the soil skeleton resulting from the change in the effective stresses at each point in the domain being analysed. In the present investigation, direct use of the constitutive model for the pore water is avoided and is replaced by an empirical law traditionally appreciated as Darcy's law.

Both the constitutive model for soils and the Darcy's law require us to specify the material parameters needed in describing the mechanical characteristics of the soil skeleton and the pore water flow. These parameters should primarily be obtained either from the laboratory tests or from the in-situ tests purposely designed for the use of specifying each of the parameters needed. However, the use of traditionally available methods of soil testing is much more practical provided that test results are

properly interpreted. This paper presents some methods of correcting the field vane strength and unconfined compression strength to convert them into theoretically interpretable values from which soil parameters needed in the constitutive model are to be derived. The deformation and stability of soft clay foundations observed in the field will then be successfully analysed by employing thus obtained parameters.

2 CONSTITUTIVE PARAMETERS

The parameters used in the constitutive model proposed by Sekiguchi and Ohta (1977) are listed in Table 1. The constitutive model proposed by Sekiguchi and Ohta (1977) is essentially an extension of a model originally developed by Shibata (1963). Shibata's model was then expanded and sophisticated by Hata and Ohta (1969), Ohta (1971) and Sekiguchi and Ohta (1977). They took into consideration some of the important findings obtained by Shibata and Karube (1965) and Karube and Kurihara (1966) through a series of very careful experiments on the negative dilatancy (contractancy) of clays. The development of the constitutive model went forward along the lines of a study in which the major object was the dilatancy characteristics of clays.

It should be mentioned that this model happened to be essentially identical in its mathematical expression with the Cam Clay model proposed by Roscoe, Schofield and Thurairajah (1963) although the basic

Table 1. Parameters needed in the model proposed by Sekiguchi and Ohta (1977)

① soil	consolidation	λ	$\lambda = 0.4343C_c$ (C_c :compression index)	
		κ	$\kappa = 0.4343C_s$ (C_s :swelling index)	
	contractancy	D	defined by Shibata (1963)	
		Λ	$\Lambda = 1 - \kappa/\lambda$ (Λ :irreversibility ratio)	
	parameters derived from λ, κ, D	M	$M = \dfrac{\lambda - \kappa}{(1+e_0)D} = \dfrac{6\sin\phi'}{3-\sin\phi'}$ M :critical state parameter ϕ' : ϕ' from triaxial compression	
symbols used only for convenience		p'	$p' = \dfrac{1}{3}\sigma'_{ij}\delta_{ij}$ (p' :effective mean principal stress, δ_{ij} :Kronecker's delta)	
		q	$q = \sigma'_1 - \sigma'_3$ (q :principal stress deviator, σ'_1, σ'_3 :max and min effective principal stresses)	
		s_{ij}	$s_{ij} = \sigma'_{ij} - p'\delta_{ij}$ (s_{ij} :deviatoric stress δ_{ij} :Kronecker's delta)	
		η^*	$\eta^* = \sqrt{\dfrac{3}{2}\left(\dfrac{s_{ij}}{p'} - \dfrac{s_{ij0}}{p'_0}\right)\left(\dfrac{s_{ij}}{p'} - \dfrac{s_{ij0}}{p'_0}\right)}$ .normalized shear stress	
② stress history		σ'_{v0}	preconsolidation pressure	
		K_0	$K_0 = \sigma'_{h0}/\sigma'_{v0}$ (K_0 :at completion of virgin consolidation) σ'_{v0} : preconsolidation pressure (in vertical direction) σ'_{h0} : preconsolidation pressure (in horizontal direction)	
symbols used only for convenience		η_0	$\eta_0 = \dfrac{3(1-K_0)}{1+2K_0}$ (η_0 :normalized shear stress $\dfrac{q_0}{p'_0}$ at virgin consolidation)	
		β	$\beta = \sqrt{3}\eta_0\Lambda/2M$	
③ current stress state		σ'_{vi}	current in-situ effective overburden stress	
		K_i	$K_i = \sigma'_{hi}/\sigma'_{vi}$ (K_i :current coefficient of earth pressure at rest) σ'_{vi} :current effective vertical stress σ'_{hi} :current effective horizontal stress	
		p_{wi}	p_{wi} :current in-situ pore water pressure	
notes			σ'_{ij} : ij component of effective stress σ' subscript 0 : at completion of virgin consolidation subscript i : at initial state prior to the on-going stress changes	

assumptions employed in developing these two models are totally different from each other. The Cam Clay model, as widely known, is based on an assumption about the energy dissipated during shear, while the model employed here is based on an empirical equation of negative dilatancy (contractancy) originally found by Shibata (1963) through a series of experiments in which clay specimens were sheared keeping the effective mean principal stress p' constant during shear. The model proposed by Sekiguchi and Ohta (1977) may be distinguished from the Original Cam Clay model when it is applied to K₀-consolidated clays, see Ohta and Sekiguchi (1979).

The failure state of soils is described in terms of strain increment as the state where infinite plastic shear strain increment does exist. This means it is the state where no plastic volumetric strain increment exists. The undrained conditions are described in terms of strain increment as the state where the summation of elastic and plastic volumetric strain increments is zero. The plane strain state is described as zero intermediate principal strain increment, while the axi-symmetric state as two identical principal strain increments in the horizontal direction. Applying the constitutive model proposed by Sekiguchi and Ohta (1977) to these conditions, Ohta, Nishihara and Morita (1985) derived the theoretical expressions of the undrained strength to be obtained by different types of testing procedures as seen in Table 2.

Table 2. Undrained strength to be obtained from various testing methods (after Ohta, Nishihara and Morita, 1985, test data reported by Ladd, 1973)

$$\frac{S_u}{\sigma'_{vi}} = \frac{OCR^\Lambda(1+2K_0)\,M\exp(-\Lambda)}{3\sqrt{3}(\cosh\beta - \sinh\beta\cos2\theta)} \quad (1)$$

$$\frac{S}{\sigma'_{vi}} = \frac{OCR^\Lambda(1+2K_0)\,M\exp(-\Lambda)}{3\sqrt{3}(\sqrt{\cosh^2\beta - \sinh^2\beta\cos^2 2\omega} - \sinh\beta\sin2\omega)} \quad (2)$$

$$M = \frac{6\sin\phi'}{3-\sin\phi'} \qquad \Lambda = 1 - Cs/Cc \qquad n_0 = \frac{3(1-K_0)}{1+2K_0} \qquad \beta = \frac{\sqrt{3}\,n_0\Lambda}{2M}$$

Type of test	Reduced equation for specified test on normally consolidated clay	Blue marine clay PI=20 ϕ'=33° K0=0.5 measured	predicted
Ko-consolidated Plane strain Comp. KoPUC (A)	$\dfrac{S_u}{\sigma'_{vo}} = \dfrac{(1+2K_0)\,M\exp(-\Lambda)}{3\sqrt{3}(\cosh\beta - \sinh\beta)}$	0.34	0.347
Ko-consolidated Triaxial Comp. KoUC	$\dfrac{S_u}{\sigma'_{vo}} = \dfrac{1+2K_0}{6}\,M\exp\left(\dfrac{\Lambda n_0}{M} - \Lambda\right)$	0.33	0.318
Shear Box Test SBT	$\dfrac{S}{\sigma'_{vo}} = \dfrac{(1+2K_0)\,M\exp(-\Lambda)}{3\sqrt{3}}$	—	0.239
Direct Simple Shear DSS (D)	$\dfrac{S}{\sigma'_{vo}} = \dfrac{(1+2K_0)\,M\exp(-\Lambda)}{3\sqrt{3}\cosh\beta}$	0.20	0.224
Ko-consolidated Plane strain Ext. KoPUE	$\dfrac{S_u}{\sigma'_{vo}} = \dfrac{(1+2K_0)\,M\exp(-\Lambda)}{3\sqrt{3}(\cosh\beta + \sinh\beta)}$	0.19	0.165
Ko-consolidated Triaxial Ext. KoUE (P)	$\dfrac{S_u}{\sigma'_{vo}} = \dfrac{1+2K_0}{6}\,M\exp\left(-\dfrac{\Lambda n_0}{M} - \Lambda\right)$	0.155	0.135
Field Vane FV	$\dfrac{S_h}{\sigma'_{vo}} = \dfrac{(1+2K_0)\,M\exp(-\Lambda)}{3\sqrt{3}}$ $\dfrac{S_v}{\sigma'_{vo}} = \dfrac{1+2K_0}{3\sqrt{3}}\sqrt{\left(\dfrac{M}{\Lambda}\dfrac{P}{P_0}\ln\dfrac{P}{P_0}\right)^2 - \left(1-\dfrac{P}{P_0}\right)^2 n_0^2}$	0.19	0.182

Equations (1) and (2) in Table 2 give the undrained strength of an anisotropically consolidated clay as a function either of the principal stress direction at failure or of the inclination of the slip surface. These equations are reduced to the equations shown in the lower rows in Table 2, when the stress and geometrical boundaries represent each of the listed testing methods.

In the right hand columns, the experimentally obtained values of undrained strength of Boston Blue Clay reported by Ladd (1973) are compared with values calculated by using the equations listed in the central column. The parameters reported by Ladd are also shown. Theoretical values are in good agreement with experimental values. Since the values of parameters are already specified by Ladd (1973), it is not possible to the author to intentionally adjust the parameters aiming at obtaining the closer agreement between the experimental and theoretical values.

As seen in Table 2, the undrained strength of K_0-consolidated clay is a function of the principal stress direction θ at failure and hence of the inclination ω of the slip surface. This means that the undrained strength mobilized along the slip surface depends on the inclination of the slip surface. The undrained strength mobilized on a fractional segment of the slip surface changes its value from segment to segment depending on the inclination of the segment. If the geometry of the slip surface is specified, it is possible to derive the theoretical value of the representative strength (design strength) mobilized in average along the slip surface. In the case of a circular slip surface which has right-left symmetry about a vertical line (such as a circular slip surface caused by a strip load applied on a part of the horizontal ground surface), the ratio s_u/σ'_{vo} of undrained design strength s_u to preconsolidation pressure σ'_{vo} ranges in a very narrow band between 0.24 and 0.26 depending on the value of plasticity index, see Ohta , Nishihara and Morita (1985). In calculating the value of the theoretically expected 'design strength', the material parameters M, Λ and K_o, are estimated from the plasticity index by using widely accepted empirical equations which were summarized by Iizuka and Ohta (1987).

In interpreting any of laboratory tests for undrained strength, the following 5 effects must be carefully considered:

(a) stress release (effective stress change caused by taking undisturbed samples from deep in the ground to the air, from anisotropic in-situ effective stress state (ideal samples) to isotropic stress state (perfect samples))

(b) sample disturbance during sampling, handling and trimming processes

(c) shearing rate (loading rate) in the test relative to the shearing rate observed in the actual failure of the soft foundations

(d) incompletely undrained conditions arising from partial drainage, suction or migration of the pore water

(e) anisotropy of undrained strength (as apparently seen in Table 2, the undrained strength of an anisotropically consolidated clay is a function of the principal stress direction at failure and hence of the inclination of the slip surface; this requires a correction factor to convert the strength obtained from a particular type of laboratory test into the average strength (design strength) that is expected to be mobilized, in average, along the whole portion of the slip surface in the field)

The unconfined compression strength is accepted in some countries including Japan as an inexpensive substitute for Ko-consolidated undrained triaxial compression strength. In the author's opinion, unconfined compression strength has to be adequately corrected, when it is used as the substitute, to eliminate the effect of the sample disturbance (Ohta, Nishihara, Iizuka, Morita, Fukagawa and Arai, 1989), the effect of the stress release from anisotropic in-situ effective stress state (ideal sample) to isotropic stress state (perfect sample)(Skempton and Sowa, 1963, Ladd

and Lambe, 1964, Ohta, Nishihara, Iizuka, Morita, Fukagawa and Arai, 1989), lack of confining pressure (Nakase, Katsuno and Kobayashi, 1972) and the effect of the rate of shear (Bjerrum, 1972).

Suppose we have the data of undrained strength of undisturbed clay specimens obtained from K_o-consolidated undrained triaxial compression tests. Suppose the loading rate during the tests was so slow that there is no need to apply the correction factor μ_R for rate effect. The ratio μ_A of 'design strength' to K_oUC strength can be theoretically calculated, provided the material parameters M, Λ and K_o are known. This ratio μ_A should be applied to the K_oUC strength when we need to convert K_oUC strength to the 'design strength' in the case of a circular slip surface which has right-left symmetry about a vertical line. The correction factor μ_A for strength anisotropy thus obtained is plotted against the plasticity index in Fig. 1.

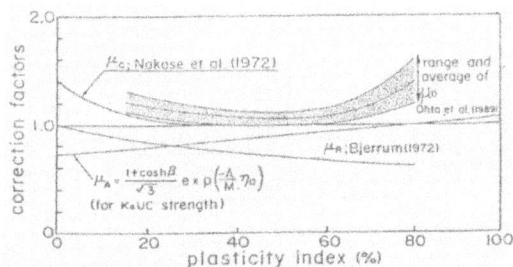

Figure 1. Correction factors for unconfined compression strength (after Ohta, Nishihara, Iizuka, Morita, Fukagawa and Arai, 1989)

The correction factor μ_S for stress release theoretically proved to be unity ($\mu_S = 1$), see Ohta, Nishihara, Iizuka, Morita, Fukagawa and Arai (1989). It is also experimentally shown to be unity by Skempton and Sowa (1963) and Ladd and Lambe (1964). The correction factors for rate effect (μ_R) and for confining pressure (for incomplete drainage, for partial suction)(μ_C) are also plotted in Fig. 1. These two correction factors are respectively proposed by Bjerrum (1972) and by Nakase, Katsuno and Kobayashi (1972). The use of Bjerrum's correction factor μ_R in correcting the unconfined compression strength may be justified since the time required to reach the failure state is more or less the same in the field vane tests as in the unconfined compression tests. A band in Fig. 1 indicates the probable range of correction factor for sample disturbance (μ_D). The band was estimated by Ohta, Nishihara, Iizuka, Morita, Fukagawa and Arai (1989) based on the experimentally obtained negative pore water pressure remaining in the undisturbed samples collected from 22 sites distributed all over Japan.

Suppose we have data of the unconfined compression tests on undisturbed soft clay samples. By

400

multiplying μ_S μ_R μ_C μ_D by half of the unconfined compression strength $q_u/2$, we get the value of K_0UC strength converted from $q_u/2$. Since we have the theoretical equation for K_0UC strength as seen in Table 2, the value of K_0UC strength converted from $q_u/2$ will provide the information about the material parameters. This implies that q_u values can be a source of information useful in estimating the material parameters needed in the constitutive model proposed by Sekiguchi and Ohta (1977). If we need the 'design strength' for a symmetric circular slip surface, the correction factor μ_A for strength anisotropy should further be multiplied to the K_0UC strength converted from $q_u/2$.

When in-situ tests for undrained strength are to be interpreted, care should be taken in correcting the effects of

(a) disturbance caused by the placement of the measuring device in the soil deposit
(b) shearing rate effect
(c) partial drainage
(d) failure mode/anisotropy of undrained strength (in-situ tests should be analysed as boundary value problems, hence the value of test data obtained should adequately be converted into the 'design strength' taking into account both the failure mode seen in an in-situ test and the geometry of the slip surfaces expected to exist in the failure of foundations).

Fukagawa, Fahey and Ohta (1990) and Ohta, Iizuka, Nishihara, Fukagawa and Morita (1991) analysed the pressuremeter tests and concluded that the in-situ pressuremeter tests should be carried out much faster than they are currently performed. Current rate of loading seems to be too slow to minimize the effect of partial drainage of the pore water around the pressuremeter. Test data obtained from fast loading tests have to be corrected by multiplying the correction factor for rate effect. Test data obtained from slow loading tests have to be corrected by multiplying a correction factor for effect of partial drainage. Since the possible range of the permeability of wide variety of soil is much wider than the possible range of viscous response of the loaded subsoil, the author feels that the correction for rate effect seems to be easier to apply, see Fig. 5.

3 SITES INVESTIGATED

A polder was made in an area of the shallow inland sea located at the place called Kasaoka in the western part of Japan during a period from March 1976 through August 1977 for a purpose of agricultural use. The dewatered area of 861 ha was protected either by an island or by the dike from the flooding as seen in Fig. 2. At the western side of the polder, there was a reclaimed land developed for the industrial use.

The old sea-bed in the polder was covered by a soft clay layer of 11m in thickness. In the polder, there were three sites interesting from the geotechnical point of view. During the process of excavating the network of drainage channels in the polder, a slope failure unexpectedly took place in the summer of 1989 at a location indicated by a small circle named "channel" in Fig. 2. This site is called site C hereafter. A pre-loading embankment was placed in 1991 on the soft clay foundation to eliminate the long-term settlement of a stockyard facility at a location indicated by "stockyard" in Fig. 2. This site is called site S hereafter. A small airport with an 800m runway was completed at a location indicated by "airport" in Fig. 2. This site is called site A hereafter. The settlement of the pre-loading embankment was analysed by Ohta, Nishihara, Iizuka and Morita (1992). Undisturbed samples of the soft clay were taken at sites A and C.

The results of the tests carried out on the soft clay at site A and site C are plotted in Figs. 3 in which there was no significant difference between the data at these two sites. This implies that the soft clay layer was so uniform that the material parameters should be the same regardless of the location in the polder. The slope failure at site C took place when the drainage channel was excavated to a depth of 2m. This failure was probably triggered by tentatively placing the excavated material on the ground surface close to the channel. It was believed that the deepest part of the slip circle reached a depth of about 5m. The in-situ tests were carried out after the slide of the slope and hence were somewhat affected by the slide as typically seen in Fig. 3 (h).

At site C, a series of very careful field vane tests were carried out together with dilatometer tests and pressuremeter tests of pre-boring type. In Fig. 3 (e), plotted is the undrained strength obtained from the standard (vane height/vane width=H/B=2) field vane tests. Since the vane torque T is interpreted by using an equation

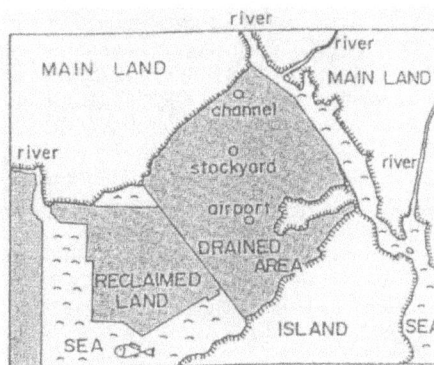

Figure 2. Site investigated (after Ohta, Nishihara, Iizuka and Morita, 1992)

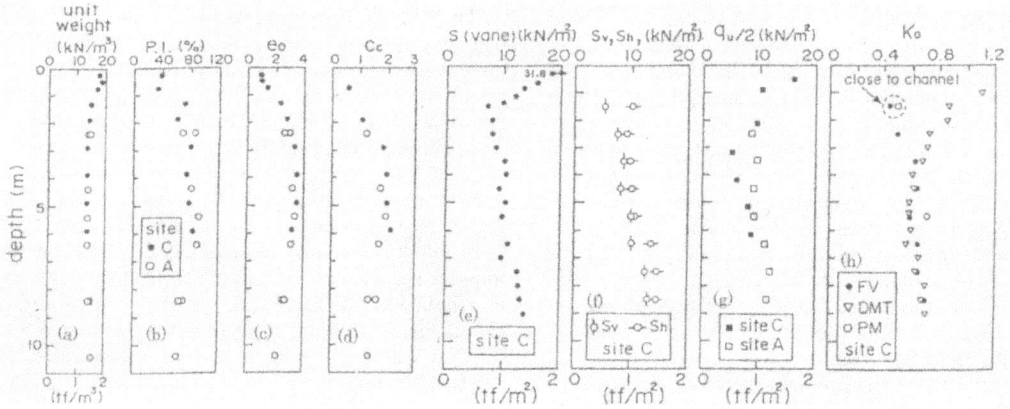

Figure 3. In-situ and laboratory test results (after Ohta, Nishihara, Iizuka and Morita, 1992)

$$\frac{2T}{\pi B^3} = \frac{H}{B}s_v + \frac{1}{3}s_h \qquad (1),$$

undrained strength s_v mobilized on the vertical-cylindrical side shear surface can be separated from undrained strength s_h mobilized on horizontal top and bottom shear planes provided the data of a series of vane tests with different H/B are available as explained by Cadling and Odenstad (1950), see Fig. 4. The author carried such vane tests with different H/B and obtained s_v and s_h summarized in Fig. 3 (f).

Undisturbed samples were taken from sites A and C, and subjected to unconfined compression tests. Half of the unconfined compression strength $q_u/2$ is plotted against the depth in Fig. 3 (g). The undrained strength $s_u(K_0UC)$ to be obtained from K_0UC tests can be estimated by multiplying correction factors $\mu_S \mu_R \mu_C \mu_D$ to $q_u/2$ as explained already, see Fig. 1. The undrained strength $s_u(IUC)$ is theoretically obtained by substituting $K_0=1$ into the theoretical equation of $s_u(K_0UC)$ in Table 2. The sign $s_u(IUC)$ denotes the undrained strength to be obtained from undrained compression test on an isotropically consolidated specimen. The ratio of $s_u(K_0UC)/s_u(IUC)$ is calculated by using the theoretical equations for $s_u(K_0UC)$ and $s_u(IUC)$ together with an empirical relation $M=1.75\Lambda$ proposed by Karube (1975). It is interesting to see that the ratio $s_u(K_0UC)/s_u(IUC)$ happens to be very close to unity (from 0.989 through 1.023 for K_0 values ranging between 0.5 and 1.0). This coincidence of $s_u(K_0UC)$ and $s_u(IUC)$ theoretically justifies the traditional usage of $s_u(IUC)$ as a substitute of $s_u(K_0UC)$ in engineering practice. This coincidence also indicates that $q_u/2$ multiplied by correction factors $\mu_S \mu_R \mu_C \mu_D$ equals to $s_u(IUC)$.

Substituting $K_0=1$ into $s_u(K_0UC)$ in Table 2, we get $s_u(IUC)$ as

$$\frac{s_u}{\sigma'_{vo}} = \frac{1}{2}M\exp(-\Lambda) : s_u(IUC) \qquad (2)$$

The vane shear strength s_h mobilized on the horizontal shear plane is

$$\frac{s_h}{\sigma'_{vo}} = \frac{1+2K_o}{3\sqrt{3}}M\exp(-\Lambda) :s_h(vane) \qquad (3)$$

as listed in Table 2. Then

$$\frac{s_h/\sigma'_{vo}}{s_u(IUC)/\sigma'_{vo}} = \frac{2(1+2K_o)}{3\sqrt{3}} \qquad (4)$$

Since we now know the values of s_h from vane tests and $s_u(IUC)$ from unconfined compression tests, we can calculate the left hand side of Eq. (4). K_0-values thus calculated from Eq. (4) are plotted in Fig. 3 (h). As seen in the figure, thus estimated K_0-values (solid circles) are in surprisingly good agreement with the K_0-values obtained from dilatometer tests and pressuremeter tests, with exception of two data points circled by dotted curve. These are the data measured 15 months after the slope failure at a location very close (2m) to the failure surface in the field, and hence are very likely to be affected by the lateral stress release due to the preceding failure of

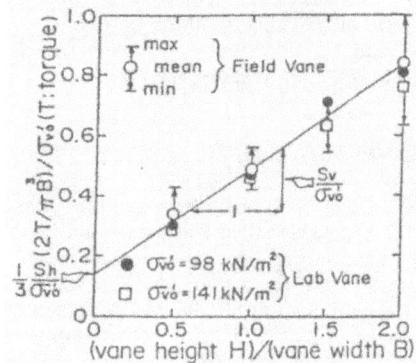

Figure 4. Separation of vane strength on vertical and horizontal planes (after Ohta, Nishihara, Iizuka and Morita, 1992)

402

the excavated channel slope. It should be noted here that the correction factor for the rate effect is cancelled when we calculate the ratio $s_h/ s_u(IUC)$, since both the vane shear test and the unconfined compression tests are performed within the same order of time period and hence the correction factor for the rate effect is likely to be the same.

It is believed that the deepest part of the slip surface of the channel slope failure reaches a depth of about 5m. The average of the K_o-values shown by solid circles in Fig. 3 (h) obtained by the use of Eq. (4) is 0.6 at the middle depth of the soft clay layer (K_o=0.6). The ratio s_v/s_h estimated from Fig. 3 (f) is 0.84 in average (s_v/s_h=0.84). It is possible to calculate the value of M from the ratio s_v/s_h together with K_o=0.6 and the Karube's relation M=1.75Λ. This procedure is not simple since the value of the effective mean principal stress p' at failure in the vane test has to be calculated, see Table 2. The value of M obtained from the ratio s_v/s_h is 1.09 (M=1.09, φ'=27.5 degrees). Then, the Karube's relation leads to Λ=M/1.75 =0.62.

Fig. 4 shows the results both of laboratory and field vane tests. The data points represented by solid circles and open squares are the results of the laboratory vane tests on one-dimensionally reconsolidated clay specimens. The arrows indicating the range of values with the mean value represented by open circles are the data obtained from the field vane. Since the ground is still under consolidation, the effective vertical stress σ'_{vo} is not clearly known. The vane torque T on the vertical axis of Fig. 4 is divided by the σ'_{vo} obtained by 1-D FEM (dotted curve in Fig. 6). Together with the laboratory vane data, the open circles more or less overlap the laboratory vane data and form a straight line, as they should, implying that the effective vertical stress σ'_{vo} estimated by FEM was close to the actual value.

Fig. 5 shows how the undrained strength experimentally obtained by vane and shear box tests is reduced as the rate of shear diminishes (time to failure increases). The data points shown by the open circles and open squares represent the strength of re-molded-reconsolidated samples. As pointed out by Torstensson (1977), the time to failure in the field is generally of the order of 1000 times longer than the time to failure either in vane tests or in shear box tests. The experimentally obtained value of s(vane)/σ'_{vo} (field vane) at time to failure of 10 minutes is 0.36, while the values of s(vane)/σ'_{vo} at time to failure of 10 000 minutes is estimated by extrapolation of the experimental data as 0.22. This makes the correction factor for rate effect to be 0.61. Since Bjerrum's correction factor is 0.62 (for clays of PI=80%) and is in agreement with the experimental results as seen in Fig. 5, let us use Bjerrum's factor in this investigation. In Fig. 5, the rate effect in the shear box test is also shown in comparison with Bjerrum's factor.

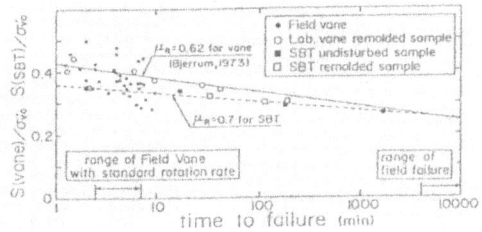

Figure 5. Rate effects in vane and shear box test (after Ohta, Nishihara, Iizuka and Morita, 1992)

4 INITIAL STRESS STATE

In analysing the behaviour of soft clay foundations, it is essential to estimate the in-situ effective stress state prior to the construction works. Fig. 6 shows the effective overburden pressure before and long after the draining of the sea-bed. Open circles and solid circles are the effective vertical stresses obtained from oedometer tests on the specimens sampled before the placement of embankment and excavation of the channel. These data points are plotted between two solid lines representing the effective overburden pressure before and long after dewatering of the polder. At a shallow depth the data points are close to the effective overburden pressure long after dewatering of the polder implying that the excess pore water pressure has quickly dissipated at a depth close to the drainage boundary. This indicates that the soft clay ground is still on the process of consolidation especially at the mid-depth of the soft clay layer.

Substituting M=1.09, Λ=0.62 and K_o=0.6 into Eq. (3), we get s_h/σ'_{vo}=0.25. Since we have already known the values of s_h as seen in Fig. 3 (f), the val-

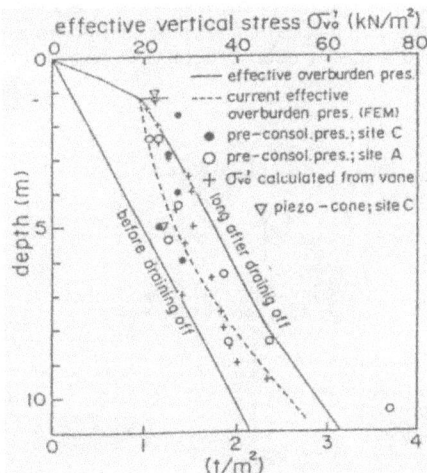

Figure 6. Effective vertical stress prior to embankment and excavation (after Ohta, Nishihara, Iizuka and Morita, 1992)

403

ues of the effective overburden pressure σ'_{vo} can be calculated as shown by data points represented by + in Fig. 6 where the effective overburden pressure σ'_{vo} directly obtained from piezocone tests are also plotted as a triangle ∇. The dotted curve in Fig. 6 represents the effective vertical stress distribution computed by 1-D FEM using the soil parameters specified through the process introduced in this paper. Since all the data points are located around the dotted curve, the material parameters used in the analysis seem to be reasonable. In order to further confirm the appropriateness of the material parameter specification, the author wanted to compare the computed rate of settlement of the sea-bed with the monitored settlement record after dewatering. However, unfortunately, there was no settlement record taken after the completion of dewatering.

5 FAILURE OF THE CHANNEL SLOPE

At site C, a slide took place during the excavation work of a channel as shown in Fig. 7. It is reported that the slope failed when the excavation depth reached about 2m, about a week after they started to excavate. The deepest part of the slip circle was believed to be at a depth of 5m. Eq. (2) in Table 2 is rewritten in a form shown in Fig. 7. The shear strength s_o along the horizontal plane ($\omega = 0$) is theoretically proved to be identical with the average strength for a circular slip surface generated by placing a low embankment on a flat ground surface, see Ohta, Nishihara and Morita (1985). However, in the case shown in Fig. 7, the average strength mobilized along the slip surface is slightly higher than s_o because of the lack of slip surface at the portion shown by the dotted curve in Fig. 7.

The factor of safety calculated from the strength based either on the field vane tests or on the unconfined compression tests is shown in Fig. 8. Apparently the use of raw data of field vane and uncon-

fined compression strength (open circles and open squares) leads to a factor of safety higher than 1.0 at the excavation depth of 2m indicating that some strength correction has to be applied to the raw data of strength. The use of strength corrected in a way explained in the previous sections (solid circles and solid squares) seems to give reasonable factors of safety as seen in Fig. 8. It should be noted here that s_h value obtained from field vane tests must theoretically be equal to the strength obtained from shear box tests (see Table 2) and to s_o in Fig. 7 as well. Since the correction is consistent with the constitutive model used and the parameters obtained, it may be concluded that the whole story introduced in this paper is in accordance with the observed failure. The critical slip surfaces obtained by the stability analysis (solid circle and solid square at a depth of excavation of 2m in Fig. 8) are found to be very close to the circle observed in the field and schematically shown in Fig. 7.

Fig. 9 shows the correction factors back-calculated from the failure records observed at various construction sites. Open circles (Ladd et al., 1977) are the data points back-calculated from the strength obtained by the field vane tests, while solid circles are those back-

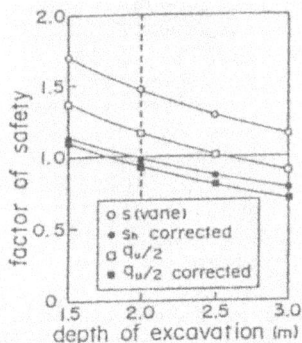

Figure 8. Factor of safety for the channel slope failure (after Ohta, Nishihara, Iizuka and Morita, 1992)

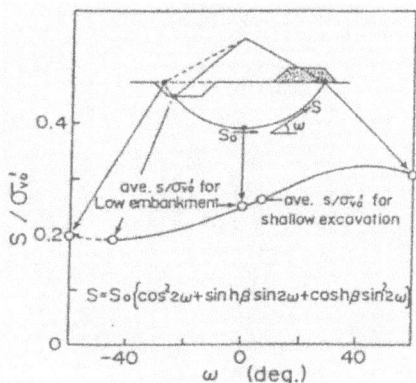

Figure 7. Average strength mobilized on the slip surface at the channel slope(after Ohta, Nishihara, Iizuka and Morita, 1992)

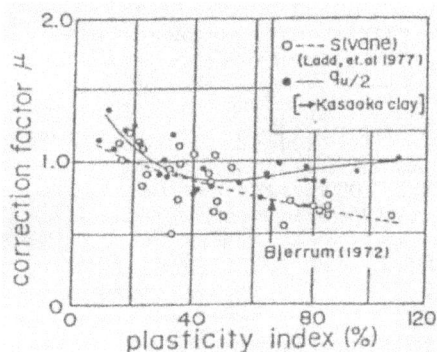

Figure 9. Correction factor back-calculated from the actual failure records(after Ohta, Nishihara, Iizuka and Morita, 1992)

calculated from the unconfined compression strength. The data represented by the solid circles are all collected by Ohta, Nishihara, Iizuka and Morita (1992) from the failures observed in Japan. The dotted curve was proposed by Bjerrum (1972) and the solid curve is drawn by Ohta, Nishihara, Iizuka and Morita (1992) in such a way that it passes through the middle of the data points shown by solid circles. The failure at site C in Kasaoka is represented in Fig. 9 by a plot specified with a horizontal arrow. The plot for the failure at Kasaoka is not far from the solid curve in Fig. 9 indicating that the failure at site C was just one of the common failures.

The solid curve shown in Fig. 9 is only an empirical curve and is not theoretically supported. Ohta, Nishihara, Iizuka, Morita, Fukagawa and Arai (1989) obtained a theoretical curve that locates in the position lower than the solid curve shown in Fig. 9 by an amount 0.1-0.2 of the correction factor. They assumed fully undrained conditions at the sites during loading while most of the actual failures take place under partially drained conditions. Asaoka (personal correspondence) mentioned that the strength gain due to partial drainage during embankment loading is generally equivalent to 0.1 in terms of correction factor in Fig. 9. This statement seems to have derived from his in-depth research on the reliability of stability analysis in geotechnical engineering. The author feels that the difference between the back-calculated correction factor typically shown by the solid curve in Fig. 9 and the curve theoretically derived based on the fully undrained conditions may arise from the on-going consolidation due to partial drainage during construction works.

6 SETTLEMENT OF PRELOAD

At site S, an embankment was placed as a preload to strengthen the soft clay foundation of the stockyard facilities. The height of the fill was different from place to place as shown in Fig. 10 (a). Since the loading rate at each part of the fill was also different as seen in Fig. 10 (b), the deformation of the soft clay foundation was not uniform. Such 2-dimensional consolidation process is analysed by employing a soil-water coupling finite element programme called DACSAR (Deformation Analysis Considering Stress Anisotropy and Reorientation) developed by Iizuka and Ohta (1987). An elasto-viscoplastic constitutive model proposed by Sekiguchi and Ohta (1977) was incorporated into the DACSAR programme.

The material parameters are those obtained in the previous sections based on the field vane tests and unconfined compression tests. In addition to those, the results of oedometer tests are also used. In specifying all the material parameters needed in the analysis, some empirical procedures recommended by Iizuka and Ohta (1987) are complementally applied. The vertical distributions of material parameters are summarized in Fig. 11. The permeability of the soft clay immediately beneath the embankment was improved by installing sand drains, the effect of which is evaluated in the computation by assuming the permeability 20 times higher than the adjacent natural clay.

The settlement records of the different part of the embankment are compared in Fig. 12 with those computed by the DACSAR programme. The computed results seem to be in accordance with the settlements monitored at the different part of the fill. This implies that the computer simulation of 2-dimensional consolidation process at site S is successfully performed by using the estimated material parameters, mostly specified based on the field vane tests, unconfined compression tests and oedometer tests. In judging whether the computer simulation was successful or not, the lateral movement of the subsoil as well as the change in pore water pressure

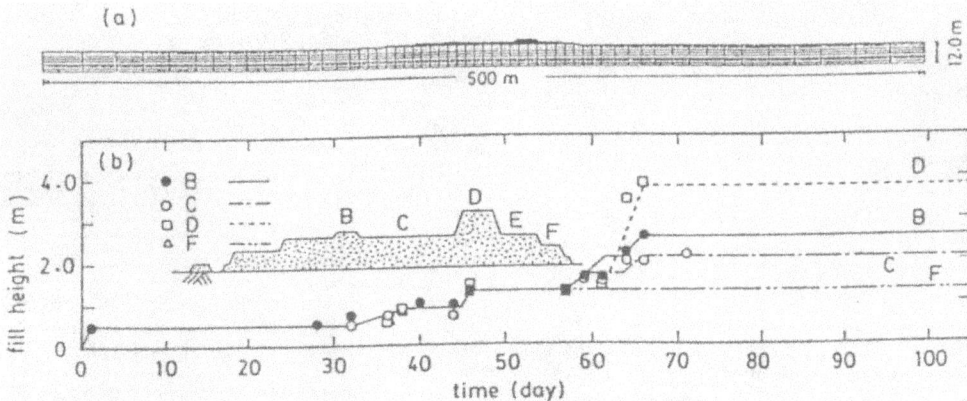

Figure 10. Finite element modeling of the preload (a) Mesh formation and (b) Loading rate (after Ohta, Nishihara, Iizuka and Morita, 1992)

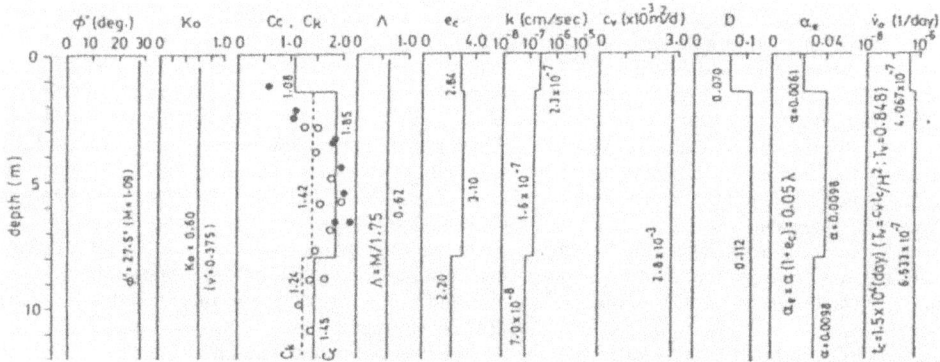

Figure 11. Vertical distributions of material parameters used in the computations (after Ohta, Nishihara, Iizuka and Morita, 1992)

are needed. However, regretfully, those data were not available and could not be compared with the computed results. Since no attempt was made to adjust the material parameters (except the overall permeability of the sand drained area) aiming at better agreement between computed and monitored settlement, the agreement seen in Fig. 12 may justify the use of field vane in estimating the material parameters needed in the constitutive modeling.

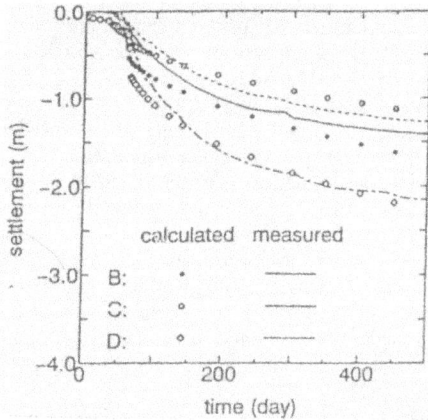

Figure 12. Computed and monitored settlement (after Ohta, Nishihara, Iizuka and Morita, 1992)

7 CONCLUSIONS

The use of field vane data in specifying the material parameters needed in the constitutive model proposed by Sekuguchi and Ohta (1977) was introduced in this paper. In the interpretation of the test data, some theoretical reasoning derived from the constitutive model was used. Material parameters thus specified were found to be in good agreement with the failure of a slope and with the settlement of an embankment. This leads to the conclusion that the use of field vane data in specifying the material pa-

rameters is promising from the viewpoint of engineering practice. It may also be concluded that the use of a constitutive model has reached the level of practical applications. In this paper, the effect of partial drainage during the field vane tests was not considered since the tests are usually carried out very quickly. The effect of insertion of the vane blade into the soft clay ground was also ignored in this paper. Whether this is appropriate or not is still open to question.

REFERENCES

Bjerrum, L. (1972): Embankments on the soft ground, Performance of Earth and Earth-Supported Structures, ASCE Specialty Conference, Vol.2, pp.1-54

Fukagawa, R., Fahey, M. and Ohta, H. (1990): Effect of partial drainage on pressuremeter test in clay, Soils and Foundations, Vol. 30, No. 4, pp.134-146

Hata, S. and Ohta, H. (1969): On the effective stress paths of normally consolidated clays under undrained shear, Proc. JSCE, No. 162, pp.21-29 (in Japanese)

Iizuka, A. and Ohta, H. (1987): A determination procedure of input parameters in elasto-viscoplastic finite element analysis, Soils and Foundations, Vol. 27, No. 3, pp.71-87

Karube, D. (1975): Non-standardized triaxial testing methods and their problems, Proc. 20th Symposium on Soil Engineering, JSSMFE, pp.45-60 (in Japanese)

Karube, D. and Kurihara, N. (1966): Dilatancy and shear strength of saturated remoulded clay, Trans. JSCE, No. 135

Ladd, C. C. (1973): Discussion, Main Session 4, Proc. 8th ICSMFE, Moscow, Vol. 4.2, pp.108-115

Ladd, C. C., Foott, R., Ishihara, K., Schlosser, F. and Poulos, H. G. (1977): Stress-deformation and strength characteristics, SOA Report, Proc. 9th ICSMFE, Tokyo, Vol. 2, pp.421-494

Ladd, C. C. and Lambe, T. W. (1964): The strength of "undisturbed" clay determined from undrained tests, ASTM SPT 361, pp.342-371

Nakase, A., Katsuno, M. and Kobayashi, M. (1971): Unconfined compression strength of sandy clay, Report of Port and Harbour Research Institute, Vol. 11, No. 4, pp.83-102 (in Japanese)

Ohta, H. (1971): Analysis of deformation of soils based on the theory of plasticity and its application to settlement of embankments, D. Eng. Thesis, Kyoto University

Ohta, H., Iizuka, A., Nishihara, A., Fukagawa, R. and Morita, Y. (1991): Design strength Su derived from pressuremeter tests, Proc. 7th Int. Conf. Computer Methods and Advances in Geomechanics, Cairns, Vol. 1, pp.273-278

Ohta, H., Nishihara, A. and Morita, Y. (1985): Undrained stability of Ko-consolidated clays, Proc. 11th ICSMFE, San Francisco, Vol. 2, pp.613-616

Ohta, H., Nishihara, A., Iizuka, A., Morita, Y., Fukagawa, R. and Arai, K. (1989): Unconfined compression strength of soft aged clays, Proc. 12th ICSMFE, Rio de Janeiro, Vol. 1, pp.71-74

Ohta, H. and Sekiguchi, H. (1979): Constitutive equations considering anisotropy and stress reorientation in clay, Proc. 3rd Int. Conf. Numerical Methods in Geomechanics, Aachen, pp.475-484

Roscoe, K. H., Schofield, A. N. and Thurairajah, A. (1963): Yielding of clays in state wetter than critical, Geotechnique, Vol. 13, No. 3, pp. 211-240.

Sekiguchi, H. and Ohta, H. (1977): Induced anisotropy and time dependency in clays, Proc. 9th ICSMFE, Speciality Session 9, Tokyo, pp.229-238

Shibata, T. (1963): On the volume change of normally consolidated clays, Disaster Prevention Research Institute Annuals, Kyoto University, No. 6, pp.128-134 (in Japanese)

Shibata, T. and Karube, D. (1965): Influence of the variation of the intermediate principal stress on the mechanical properties of normally consolidated clays, Proc. 6th ICSMFE, Montreal, Vol. 1, pp.359-363

Skempton, A. W. and Sowa, V. A. (1963): The behaviour of saturated clays during sampling and testing, Geotechnique, Vol. 13, No. 4, pp.269-290

Torstensson, B. A., (1977): Time-dependent effects in the field vane test, Proc. Int. Symp. On Soft Clay, Bangkok, pp.387-397

407

Soft Ground Engineering in Coastal Areas, Tsuchida et al. (eds)
© 2003 Swets & Zeitlinger, Lisse, ISBN 90 5809 613 0

New improving method for the bearing capacities of steel pipe piles installed by water jet vibratory technique

A. Uezono & K. Takahashi
Port and Airport Research Institute, Japan

H. Yamashita
Nippon Steel Corporation, Japan

S. Nishimura
Fugro Geoscience Co., Ltd., Japan

ABSTRACT: At the port of Fushiki Toyama a quay was constructed over 14m of water. This quay was built on steel pipe piles installed using water jet vibratory technique (JV). However static and dynamic loading tests performed on the JV piles showed deficiencies in the bearing capacity. To increase the bearing capacity, a new technique for improving soil using cement milk in place of water (CJV method) was developed. The results of the static and dynamic loading tests on the CJV piles showed that the designed bearing capacity was achieved.

1 INTRODUCTION

To handle the increasing amount of container cargos and large ships at the port of Fushiki Toyama, a new multi-purpose international terminal is being built with a quay over 14m of water in the Shinminato district.

As we can see in Figure 1, this quay is a pier structure in which caissons are used to anchor it in place. There are 47m-long, φ1200mm steel pipe piles arranged in four rows having a supporting layer 45m below the water surface.

The original construction method for the steel pipe piles was the water jet-vibratory hammer method (hereafter referred to as the JV method) that was used in consideration of local ground conditions and to reduce noise and vibration in the vicinity. However, there is still much that is unknown about the bearing capacity of JV piles. Therefore, in the construction of the steel pipes in the present project, static and dynamic loading tests were conducted on piles installed with the JV method to gain an accurate understanding of the bearing capacity. Based on the results of the JV loading tests, we decided to make some improvements to increase the bearing capacity.

Various improvement methods were considered based on ease of construction and economy. As a result, it was decided to inject cement milk instead of spraying jet water in the JV method (hereafter called the CJV method) to improve the surrounding ground. An experimental improvement was then conducted.

The improvements were subjected to static and dynamic loading tests. As a result, we found that the CJV method was able to provide the designed bearing capacity.

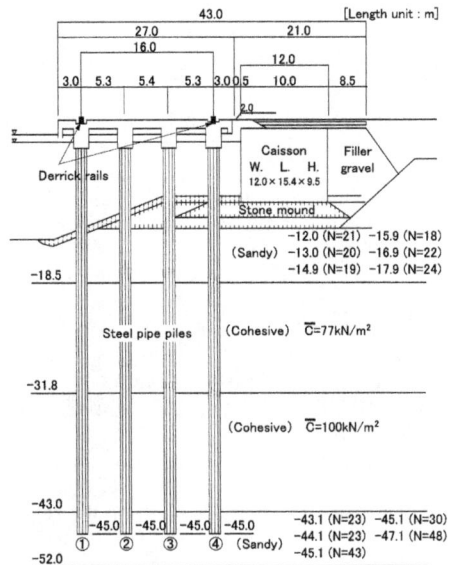

Figure 1. Structure of the pier

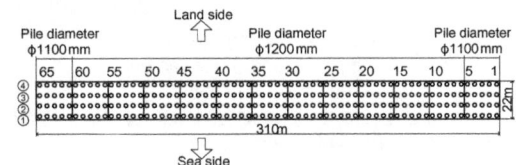

Figure 2. Disposition of piles

2 DESIGN AND CONSTRUCTION OF THE PIER STEEL PIPE PILES USING THE JV METHOD

2.1 *Design of the pier steel pipe piles*

Table 1 and Table 2 show, respectively, the specifications for the pier steel pipe piles for the quay over 14m of water and the bearing capacity required in the design. The design bearing capacities are shown for piles in the first and fourth rows, which would support large loads.

Table 1. Specifications of the pier piles

Pile type	Shape of toe	Diameter (mm)	Wall thickness (mm)		Length (m)	Nominal tensile strength (N/mm²)
			Upper part	Lower part		
Steel Pipe pile	Closs ribs or open	1200 or 1100	14□`17	12	47.0	490

Table 2. Bearing capacities required in the design

NO.	Dead load (kN)	× safety factor: 2.5 (kN)	Seismic load (kN)	× safety factor: 2.0 (kN)	Maximum required capacity (kN)
Row ①	3970	9930	4480	8960	9930
Row ④	3600	9000	4640	9280	9280

The bearing capacity of the steel pipe piles is ensured by the shaft friction and the pile toe capacity. The design bearing capacity of driven piles in port structures is calculated as follows (Technical Standards and commentaries for Port and Harbor Facilities in Japan 1995):

Shaft friction R_f (kN) is expressed as follows.

$$R_f = \Sigma C_a \cdot A_c + \Sigma 2N \cdot A_s \qquad (1)$$

Here,

C_a: adhesive strength of the cohesive soil
($C_a = 3.4 \cdot Z + 0.7$ $Z=0$ at -2.6m, 100kN/m² or less)
N: N-value of sandy soil
A_c: peripheral area of the piles abutting the cohesive soil
A_s: peripheral area of the piles abutting the sandy soil

The pile toe capacity is expressed as follows.

$$R_s = 300\alpha \cdot N \cdot A_p \qquad (2)$$

Here,

N: N-value of the soil at the pile toe
A_p: area of the pile toe with closed end
α: plugging ratio of the pile toe

Ultimate bearing capacity R(kN) can be considered with the following formula:

$$R = R_f + R_s \qquad (3)$$

Because the JV method disturbs the surrounding ground during pile installation, it reduces the shaft friction of the JV piles used for calculating the design bearing capacity based on the experimental values, as shown in Table 3(Vibratory Hammer Research Association, 1993).

The ground conditions used in the design were those shown in Figure 1. Table 4 shows the calculated safety ratios for ultimate and design bearing capacity of the fourth row of piles constructed with the JV method under these conditions.

In the present project, the first and fourth rows of piles supported large design loads because the derrick will be built on these piles. Therefore, to ensure the plugging ratio, cross ribs were attached to the pile toes.

Table 3. Shaft resistance for JV pile in design

Soil type	Shaft friction stress in design		
	Case of limited pressure and quantity of water (kN/m²)	Case of unlimited pressure and quantity of water (kN/m²)	Case of grouting after installing (kN/m²)
Sandy soil [Reduction ratio to driven pile]	$2N \leqq$ (100) [1.0]	$1N (\leqq 50)$ [0.5]	$2N (\leqq 100)$ [1.0]
Silty soil [Reduction ratio to driven pile]	$0.5C$ or $5N$ ($\leqq 150$) [0.5]	$0.2C$ or $2N$ ($\leqq 100$) [0.2]	C or $10N$ ($\leqq 150$) [1.0]

Table 4. Design values of bearing capacity (4th row)

Shaft resistance	Sand layer (L=8.5m) Ru①	736 kN [Reduction ratio: 0.5]	
	Silt layer(L=24.5m) Ru②	1616 kN [Reduction ratio: 0.2]	
Toe capacity Ru③		12204 kN [Plugging ratio: 1.0, N=36]	
Total capacity Ru①+Ru②+Ru③		14556 kN	
Safety ratio	Dead load (2.5)	4.04	
	Seismic load (2.0)	3.13	

2.2 *JV construction method*

The pile installation method using the JV method is shown in Figure 3. The pile penetrated the sandy ground with just a 15MPa water jet from the sea floor to create a foundation for the piles, which were set in place an average of 22m below the water surface, where they were self-standing. From the cohesive soil below that, a vibratory hammer was used in conjunction with the water jet, whose jet pressure was reduced to 2Mpa, to drive the piles to a depth 39 m below the water surface, which is 5D (D: pile diameter of 1200 mm) above the design pile toe depth. For the remaining 5D spaces, only the vibratory hammer was used for the pile installation.

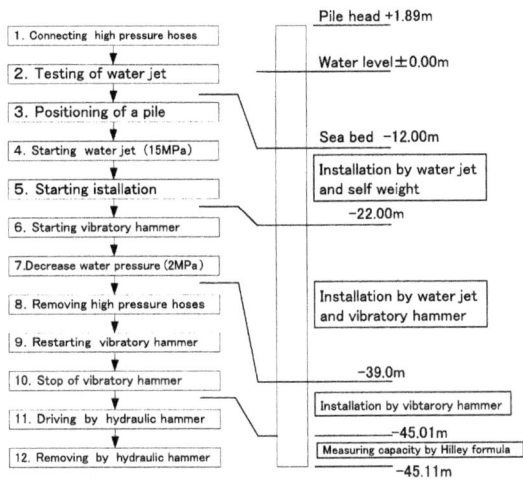

Figure 3. Construction order of JV method

3 LOADING TESTS CONFIRM THE BEARING CAPACITY OF THE JV METHOD

3.1 *Summary of loading test*

The static loading test was conducted on Pile 3-59 (third row) 40 days after 3-59 was installed, and the adjoining two piles (3-58 and 3-60) were used as reaction piles, as shown in Figure 4. The specifications of the test piles are shown in Table 5.

As shown in Figure 5, four hydraulic jacks were used for the loading. Load cells were mounted on the reaction piles to control the load and to prevent any accidents arising from the eccentricity of the load. Displacement due to pile settlement was measured at the two shaft sections (end of pile and sea floor), while the strain was measured at the seven shaft sections shown in Figure 6. The experiment was performed so that the amount of settlement displacement of the pile toe would be equal to about 10% of the pile diameter.

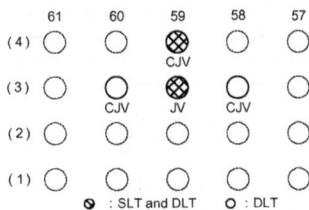

Figure 4. Location of the test piles

In order to compare the static and dynamic loading tests and use the results in the construction administration, the piles used on the static loading test were also used in the dynamic loading test, which was conducted two weeks later.

Table 5. Specifications of the test piles

NO.	Shape of toe	Diameter (mm)	Wall thickness (mm)	Length (m)	Total length (m)	Nominal tensile strength (N/mm²)
3-59	Cross ribs	1200	15 (Upper) 12 (Lower)	22.0 25.0	47.0	490

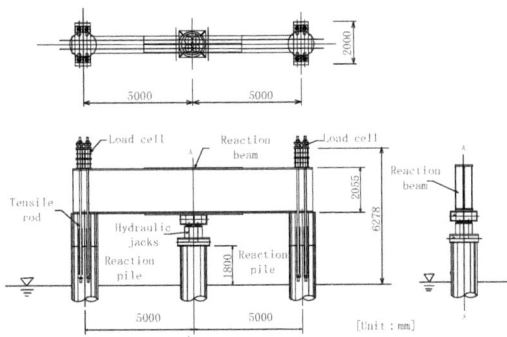

Figure 5. Static loading test equipment

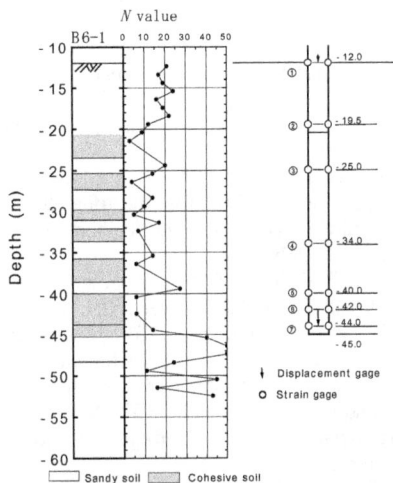

Figure 6. Soil type and pile location with measuring points

In the dynamic loading test, a strain gauge and an acceleration gauge were attached near the pile head. The behavior of the pile while it was being driven into place by impacting the pile head with a hydraulic hammer was analyzed based on one-dimensional stress-wave analysis. The results were used to evaluate the bearing capacity of the pile. A schematic drawing of the dynamic loading test is shown in Figure 7.

3.2 *Result of the static loading test*

Although the maximum load for the static loading test was set at 6500kN, displacement increased while the load was still low, so the experiment was terminated when the load was 4350kN.

411

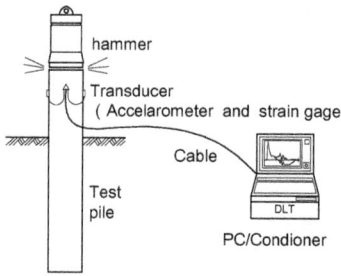

Figure 7. Dynamic loading test equipment

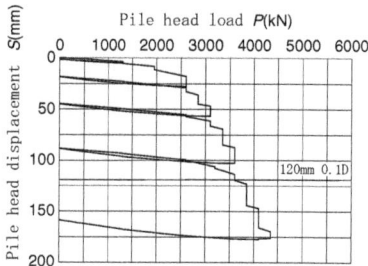

Figure 8. Pile head load-displacement curve (JV)

The load-settlement curve for this experiment is shown in Figure 8.

a) First and second limit loads
The log$P$-log$S$ relationship for the pile head is shown in Figure 9. Here we can see that the first limit load was 2660kN.

The second limit load of the experiment was derived as follows: Following the vertical loading test method and associated guidelines for piles meeting the standards of the Japanese Geotechnical Society (1993), it was determined to be 4350kN which is the maximum load on the pile head, or 3700kN which is the bearing capacity when settlement reached 10% of the pile diameter.

Figure 10 shows the values of axial force calculated from the strain gauge measurements. Here we can see that the shaft friction for the second limit load (4350kN) was 1480kN.

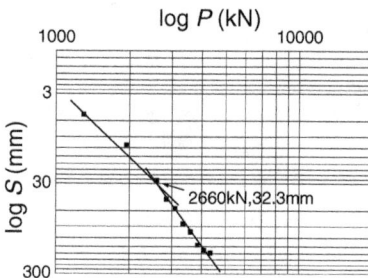

Figure 9. log$P$- log$S$ relationship (JV)

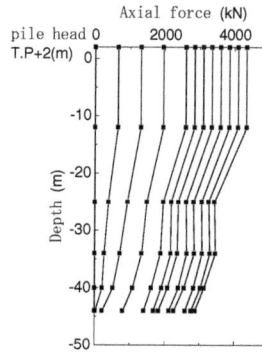

Figure 10. Distribution of axial forces (JV)

b) Shaft friction
Table 6 and Figure 11 show the calculated values of maximum shaft friction for each layer. Table 6 also lists the N-value of the ground. In addition, Figure 12 shows a comparison of the shaft friction obtained from the static loading test and the design shaft friction with and without reduction for JV piles, which was used in the calculation of the design bearing capacity that was reported in 2.1.

The results of the static loading tests shown in Figure 12 tend to agree roughly with the design shaft friction values in 2.1.2 that take reduction into account.

c) Pile toe capacity
Figure 13 shows the displacement at the pile toe and the toe axial force as indicated by the measurements of the gauges at the toe. Here we can see that the toe capacity was 2546 kN for the transmitted second

Table 6. Maximum shaft friction stress

Depth (m)	Soil type	Stress τ (kN/m^2)	N value
-12.0 ～ -19.5	Fine sand	17	18.8
-19.5 ～ -25.0	Silty clay	17	10.7
-25.0 ～ -34.0	Silt	6	10.1
-34.0 ～ -40.0	Silty fine sand	15	15.7
-40.0 ～ -42.0	Clayey silt	23	6.0
-42.0 ～ -44.0	Clayey silt	23	14.0
-44.0 ～ -45.0	Silty fine sand	23	14.0

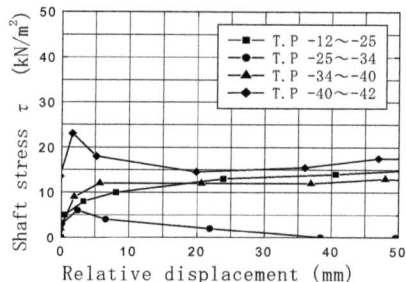

Figure 11. Shaft stress - relative displacement curves

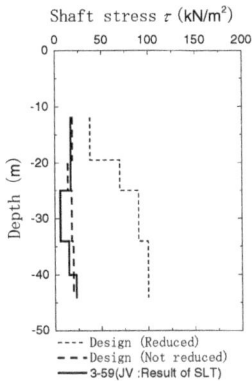

Figure 12. Comparison between design and measured values of shaft friction stress

Table 7. Result of Dynamic loading test 3-59(JV)

Soil type	Thickness of layer (m)	Shaft stress τ (kN/m^2)
Fine sand	8.7	17
Silty clay	17.9	15
Silt		
Silty fine sand	1.4	15
Clayey silt	3.8	20
Clayey silt	1.5	20
Shaft resistance (kN)		2064
Toe resistance (kN)		875

limit load (when settlement reached 10% of the pile diameter), and 2870kN for the transmitted second limit load (when the maximum load of 4350 kN was applied).

While it was assumed that the pile toe was completely plugged (the plugging ratio was 1.0) when cross ribs were attached, the results of the loading tests showed a bearing capacity of only 1/4 of the expected value. A possible reason for this discrepancy was that the cross rib effect was not manifested well due to the alternation of the toe foundation strata, an excessively small N-value of the supporting layer for a large diameter pile, or other factors.

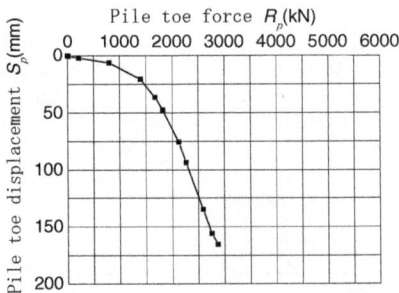

Figure 13. Pile toe force Rp - displacement Sp curves

3.3 Result of the dynamic loading test

The dynamic loading test showed an ultimate bearing capacity of 2939kN (shaft friction of 2060kN + toe capacity of 880kN).

4 PROPOSED METOD FOR IMPROVING BEARING CAPACITY

4.1 Investigation of improvement method

Three main methods were considered for improving the bearing capacity:

A: Increasing shaft friction by injecting chemicals, etc., into the ground adjoining the piles
B: Increasing toe friction by hardening the ground around the pile toes
C: Increasing shaft friction by injecting cement instead of jet water of the JV method (CJV method).

Given the considerations at the construction site were mainly for ease of construction and economy, we elected to go with the CJV method (C).

The procedure for the CJV method is shown in Figure 14. However, it should be noted that the present experiment marks the first time this procedure has been used.

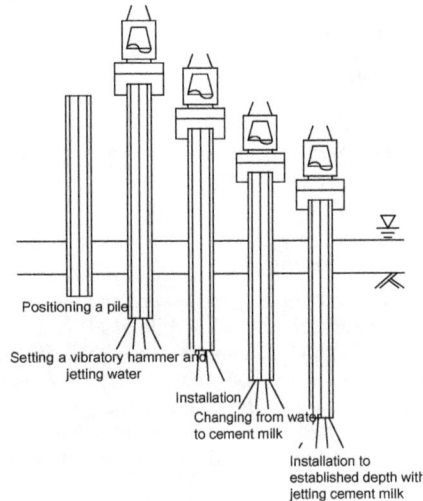

Figure 14. Procedure of the CJV method

4.2 Concept of improvement using the CJV method

In the JV method, the ground adjoining the piles is disturbed by jet water and a vibratory hammer. The CJV method is based on the concept of injecting cement milk into the ground adjoining the piles, instead of jet water, to restore the adjoining foundation of the piles to its original strength and create a cement-hardened foundation that should allow for the installation of piles with larger diameters.

413

Since this was an untested method, it was necessary to determine the amount of cement to be injected. The target cement injection volume was set at the 300kg per cubic meter of soil used in the similar steel pipe soil cement pile installation method ("GANTETSU",1995). Improvements were to be made in an area encompassed by pile radius plus 30cm, including inside the pipe (see Fig. 15).

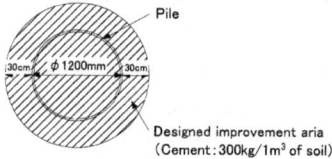

Figure 15. Designed improved aria for CJV method

4.3 Construction method for the CJV method

The CJV method involved spraying cement milk into a cohesive soil layer 25m or deeper below the water surface. As with the main construction, the upper sandy ground was penetrated down to -22m with just a 15MPa water jet, where they were self-standing. Between there and -25m, a vibratory hammer was used in conjunction with the water jet, whose jet pressure was reduced by 2MPa. Below that (down to the bottom depth of -45m), a vibratory hammer was used in conjunction with cement milk, which was used instead of jet water and sprayed at a pressure of 4MPa.

5 RESULTS OF CORROBORATIVE TESTS ON BEARING CAPACITY DERIVED FROM THE CJV METHOD

5.1 Summary of loading tests

Static and dynamic loading tests were conducted on Pile 4-59 (same specifications as Pile 3-59) installed with the CJV method. A comparison with 3-59 (installed with the JV method near 4-59) showed an improvement effect with the CJV method.

Additionally, a dynamic loading test was conducted on Piles 3-58 and 3-60, which were installed with the CJV method.

The static loading test was conducted 35 days after pile installation, while the dynamic loading test was conducted 50 days after installation.

5.2 Results of loading tests

Figure 16 shows the relation between load and displacement, as derived from the static loading test. The maximum settlement at the time of maximum load (13500kN) was 190.6mm.

Figure 17 shows the logP-logS relationship for the pile head. Here we can see that the first limit load was 10300kN.

The second limit load was 12,000kN (toe capacity of 2,160kN + shaft friction of 9,840kN) when displacement of the pile toe reached 10% of the pile diameter, and 13,500kN (toe capacity of 2,450kN + shaft friction of 10,960kN) when the maximum load was applied.

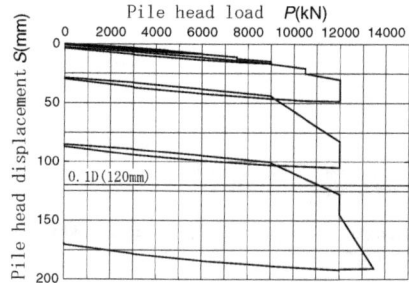

Figure 16. Pile head load – displacement curve (CJV)

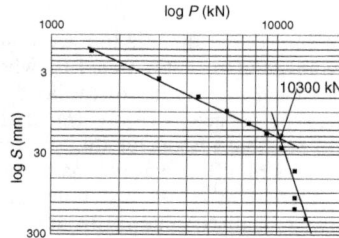

Figure 17. logP - logS relationship (CJV)

A comparison with the results of the (unimproved) JV method is shown in Table 8. Compared to the JV method, the bearing capacity of the piles installed with the CJV method for the second limit load was about 8,000kN higher.

Table 8. Comparison of capacities between JV and CJV

Values		CJV NO.4-59	JV NO.3-59
Maximum load (kN)		13500	4350
Maximum displacement (mm)		190.6	176.2
1st limit load (kN)	logP - logS	10300	2660
2nd limit load (kN)	0.1D	12000	3700
	Maximum	13500	4350

It should also be noted that the total bearing capacity of 12,368kN of Pile 4-59 (10,400kN shaft friction + 1,968kN toe capacity) derived from dynamic loading test adequately met the design bearing capacity as shown in Table 9.

5.3 Analysis of the results of bearing capacity improvement

Figure 18 shows a comparison of the shaft friction stress and toe bearing stress for each layer for the JV

Table 9. Result of DLT (CJV)

Soil type	Thickness of layer (m)	Shaft stress τ (kN/m²)		
		4-59	3-58	3-60
Fine sand	8.65	102	102	82
Silty soil	4.35	102	71	122
Silty soil	13.55	66	51	31
Fine sand	1.4	87	31	36
Silty clay	3.8	87	71	61
Silt	1.45	71	92	82
Shaft resistance (kN)		10400	8754	7634
Toe resistance (kN)		1968	2416	2416
Total resistance (kN)		12368	11170	10050

and CJV methods obtained from strain gauge measurements. The figure indicates that even in the sandy ground in the top section, where cement milk was not injected, there was improvement of about 6 to 8 fold. In the cohesive layer beneath that (-25 to 34m) the improvement was about 13 fold, and 3 to 6.5 fold in the layer from –34 to -45m.

However, there was a very noticeable lack of improvement in the toe friction. These results prove that the injection of cement milk helped to greatly improve shaft friction and achieve the design bearing capacity. It was also found that the injection of 300kg of cement per cubic meter of soil was enough to provide the desired bearing capacity.

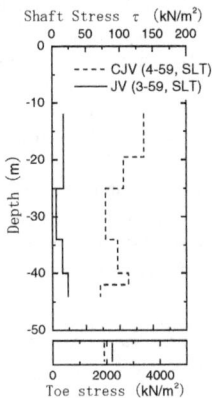

Figure 18. Comparison between JV and CJV method

6 A COMPARISON OF THE RESULTS OF THE STATIC AND DYNAMIC LOADING TESTS

6.1 *JV piles*

Table 10 shows a comparison between the results of the static and dynamic loading tests. The results for the static loading tests were those that were recorded at the first limit load (2660kN) and the second limit

load when the displacement of the toe was 10% (120mm). Figure 19, on the other hand, shows a comparison of the load-settlement curve obtained from the static loading test results and the load-settlement curve that was derived from signal matching analysis of the dynamic loading test results.

These results indicate that the confirmed bearing capacity during the dynamic loading test was lower than the bearing capacity at the second limit load in the static loading test, or rather was roughly equivalent to the bearing capacity at the first limit load. The reason is that the amount of displacement of the pile head that occurred during the dynamic loading test was 21 mm (penetration of 10mm, rebound of 11mm) which was much smaller than the 120mm in the static loading test, resulting in a failure to manifest bearing capacity.

In addition, the load-settlement curves in Figure 15 indicate that the stiffness of the pile head displacement was greater in the dynamic loading test (DLT) than that in static loading test (SLT). This was because the DLT was conducted after the SLT, so the DLT was the beneficiary of the loading hysteresis. In addition, the penetration rate was faster during the DLT, and rebound was greater due to dependence on the strain rate of the ground.

Table 10 shows that there were differences in the bearing capacity ratio of the pile toe and shaft in the SLT and DLT. Among the likely reasons were the skew in the DLT analysis, and the errors in the axial force measurements induced by the residual stress that occurred during the pile installation by the vibratory hammer.

Table 10. Comparison of test results between DLT and SLT(JV)

	SLT 3-59		DLT 3-59
	1st limit load (Yield)	2nd limit load (Ultimate)	Estimated static resistance
Shaft resistance (kN)	1227	1228	2064
Toe resistance (kN)	1443	2546	875
Total resistance (kN)	2660	3774	2939

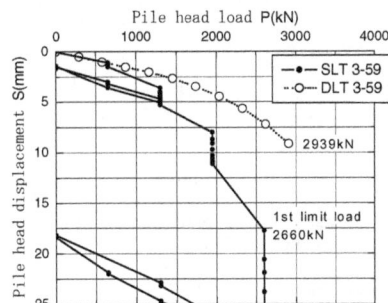

Figure 19. Comparison of Pile head force – displacement curves between SLT and DLT(JV)

6.2 CJV piles

Table 11 and Figure 20 show comparisons between the SLT and DLT results for the CJV method. A dynamic loading test was conducted on three CJV piles: 4-59 (which was used in the static loading test), and 3-58 and 3-60, which were installed in the same way as 4-59.

As with the JV method, the DLT results were roughly equivalent to the results at the first limit load in the static loading test. Furthermore, the load-settlement curve shows that the stiffness of the pile head displacement was greater in the dynamic loading test. However, the difference in stiffness between the DLT and SLT results for the CJV method was less than that for the JV method. This was probably because of differences in the shaft friction resulting from improvements made to the shaft foundation in the CJV method. We also found that the stiffness of Pile 4-59, which was subjected to the static loading tests, was greater than those of Piles 3-58 and 3-60, which was not subjected to the static loading tests, due to the loading hysteresis.

Table 11. Comparison of test results between DLT and SLT(CJV)

	SLT 4-59		DLT 4-59	DLT 3-58	DLT 3-60
	1st limit load	2nd limit load	Estimated static resistance (Signal matching analysis)		
Shaft resistance (kN)	9809	9837	10400	8754	7634
Toe resistance (kN)	491	2163	1968	2416	2416
Total resistance (kN)	10300	12000	12368	11170	10050

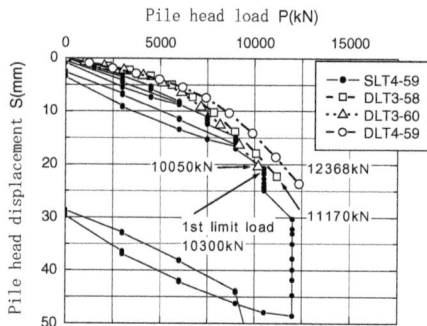

Figure 20. Comparison of Pile head force – displacement curves between SLT and DLT(CJV)

6.3 Comparison between the SLT and DLT results

Comparisons between the static loading and dynamic loading test results of the JV and CJV piles led us to make the following conclusions:

Signal matching analysis results of the dynamic loading test were roughly equivalent to the bearing capacity at the first limit load in the static loading test.

The stiffness of the pile head displacement obtained from the dynamic load-settlement curve estimated from the DLT results were much higher than that obtained from the static loading tests.

It is our hope that more data will be collected to help to resolve issues about the correlation between the static loading and dynamic loading tests, such as differences in the bearing capacity ratio of pile sections.

7 CONCLUSIONS

Static and dynamic loading tests were conducted on pier steel pipe piles that were installed with the JV method, and piles that were given added bearing capacity by the CJV method. The main confirmed results were as follows:

The shaft friction of the JV piles tended to roughly coincide with the design value proposed here which takes reduction for the JV piles into account.

In this experiment which marked the first time the CJV method has been used, the loading tests confirmed that the design bearing capacity was achieved.

Improvements made to the shaft friction stress (reinforcement) amounted to a 6 to 8 fold increase in strength in the sandy ground, and a 3 to 13 fold increase in the strength of the cohesive ground.

The dynamic loading test was found to be a suitable means of confirming the bearing capacity near the first limit load in the static loading test.

While the JV method is the preferred method to use in environments where noise and vibration must be suppressed, there is still much that is unknown about the bearing capacity. However, loading tests conducted on the JV method in the present study have provided much important information about the bearing capacity.

In addition, the CJV method proposed here produced desired improvement results, revealing that it is a simple and economical on-site construction method.

In the future, there will undoubtedly be more opportunities to make construction methods more environmentally-friendly, and examples of applications of such methods will likely increase. If that happens, we hope that corresponding advances will be made in analyzing improvement mechanisms and establishing design methods.

ACKNOWLEDGMENTS

We would like to express our gratitude to Professor Kusakabe of Tokyo Institute of Technology, Professor Zen of Kyushu University, and Mr. Kikuchi, Chief researcher of Port and Airport Research Institute, for their assistance with this study.

REFERENCES

The Japan Port and Harbour Association: Technical Standards and Associated Guidelines for Port Facilities. 1995

Vibratory Hammer Research Association: Construction techniques using the vibratory hammer. 1993

The Japan Institute of Construction Engineering: General Civil Engineering Methods, Technical Corroboration and Report, The simultaneous installation of buried steel alloy pipe piles"GANTETSU". 1995

Japanese Geotechnical Society: Standards of the Japanese Geotechnical Society for vertical loading tests on piles and associated guidelines. 1993.

Soft Ground Engineering in Coastal Areas, Tsuchida et al. (eds)
© 2003 Swets & Zeitlinger, Lisse, ISBN 90 5809 613 0

Study on design method of suction foundation using model tests

H. Yamazaki, Y. Morikawa & F. Koike
Port and Airport Research Institute, Yokosuka, Japan

ABSTRACT: A new type of foundations called Suction Foundation has been developed for offshore structures. The foundations are made by penetration of cylindrical caissons into seabed with suction force. The penetration gives the foundations some advantages such as increase in both the lateral resistance and bearing capacity. In addition, when a soft soil layer exists over a stiff soil layer, the suction foundations can be constructed without any soil improvements because the caissons have penetrated into the stiff layer underlying the soft layer, although conventional foundations such as mound type foundations usually necessitate soil improvements. This paper describes the design method of the suction foundations for quaywalls, and also reports the shaking table test results on quaywalls.

1 INTRODUCTION

The suction foundations are made by penetration of bottom-open caissons into seabed by use of suction force as shown in Figure 1 (Zen et al 1998). The penetration enhances the horizontal resistance of the foundations because passive earth pressures act the foundation caissons. In case that soft soil layers exists near the surface, the foundations can obtain sufficient bearing capacity, without any soil improvements, by the penetration into the stiff soil layer underneath the soft soil layers. The foundations are anticipated to use for offshore structures subjected to large horizontal forces such as breakwaters and quaywalls that are constructed on soft soil layers overlying stiff layers.

In this paper, a design method for suction foundations are described in respect of use for quaywalls, and shaking table tests results on quaywalls are examined to validate the design method.

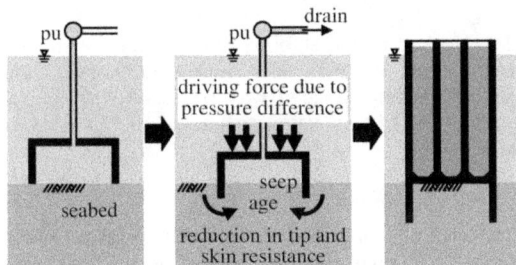

Figure 1. Installation of suction foundation

2 DESIGN METHOD FOR QUAYWALLS

Figure 2 shows quaywalls using suction foundations. There are two types of suction foundations, a united type and a separate type, in regard to the construction of their superstructures. In the united type suction foundations, the superstructures are constructed together with the foundation caissons. However, in the separate type ones, the superstructures are constructed separately from the foundation caissons. The choice of the types depends on work condition, soil condition and so on.

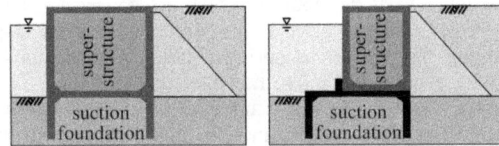

Figure 2(a). United type   Figure 2(b). Separated type

Figure 3 shows forces acting on quaywalls during an earthquake, where P_{wd} = dynamic water pressure, W'_1, H_{w1}, P_{w1}, P_{ah1} and P_{av1} = submerged weight, inertial force, residual water pressure, and horizontal and vertical component of the active earth pressure on the superstructure. W'_2, H_{w2}, P_{w2}, P_{ah2} and P_{av2} = submerged weight, inertial force, residual water pressure, and horizontal and vertical component of the active earth pressure on the foundation caisson. P_{ph} and P_{pv} = horizontal and vertical component of the subgrade reaction on the front wall of the foundation caisson. Q, q_1 and q_2 = shearing resistance and vertical subgrade reactions at the bottom of the foundation caisson.

The above external forces are estimated with Westagaard's formula, earth pressure formula, etc. using a design seismic coefficient. In the estimate of the inertial force acting on the foundation caisson below the seabed surface level, the seismic coefficient is linearly reduced to zero from the seabed surface to 10m deep. In the estimate of the horizontal active earth pressure below the seabed surface level in the land side, the earth pressure coefficient is assumed 0.5 and the overburden pressure is assumed the constant equal to that at the seabed surface level. Therefore, the active earth pressure distribution in the vertical direction is constant below the seabed surface level. The vertical component of the active earth pressure is obtained by multiplying the horizontal component by the frictional coefficient of the foundation caisson wall.

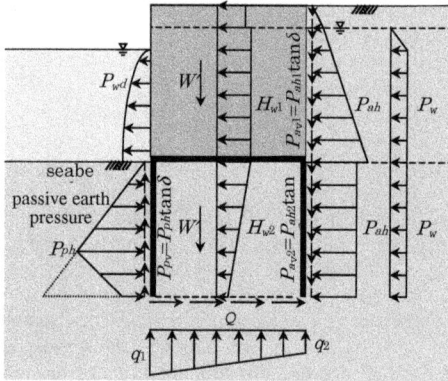

Figure 3. Forces acting on quay wall

The subgrade reactions P_{ph}, q_1, q_2 and the shear resistance Q at the bottom, which are caused by the aforementioned external forces, are calculated by assuming the subsoil as spring. The horizontal subgrade reaction P_{ph} must, however, be less than the passive earth pressure under an ultimate stress condition, which is obtained by an earth pressure formula. The vertical component of the subgrade reaction P_{pv} is calculated by multiplying the subgrade reaction P_{ph} by the frictional coefficient.

The stability of suction foundations against the aforementioned forces are analyzed in respect to the bottom of the foundation caisson as shown in Figure 4. The safety factors for the sliding, overturning and the allowable bearing capacity are given by the following equations.

$$V = F_v + W_1' + W_2' \tag{1}$$

$$R_h = \mu_b V \tag{2}$$

$$F = \frac{M_r}{M_d} : \text{overturning,} \tag{3}$$

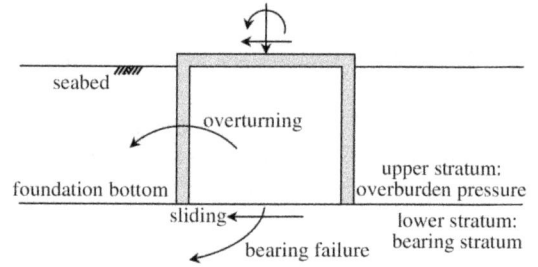

Figure 4. Failure modes

$$F = \frac{R_h}{Q} : \text{sliding} \tag{4}$$

$$q_a = \frac{\beta \gamma_1 B N_\gamma + \gamma_2 d N_q}{F} + \gamma_2 d : \text{sandy soil} \tag{5}$$

$$q_a = \frac{N_c c_o}{F} + \gamma_2 d : \text{clayey soil} \tag{6}$$

where Fv = total vertical load acting on the foundation caissons, μ_b = frictional resistance at the bottom of the foundation caissons, M_d = overturning moment respected to the toe of the foundation caissons, B = width of the foundation caissons, d = penetration depth, b = shape factor that is 0.5 for continuous foundations, γ_1 and γ_2 = submerged unit weights of soil under and over the bottom of the foundation caissons, c_o = cohesion of soil at the bottom. When the foundation caissons are a cylindrical shape, the width B is obtained by converting the circular shape to the rectangular shape.

The safety factors under ordinary condition are 1.2 for both overturning and sliding and 2.5 for bearing capacity. Under unordinary condition such as an earthquake, they are 1.1 for overturning, 1.0 for sliding and 1.5 for bearing capacity. The stability analysis using Bishop's method should be, also, conducted on inclined loads.

For the separated type foundations, the stability of their superstructures should be analyzed on both the sliding and overturning on the top of the foundation caissons.

3 SHAKING TABLE TESTS

The method of the stability analysis for the suction foundations of quaywalls during earthquakes described in the chapter 2 was examined with the results of model tests using a shaking table (Yamazaki et al 2000).

3.1 Outline of the tests

The shaking table tests were conducted on model quaywalls using suction foundations in a shaking box as shown in Figure 5 and Table 1.

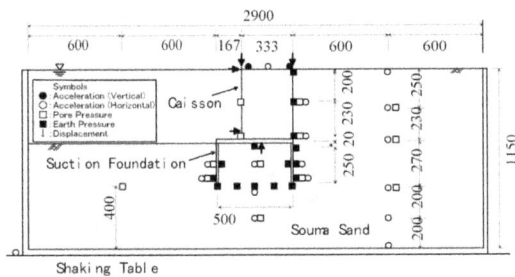

Figure 5. Example of shaking table tests

Table 1. Test case

Case	Density of ground	Unit weight of caisson γ_1(kN/m³)	Penetration Depth(m)	Penetration Depth/Diameter	Caisson width (m)	Seismic wave	Remarks
D1	Dense	15.3	0.25	0.5	0.33	Kobe	Caisson not filled
D2	Dense	22.8	0.25	0.5	0.33	Kobe	
D3	Loose	22.4	0.25	0.5	0.33	Kobe	
D4	Dense	22.7	0.125	0.25	0.33	Kobe	
D5	Dense	22.7	0	0	0.33	Kobe	
D6	Loose	22.4	0	0	0.33	Hachinohe	
D7	Loose	22.4	0.25	0.5	0.33	Hachinohe	
D8	Loose	22.1	0.25	0.5	0.50	Hachinohe	Large caisson width

The model suction foundations were hollow cylinders whose tops were closed but bottoms were open. The rectangular model caissons as superstructures were set on the foundation caissons. Semi-models of the superstructures and the foundation caissons (cylinders) were placed as dummy in both sides of the model in order to prevent the wall of the shaking box from affecting the behaviors of the model for the measurements. The model suction foundation had a diameter of 50cm. A scaling factor in the modeling was 1/30, as the diameter of the prototype was supposed 15m. According to the similitude of the modeling, a displacement of 1mm of the model corresponds to 16mm of the prototype. Tests were conducted on three kinds of penetration depth of the suction foundations, which were 25cm, 12.5cm and 0cm. The test cases of the penetration depth of 0cm had caissons (superstructures) whose bottoms had plates that were top plates of the suction foundations as shown in Figure 5. The model superstructures had a rectangular shape with 50cm in height and 33cm in width. Inside the caissons was filled with sand. However, two test cases were conducted with different caissons to examine the effect of the caisson type. One was conducted with the caisson of 50cm in width, the other with the caisson not filled with the sand.

Model grounds were made with the Souma No.6 sand. The sand has physical properties that are the particle density of $2.651 g/cm^3$, the maximum void ratio of 1.303 and the minimum void ratio of 0.840. Two types of model ground were made. One was a loose deposit having a relative density of 40% and the other a dense deposit having a relative density of 60%. Triaxial test results on the sand shown that the frictional angles of both the loose specimen and the dense specimen were, almost same each other, 38degrees. However, other mechanical properties such as a stress-strain relationship were different from each other. For instance, liquefaction resistances, which were shear stress ratios causing double amplitude axial strain of 5% at 20 cyclic loads, obtained from cyclic triaxial test results were 0.25 for the loose specimen (Dr=45%) and 0.4 for the dense specimen (Dr=80%).

The model suction foundations were made by hand without a suction force although the actual implementation in a field used the suction force, because the aim of the tests was not for the implementation by the suction force but the stability of suction foundations.

The shaking table tests were performed by a stage test where the amplitude of an incident seismic wave is increased from 50gal to 1000gal with increments of 50gal or 100gal. The employed seismic waveform as an incident seismic motion was Kobe wave and Hachinohe wave. The time scale of the wave data is compressed to 0.078 times the original because of the similitude of the modeling.

3.2 Test results

Only displacements of the front of the crest of the superstructures are described in this paper, although measurements were variously conducted on earth pressures, pore water pressures and the other displacements, etc. However, it is especially commented on the pore water pressures with related to liquefaction of the model ground. In all the tests, liquefaction is not observed from pore water pressure measurements, even in loose deposited model grounds. Such pore pressure behaviors are different from that reported by other researchers. The reason why the deposits did not liquefy is not found.

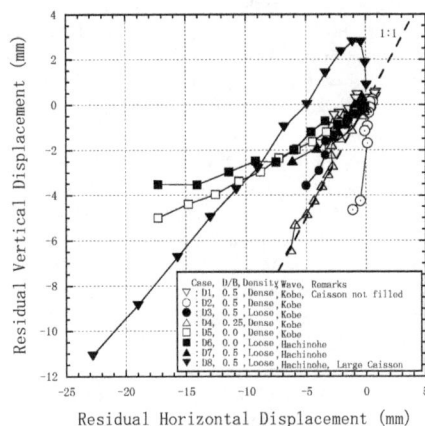

Figure 6. Relationship between horizontal displacements and vertical displacements

Figure 6 shows the relationship between the residual vertical displacements and the residual horizontal ones at the crown. It is found from Figure 6 that the horizontal displacement is larger in the case of the penetration of 0cm than that of the penetration of 25cm. This means that the penetration enhances the stability of quaywalls against the sliding. The horizontal displacements of the loose deposits are larger than that of the dense one when the penetration depth is the same. This difference comes from the difference of the stress-strain relationship between the dense deposit and the loose one.

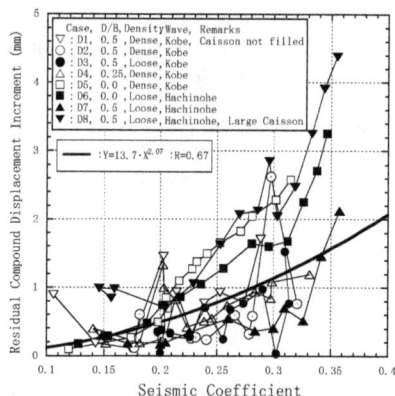

Figure 7. Relationship between displacements and seismic coefficients

Figure 7 shows the relationship between the increments of residual compounded displacement increments and the seismic coefficients. The residual compounded displacement is defined as the magnitude of the vector that is the sum of the vertical displacement increment vector and the horizontal one. The displacement increment is the residual displacement caused in each stage. The seismic coefficient is obtained with Noda-Uwabe's equations using maximum acceleration at the ground surface behind the caisson (Noda et al 1975). In Figure 7 is drawn the curve which approximates the plots by the least square method using a power function and the correlation factor is 0.67. It is found from Figure 7 that the displacement increment increases with increase of the seismic coefficient and that the displacement increment in the penetration of 0cm is larger than that in the penetration of 25cm at the same seismic coefficient. Therefore, the seismic resistance becomes larger in the case of the large penetration depth than that of the small one.

Figure 8 shows the relationship between the compounded residual displacement increments and risk factors. The risk factor is defined by the ratio of the measured seismic coefficient to the critical seismic coefficient. The critical seismic coefficient is the seismic coefficient causing failure to the quaywalls and is obtained by using the design method proposed in the chapter 2. In Figure 8 is also drawn the curve which approximates the plots by the least square method using a power function and the correlation factor is 0.85. It is found from Figure 8 that the correlation between the residual compounded displacement increments and the risk factors becomes much better than that in Figure7. And when the risk factor exceeds around 1.0 the displacement increment becomes large. Therefore, it is thought that the test results support the proposed design method in the chapter 2.

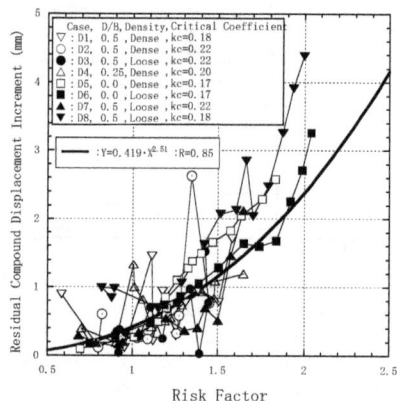

Figure 8. Relationship between displacements and risk factors

4 CONCLUDING REMARKS

The design method of the suction foundations used for quaywalls and shaking table tests using small models are presented. The test results have validated the design method. The suction foundations are anticipated to widely use for the foundations of many kinds of offshore structures as well as quaywalls.

REFERENCES

Noda, S., Uwabe, T. and Chiba, T., 1975. Relation between seismic coefficient and ground acceleration for gravity wall, Report of the Port and Harbour Research Institute, Vol.14, No.4, 67-111 (in Japanese).
Yamazaki, H. and Takahashi, K. 2000. Suction Foundation and case history of its application to the breakwaters. Foundation engineering & Equipment, Vol.28, No.1, 74-76 (in Japanese).
Zen, K., Yamazaki, H. and Maeda, K. 1998. Case history on the penetration of caisson-type foundations into seabed by use of suction force. Journal of JSCE, No.603/III-44, 21-34 (in Japanese).

Author index

For Product Safety Concerns and Information please contact our EU
representative GPSR@taylorandfrancis.com
Taylor & Francis Verlag GmbH, Kaufingerstraße 24, 80331 München, Germany